Corrosion Resistance of Steels against Inorganic Acid

Edited by
Michael Schütze, Thomas Ladwein
and Roman Bender

Corrosion Resistance of Steels against Inorganic Acids

DECHEMA

WILEY-VCH

WILEY-VCH Verlag GmbH & Co. KGaA

Editors

Prof. Dr.-Ing. Michael Schütze
DECHEMA-Forschungsinstitut
Chairman of the Executive Board
Theodor-Heuss-Allee 25
60486 Frankfurt am Main
Germany

Prof. Dr. Thomas Ladwein
Aalen University of Applied Sciences
Surface and Materials Science
73428 Aalen
Germany

Dr. rer. nat. Roman Bender
Chief Executive of GfKORR e. V.
Society for Corrosion Protection
Theodor-Heuss-Allee 25
60486 Frankfurt am Main
Germany

Cover Illustration
Source: DECHEMA-Forschungsinstitut, Frankfurt (Main), Germany

Library of Congress Card No.: Applied for.

British Library Cataloguing-in-Publication Data:
A catalogue record for this book is available from the British Library.

Bibliographic information published by Die Deutsche Bibliothek
Die Deutsche Bibliothek lists this publication in the Deutsche Nationalbibliografie; detailed bibliographic data is available in the Internet at <http://dnb.ddb.de>.

Printed in the Federal Republic of Germany
Printed on acid-free paper

Typesetting Kühn & Weyh, Satz und Medien, Freiburg
Printing Strauss GmbH, Mörlenbach
Binding Strauss GmbH, Mörlenbach
Cover Design Graphik-Design Schulz, Fußgönheim

ISBN: 978-3-527-33520-6

Contents

Preface

Practically all industries face the problem of corrosion – from the micro-scale of components for the electronics industries to the macro-scale of those for the chemical and construction industries. This explains why the overall costs of corrosion still amount to about 2 to 4 % of the gross national product of industrialized countries despite the fact that zillions of dollars have been spent on corrosion research during the last few decades.

Much of this research was necessary due to the development of new technologies, materials and products, but it is no secret that a considerable number of failures in technology nowadays could, to a significant extent, be avoided if existing knowledge were used properly. This fact is particularly true in the field of corrosion and corrosion protection. Here, a wealth of information exists, but unfortunately in most cases it is scattered over many different information sources. However, as far back as 1953, an initiative was launched in Germany to compile an information system from the existing knowledge of corrosion and to complement this information with commentaries and interpretations by corrosion experts. The information system, entitled "DECHEMA-WERKSTOFF-TABELLE" (DECHEMA Corrosion Data Sheets), grew rapidly in size and content during the following years and soon became an indispensable tool for all engineers and scientists dealing with corrosion problems. This tool is still a living system today: it is continuously revised and updated by corrosion experts and thus represents a unique source of information. Currently, it comprises more than 12,000 pages with approximately 110,000 corrosion systems (i.e., all relevant commercial materials and media), based on the evaluation of over 100,000 scientific and technical articles which are referenced in the database.

Last century, an increasing demand for an English version of the DECHEMA-WERKSTOFF-TABELLE arose in the 80s; accordingly the first volume of the DECHEMA Corrosion Handbook was published in 1987. This was a slightly condensed version of the German edition and comprised 12 volumes. Before long, this handbook had spread all over the world and become a standard tool in countless laboratories outside Germany. The second edition of the DECHEMA Corrosion Handbook was published in 2004. Together the two editions covered 24 volumes.

The present book compiles all information on the corrosion behaviour of Steels against inorganic acids that was compiled in the volumes of the corrosion handbook. This compilation is an indispensable tool for all engineers and scientists dealing with corrosion problems of Steels in contact with inorganic acids.

Steel is one of the most widely used construction materials with more than 1.3 billion tons produced each year. Buildings, industrial plants, machines, tools, pipelines, vessels and tanks are only a few of its applications in our daily life. Steel is an alloy made of iron and additional elements like carbon, chromium, manganese, vanadium and tungsten, and its quality, ductility, hardness and strength vary with the amount of the alloying element.

As steels corrode in various atmospheres, in water and in soil its corrosion resistance against the four most common chemicals and inorganic acids – hydrochloric, nitric, sulfuric and phosphoric acid – is essential and a crucial financial factor for many industries. These acids are present in nearly every industrial production process such as metal manufacturing but also explosives, food, dyes, leather, paper and fertilizers, to name only a few.

Understanding how to strengthen the corrosion resistance of steels as reaction, transport and storage devices against these omnipresent and aggressive acids is key for all industries involved. This book is therefore a must-have for all mechanical, civil and chemical engineers, material scientists and chemists working with steel or acidic media.

This handbook highlights the limitations of Steels in inorganic acids and provides vital information on corrosion protection measures.

Corrosion is a complex phenomenon that depends on a number of parameters, related to both the environment and the metal.

The chapters are arranged by the agents leading to individual corrosion reactions, and a vast number of steels are presented in terms of their behaviour in these agents. The key information consists of quantitative data on corrosion rates coupled with commentaries on the background and mechanisms of corrosion behind these data, together with the dependencies on secondary parameters, such as flow-rate, pH, temperature, etc. This information is complemented by more detailed annotations where necessary, and by an immense number of references listed at the end of each chapter.

An important feature of this handbook is that the data was compiled for industrial use. Therefore, particularly for those working in industrial laboratories or for industrial clients, the book will be an invaluable source of rapid information for day-to-day problem solving. The handbook will have fulfilled its task if it helps to avoid the failures and problems caused by corrosion simply by providing a comprehensive source of information summarizing the present state-of-the-art. Last but not least, in cases where this knowledge is applied, there is a good chance of decreasing the costs of corrosion significantly.

Finally the editors would like to express their appreciation to Gudrun Walter of Wiley-VCH for their valuable assistance during all stages of the preparation of this book.

Michael Schütze, Thomas Ladwein and Roman Bender

How to use the Handbook

The Handbook provides information on the chemical resistance and the corrosion behavior of steels in inorganic acids.

The user is given information on the range of applications and corrosion protection measures.

Research results and operating experience reported by experts allow recommendations to be made for the selection of steels and to provide assistance in the assessment of damage.

The objective is to offer a comprehensive and concise description of the behavior of steels in contact with a particular acid.

The information on resistance is given as text, tables, and figures. The literature used by the authors is cited at the corresponding point. There is an index of materials as well as a subject index at the end of the book so that the user can quickly find the information given for a particular keyword.

The Handbook is thus a guide that leads the reader to steels that have already been used in certain cases, that can be used or that are not suitable owing to their lack of resistance.

The resistance is coded with three evaluation symbols in order to compress the information. Uniform corrosion is evaluated according to the following criteria:

Symbol	*Meaning*	*Area-related mass loss rate*		*Corrosion rate*
		g/(m² h)	g/(m² d)	mm/a
+	resistant	≤ 0.1	≤ 2.4	≤ 0.1
⊕	fairly resistant	> 0.1 to ≤ 1.0	> 2.4 to ≤ 24.0	> 0.1 to ≤ 1.0
–	not resistant	> 1.0	> 24.0	> 1.0

The evaluation of the corrosion resistance of metallic materials is given

- for uniform corrosion or local penetration rate, in: mm/a
- or if the density of the material is not known, in: g/(m² h) or g/(m² d).

Pitting corrosion, crevice corrosion, and stress corrosion cracking or non-uniform attack are particularly highlighted.

The following equations are used to convert mass loss rates, x, into the corrosion rate, y:

from x_1 into $g/(m^2\,h)$

$$\frac{x_1 \cdot 365 \cdot 24}{\rho \cdot 1{,}000} = y\ (mm/a)$$

from x_2 into $g/(m^2\,d)$

$$\frac{x_2 \cdot 365}{\rho \cdot 1{,}000} = y\ (mm/a)$$

where

x_1: value in $g/(m^2 h)$
x_2: value in $g/(m^2 d)$
ρ: density of material in g/cm^3
y: value in (mm/a)
d: days
h: hours

In those media in which uniform corrosion can be expected, if possible, isocorrosion curves (corrosion rate = 0.1 mm/a) are given.

Unless stated otherwise, the data was measured at atmospheric pressure and room temperature.

The resistance data should not be accepted by the user without question, and the materials for a particular purpose should not be regarded as the only ones that are suitable. To avoid wrong conclusions being drawn, it must be always taken into account that the expected material behavior depends on a variety of factors that are often difficult to recognize individually and which may not have been taken deliberately into account in the investigations upon which the data is based. Under certain circumstances, even slight deviations in the chemical composition of the medium, in the pressure, in the temperature or, for example, in the flow rate are sufficient to have a significant effect on the behavior of the materials. Furthermore, impurities in the medium or mixed media can result in a considerable increase in corrosion.

The composition or the pretreatment of the material itself can also be of decisive importance for its behavior. In this respect, welding should be mentioned. The suitability of the component's design with respect to corrosion is a further point which must be taken into account. In case of doubt, the corrosion resistance should be investigated under operating conditions to decide on the suitability of the selected materials.

Warranty disclaimer

This book has been compiled from literature data with the greatest possible care and attention. The statements made in this book only provide general descriptions and information.

Even for the correct selection of materials and correct processing, corrosive attack cannot be excluded in a corrosion system as it may be caused by previously unknown critical conditions and influencing factors or subsequently modified operating conditions.

No guarantee can be given for the chemical stability of the plant or equipment. Therefore, the given information and recommendations do not include any statements, from which warranty claims can be derived with respect to DECHEMA e.V. or its employees or the authors.

The DECHEMA e.V. is liable to the customer, irrespective of the legal grounds, for intentional or grossly negligent damage caused by their legal representatives or vicarious agents.

For a case of slight negligence, liability is limited to the infringement of essential contractual obligations (cardinal obligations). DECHEMA e.V. is not liable in the case of slight negligence for collateral damage or consequential damage as well as for damage that results from interruptions in the operations or delays which may arise from the deployment of this book.

Hydrochloric Acid

Unalloyed and low-alloyed steels/cast steel

Iron reacts with hydrochloric acid to produce hydrogen according to the reaction

Eq. 1 $Fe + 2\,HCl \rightarrow FeCl_2 + H_2$

Depending on the concentration and temperature of the acid, the removal rates lie between 2 and 20 mm/a (78 and 787 mpy). As the content of iron chloride in the solution increases the corrosion rate decreases; however, hydrochloric acid cannot be used without a suitable inhibitor to clean the iron and steel surfaces. Only after effective inhibitors had been developed could hydrochloric acid be used to pickle steel. With more advances in the development of inhibitors and pickling bath monitoring, hydrochloric acid pickling baths, which dissolve the rolling scale and annealing scale on steel much better than sulphuric acid, have increasingly displaced the previously used sulphuric acid pickling baths in recent years.

The following requirements are specified for the inhibitors:

- good solubility in hydrochloric acid
- simple dosability
- stability at the application temperature over longer operating periods
- insensitive to hydrogen
- reduction of hydrogen absorption into steel
- no reaction with the iron(II) and iron(III) ions produced during pickling
- no significant lengthening of the pickling time
- simple rinsing off from the surface
- no problems in the further processing of the pickled material

In addition, they should, of course, be as inexpensive as possible.

Most of the inhibitors used in hydrochloric acid pickling baths for steel and which have been tested in practice are nitrogen-containing organic compounds. They are mostly based on the following substances:

- acetylene derivatives
- pyridine derivatives
- urotropine and its derivatives
- thiourea and its derivatives
- quaternary ammonium compounds
- alkylamines and arylamines
- aldehydes (e.g. formaldehyde)
- ketones (e.g. cyclohexanone)

Acid inhibition is a classical field of corrosion protection engineering and therefore has been extensively investigated in the older literature [1].

In recent literature, numerous reports have been published of investigations and developments of new compounds that are suitable as inhibitors for the reaction of unalloyed steels in hydrochloric acid [2–6]. Attempts are often made to correlate the relationship between the effect and the structure of the compound.

Indeed, it is possible that the substances have differing efficiencies in hydrochloric acid and sulphuric acid [7–9]. There are even cases in which they act as an effective inhibitor against metal corrosion and against hydrogen absorption by the steel in hydrochloric acid, but are less effective or even promote corrosion in sulphuric acid. An example of this is given in Table 1 [9].

Inhibitor concentration mmol/l	*Inhibition efficiency (in HCl) %*			
	1 M HCl		*0.5 M H_2SO_4*	
	mass loss	*gas evolution*	*mass loss*	*gas evolution*
ortho-Anisidine				
20	62.7	62.1	21.5	22.9
50	68.6	70.2	39.6	38.1
75	80.2	80.3	46.9	50.9
100	87.3	85.9	66.9	61.8
meta-Anisidine				
20	70.5	71.2	– 21.9	– 10.0
50	81.1	80.8	3.1	1.4
75	86.6	84.3	24.2	38.1
100	90.0	89.9	58.1	63.6
para-Anisidine				
20	84.5	81.8	– 51.5	– 39.0
50	91.4	90.4	– 45.8	– 33.1
75	95.0	93.9	– 37.7	– 16.1

Table 1: Inhibition efficiency of anisidine isomers on the corrosion of unalloyed steel in hydrochloric acid and in sulphuric acid [9]

For practical applications, tested products of well-known manufacturers should be used because they are based on many years of experience and competent customer services can be expected.

The study in [10] reports in detail on the removal of deposits and corrosion products on pipelines and plant components. In chemical cleaning, in addition to hydrofluoric acid, hydrochloric acid, in particular, plays a decisive role because it dissolves many deposits that originate from water-carrying systems, e.g. boiler scale, rust or magnetite, but not silicic acid or silicates. Figure 1 and Figure 2 show the dissolution rates of magnetite and natural rust on unalloyed steel in solutions of hydrochloric acid compared to other acid solutions. After hydrofluoric acid, hydrochloric acid is the most effective. However, it is not recommended to use hydrochloric acid to clean stainless steel components, even in combination with effective corrosion inhibitors, because there is a risk of local corrosion due to residual chloride ions.

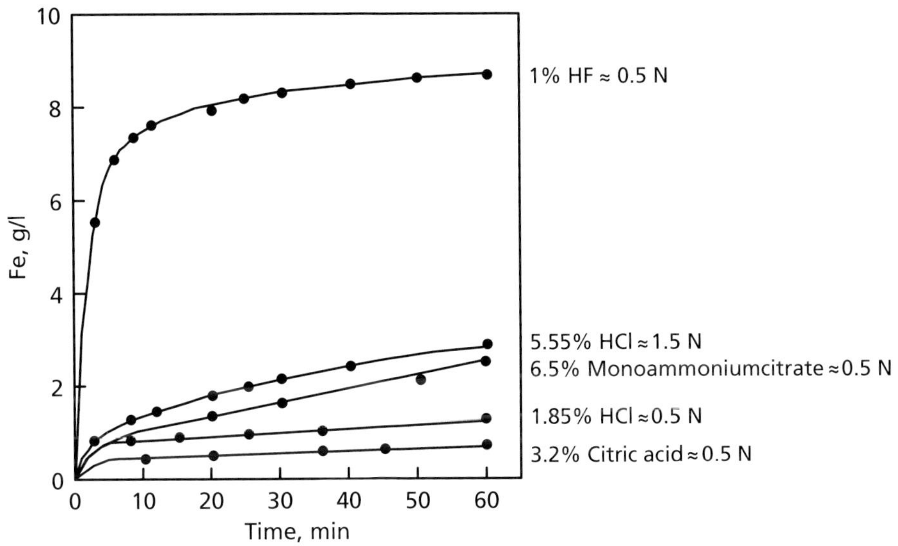

Figure 1: Dissolution of magnetite (Fe_2O_3) in various acids at 343 K (70 °C) [10]

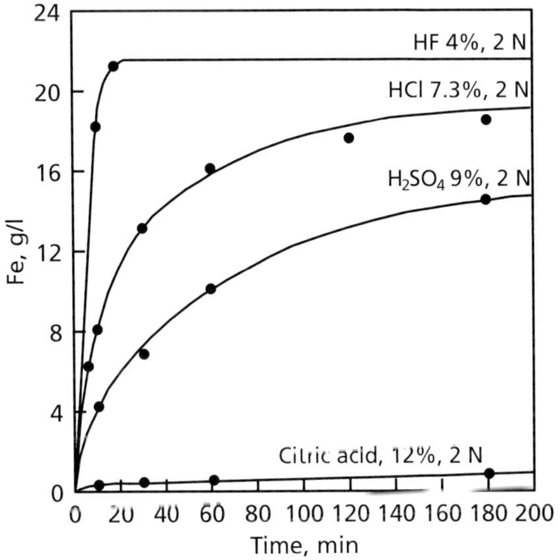

Figure 2: Dissolution of natural rust in various acids at 333 K (60 °C) [10]

The hydrolysis of chlorinated hydrocarbons is already catalysed at room temperature by rusting steel and by rust. The thus liberated hydrochloric acid is very corrosive [11]. If the action of external energy (heat, light) is excluded, a good hydrolysis resistance can be generally assumed. This was confirmed by tests with 50 ml H_2O and chlorinated hydrocarbons (CCl_4, $CHCl_3$, CH_2Cl_2, C_2HCl_3, C_2Cl_4) at 323 K (50 °C) stored in the dark in glass-stoppered flasks. After one week, the aqueous phase had the same pH and Cl^- ions could not be detected. When strips of sheet (35 × 70 × 1 mm from St 34) were added, then after 2 days there was strong rust formation with noticeable hydrolysis and formation of hydrochloric acid. Table 2 gives the corresponding corrosion rates.

Medium	g/m^2	mm/a (mpy)
CCl_4	190	4.4 (173)
$CHCl_3$	24.2	0.56 (22)
CH_2Cl_2	18.6	0.43 (17)
C_2HCl_3	35.0	0.82 (32)
C_2Cl_4	9.3	0.22 (8.7)

Table 2: Corrosion rates of unalloyed steel as a result of hydrolytically produced HCl in various chlorinated hydrocarbons [11]

These corrosion rates are much too high for normal rusting in H_2O.

The hydrolytic formation of hydrochloric acid in aqueous chlorinated hydrocarbons in the presence of unalloyed steel is also confirmed in [12]. In a two-component mixture of carbon tetrachloride and water in a 5 : 2 ratio, the corrosion rates of samples of unalloyed steel DIN-Mat. No. 1.0402 (UNS G10200, C22) were measured at the boiling point in darkened rooms for test periods of 8–30 days in the three phases listed in Table 3.

organic phase	0.7 mm/a (27.5 mpy)
aqueous phase	9.0 mm/a (354 mpy)
vapor phase	9.0 mm/a (354 mpy)

Table 3: Corrosion rates of steel C 22 in the two-component system CCl_4/H_2O at the boiling point [12]

These results were obtained by means of a radionuclide method in which the mass loss was calculated from the activity increase in the corrosion media caused by the corrosion products from a reactor-activated material sample.

The authors of [13] investigated the problem of stress corrosion cracking of high-strength carbon steels in hydrochloric acid solutions. The compositions and tensile strengths of the investigated steels are given in Table 4.

Steel	C %	Mn %	Si %	P %	S %	Cr %	Mo %	V %	Rm MPa
J-55	0.42	1.07	0.37	0.029	0.019	0.13	0.01	–	593
C-75	0.37	1.48	0.29	0.014	0.016	0.14	0.07	–	753
N-80 DIN-Mat. No. 1.0564	0.33	1.34	0.29	0.011	0.013	0.16	0.05	–	746
P-105 DIN-Mat. No. 1.0670	0.40	1.59	0.33	0.012	0.012	0.57	0.20	0.09	936

Table 4: Chemical compositions and tensile strengths (Rm) of the investigated high-strength steels [13]

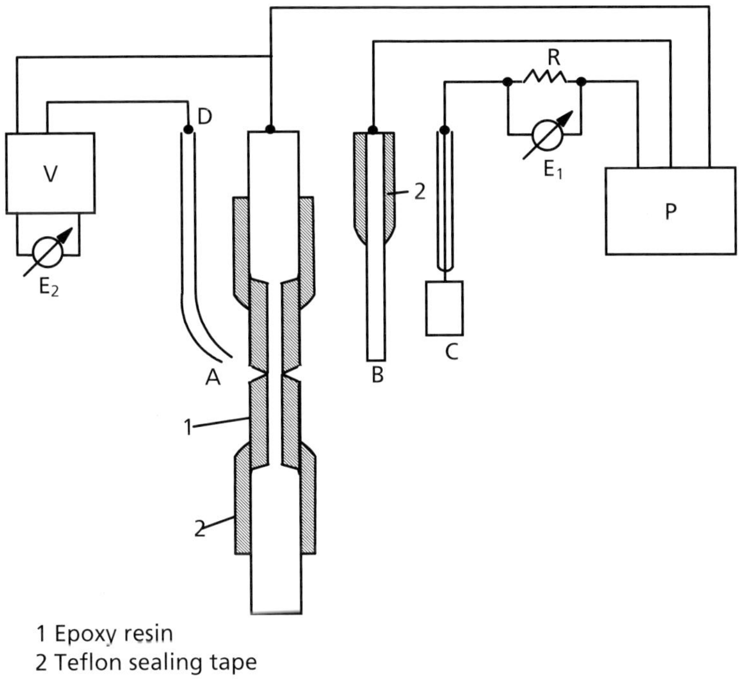

Figure 3: Diagram of the set-up of the equipment for electrochemically controlled SCC testing [13]
A – strained microtension specimen
B – unloaded macroelectrode
C – Pt counterelectrode
D – SCE reference electrode
R – resistance
E_1 and E_2 – recorders
V – voltmeter
P – potentiostat

The investigations were carried out on round tensile samples with a diameter of 3 mm in the test direction that could be polarised in an electrolysis cell and subjected to tensile loading (Figure 3). The incubation time, crack advancing speed and fracture time were derived from the course of the current density/potential curves. As expected, the fracture time decreased with increasing voltage, temperature and acid concentration. The fracture time also depended in a characteristic manner on the potential of the samples (Figure 4). This dependency results from the interaction of iron dissolution at the crack tip in the anodic potential range and on hydrogen-induced crack formation in the cathodic potential range. The influence of anodic dissolution was the major factor.

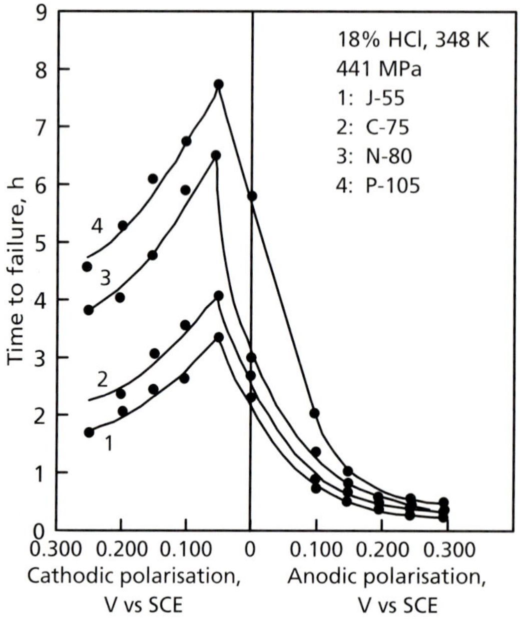

Figure 4: Time to fracture depending on the polarisation in 18 % HCl at 348 K (75 °C) and 441 MPa [13]

In reference [14], the influence of molybdenum additions of up to 1 % on the corrosion behavior of highly pure carbon steels with up to 0.4 % C was investigated in 5, 10, 15 and 20 % HCl at ambient temperature. The current density/potential curves of the samples were recorded. The Stern-Geary equation

Eq. 2 $$i_{cor} \frac{\beta_a \beta_c}{\beta_a + \beta_c} \cdot \frac{1}{2,3 \cdot R_p}$$

where β_a and β_c are the anodic and cathodic Tafel constants, R_p is the slope of the polarisation resistance lines and i_{cor} is the corrosion current density, was used to calculate the surface-related removal rate in mpy (1 mpy = 0.0254 mm/a).

The electrochemical principles are explained in [47] in an understandable manner.

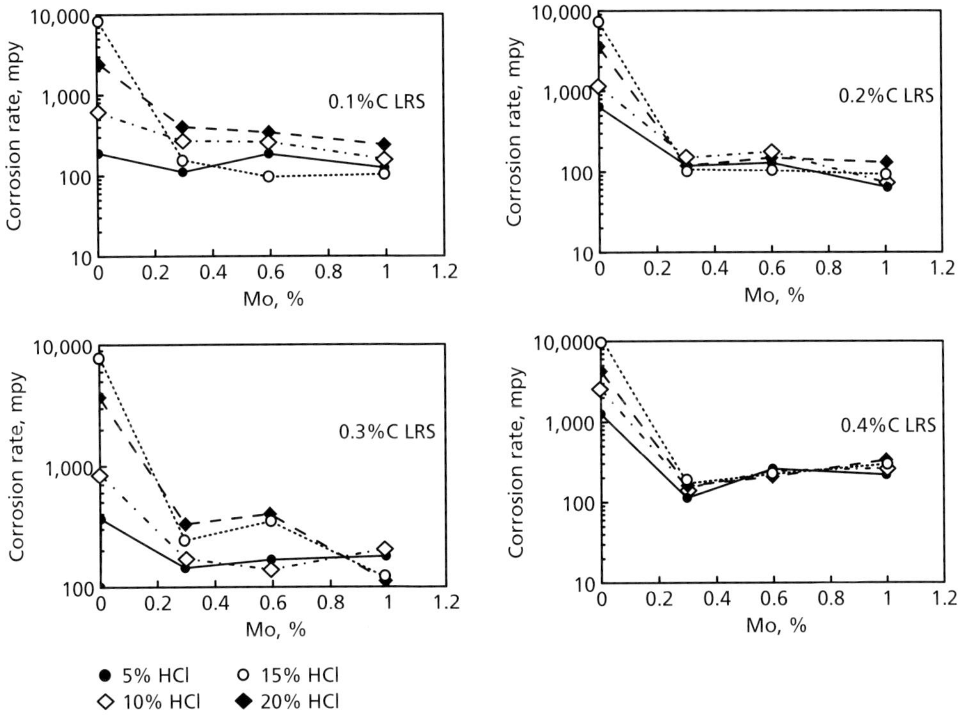

Figure 5: Corrosion rate of low residual steels (LRS) with differing carbon contents depending on the Mo content in various HCl solutions (1 mpy = 0.0254 mm/a) [14]

Figure 5 gives the corrosion rate of the steels with differing carbon (C) contents depending on the molybdenum content in various HCl solutions. Independent of the C content, the corrosion rate deceased by up to approx. 0.3 % Mo as the acid concentration increased. It attained values of up to two powers of ten. No statement is made concerning the extent to which this favorable effect was influenced by Si, Mn, P, S etc. within the usual limits.

Corrosion protection in oil refineries

The corrosion of condensers, columns and connecting systems by overhead streams from the distillation of crude oil is a major problem in refineries.

Reference [15] discusses corrosion in the head region of fractionating columns used for oil distillation. The main cause is the generation of hydrochloric acid by hydrolysis of $MgCl_2$ and $CaCl_2$

Eq. 3 $$MgCl_2 + H_2O \rightarrow 2\ HCl + MgO$$

Eq. 4 $$CaCl_2 + H_2O \rightarrow 2\ HCl + CaO$$

and by absorption of HCl in the condensation water in the head region. In the region of the preheater, where the hydrolysis occurs, corrosion is low because water is not present as a liquid. The NaCl present in large amounts in crude oil is stable and only hydrolyses to a small extent in the preheater stage.

Because of the importance of this problem in oil refineries, the NACE (National Association of Corrosion Engineers) questioned many operators with regard to this problem with the aim of finding out how to deal with this type of corrosion. Possible measures are desalting of the crude oil and neutralisation with sodium hydroxide or sodium carbonate in the preheater circuit. Furthermore, ammonia and neutralising amines can be added to the heads of the columns to increase the pH value from less than 1 to 6–7. Because the neutralisation products are often solids that can also be corrosive and cause blockages, they are flushed out with water. Amine-based inhibitors are also added in the head region. These inhibitors are more effective in neutral environments than in acidic ones. Therefore, they are usually added together with neutralisers.

There are particularly aggressive corrosion conditions in the head region of the fractionating columns used to distil crude oil. The nitrogen and sulphur compounds present in crude oil decompose at high temperatures to form corrosive mercaptans and hydrogen sulphide. Any chlorides present are hydrolysed to hydrogen chloride. Hydrogen sulphide and hydrogen chloride tend to become enriched in the head region of the fractionating columns together with hydrocarbons and heated steam. Therefore, in the head region, corrosion occurs wherever water can condense. This preferentially absorbs the HCl arriving in the head region to form hydrochloric acid. The reaction is additionally complicated by the presence of H_2S in the head region where iron sulphide is deposited from the reaction between H_2S and soluble iron chloride, which is produced by the corrosion of the steel construction by HCl.

Nitrogen compounds, such as quaternary ammonium compounds, amines, amino salts and heterocyclic compounds, are used as corrosion inhibitors in the oil industry.

On the basis of several cases of damage in the head region of such crude oil processing plants, it is demonstrated in [16] that water-soluble inhibitors can be used to avoid these corrosion problems. These water-soluble inhibitors are more favorable than oil-soluble ones because if the temperature drops below the dew point of water, they are directly active in the critical HCl-containing condensates. Inhibitors based on modified fatty acids have also proved suitable in this field [17, 18].

The work in [19] aimed to investigate the behavior of 3-amino-1,2,4-triazole (ATR), 2-amino-thiazole (ATH) and 2,6-diamino-pyridine (DAP) as corrosion inhibitors for carbon steel exposed to a kerosine-water mixture (10 vol% water) with 3 ppm HCl and 800 ppm H_2S/per day at a pH value of 6 to 6.5 and a temperature of 328 to 333 K (55 to 60 °C) using ammonia as a neutralising agent. The three heterocyclic compounds were selected on the basis of their thermal stability up to approx. 573 K (300 °C), which facilitates the formation of a coating on the metal and the possibility of forming stable complexes with the surface of the carbon steel.

The three substances were used in the technically pure state (> 95 %) without purification. They are soluble in aromatic and aliphatic hydrocarbons and partly soluble in water. In the amounts used, there is no significant emulsification.

The investigated steel had the composition: 0.2 % C, 0.7 % Mn, 0.04 % P, 0.04 % S. Table 5 gives the results after a test period of 50 h. For an addition of 10 ppm of all three substances, the highest inhibition efficiency was 82.7–88.2 %. Higher additions lowered the inhibition efficiency.

The effectiveness of the organic compounds (ATR, ATH, DAP) was also investigated using air as the carrier gas because some commercially available inhibitors are also recommended for systems that contain oxygen in addition to H_2S, HCl and cyanides. For additions of 10 ppm and a test duration of 100 h, the inhibition efficiencies were 60.3 % for ATR, 55.8 % for ATH and 50.7 % for DAP.

Reference [19] gives the results for additions of 10 ppm after various test periods with nitrogen as the carrier gas along with ammonia as the neutralising agent and H_2S. Furthermore, octylamine was added in order to test the behavior of the three compounds in the presence of amines. Octylamine has a high boiling point compared to other aliphatic amines that are sometimes used to control the pH value.

The results show that the addition of octylamine alone gave a low inhibition efficiency, particularly after a test duration of 100 h. Without the addition of octylamine, the three investigated substances exhibited a high inhibition efficiency of approx. 90 % after 50 h, which dropped to approx. 70 % after 400 h. On adding 10 ppm octylamine, the values after an exposure time of 100 h were lower than these. The best inhibition efficiency was exhibited by ATH. The investigation showed that the three substances formed Fe(II) and Fe(III) complexes that form protective layers on the steel surface. They are suitable as inhibitors for fractionating columns in the oil industry.

It is known that the combustion of polyvinylchloride (PVC) produces hydrogen chloride which condenses together with water vapor as hydrochloric acid that can lead to strong corrosion of metallic objects. In reference [20], it is shown that the thermal decomposition of freones (hydrofluorocarbons, HFC), which can also contain chlorine, and also of halones (halogenated hydrocarbons), which in addition to fluorine also contain chlorine and particularly bromine, can lead to the formation of not only hydrochloric acid but also hydrofluoric and hydrobromic acids. Iodine-containing HFCs can also lead to the formation of hydriodic acid.

These compounds are used as extinguishing agents for fires. A massive dosage of halons in the event of a fire produces a cloud of halogen hydracid gas with a relatively high density around the source of the fire. This blocks off any oxygen from reaching the fire so that it stagnates after a short period. The halogen hydracid liberated from the extinguishing agents are deposited on metal parts, such as iron, and can thus attack them to form the corresponding iron halogen salt. Any further processes depend on the relative humidity. Figure 6 gives the results of laboratory tests in which the surface of unalloyed steel was exposed for 1 hour to the vapor of concentrated halogen hydracids followed by 95 hours exposure to an acid-free atmosphere at various levels of relative humidity. Depending on the type of acid, there was a more or less strong corrosion for the different relative humidities. The corrosive attack

Test	**Additive**	**3-Amino-1,2,4-triazole**		**2-Amino-thiazole**		**2,6-Diamino-pyridine**	
		Corrosion rate mm/a (mpy)	**Inhibitor efficiency %**	**Corrosion rate mm/a (mpy)**	**Inhibitor efficiency %**	**Corrosion rate mm/a (mpy)**	**Inhibitor efficiency %**
1	None	0.295 (11.5)	–	0.295 (11.5)	–	0.295 (11.61)	–
2	Ammonia	0.113 (4.5)	–	0.113 (4.5)	–	0.113 (4.45)	–
3	Ammonia + 10 ppm inhibitor	0.017 (0.7)	84.7	0.013 (0.51)	88.2	0.020 (0.79)	82.7
4	Ammonia + 20 ppm inhibitor	0.018 (0.71)	84.0	0.025 (0.98)	77.6	0.030 (1.18)	73.1
5	Ammonia + 30 ppm inhibitor	0.033 (1.3)	70.9	0.024 (0.94)	79.1	0.051 (2.01)	55.1
6	Ammonia + 40 ppm inhibitor	0.048 (1.9)	57.1	0.029 (1.14)	74.7	0.058 (2.28)	48.7

The calculation of the inhibitor efficiency in tests 3–6 is based on the results from test 2

Table 5: Corrosion rates and inhibition efficiencies for steel in the presence of 3-amino-1,2,4-triazole, 2-amino-thiazole and 2,6-diamino-pyridine after an exposure time of 50 h [19]

was the strongest for hydrochloric acid and hydrobromic acid. In comparison, hydrofluoric acid gives a corrosion maximum only at 90 % relative humidity, which is significantly lower than the corrosion values for HCl and HBr.

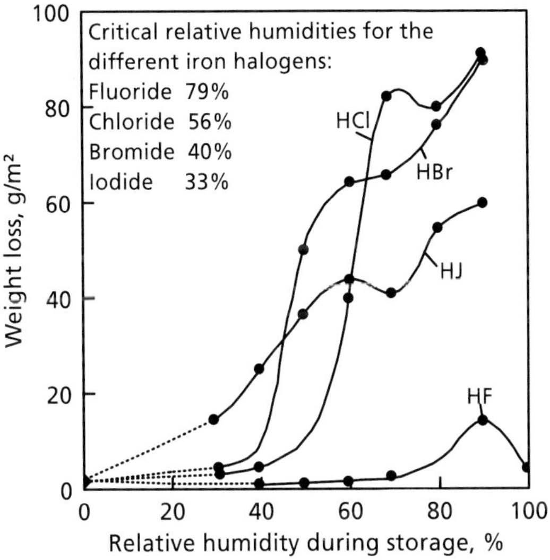

Figure 6: Corrosion experiment with unalloyed steel in an acidic atmosphere. Firstly infection for one hour in acidic vapors above solutions of the concentrated acid. Then storage for 95 hours in an acid-free atmosphere at various relative humidities [20]

The work in [21] describes chemical cleaning without interrupting operations (on-stream cleaning) in heat-exchangers and other plant sections in an oil refinery. This method, introduced at the start of the 1960s, injects inhibited concentrated hydrochloric acid into the flow of cooling water in sufficient quantities to obtain an acid concentration of 5 to 10 % at the outlet. The exact procedure used in nine cases and the corresponding improvements in the results obtained with this type of cleaning are described. Apart from savings of heating oil costs, advantages include lower operating downtimes and lower maintenance costs.

No information is given on the inhibitor used. However, it is mentioned that hydrochloric acid should not be used for equipment made of aluminium and stainless steels. Therefore, it can be concluded that the inhibitor used is only effective for carbon steel.

Unalloyed and low-alloy cast iron

The behavior of cast iron in hydrochloric acid can be compared to that of unalloyed steels, so that the information given in Section Unalloyed and low-alloyed steels is also largely applicable to these materials.

High-alloy cast iron
Silicon cast iron

Grades of cast iron alloyed with silicon (approx. 14.5 % Si) with low contents of Mo and Cu have good resistance at room temperature to hydrochloric acid of all concentrations [22].

Austenitic cast iron (and others)

The austenitic, high-nickel cast-iron grades (Ni-Resist®, ASTM A 436) exhibit a certain resistance to very dilute non-aerated hydrochloric acids. However, aerated solutions or higher temperatures lead to strong attack [22].

Ferritic chromium steels with < 13 % Cr

Constructional steels with up to 10 % chromium

Chromium often serves as an alloying additive in carbon steels that are used for heat-exchanger tubes in steam generators. To maintain optimum heat-exchange conditions, the pipes must be regularly cleaned with a dilute acid solution with or without inhibitors.

Although hydrofluoric acid is expensive and difficult to handle, it does have certain advantages in acid washing compared to other acids, particularly with regard to the dissolution of silicate compounds. Because little was known about the influence of chromium on the corrosion behavior of steel in contact with hydrofluoric acid, this was studied in [23] in a comparative investigation with other acids, particularly hydrochloric acid.

The mass losses were determined by solution analysis and from polarisation curves. Table 6 gives the compositions of the investigated steels with increasing chromium and molybdenum contents.

	Corresponding DIN-Mat. No.	*C*	*Si*	*Mn*	*P*	*V*	*Cr*	*Mo*
Armco iron		0.012	trace	0.017	0.025	0.005	0	0
ASTM A 106	1.0256, 1.0481	0.16	0.17	0.89	0.024	0.027	0	0
ASTM A 335 (P 11)		0.13	0.61	0.51	0.017	0.019	1.35	0.54
ASTM A 335 (P 22)	1.7375, 1.7380	0.13	0.34	0.46	0.021	0.017	2.28	0.93
ASTM A 335 (P 5)	1.7362	0.11	0.33	0.51	0.015	0.020	4.82	0.52
ASTM A 335 (P 9)	1.7386	0.15	0.25 – 1.0	0.3 – 0.6	0.03	0.03	10	0.9 – 1.1
SAE 430	1.4016	0.12	1.0*	1.0*	0.025	0.03	17	0

Table 6: Composition of the investigated steels, mass% [23]

The acids listed in Table 7 were used in the comparative investigations at 318 K (45 °C). The acid concentrations of 1.7 M HF (3 %) or 1.4 M HCl (5 %) correspond to the concentrations that are generally used for cleaning with these acids.

Table 7 compares the corrosion rates obtained by solution analysis and from the polarisation curves. From the results, which are all in good agreement, it can be concluded that dilute hydrochloric acid is advantageous for the cleaning of heat-exchangers made of chromium steels because of the considerably lower corrosion rate compared to that obtained in hydrofluoric acid. However, it is worthwhile reducing the corrosion rate further by adding a suitable inhibitor (see Section Unalloyed and low-alloyed steels); this is also what is done in practice.

Figure 7 and Figure 8 give the mass losses depending on the time for hydrofluoric acid and hydrochloric acid. In the case of hydrofluoric acid, the addition of chromium to the steel results in a considerable increase of the mass loss compared to the chromium-free steel. Hydrochloric acid exhibits the reverse tendency:
the mass loss decreases with increasing chromium content. The result from steel ASTM A 355 P22 with 2.28 % Cr is anomalous. The reason for this is probably thc relatively high molybdenum content of 0.93 %.

	1.4 M HCl		*1.7 M HF*	
	■	◆	■	◆
Armco iron	1.2	2.0	1.3	–
ASTM A 106	2.1	1.8	1.8	3
ASTM A 335 (P 11)	0.4	0.3	11	10
ASTM A 335 (P 22)	0.9	1.2	58	40
ASTM A 335 (P 5)	0.3	0.3	> 26	25
ASTM A 335 (P 9)	0.5	0.5	30	30
SAE 430	0.4	–	37	–

■ From solution analysis; ◆ From polarisation curves

Table 7: Corrosion rates of alloys, mg/cm^2 h [23]

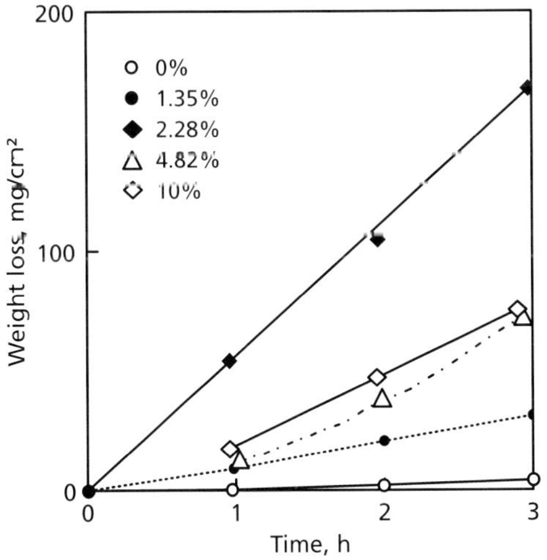

Figure 7: Mass losses of steels with differing chromium contents as a function of the exposure time in 1.7 M HF at 318 K (45 °C) [23]

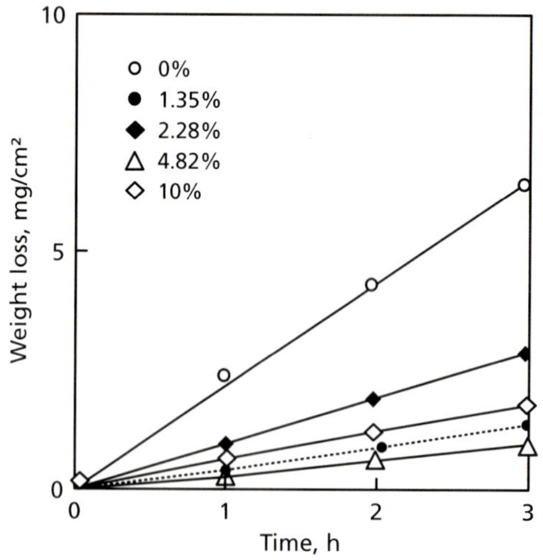

Figure 8: Mass losses of steels with differing chromium contents as a function of the exposure time in 1.4 M HCl at 318 K (45 °C) [23]

Constructional steels with up to 12 % chromium

High-strength tempering steels contain not only an increased amount of carbon, but also chromium alone or in combination with nickel as a significant alloying element. Such tempering steels are used for high-strength screws, for example. Because of the material state (high material strength with limited toughness) and because of the operational loading (transfer of high tensioning forces and superimposed operating forces), these materials are susceptible to hydrogen-induced cracking if there is sufficient atomic hydrogen present in the environment. To prevent this, these screws are usually protected from corrosion by a galvanic coating. However, to ensure sufficient adhesion of these coatings, a bare, active metal surface is necessary and this can be obtained by acid pickling. It is known that, in the event of incorrect pickling and possibly during the subsequent galvanic coating, hydrogen is able to penetrate the high-strength screws in such amounts as to produce irreversible cracking.

This problem is studied in detail in [24]. Table 8 lists the investigated steels. Screws with different surface states were pickled in 15 % HCl at room temperature for different periods of time, and without or after increasing exposure times in creep tests for which the sample was pretensioned to 90 % of its 0.2 % yield strength.

Figure 9 shows the influence of the exposure time on the creep behavior of pickled M8 screws made of tempering steel 35B2 (1.5511) after 8 min pickling time in 15 % HCl at room temperature. As the exposure time increased, there was a continuous increase in the critical tensile strength (tensile stress) R_{mcrit}, below which there was no longer any delayed hydrogen-induced cracking. After 138 h there was no longer a negative effect as a result of continuous hydrogen effusion.

Material	DIN-Mat. No.			C	Si	Mo	P	V	Cr	Ni	Mo	Al	B
35B2	1.5511	standard analysis values	min.	0.32	0.15	0.50	–	–	–	–	–	–	0.005
			max.	0.40	0.40	0.80	0.035	0.035	–	–	–	–	0.008
		actual value		0.37	0.20	0.71	0.019	0.021	0.12	0.03	< 0.01	0.05	0.0023
34Cr4	1.7033	standard analysis values	min.	0.30	0.15	0.60	–	–	0.90	–	–	–	–
			max.	0.37	0.40	0.90	0.035	0.035	1.20	–	–	–	–
		actual value		0.34	0.11	0.82	0.025	0.009	0.92	0.14	0.05	0.02	0.0014
34CrMo4 G34CrMo4	1.7220	standard analysis values	min.	0.30	0.15	0.50	–	–	0.90	–	0.15	–	–
			max.	0.37	0.40	0.80	0.035	0.035	1.20	–	0.30	–	–
		actual value		0.34	0.09	0.71	0.026	0.021	1.00	0.10	0.16	0.026	0.0016
30CrNiMo8	1.6580	standard analysis values	min.	0.26	0.15	0.30	–	–	1.80	1.80	0.30	–	–
			max.	0.33	0.40	0.60	0.035	0.035	2.20	2.20	0.50	–	–
		actual value		0.31	0.34	0.45	0.016	0.019	1.96	1.93	0.37	0.02	< 0.001

Table 8: Chemical composition (mass%) of the investigated tempering steels [24]

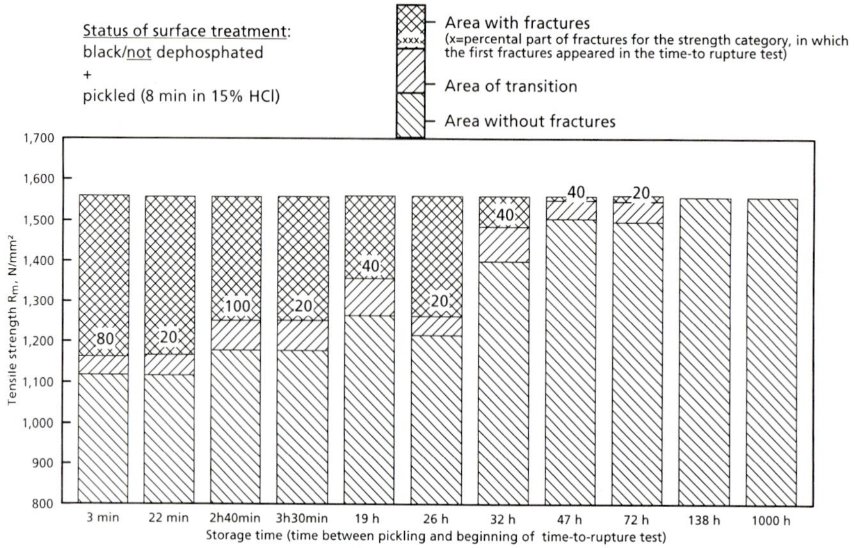

Figure 9: Influence of the exposure time on the creep behavior of pickled M8 screws made of tempering steel 35B2 [24]

The effect of pickling time on the creep behavior of M8 screws made of tempering steel 35B2 is shown in Figure 10. For a pickling time of 1 min, no negative effect of hydrogen can be seen; however, it can be seen above a pickling time of 4 min.

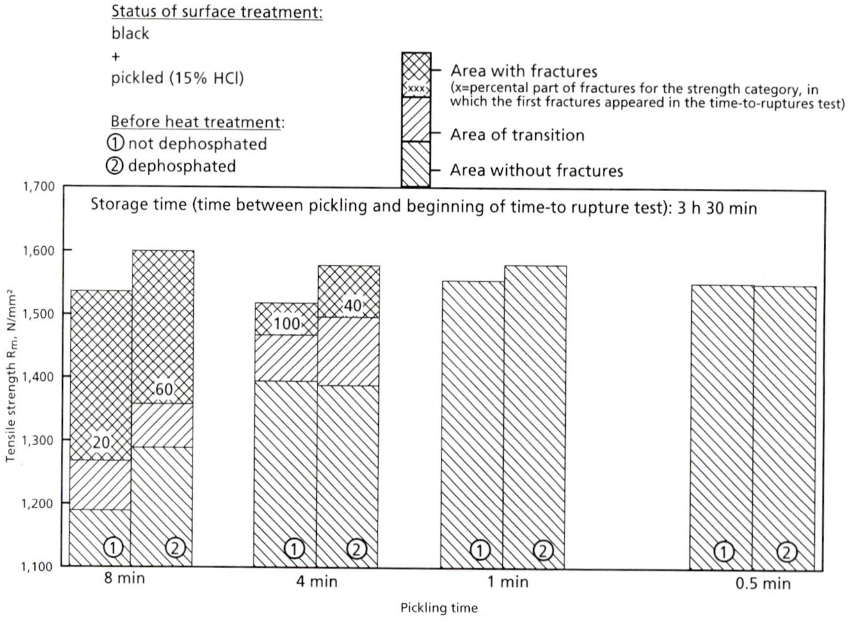

Figure 10: Influence of the pickling time on the creep behavior of pickled M8 screws made of tempering steel 35B2 [24]

To avoid delayed hydrogen-induced crack formation of high-strength screws it is important to keep the pickling time as short as possible and/or to store the screws for a long time before applying a galvanic zinc coating to protect against corrosion because the zinc layer acts as an effusion barrier. However, corrosion protection of high-strength screws is necessary because, depending on the aggressiveness of the surrounding conditions, hydrogen-induced cracking can also occur in rusting screws.

As can be seen from the analysis in Table 8, steel 34Cr3 with 0.009 % S has an extremely low sulphur content. Under otherwise identical test conditions, this steel proved to be the least susceptible to hydrogen-induced cracking. This is attributed to the fact that, for this low content of manganese sulphide, pickling with hydrochloric acid produces hardly any hydrogen sulphide, which is known to act as a promotor for hydrogen absorption into steel.

Ferritic chromium steels with ≥ 13 % Cr

The work in reference [25] reports on experience gained with 17 % ferritic, stainless chromium steels in an oil refinery. Advantages compared to 18/8-CrNi-steels are the lower thermal expansion, the higher thermal conductivity and the insensitivity to transgranular stress corrosion cracking. A disadvantage is the limited resistance to acidic, chloride-containing corrosive media that occur in some sections of the refinery.

Therefore, it appeared promising to use stainless ferritic steels with higher chromium contents that also contained molybdenum, known as superferrites, that were developed in the 1970s. The behavior of the steels, listed in Table 9, that were used as heat-exchanger tubes in various sections of the oil refining process was investigated.

The evaluation of the practical behavior proved this extended application possibility for the high chromium steels. Because the tendency of embrittlement at 748 K (475 °C) increases as the chromium content increases and at even higher temperatures embrittlement increases due to the precipitation of intermetallic phases, its use is limited to temperatures of up to 700 K (427 °C).

The following tables give the results of comparative laboratory tests with austenitic chromium-nickel steels and ferritic chromium steels. Table 10 shows the excellent resistance to pitting corrosion, and Table 11 shows the resistance to crevice corrosion compared to the austenitic chromium-nickel steels X5CrNi18-10 (SAE 304, DIN-Mat. No. 1.4301) and X5CrNiMo17-12-2 (SAE 316, DIN-Mat. No. 1.4401). As expected, the superferrites exhibited low corrosion rates of < 0.15 mm/a (< 6 mpy) in the Huey test (ASTM A 262 – Practice C).

Element/ UNS No.:	SAE 430 (S43000)	SAE 439 [1] (S43025)	SAE 444 [2] (S44400)	26-1S [3] (S44626)	E-BRITE® [4] (S44627)	29-4 (S44700)	29-4-2 (S44800)
Carbon	0.07	0.03	0.02	0.02	0.002	0.005	0.005
Nitrogen	0.025	0.020	0.025	0.025	0.010	0.013	0.013
Chromium	17	18	18	26	26	29	29
Molybdenum	–	–	2	1	1	4	4
Nickel	–	0.20	0.20	0.25	0.10	0.10	2
Manganese	0.45	0.35	0.35	0.30	0.05	0.05	0.05
Silicon	0.45	0.30	0.15	0.30	0.25	0.10	0.10
Phosphorus	0.040	0.020	0.020	0.030	0.010	0.015	0.015
Sulphur	0.010	0.010	0.010	0.010	0.010	0.010	0.010
Titanium	–	0.80	0.30	0.50	–	–	–
Niobium	–	–	0.35	–	0.10	–	–

(1) ASTM XM-8
(2) ASTM 18-2
(3) ASTM XM-33
(4) ASTM XM-27

Table 9: Composition of the stainless steels, mass% [25]

Alloy	U_{SCE}, mV		
	pH 10	pH 6	pH 2
SAE 304	+40	−50	−50
SAE 316	+120	+10	−20
ASTM XM-27	+400	+420	+430
29-4	+1020	+940	+880
29-4-2	+990	+990	+860

Table 10: Critical pitting corrosion potentials of the steels in saturated NaCl solution at 38 °C (311 K) [25]

Alloy	Critical crevice corrosion temperature	
	K	°C
SAE 304	270.5	< −2.5
SAE 316	275.5	2.5
18-2	275.5	2.5
ASTM XM-27	293–298	20–25
Inconel® 625	< 323	< 50
29-4	323	50
29-4-2	323	50
Alloy C	338–347	65–74
Titanium	350	77

Table 11: Resistance of the steels to crevice corrosion [25]

Alloy		1 % HCl		10 % H_2SO_4	
		mm/a	mpy	mm/a	mpy
SAE 304		81	3,189	400	15,748
SAE 316		71	2,795	22	866
SAE 430		1,500	59,055	6,400	251,969
18-2		850	33,465	2,400	94,488
ASTM XM-27	active range	2,000	78,740	3,400	133,858
	passive range	0.7	28		
29-4	active range	500	19,685	1,300	51,181
	passive range	0.2	8		
29-4-2		0.2	8	0.2	8

Table 12: Corrosion rates in a boiling 1 % HCl and 10 % H_2SO_4 solution for austenitic and ferritic steels [25]

Table 12 gives the corrosion rate in a boiling 1 % HCl and 10 % H_2SO_4 solution for austenitic and ferritic steels. With the exception of superferrite 29-4-2, high corrosion rates are exhibited by austenitic CrNi steels, and particularly by the ferritic Cr steels. Therefore, they are not suitable for use in reducing mineral acids, particularly at elevated temperatures. However, steel 29-4-2 was still resistant in 1.5 % HCl and 2.5 % H_2SO_4 from room temperature to the boiling point.

Furthermore, in crevice corrosion tests in acidic chloride-containing solutions, which occur in certain refinery processes, the superferrite 29-4-2 proved to be very resistant after four years in a condenser.

In summary, the behavior of superferrite 29-4-2 is evaluated as follows: on account of the combination of resistance to stress corrosion cracking, chloride corrosion and acid corrosion, this steel is particularly suitable for demanding conditions in heat-exchangers, such as those found in the main systems of petroleum distillation units.

Ferritic/pearlitic-martensitic steels

Reference [26] investigated the relationship between heat-treated state and corrosion or erosion-corrosion behavior of stainless tempering steel X17CrNi16-2 (SAE 431, DIN-Mat. No. 1.4057) in solutions with 0.1 M HCl, 0.05 M H_2SO_4 and 3 % NaCl at 295 K (22 °C). The composition of the steel was 0.14 % C, 0.36 % Si, 0.41 % Mn, 0.011 % P, 0.01 % S, 16.42 % Cr, 2.24 % Ni. After solution annealing for one hour at 1,333 K (1,060 °C) and quenching in oil, the samples were each subjected to two-hour tempering treatment between 473 and 1,023 K (200 and 750 °C) with air quenching.

The measured Vickers microhardness depending on the heat-treatment for the microstructural components martensite, ferrite and residual austenite are shown in Figure 11. Above approximately 823 K (550 °C) there was a steep drop in hardness as a result of martensite decomposition due to the precipitation of chromium carbide. The conversion of residual austenite at 673 K (400 °C) produced a secondary hardening. Due to the chromium carbide precipitation, there was a depletion of chromium in the vicinity of the precipitation centres that greatly reduced the corrosion resistance. Figure 12 shows the mass loss rates in solutions with 0.1 M HCl and 0.05 M H_2SO_4, which indicates the large dependence of the corrosion behavior on the tempering temperature.

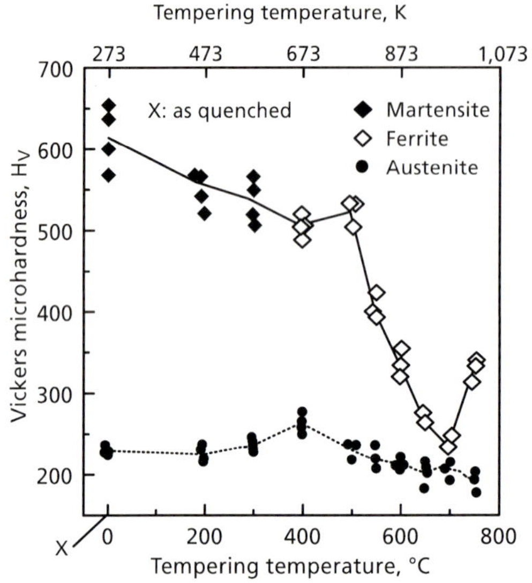

Figure 11: Vickers microhardness depending on the heat treatment for the microstructural components martensite (◆), ferrite (◇) and residual austenite (●) after cooling from 1,333 K (1,060 °C) (x = without further heat treatment) [26]

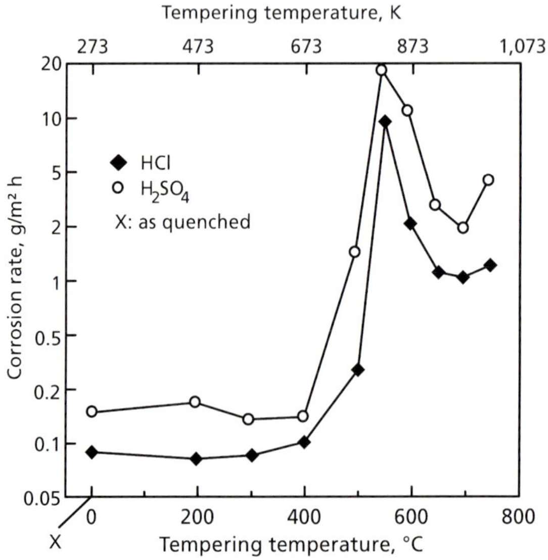

Figure 12: Results of the corrosion tests in 0.05 M H_2SO_4 (○) and 0.1 M HCl (◆) solutions at 295 K (22 °C) after 24 h (x = without further heat treatment) [26]

Figure 13 gives the result of the vibration cavitation test in a 0.1 M HCl solution at a frequency of 20 kHz and an amplitude of 50 µm depending on the test duration. As already observed for the corrosion rate, above approximately 823 K (550 °C) there was a considerable increase in the mass loss which was caused by the combination of a drop in hardness and lowering of the corrosion resistance. A quantitatively similar result was also found in the 0.05 M H_2SO_4 solution. The mass loss in the 3 % NaCl solution was lower by approximately one third.

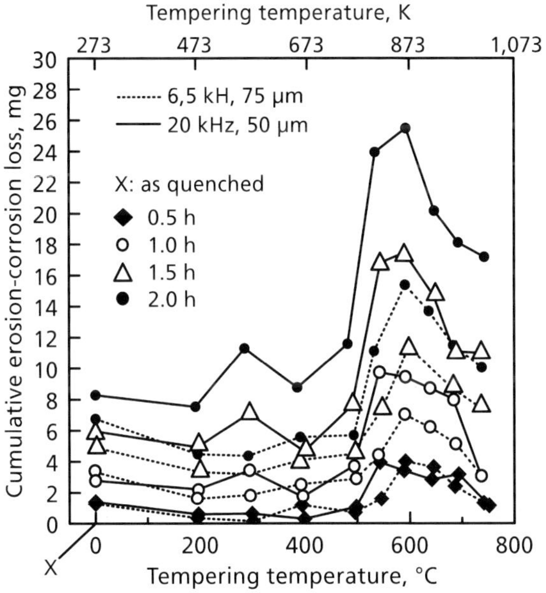

Figure 13: Result of the vibration cavitation test in a 0.1 M HCl solution at a frequency of 20 kHz and an amplitude of 50 µm depending on the test duration after 0.5 (◆), 1.0 (○), 1.5 (△) and 2.0 (●) h (x = without further heat treatment) [26]

Figure 14 shows the mass loss in the vibration cavitation test in the non-corrosive medium ethylene glycol. Because of the secondary hardening caused by the decomposition of residual austenite at 673 K (400 °C), there is a minimum in the mass loss. Above approx. 823 K (550 °C), an increase starts, although it is more than one power of ten less than in the 0.1 M HCl solution. Therefore, there is a strong synergetic effect for erosion corrosion depending on the hardness and corrosion resistance.

The investigated steel contained 0.14 % C, which is somewhat less than the carbon content of 0.15–0.23 % specified for X17CrNi16-2 (SAE 431, DIN-Mat. No. 1.4057). The dependence of the corrosion and erosion-corrosion behavior on the heat treatment would have been even more pronounced if the investigated steel had had a higher carbon content.

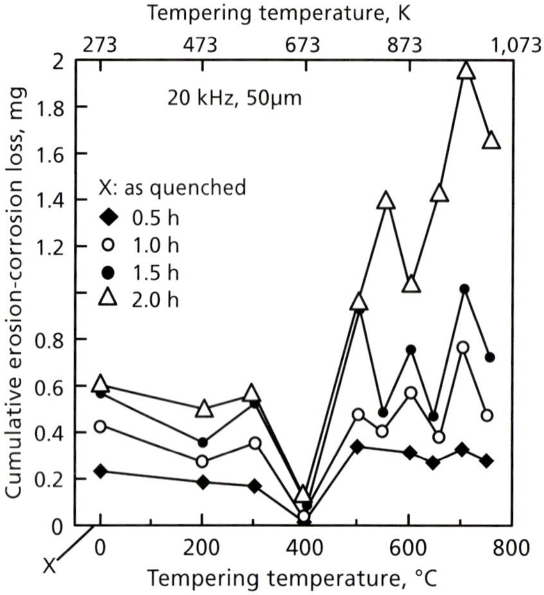

Figure 14: Mass loss in the vibration cavitation test in the non-corrosive medium ethylene glycol after 0.5 (◆), 1.0 (○), 1.5 (●) and 2.0 (△) h [26]

Ferritic-austenitic steels/duplex steels

Since the beginning of the 1970s, stainless duplex steels based on 25Cr-5Ni have been used as pipes in oil boreholes because of their approximately two-fold higher 0.2% yield strength compared to stainless austenitic chromium-nickel steels and because of their good resistance to chloride corrosion (pitting, crevice and stress corrosion cracking). In those cases in which hot concentrated inorganic acids, especially hydrochloric acid, are used to increase oil production, the corrosion resistance of the duplex steel is insufficient. The acid must be inhibited in order to protect the pipes made of this steel from premature failure as a result of corrosion. Since the proven inhibitors for low-alloyed steels are not sufficiently effective for duplex steels, new inhibitors had to be developed [27].

Reference [27] reports on investigations with a steel having the composition 0.03 % C, 0.33 % Si, 1.51 % Mn, 0.013 % P, 0.0045 % S, 23.3 % Cr, 6.0 % Ni, 3.0 % Mo, 0.16 % N (DIN-Mat. No. 1.4462, X2CrNiMoN22-5-3, 2205). The mass loss rates were determined along with the polarisation resistance, which was measured at certain intervals during these tests. The current density/potential curves were recorded in separated tests.

Condensation products of aromatic aldehydes and amines were tested as inhibitors. In order to increase their effectivity, surfactants were added that improved the distribution of the inhibitors and the wettability of the metal surface. Finally, potassium iodide was also tested as an inhibitor. The following pure substances were used:

Aldehydes

trans-cinnamaldehyde (TCA)
benzaldehyde (BAL)
salicylaldehyde (SAL)

Amines

aniline (PHA)
benzylamine (BAM)

Surfactants

N-dodecylpyridine chloride (DDPC)
N-dodecylquinoline bromide (DDQBr)
N-denzylquinoline chlorid (BZQC)

Auxiliary inhibitor

potassium iodide (KI)

The listed aldehydes and amines were tested individually as well as in mixtures for their inhibition efficiency. Their reaction products were obtained after refluxing the mixtures for 3 hours.

	Corrosion rate $mg/cm^2\ h$	**Inhibition efficiency** %
HCl 20 %	144.1	–
+ TCA-PHA	6.55	95.45
+ TCA-BAM	2.65	98.16
+ BAL-PHA	75.7	47.50
+ BAL-BAM	58.8	59.20
+ SAL-PHA	84.2	41.57
+ SAL-BAM	102.8	28.66

Table 13: Corrosion rates of the duplex steel and the inhibition efficiencies for the various combinations of substances with an inhibitor addition level of 0.2 % [27]

	Corrosion rate $mg/cm^2\ h$	**Inhibition efficiency** %
TCA-PHA 0.2 %	6.55	95.45
TCA-PHA 0.2 % + DDPC 0.2 %	2.21	98.47
TCA-PHA 0.2 % + KI 0.2 %	1.28	99.11
TCA-PHA 0.2 % + DDPC 0.2 % + KI 0.2 %	0.21	99.85

Table 14: Corrosion rates of the duplex steel and the inhibition efficiencies in 20 % HCl at 363 K (90 °C) [27]

	Corrosion rate mg/cm² h	*Inhibition efficiency (in HCl) %*
TCA-PHA + DDPC + KI	0.21	99.85
TCA-BAM + DDPC + KI	0.20	99.86
BAL-PHA + DDPC + KI	0.71	99.50
BAL-BAM + DDPC + KI	1.30	99.10
SAL-PHA + DDPC + KI	0.66	99.54
SAL-BAM + DDPC + KI	0.70	99.51

Table 15: Corrosion rates of the duplex steel and the inhibition efficiencies in 20 % HCl at 363 K (90 °C) [27]

	Corrosion rate mg/cm² h	*Inhibition efficiency (in HCl) %*
TCA-PHA 0.4 %	1.98	98.63
TCA-PHA 0.4 % + DDPC 0.2 % + KI 0.2 %	0.093	99.94

Table 16: Corrosion rates of the duplex steel and the inhibition efficiencies with an addition of 0.4 % [27]

	Corrosion rate mg/cm² h	*Inhibition efficiency (in HCl) %*
BzQC 0.2 %	0.88	99.39
BzQC 0.2 % + TCA-PHA 0.4 % + KI 0.2 %	0.20	99.86
DDQBr 0.2 %	59.62	58.60
DDQBr 0.2 % + TCA-PHA 0.4 % + KI 0.2 %	0.10	99.93
DDPC 0.2 %	13.10	91.00
DDPC 0.2 % + TCA-PHA 0.4 % + KI 0.2 %	0.093	99.94

Table 17: Corrosion rates of the duplex steel and the inhibition efficiencies with different additives [27]

Table 13 to 17 give the mass loss rates and the inhibition efficiencies for the different material combinations as well as the additional effects due to the surfactants and the auxiliary inhibitor KI. Efficiencies of > 99 % were obtained for a series of substance combinations with additions of the surfactant DDPC and of KI at an individual level of 0.2 % (Table 15).

Austenitic CrNi steels

The austenitic standard steels (SAE series 300) are already attacked by dilute hydrochloric acid. The isocorrosion curve for a removal rate of 0.1 mm/a (4 mpy), e.g. for the steel SAE 316 L, lies below an acid concentration of 2.5 % at room temperature. These steels can undergo both pitting corrosion as well as stress corrosion cracking in acids whose concentration is not sufficient for general surface corrosion [22].

High-temperature corrosion in combustion gases at elevated temperatures

Vapors of vinyl chloride monomers are extremely hazardous to health and must therefore be made as ineffective as possible. Thermal oxidation is the most suitable method for this. However, this produces HCl and thus creates corrosion problems. If the temperature of the combustion gas is less than 393 to 423 K (120 to 150 °C), then equipment made of carbon steel can be protected against corrosion by means of linings made of rubber, plastics, graphite etc. At higher temperatures and in equipment used for heat transfer, metallic materials have to be used. In reference [28], a laboratory set-up and sample exposure tests in an industrial plant were used to find out whether stainless steels SAE 304 (X5CrNi18-10, DIN-Mat. No. 1.4301) and SAE 316 (X5CrNiMo17-12-2, DIN-Mat. No. 1.4401) are economically viable materials for heat exchangers in such combustion plants.

The samples were evaluated gravimetrically and visually after 50 h. At temperatures of up to 773 K (500 °C) and with 6,000 ppm HCl, steel 1.4301 exhibited very low uniform surface corrosion rates of < 0.1 mm/a (< 4 mpy). The authors concluded that the standard stainless steels 1.4301 and 1.4401 can be used for heat exchangers if the process conditions are adjusted correspondingly. The testing equipment and operating units are described.

In mixtures of hydrochloric acid and nitric acid, the corrosion of materials and also the inhibiting effects are difficult to predict because reactions between the acids must be expected. Thus, at certain quantitative ratios of the acid mixture, it is possible that nitrosyl chloride is produced, which strongly attacks chromium-nickel steels with generation of a lot of heat. In reference [29], the temperature increase of mixtures of HCl-HNO_3 acids was used to characterise the corrosion of the steel SAE 304 (X5CrNi18-10, DIN-Mat. No. 1.4301).

For better reproducibility, samples of the same size were activated before the tests by contact with a zinc anode in 0.5 M H_2SO_4. They were then exposed to a constant quantity of an acid mixture in an open beaker. The temperature profile was measured with a thermometer over the course of one hour.

Figure 15 gives the temperature-time curve in 6 M mixtures of HCl-HNO_3 with differing mixing ratios. The rapid and large temperature increase for the mixture 40 HCl + 60 HNO_3 indicates a particularly high aggressiveness. In mixtures with a ratio of 20 HCl + 80 HNO_3 there was no temperature increase because the steel is passive. Figure 16 shows the result for a constant mixing ratio of 60 : 40 and variable mixture concentrations of 1 to 8 M. As expected, the aggressiveness increases as the concentration of the mixture increases.

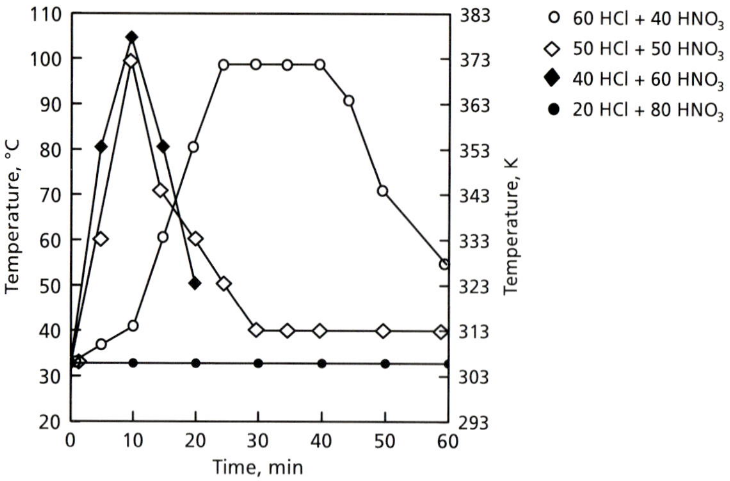

Figure 15: Temperature profile depending on the time for SAE 304 in mixtures of 6 M HCl with 6 M HNO_3 at various mixing ratios (○ 60 HCl : 40 HNO_3, ◇ 50 HCl : 50 HNO_3, ◆ 40 HCl : 60 HNO_3 and ● 20 HCl : 80 HNO_3) [29]

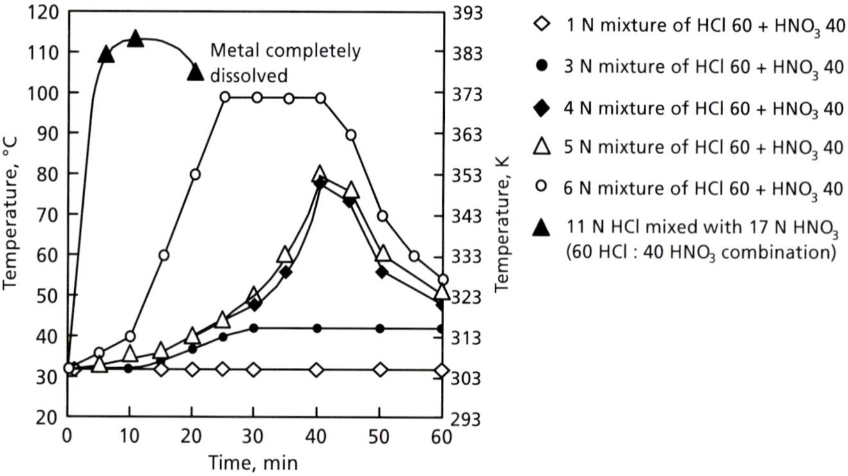

Figure 16: Temperature profile depending on the time for SAE 304 at a constant mixing ratio of 60 HCl : 40 HNO_3 and variable concentration of the mixture from 1 to 8 M [29]

A mixture of concentrated HCl and HNO_3 in a 60 : 40 ratio dissolved the sample with a large temperature increase of the solution after 20 min. At a 6 M mixture concentration and a mixing ratio of 60 HCl : 40 HNO_3, thiourea and hexamines gave a large inhibition (Figure 17 and Figure 18). At the critical ratio of 40 HCl : 60 HNO_3, the addition of hexamine up to 4 % did not produce any inhibition effect, instead it promoted the reaction (Figure 19). Therefore, there are complicated effects for acid mixtures (mixed acids) that can only be elucidated by means of tests.

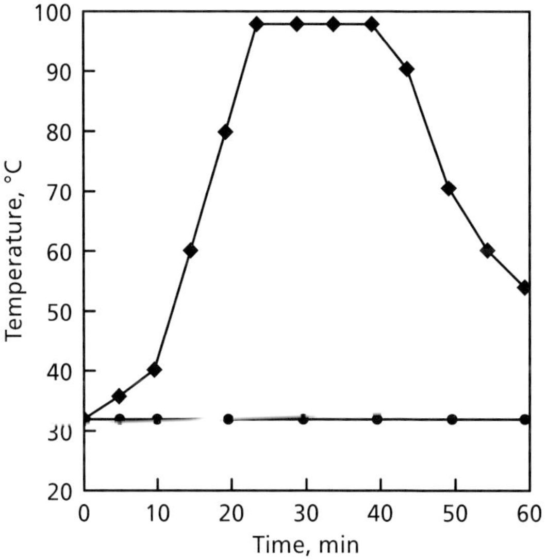

Figure 17: Temperature profile depending on the time for SAE 304 in 6 M mixtures of $HCl-HNO_3$ (60 HCl : 40 HNO_3) with (●) and without addition (◆) of 1 % thiourea [29]

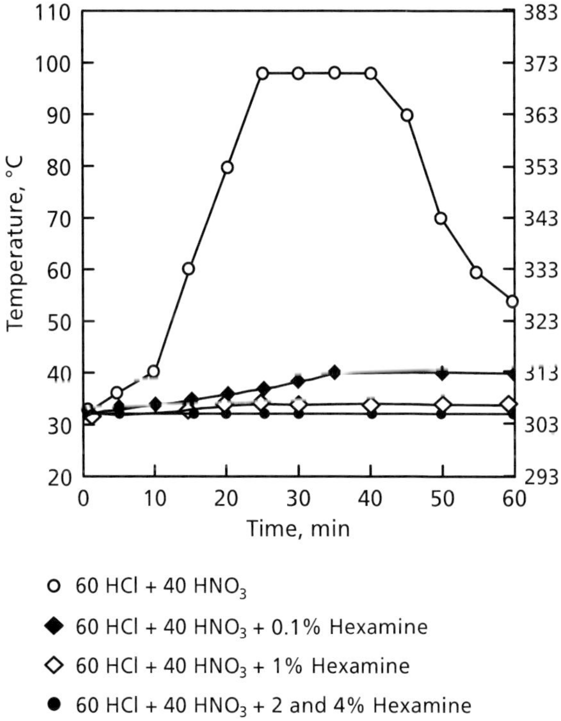

Figure 18: Temperature profile depending on the time for SAE 304 in 6 M mixtures of $HCl-HNO_3$ (60 HCl : 40 HNO_3) with and without addition of various amounts of hexamine [29]

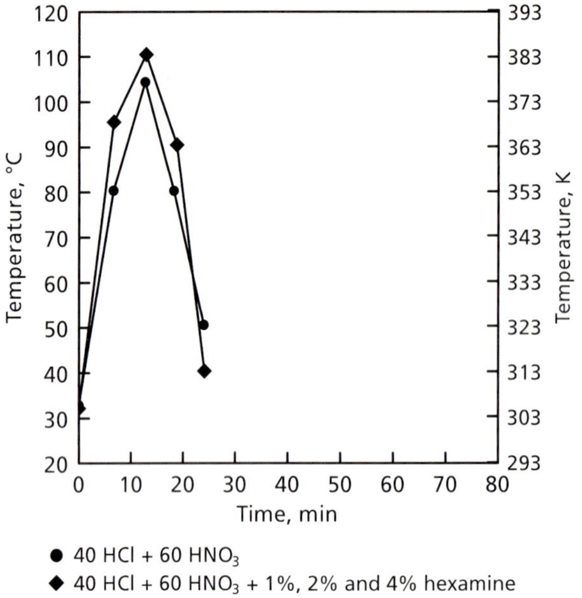

Figure 19: Temperature profile depending on the time for SAE 304 in 6 M mixtures of HCl-HNO_3 (40 HCl : 60 HNO_3) with and without addition of various amounts of hexamine [29]

Atmosphere in a swimming pool hall

The collapse of a ceiling in a swimming pool hall in Ulster/Switzerland in 1985 resulted in twelve fatalities. The ceiling was suspended by wires made of stainless austenitic steel SAE 304 (X5CrNi18-10, DIN-Mat. No. 1.4301) with a diameter of 10 mm [30]. Because of the chlorine-containing atmosphere in the swimming pool hall, there was stress corrosion cracking and this caused the catastrophe. The occurrence of stress corrosion cracking in stainless austenitic steels at elevated temperatures > 353 K (> 80 °C) is well known. The stress corrosion cracking occurs in the passive state without recognizable general corrosive attack. Because the hall temperature was well below this temperature, it was suspected that the stress corrosion cracking occurred in the active state because this is possible from the action of HCl-containing media at ambient temperature. As a result of the accident in Ulster, a group of researchers investigated active stress corrosion cracking of stainless austenitic steels [31].

Table 18 lists the investigated steels. Apart from the basic steel SAE 304 (1.4301), that was damaged in Ulster, more highly alloyed austenitic steels, a semi-austentic steel and a chromium steel (SAE 443, 1.4522) were investigated. Round rods, some of which had been cold-formed, were used as test objects. They had a neck that was electropolished to remove the surface hardening resulting from mechanical processing. The tests were carried out at constant load. The test solutions were mixtures of HCl-NaCl, based on reference [48].

Material		*DIN-Mat. No.*	*C*	*Si*	*Mn*	*P*	*V*	*Cr*	*Mo*	*Ni*	*N*	*Ti*	*Nb*	*Cu*
X2CrNiMnMoNbN23-17-6-3		1.3974	0.022	0.21	6.19	0.02	0.003	23.66	2.79	15.58	0.448	–	0.21	–
X5CrNi18-10	SAE 304	1.4301	0.035	0.71	1.65	0.025	0.019	18.31	0.45	8.79	0.06	–	–	–
X4CrNi18-12	SAE 308	1.4303	0.028	0.42	1.35	0.029	0.002	18.10	0.82	10.80	–	–	–	–
X2CrNiMo18-14-3	cf. SAE 316 L	1.4435	0.017	0.25	1.76	0.019	0.001	17.58	2.82	14.18	0.08	–	–	–
X2CrNiMoN17-13-5	SAE 317 LMN	1.4439	0.015	0.14	1.51	0.020	0.003	17.69	4.19	13.68	0.160	–	–	–
X2CrNiMoN22-5-3	2205	1.4462	0.024	0.44	1.71	0.023	0.003	22.14	3.11	5.43	1.11	–	–	–
X2CrMoNb18-2	SAE 443	1.4522	0.010	0.38	0.38	0.015	0.003	18.30	2.15	0.13	0.008	–	0.32	–
X1NiCrMoCu25-20-5	SAE 904 L	1.4539	0.018	0.10	1.72	0.019	0.003	20.68	4.59	24.44	0.07	–	–	1.35
X6CrNiMoTi17-12-2	SAE 316 Ti	1.4571	0.046	0.44	1.43	0.025	0.003	16.89	2.12	11.54	0.011	0.48	–	–

Table 18: Composition of the investigated steels, mass% [31]

Table 20 gives the result of steel SAE 304 (1.4301) depending on the electrolyte composition for a constant Cl^- ion concentration of 1.5 mol/l. It was found that the endurance of the samples to fracture was very dependent on the HCl content because general attack as well as the stress corrosion cracking are both dependent on it. This is also shown by the results in Table 21. Here, the HCl concentration was varied from 0.01 to 1.0 mol/l in a saturated NaCl solution at 298 K (25 °C).

Composition of the medium mol/l		***Time to fracture h***	***Type of corrosion***		
HCl	***NaCl***		***surface corrosion***	***SCC***	***pitting***
1.5	0	25	yes	yes	no
1.0	0.5	35	yes	yes	no
0.3	1.2	188	yes	yes	no
0.1	1.4	1,011	yes	no	no
0	1.5	> 4,000	no	no	yes

Table 20: Results from steel SAE 304 (1.4301) depending on the composition of the test solution for a constant concentration of Cl^- ions of 1.5 mol/l [31]

HCl concentration mol/l	***Time to fracture h***	***Type of corrosion***		
		surface corrosion	***SCC***	***pitting***
1.0	348		weak	
0.3	507		yes	
0.1	887	yes	yes	no
0.03	2,537		no	
0.01	> 4,000		no	

Table 21: Influence of the HCl concentration on the stress corrosion cracking behavior of steel SAE 304 (1.4301) in a saturated NaCl solution at 298 K (25 °C) for a loading level of 0.2 % of the yield strength [31]

In agreement with earlier publications [49], the appearance of stress corrosion cracking in the active state is most strongly pronounced in a solution with 1.0 mol/l HCl and 0.5 mol/l NaCl. Therefore, it was used as the standard solution in further tests. Figure 20 gives the endurance times in the standard solution at 323 K (50 °C) for the individual 0.2% yield strength of the steels. Figure 21 gives these times for a constant load of 355 N/mm^2 for samples with 30 % cold deformation. The results show that stress corrosion cracking in the active state is always accompanied by a more or less pronounced general corrosion and that the endurance time is largely dependent on the composition of the steel, like passive stress corrosion cracking of these materials (the nickel content is decisive).

The authors conclude that the type of active stress corrosion cracking in HCl-NaCl solutions does not correspond to that observed in the damaged materials at Ulster where the stress corrosion cracks initiated at the corrosion pits; however, in these investigations, this type of corrosion only occurred in the pure NaCl solution without stress corrosion.

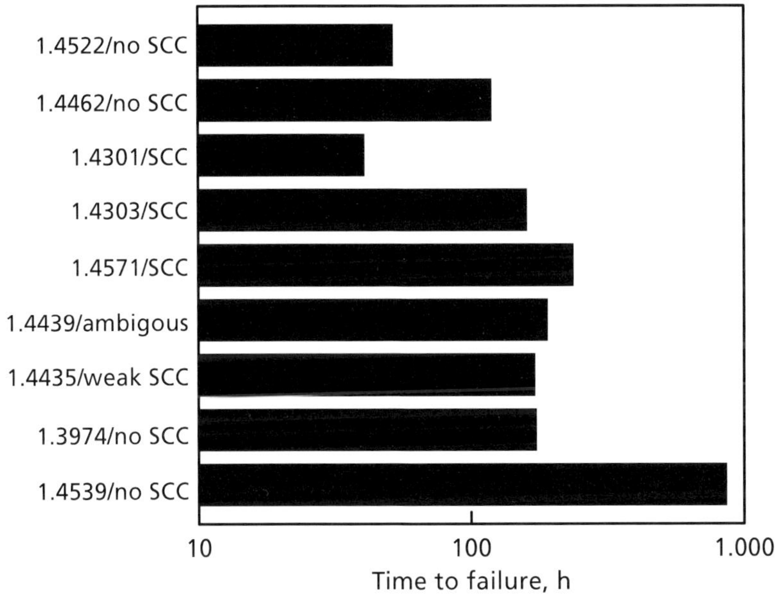

Figure 20: Endurance times in the standard solution at 323 K (50 °C) at the individual 0.2 % yield strength limit of the steels [31]

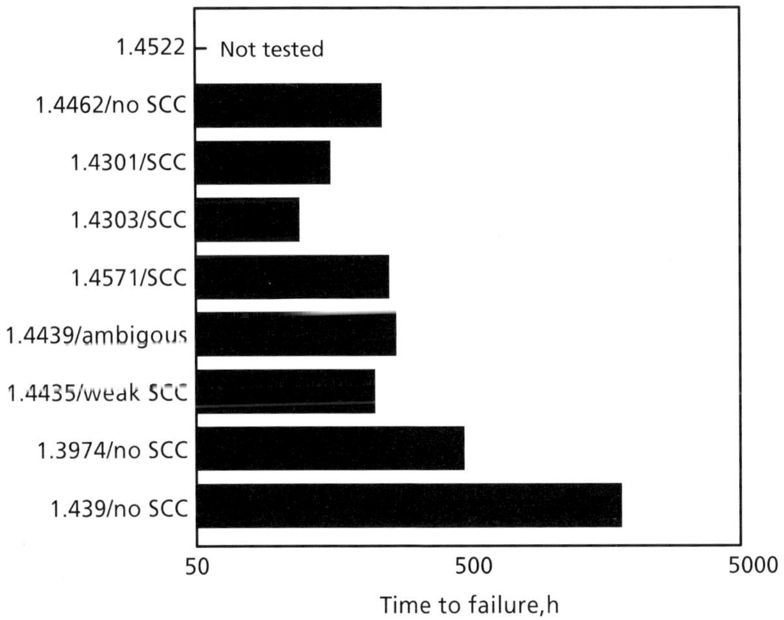

Figure 21: Endurance times of the steels for constant load of 355 N/mm^2, standard solution at 323 K (50 °C), 30 % cold deformation [31]

Measures to avoid the risk of stress corrosion cracking of stainless steels in swimming pool halls include the use of materials that have improved passivation properties or that have a more stable passive layer and increased resistance to pitting corrosion. These include austenitic steels with increased contents of molybdenum and nickel (e.g. DIN-Mat. No. 1.4439 (SAE 317 LMN, X2CrNiMoN17-13-5), 1.4539 (SAE 309 L, X1NiCrMoCu25-20-5), 1.4529 (UNS N08926, X1NiCrMoCuN25-20-7)), a ferritic-austenitic steel DIN-Mat. No. 1.4462 (2205, X2CrNiMoN22-5-3) with enhanced SCC resistance as well as nickel-based alloys [30]. According to Table 10 of the Building Inspectorate Approval Z-30.3-6, only the materials 1.4539, 1.4529, 1.4565 (X2CrNiMnMoNbN25-28-5-4) and 1.4547 (UNS S31254, X1CrNiMoCuN20-18-7) are qualified [32] for use as load-bearing components exposed to atmospheres in a swimming pool hall that are not cleaned regularly and upon which highly aggressive critical deposits can form.

In reference [33], the influence of the pH value on the stress corrosion cracking behavior of stainless austenitic steels 1.4300 (X 12 CrNi18 8) and 1.4301 (SAE 304, X5CrNi18-10) was investigated in the system NaCl-Na_2SO_4 + HCl with the aim of optimising the pH value. The standard solution had a boiling point of 381 K (108 °C), and contained 4.8 M NaCl, 0.15 M Na_2SO_4 and had a pH value of approx. 1 (3 ml conc. HCl/l). The solution was deaerated by continuous purging with nitrogen. The samples were tensile sheet specimens that were loaded to 70 % of their 0.2 % yield strength.

The potential-time profile during the stress corrosion cracking test of a sample of steel SAE 304 (corresp. to DIN-Mat. No. 1.4301, 18.32 % Cr, 8.48 % Ni, 0.19 % Mo, 0.07 % C, 1.68 % Mn, 0.020 % P, 0.004 % S, 0.51 % Si) is given in Figure 22. The

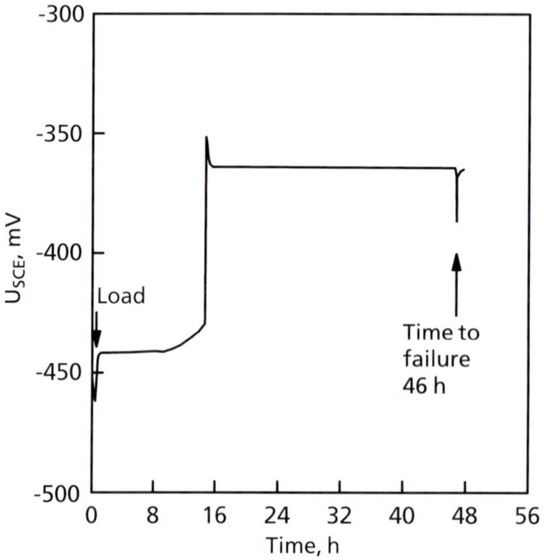

Figure 22: Potential-time profile of stress cracking samples of steel 304 in 4.9 M NaCl, 15 M Na_2SO_4, pH approx. 1 at 380 K (107 °C) [33]

sample was activated during use with a corrosion potential of approx. $U_H = -0.2$ V. When the load was applied there was a short-term drop in the potential, then the active starting potential was maintained over many hours with liberation of hydrogen on the sample. After 16 h and within a period of 3–5 min, the potential rose by approx. 0.2 V, with concomitant suppression of the hydrogen evolution. This indicates that the steel has become more or less passivated. After 46 h, the sample fractured due to stress corrosion cracking. The positive potential jump is a surprising result for a chloride-containing electrolyte; however, it does show that the stress corrosion cracking also occurs in this acid system with formation of an almost continuous protective layer, namely it is in a state that is almost passive and not active. The presence of Na_2SO_4 is probably significant.

Whereas stress corrosion cracking has been investigated in detail for highly concentrated chloride solutions, there are only a few investigations on stress corrosion cracking in relatively weak chloride-containing solutions at elevated temperatures, although it is exactly this type of corrosive conditions that frequently occurs in practice. This problem is dealt with in [34]. U-bending specimens of the materials 1.4301, 1.4306, 1.4401, 1.4406 (see Table 48), Carpenter® 20 Cb-3 (corresp to. DIN-Mat. No. 2.4660), Incoloy® 825 (corresp to. DIN-Mat. No. 2.4858), Haynes® 20 Mod. alloy; Hastelloy® alloy G (corresp to. DIN-Mat. No. 2.4618); Hastelloy® alloy C-276 (corresp to. DIN-Mat. No. 2.4819, 2.4887) were exposed for up to 30 days in an autoclave at 414 K (141 °C) to 4 % NaCl solutions that had been acidified with phosphoric acid, acetic acid and hydrochloric acid to a pH value of ~ 2. None of the investigated materials exhibited stress corrosion cracking in the pure NaCl solution. On the other hand, there was pronounced stress corrosion cracking of all four 18/8CrNi-based steels in all three acidified solutions. In contrast, steels containing higher amounts of nickel and nickel-based alloys Carpenter® 20 Cb-3 to Hastelloy® alloy C-276 were free of stress corrosion cracking. The results confirmed the finding that as the nickel content increases, the resistance to stress corrosion cracking in chloride-containing media is improved. This is also known from tests in highly concentrated, hot chloride solutions.

SAE	DIN-Mat. No.	Designation acc. to DIN EN 10028 T7	Designation acc. to EU 88
304	1.4301	X5CrNi18-10	X6 CrNi 18 10
304 L	1.4306	X2CrNi19-11	X3 CrNi 18 10
316	1.4401	X4CrNiMo17-12-2	X6 CrNiMo 17 12 2
316 LN	1.4406	X2CrNiMoN17-11-2	X3 CrNiMoN 17 12 2

Table 22: Designation of the materials according to various standards

As already known, crack formation in the U samples of the 18/8Cr/Ni steels can be stopped by contact with less noble materials (carbon steel, copper, nickel) and accelerated by contact with more noble materials with higher nickel contents.

The results agree with practical experience according to which pipes made of DIN-Mat. No. 1.4401 (SAE 316) failed in less than 3 months in a hot medium containing 1 to 2 % NaCl and approx. 7 % acetic acid as a result of stress corrosion cracking.

Sulphuric acid in all concentrations is widely used in the chemical industry. The corrosion behavior of stainless steels in pure sulphuric acid solutions has been investigated in detail. However, it is often contaminated with other acids. The effect on the corrosion behavior of materials is difficult to predict. Reference [35] reports an investigation of how the corrosion behavior of steel 1.4301 (SAE 304) changes if small amounts of hydrochloric acid and nitric acid are added to sulphuric acid.

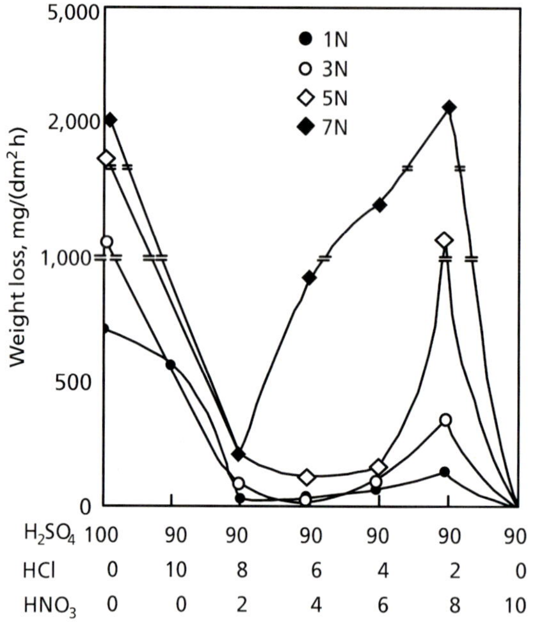

Figure 23: Influence of the acid concentration on the corrosion rate of steel SAE 304, 1 h test duration [35]

Starting from the mass loss rates in pure sulphuric acid solutions of samples that had been previously activated by contact with zinc in a 0.5 M H_2SO_4 solution, the mass loss rates were then determined in the acid mixtures given in Figure 23. The total acid concentration was kept constant between 1 M and 7 M. It can be seen that the corrosion rate decreases up to the mixing ratio H_2SO_4 : HCl : HNO_3 = 90 : 2 : 8, and strongly increases for a ratio 90 : 2 : 8, particularly for the more highly concentrated 5 N and 7 N solutions and drops to zero for all concentrations of the HCl-free composition H_2SO_4 : HNO_3 = 90 : 10, i.e. the steel is passivated in these solutions. The most critical composition was 90 : 2 : 8. The dependence of the corrosion rate on the composition is caused by the interaction between activating and passivating effects of the acids, the result of which cannot be predicted.

Further test results show that corrosion rates in the 1 M acid solution can considerably increase according to the mixing ratio with increasing test time and temperature. Under these test conditions, the highest corrosion rates were found for the mixing ratio H_2SO_4 : HCl : HNO_3 = 90 : 2 : 8.

The work described in [36] concerned stress corrosion cracking (in HCl) of stainless austenitic steels in the active state at ambient temperature. The materials given in Table 23 were tested both as sheets in electrochemical studies and as round tensile specimens in slow tensile tests (CERT test). The test solution contained HCl and NaCl with $c(Cl^-) = 1.5$ mol/l and $c(H^+) = 1$ mol/l. Tests were also carried out with steel 1.4301 in solutions with lower acid concentrations of 0.1 and 0.01 mol/l. The influence of martensite was also tested in separate experiments for steel 1.4301. For this, the specimens were cooled in liquid nitrogen.

No *	DIN-Mat. No. (abbreviation)	SAE	C	Si	Mn	P	V	Cr	Ni	Mo
1a, b	1.4301 (X5CrNi18-10)	304	0.07	1.00	2.00	0.045	0.030	20.0	10.5	–
2a, b	1.4571 (X6CrNiMoTi17-12-2)	316 Ti	0.10	1.00	2.00	0.045	0.030	18.0	13.5	2.5
3a	1.4439 (X2CrNiMoN17-13-5)	317 LMN	0.04	0.41	1.15	0.030	0.018	17.1	13.8	4.15
3b	1.4439	317 LMN	0.01	0.44	0.92	0.020	0.007	16.6	13.2	4.15
4a	1.4558 (X2NiCrAlTi32-20)	N08800	0.07	0.54	1.01	0.012	0.002	20.0	31.1	–
4b	1.4558	N08800	0.01	0.31	0.59	0.009	0.003	22.5	33.1	–

* a) Sheet specimens, b) tensile specimens

Table 23: Designation and chemical composition (mass%) of the investigated materials [36]

The measured current density/potential curves show that martensite has extended the active range somewhat, and has increased activation current density with a shift to more positive potentials.

The strain rate in the CERT test was 10^{-6} 1/s, the samples were elongated until they fractured. At $c(H^+) < 0.1$ mol/l there were no transgranular cracks like those found for steel 1.4301. In the standard solution, steels SAE 304 and 316 (1.4301 and 1.4401) exhibited approximately the same types of corrosion and crack progression. In the martensitic variant of steel SAE 304, in addition to the transgranular cracks running perpendicularly from the surface, there was also corrosion in the longitudinal direction, which is due to attack of the martensite. There are only transgranular notches in steels SAE 317 LMN (1.4439) and 1.4558. The resistance of the materials correlates with that in hot $MgCl_2$ solutions. For these systems, the CERT test does not provide any advantages compared to tests with a constant load.

Comparative studies of stress corrosion cracking using the CERT test and CCT tests in 1 M HCl at ambient temperature on steel SAE 304 (1.4301) were carried out in reference [37]. Under these corrosive conditions and for both type of test, when a relative elongation of 0.006 % of the test length was attained there was a permanent local destruction of the passive layer and cracking started immediately.

Phosphoric acid is the main component in the manufacture of phosphate fertiliser. During manufacture of phosphoric acid, the phosphate rock is dissolved by addition of hydrochloric acid. This leads to contamination with chloride ions, which initiate corrosion damage of stainless steels. Thus, e.g. steel SAE 304 (1.4301) exhibits pitting in a solution of 3.3 M H_3PO_4 + 0.1 M HCl. To avoid corrosion damage in phosphoric acid-chloride systems, guanidines and malonamides are added as inhibitors. In reference [38] some substituted dithiomalonamides were investigated for their inhibition efficiency towards steel 1.4301 in 3.3 M H_3PO_4 + 0.1 M HCl.

The structures of the investigated compounds were:

(I) 1,5-diphenyl-2,4-dithiomalonamide
(II) 1,5-di-p-methylphenyl-2,4-dithiomalonamide
(III) 1,5-dimethoxyphenyl-2,4-dithiomalonamide
(IV) 1,5-di-p-chlorophenyl-2,4-dithiomalonamide

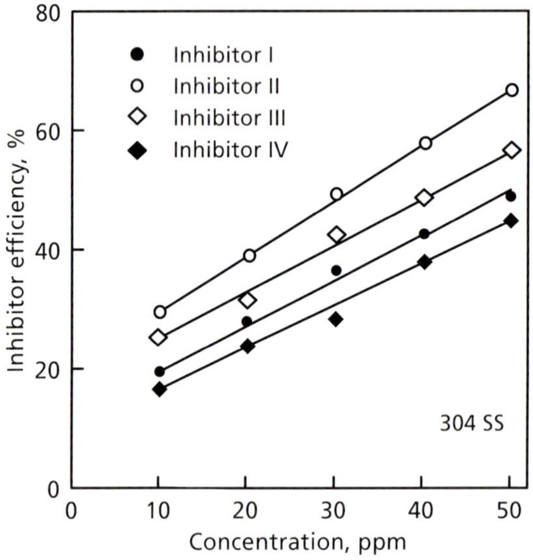

Figure 24: Inhibition action in H_3PO_4/HCl solutions [38]

Potentiodynamic current density/potential curves were recorded at 298 K (25 °C). The corrosion current densities, obtained by extrapolation of the anodic and cathodic polarisation curves, were used to calculate the inhibition efficiency. The results are summarised in Figure 24. All compounds exhibited a linear dependency for additions of 10 to 50 ppm. Compound II had an efficiency of 65 % at the highest addition of 50 ppm. The authors accounted for the differing efficiencies of the compounds on the basis of their structure using Auger electron spectra of the steel samples.

Chemical cleaning in the sugar industry

In former times in the sugar industry, unalloyed steels were used almost exclusively for pipes in evaporator units; however, from 1967 onwards stainless ferritic chromium steels and austenitic chromium-nickel steels were used increasingly as well.

Here too, it is necessary to regularly remove deposits. The most effective results were obtained with dilute hydrochloric acid. Since the steels are active in these solutions and are consumed at a high rate, they cannot be used without inhibitors [39].

Figure 25 demonstrates how the dissolution rate of materials that lie within the active potential range can be reduced in acids by inhibitors. Table 24 gives the results for the unalloyed steel St 35 (SAE 1010, DIN-Mat. No. 1.0308), the ferritic 17% chromium steel SAE 430 (DIN-Mat. No. 1.4016, (X6Cr17) and the austenitic 18/8 CrNi steel SAE 321 (DIN-Mat. No. 1.4541, X6CrNiTi18-10) in 5 % HCl solution with and without addition of 2.5 g/l of the commercially available inhibitors LITHSOLVENT® (Keller and Bohacek, Düsseldorf), that are used in the sugar industry. The mass loss rates were greatly reduced for all steels and temperatures by the three inhibitors used. Thus, at 373 K (100 °C) LITHSOLVENT® EB-II gave the following efficiencies for the three steels:

St 35 (SAE 1010)	99.1 %
1.4016 (SAE 430)	99.5 %
1.4541 (SAE 321)	98.4 %

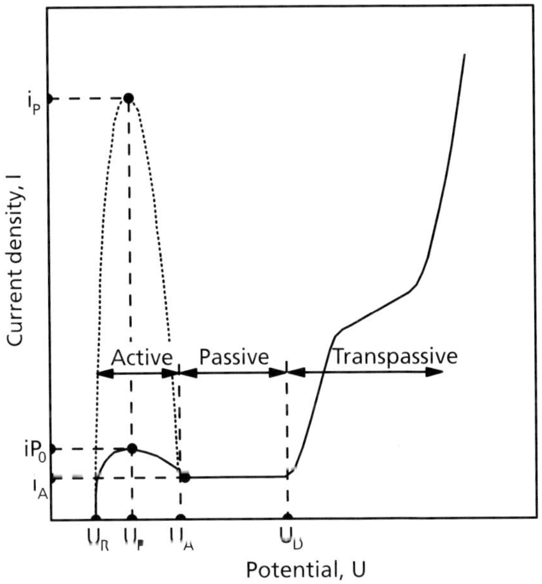

Figure 25: Influence of inhibitors on the corrosion current density in the active range (schematic) [39]
iP_0 = inhibition current density
i_A = passive dissolution current density
i_P = passivating current density
U_R = resting potential
U_P = passivating potential
U_A = activating potential
D_U = breakdown potential

The tests were carried out in static solutions. A decrease in the efficiency must be expected for higher flow rates of the solution. For the relatively low flow rates used in chemical cleaning, however, no great influence on the inhibition effect of the inhibitors can be expected from any movement. In addition, the report gives a detailed description of the execution of chemical cleaning in the sugar industry.

Inhibitor	*Temp. K (°C)*	*Duration h*	*Area-related mass loss rate, g/m² h*		
			St 35 (SAE 1010)	*1.4016 (SAE 430)*	*1.4541 (SAE 321)*
without	313 (40)	1	65	117	1.5
	333 (60)	1	580	440	5.0
	353 (80)	1	800	760	27.0
	373 (100)	1	600	1000	63.0
LITHSOLVENT® EB-II	313 (40)	6	0.36	0.17	0.2
	333 (60)	6	0.80	0.42	0.2
	353 (80)	4	1.5	1.9	0.4
	373 (100)	2	5.0	5.1	1.0
LITHSOLVENT® EB	313 (40)	6	0.5	0.2	0.2
	333 (60)	6	1.3	1.4	0.25
	353 (80)	6	2.2	3.1	0.3
	376 (103)	3	7.5	9.5	3.2
LITHSOLVENT® HEN	293 (20)	6	0.15	0.15	0.04
	313 (40)	6	0.7	0.6	0.15
	333 (60)	6	2.0	1.0	0.32

$0.9\ g/m^2 \times h = 1\ mm/a = 39.4\ mpy$

Table 24: Area-related mass loss rates from chemical cleaning of boiler tubes with hydrochloric acid – with and without addition of inhibitors [39]

Mass loss measurements and current density/potential curves were used to investigate the inhibition efficiency of potassium iodide in hydrochloric acid on the four stainless steels listed in Table 51 [40]. Table 26 gives the results in 1 M HCl at 303 K (30 °C) in dependence on the inhibitor concentration. Except for steel SAE 304 (1.4301), efficiencies of > 93 % were obtained for high and intermediate inhibitor concentrations. Table 27 gives the mass loss rates in 1 M HCl + 0.1 M KI in dependence on the temperature. Even at elevated temperatures, there is a high efficiency. Thus from the mass loss rates for all four steels, an efficiency of > 98 % was calculated at the highest temperature of 353 K (80 °C).

	Corresponds to	*Cr*	*Ni*	*Mo*	*C*
SAE 304	DIN Mat. No. 1.4301	18.20	8.12	–	0.08
SAE 316	DIN Mat. No. 1.4401; 1.4436	16.18	10.14	2 – 3	0.08
SAE 430	DIN Mat. No. 1.4016	14.18	–	–	0.12
SAE 440		16.18	–	0.75	0.60

Table 24: Composition of the investigated stainless steels, mass% [40]

KI M	SAE 304 corrosion rate mg/cm² h	SAE 304 efficiency %	SAE 316 corrosion rate mg/cm² h	SAE 316 efficiency %	SAE 430 corrosion rate mg/cm² h	SAE 430 efficiency %	SAE 440 corrosion rate mg/cm² h	SAE 440 efficiency %
–	0.036	–	0.064	–	0.580	–	0.260	–
0.1	0.011	69.4	0.023	64.0	0.018	96.9	0.015	94.22
10^{-2}	0.012	66.6	0.002	96.8	0.019	96.7	0.031	88.0
5×10^{-3}	0.012	66.6	0.004	93.7	0.036	93.8	0.060	76.9
10^{-3}	0.015	58.3	0.009	85.9	0.140	75.8	0.130	50.0
10^{-4}	0.015	58.3	0.009	85.9	0.230	60.3	0.150	42.3

Table 26: Influence of various potassium iodide (KI) concentrations on the corrosion rate of the stainless steels in 1 M HCl at 303 K (30 °C) [40]

Temperature K (°C)	SAE 304 corrosion rate, mg/cm² h		SAE 316 corrosion rate, mg/cm² h		SAE 430 corrosion rate, mg/cm² h		SAE 440 corrosion rate, mg/cm² h	
	HCl	HCl + KI	HCl	HCl + KI	HCl	HCl + KI	HCl	HCl + KI
303 (30)	0.036	0.011	0.064	0.023	0.58	0.018	0.26	0.015
313 (40)	0.27	0.019	0.250	0.008	3.50	0.040	2.60	0.020
323 (50)	0.40	0.018	0.500	0.016	10.50	0.060	6.20	0.020
333 (60)	0.52	0.020	0.700	0.021	17.50	0.600	14.00	0.060
343 (70)	1.05	0.024	1.500	0.021	32.00	0.200	23.00	0.160
353 (80)	1.53	0.025	2.800	0.022	41.00	0.520	37.00	0.460

Table 27: Influence of the temperature on the corrosion rate of the stainless steels in 1 M HCl with 0.1 M potassium iodide [40]

The determination of the current density from the Tafel lines in the current density/potential curves gave comparable results to those given in Table 26.

Furthermore, the investigations show that the addition of very small quantities of 10^{-4} M KI in combination with other known inhibitors (such as: hexamine, quinoline and thiourea) in 1 M HCl at 303 K (30 °C) gave a considerable improvement in the efficiency for the steels SAE 304, SAE 316 and SAE 430. Therefore, there is a synergistic effect.

As a matter of principle, it is not recommended to use hydrochloric acid, whether with or without inhibitor, to clean components made of stainless austenitic steels.

Decomposition of chlorinated hydrocarbons

Halogenated hydrocarbons are frequently used to clean metallic objects. Anhydrous chlorinated hydrocarbons exhibit a completely inactive corrosion behavior at ambient temperature. In the presence of small quantities of moisture, however, they can be hydrolysed to form HCl and thus become aggressive [41–43].

Table 27 [44] shows how the corrosion rate of steel 1.4541 (SAE 321, X6CrNiTi18-10) is increased by HCl in dichloroethane. Thus the corrosion rate in the dichloroethane-HCl system (the mixture was made with 35 % HCl) with a content of 5 % HCl, is approx. 20 times greater than in an aqueous solution of the same concentration (5.16 : 0.26 mm/a). Table 28 gives the temperature-dependent corrosion rates of steel 3 (unalloyed carbon steel) and the steel 1.4541 in dichloroethane with 0.11 % water. The corrosion rate increases 4- to 5-fold at 293 to 353 K (20 to 80 °C) for differing initial values. At even higher temperatures, the corrosion rate increases much more strongly.

Dichloroethane-HCl			Aqueous HCl solution			0.1% HCl solution in dichloroethane		
proportion of HCl %	corrosion rate mm/a	corrosion rate mpy	proportion of HCl %	corrosion rate mm/a	corrosion rate mpy	proportion of dichloroethane %	corrosion rate mm/a	corrosion rate mpy
0.00	0.004	0.16	–	–	–	11.0	0.066	2.6
0.01	0.015	0.6	0.01	–	–	55.0	0.093	3.7
0.05	0.122	4.8	0.06	–	–	75.0	0.108	4.3
0.10	0.980	38.6	0.11	0.002	0.1	99.7	0.980	38.6
1.00	1.860	73.2	0.33	0.002	0.1	–	–	–
5.00	5.160	203.1	5.00	0.26	10.2	–	–	–

Table 27: Change of the corrosion rate of steel 1.4541 (SAE 321, X6CrNiTi18-10) by HCl in dichloroethane (temperature 293 K (20 °C), test period 360 h) [44]

Temperature, K (°C)	*Steel 3 **	*1.4541*
293 (20)	0.011 (0.43)	0.003 (0.12)
303 (30)	0.011 (0.43)	0.006 (0.24)
313 (40)	0.019 (0.75)	0.006 (0.24)
323 (50)	0.019 (0.75)	0.006 (0.24)
333 (60)	0.028 (1.10)	0.006 (0.24)
343 (70)	0.033 (1.30)	0.011 (0.43)
353 (80)	0.051 (2.01)	0.011 (0.43)

* unalloyed carbon steel

Table 28: Corrosion rate in mm/a (mpy) of the two steels in dichloroethane with 0.11 % water in dependence on the temperature (test duration 12 h) [44]

The cleavage of HCl at elevated temperatures can also initiate pitting corrosion and stress corrosion cracking in stainless steels.

δ-Ferrite attack by inhibited 7–10 % HCl at 393 K (120 °C)

In paper mills, white-liquor clarifiers are used to separate the solid lime sludge ($CaCO_3$) from the white liquor. White liquor is a strong base containing NaOH (approx. 100 g/l), Na_2S (30 g/l), Na_2CO_3 (25 g/l), Na_2SO_5 (10 g/l), $Na_2S_2O_3$ (40 g/l) and undissolved $CaCO_3$ (170 g/l) at 363 K (90 °C). At certain time intervals, it is necessary to clean the filters with dilute 7 to 10 % HCl (inhibited) at 393 K (120 °C) [45]. After cleaning 27 times in a period of seven months, a crack approx. 1 m long appeared in the longitudinal seam of a feed pipe made of austenitic steel 1.4306 (SAE 304 L, GX2CrNiN18-9, X2CrNi19-11) containing 0.024 % C, 18.5 % Cr, 9.3 % Ni. An inspection showed that a shut-off valve was leaking so that acid was able to enter the feed pipe during cleaning. Metallographic investigations showed that there was a pronounced selective corrosion of the δ-ferrite by the inhibited hydrochloric acid and that this had led to the damage. The longitudinal seam had been welded without a filler and thus had the same composition as the pipe material.

The corrosion of the welded material itself was low in circumferential assembly seams welded with a molybdenum containing weld filler (1.7 and 2.6 % Mo). However, on the melt line to the substrate material there was enhanced attack as a result of increased δ-ferrite formation. The boiler itself, made of steel SAE 316 (1.4401), was not attacked except for slight pitting on the air/white liquor interface.

This damage shows that the inhibitor was able to provide sufficient corrosion protection for the austenitic microstructure, but not for the δ-ferrite.

Austenitic CrNiMo(N) and CrNiMoCu(N) steels

The behavior of molybdenum-alloyed austenitic standard steels in hydrochloric acid is essentially the same as that of the molybdenum-free grades. Some of the highly alloyed austenitic CrNiMoN steels (superaustenites), such as 254 SMO (UNS S31254, corresp to. DIN-Mat. No. 1.4547) N08367, N08926, can be used in hydrochloric acids with concentrations below approximately 3 %. Alloy 654 SMO (UNS S32654, corresp to. DIN-Mat. No. 2.4652) with nominal 7.3 % Mo can also be used at room temperature at an acid concentration of approx. 8 %. Figure 26 gives the isocorrosion curves for a corrosion rate of 0.1 mm/a (3.94 mpy) for a series of these materials [22].

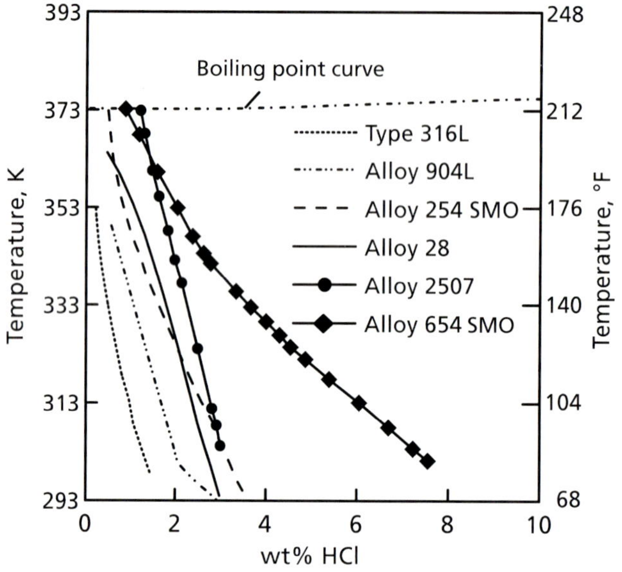

Figure 26: Isocorrosion diagram (0.1 mm/a (3.94 mpy)) for various austenitic steels in hydrochloric acid [22, p. 18]

Special iron-based alloys

In reference [46], the corrosion behavior of an austenitic cast alloy 18Cr-37Ni was investigated in alkaline, acidic and neutral solutions. Such austenitic steels and cast grades are suitable for nuclear power stations because of their enhanced resistance against stress corrosion cracking. The composition of the investigated cast alloy is given in Table 29. It was investigated in NaOH, H_2SO_4, HCl, NaCl and seawater in the cast state and after a homogenisation annealing for 100 and 400 h at 1,173 K (900 °C). The mass loss rates at 289 K (16 °C) and at 373 K (100 °C) are summarised in Table 30 and Table 31. As expected, the corrosion rate in the acid solutions was high. The homogenisation treatment improved the corrosion resistance in these media.

C	Mn	Si	Cr	Ni	Mo	V	P	Fe
0.1	0.71	1.25	19.7	36.6	0.026	0.016	0.010	rest

Table 29: Chemical composition of the investigated alloy

Heat treatment	*NaOH*			*H_2SO_4*			*HCl*			*NaCl*		
	5 %	***10 %***	***15 %***	***5 %***	***10 %***	***15 %***	***5 %***	***10 %***	***15 %***	***5 %***	***10 %***	***15 %***
Initial state	0.025 (0.98)	0.04 (1.57)	0.07 (2.76)	1.075 (42.32)	1.25 (49.21)	2.9 (114.17)	0.85 (33.46)	1.35 (53.15)	1.9 (74.8)	0.05 (1.97)	0.02 (0.79)	0.02 (0.79)
100 h, 1,173 K (900 °C)	0.0085 (0.33)	0.01 (0.39)	0.0225 (0.89)	0.85 (33.46)	1.125 (44.29)	2.9 (114.17)	0.83 (32.68)	1.125 (44.29)	1.2 (47.24)	0.055 (2.17)	0.04 (1.57)	0.04 (1.57)
400 h, 1,173 K (900 °C)	0.0025 (0.1)	0.0075 (0.3)	0.0075 (0.3)	0.65 (25.59)	1.125 (44.29)	2.35 (92.52)	0.475 (18.7)	0.4 (15.75)	0.525 (20.67)	0.005 (0.20)	0.015 (0.59)	0.015 (0.59)

Table 30: Corrosion rate at 289 K (16 °C), mm/a (mpy) [46]

Heat treatment	*H_2SO_4*	*HCl*	*NaCl*	*Seawater*
untreated	1,129.1 (44,453)	53.9 (2,122)	0.26 (10.24)	0.20 (7.87)
100 h, 1,173 K (900 °C)	1,152.8 (45,386)	119.2 (4,693)	0.06 (2.36)	0.0025 (0.10)
400 h, 1,173 K (900 °C)	1,283.4 (50,528)	121.2 (4,772)	0.21 (8.27)	0.005 (0.20)

Table 31: Corrosion rate at 373 K (100 °C), mm/a (mpy) [46]

Bibliography

[1] DECHEMA-WERKSTOFF-TABELLE
Salzsäure
(Hydrochloric Acid) (in German)
DECHEMA e.V., Frankfurt a.M. (1976)

[2] Hanna, F.; Sherbini, G. M.; Barakat, Y.
Commercial fatty acid ethoxylates as corrosion inhibitors for steel in pickling acids
Br. Corros. J. 24 (1989) 4, p. 269–272

[3] Trabanelli, G.; Zucchi, F.; Brunoro, G.; Rochini, G.
Corrosion inhibition of carbon and low alloy steels in hot hydrochloric acid solutions
Br. Corros. J 27 (1992) 3, p. 213–217

[4] Zuchi, F.; Trabanelli, G.; Brunoro, G.
Iron corrosion inhibition in hot 4 M HCl solution by t-cinnamaldehyde and its structure-related compounds
Corros. Sci. 36 (1994) 10, p. 1683–1690

[5] Horvath, T.; Kalman, E.; Kutsan, G.; Rauscher, A.
Corrosion of mild steel in hydrochloric acid solutions containing organophosphonic acids
Br. Corros. J. 29 (1994) 3, p. 215–218

[6] Ajmal, M.; Miden, A. S.; Quraishi, M. A.
2-Hydrazino-6-methyl-benzothiazole as an effective inhibitor for the corrosion of mild steel in acidic solutions
Corros. Sci. 36 (1994) 1, p. 79–84

[7] Rengamani, S.; Vasudevan, T.; Venkatakrishna Iyer, S.
Beta-phenylamine as an inhibitor for corrosion of mild steel in acidic solutions
Indian J. Technol. 31 (1993) 7, p. 519–524

[8] Rengamani, S.; Muralidharan, S. et al.
Inhibiting and accelerating effects of aminophenols on the corrosion and permeation of hydrogen through mild steel in acid solutions
J. Appl. Phys. (London) 24 (1994) 4, p. 355–360

[9] Rengamani, S.; Muralidharan, S.; et al.
Influence of anions on the performance of anisidines as inhibitors for the corrosion of mild steel in acidic solutions
Indian J. Chem. Technol. 1 (1994) 3, p. 168–174

[10] Reimann-Dubbers, V.
Entfernung von Ablagerungen und Korrosionsprodukten aus Rohrleitungen und Anlagenteilen
(Elimination of deposits of corrosion products from pipes and plant components) (in German)
3 R International 24 (1985) 3, p. 133–144

[11] Schwenk, W.
Untersuchungen über die Hydrolyse von C-Cl- und C-S-Bindungen durch rostenden Stahl
(Investigations on hydrolysis of C-Cl- and C-S-bonds by rusting steel) (in German)
Werkst. Korros. 33 (1982) 10, p. 551–553

[12] Vehlow, J.
Korrosion von Metallen in halogenierten Kohlenwasserstoffen
(Corrosion of metals in halogenated hydrocarbons) (in German)
Ergebnisse des Forschungs- und Entwicklungsprogramms "Korrosion und Korrosionsschutz" 1 (1980), p. 175–177
(Results of the research and development program "Corrosion and corrosion protection")
Korrosion verstehen – Korrosionsschäden vermeiden
(ed. Gräfen, H.: Rahmel, A.) vol. 1, p. 267–276
Verlag Irene Kuron, Bonn (1994)

[13] Radovici, O.; Popa, M. V.
The electrochemical study of stress corrosion cracking of some high strength carbon steels
CORROSION-NACE 37 (1981) 8, p. 443–449

[14] Al-Hajji, J. N.; Nawwar, A. M.
Corrosion behavior of molybdenum alloyed low residual carbon steels in HCl solutions
Br. Corros. J. 26 (1991) 2, p. 127–132

[15] Merrick, R. D.; Auerbach, T.
Crude Unit Overhead Corrosion Control
Mater. Performance22 (1983) 9, p. 15–21

[16] French, E. C.; Fahey, W. F.
Water soluble filming inhibitor system for corrosion control in crude unit overheads
Mater. Performance 22 (1983) 9, p. 9–14

[17] Badran, B. M.; Abdel Fattah, A. A.; Abdul Azim, A. A.
New corrosion inhibitors based on fatty materials – I. Epoxidized fatty materials modified with aliphatic amines
Corros. Sci. 22 (1982) 6, p. 513–523

[18] Badran, B. M.; Abdel Fattah, A. A.; Abdul Azim, A. A.
New corrosion inhibitors based on fatty materials – II. Epoxidized fatty materials modified with aromatic amines
Corros. Sci. 22 (1982) 6, p. 525–536

[19] Al-Kharafi, F. M.; Al-Hajjar, F. H.
Control of the corrosion of carbon steel in crude oil distillation units
Br. Corros. J. 25 (1990) 3, p. 209–212

[20] Bresel, Ake
Korrosionsschäden an Metall infolge von Halonen und Fluorkohlenwasserstoffen
(Corrosion damages on metal as a consequence of halones and fluorocarbons) (in German)
Maschinenmarkt, Würzburg 87 (1981) 43, p. 888–890

[21] Axsom, J. F.
On-stream cleaning of refineries pays off
Oil and Gas Journal 71 (1973), p. 106

[22] Materials Selector for Hazardous Chemicals, vol. 3
Hydrochloric Acid, Hydrogen Chloride and Chlorine
MIT Publication MS-3, Elsevier Science, 1999, p. 15

[23] Frignani, A.; Monticelli, C.; Brunoro, G.
Influence of chromium content on corrosion resistance of low alloyed chromium steels in hydrofluoric acid and in other acid environments
Br. Corros. J. 22 (1987) 3, p. 190–194

[24] Kloos, K. H.; Landgrebe, R.; Speckhardt, H.
Untersuchungen zur wasserstoffinduzierten Rißbildung bei hochfesten Schrauben aus Vergütungsstählen
(Investigations on the hydrogen-induced crack formation in high-strength screws of tempering steels) (in German)
VDI-Z 127 (1985) 19, p. S92–S102

[25] Moller, G. E.; Franson, I. A.; Nichol, T. J.
Experience With Ferritic Stainless Steel In Petroleum Refinery Heat Exchangers
Mater. Performance 20 (1981) 4, p. 41–50

[26] Ogino, K.; Hida, A.; Kishima, S.; Kumanomido, S.
Susceptibility of type 431 Stainless Steel to erosion-corrosion by vibration cavitation in corrosive media
Wear 116 (1987) 3, p. 299–307

[27] Zucchi, F.; Trabanelli, G.; Brunoro, G.
Corrosion inhibition of duplex steel in hot HCl solutions
Eur. Congr. Corros. 9 (1989), p. 1–6

[28] Machida, M.; Nakajima, S.
Corrosion prevention for incineration treatment plant of vinyl chloride number
AIChE Meeting Porg., Pachec (1977), p. 1306–1313

[29] Gupta, S.; Kumar, Y.; Sanyal, D. B.; Pandey, G. N.
The exothermic reaction of stainless steel (SAE 304) in mixtures of HCl and HNO_3, and the inhibition of its corrosion
Corros. Prev. Control 30 (1983) 1, p. 11–14

[30] Nürnberger, Ulf
Korrosion und Korrosionsschutz im Bauwesen, 1. ed., vol. 2
(Corrosion and corrosion protection in civil engineering) (in German)
Bauverlag, Wiesbaden und Berlin, 1995, p. 997

[31] Hirschfeld, D.; Busch, H.; Stellfild, I.; Arlt, N.; Michel, E.; Grimme, D.; Steinbeck, G.
Stress corrosion cracking behavior of stainless steels with respect to their use in architecture, part 1: corrosion in the active state
Steel Research 64 (1993) 8/9, p. 461–465

[32] Allgemeine bauaufsichtliche Zulassung
(General permission of the building inspection) Z-30.3-6 of August 3, 1999
Bauteile und Verbindungselemete aus nichtrostenden Stählen
(Components and joining elements of stainless steels) (in German)
Deutsches Institut für Bautechnik, Berlin
Offprint 862 of the Informationsstelle Edelstahl Rostfrei, Düsseldorf 2000

[33] Vaccaro, F. P.; Hehemann, R. F.; Troiano, A. R.
Stress Corrosion Cracking of Austenitic Stainless Steel in an Acidified Chloride Solution
Corrosion-Nace 36 (1980) 10, p. 530–537

[34] Asphahani, A. I.
Effect of Acids on the Stress Corrosion Cracking of Stainless Materials in Dilute Chloride Solutions
Mater. Performance19 (1980) 11, p. 9–14
[35] Vajpeyi, M.; Gupta, S.; Pandey, G. N.
Corrosion of stainless steel (SAE 304) in H_2SO_4 contaminated with HCl and HNO_3
Corros. Prev. Control32 (1985) 5, p. 102–104
[36] Liu, X.; Wu, Y.; Dahl, W.; Schwenk, W.
Untersuchungen der Spannungsrißkorrosion austenitischer Stähle durch saure Chloridlösungen bei niedrigen Temperaturen
(Investigation of stress corrosion cracking of austenitic steels in acid chloride solutions at low temperatures) (in German)
Werkst. Korros. 44 (1993) 5, p. 179–186
[37] Davies, D. E.; Dennison, J. P.; Odeh, A. A.
The assessment of stress corrosion damage in austenitic stainless steel by measurements of cracks formed during constant strain rate and constant load tests in 1 M HCl
Corros. Sci. 24 (1984) 11/12, p. 953–964
[38] Kumar, A.; Singh, M. M.
Substituted dithiomalonamides as inhibitor for the corrosion of SAE 304 SS in phosphoric acid-hydrochloric acid mixture
Anti-Corrosion 40 (1993) 12, p. 4–7
[39] Heppner, H.-J.
Korrosionsverhalten chemisch beständiger Stähle in der Zuckerindustrie unter besonderer Berücksichtigung der chemischen Reinigung
(Corrosion behavior of chemically resistant steels in the sugar industry under special consideration of the chemical cleaning) (in German)
Z. Zuckerind. 25 (1975) 11, p. 622–626
[40] Sanad, S. H.; Ismail, A. A.; Mahmoud, N. A.
Efecto inhibidor del ioduro potasico sobre las corrosion del acero inoxidable en contacto con soluciones de acido clorhidrico
Revista de Metalurgia (Madrid) 28 (1992) 2, p. 89–97
[41] Schober, J.; Sandmann, H.; Kaufmann, W.
Korrosionsverhalten rostfreier Stähle gegenüber halogenhaltigen Lösungsmitteln
(Corrosion behavior of stainless steels in halogen containing solutions) (in German)
Werkst. Korros. 33 (1982) 7, p. 404–410
[42] DECHEMA-WERKSTOFF-TABELLE
32. Ergänzungslieferung
Chlorkohlenwasserstoffe – Chlorethane
(Chlorinated Hydrocarbons – chloroethanes) (in German)
DECHEMA e.V., Frankfurt (1993)
[43] DECHEMA-WERKSTOFF-TABELLE
33. Ergänzungslieferung
Chlorkohlenwasserstoffe – Chlormethane
(Chlorinated Hydrocarbons – chloromethanes) (in German)
DECHEMA e.V., Frankfurt (1993)
[44] Ovchiyan, V. N.; Abalyan, N. P.; Buldukyan, A. M.; Solunkina, G. P.; Dzhanumov, A. N.
Corrosion of steels in acid mixtures based on dichloroethane
The Soviet Chemical Industry 47 (1971) 10, p. 748–749
[45] Karjalainen, L. P.; Suutala, N.
Corrosion Damage in White Liquor Filter Piping – Korrosion in Filterrohren für Weißlauge
Praktische Metallographie 20 (1983) 7, p. 360–364
[46] Ahmad, Z.; Davami, P.
The Corrosion Resistance of 18 Cr-37 Ni Austenitic Steel in Alkaline, Acid and Neutral Solutions
Boshoku Gijutsu 29 (1980) 12, p. 595–601
[47] Heitz, E.; Henkhaus, R.; Rahmel, A.
Korrosionskunde im Experiment: Untersuchungsverfahren-Meßtechnik-Aussagen, 2. edition
(Corrosion science in the experiment: Investigation methods – measurement-technique – conclusions) (in German)
VCH Verlagsgesellschaft, Weinheim (1990)
[48] Yakovleva, L. A.; Vakulenko, L. I.; Vdovenko, I. D.; Lisogor, A. I.; Kalinyuk, N. N.; Novitskaya, G. N.
Corrosion disintegration of α-brass in hydrochloric and sulfamic acid solutions
Ukrainskii Khimicheskii Zhurnal 53 (1987) 7, p. 709–711
[49] Saber, T. M. H.; Tag El Din, A. M. K.
Dibutyl thiourea as corrosion inhibitor for acid washing of multistage flash distillation plant
Br. Corros. J. 27 (1992) 2, p. 139–143

Mixed Acids

Unalloyed and low alloy steels/cast steel

Formerly, unalloyed iron was *used extensively for handling nitrating acid (H_2SO_4 + HNO_3), particularly for storage at room temperature [1]. It was decisive that moisture was avoided and that the steel was in a stress-free state in order to avoid the often observed stress corrosion cracking caused by cold deformation on weldments, and which occurred, for example, in nitrating acid tanks containing 9 parts conc. HNO_3 + 1 part conc. H_2SO_4 after only 14 days [2,3]. Figure 1 shows a resistance diagram of unalloyed iron in mixed acids consisting of HNO_3 + H_2SO_4 + H_2O. It can be seen that higher sulfuric acid concentrations improve the resistance, while it is diminished with increasing amounts of water (> 20% H_2O) [4].*

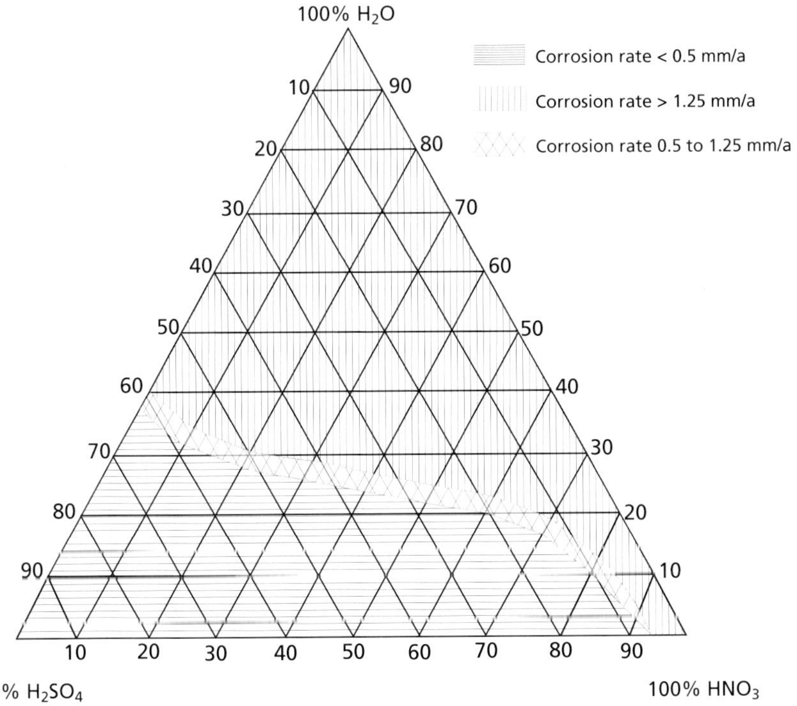

Figure 1: Behavior of unalloyed steel in various mixtures of HNO_3 and H_2SO_4 at room temperature [4]

Steel grade	*Approx. analysis*	*Corres-ponding DIN-Mat. No.*	*Testing time h*	*Corrosion rate at 283–300 K mm/a (mpy)*	*Testing time h*	*Corrosion rate at 393–423 K mm/a (mpy)*
SAE 1020 (cold rolled)	0.20 % C, 0.45 % Mn	1.0402 1.0414	48	0.025 (1)	8	0.60 (23.5)
SAE 303	18 % Cr, 9 % Ni, 0.15 % C	1.4305	48	0.025 (1)	8	0.20 (7.9)
SAE 304	19 % Cr, 9.5 % Ni, 0.08 % C	1.4301	48	0.015 (0.6)	8	2.0 (79)
SAE 321	18 % Cr, 9.5 % Ni, 0.10 % C stabilised with Ti	1.4541	48	0.038 (1.5)	8	2.0 (79)
SAE 347	18 % Cr, 10.5 % Ni, 0.10 % C stabi-lised with Nb	1.4550	48	0.038 (1.5)	8	0.60 (23.5)
SAE 416	13 % Cr, 0.15 % C	1.4005	48	0.008 (0.3)	8	0.08 (3.2)
SAE 4130	0.31 % C, 1.0 % Cr, 0.20 % Mo	1.7218 1.7220	48	0.100 (4)	8	3.8 (150)
SAE 8630	0.30 % C, 0.55 % Ni, 0.50 % Cr, 0.20 % Mo	1.6545	48	0.100 (4)	8	3.8 (150)
SAE 9130	0.10 % C, 3.25 % Ni, 1.23 % Cr, 0.12 % Mo		48	0.100 (4)	8	7.5 (295)
cast iron	3.0–4.5 % C		48	7.5 (295)	8	75 (2,950)
Duriron	14.5 % Si, 0.80 C		48	0.056 (2.2)	8	0.50 (20)

Table 1: Behavior of steels in anhydrous nitrating acid with 88 % HNO_3 (95 %) + 12 % H_2SO_4 (with 20 % SO_3) [5]

It can be seen in Table 1 that all steels, with the exception of grey cast iron, are resistant at room temperature with corrosion rates of < 0.1 mm/a (< 3.94 mpy), but not at relatively high temperatures of 393–423 K (120–150 °C). Here only the ferritic steel SAE 416 with 13 % chromium (0.08 mm/a / 3.15 mpy) as well as SAE 303 (0.2 mm/a) still have good resistance [5]. Grey cast iron exhibits bursting in mixtures of 70 % H_2SO_4 + 30 % HNO_3. In mixtures with 70–95 % H_2SO_4 + 5–30 % HNO_3, corrosion rates between 0.2 and 0.4 mm/a (7.87 and 15.7 mpy)were found at room temperature [6]. It is advantageous to use pearlitic cast iron containing 3 % nickel to heat these mixed acids since this allows a lower wall thickness as a consequence of the higher strength [7].

The most resistant steels in mixtures of 40 % H_2SO_4 + 40 % HNO_3 + 20 % H_2O are those with higher carbon contents, as shown in Table 2 [8]. Further corrosion data on the behavior of unalloyed steel and cast iron are given in Table 3. It can be seen from the values that up to a HNO_3 concentration of 30 %, unalloyed steel has quite good resistance and cast iron has less favorable behavior as long as the mixed acids are anhydrous in both cases. In 5 % HNO_3 + 70 % H_2SO_4 and 25 % water, the corrosion rates rapidly increase; steel is no longer suitable. Only at H_2SO_4 contents of < 10 % and HNO_3 contents of > 70 % does steel show very low corrosion rates again.

% C	Corrosion rates, g/m^2 d
0.95	7.2
0.87	8.0
0.55	8.0
0.35	8.4
0.20	12
0.08	13
0.02	continuous H_2 evolution

Table 2: Influence of the carbon content on the attack of steel in mixtures of 40 % H_2SO_4 + 40 % HNO_3 + 20 % H_2O at 298 K (25 °C) [8]

For electrochemical investigations of the system iron-nitrating acid, see also [9].

Formerly, nitrating acid (H_2SO_4 + HNO_3) was produced in vats made of wrought iron (also of aluminium), which were equipped with a stirrer and a cooling system. Instead of cooling coils in the vat or a cooling jacket, it is better to use external coolers made of silicon cast iron (14.5 % Si) or aluminium. The lid of the mixing vat is made of aluminium because the evolving nitrous gases attack unalloyed steel too strongly. Pumps are made of silicon cast iron with at least 14.5 % Si or of austenitic 18/10 CrNiMo steel (1.4401 etc.) or austenitic special steel [10]

The nitration of toluene is carried out in stirred vats made of wrought iron [11]. Mixtures of 55 % H_2SO_4 + 25 % HNO_3 + 20 % H_2O are used for mononitrotoluene, while mixtures of 65 % H_2SO_4 + 30 % HNO_3 + 5 % H_2O are used for dinitrotoluene; temperatures do not exceed 323 K (50 °C).

Nitrating centrifuges for the manufacture of nitrocellulose are made of steel, or even better of 18/10 CrNi steel. Mixed acids are used that contain 16 to 18 % water and a H_2SO_4 : HNO_3 ratio of 2.5 : 1 [12].

Carbon steel is not resistant in all mixtures of nitric and hydrofluoric acids, even at very low concentrations (1.5 and 0.5 %) [52].

In test mixtures of 10 % HNO_3 + 60 % H_2SO_4, C steel is not resistant at 333 and 353 K (60 and 80 °C) nor in a 50 : 50 mixture at 333 K (60 °C) [52].

Nitrating acid*)			
		corrosion rates, mm/a (mpy)	
% H_2SO_4	% HNO_3	cast iron	steel
95	5	0.25 (9.8)	0.07 (2.8)
90	10	0.25 (9.8)	0.13 (5.1)
85	15	0.16 (6.3)	0.10 (3.9)
80	20	0.15 (5.9)	0.12 (4.7)
75	25	0.14 (5.5)	0.10 (3.9)
70	30	0.14 (5.5)	0.13 (5.1)
72.5	27	–	0.42 (16)
70	5	1.31 (51)	unsuitable
65	10	0.36 (14)	0.75 (30)
60	23.5	–	0.30 (12)
60	15	0.36 (14)	0.46 (18)
55	20	0.31 (12)	0.43 (17)
53	46	–	0.08 (3.1)
50	25	0.30 (11.8)	0.36 (14.2)
45	30	0.29 (11.4)	0.35 (13.8)
10**)	40	unsuitable	unsuitable
10**)	70	–	0.04 (1.6)
1***)	99	–	0.05 (2)
3***)	97	–	0.025 (1)
4***)	96		0.015 (0.6)

*) balance water
**) Steel with 0.03 % C, 0.07 % Si, 0.019 % P, 0.024 % S, 0.059 % Cu, 0.11 % Ni, 0.145 % Mn
***) Steel SAE 1020 (0.18–0.23 % C, 0.30–0.50 % Mn) in smoking HNO_3 with 16 % NO_2 and the given amounts of H_2SO_4

Table 3: Behavior of unalloyed steel and cast iron in various ratios of nitrating acid at room temperature [13]

In a mixture of 50 % nitric acid and 20 % oxalic acid, C steel is not resistant at 343 K (70 °C) [52].

For the pickling of C steel, a mordant is recommended that consists of (275–50) g/l H_2SO_4 + (150–200) g/l HCl + (40–80) g/l Fe^{2+} for a pickling temperature of 323–363 K (50–90 °C) [14]. Noticeable pickling effects are already observed after 1 min. Therefore, steel has absolutely no resistance in this mixed acid.

A low-alloy Si steel with 3.15 % Si is post-treated by a steel manufacturer in mixed acids based on 13 % hydrochloric acid with additional nitric and hydrofluoric acids. Three different mordants were tested in the laboratory: 13 % HCl, 13 % HCl + 2 % HNO_3 and 13 % HCl + 3 % HF, all at 333 K (60 °C) [15]. While all mordants gave a good pickling effect, the mordant with 13 % HCl + 3 % HF is favored and recommended because the attack on the steel, and thus the loss of metal, is the lowest.

The strong attack of C steel by mixtures of sulfuric and hydrochloric acids can be noticeably decreased by inhibitors based on aniline and its derivatives [16]. However, in the case of the composition 3 M HCl + 3 M HNO_3, only an inhibiting effect of approx. 45 to 67 % is to be expected.

Unalloyed cast iron and low-alloy cast iron

The behavior of cast iron in hydrochloric acid can be compared to that of unalloyed steels, so that the information given in Section Unalloyed and low alloyed steels/cast steel can be essentially transferred.

High-alloy cast iron

Silicon cast iron

Silicon cast iron with more than 14.5 % Si has excellent resistance in dilute mixtures of sulfuric and nitric acids, in which the former acid predominates. However, this material should not be used for smoking mixed acids [17]. Distillation columns for waste acid with 5–6 % nitric acid, rest sulfuric acid and water are also made of silicon cast iron, as well as coolers used to recover nitrating material from waste acid [18]. Pumps made of silicon cast iron, in which abrupt temperature changes were avoided, were operated continuously for three years at 333 K (60 °C) in sulfuric and nitric acid mixtures of very different compositions. Acids with 65 % H_2SO_4 + 2 % HNO_3 (rest water) do not have an observable effect on silicon cast iron at 423 K (150 °C). However, in Pauling towers, which operate under these conditions, a material is required which is tailored to the construction [19]. Ventilators made of silicon cast iron for mixed acid vapors were still in a good state after 20 years [17].

Ferritic chromium steels with < 13 % Cr

All ferritic chromium steels are strongly attacked, even in very dilute acid mixtures. High-chromium superferrites exhibit better behavior.

Immersion tests in mixtures of sulfuric and nitric acids as well as phosphoric and nitric acids were carried out with a 9Cr-1Mo steel (cf. DIN-Mat. No. 1.4903) in 1 N HNO_3 with additions of 0.01, 0.1 and 1 N solutions of the two other acids (sulfuric and phosphoric acids) at 308, 333 and 372 K (35, 60 and 99 °C) [20]. As shown in Table 4, the steel is not resistant under all the above conditions, particularly in the

1N solutions, and extremely high corrosion rates were determined at 372 K (99 °C). Additionally, in all samples tested, the region of the weld seam was more strongly attacked than the matrix. The attack in the mixtures of phosphoric and nitric acids was much less than in the mixtures of sulfuric and nitric acids; however, there was no resistance here either because corrosion rates of > 1 mm/a (> 39.4 mpy) were measured, even in very dilute solutions.

Temperature K	*9Cr-1Mo steel (cf. DIN-Mat. No. 1.4903)*	*upper side / HAZ*	*weld root side / HAZ*	*Uniform surface corrosion*
		1.0 M H_2SO_4 + 1 % HNO_3		
308	1,831	2,227	2,784	1,442
333	6,172	8,164	9,105	6,296
372	41,430	4,220	47,860	19,950
		0.1 M H_2SO_4 + 1 % HNO_3		
308	1,552	1,857	1,920	1,242
333	2,848	4,411	4,924	3,380
372	6,367	6,532	6,686	3,464
		0.01 M H_2SO_4 +1 % HNO_3		
308	988	1,008	1,137	771
333	1,288	1,920	2,212	1,644
372	2,473	2,726	2,870	1,790
		1.0 M H_3PO_4 + 1 % HNO_3		
308	162.1	456.0	708.0	236.4
333	188.9	654.0	900.0	282.9
372	220.5	738.0	990.0	325.9
		0.1 M H_3PO_4 + 1 % HNO_3		
308	30.80	108.0	198.0	34.70
333	34.80	126.0	228.0	44.11
372	44.59	192.0	282.0	51.18
		0.01 M H_3PO_4 + 1 % HNO_3		
308	23.57	60.0	126.0	29.99
333	27.56	108.0	150.0	32.94
372	32.43	132.0	204.0	36.47

Table 4: Effect of temperature on corrosion of 9Cr-1Mo steel in sulfuric acid and phosphoric acid based solutions, corrosion rate in g/m^2 d [20]

The oxide layers of only 0.1 μm thickness produced by hot deformation on ferritic steels of type E409 (13Cr-0.01 Ti) must be removed because they affect both the corrosion resistance of components as well as the service life of the presses used for deformation. Therefore, an efficient mordant was developed for a coil annealing line (CAL) with a directly integrated pickling line. This had to be able to carry out the pickling and the passivation processes in the shortest possible time (corresponding to the coil advance speed) [21]. If the ferritic steel is treated electrolytically in approx. 15 % nitric acid, then the oxides are either not completely removed or only after a longer time. However, if hydrochloric acid is added to this mordant, then very good effects are obtained in very short times. The optimum conditions are given as: 100–200 g/l HNO_3; 2–7 g/l HCl at 328 K (55 °C) and with a current density of 25–30 A/dm^2 (under these conditions the oxide layers are not completely removed in the immersion bath i.e. without the use of a current, even after 60 seconds). If the HCl concentration is too high, then over-pickling occurs in the form of large deep pits in the steel itself whilst the oxide layer is not completely removed. This starts above additions of 10 g/l HCl, which is why 5 g/l was specified as the optimum amount for the in-line pickling and passivation process.

Ferritic chromium steels with ≥ 13 % Cr

The resistance diagrams for 17 % chromium steel in mixtures of nitric and sulfuric acids at 333 and 353 K (60 and 80 °C) are given in Figure 2 [22]. The steel can be used in the concentration ranges I and II.

13 % Cr steel is not resistant, even in relatively dilute mixtures of nitric and hydrofluoric acids (see also Table 16) [52]. The same also applies to mixtures of nitric and sulfuric acids. Corrosion rates between 0.1 and 1.0 mm/a (4–40 mpy) were determined in a mixture of 50 % HNO_3 + 50 % H_2SO_4 at 333 K (60 °C) [52].

For the behavior of ferritic chromium steels with 17 % Cr or 17 % Cr + 1 % Mo, see Table 8 [43].

In the case of the ferritic steels, the information given in Table 1 from [5] for the steel SAE 416 is an exception. All ferritic steels with 12 to 18 % Cr, even with additions of Mo, are not regarded to be as resistant towards mixtures of H_2SO_4 and HNO_3, particularly at elevated temperatures. Better behavior is exhibited by cast alloys with very high chromium contents as well as additions of Ni or Al (as shown in Table 5), but, as a rule, only up to 323 K (50 °C) [23]. However, with regard to higher temperatures, this information appears to be too optimistic [24].

For information on the behavior of ferritic stainless steels in mixtures of sulfuric and nitric acids (nitrating acids) see also Table 23.

13 % Cr steel is not resistant, even in relatively dilute mixtures of nitric and hydrofluoric acids (see also Table 21) [52]. The same also applies to mixtures of nitric and sulfuric acids. In a mixture of 50 % HNO_3 + 50 % H_2SO_4 at 333 K (60 °C), corrosion rates between 0.1 and 1.0 mm/a (4 and 40 mpy) were determined (see also Table 20) [52].

Approx. analysis, %	% H_2SO_4	% HNO_3	% H_2O	K	Corrosion rate	
					mpy	mm/a
25-36 Cr, 0.5 C	88.7	4.7	6.6	323	< 4.7	< 0.12
	88.7	4.7	6.6	393	< 47	< 1.2
	77	12	11	323	< 4.7	< 0.12
28-30 Cr	58	40	2	293	< 4.7	< 0.12
	58	40	2	343	< 47	< 1.2
	25	9	64	293	< 4.7	< 0.12
	20	15	65	293	< 4.7	< 0.12
	20	15	65	343	< 47	< 1.2
	17	62	21	353	unsuita-ble	unsuita-ble
	3	3	94	293	< 4.7	< 0.12
	2.5	97.5	0	293	< 4.7	< 0.12
	2.5	97.5	0	343	< 47	< 1.2
28-30 Cr, 2-3 Mo	58	40	2	293–343	< 4.7	< 0.12
	58	40	2	373	< 118	< 3
	20	15	65	293	< 4.7	< 0.12
	20	15	65	343	< 47	< 1.2
	2.5	97.5	0	293–343	< 4.7	< 0.12
	2.5	97.5	0	373	< 47	< 1.2
28 Cr, 10 Ni, 2 Mo	57	9.5	33.5	293–343	< 4.7	< 0.12
28 Cr, some Al	75	25	0	323	< 4.7	< 0.12
	70	10	20	323	< 4.7	< 0.12
	20	15	65	323	< 4.7	< 0.12
	15	5	80	134	unsuita-ble	unsuita-ble

Table 5: Behavior of cast high-alloy chromium steel in mixtures of sulfuric and nitric acids [5]

A superferrite alloyed with ruthenium and containing approx. 29Cr, 4Mo, 2Ni (cf. DIN-Mat. No. 1.4575) exhibited improved corrosion behavior in pure sulfuric acid and in a mixture with phosphoric acid, as long as the concentration of ruthenium was > 0.02 %, preferably 0.2 % [25]. In 10 % boiling H_2SO_4, this superferrite exhibited a corrosion rate of only 0.001 mm/a (0.04 mpy). After exposure in a phosphoric acid plant, the corrosion rate was 0.01 mm/a (0.4 mpy). The composition of the mixed acid was: 42.4 % P_2O_5 (58 % H_3PO_4) + 3 % H_2SO_4 + 1.2 % F^- + 400 mg/l Cl^-, at an operating temperature of 353 K (80 °C).

Very dilute mixed acids can be formed in condensates which arise from the combustion of natural gases [26]. The primary products correspond to the usual gaseous components and other impurities from the combustion: carbonic, nitric, hydrochloric and sulfuric acids. The latter is not only produced by the SO_2 brought in by the industrial air, but also from the natural gas itself, for which a maximum of 120 mg/m^3 is permissible. Sulfuric acid can form from SO_3 in the dew-point region at 393–423 K (120–150 °C) and be concentrated via soot and fly ashes. Hydrochloric acid is introduced in chlorine-containing household products in the combustion air. Unfavorable designs, which, for example, allow draining of the condensate into hot parts of the plant, lead to local increases in the concentration and thus to an increased risk of corrosion, likewise fluctuations close to the dew point. On the other hand, if the temperature falls noticeably below the dew-point, this leads to dilution effects and thus reduces the corrosive attack. Although the acid concentration is in the very low range of 0.1 %, typical pH values of the condensate are 2.5-4. Investigations using the amount of metal ions entering the solution as a function of the acidity of natural gas condensates prove that the corrosion of a ferritic steel based on 18Cr2Mo (1.4521) is primarily determined by sulfuric and nitric acids. According to metallographic examinations, no pitting corrosion was observed in spite of the presence of HCl or chlorides. The amount of metal ions in the condensates of heat exchangers allows an estimation of the corrosion rate with reference to the surface of the heat exchanger. According to this, it is < 0.001 mm/a (< 0.039 mpy). Such low values do not indicate any damage to the components.

Ferritic-austenitic steels/duplex steels

Duplex steels with higher Cr and Mo contents exhibit relatively good resistance in dilute acid mixtures, with the exception of HNO3 + HF.

Nitric and hydrochloric acids

Ferralium® (cf. 1.4507) has absolutely no resistance in dilute hydrochloric acid at 353 K (80 °C). If nitric acid is added to the hydrochloric acid, then the steel becomes corrosion resistant above 5 % HNO_3 and remains so up to a concentration of 60 % HNO_3 (see also Figure 2). This is also applies to a mixtures of 9 % HNO_3 and up to 20 % HCl at a temperature of 325 K (52 °C). The steel is still resistant for additions of up to 16 % HCl [27].

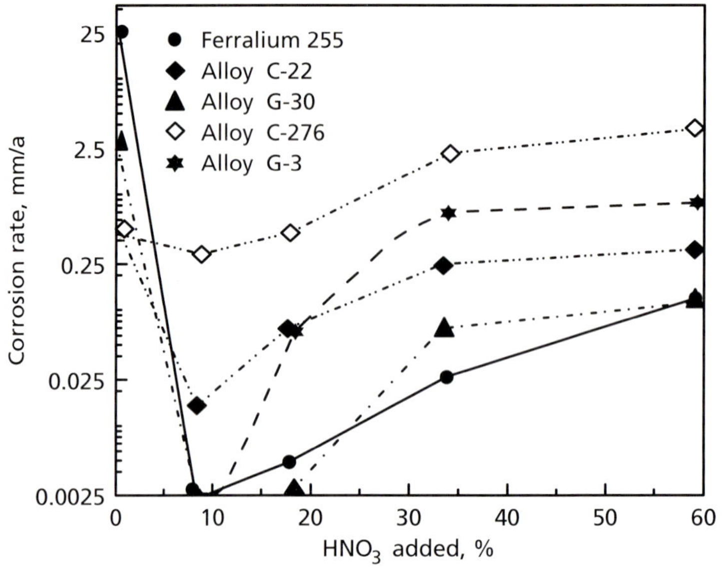

Figure 2: Influence of HNO_3 on the corrosion of various alloys in a 4 % HCl solution at 333 K (80 °C) [??]

Nitric and hydrofluoric acids

In mixed acids with up to 20 % HNO_3 + 5 % HF, duplex steels are not resistant, even at room temperature (see also Table 20).

In dilute nitric acid, the addition of fluorides in the form of hydrofluoric acid and chlorides have a different effect on the resistance of austenitic-ferritic steels. This was demonstrated by investigations in 10 M HNO_3 with 0.01 or 0.1 M/l HF and 5 g/l chlorides at 293 and 323 K (20 and 50 °C) [28]. The generally higher-chromium AF steels are obviously more resistant than austenitic CrNi steels, cf. Table 6 (08Ch22N6T and 03Ch23N6). However, the term "resistant" can only be used for the case of very low HF additions (0.01 M/l). Even in the presence of an HF concentration of 0.1 M/l (which corresponds to approx. 2 g/l), the corrosion rates increased, except for 03Ch23N6, at 293 K (20 °C) to 0.3 up to >1 mm/a (12 up to >40 mpy). The addition of chlorides only slightly increased the corrosion rate. It is interesting that, in some cases, the corrosion rates in the mixed acid were higher without additional chlorides than in the mixed acid (0.01 M/l) with 5 g/l chlorides. This obviously inhibiting effect can be explained by a competing process of chloride adsorption.

Steel	*293 K*				*323 K*			
	10 M HNO_3				*10 M HNO_3*			
	pure	*+ HF*	*+ Cl^-*	*+ HF and Cl^-*	*pure*	*+ HF*	*+ Cl^-*	*+ HF and Cl^-*
12Ch18N10T (cf. DIN-Mat. No. 1.4541)	0.018	0.12[a)]/0.77[b)]	0.014	0.052	0.020	0.48/5.08	0.036	0.16
08Ch22N6T (22 Cr, 6 Ni, Ti)	0.008	0.067/0.47	0.013	0.027	0.022	0.26/1.29	0.037	0.11
08Ch21N6M2T (21 Cr, 6 Ni, 2 Mo, Ti)	0.008	0.060/0.42	0.016	0.038	0.030	0.39/1.18	0.042	0.14
03Ch23N6 (23 Cr, 6 Ni, Mn)	0.010	0.023/0.093	0.011	0.033	0.016	0.16/0.68	0.018	0.087
08Ch18G8N2T (18 Cr, 8 Mn, 2 Ni, Ti)	0.020	0.18/1.16	0.020	0.050	0.034	0.058/7.15	0.058	0.19

Table 6: Mean corrosion rates (mm/a) of steels in 10 M HNO_3, pure and with added hydrofluoric acid (0.01[a)] or 0.1[b)] M/l) and chloride (5 g/l) [28]

Nitric and sulfuric acids

Duplex steels are resistant in 10% HNO_3 + 60% H_2SO_4 up to 353 K (80 °C) (Table 16), likewise in 30% HNO_3 + 40% H_2SO_4 and 50% HNO_3 + 20% H_2SO_4 up to 353 K (80 °C).

In a mordant of sulfuric and hydrofluoric acids (approx. 12% + approx. 3.5%), the corrosive removal was approximately 5 µm after 10 hours at room temperature. This gives a corrosion rate of approx. 5 mm/a (197 mpy), i.e. the steel is not resistant [53].

A mixed acid containing 30% H_3PO_4 and 3% H_2SO_4 along with small amounts of fluoride ions is formed during the production of phosphoric acid. Exposure tests on Ferralium® (cf. 1.4507) in an ageing tank at 339 K (66 °C) gave a corrosive removal of 1.7 mpy or 0.04 mm/a [29].

The duplex steel UR 52 N (25Cr, 6.5 Ni, 3 Mo, 0.18 N and 1.5 Cu) was tested in a phosphoric acid plant with regard to its corrosion resistance and abrasion in comparison with other stainless steels [30]. In the reactor, 98% sulfuric acid at 363 K (90 °C) was added to the phosphate ore to produce an approx. 30% phosphoric acid with approx. 2% residual sulfuric acid as well as impurities of hydrofluoric acid and up to 2,000 ppm chlorides. This corrosive mixed acid is also very abrasive because it contains 30% fine gypsum. In order to determine the resistance under these conditions, samples were attached to stirrers and were tested at a rotation speed of approx. 8 m/s for 7 months.

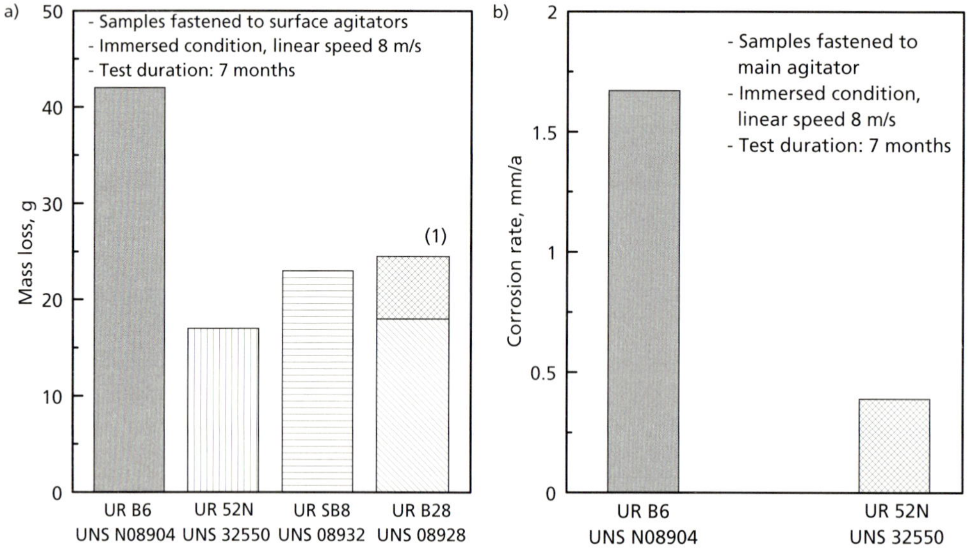

Figure 3: Corrosion tests in a Rhone-Poulenc plant [30]; (1) shorter test period

As shown in Figure 3, the duplex steel UR 52N exhibited the best behavior of all steels, including the steels with higher Mo and Ni contents UR B6 (cf. DIN-Mat. No. 1.4539, X1NiCrMoCu25-20-5) and the high-chromium UR B28 (cf. DIN-Mat. No. 1.4563, X1NiCrMoCu31-27-4). This is attributed to the higher stability and faster regeneration of the passivating layer after abrasion of the solids immediately after impact. Here, the higher chromium content does not appear to be the decisive factor (cf. also the behavior of UR B28 with 28 % Cr). Instead, it appears that nitrogen as well as copper have a positive effect. It can also be seen, however, that the corrosion rate of approx. 0.3 mm/a (12 mpy) is much better than that found for the other steels, particularly for UR B6. Although this value is still too high in the sense of the classical definition of good corrosion resistance, such a corrosion rate is acceptable in industrial practice and in view of the thick dimensions of the stirrers.

In a further study, the behavior of duplex and high-alloy superaustenites in mixed acids containing phosphorus was investigated at several phosphoric acid producers [31]. The composition of the individual process solutions varied, less with regard to the phosphoric acid content, which was 30–31.85 % P_2O_5, but rather with regard to the amount of sulfuric acid (1.4–3.2 %) and chlorides (400–650 mg/l). The content of hydrofluoric acid is not given; however, 1.5 % HF was used in laboratory simulations. Relatively low amounts of hydrofluoric acid can markedly increase the corrosion, as shown in Figure 4, especially in the case of steel UR B6.

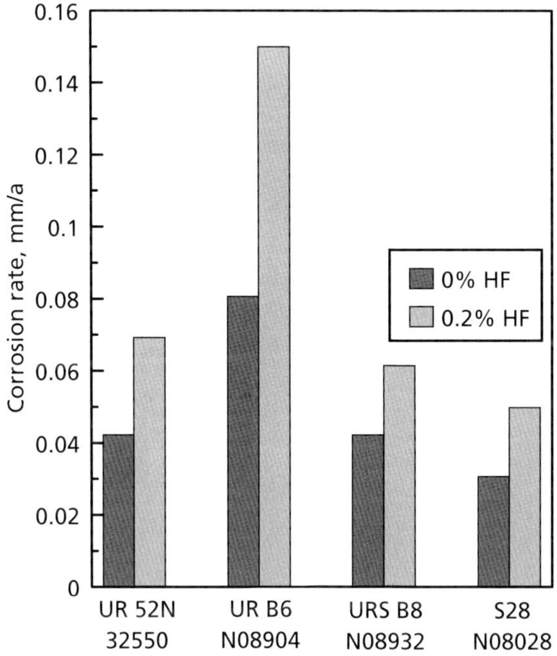

Figure 4: Effect of hydrofluoric acid on the corrosion rate of stainless steels in phosphoric acid at 353 K (80 °C) [31]
(UR 52N: 25 Cr, 6.5 Ni, 3 Mo, 1.5 Cu, 0.17 N; UR SB8: cf. DIN-Mat. No. 1.4529; UR B28: DIN-Mat. No. 1.4563; UR B6: DIN-Mat. No. 1.4539)

The influence of sulfuric acid is similarly pronounced, as shown in Figure 5. In both cases, the relatively high amount of chromium obviously has a positive effect. Even the duplex steel shows a slight increase in the corrosion rate when the sulfuric acid content is doubled. If the temperature is increased to 383 K (110 °C), then the corrosion rate increases by a factor of 3, Figure 6. While the steel UR B6 is no longer recommended for use under these conditions, the remaining high-chromium steels that have a corrosion rate of 0.15 mm/a (6 mpy) still have good resistance, along with the duplex steel. Finally, the influence of the abrasion was tested in the laboratory: the synthetic phosphoric acid solution contained 0.1 mm SiC particles as the abrasive agent.

Figure 7 shows that the corrosion rates are drastically increased and a steel such as UR B6 (DIN-Mat. No. 1.4539) is no longer suitable because the corrosion rate is > 1 mm/a (> 40 mpy). Although the superaustenite UR B8 and the duplex steel UR B52 show much better behavior, the observed values of 0.6 mm/a (24 mpy) were still too high. Since the laboratory conditions were certainly enhanced, the results must be taken as differentiation tests. In practice, the two steels exhibit good abrasion behavior, as confirmed in the practical exposure tests reported in this article (see below).

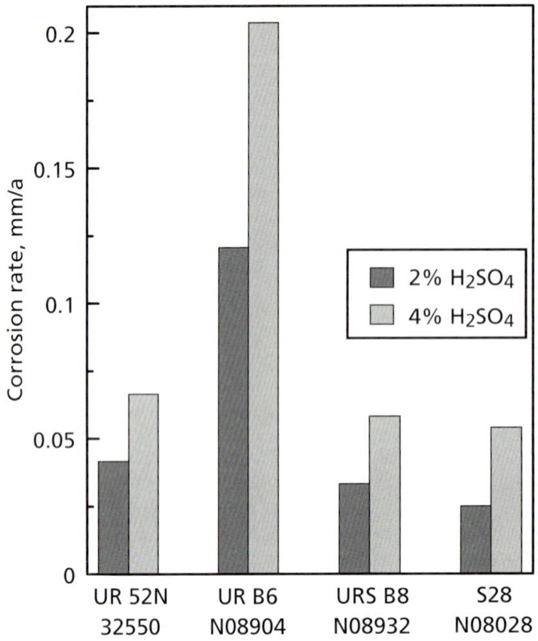

Figure 5: Effect of excess of sulfuric acid on the corrosion rate of stainless steels in phosphoric acid at 353 K (80 °C) [31]

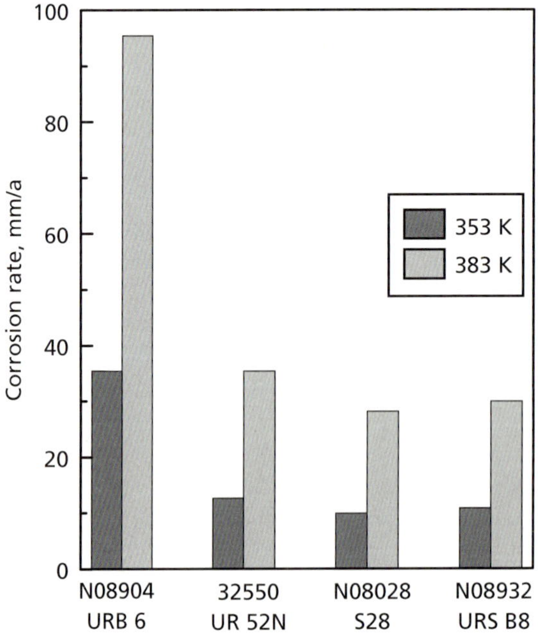

Figure 6: Effect of temperature on the corrosion rate of stainless steels in phosphoric acid [31]

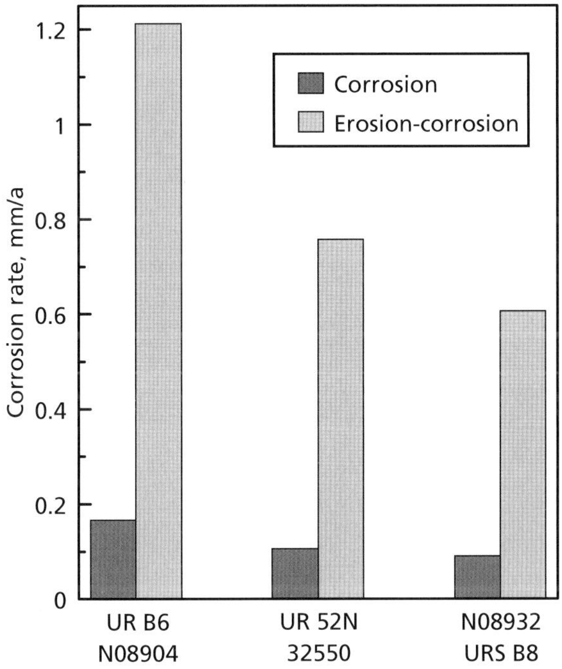

Figure 7: Effect of erosion by solid particles in suspension in phosphoric acid at 353 K (80 °C) [31]

Ferralium® 255 (corresp. to DIN-Mat. No. 1.4492) shows relatively good resistance up to approx. 343 K (70 °C) in a mixed acid of 50 % H_3PO_4 + 50 % HNO_3 (max. corrosion rate approx. 0.5 mm/a / 20 mpy) [32].

Austenitic CrNi steels

The molybdenum-free austenitic steels are only resistant in very dilute mixed acids and at temperatures that are only slightly above room temperature. They are strongly attacked by mixed acids based on HNO_3 + HF, which is why these mixtures are used as mordants. Corrosive attack is also very noticeable in mixtures with hydrochloric acid, which also exhibit pitting corrosion. In sulfuric acid, additions of nitric acid have a corrosion-inhibiting effect, even at higher mixing concentrations.

Nitric and hydrofluoric acids

A 18Cr10Ni steel (X5CrNi18-10, DIN-Mat. No. 1.4301) was tested at 323 K (50 °C) in a constant nitric acid concentration. The corrosion rate increased linearly with increasing amount of added hydrofluoric acid (Figure 9) [33]. The extremely high corrosion rates – the values are in the mm/a range – show that the steel is definitely not resistant. The niobium-stabilised steel X8CrNiNb16-13 (DIN-Mat. No. 1.4961,

SAE 309 SCb) exhibited similar behavior, Figure 8. The steel is only resistant at very low nitric acid and hydrofluoric acid concentrations (max. 2 M or < 0.1 M HF). In the case of constant HF concentrations of 0.05–1 M and at the boiling point of the solutions, the corrosion rate shows a minimum for 0.05–0.2 M nitric acid, and then drastically increases with increasing nitric acid concentration. Corrosion rates of 1 mm/a or >> 10 mm/a (40 or >> 400 mpy) were obtained above 4 M HNO_3. There is a certain resistance only up to 0.1 M HF and 1 M HNO_3 (0.25 mm/a / 10 mpy).

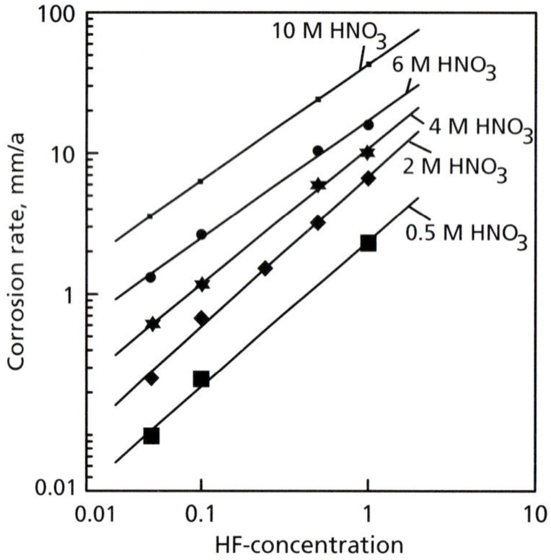

Figure 8: X8CrNiNb16-13 (SAE 309 SCb) in boiling HNO_3-HF mixtures with variable HF concentrations [33]

Figure 10 shows the mass losses at 343 K (70 °C) in a dilute acid mixture as a function of the time [34].

Under these conditions, the corrosion rate was > 10 mm/a (> 400 mpy). In these investigations, it was also found that the corrosion rate in the presence of iron already present in the solution was lower (important for the pickling process). Thus, the corrosion rate in a mixture of 3.5 M HNO_3 + 2 M HF at 323 K (50 °C) is decreased by a half.

In a large-scale trial, Cr/Ni steel pipes were completely descaled after 30 min in a mordant of 12–15 % HNO_3 + 1–3 % HF at max. 333 K (60 °C) after they had been subjected to austenitising annealing at 1,323 K (1,050 °C) for approx. 20 min. In contrast, the pipes of the same composition that were subjected to stabilising annealing at 1,223 K (950 °C)/3 h were not completely free of scale after several hours [35]. In mixtures of HCl and HNO_3, the scale can be completely removed, but this leads to very roughened surfaces and to pickling pores.

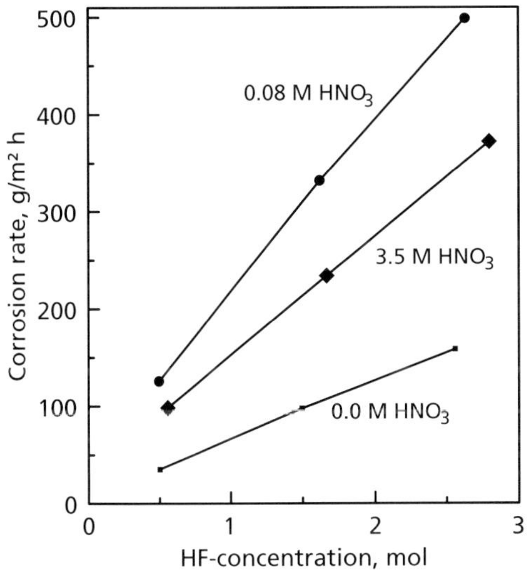

Figure 9: Corrosion of DIN-Mat. No. 1.4301 (SAE 304) at 323 K (50 °C) in HF and HNO_3-HF solutions; test duration 90 min [33]

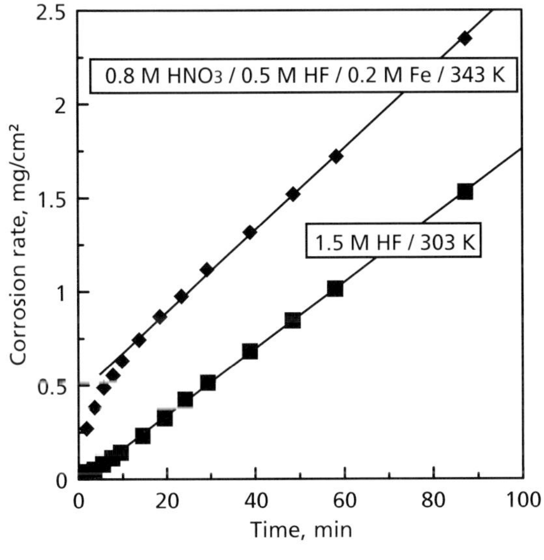

Figure 10: Mass loss as a function of the time for steel SAE 304 (DIN-Mat. No. 1.4301) after exposure to a HNO_3/HF solution [34]

Sulfuric, hydrochloric and nitric acids

Sulfuric acid is used in the production of many organic substances in the chemical industry, but is often contaminated with hydrochloric and nitric acids in the process. With this in mind, the behavior of SAE 304 (DIN-Mat. No. 1.4301) was tested in 1–7 N mixed acids, whereby the proportion of sulfuric acid was always very high, i.e. 90 % [36]. Figure 11 shows that the addition of HCl and HNO_3 decreased the corrosion rate, that is up to a mixing ratio of 90:4:6 – with the exception of the most highly concentrated 7 N acid mixture. The maximum corrosion rate occurred in the mixing range 90:2:8, which is due to the dominating influence of the oxidising nitric acid. As the corrosion rates show, only a certain corrosion resistance was obtained, as a minimum, in 1 N acid mixtures (20 mg/dm^2 h = 2 mm/a = 80 mpy). The steel was not resistant at all other concentrations.

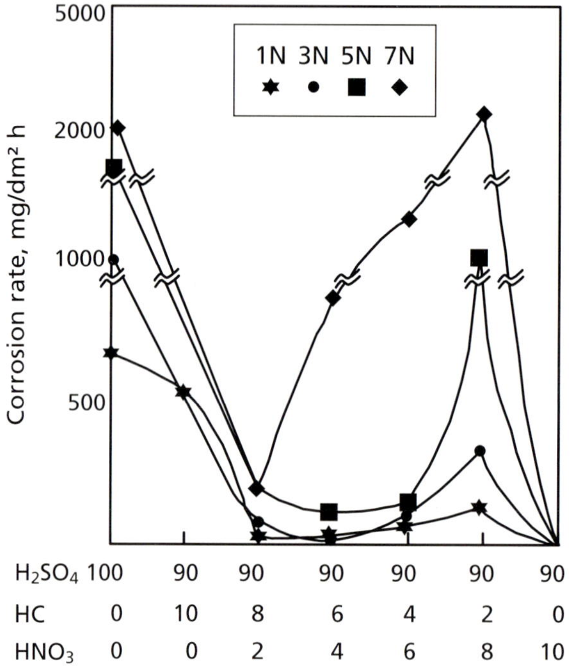

Figure 11: Influence of the acid concentration on the corrosion of SAE 304 (DIN Mat. no.1.4301) [36]

The pronounced influence of the temperature on the corrosion rate is shown in Figure 12.

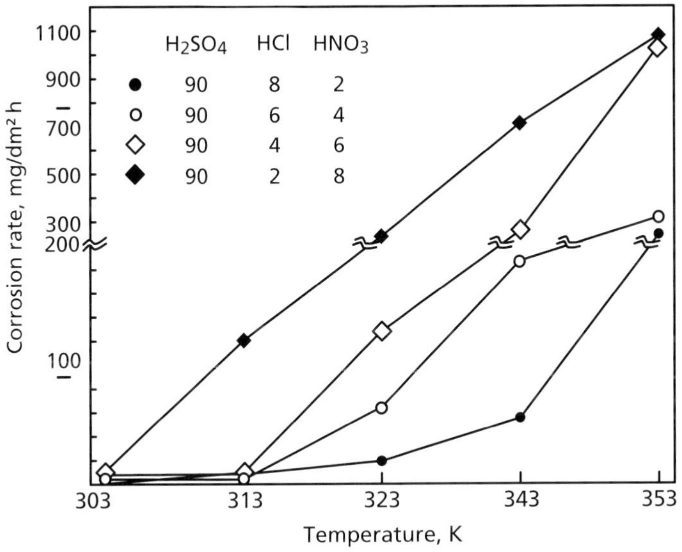

Figure 12: Influence of the temperature on the corrosion of SAE 304 (DIN-Mat. No. 1.4301) in ternary mixtures of HCl, H_2SO_4 and HNO_3 [36]

Nitric and sulfuric acids

At room temperature, all the molybdenum-free austenitic steels are resistant to mixed acids of all mixing ratios; however, considerable corrosion can occur in the vapor phase at moderate temperatures, as shown in Table 7 [37].

Corrosion products, such as iron(III) sulfate and chromium(III) sulfate, as well as admixtures of chlorides (130 mg/l) do not have any influence in a mixture of 85 % HNO_3 + 13.5 % H_2SO_4 at 316 K (43 °C). Also nitrogen and dry air do not have a negative effect on the resistance. Humid air in the already aggressive vapor phase of the mixed acids doubled the corrosion rate [37]. Investigations of steel SAE 304 in boiling mixtures of nitric and sulfuric acids, whose total acidity did not exceed 2 N, showed that the sulfuric acid content had an activating influence and that as the sulfuric acid proportion increased there were four zones of attack: passive zone, transition zone, active and very active zones with corresponding increases in the corrosion rates of 1 to 1,000 g/m² d (0.04 to 40 mm/a) [38].

In the technical nitration process, austenitic steels are used for centrifuges, vats, pumps, valves, pipelines etc. The production of nitro-glycerine is carried out with nitrating acids consisting of 70 % H_2SO_4 + 15 % HNO_3 + 15 % H_2O, as well as 55 % H_2SO_4 + 45 % HNO_3 in equipment made of polished 18/8 CrNi steel ; the pumps and pipelines are also made of austenitic steel. The mirror finish of the tanks used as nitrating vessels, which cannot have any welding seams or stop-cocks, is carried out so that nitro-glycerine (blasting oil) residues cannot adhere anywhere. The separators and scrubbers are also made of austenitic steel. The raw nitro-glycerine still contains approx. 65 % HNO_3 and 0.3 % H_2SO_4. The first scrubber, which operates with a soda solution, has a control lid made of polyethylene [39]. Similarly, all the more highly loaded tanks and equipment used for the nitration of cellulose are

% H_2SO_4	Corrosion rate for a total acidity of						
	90 %	93 %	95 %	97 %	98.5 %	100 %	Phase
0	0.075 (3)	0.05 (2)	0.675 (27)	0.025 (1)	–	8.2 (323)	liquid
3	0.025 (1)	0.025 (1)	0.025 (1)	0.025 (1)	–	0.70 (28)	liquid
6	0.025 (1)	0.025 (1)	0.025 (1)	0.025 (1)	–	0.58 (23)	liquid
13.5	–	–	–	0.025 (1)	0.025 (1)	–	liquid
0	0.025 (1)	0.3 (12)	0.5 (20)	1.8 (71)	–	3.9 (154)	vapor
3	0.05 (2)	0.9 (35)	1.1 (43)	1.2 (47)	–	3.5 (138)	vapor
6	0.025 (1)	0.075 (3)	1.5 (59)	3.1 (122)	–	2.7 (106)	vapor
13.5	–	–	–	1.6 (63)	2.6 (102)	2.4 (95)	vapor
13.5	–	–	–	8.3*) (327)	–	–	vapor
0**)	0.0 (0)	–	0.075 (3)	–	–	1.1 (43)	vapor
3**)	0.0 (0)	–	0.025 (1)	–	–	0.5 (20)	vapor
6**)	0.0 (0)	–	0.05 (2)	–	–	0.5 (20)	vapor

*) Sample sensitised
**) Tests at 298 K (25 °C)

Table 7: Corrosion rates in mm/a (mpy) of Nb-stabilised austenitic CrNi steel (SAE 347, cf. DIN-Mat. No. 1.4550) in mixtures of H_2SO_4 and HNO_3 in the liquid and vapor phases at 316 K (43 °C) after 168 h [37]

made of austenitic CrNi steel. This also applies to the nitration of 2-methylnaphtaline at 273 to 303 K (0 to 30 °C) with a mixed acid consisting of 55 % H_2SO_4 + 25 % HNO_3 + 0 % H_2O and for the production of propyl nitrate from a mixed acid with 68 % H_2SO_4 + 20 % HNO_3 + 2 % H_2O at 273 to 278 K (0 to 5 °C) [40]. The austenitic special steel Worthite® (20 Cr, 24 Ni, Mo) has proven very suitable for particularly loaded components used in the production of picric acid from phenylsulfonic acid at 373 K (100 °C) using a mixed acid of 48 % H_2SO_4 + 37.5 % HNO_3 + 0.5 % oxide + phenylsulfonic acid. After eight nitrations, a corrosion rate of 0.015 mm/a (0.6 mpy) was found [41]. This also applies to other special austenites, such as Carpenter® 20 Cb-3 (cf. DIN-Mat. No. 2.4660), which are always superior to the usual austenitic steels under the same conditions and which are especially suitable for valves and pumps [42].

Laboratory tests were carried out by a stainless steel manufacturer with regard to the behavior of stainless steels in mixtures of sulfuric and nitric acids of very different mixing ratios and at temperatures (from room temperature to > 373 K (> 100 °C)) [43]. The individual evaluations are given in Table 8.

Mixed acids, mass %	Temperature, K	Material			
		CROFER®		CRONIFER®	
		DIN-Mat. No. 1.4016, 1.4510*	DIN-Mat. no.1.4113**	***	****
99 % HNO_3 + 1 % H_2SO_4	293		0	0	0
	323		0	0	0
90 % HNO_3 + 10 % H_2SO_4	293		0	0	0
	323		0	0	0
50 % HNO_3 + 50 % H_2SO_4	293		0	0	0
	323	2	1	0	0
	363	3	2	1	1
	393	3	3	2	2
38 % HNO_3 + 60 % H_2SO_4 + 2 % H_2O	293		0	0	0
	323	3	1	0	0
25 % HNO_3 + 75 % H_2SO_4	323	2		1	0
	363	3		1	1
	430	3		3	3
15 % HNO_3 + 20 % H_2SO_4 + 65 % H_2O	293		0	0	0
	323	3	1	0	0
	353	3	2	1	0
10 % HNO_3 + 70 % H_2SO_4 + 20 % H_2O	323	3	1	0	0
	363	3	3	1	0
	441	3	3	3	3
5 % HNO_3 + 30 % H_2SO_4 + 65 % H_2O	363	3	1	0	0
	383	3	3	0	0
5 % HNO_3 + 15 % H_2SO_4 + 80 % H_2O	407	3	2	1	1
3 % HNO_3 + 30 % H_2SO_4 + 67 % H_2O	293		0	0	0
	323		1	0	0
1 % HNO_3 + 2 % H_2SO_4 + 97 % H_2O	boiling	3		2	0

* = CROFER® 1700 / 1700Ti
** = CROFER® 1701
*** = CRONIFER® materials 1808 (~ 1.4300), 1809 (~ 1.4301), 1809 NC (~ 1.4306), 1809 NC N (~ 1.4311), 1809 Ti (~ 1.4541), 1809 Nb (~ 1.4550)
**** = CRONIFER® materials 1810 (~ 1.4401), 1810 NC (~ 1.4404), 1810 NC N (~ 1.4406), 1810 Ti (~ 1.4571), 1810 Nb (~ 1.4580), 1812 (~ 1.4436), 1812 NC (~ 1.4435), 1812 NC N (~ 1.4429), 1812 Ti (~ 1.4573), 1812 Nb (~ 1.4583),1820 Ti (~ 1.4506), 1820 Nb (~ 1.4505)

Table 8: Corrosion resistance of stainless steels in mixtures of sulfuric and nitric acids as well as water [43].
Corrosion rates: 0 = < 0.11 mm/a; 1 = 0.11–1.1 mm/a; 2 = 1.1–11 mm/a; 3 = > 11 mm/a
Corrosion rates: 0 = < 4.3 mpy; 1 = 4.3–43 mpy; 2 = 43–433 mpy; 3 = > 433 mpy

Another manufacturer has provided similar data for nitrating acids (cf. Table 23).

In the search for an optimum mixed acid to dissolve plutonium from waste and fragments of ceramic reactor fuel elements, a stainless steel of type SAE 304 was considered as a possible vessel material [44]. Correspondingly, the material was investigated in mixed acids in the concentration range 0–100 % at 323, 348 and 373 K (50, 75 and 100 °C). The corrosion rates were determined gravimetrically and also by atomic absorption spectrometry of the alloying components Fe, Cr and Ni that had been dissolved.

Composition HNO_3 / H_2SO_4		Exposure time h	mg/dm^2	Iron (dissolved) mg	Chromium (dissolved) mg	Nickel (dissolved) mg	Corrosion rate mpy	Corrosion rate (mm/a)
				323 K (50 °C)				
0	100	25	3.45				18.30	0.465
		50	6.05				16.00	0.406
		75	8.64				15.25	0.387
		100	10.36	1.79	0.13	0.37	13.73	0.349
10	90	25	13.38				70.94	1.80
		50	22.45				59.50	1.51
		75	29.36				51.86	1.32
		100	37.91	6.24	1.29	0.31	50.23	1.28
15	85	25	18.41	3.43	0.28	0.48	96.12	2.44
		50	28.07	4.73	0.95	0.712	74.38	1.89
		75	39.73	6.48	1.58	0.95	70.17	1.78
		100	51.42	8.04	2.24	1.29	68.14	1.73
20	80	25	20.30	3.99	0.09	0.56	107.60	2.73
		50	36.71	6.78	0.64	0.91	97.26	2.47
		75	44.91	7.75	1.40	1.04	79.33	2.02
		100	57.44	9.34	2.53	1.17	76.09	1.93
35	65	25	17.27				91.54	2.33
		50	30.23				80.10	2.04
		75	41.03				72.46	1.84
		100	48.37	7.92	2.42	0.69	64.09	1.63
50	50	25	16.41	2.81	0.48	0.44	86.96	2.21
		50	25.48	4.13	1.09	0.59	67.51	1.72
		75	34.17	5.35	1.67	0.73	60.26	1.53
		100	34.96	5.37	2.15	0.76	46.33	1.18
65	35	25	6.48				35.76	0.908
		50	12.96				34.33	0.872
		75	18.14				32.04	0.814
		100	22.46	3.32	1.41	0.37	29.75	0.756

Table 9: Results of corrosion tests of SAE 304 in mixtures of nitric and sulfuric acids [44]

Table 9: Continued

Composition HNO$_3$ / H$_2$SO$_4$		*Exposure time h*	*mg/dm^2*	*Iron (dissolved) mg*	*Chromium (dissolved) mg*	*Nickel (dissolved) mg*	*Corrosion rate mpy*	*Corrosion rate (mm/a)*
				323 K (50 °C)				
90	10	25	3.88	0.67	0.13	0.08	21.45	0.545
		50	4.32	0.74	0.15	0.09	11.44	0.291
		75	4.75				8.39	0.213
		100	5.18	0.79	0.28	0.13	6.87	0.175
100	0	25	118.30				627.00	15.93
		50	210.30	32.68	11.01	4.09	557.00	14.15
		75	279.80	41.67	17.17	4.67	494.30	12.55
		100	323.90	46.20	22.05	5.18	427.10	10.85
				348 K (75 °C)				
0	100	25	10.36	1.99	0.06	0.31	54.92	1.40
		50	19.87				52.63	1.34
		75	26.34				46.53	1.18
		100	30.23	5.12	0.896	0.85	40.05	1.02
10	90	25	70.39				373.00	9.47
		50	127.40				337.50	8.57
		75	154.60				273.10	6.94
		100	187.90	31.06	7.83	3.61	248.80	6.32
15	85	25	80.32	15.38	0.86	1.99	425.70	10.81
		50	157.60	28.76	3.54	3.50	417.60	10.61
		75	217.20	36.52	7.65	5.13	383.70	9.75
		100	260.60	41.43	12.12	5.33	345.30	8.77
20	80	25	95.87	18.25	1.22	2.39	508.00	12.90
		50	189.20	32.41	6.04	4.51	501.20	12.73
		75	242.70	38.10	10.39	5.34	428.60	10.89
		100	279.40	44.64	12.94	5.69	370.20	9.40
35	65	25	84.21				446.30	11.34
		50	160.60				425.70	10.81
		75	215.90				381.40	9.69
		100	247.70	39.94	12.20	4.30	328.20	8.34
50	50	25	82.47	16.01	0.94	2.10	437.00	11.10
		50	128.70	21.69	3.99	3.70	340.90	8.66
		75	162.80	26.65	5.01	4.29	287.60	7.31
		100	182.80	27.45	9.18	4.79	242.30	6.15
65	35	25	32.39	5.32	0.83	1.25	171.60	4.36
		50	50.58				149.90	3.81
		75	73.41				129.70	3.29
		100	88.09	15.89	4.67	2.57	116.70	2.96

Table 9: Results of corrosion tests of SAE 304 in mixtures of nitric and sulfuric acids [44]

Table 9: Continued

Composition HNO_3 / H_2SO_4		Exposure time h	mg/dm²	Iron (dissolved) mg	Chromium (dissolved) mg	Nickel (dissolved) mg	Corrosion rate mpy	Corrosion rate (mm/a)
				373 K (100 °C)				
0	100	25	17.27				91.54	2.32
		50	28.50				75.52	1.92
		75	37.39				66.05	1.68
		100	43.62	7.59	0.92	1.27	57.78	1.47
10	90	25	239.20	45.32	3.55	6.59	1267.80	32.20
		50	433.10	74.52	16.05	8.93	1147.60	29.15
		75	622.70	102.50	27.25	12.23	1099.90	27.94
		100	758.80	120.90	35.67	15.64	1005.50	25.54
15	85	25	288.00	46.82	9.54	8.94	1526.00	38.76
		50	601.60	101.90	18.94	16.58	1593.00	40.46
		75	881.40	139.20	40.82	19.49	1556.00	39.52
		100	1037.00	159.80	55.27	19.58	1374.00	34.90
20	80	25	323.00	59.09	6.21	8.15	1711.00	43.46
		50	660.70	109.80	26.01	14.23	1750.00	44.45
		75	919.40	147.30	43.01	18.09	1624.00	41.25
		100	1096.00	173.40	55.09	19.80	1452.00	36.88
35	65	25	290.20	51.74	6.65	7.73	1538.00	39.07
		50	582.90	100.00	19.04	13.37	1544.00	39.22
		75	783.80	123.90	26.23	17.97	1384.00	35.15
		100	972.50	139.80	61.70	19.82	1288.00	32.72

Table 9: Results of corrosion tests of SAE 304 in mixtures of nitric and sulfuric acids [44]

The concentrations of mixed acids, test durations, weight losses and the alloying elements that had been dissolved are listed in Table 9. As can be seen from the last column, even at the lowest testing temperature of 323 K (50 °C) the corrosion rates were considerable and generally lay well above 1 mm/a (40 mpy). There was a certain resistance only in 100 % H_2SO_4 (0.35–0.45 mm/a / 14–18 mpy) as well as in the mixed acid 90 % HNO_3 + 10 % H_2SO_4 (approx. 0.2–0.5 mm/a / 8–20 mpy).

At the interface between a corrosive liquid medium and the gas phase, there can be increased corrosion due to oxygen uptake (water-line corrosion). The behavior of a 18/10 CrNi steel of type SAE 321 (cf. DIN-Mat. No. 1.4541) was investigated in sulfuric acid as well as in a mixed acid with additional nitric acid [45]. The aim was to test whether nitric acid had an inhibiting effect, and if applicable, above which concentration.

Table 10 shows the very high corrosion rates in pure sulfuric acid of both the completely immersed samples (shorter, column 3) as well as of the half-immersed (interface) sample (column 4). The values for the samples simulating the interface (column 4) are mostly higher and thus support the visually observed attack above the immersed surface. The main aim of the investigations was to check the known inhi-

(100 % relative humidity, 8 d)			
Acid concentration		Weight loss of the samples mg/dm^2, test duration 8 d	
H_2SO_4 (N)	HNO_3 (vol%)	5 × 2.5 (cm)*	15 × 2.5 (cm)*
5	0	2550	2670
12	0	6760	8300
20	0	3010	7560
27	0	2610	4480
34	0	4100	2450
5	1	4.0	16.6
12	1	0.2	1.4
20	1	1.2	1.4
27	1	0.8	0.8
34	1	2.0	4.6
5	0.01	4880	7950
12	0.05	5720	9040
20	0.1	5370	8110
27	0.5	3.6	3.0
34	1.0	2.2	4.2

* = Sample dimensions

Table 10: Corrosion of SAE 321 on and above the level of H_2SO_4, with and without additional HNO_3 [45]

bition of sulfuric acid corrosion of stainless steels resulting from the addition of nitric acid for the case of **interface corrosion**. As shown in Table 10, nitric acid already has a pronounced inhibiting effect above a concentration of 1 % HNO_3, for both the completely submerged samples as well as for the half-immersed samples. The corrosive attack decreases by more than a thousand-fold from several mm/a to values of < 0.01 mm/a (< 0.4 mpy). Finally, the critical concentration of nitric acid was determined above which the corrosive attack is markedly decreased. It can be seen from Table 10 that this is already the case above 0.5 vol% HNO_3, and that 0.1 % or less are not sufficient to passivate the steel. This also applies to the half-immersed state, namely for interface corrosion. The critical nitric acid concentration for effective inhibition of 6 M H_2SO_4 lies between 0.5 % and 0.1 % HNO_3.

Nitric and hydrochloric acid

The behavior of SAE 304 in vapors of mixed acids of different concentration and at temperatures up to 321 K (48 °C) was investigated in the laboratory. On the basis of preliminary trials in 6 M solutions, a mixing ratio of 40 HCl : 60 HNO_3 was used because the thermometric curves indicated the strongest corrosion for this ratio. Table 11 shows that stainless steel is also already strongly corroded in such vapors above 1 M mixtures (37.6 mg/dm^2 h = approx. 4 mm/a corrosion rate) [46]. In mixtures of hydrochloric and nitric acids in the concentration range 40–60 %, corrosive attack can be reduced by a factor of 10 by inhibitors, such as hexamine and thiourea (see Figure 13) [47].

Concentration of HCl and HNO_3, M	*Mass loss mg/dm^2 h*	*Final temperature K*	*Temperature difference K*
1	37.6	306	1.0
3	58.4	311,1	6.1
6	127.0	314,8	9.8
9	212.0	317,7	12.7
conc. solution*	754.4	321	16.0

* concentrated solution of 11.5 M HCl and 17 M HNO_3

Table 11: Mass loss and temperature increase during the reaction between steel SAE 304 (DIN-Mat. No. 1.4301) and vapors of HCl and HNO_3 (40 HCl : 60 HNO_3) (initial temperature 305 K (32 °C)) [46]

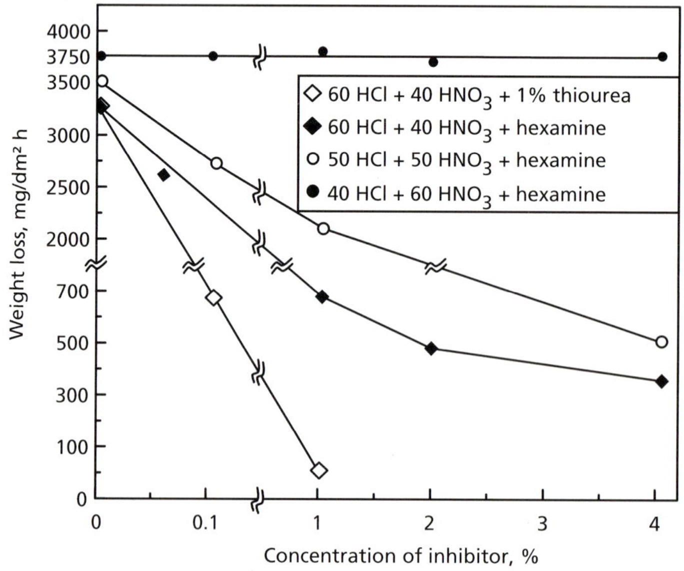

Figure 13: Influence of hexamine on the corrosion of SAE 304 in 6 M mixtures of HCl and HNO_3 [47]

Phosphoric and sulfuric acids

For electrolytic polishing of stainless steels of the group 18Cr10Ni (DIN-Mat. No. 1.4301, 1.4541 etc.), mixtures of phosphoric acid (10–50 %) and sulfuric acid (10–75 %) are used (very shiny surfaces are obtained), also with added acetic acid (up to 25 %). The electrochemical polarisation curves shown in Figure 14 and Figure 15 show that in all solutions the steel has a pronounced anodic active range (high current densities and thus higher corrosion rates), which is why it cannot be used as a construction material for vessels and components in these mixed acids. It can be used in the presence of acetic acid if the steel is brought into the passive range (cf. Figure 15), namely, by means of anodic protection (current densities of 10 µA/cm^2 correspond to a corrosion rate of approx. 0.1 mm/a / 4 mpy) [48].

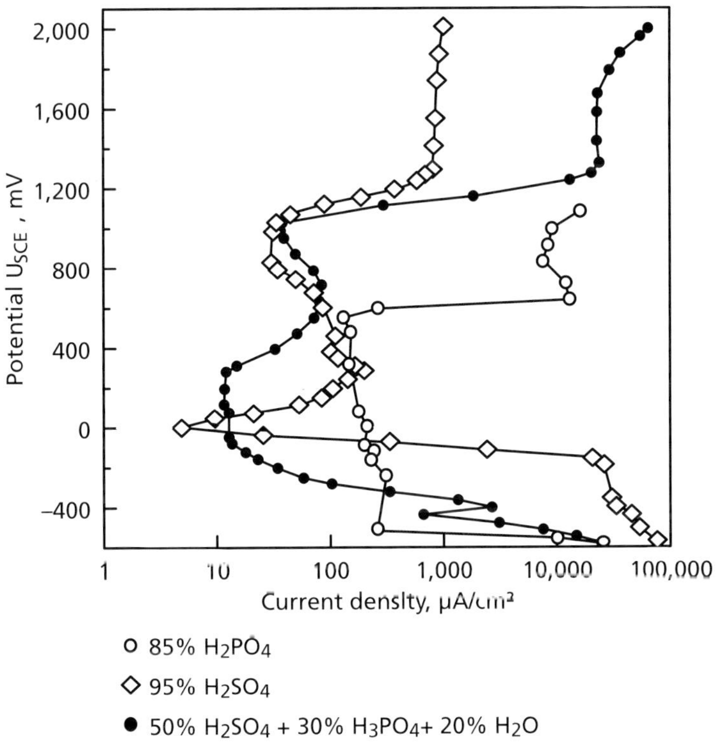

Figure 14: Polarisation behavior of SAE 304 in mixtures of sulfuric and phosphoric acids [48]

During the production of phosphoric acid, a mixture of 30 % H_3PO_4 and 3 % H_2SO_4 is produced that also contains a small amount of fluoride ions. Exposure tests of Haynes® 20 Mod. alloy (UNS N08320; 26Ni, 22Cr, 5Mo) in an ageing tank at 339 K (66 °C) gave a corrosion rate of 1.7 mpy or 0.04 mm/a [29].

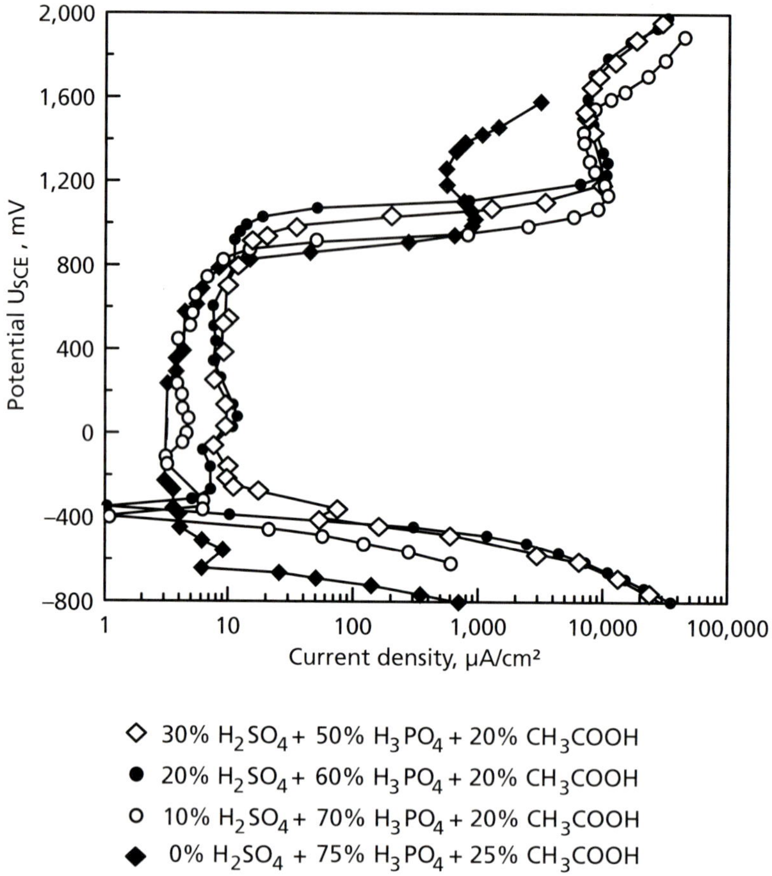

Figure 15: Polarisation behavior of SAE 304 in phosphoric acid + sulfuric and acetic acid [48]

A mixture of hydrochloric and phosphoric acids is used to remove salt deposits in extraction plants [49]. In order to test the resistance of various CrNi or CrNiMo steels, corrosion tests were carried out in 83–93 % H_2SO_4 with the addition of different amounts of 28 % H_3PO_4 (20 % P_2O_5) at temperatures of 333 and 353 K (60 and 80 °C). Table 12 gives the mixing ratios of the acids, test temperatures and corrosion rates after 1–60 days test duration.

Concentration of H_2SO_4 in the solution %	*Mixing ratio H_2SO_4:EPA:H_2O*	*Temperature of the solution K*	*Material*	*Corrosion rate 10^{-6} g/cm^2 h Exposure time of up to 60 days*					
				1 d	*3 d*	*7 d*	*14 d*	*28 d*	*60 d*
93	1:0.22:0	333	X18H10T	12	7	7	13	15	13
			EI-943	11	13	10	14	14	12
87	1:0.077:0.032	333	X18H10T	6	5	5	15	14	15
			EI-943	14	17	10	11	8	9
83	1:0.124:0.032	333	X18H10T	53	52	3	181	14	10
			EI-943	49	24	11	8	5	4
93	1:0.032:0	353	X18H10T	18	48	76	95	87	93
			EI-943	33	84	97	92	80	77
87	1:0.077:0.032	353	X18H10T	29	103	106	120	132	118
			EI-943	56	172	122	106	105	74
83	1:0.124:0.032	353	X18H10T	4	21	23	40	89	83
			EI-943	5	13	14	16	20	24
93	1:0.032:0	333	X18H10T	41	97	65	50	31	21
			EI-943	33	125	60	40	25	17
93	1:0.075:0	333	X18H10T	24	64	56	37	24	26
			EI-943	20	106	61	38	24	20
96	1:0:0	353	iron	0.00156	0.000897	0.000578	0.000546	0.0051	0.0044
93	1:0.032:0	333	X18H10T	7	4	5	7	7	7
			EI-943	10	6	1	1	1	1

EI-943 corresponds to 06ChN28MDT (Cr 22-25, Ni 26-29, Cu 2.5-3.5, Mo 2.5-3, Ti 0.5-0.9, Mn-0.8, bal. Fe)
X18H10T corresponds to approx. X10CrNiTi18 9 (DIN-Mat. No. 1.4541)

Table 12: Corrosion rates of various steels in mixed acids of sulfuric acid and extraction phosphoric acid (EPA) [49]

According to the results, there is an increase in the corrosion as the amount of sulfuric acid decreases and at higher temperatures. However, all steels are definitely resistant under the given conditions. Thus, extraction tanks can be made of CrNi steels.

Phosphoric and hydrochloric acids

The electrochemical behavior of SAE 304 in 14 M H_3PO_4 with additions of 1,000–20,000 ppm (mg/l) HCl was tested at 298, 308 and 318 K (25, 35 and 45 °C) [50]. The current density/voltage curves (not shown here) show that as the HCl concentration increases the passivating current density strongly increases and the passive range is narrowed. According to these results, spontaneous passivation is only possible at 1,000 ppm HCl. At 10,000 ppm HCl, the temperature influence is so great that even a temperature increase of 10 K increases the passivating current by 2 powers of ten so that there is no resistance.

Boric, sulfuric and hydrochloric acids

A Cr18Ni10T steel (cf. DIN-Mat. No. 1.4541) was tested in mixed solutions of hydrochloric and sulfuric acids with dichloroethane (DCE) [51]. In mixtures of DCE with hydrochloric acid (with 35 % hydrochloric acid), the addition of up to 5 % HCl gave a strong increase in the corrosion rate at room temperature after a test period of 360 h (Table 13). Even additions of only 0.10 % HCl give a corrosion rate of approx. 1 mm/a (40 mpy).

DCE – HCl			Aqueous HCl solution			0.1 % HCl solution in DCE		
Amount of HCl %	Corrosion rate		Amount of HCl %	Corrosion rate		Amount of DCE %	Corrosion rate	
	mm/a	mpy		mm/a	mpy		mm/a	mpy
0.00	0.004	0.2	0.01	–	–	11.0	0.006	0.2
0.01	0.015	0.6	0.06	–	–	55.0	0.093	3.7
0.05	0.122	4.8	0.11	0.002	0.1	75.0	0.108	4.3
0.10	0.980	38.6	0.33	0.002	0.1	99.7	0.98	38.6
1.00	1.860	73.2	5.00	0.26	10.2			
5.00	5.160	203.1						

Table 13: Corrosion rate of a Cr18Ni10T steel (cf. DIN-Mat. No. 1.4541) in DCE/aqueous hydrochloric acid solution; at 293 K (20 °C), after 360 h [51]

Therefore, corrosive attack is noticeably higher than in a DCE-free aqueous hydrochloric acid solution of the same concentration, cf. column 4. In a 0.1 % HCl solution, the corrosion rate also increases with increasing DCE content. While the steel is still resistant with a corrosion rate of 0.1 mm/a (4 mpy) in up to 75 % DCE, the corrosion rate is approx. 1 mm/a (40 mpy) in a solution with 99.7 % DCE. If sulfuric acid (80 %) is added to a DCE-0.1 % HCl mixture in amounts of 0.1 to 0.8 %, then the corrosion rate decreases from 0.98 (cf. Table 14) to 0.11 mm/a (38.6 to 4.33 mpy) for 0.8 % sulfuric acid. This also applies to USt 3 (DIN-Mat. No. 1.0333).

Amount of H_2SO_4, %	USt 3 (DIN-Mat. No. 1.0333)	Cr18Ni10T steel (corresp. to DIN-Mat. No. 1.4541)
0.1	0.93	0.75
0.2	0.87	0.72
0.3	0.49	0.45
0.5	0.44	0.13
0.8	0.16	0.11

Table 14: Corrosion rates of steels in DCE-HCl (0.1 %) solution with different amounts of H_2SO_4; 293 K (20 °C), 96 h [51]

Austenitic CrNiMo(N) and CrNiMoCu(N) steels

Molybdenum-containing austenitic steels exhibit a relatively good resistance in mixed acids, except for mixtures of nitric and hydrofluoric acids as well as sulfuric and hydrochloric acids of higher concentrations. It noticeably increases with increasing Mo and Cr content. The best behavior is exhibited by the so-called "superaustenites".

Several high-alloy special/superaustenites are listed below; some, especially foreign grades, are given with their trade names. To aid understanding, these less well-known grades, some of which have only been developed in recent years, are summarised with their approximate composition and material number in Table 15.

Name	Mat. no.	Cr	Ni	Mo	Cu	N
UR B6	1.4539	20	25	4.5	1.2	
UR B28, Sanicro® 28	1.4563	27	31	3.5	1.2	
UR B8, HV9A, 1925hMo	1.4529	21	25	6.5	1	0.20
254 SMO®	1.4547	20	18	6	0.7	0.20
Haynes® 20 Mod. alloy	(UNS N08320)	22		4.5		
Nicrofer® 3127 hMo	1.4562	31	27	6.5	1.3	0.20
Nicrofer® 3033	1.4591	31	33	1.6	0.6	0.40
Remanit® 4565 (5–7Mn)	1.4565	22–25	16–18	4–6	3.3	0.50
AF UR 52 N	(1.4462)	25	6.5	3	1.5	0.18
Ferralium® 255	(1.4507)	26	5.5	3	1.8	0.20

Table 15: Approximate compositions of high-alloy special stainless steels

Nitric and hydrofluoric acids

In the concentration range up to 20 % HNO_3 and up to 4 % HF and at temperatures from 298 to 353 K (25 to 80 °C), all CrNiMo steels have corrosion rates >1 mm/a (40 mpy) and are thus not resistant, as shown in Table 16 [52]. Only those steels containing higher amounts of chromium and molybdenum exhibit a certain resistance at temperatures up to 303 K (30 °C).

Pickling tests in mordants of the composition 5–13 % HNO_3 + 1–10 % HF with the steels X6CrNiMoTi17-12-2 (DIN-Mat. No. 1.4571) and X2CrNiMoN17-13-5 (DIN-Mat. No. 1.4439) over a period of a few hours gave pickling removal of up to 10 μm, that is, corrosion rates >> 10 mm/a (>> 400 mpy) [53].

In 20 % HNO_3 with 1–3 % HF at 323 K (50 °C), the steel X2CrNiMo17-12-2 (DIN-Mat. No. 1.4404) is not resistant (corrosion rate 7.6–29 mm/a (300–1,150 mpy)) [33].

During the development of the very highly alloyed superaustenite Nicrofer® 3033 (alloy 33, X1CrNiMoCuN33-32-1, DIN-Mat. No. 1.4591), comparative investigations with the nickel materials were carried out in mixtures of nitric and hydrofluoric acids [54]. This material was especially developed for use in highly oxidising media, particularly for highly concentrated sulfuric acid. Therefore, it contains an unusually high amount of chromium, only a little molybdenum, but nitrogen to stabilise the austenitic microstructure.

Concentration HNO$_3$, %		**1.5**	**10**	**15**	**15**	**15**	**20**	**20**	**20**	**30**
Concentration HF, %		**0.5**	**3**	**4**	**4**	**4**	**4**	**4**	**4**	**0.1**
Temperature, K	***DIN-Mat. No.***	**353**	**343**	**313**	**323**	**333**	**298**	**303**	**338**	**BP**
13 % Cr steel		2	2				2		2	2
18-10	1.4301	2	2				2		2	2
17-12-2.5	1.4436	2	2	2	2	2	2	2	2	2
18-13-3	1.4438	2	2				2		2	2
17-14-4	1.4439	2	2				2		2	2
904L	1.4539	1	2				1	1	2	2
Sanicro® 28	1.4563								2	
254 SMO®	1.4547		2	2	2	2	1	1		
654 SMO®	1.4652						0	0		
SAF® 2304	1.4362			2	2	2	2	2		
2205	1.4462			2	2	2	1	1		
SAF® 2507	(1.4410)						0	1		
titanium		2	2				2		2	2

grading: 0 = < 0.1 mm/a (4 mpy); 1 = 0.1–1.0 mm/a (4–40 mpy); 2 = > 1 mm/a (> 40 mpy)
BP: boiling point

Table 16: Corrosion rates of CrNiMo steels in mixtures of nitric and hydrofluoric acids [52]

Material	***2 mol/l HNO$_3$***			***0.25 mol/l HF***			
	+ 0 mol/l HF	***+ 0.5 mol/l HF***	***+ 2 mol/l HF***	***+ 6 mol/l HNO$_3$***	***+ 9 mol/l HNO$_3$***	***+ 12 mol/l HNO$_3$***	***+ 15 mol/l HNO$_3$***
Nicrofer® 3127 LC, alloy 28, DIN-Mat. No. 1.4563	< 0.01 (< 0.4)	6.26 (246.5)	22.6 (890)	1.05 (41.3)	1.94 (76.4)	3.62 (143)	5.95 (234)
Nicrofer® 6030, alloy 690 DIN-Mat. No. 2.4642	< 0.01 (< 0.4)	0.66 (26)	6.91 (272)	1.59 (62.6)	2.15 (84.6)	5.11 (201)	8.09 (319)
alloy G- 30, DIN-Mat. No. 2.4603	< 0.01 (< 0.4)	0.31 (12.2)	1.32 (52)	0.53 (20.9)	1.58 (62.2)	2.61 (102)	4.89 (193)
Nicrofer® 3033, alloy 33 DIN-Mat. No. 1.4591	< 0.01 (< 0.4)	0.26 (10.2)	1.30 (51)	0.29 (11.4)	0.73 (28.7)	1.81 (71.3)	3.36 (132)

Table 17: Corrosion rate of Nicrofer® 3033 (DIN-Mat. No. 1.4591) and other materials in nitric acid + hydrofluoric acid at 363 K (90 °C) in dependence on the acidity, corrosion rate in mm/a (mpy) [54]

Compared to other materials with high amounts of chromium, this material showed the best resistance. Moderate resistance was only observed in the mixture 2 mol/l HNO_3 + 0.5 mol/l HF. The behavior in mixed acids up to 323 K (50 °C) is much more favorable, as shown in Table 18. Alloy 33 shows very good resistance in all mixed acids. Apart from one exception, the corrosion rates were ≤ 0.1 mm/a (≤ 4 mpy).

Material	**298 K**			**323 K**		
	+3% HF	**+5% HF**	**+7% HF**	**+3% HF**	**+5% HF**	**+7% HF**
SAE 316 Ti	3.33	6.20	5.68	17.3*	24.4*	33.45*
alloy 28	0.03	0.04	0.06	0.18	0.29	0.41
alloy 33	0.01	0.01	0.02	0.08	0.11	0.17

* Test period 7 days

Table 18: Corrosion test in 20% HNO_3 with additions of HF; immersion tests over 3 × 7 days; corrosion rate in g/m^2 h [54]

Nitric and sulfuric acids

Resistance diagrams for austenitic 18/10 CrNi and 18/10 CrNiMo steels are shown in Figure 16 [55].

The temperature-dependent behavior of austenitic CrNi and CrNiMo steels in mixtures of 50–80% sulfuric acid and 20% nitric acid is given in Figure 17 [22]. There are only a few differences between both types of austenites (with and without Mo): there is good resistance at temperatures up to 353 K (80 °C) in up to 60% H_2SO_4 + 20% HNO_3. The steels containing molybdenum are only resistant up to 333 K (60 °C) in 80% H_2SO_4 + 20% HNO_3, while the molybdenum-free steels are only resistant up to 323 K (50 °C).

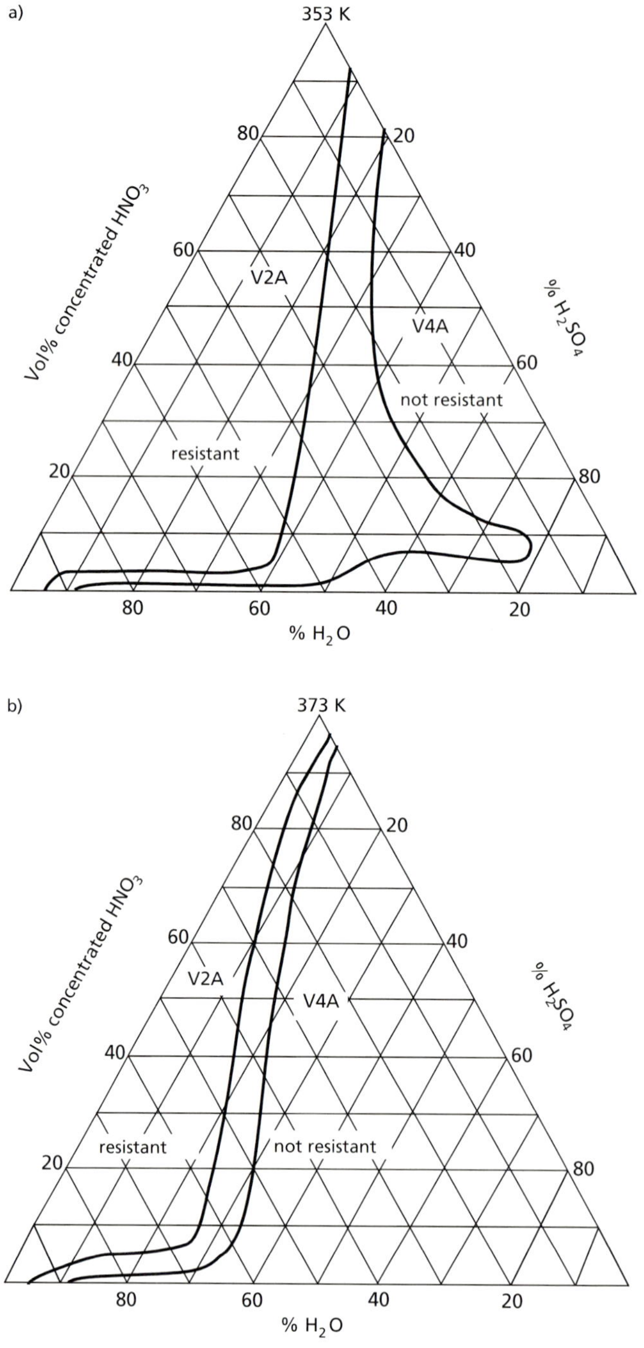

Figure 16: Resistance limits (maximum corrosion rates 0.1 mm/a) of austenitic steels in mixtures of sulfuric and nitric acids; a) at 353 K (80 °C), b) at 373 K (100 °C) [55]

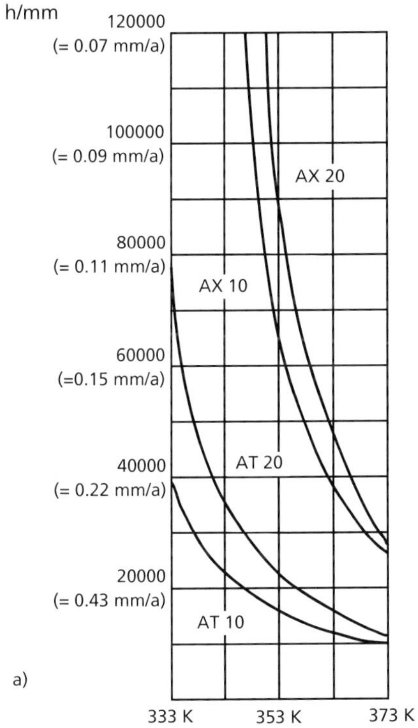

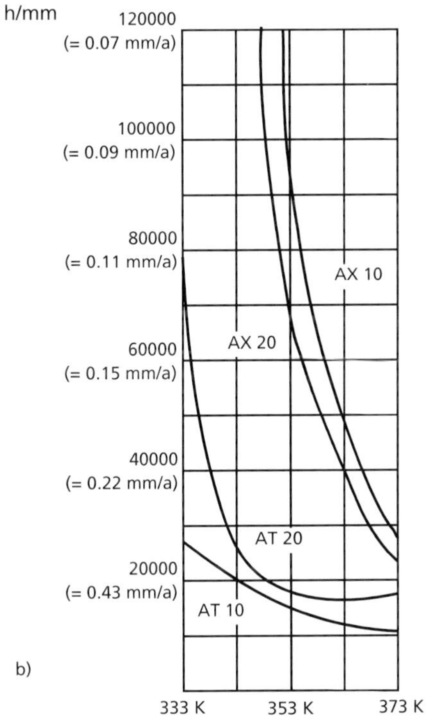

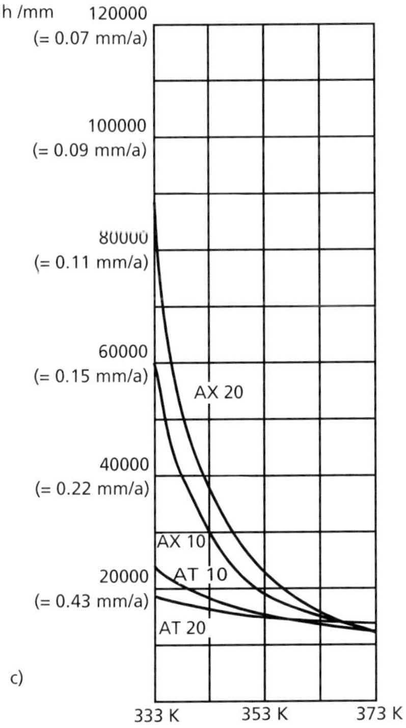

Figure 17: Temperature-dependent resistance of steels AX10, AT10, AX 20, and AT 20 in mixed acids; a) 50 vol% H_2SO_4 + 20 vol% HNO_3 + 30 vol% H_2O; b) 60 vol% H_2SO_4 + 20 vol% HNO_3 + 20 vol% H_2O; c) 80 vol% H_2SO_4 + 20 vol% HNO_3 [22]
AX 10 (corresp. to DIN-Mat. No. 1.4301; X5CrNi18-10); AX 20 (corresp. to DIN-Mat. No. 1.4401; X5CrNiMo17-12-2); AT 10 (corresp. to DIN-Mat. No. 1.4541; X6CrNiTi18-10) and AT 20 (corresp. to DIN-Mat. No. 1.4571; X6CrNi-MoTi17-12-2)

The resistance of X5CrNiMo17-12-2 (SAE 316, DIN-Mat. No. 1.4401) in anhydrous mixed acids of 67 % HNO_3 and conc. H_2SO_4 as a function of the temperature is given in Figure 18. The molybdenum-containing steel is resistant in all mixing ratios up to 333 K (60 °C); but at 353 K (80 °C) only above 90 vol% HNO_3 and less than 10 % H_2SO_4.

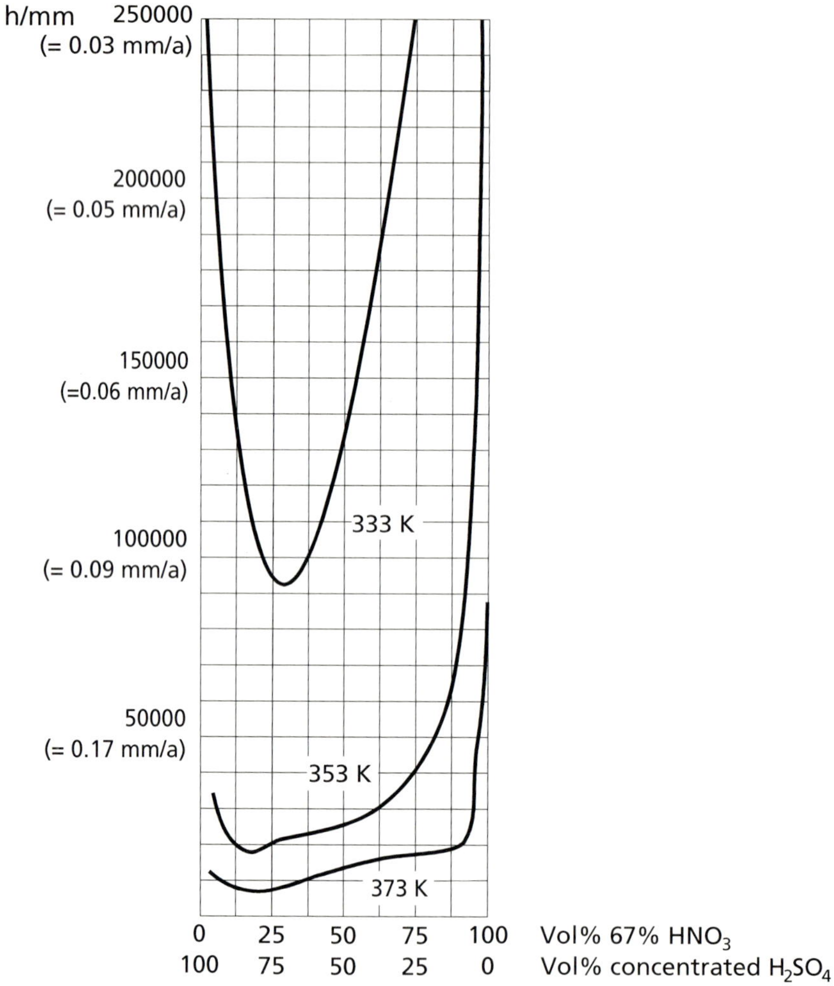

Figure 18: Temperature-dependent corrosion resistance of steel X5CrNiMo17-12-2 (SAE 316, DIN-Mat. No. 1.4401) in anhydrous mixed acids of $HNO_3 + H_2SO_4$ [22]

Corrosion rates of various austenitic steels alloyed with copper, silicon and nitrogen, are shown in Table 19. Silicon and molybdenum obviously have a positive influence on the resistance.

Approximate composition, %	H_2SO_4, vol%	HNO_3, vol%	H_2O, vol%	Temperature, K	Corrosion rate, mm/a (mpy)
18 Cr, 8 Ni; 18 Cr, 12 Ni, 2.5 Mo	75	25	0	363	0.1–1 (4–40)
18 Cr, 8 Ni; 18 Cr, 12 Ni, 2.5 Mo	75	25	0	430	unsuitable
18 Cr, 8 Ni	70	10	20	363	0.1–1 (4–40)
18 Cr, 12 Ni, 2.5 Mo	70	10	20	363	0.1 (4)
18 Cr, 8 Ni; 18 Cr, 12 Ni, 2.5 Mo	70	10	20	430	unsuitable
18 Cr, 8 Ni	58	40	2	383	unsuitable
18 Cr, 8 Ni; 18 Cr, 12 Ni, 2.5 Mo	50	50	0	363	0.1–1 (4–40)
18 Cr, 8 Ni; 18 Cr, 12 Ni, 2.5 Mo	30	5	65	363	0.1 (4)
18 Cr, 8 Ni	30	5	65	383	0.1–1 (4–40)
18 Cr, 12 Ni, 2.5 Mo	30	5	65	383	0.1 (4)
18 Cr, 8 Ni; 18 Cr, 12 Ni, 2.5 Mo	15	5	80	407	1 (40)
18 Cr, 8 Ni	2	1	97	boiling point	10 (400)
18 Cr, 12 Ni, 2.5 Mo	2	1	97	boiling point	0.1 (4)
18 Cr, 8 Ni	60*	20*	20*	80	0.18 (7)
23 Cr, 4 Ni, 0.29 N	60*	20*	20*	80	0.08 (3)
23 Cr, 5 Ni, 0.28 N	60*	20*	20*	80	0.07 (2.7)
18 Cr, 8 Ni	50*	20*	30*	80	0.18 (7)
23 Cr, 4 Ni, 0.29 N	50*	20*	30*	80	0.07 (2.7)
23 Cr, 5 Ni, 0.28 N	50*	20*	30*	80	0.05 (2)

*) According to W. Tofaute [56]

Table 19: Behavior of austenitic steels in mixtures of $H_2SO_4 + HNO_3$ [19]

Investigations in mixtures with up to 20 % HNO_3 showed resistance to sulfuric acid in the entire range of concentrations up to approx. 323 K (50 °C) [52]. Corrosion rates of up to 1 mm/a (40 mpy) are found only at temperatures above 333–353 K (60–80 °C). An individual evaluation is given in Table 20.

The steel SAE 316 L (DIN-Mat. No. 1.4404) is resistant at room temperature in a nitrating acid of the composition 75 % H_2SO_4 + 25 % HNO_3, as shown by electrochemical investigations [57]. It has better behavior than SAE 316 (DIN-Mat. No. 1.4401), because it has a low C content and a slightly higher Cr content and is thus not sensitised.

The behavior of austenitic Mo steels, such as X5CrNiMo17-12-2 (DIN-Mat. No. 1.4401), X6CrNiMoTi17-12-2 (DIN-Mat. No. 1.4571), X5CrNiMo17-13-3 (DIN-Mat. No. 1.4436) and X5NiCrMoCuTi20-18 (DIN-Mat. No. 1.4506), in various mixed acids is given in Table 8.

Concentration HNO_3	Concentration H_2SO_4	Temperature K	18-10 (DIN-Mat. No. 1.4301)	17-12-2.5 (DIN-Mat. No. 1.4436)	18-13-3 (DIN-Mat. No. 1.4438)	17-14-4 (DIN-Mat. No. 1.4439)	904L (DIN-Mat. No. 1.4539)	Sanicro® 28 (DIN-Mat. No. 1.4563)	254 SMO® (DIN-Mat. No. 1.4547)	654 SMO® (DIN-Mat. No. 1.4652)	SAF® 2304 (DIN-Mat. No. 1.4362)	SAF® 2205 (DIN-Mat. No. 1.4462)	SAF® 2507 (DIN-Mat. No. 1.4410)	Titanium
1	5	298	0	0	0	0	0							0
1	5	323	0	0	0	0	0							1
1	10	298	0	0	0	0	0							1
1	10	353	1	0	0	0	0							2
1	17	373	1											
1	95	323	1	0	0	0	0							2
1	99	308	0	0	0	0	0							1
3	10	298	0	0	0	0	0							0
3	10	353	1	0	0	0	0							1
3	50	298	0	0	0	0	0							0
5	20	298	0	0	0	0	0							0
5	20	323	0	0	0	0	0							0
5	60	298	0	0	0	0	0							0
5	60	323	0	0	0	0	0							0
5	60	353		1	1	1	1							2
7	17	373	0											
10	60	333	0	0	0	0	0	0			0	0	0	1
10	60	353	1	1	1	1	1	0			0	0	0	2
10	80	323	0	0	0	0	0							1
10	90	308	0	0	0	0	0							2

Key: 0 = < 0.1 mm/a (4 mpy); 1 = 0.1–1.0 mm/a (4–40 mpy); 2 = > 1 mm/a (> 40 mpy)

Table 20: Corrosion rates of CrNiMo steels in mixtures of nitric and sulfuric acids [52]

Table 20: Continued

Concentration HNO_3	*Concentration H_2SO_4*	*Temperature K*	*18-10 (DIN-Mat. No. 1.4301)*	*17-12-2.5 (DIN-Mat. No. 1.4436)*	*18-13-3 (DIN-Mat. No. 1.4438)*	*17-14-4 (DIN-Mat. No. 1.4439)*	*904L (DIN-Mat. No. 1.4539)*	*Sanicro® 28 (DIN-Mat. No. 1.4563)*	*254 SMO® (DIN-Mat. No. 1.4547)*	*654 SMO® (DIN-Mat. No. 1.4652)*	*SAF® 2304 (DIN-Mat. No. 1.4362)*	*SAF® 2205 (DIN-Mat. No. 1.4462)*	*SAF® 2507 (DIN-Mat. No. 1.4410)*	*Titanium*
13	16	373	0											
20	80	293	0	0	0	0	0							
20	80	333	1	0	0	0	0							2
20	80	373	1	1	1	1	1							2
25	15	373	0											
30	20	353	1	0	0	0	0							1
30	40	353	1	[illegible]	1	1	1	0			0	0	0	1
30	70	308	0	0	0	0	0							1
47	14	373	0											
50	20	353	1	1	1	1	0	0	0		0	0	0	1
50	50	333	1	1	0	0	0							0
54	67	348	0											
54	67	BP	2											
54	95	333	1											
56	14	373	0											
65	35	308	0	0	0	0	0							0
90	10	308	0	0	0	0	0							0

Key: 0 = < 0.1 mm/a (4 mpy); 1 = 0.1–1.0 mm/a (4–40 mpy); 2 = > 1 mm/a (> 40 mpy)
BP: boiling point

Table 20: Corrosion rates of CrNiMo steels in mixtures of nitric and sulfuric acids [52]

Nitric and phosphoric acids

In plants of the food industry, residues that occur are fats, carbohydrates and proteins and, in particular, salts. These residues cannot be completely removed by neutral or alkaline cleaning agents. Therefore, acidic cleaning solutions are used that are very dilute solutions of 1–2% acids based on HNO_3 with small amounts (2–10%) of H_3PO_4. An investigation was carried out to find optimum cleaning solutions which gave a good cleaning effect but did not attack the stainless steel X5CrNiMo17-12-2 (DIN-Mat. No. 1.4401) that is generally used for this type of plant, taking account of the fact that the cleaning temperatures can be up to 373 K (100 °C) or even 413 K (140 °C) in pressurised tanks [58]. Figure 19 shows a plot of the corrosion rate as a function of the concentration of both acids at 363 K (90 °C) for all proportions of H_3PO_4 from 0 to 5%. In pure HNO_3, the corrosion rate decreases with increasing concentration (up to 10%). In solutions with up to 3 parts H_3PO_4, a maximum corrosion rate was found at an acid concentration of about 4%, and above 4 parts H_3PO_4, the corrosion rate was very low, independent of the acid concentration. At 413 K (140 °C), the corrosive attack was much stronger, as shown

Concentration, %	
HNO_3	H_3PO_4
● 100	-
◆ 98	2
○ 97	3
◇ 96	4
■ 95	5

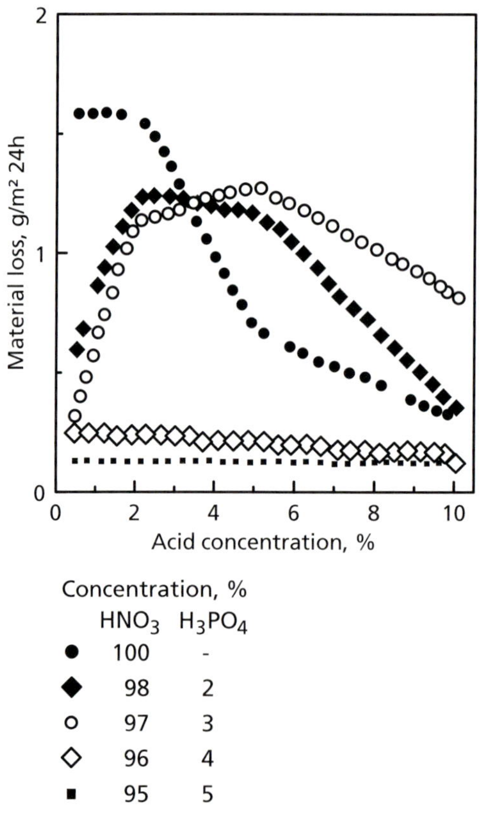

Figure 19: Material loss (g/m² 24 h at 363 K (90 °C)) as a function of the acid concentration for stainless steel sheets immersed in nitric/phosphoric acid mixtures [58]

in Figure 20. For the cleaning solutions used in practice, up to approx. 5 % acid concentration and with 5–10 parts H_3PO_4, the corrosion rates are still low (corresponding to corrosion rates of less than 0.05 mm/a / 2 mpy).

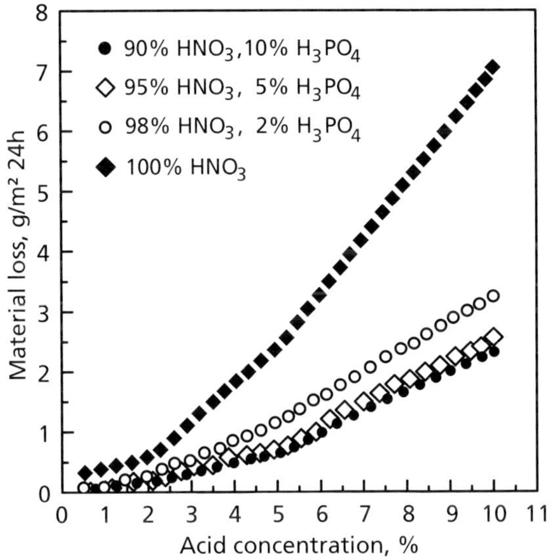

Figure 20: Material loss (g/m^2 24 h at 413 K (140 °C)) as a function of the acid concentration of stainless steel sheets immersed in nitric/phosphoric acid mixtures [58]

Phosphoric and sulfuric acids

During the digestion of phosphates to produce phosphoric acid, mixtures are formed that contain phosphoric and sulfuric acids as well as small amounts of hydrofluoric and hydrofluosilicic acids. The resistance of three stainless steels in several different solutions, to which additional varying amounts of chlorides were added, was investigated by means of immersion tests (Table 21).

It can be seen from the values given in the table that a binary mixed acid with 24 % P_2O_5 and 10 % H_2SO_4 produced strong corrosion (approx. 2.5 mm/a (98 mpy)) of the CrNiMo steel SAE 316 L (X2CrNiMo17-12-2, DIN-Mat. No. 1.4404), while the steel alloyed with more molybdenum HV-9A (X1NiCrMoCuN25-20-7, DIN-Mat. No. 1.4529) exhibited better behavior (approx. 0.5 mm/a (19.7 mpy)); however, it is still not sufficiently resistant. In contrast, no corrosion was found for the high-chromium steel Sanicro® 28 (X1NiCrMoCu31-27-4, DIN-Mat. No. 1.4563) under the given conditions. The influence of up to 0.5 % hydrofluosilicic acid is of no importance; however, hydrofluoric acid accelerates corrosion above 0.5 %. Chlorides increase the corrosion rate above a concentration of only 500 ppm, particularly in the presence of hydrofluoric acid. The steel SAE 316 L (DIN-Mat. No. 1.4404) failed, even HV-9A (1.4529) can no longer be used (> 1 mm/a (> 40 mpy)). The lowest corrosion rate was exhibited by the steel Sanicro® 28 (DIN-Mat. No. 1.4563) with approx. 0.5 mm/a (20 mpy) [59].

				Corrosion rate, mg/dm² d		
Medium	**Cl, ppm**	**H_2SiF_6, %**	**HF, %**	**SAE 316 L**	**HV-9A**	**Sanicro® 28**
24 % P_2O_5 + 10 % H_2SO_4	–	–	–	674.0	119.5	–
	100	–	–	682	214.2	–
	200	–	–	786	260.7	–
	500	–	–	25,430	367.0	105.3
	1,000	–	–	–	–	217.0
	10,000	–	–	–	–	689.0
	–	0.125	–	645	150.5	–
	–	0.25	–	685	180.1	–
	–	0.50	–	707	191.6	85.4
	–	–	0.125	805	163.0	–
	–	–	0.25	881	198.5	–
	–	–	0.50	1,100	216.3	101.2
	100	0.125	0.125	2,549	–	–
	500	0.50	0.50	39,080	375.1	125.3

Table 21: Corrosion of SAE 316 L, HV-9A and Sanicro® 28 at 358 K (85 °C) [59]

This behavior of the 18Cr10NiMo steels in mixtures based on phosphoric acid can be explained electrochemically [60]. Figure 21 shows that, as the Mo content increases, the passivating current densities and thus the anodic dissolution range strongly decrease so that in the case of >4 % Mo spontaneous passivation occurs. Thus, steels with higher amounts of Mo are resistant in contaminated phosphoric acid, even in the presence of chlorides.

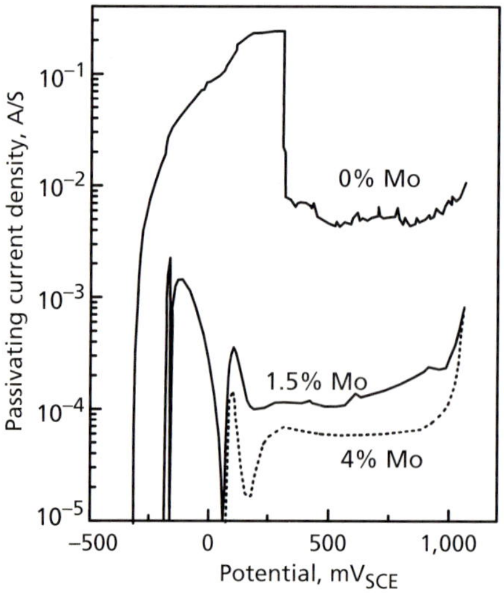

Figure 21: Influence of molybdenum on the anodic behavior of 18Cr10Ni steels in 40 mass % H_3PO_4 + 4 mass % H_2SO_4 + 3,000 ppm Cl^- + 24 g/l SiC at 333 K (60 °C) [60].
S = electrode surface area = 654 mm²

Because of the relatively good behavior in phosphoric acid contaminated with approx. 1–4 % sulfuric acid, the steels X1NiCrMoCu31-27-4 (DIN-Mat. No. 1.4563), X1NiCrMoCuN25-20-7 (DIN-Mat. No. 1.4529) and X1NiCrMoCu25-20-5 (DIN-Mat. No. 1.4539) are preferentially recommended for the individual process steps given in Table 22 [61].

Plant component	***Construction material***
Reaction stage	
Reaction tanks	concrete, lined with rubber and graphite bricks
Reactor turbine agitators	N08904 and N08028 (cf. 1.4539 and 1.4563)
Digestor agitator	N08904 and S31803 (cf. 1.4539 and 1.4462)
Slurry pumps	N08904 and N08028 and CD4MCu (cast duplex steel)
Pipelines	rubberised C steel, PP, HDPE
Filtration stage	
Filter frame	painted C steel, N08904, S31600, S31803 (cf. 1.4539, 1.4401, 1.4462)
Tilting pans	N08904, S31254, S31803 (cf. 1.4539, 1.4547, 1.4462
Pipelines	rubberised C steel
Concentration stage	
Heat exchanger	shell: C steel; pipes: graphite, N08028 (1.4563)
Pumps	N08028 and CD4MCu (cast components)
Evaporator	body: rubberised steel, lined with graphite bricks; pipes: graphite
Storage tanks	rubberised steel (natural and neoprene)
Pipelines	rubberised steel, PP

Table 22: Chemical apparatus and construction materials in a typical phosphoric acid processing plant [61]

The steel Z 2 NCDU 25-20 (corresp. to DIN-Mat. No. 1.4539) has been used to transport contaminated phosphoric acid in tankers, in the form of tubular heating coils with a total length of 7,000 m, that are obviously used to keep the acid warm during transport in extremely cold regions of Russia. The mixed acid contained: 70 % P_2O_5 (almost 100 % H_3PO_4), <5 % H_2SO_4, <100 ppm HF and <600 ppm Cl^-, the maximum temperature was 358 K (85 °C). Under these conditions, the corrosion rate was max. 0.1 mm/a (4 mpy) [62].

The process solution generally used in Great Britain for chemical brightening of aluminium is a mixture of 80 % H_3PO_4 + 15 % H_2SO_4 + 5 % HNO_3 with low additions of copper and nickel ions as inhibitors. Higher amounts of sulfuric acid instead of phosphoric acid are also possible. The process temperature lies at 368–378 K (95–105 °C). Stainless steels (no exact details, probably based on 18/10 CrNiMo) can be used for heating pipes in the process tanks and for heating the brightener baths because their resistance is based on the presence of nitric acid: if the concentration of HNO_3 falls below 2 %, then the heating pipes, in particular, are attacked by pitting corrosion [63].

Electrochemical investigations were used to study the behavior of austenitic stainless steels SAE 304 (DIN-Mat. No. 1.4301) and SAE 316 (DIN-Mat. No. 1.4401) in 14 M phosphoric acid with additions of hydrochloric acid up to 20,000 mg/l [63]. Anodic polarisation curves show that, as the HCl concentration increases, both the attack is markedly increased in both the active and the passive ranges and that the pitting corrosion potential is decreased. These effects are more noticeable for the Mo-free steel SAE 304. For the Mo-containing steel SAE 316, the minimum concentration of HCl that triggers pitting corrosion is much higher. According to these results, Mo improves the resistance to local corrosion in mixtures of phosphoric and hydrochloric acids to the same extent as in phosphoric acid with additional chlorides (see also above under phosphoric acid production). In all cases, at increased HCl contents and higher temperatures, the special stainless steels containing high amounts of Mo (e.g.: X1NiCrMoCu32-28-7, DIN-Mat. No. 1.4562) or even nickel alloys (e.g.: NiCr22Mo9Nb, DIN-Mat. No. 2.4856; NiCr23Mo16Al, DIN-Mat. No. 2.4605 etc.) are the most resistant.

All CrNiMo steels are resistant in a mixture of 50 % nitric acid and 20 % oxalic acid up to 343 K (70 °C) [52].

Austenitic CrNiMoCu(N) steels

Sulfuric and nitric acids

In boiling mixed acids with 12.5–47.1 % H_2SO_4 and 21.5–55.9 % HNO_3, the steel 18Cr10NiMo (X5CrNiMo17-12-2, DIN-Mat. No. 1.4401) exhibits corrosion rates of > 1 mm/a (> 40 mpy) and is thus not resistant [64].

In the three-component mixture of 72 % H_2SO_4 + 8 % HNO_3 + 4 % HF, the superaustenite Nicrofer® 3127 hMo (alloy 31, DIN-Mat. No. 1.4562) exhibits excellent resistance at 327 K (54 °C) with corrosion rates of approx. 0.02 mm/a (0.79 mpy), likewise the nickel alloy G-30, cf. also Figure 22 [67].

Figure 22 shows the behavior of the superaustenite Nicrofer® 3127 hMo (DIN-Mat. No. 1.4562) in further inorganic acids. In all cases, noticeably lower corrosion rates were found compared to alloy G-30 (NiCr30FeMo, DIN-Mat. No. 2.4603). Nicrofer® 3127 hMo is resistant in 30 % hydrochloric acid at room temperature, and in 5 % HCl up to 323 K (50 °C). In approx. 75 % phosphoric acid (54 % P_2O_5) and up to 393 K (120 °C), the corrosion rate was only 0.05 mm/a (2 mpy). In the test solution according to ASTM G-28 A (a test to determine the resistance to intergranular corrosion) that was carried out in 50 % H_2SO_4 + 42 g/l $Fe_2(SO_4)_3$ x 9 H_2O, the corrosion rate was 0.15 mm/a (6 mpy).

KRUPP THYSSEN NIROSTA [65] has published information for its entire range of stainless steel products with regard to their resistance in mixed acids based on sulfuric and nitric acids (nitrating acids) (Table 23).

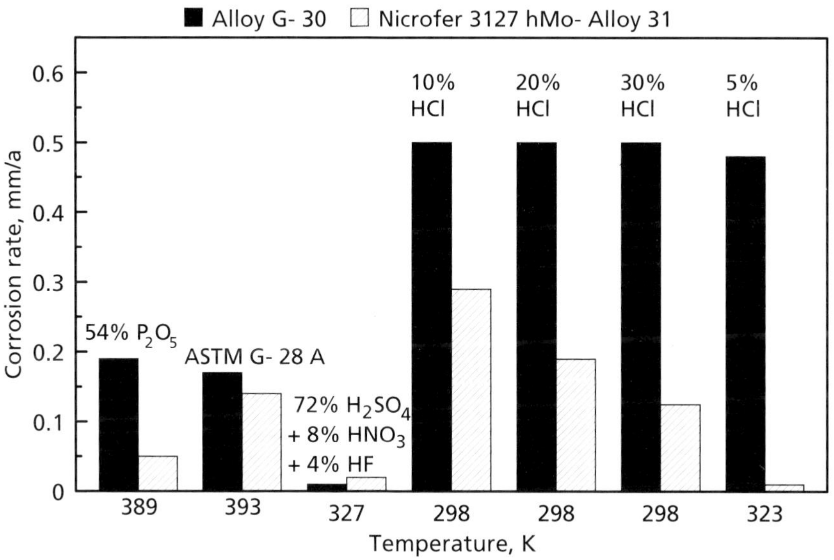

Figure 22: Comparison of the corrosion rates of the superaustenite Nicrofer® 3127 hMo (DIN-Mat. No. 1.4562) with the nickel alloy G-30 in various acids and at different temperatures [67]

Attacking agent	*Concentration*	*Temperature*	*Group 1*	*Group 2*	*Group 3*	*Group 4*	*Group 5*	*1.4465*	*1.4539*	*1.4565*
Nitrating acid	50 % H_2SO_4 + 50 % HNO_3	323 K	3	2	1	0	0	0	0	0
Nitrating acid	50 % H_2SO_4 + 50 % HNO_3	363 K	3	3	2	1	1			
Nitrating acid	50 % H_2SO_4 + 50 % HNO_3	393 K	3	3	3	2	2			
Nitrating acid	75 % H_2SO_4 + 25 % HNO_3	323 K	3	2	1	1	0	0	0	0
Nitrating acid	75 % H_2SO_4 + 25 % HNO_3	363 K	3	3	1	1	1			
Nitrating acid	75 % H_2SO_4 + 25 % HNO_3	430 K	3	3	3	3	3			
Nitrating acid	20 % H_2SO_4 + 15 % HNO_3	323 K	3	3	1	0	0	0	0	0
Nitrating acid	20 % H_2SO_4 + 15 % HNO_3	353 K	3	3	2	1	0	0	0	0

0 = Rate of mass loss < 0.1 g/m^2 h, corresponding to a corrosion rate < 0.11 mm/a (< 4.3 mpy)
1 = Rate of mass loss 0.1–1.0 g/m^2 h, corresponding to a corrosion rate 0.11–1.1 mm/a (4.3–43 mpy)
2 = Rate of mass loss 1.0–10.0 g/m^2 h, corresponding to a corrosion rate 1.1–11 mm/a (43–433 mpy)
3 = Rate of mass loss > 10.0 g/m^2 h, corresponding to a corrosion rate > 11.0 mm/a (> 433 mpy)

Table 23: Resistance of NIROSTA® steels in mixtures of sulfuric and nitric acids (nitrating acid) [65]. See Table 24 for the classification of groups

Table 23: Continued

Attacking agent	*Concentration*	*Temperature*	*Group 1*	*Group 2*	*Group 3*	*Group 4*	*Group 5*	*1.4465*	*1.4539*	*1.4565*
Nitrating acid	70 % H_2SO_4 + 10 % HNO_3	323 K	3	3	1	0	0	0	0	0
Nitrating acid	70 % H_2SO_4 + 10 % HNO_3	363 K	3	3	3	1	0	0	0	0
Nitrating acid	70 % H_2SO_4 + 10 % HNO_3	441 K	3	3	3	3	3			
Nitrating acid	30 % H_2SO_4 + 5 % HNO_3	363 K	3	3	1	0	0	0	0	0
Nitrating acid	30 % H_2SO_4 + 5 % HNO_3	383 K	3	3	2	1	0	0	0	0
Nitrating acid	15 % H_2SO_4 + 5 % HNO_3	407 K	3	3	2	1	1			
Nitrating acid	2 % H_2SO_4 + 1 % HNO_3	boiling	3	3	2	2	0	0	0	0

0 = Rate of mass loss < 0.1 g/m^2 h, corresponding to a corrosion rate < 0.11 mm/a (< 4.3 mpy)
1 = Rate of mass loss 0.1–1.0 g/m^2 h, corresponding to a corrosion rate 0.11–1.1 mm/a (4.3–43 mpy)
2 = Rate of mass loss 1.0–10.0 g/m^2 h, corresponding to a corrosion rate 1.1–11 mm/a (43–433 mpy)
3 = Rate of mass loss > 10.0 g/m^2 h, corresponding to a corrosion rate > 11.0 mm/a (> 433 mpy)

Table 23: Resistance of NIROSTA® steels in mixtures of sulfuric and nitric acids (nitrating acid) [65]. See Table 24 for the classification of groups

Group 1 DIN-Mat. No.	*Group 2 DIN-Mat. No.*	*Group 3 DIN-Mat. No.*	*Group 4 DIN-Mat. No.*	*Group 5 DIN-Mat. No.*
1.4000			1.4301	
1.4002			1.4303	
1.4003	1.4016		1.4306	1.4401
1.4006	1.4057		1.4307	1.4404
1.4021	1.4120		1.4310	1.4429
1.4024	1.4305		1.4311	1.4435
1.4028	1.4427		1.4315	1.4436
1.4031	1.4509		1.4318	1.4438
1.4034	1.4510		1.4541	1.4439
1.4313	1.4511		1.4544	1.4462
1.4512	1.4520	1.4113	1.4546	1.4561
1.4589	1.4521	1.4568	1.4550	1.4571

Table 24: Classification of groups for the materials listed in Table 23

Good resistance was found for most of the molybdenum-containing steels in several mixtures, some even at elevated temperatures, particularly the special grades 1.4465, 1.4539 and 1.4565. The molybdenum-free austenites (Group 4) are less resistant; the ferritic stainless steels are not resistant (Groups 1 and 2).

Another stainless steel manufacturer gives the resistance of its products (ferritic, austenitic steels with and without molybdenum as well as austenitic-ferritic and special stainless steels) in mixtures of nitric and sulfuric acids, as summarised in Table 25 [66]

Concentration of the mixed acid	***Temperature, K***	***Material***					
		1.4000, 1.4002, 1.4006, 1.4021, 1.4028, 1.4104, 1.4034, 1.4110, 1.4112, 1.4528	***1.4016, 1.4057, 1.4122***	***1.4405, 1.4542, 1.4548***	***1.4301, 1.4304, 1.4306, 1.4361, 1.4544, 1.4546, 1.4550***	***1.4401, 1.4404, 1.4435, 1.4436, 1.4571***	***1.4439, 1.4460, 1.4462, 1.4465, 1.4466, 1.4467, 1.4539***
1 % HNO_3 + 2 % H_2SO_4	boiling	3	3	2	2	0	–
1 % HNO_3 + 5 % H_2SO_4	298	–	–	–	0	0	0
1 % HNO_3 + 5 % H_2SO_4	323	–	–	–	0	0	0
1 % HNO_3 + 10 % H_2SO_4	298	–	–	–	0	0	0
1 % HNO_3 + 10 % H_2SO_4	353	–	–	–	1	0	0
1 % HNO_3 + 95 % H_2SO_4	323	–	–	–	1	0	0
1 % HNO_3 + 99 % H_2SO_4	308	–		–	0	0	0
3 % HNO_3 + 10 % H_2SO_4	298	–	–	–	0	0	0
3 % HNO_3 + 10 % H_2SO_4	353	–	–	–	1	0	0
5 % HNO_3 + 15 % H_2SO_4	407	3	3	2	1	1	–
5 % HNO_3 + 20 % H_2SO_4	323	–	–	–	0	0	0

0 = Rate of mass loss < 0.1 g/m^2 h, corresponding to a corrosion rate < 0.11 mm/a (< 4.3 mpy)
1 = Rate of mass loss 0.1–1.0 g/m^2 h, corresponding to a corrosion rate 0.11–1.1 mm/a (4.3–43 mpy)
2 = Rate of mass loss 1.0–10.0 g/m^2 h, corresponding to a corrosion rate 1.1–11 mm/a (43–433 mpy)
3 = Rate of mass loss > 10.0 g/m^2 h, corresponding to a corrosion rate > 11.0 mm/a (> 433 mpy)

Table 25: Chemical resistance of stainless steels in mixtures of nitric and sulfuric acids [66]

Table 25: Continued

Concentration of the mixed acid	*Temperature, K*	*Material*					
		1.4000, 1.4002, 1.4006, 1.4021, 1.4028, 1.4104, 1.4034, 1.4110, 1.4112, 1.4528	**1.4016, 1.4057, 1.4122**	**1.4405, 1.4542, 1.4548**	**1.4301, 1.4304, 1.4306, 1.4361, 1.4544, 1.4546, 1.4550**	**1.4401, 1.4404, 1.4435, 1.4436, 1.4571**	**1.4439, 1.4460, 1.4462, 1.4465, 1.4466, 1.4467, 1.4539**
5 % HNO_3 + 30 % H_2SO_4	363	3	3	1	0	0	–
5 % HNO_3 + 30 % H_2SO_4	383	3	3	2	1	0	–
5 % HNO_3 + 60 % H_2SO_4	323	–	–	–	0	0	0
5 % HNO_3 + 60 % H_2SO_4	353	–	–	–	–	1	1
10 % HNO_3 + 60 % H_2SO_4	333	2	2	–	0	0	0
10 % HNO_3 + 60 % H_2SO_4	353	2	2	–	1	0	1
10 % HNO_3 + 70 % H_2SO_4	323	3	3	1	0	0	–
10 % HNO_3 + 70 % H_2SO_4	363	3	3	3	1	0	–
10 % HNO_3 + 70 % H_2SO_4	441	3	3	3	3	3	–
10 % HNO_3 + 80 % H_2SO_4	323	–	–	–	0	0	0
10 % HNO_3 + 80 % H_2SO_4	353	–	–	–	1	1	1
10 % HNO_3 + 90 % H_2SO_4	308	–	–	–	0	0	0
15 % HNO_3 + 20 % H_2SO_4	323	3	3	1	0	0	–
15 % HNO_3 + 20 % H_2SO_4	353	3	3	2	1	0	–
20 % HNO_3 + 80 % H_2SO_4	293	0	0	–	0	0	0

0 = Rate of mass loss < 0.1 g/m^2 h, corresponding to a corrosion rate < 0.11 mm/a (< 4.3 mpy)
1 = Rate of mass loss 0.1–1.0 g/m^2 h, corresponding to a corrosion rate 0.11–1.1 mm/a (4.3–43 mpy)
2 = Rate of mass loss 1.0–10.0 g/m^2 h, corresponding to a corrosion rate 1.1–11 mm/a (43–433 mpy)
3 = Rate of mass loss > 10.0 g/m^2 h, corresponding to a corrosion rate > 11.0 mm/a (> 433 mpy)

Table 25: Chemical resistance of stainless steels in mixtures of nitric and sulfuric acids [66]

Table 25: Continued

Concentration of the mixed acid	*Temperature, K*	*Material*					
		1.4000, 1.4002, 1.4006, 1.4021, 1.4028, 1.4104, 1.4034, 1.4110, 1.4112, 1.4528	***1.4016, 1.4057, 1.4122***	***1.4405, 1.4542, 1.4548***	***1.4301, 1.4304, 1.4306, 1.4361, 1.4544, 1.4546, 1.4550***	***1.4401, 1.4404, 1.4435, 1.4436, 1.4571***	***1.4439, 1.4460, 1.4462, 1.4465, 1.4466, 1.4467, 1.4539***
20 % HNO_3 + 80 % H_2SO_4	333	1	1	–	1	0	0
20 % HNO_3 + 80 % H_2SO_4	373	–	–	–	1	1	1
25 % HNO_3 + 75 % H_2SO_4	323	3	2	1	1	0	–
25 % HNO_3 + 75 % H_2SO_4	363	3	3	1	1	1	–
25 % HNO_3 + 75 % H_2SO_4	430	3	3	3	3	3	–
30 % HNO_3 + 20 % H_2SO_4	353	–	–	–	1	0	0
30 % HNO_3 + 40 % H_2SO_4	353	–	–	–	1	1	1
30 % HNO_3 + 70 % H_2SO_4	308	–	–	–	0	0	0
50 % HNO_3 + 20 % H_2SO_4	353	–	–	–	1	1	1
50 % HNO_3 + 50 % H_2SO_4	323	3	2	1	0	0	–
50 % HNO_3 + 50 % H_2SO_4	363	3	3	2	1	1	–
50 % HNO_3 + 50 % H_2SO_4	393	3	3	3	2	2	–
65 % HNO_3 + 35 % H_2SO_4	308	0	0	–	0	0	0
90 % HNO_3 + 10 % H_2SO_4	308	–	–	–	0	0	0

0 = Rate of mass loss < 0.1 g/m^2 h, corresponding to a corrosion rate < 0.11 mm/a (< 4.3 mpy)
1 = Rate of mass loss 0.1–1.0 g/m^2 h, corresponding to a corrosion rate 0.11–1.1 mm/a (4.3–43 mpy)
2 = Rate of mass loss 1.0–10.0 g/m^2 h, corresponding to a corrosion rate 1.1–11 mm/a (43–433 mpy)
3 = Rate of mass loss > 10.0 g/m^2 h, corresponding to a corrosion rate > 11.0 mm/a (> 433 mpy)

Table 25: Chemical resistance of stainless steels in mixtures of nitric and sulfuric acids [66]

Sulfuric and hydrochloric acids

In dilute mixtures with 1.8–6.8 % H_2SO_4 and 0.45–1.7 % HCl, a steel of type 1.4401 is absolutely not resistant at 353 K (80 °C) with corrosion rates of 2–15 mm/a (80–590 mpy).

The common factor in all processes used to produce pulp and paper from wood chippings is that the bleaching solutions based on sulfuric acid also contain chlorides (from the wood material as a result of marine transportation as well as from the process water used), so that stainless steels are mainly endangered by local corrosion, i.e. pitting corrosion, crevice corrosion and stress corrosion cracking because, in addition, the temperature can rise up to 343 K (70 °C). In order to produce very white paper, the fibres must be bleached in oxidising chemicals, such as chlorine, chlorine dioxide, hypochlorite and hydrogen peroxide (so-called C and D stages). This means the formation of hydrochloric acid in the bleaching solution, which is thus a mixed acid. In the subsequent scrubbers and filters, the conditions are similarly corrosive. The conditions are even more corrosive if, because of the dioxin problem, the more environmentally friendly chlorine dioxide is used: the bleaching solution is more strongly oxidising and the process is carried out at even higher temperatures. In the past, because of these very corrosive conditions and because of warranty reasons and low maintenance costs, highly alloyed nickel materials that were also relatively expensive were used to be on the safe side. After laboratory investigations and extensive pilot and exposure tests, the more economically viable superaustenites with higher chromium and molybdenum contents are now being used more and more frequently. The special stainless steel Nicrofer® 3127 hMo – alloy 31 (DIN-Mat. No. 1.4562) represents an alternative, as exposure tests in the bleaching stage and scrubber regions show [67]. According to these tests, it exhibits a better performance in some cases than nickel alloys because, according to electrochemical studies in acid solutions, it has an extensive passive range and a very high transpassivation potential. This is attributed to the unusually high chromium content of 27 % Cr coupled with a high molybdenum content (6.5 % Mo). A wide range of stainless steels and nickel alloys were exposed on test racks in a bleaching plant of a foreign paper and pulp manufacturer. The materials listed in Table 26 were tested for 136 days under the process conditions: 348 K (75 °C), 900 ppm chlorides, pH value 3.0–5.0.

It can be seen from the table, that uniform surface corrosion is very low for all steels and nickel alloys and that local corrosion was observed for most steels and even a few nickel alloys. Alloy 31 (X1NiCrMoCu32-28-7, DIN-Mat. No. 1.4562) exhibited the best behavior along with the Mn-containing steel X2CrNiMnMoNbN25-18-5-4 (DIN-Mat. No. 1.4565). Even the superaustenite containing a high amount of Mo X1NiCrMoCuN25-20-7 (DIN-Mat. No. 1.4529) exhibited slight etching. Of the nickel alloys, alloy 625 (NiCr22Mo9Nb, DIN-Mat. No. 2.4856), a material generally used for critical applications, did not exhibit very favorable behavior as noticeable etching was observed. Only alloy C-276 (NiMo16Cr15W, DIN-Mat. No. 2.4819) obviously has unlimited suitability; it exhibited the lowest corrosion rate and total absence of any local attack. The commercial stainless steel 316 L (X2CrNiMo17-12-2,

Material	Mat. No.	PRE *)	Corrosion rate		Observations
			mm/a	mpy	
Alloy 316 L	1.4435	26	0.0062	0.24	pitting up to 2 mm
Cronifer® 1925 hMo	1.4529	47	0.0028	0.11	slight attack
Alloy 24	1.4565	>40	0.0039	0.15	no attack
Nicrofer® 3127 hMo – alloy 31	1.4562	53	0.0060	0.24	no attack
Nicrofer® 3127 LC – alloy 28	1.4563	39	0.0051	0.20	slight attack
Nicrofer® 4823 hMo – alloy G-3	2.4619	46	0.0083	0.33	slight attack
Nicrofer® 5716 hMoW – alloy C-276	2.4819	69	0.0030	0.12	no attack
Nicrofer® 5923 hMo – alloy 59	2.4505	75	0.0051	0.20	slight attack
Nicrofer® 6020 hMo – alloy 625	2.4856	51	0.0092	0.36	uniform surface corrosion
Nicrofer® 6616 hMo – alloy C-4	2.4610	68	0.0064	0.25	

*) Pitting Resistance Equivalent: % Cr + 3.3 × % Mo + 30 × % N

Table 26: Results of the exposure of various high-alloy steels and nickel alloys on test racks in the bleaching solution of a paper and pulp plant after 136 days at 348 K (75 °C) [67]

DIN-Mat. No. 1.4404) is definitely unsuitable for such plant regions due to strong local corrosion and pitting depths of up to 2 mm.

Similar results were found in a Scandinavian plant. The conditions in the chlorine dioxide stage were: 200 ppm active chlorine, pH value 2–3, temperature 323–333 K (50–60 °C). After an exposure time of 174 days and compared to other tested steels, the superaustenite Nicrofer® 3127 hMo – alloy 31 (DIN-Mat. No. 1.4562) exhibited very low uniform corrosion without any local attack. In contrast, the steel X6CrNiMoTi17-12-2 (DIN-Mat. No. 1.4571) exhibited pitting corrosion [67]. This behavior was confirmed in a further plant, located in South America, under the following conditions: pH value 3.4–4.5 (1.9 at times) and a temperature of 345–351 K (72–78 °C). After a test period of 184 days, the superaustenites Nicrofer® 3127 hMo and Cronifer® 1925 hMo – alloy 926 (X1NiCrMoCuN25-20-7, DIN-Mat. No. 1.4529) – both highly alloyed with Mo (6.5 %) – exhibited the lowest corrosion rates with 0.034 and 0.011 mm/a (1.339 and 0.433 mpy), respectively, and no local corrosion. Although the nickel alloys that were also tested, did not exhibit local corrosion, they did show quite considerable general corrosion at a rate of 0.15 mm/a (6 mpy). Therefore, when selecting materials for use under the given operating conditions, the plant engineer does not necessarily need to use nickel materials because the less expensive superaustenites are also very resistant and the very low corrosion rates also promise a higher service life of the plant and its components [67]. Finally, both superaustenites were tested in a North American plant along with other stainless steels and nickel alloys over a period of 60–128 days in the intake region of a scrubber under the following conditions: 20–200 ppm chlorine gas, 600 ppm chloride,

300 ppm chlorate, pH value 6–7. While the steel SAE 317 L (X2CrNiMo18-15-4, DIN-Mat. No. 1.4438) did indeed exhibit lower corrosion rates, it also exhibited marked crevice corrosion in all stages, with attack depths of 0.4, 0.6 and even 1.25 mm. The nickel materials were free of local corrosion; however, they showed relatively high uniform corrosion rates of 0.5 and 0.3 mm/a (20 and 12 mpy). The results of exposure tests in the scrubber filtrate are given in Table 27.

Material	*Chemical composition, mass %*					*Corrosion rate mm/a (mpy)*	*Crevice corrosion*	
	Cr	*Ni*	*Mo*	*N*	*other*		*Max. depth mm*	*No. of sites attacked*
UNS S31703 (DIN-Mat. No. 1.4438)	18.3	14.5	3.47	0.09		< 0.0025 (< 0.1)	0.125	3/40
UNS S31254 (DIN-Mat. No. 1.4547)	20.1	18.05	6.10	0.19	Cu 0.78	< 0.0025 (< 0.1)	0.025	3/40
UNS S31254 (DIN-Mat. No. 1.4547)	19.7	18.2	6.27	0.20	Cu 0.16	< 0.0025 (< 0.1)	0.325	3/40
UNS N08367	20.3	24.7	6.3	0.19		< 0.0025 (< 0.1)	0.425	3/40
Cronifer® 1925 hMo (DIN-Mat. No. 1.4529)	20.7	25.05	6.18	0.18	Cu 0.86	0 (0)	0	0
Nicrofer® 3127 hMo (DIN-Mat. No. 1.4562)	26.8	31.7	6.5	0.2	Cu 1.2	0.0025 (0.1)	0	0
UNS N08026	24.9	33.6	5.2	0.12	Cu 2.9	< 0.0025 (< 0.1)	0.0125	3/40
UNS N06030 (DIN-Mat. No. 2.4603)	28.6	bal.	5.2		W 2.3	< 0.0025 (< 0.1)	0	0
UNS N10276 (DIN-Mat. No. 2.4819)	15.7	bal.	15.7		W 3.4	0.4725 (19)	0	0
UNS N06022 (DIN-Mat. No. 2.4602)	20.5	bal.	13.1		W 2.8	0.38 (15)	0	0

Table 27: Behavior of various alloys and steels in a paper plant and pulp plant; after 2 months' exposure in the scrubber filtrate [67]

With regard to the observed crevice corrosion, the maximum attack depths are given in column 8. Column 9 gives the number of sites attacked by crevice corrosion from a total of 40 artificially generated crevices. The corrosion rates given in column 7 are generally extremely low. The overall result shows that, once again, the super-austenites show excellent behavior because they are also free of any local corrosion [67].

In flue gas desulfurisation plants (FGDP) of hard coal- and brown coal-fired power stations as well as in waste incineration plants (WIPs), the corrosion conditions in the plants operating with the wet scrubbing process are very demanding [70]. Scrubber solutions are formed which are very dilute mixtures of sulfuric and hydrochloric acids (cf. also A 29). Commercial stainless steels based on 18/10CrNi, even those with 3–4 % Mo, fail due to pitting and crevice corrosion. Only superaustenites containing higher amounts of chromium and/or molybdenum can be used in some areas of a FGDP or WIP. In connection with the qualification of high-alloy nickel materials for corrosively loaded plant regions, namely the raw gas intake, the pre-scrubber and the quencher, the superaustenites Cronifer® 1925 hMo (DIN-Mat. No. 1.4529) and Nicrofer® 3127 hMo (DIN-Mat. No. 1.4562) were studied. The following practice-based testing medium was used as a FGD-simulating solution: H_2SO_4 with pH 1 + 3 % Cl^- + 0.05 % F^- + 15 % FGD gypsum ($CaSO_4$, contaminated). This solution is suitable to differentiate, in an acceptable time period (max. 3 × 30 days), between resistant and non-resistant materials with high contents of nickel, chromium and molybdenum. With regard to the resistance towards crevice corrosion, which was also tested, an additional time-lapse effect is obtained mounting a slotted Teflon block (i.e. a 12-fold slotted Teflon disc) on every surface of the sample with a fixed torque. Temperatures between 333 and 353 K (60 and 80 °C) were used in accordance with practical conditions.

Material	*DIN Mat. No.*	*Temperature K*	*Break-through potential mV*	*Repassi-vation potential mV*	*Corrosion rate*		*Observations*
					mm/a	*mpy*	
Cronifer® 1925 hMo	1.4529	333	1,150	1,065	< 0.001	< 0.04	no attack
		353	575	< 375	approx. 0.7	approx. 28	pitting corrosion
Nicrofer® 3127 hMo-alloy 31	1.4562	333	1,140	995	< 0.001	< 0.04	no attack
		353	1,110	935	< 0.001	< 0.04	no attack

Table 28: Corrosion behavior of the materials Cronifer® 1925 hMo (DIN-Mat. No. 1.4529) and Nicrofer® 3127 hMo – alloy 31 (DIN-Mat. No. 1.4562) in the simulated FGD solution: H_2SO_4 with pH 1 + 3 % Cl^- + 0.05 % F^- + 15 % FGD gypsum ($CaSO_4$, contaminated, aeration, stirring); determination of the breakthrough potential using the potentiostatic method, the corrosion rate was determined after a test period of 10 days [70]

The experiments carried out under electrochemically controlled conditions showed that after 10 days, the steel X1NiCrMoCuN25-20-7 (UNS N08926, DIN-Mat. No. 1.4529) is resistant up to the process temperatures of 333 K (60 °C); however, it already exhibited strong uniform and pitting corrosion above 353 K (80 °C) with a corrosion rate of 0.7 mm/a (28 mpy). This is explained by the marked decrease of the breakthrough potential as well as the repassivation potential. In contrast, Nicrofer® 3127 hMo is much more stable due to its unusually high chromium content of 27 %. The breakthrough and repassivation potential remained at a high level, even at 353 K (80 °C). Local corrosion was not observed, the corrosion rates of 0.001 mm/a (0.04 mpy) were hardly measurable. Similar results were obtained for a test in an enhanced

FGD medium containing 7% Cl^-, i.e. with more than double the chloride content of the otherwise comparable testing medium.

In the very stringent test ASTM G-28 (method B), which is actually used only for the highest alloyed NiCrMo alloys to determine their resistance to local corrosion, the high-alloy superaustenites are not resistant, as shown in Table 29 [68].

Material	*Corrosion rate*	
	mm/a	*mpy*
Cronifer® 1925 hMo, DIN-Mat. No. 1.4529	80	3,150
Nicrofer® 3127 LC, DIN-Mat. No. 1.4563	80	3,150
Nicrofer® 3127 hMo, DIN-Mat. No. 1.4562	4.7	185

Table 29: Test of Nicrofer® 3127 hMo – alloy 31 and reference materials according to ASTM G 28, method B, in boiling solution of 23% H_2SO_4 + 1.2% HCl + 1% $FeCl_3$ + 1% $CuCl_2$ [68]

Nevertheless, special stainless steel Nicrofer® 3127 hMo, extremely highly-alloyed with 27% Cr and 6.5% Mo, exhibits much more favorable behavior than the other two special stainless steels. Corresponding behavior and, if applicable, resistance can be expected at lower concentrations or lower temperatures of the mixed acid.

Phosphoric and sulfuric acids

The stainless steel SAE 904 L (X1NiCrMoCu25-20-5, DIN-Mat. No. 1.4539) exhibited good resistance in a highly concentrated phosphoric acid with additions of sulfuric and hydrofluoric acids up to 318 K (45 °C) or even up to 373 K (100 °C) in a chloride-free medium, while steel SAE 316 L (X2CrNiMo17-12-2, DIN-Mat. No. 1.4404) failed, Figure 23 [69]

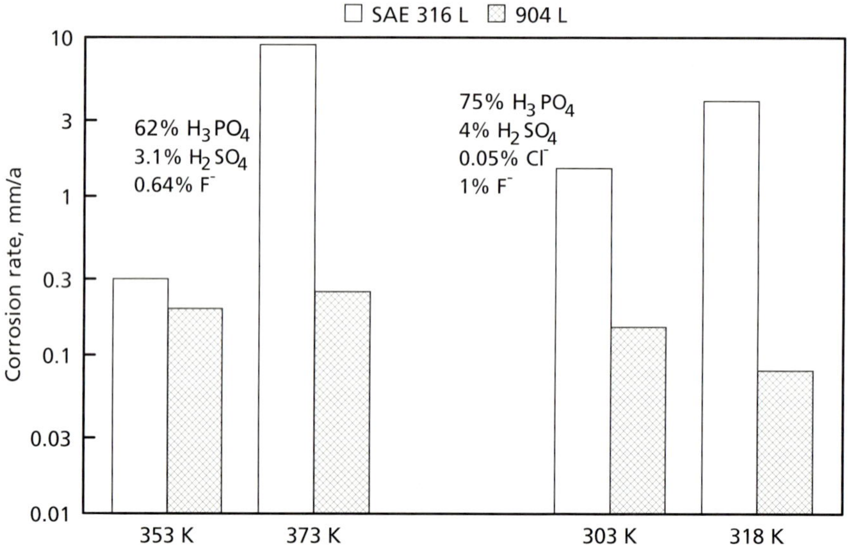

Figure 23: Corrosion rate of SAE 316 L (DIN-Mat. No. 1.4404) and 904 L (DIN-Mat. No. 1.4539) in impure phosphoric acid [69]

The superaustenites UR B6 (X1NiCrMoCu25-20-5, DIN-Mat. No. 1.4539), Sanicro® 28 (X1NiCrMoCu31-27-4, DIN-Mat. No. 1.4563), and UR SB8 (cf. DIN-Mat. No. 1.4529) were tested in comparison with the austenitic-ferritic steel UR 52N (25 Cr, 6.5 Ni, 3 Mo, 1.5 Cu, 017 N) in phosphoric acid from various production facilities (cf. also [31]). The influence of sulfuric acid, hydrofluoric acid, chlorides, temperature and abrasion were tested extensively. While the steel UR B6 was generally inferior to the austenitic-ferritic steel and cannot be entirely recommended, the superaustenites UR SB8 and S 28 exhibited equivalent behavior.

The steel Nicrofer® 3127 hMo (DIN-Mat. No. 1.4562) was tested for its suitability in the concentration stage of a phosphoric acid plant using a simulated solution under typical practical conditions, namely for a period of 3 × 7 days at 373 K (100 °C) [67]. The composition of the simulated solution was: 72 % H_3PO_4 + 4.5 % H_2SO_4 + 0.9 % hexafluorosilicic acid + 400 ppm hydrochloric acid + 1.5 % Fe_2O_3.

Material	*Corrosion rate*	
	mm/a	*mpy*
Nicrofer® 3127 LC – alloy 28 (DIN-Mat. No. 1.4563)	0.15	5.9
Nicrofer® 4823 hMo – alloy G-3 (DIN-Mat. No. 2.4619)	0.17	6.7
Nicrofer® 6020 hMo – alloy 625 (DIN-Mat. No. 2.4856)	0.25	9.8
Nicrofer® 3127 hMo – alloy 31 (DIN-Mat. No. 1.4562)	0.13	5.1

Table 30: Resistance of Nicrofer® 3127 hMo – alloy 31 (DIN-Mat. No. 1.4562) and reference materials in a simulated technical phosphoric acid solution [67]

It can be seen that the superaustenite Nicrofer® 3127 hMo – alloy 31 is superior to the nickel alloys, even the very high-alloy and versatile alloy 625 (DIN-Mat. No. 2.4856). Although the superaustenite alloy 28 (DIN-Mat. No. 1.4563) exhibited similarly good behavior with a corrosion rate of 0.15 mm/a (6 mpy), the use of alloy 31 should be given preference because it has an approx. 3 % higher Mo content and thus a higher resistance to pitting and crevice corrosion that, under the conditions of high chloride loading as well as traces of hydrochloric acid, can occur in steels and nickel alloys which are not sufficiently highly alloyed (Pitting Resistance Equivalent).

The results of further investigations of two superaustenites are summarised in Table 31 [70].

Alloy 31 (DIN-Mat. No. 1.4562) with approx. 3.5 % more Mo exhibited much better behavior than alloy 28 (DIN-Mat. No. 1.4563), which has frequently been used in the past because it is much more resistant than commercial stainless steels. However, the tests showed the limitations for the use of alloy 31 at the temperatures 393 K and 373 K (120 °C and 100 °C).

During the production of raw phosphoric acid, up to 30 % solids can be present as sulfates in the hot suspension. This gives rise to abrasive as well as corrosive conditions which markedly enhance the attack on all vessel materials. Systematic

Test medium	Temperature, K	Alloy 31 (1.4562)	Alloy 28 (1.4563)
52 % P_2O_5 + 4.5 % H_2SO_4 + 0.9 % H_2SiF_6 + 1.5 % Fe_2O_3 + 400 ppm Cl^-	353	0.02 (0.8)	0.0075 (0.3)
	393	0.78 (31)	–
52 % P_2O_5	389	0.08 (3)	1.2 (47)
30 % P_2O_5 + 2.4 % H_2SO_4 + 2.3 % H_2SiF_6 + 1 % Fe_2O_3 + 1,000 ppm Cl^-	353	0.015 (0.6)	–
54 % P_2O_5	383	0.05 (2)	1.4 (55)
54 % P_2O_5 +2,000 ppm Cl^-	383	2.04 (80)	2.3 (91)
	373	1.30 (51)	–

Table 31: Corrosion rates in mm/a (mpy) of the superaustenites alloy 28 (DIN-Mat. No. 1.4563) and alloy 31 (DIN-Mat. No. 1.4562) in practice-related phosphoric acid solutions [70]

laboratory investigations studied the effect of abrasive conditions on several high-alloy steels and nickel materials in solutions of the composition: 28 % P_2O_5 + 1,000 ppm chlorides + 2.3 % sulfate + 2 % fluorine + solids present in practice [67]. Round samples of the materials were electrochemically tested at various rotation speeds and at temperatures around 351 K (78 °C) at a range of potentials generally found in practice. It can be seen from the bar diagram in Figure 24 that alloy 904 L (X1NiCrMoCu25-20-5, DIN-Mat. No. 1.4539), which has very often been used in phosphoric acid plants as stirrers, drums etc., exhibited a corrosion rate of more than 0.5 mm/a (20 mpy) at rotation speeds (25 rpm) generally used in practice.

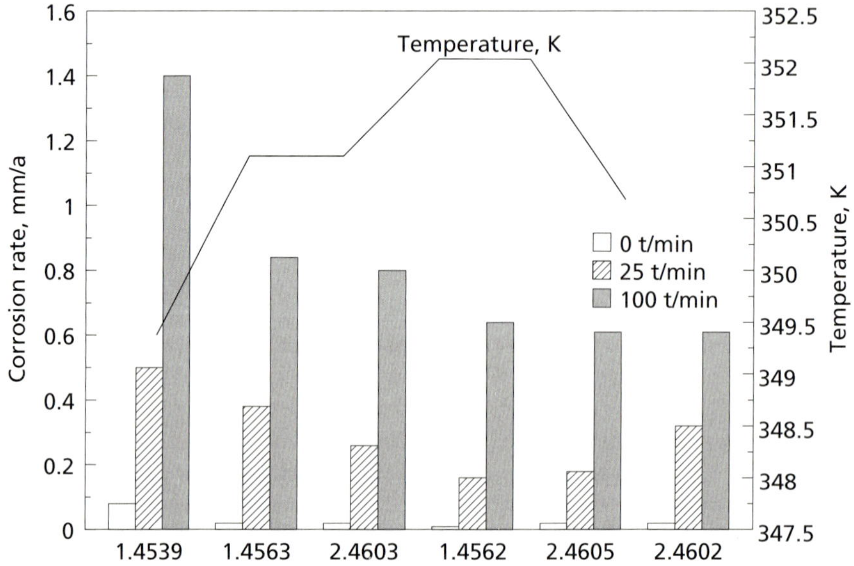

Figure 24: Corrosion under abrasive conditions; electrochemical studies of high-alloy steels and nickel materials in contaminated technical phosphoric acid at various rotation speeds [67]

At this rotation speed, alloy 31 (X1NiCrMoCu32-28-7, DIN-Mat. No. 1.4562) exhibited the most favorable behavior with corrosion rates of approx. 0.1 mm/a (4 mpy), even compared to the very highly alloyed nickel materials alloy 59 (NiCr23-Mo16Al, DIN-Mat. No. 2.4605) and alloy C22 (NiCr21Mo14W, DIN-Mat. No. 2.4602). At a very high rotation speed of 100 rpm, the corrosion rates increased to 1.4 mm/a (55.2 mpy) for alloy 904 L, while it increased to 0.6 mm/a (24 mpy) for the other materials. Such corrosion rates are acceptable in practice because the stirring arms are very thick.

Direct exposure tests with samples attached to the stirring arms of acid thickeners gave similar results: under combined corrosive and abrasive conditions, alloy 31 exhibited the best relative behavior [70].

With respect to the above, the good behavior of these superaustenites in hot sulfuric acid of low to high concentrations should be mentioned at this point. Figure 25 shows the good resistance of the steel up to a concentration of 80 % and up to a temperature of 353 K (80 °C). The conditions become extremely corrosive only above these values and indeed corrosion increases abruptly as the temperature increases [70].

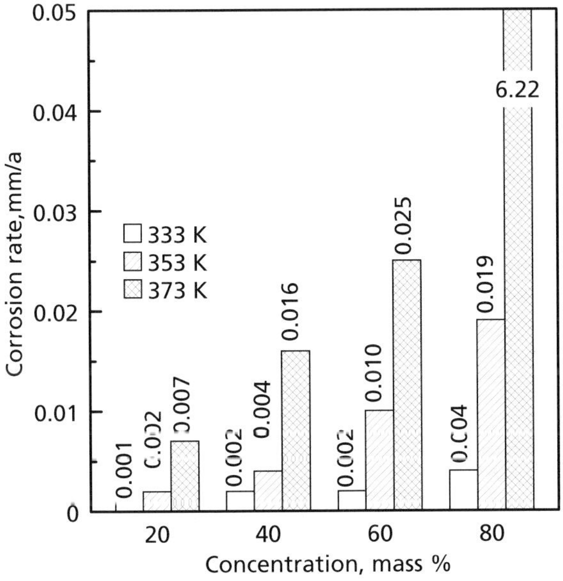

Figure 25: Corrosion rate of the superaustenite alloy 31 (DIN-Mat. No. 1.4562) in hot sulfuric acid of various concentrations and temperatures [70]

Nitric and hydrofluoric acids

A high-chromium stainless steel X1NiCrMoCu31-27-4 (DIN-Mat. No. 1.4563) exhibited greatly increasing corrosion in HNO_3-HF mixtures containing 0.25 M HF at 363 K (90 °C) in the range 6 to 14 M HNO_3. Corrosion rates between 1 and 5 mm/a (40 and 200 mpy) indicate that this material is not resistant. In alloys with varying amounts of Mo, it was shown that high Mo contents have a favorable effect, particularly for high nitric acid concentrations [33].

In boiling mixtures of nitric and hydrochloric acids as well as sulfuric acid, nitric acid has an inhibiting effect on special stainless steels alloyed with Mo, as shown in Table 32 [71].

Acid mixture	***Temperature, K***	***Corrosion rates, mpy (mm/a)***	
		Ferralium® 255	***Avesta® 254 SMO DIN Mat. no.1.4547***
6 N HNO_3	382	1.4 (0.036)	3.5 (0.089)
1 N HCl	373	11,162 (283.5)	2,935 (74.6)
		1 N HCl	
+1 N HNO_3	373	dissolved in 2 h	5.6 (0.142)
+2 N HNO_3	375		14 (0.356)
+4 N HNO_3	378		42.9 (1.01)
+6 N HNO_3	382	18.5 (0.47)	108 (2.74)
		1 N H_2SO_4	
no HNO_3	373	25.9 (0.658)	73 (1.854)
+1 N HNO_3	373	8,584 (218.0)	2.5 (0.064)
+2 N HNO_3	375	dissolved in 2 h	2.7 (0.069)
+4 N HNO_3	378		4.5 (0.114)
+6 N HNO_3	382	7 (0.178)	7.7 (0.196)

Table 32: Corrosion rates in boiling acid mixtures [71]

Because of the higher Cr content of 25 % in the duplex steel Ferralium® 255, its corrosion rate of 0.04 mm/a (1.6 mpy) in pure HNO_3 was better than that of the high-molybdenum Avesta® 254 SMO (DIN-Mat. No. 1.4547) with 0.1 mm/a (4 mpy). In contrast, the attack in 1 M HCl is extremely high for both steels. The addition of 1 M nitric acid to boiling 1 M HCl has a strongly inhibiting effect only for the high-molybdenum steel: the corrosion rate decreases from 75 mm/a (295 mpy) to 0.14 mm/a (5.5 mpy). For the addition of more highly concentrated HNO_3, the corrosive attack on the higher molybdenum steel increases again. The duplex steel is completely dissolved on addition of up to 4 M HNO_3, and became abruptly more resistant (0.5 mm/a / 19.7 mpy) on addition of 6 M HNO_3. In boiling 0.5 M H_2SO_4, the high-molybdenum stainless steel was less resistant (1.8 mm/a / 70.9 mpy) than the duplex steel Ferralium® 255 (0.65 mm/a / 25.6 mpy). Additions of up to 4 M HNO_3 decreased the corrosion rate to approx. 0.1 mm/a (4 mpy), while it increased for the duplex steel up to total dissolution.

Electroplating

More highly alloyed stainless steels are used for tanks and appliances, even for immersed bath heaters in galvanic engineering, whose baths primarily consist of very different types of mixed acids. For practical experience on the resistance of steels, such as DIN-Mat. No. 1.4571 (X6CrNiMoTi17-12-2) and DIN-Mat. No. 1.4539 (X1NiCrMoCu25-20-5) [72].

Bibliography

[1] Groggins, P. H.
Unit Process in Organic Synthesis
McGraw Hill Book Co., New York, 1952

[2] Fontana, M. G.
Stress Corrosion (monthly column)
Ind. Engng. Chem. 46 (1954) 3,
pp. 99A–102A

[3] Werner, M.; Ruttmann, W.
Korrosion an metallischen Werkstoffen
(Corrosion at metallic materials)
(in German)
Zeitschr. VDI 94 (1952), pp. 1113–1121

[4] Company publication
Nelson, G. A.
Corrosion Data Survey Emeryville
Ca. Shell Development Co.

[5] Kaplan, N.; Andrus, R. J.
Verhalten von Al-Legierungen gegenüber Mischsäure
(Behavior of Al alloys in mixed acids)
(in German)
Ind. Engng. Chem. 40 (1948),
pp. 1946–1947

[6] von der Forst, P.
Gußeisen im Apparatebau der chemischen Industrie
(Cast iron in apparatus engineering of the chemical industry) (in German)
Werkst. Korros. 10 (1959), p. 213

[7] Rabald, E.
Legierter Guß im chemischen Apparatebau
(Cast alloy in the chemical apparatus engineering) (in German)
Werkst. Korros. 7 (1956), p. 436

[8] Eddy, J.; Rohrman, F. A.
Effect of Mixed Acids upon irons and steels
Ind. Engng. Chem. 28 (1936), pp. 30–31

[9] Kaplan, N.; Andrus, R. J.
Ind. Eng. Chem. 40 (1948), pp. 1946–1947
see also: Korrosion von Metallen in roter rauchender Salpetersäure und in Mischsäuren
(Corrosion of metals in red smoking nitric acid and in mixed acids) (in German)
Werkst. Korros. (1950) 4, pp. 154–156

[10] Oknin, I. V.
Electrochemical study of the corrosion of iron in the system HNO_3-H_2SO_4-H_2O
Zhur. Fiz. Khim. 26 (1952), pp. 1057–1058
Chem. Abstr. 46 (1952), p. 6523

[11] Company publication
Technische Rundschau Sulzer 37,
pp. 15–35, 2/1955
Sulzer (Switzerland)

[12] Fierz-David, H. E.; Blangley, L.
Farbenchemie, 8. ed.
(Chemistry of colours) (in German)
Springer, Wien, 1952

[13] Thoenges, E. F.
Chem. Metallurg. Engng. 41 (1934) 10,
pp. 526

[14] Mysson, Ch.
EP 0543 729 B1 (1995)
Procede de decapage de materiau en acier doux et bain de decapage

[15] Asrar, N.; Thakur, C. P.
Pickling silicon steels in mixed acid solutions
Metal Finishing 93 (1995) 2, pp. 70–72

[16] Tandel, P. B.; Oza, B. N.
Thermometric evolution of inhibitive action of aniline and its derivatives for mild steel in binary acid mixtures
Transactions of the SAEST 33 (1998) 1,
pp. 14–17

[17] Luce, W. A.
High Silicon Irons
Chem. Engng. 61 (1954) 1, pp. 250–258

[18] Römfeld, J.
Achema-Bericht
(Achema report) (in German)
Werkst. Korros. 10 (1959), pp. 58

[19] DECHEMA-WERKSTOFF-TABELLE
45. Ergänzungslieferung "Mischsäuren"
(45th supplement "Mixed acids")
(in German)
DECHEMA e.V., Frankfurt a.M. (1961)

[20] Berchman, L. J.; Kapali, V.; Natarajan, S., Sivan, V.
General corrosion behavior of modified 9Cr-1Mo-Steel weldments in acidic environments
Br. Corros. J. 29 (1994) 2, pp. 143–146

[21] Kawabata, Y.; Owada, S. et al.
Development of Electrolytic Descaling Method for Stainless Steels Using HNO_3–HCl Acid, Conference: Process and Materials: Innovation Stainless Steel; Vol. 2, Florence Italy; 11–14 Oct. 1993
Associazione Italiana di Metallurgia, Milano, 1993, pp. 2.83–2.87

[22] Company publication
ABC der Stahlkorrosion, 1958
(ABC of steel corrosion) (in German)
Mannesmann, Düsseldorf

[23] Roesch, K.
Entwicklung und Stand des nichtrostenden Stahlgusses
(Development and conditions of stainless cast steel) (in German)
Stahl und Eisen 70 (1950), p. 602

[24] Guthmann, K.
Korrosionsbeständige Stahlsorten
(Corrosion resistant steel grades)
(in German)
Chem. Technik 15 (1942) p. 9

[25] EP 0 609 618 B1, 1997
Stainless Steel Composition
Mintek, Randburg, Transvaal, South Africa

[26] Bühler, K.
Die Rolle der Begleitwerkstoffe von Abgaskondensaten aus der Erdgasverbrennung bei der Korrosion v. Wärmetauschern
(Impurity constituents of flue gas condensates and their significance for the corrosion of heat exchangers) (in German)
Werkst. Korros. 44 (1993) 7, pp. 289–294

[27] Corrosion in Acid Mixtures
in: ASM International
Metals Handbook, 9th Edition, Vol. 13
ASM International, Metals Park, Ohio, USA, 1987, pp. 646–656

[28] Konstantina, E. V.; Lomovtsev, V. I.
Corrosion of low-nickel and nickel-free stainless steels in concentrated nitric acid with hydrofluoric acid or chloride ions
Zashch. Met. 19 (1983) 6, pp. 933–934

[29] Asphahani, A. I.
Corrosion Resistance of High Performance Alloys
Mater. Performance 19 (1980) 12, pp. 33–43

[30] Renaud, L., Bosson, J. C.; Charles, J.; Chapey, B.; Oltra, R.
Erosion-Corrosion Properties of Austenitic and Duplex Stainless Steels Conference: Duplex Stainless Steels '91, Vol 2, Bourgogne, France, 28–30 Oct. 1991
Les Editions de Physique, Courtaboeuf – Les ulis Cedex France, 1992, pp. 939–947

[31] Audouard, J. P.; Soulignac, P.
Duplex stainless steels for acid plants
Duplex Stainless Steels '91, Bourgogne, 28–30 Oct. 1991, Vol. 2
Ed. de Physique, Les Ulis, 1992, pp. 1121–1123

[32] Company publication
Langley Corrosion Resistance, GA8692/3M7481
Langley Alloys Ltd., Slough (UK)

[33] Horn, E.-M.; Manning, P. E.; Renner, M.
Korrosion nichtrostender Stähle und Nickelbasislegierungen in Salpetersäure-Flußsäure-Gemischen
(Corrosion of stainless steels and nickel-base alloys in solutions of nitric acid and hydrofluoric acid) (in German)
Werkst. Korros. 43 (1992) 5, pp. 191–200

[34] Covino, B. S.; Scalera, J. V.; Driscoll, T. J.; Carter, J. P.
Dissolution Behavior of 304 Stainless Steel in HNO_3/HF Mixtures
Metallurgical Transactions 17A (1986), pp. 137–149

[35] Neumann, G.
Chemische und mechanische Entzunderung hochlegierter Eisenwerkstoffe – unter besonderer Berücksichtigung korrosionsbeständiger Stähle
(Chemical and mechanical descaling of highly alloyed iron-base materials with special consideration of stainless steels) (in German)
Neue Hütte 35 (1990) 9, pp. 348–351

[36] Vajpeyi, M.; Gupta, S.; Dhirendra; Pandey, G. N.
Corrosion of Stainless steel (SAE 304) in H_2SO_4 contaminated with HCl and HNO_3
Corrosion Prevention and Control (1985) 10, pp. 102–104

[37] Dillon, C. P.
Corrosion of Type 347 Stainless Steel and 1100 Aluminium in strong Nitric and Mixed Nitric-Sulfuric Acids
Corrosion 12 (1956), pp. 623t–626t

[38] Krystow, P. E.; Balicki, M.
Behavior of 18-8 Stainless Steel In 2 Normal Boiling Nitric and Sulfuric Acid Mixtures
Corrosion 12 (1956), pp. 449t–454t

[39] McDonald, I. O. S.
Brit. Chem. Eng. 1 (1956), pp. 254–259

[40] Brink, J. A.; Shreve, R. N.
Nitration of 2-Methylnaphthalene
Ind. Eng. Chem. 46 (54) pp. 694–702
[41] Pratt, W. E.
Worthite
Chem. Eng. 60 (1953) 11, pp. 270
[42] Luce, W. A.
Durimet 20-Carpenter 20
Chem. Eng. 60 (1953) 12, p. 284
[43] Company publication
Cronifer – Crofer, 1965
Vereinigte Deutsche Metallwerke, Altena
[44] Viebrock, J. M.
Corrosion of Type 304 Stainless Steel in Mixed Anhydrous Nitric and Sulfuric Acids
Corrosion 25 (1969) 9, pp. 371–379
[45] Dhirenda, A.; Gupta, S.; Pandey, G. N.; Sanyal, B.
'Waterline' and 'Above Water Line' Corrosion of Stainless Steel (SAE 321) in H_2SO_4 and its Inibition
Corros. Prev. Control 34 (1987) 2. pp. 58–59
[46] Gupta, S; Vajpeyi, M; Pandey, G. N.
Determination of corrosion of stainless steel (SAE 304) by mixed vapours of HCl and HNO_3
Corros. Prev. Control, April 1986, pp. 47–50
[47] Gupta, S.; Kumar, Y.; Sanyal, D. B.; Pandey, G. N.
The exothermic reaction of stainless steel (SAE 304) in mixtures of HCl and HNO_3, and the inhibition of its corrosion
Corros. Prev. Control (1983) 2, pp. 11–14
[48] Sing, V. B., Arvind, U.
Electrodissolution of SAE stainless steel in concentrated acids leading to electropolishing
Indian Journal of Chemical Technology 2 (1995) 7, pp. 211–216
[49] Erkaev, A. U.; Namazov, S. S.; Ibragimov, I. G.
Corrosion resistance of extractors coming into contact with sulfuric acid
Khimicheskaya prommyshlennost (Moskau) 9 (1993), pp. 432–433
[50] Singh, M. M; Mukherjee, A. K.; Khare, R.
The Effect of additions of HCl on the corrosion behavior of 304 Stainless steel in concentrated H_3PO_4 at different temperatures
Bull. Electrochem. 11 (1995) 10, pp. 457–461
[51] Ovchiyan, V. N. et al.
Corrosion of steels in mixture of acids containing dichlorethane
Khim Prom. 47 (1971) 10, pp. 748–749
[52] Company publication
Corrosion Tables, 1999
AB Sandvik Steel (Sweden)
[53] Reichau, U.; Pletka, H. D.; Schütze, K. G.
Oberflächenbehandlung als Korrosionsschutzmaßnahme von nichtrostenden Stählen
(Surface treatment as corrosion protection measure of stainless steels) (in German)
Werkst. Korros. 43 (1992) 11, pp. 520–526
[54] Company publication
Rockel, M. B.; Herda, W. R.
Die Entwicklung neuer Nickelbasislegierungen an Beispielen
(The development of new nickel-base alloys by examples) (in German)
lecture Werkstoffwoche Munich 12–15 Oct. 1998
KRUPP VDM, Werdohl
[55] Rocha, H. J.
Säurebeständige Stähle. Gesichtspunkte der Bewertung ihrer Beständigkeit
(Acid-resistant steels. Criteria of the evaluation of their resistance) (in German)
Chem. Fabrik 13 (1940) 21, pp. 379–384
[56] Tofaute, W.
Nickelfreie und nickelarme rost- und säurebeständige Stähle
(Nickel free and low-nickel corrosion resistant and acid-resistant steels) (in German)
Dechema-Monographien 12 (1941) pp. 62–64
[57] Singh, I.; Bhattamishra, A. K.; Basu, D. K.
Electrochemical behavior of SAE-316 stainless steel in sulfuric/nitric acid mixtures
Anti-corrosion Methods and Materials 44 (1997) 3, pp. 200–203
[58] Puderbach, H.; Grosse-Boewig, W.
Analyse von Adsorptionsschichten auf Edelstahlblechen
(Analysis of adsorption layers on fine steel metal sheets) (in German)
Fresenius. Zeitschrift für analytische Chemie, Labor+Betriebsverfahren 319 (1984) 6/7, pp. 627–630

[59] Verma, K. M.; Ghosh, H.; Pattnaik, K. C.; Singh, R. U.
Corrosion-Erosion Characteristics of Rock Phosphate in the Manufacture of Wet Process Phosphoric Acid
Br. Corros. J. 17 (1982) 2, pp. 71–74

[60] Guenbour, A.; Faucheu, J.; Ben Bachir, A.; Dabosi, F.; Bui, N.
Electrochemical study of corrosion-abrasion of stainless steels in phosphoric acids
Br. Corros. J. 23 (1988) 4, pp. 234–238

[61] Schorr, M.
Stainless steels for corrosion control in the phosphoric acid industry
Stainless steel world 10 (1998) 3, pp. 25–29

[62] Blanchard, F.; Jollain, C.; Martin, G.
Tubes in Stainless Grades for Particularly Corrosive Conditions in the Chemical Industry
Aciers Spec. 68 (1984) 11, pp. 19–24

[63] Khare, R.; Mukherjee, A. K.; Singh, M. M
The corrosion and pitting of SAE 304 SS and SAE 316 SS in 14 M phosphoric acid containing HCl
Transactions of the SAEST 29 (1994) 2, pp. 118–125

[64] Manning, P. E.; Smith, J. D.; Nickerson, J. L.
New Versatile Alloys for the Chemical Process Industry
Mater. Performance 27 (1988) 6, pp. 67–73

[65] Company publication
Chemische Beständigkeit der NIROSTA-Stähle, 1 – 06/97
KRUPP THYSSEN NIROSTA – KRUPP, Düsseldorf

[66] Company publication
Böhler – Chemische Beständigkeit nichtrostender Böhler-Edelstähle,
AL 170 D – 1.89 – 3.000 Gl, January 1989
(Böhler – chemical resistance of stainless Böhler noble steels) (in German)
Böhler, Kapfenberg (Austria)

[67] Company publication
Rockel, M. B.
Nicrofer 3127 hMo – Alloy 31 ein neuer, hoch legierter Sonderedelstahl für die chemische Verfahrenstechnik, VDM Case History 6, March 2001
(Nicrofer 3127 hMo – Alloy 31 a new, high alloyed special stainless steels for chemial engineering) (in German)
KRUPP VDM GmbH, Werdohl

[68] Heubner, U.; Rockel, M.; Wallis, E.
Ein neuer hochlegierter Nickel-Chrom-Molybdän-Stahl für den Chemie-Apparatebau
(A new high-alloyed nickel-chromium-molybdenum steel for the chemical process industry) (in German)
Werkst. Korros. 40 (1989), pp. 418–426

[69] Nordin, S.
Stainless Special Steels for the Chemical Industry
The Chemical Engineer (1974) 11, pp. 724–726

[70] Heubner, U. et al.
Nickelwerkstoffe und hochlegierte Sonderedelstähle, 2. ed., Vol. 153
(Nickel materials and high-alloyed special stainless steels) (in German)
expert verlag, TAE Esslingen (1993)

[71] Shibad, P. R.; Das, C. M.
Kinetics of Corrosion of two Molybdenum Containing Stainless Steels and their Behavior in Mineral Acids
J. Electrochem. Soc. India 42 (1993) 4, pp. 219–224

[72] Company publication
Beständigkeitstabelle für NÜGA Sicherheitstauchbadwärmer, 2001
(Resistance table for NÜGA safety bath emersion heaters) (in German)
NÜGA Galvanotechnische Elektrowärme GmbH, Georgensgemünd

Nitric Acid

Unalloyed steels and cast steel

While carbon steel or low-alloy steel is severely attacked by dilute nitric acid, it is resistant in > 40 % concentrated nitric acid due to the formation of a protective passive layer [1].

If the concentration of nitric acid is increased to 7.9 mol/l, after a sharp increase in corrosion, the state of passivity is reached where the corrosion drops practically to zero. In the active range, e.g. at 0.5 mol/l HNO_3, the material consumption rate of iron can be reduced to 0.67 g/m^2 h by addition of about 100 g/l sulfamic acid. In current density/potential measurements, a pulsed current occurs in the range 0 to + 0.2 V_{SCE} [2].

Figure 1 shows the material consumption rate of a carbon steel in nitric acid at room temperature as a function of the HNO_3 concentration. Above 35 % HNO_3 there is a sharp drop in the material consumption rate due to the formation of a passive layer [3].

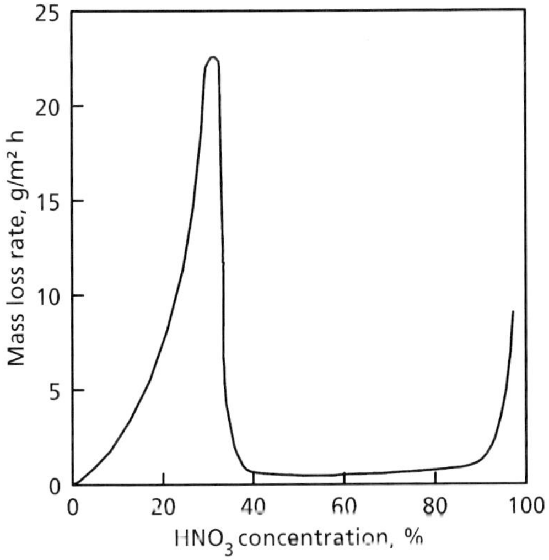

Figure 1: Dependence of the material consumption rate of C steel on the HNO_3 concentration at room temperature [3]

The corrosion of steel-10 and steel-45 in 50 % nitric acid at room temperature is increased by a factor of about 4 by 60 % deformation [4].

Dissolution of iron in 1 mol/l flowing nitric acid leads to the formation of ammonia, nitric oxide (NO) and nitrite ions (NO_2^-) in approximately equal amounts. Dinitrogen monoxide (N_2O) and nitrogen are additionally found in stationary acid [5].

Steel-20 corrodes in 1 mol/l nitric acid at room temperature at a material consumption rate of 193 g/m^2 h [6].

Hydrogen present both in soft iron and ferritic steel (Fe-0.15C-0.36Si-1.49Mn-0.017P-0.018S-0.02Ni-0.076Al) caused a further increase in corrosion in 0.5 mol/l nitric acid [7].

Normally killed steel ((%) Fe-0.013C-0.1Mn-0.005Al-0.002Si-0.065O) is resistant after deoxidation with aluminium only after careful heat treatment. Carbide-forming elements in the steel, such as, for example, titanium (Ti-stabilized steel containing 0.01 % C + 0.30 % Si + 0.50 % Mn) prevent cracking in boiling nitric acid. The resistance of low-alloy steels in boiling nitric acid generally depends largely on the structure [8].

Corrosion of iron cannot be prevented in 0.7 to 13.7 mol/l nitric acid at 300 and 333 K (27 and 60 °C) even by cathodic polarization between 30 and 40 A/cm^2. The material consumption rate of iron in 8.03 and 11.2 mol/l nitric acid at 300 K (27 °C) for example, thus increases from about 0.5 at 0 V to about 4 and 30 g/m^2 h respectively at – 2 V_{SHE} [9].

The chromized steels 3931 (Fe-0.035C-0.54Mn-0.35Ti-1.7Ni-0.1Si) and 4238 (Fe-0.06C-0.13Mn-0.44Ti-0.23Si) corrode in 65 % nitric acid at 333 K (60 °C) with a material consumption rate of 0.0026 and 0.0052 g/m^2 h [10]. The material consumption rate of the steel 4240 (Fe-0.06C-0.25Mn-0.6Nb-0.23Si) is the lowest, at < 0.00002 g/m^2 h.

Low-alloy steels cannot be inhibited effectively against corrosion in 0.3 to 2 mol/l nitric acid at room temperature by amines (monoethyl-, monomethyl- and dimethylamine) and hydrazine [11]. Benzenesulfonic acid derivatives cannot reduce the corrosion of iron in 2 mol/l nitric acid to an industrially usable degree [12].

Additions of 0.05 to 0.5 g/l potassium salts of benzoic, salicylic, oxalic, succinic, adipic and p-nitrobenzoic acid have practically no effect on corrosion of the steel St 3 (1.0333) in 1 mol/l nitric acid at room temperature [13].

While the iodides of N-decyl- and N-dodecylquinoline greatly inhibit corrosion of St 3 in sulfuric acid at 293 K (20 °C) and 353 K (80 °C), no effect was found in 0.5 and 2 mol/l nitric acid [14].

The inhibition efficiency of some substances and the material consumption rates of unalloyed steels in various concentrations of nitric acid can be seen from Table 1.

HNO_3 concentration mol/l	Temperature K (°C)	Inhibitor	Inhibitor		Material consumption rate g/m^2 h	Literature
			Concentration	Efficiency %		
0.001[1)]	303 (30)	none	–	–	0.081	[15]
		sodium petroleum sulfonate	200 ppm	36	0.052	
0.01[1)]		none	–	–	0.229	[15]
		sodium petroleum sulfonate	50 ppm	27	0.167	
			200 ppm	49	0.117	

1) Fe-0.35Mn-0.2Cu-0.025P-0.035S
2) Armco® iron
3) steel St 3
4) distillation residues obtained during ethanolamine purification

Table 1: Inhibition efficiency of various substances and material consumption rates of unalloyed steels in nitric acid

Table 1: Continued

HNO_3 concentration mol/l	Temperature K (°C)	Inhibitor	Inhibitor Concentration	Inhibitor Efficiency %	Material consumption rate g/m^2 h	Literature
0.1[1)]		none	–	–	6.25	[15]
		sodium petroleum sulfonate	50 ppm	33	4.17	
			200 ppm	83	1.04	
0.1[2)]	RT	none	–	–	59	[16]
		benzylpyridine thiocyanate	0.0044 mol/l	98.8	0.7	
		allylpyridine thiocyanate	0.0056 mol/l	98.8	0.7	
		benzyl-2-methyl pyridine rhodanide	0.0041 mol/l	99.1	0.5	
1.0	RT	none	–	–	573	[17]
		indole + thiocyanate	2.1 + 9.8 mmol/l	99.8	1.1	
			4.27 + 6.56 mmol/l	99.85	0.82	
1.0[3)]	RT	none	–	–	885	[18]
		indole + sodium sulfide	0.2 + 0.8 g/l	99.7	2.5	
2.0[3)]	298 (25)	none	–	–	1820	[19]
		[4)] + benzoquinoline thiocyanate (8:1)	2.0 + 0.25 mol/l	99.9	1.8	
3.0	RT	none	–	–	1820	[17]
		indole + thiocyanate	2.1 + 9.8 mmol/l	> 99.9	0.96	
			4.27 + 6.56 mmol/l	> 99.9	0.68	

1) Fe-0.35Mn-0.2Cu-0.025P-0.035S
2) Armco® iron
3) steel St 3
4) distillation residues obtained during ethanolamine purification

Table 1: Inhibition efficiency of various substances and material consumption rates of unalloyed steels in nitric acid

Benzotriazole and thiourea (0.010 mol/l) inhibit the corrosion of low-alloy steel (Fe-0.22C-0.79Mn-0.66Si-0.05S-0.03P) in 0.1 mol/l nitric acid at room temperature to the extent of 91 and 96 % [20]. Benzotriazole is considerably more effective than normal triazole or aminotriazole.

According to Table 2, the material consumption rate of Armco® iron (Fe-0.03C-0.10Mn-0.18Si-0.01P-0.015S), steel-45 (Fe-0.44C-0.23Cr-0.26Ni-0.69Mn-0.34Si-0.013P-0.02S) and the steels U7A (cf. 1.1520, C70U; Fe-0.70C-0.11Cr-0.09Ni-0.26Mn-0.28Si-0.013P-0.008S) and U10A (cf. 1.1545, SAE W1; Fe-0.98C-0.12Cr-0.20Ni-0.20Cu-0.26Mn-0.25Si-0.03P-0.02S) at 295 K (22 °C) in 1 mol/l nitric acid is reduced by hydrochloric acid. The greatest reduction occurs on Armco® iron [21, 22].

Material	*HNO_3*	*$HNO_3 + HCl$*		
	1 mol/l	*100 : 1*	*10 : 1*	*1 : 1*
	Material consumption rate, $g/m^2\,h$			
Armco® iron	1050	980	770	4
Steel 45	1030	960	450	8
Steel U7A	1060	950	550	19
Steel U10A	1070	1010	95	23

Table 2: Corrosion of Armco® iron and three carbon steels in 1 mol/l HNO_3 with and without additions of 1 mol/l HCl at 295 K (22 °C) [21]

The material consumption rates of Armco® iron in nitric acid/hydrochloric acid mixtures listed in Table 3 are considerably lower. Corrosion is increased by the high additions of hydrochloric acid. A halving of the material consumption rate of iron in 0.5 mol/l nitric acid is found with 1.5 mol/l hydrochloric acid [23].

HNO_3 concentration mol/l	*HCl addition mol/l*	*Material consumption rate $g/m^2\,h$*
0.5	–	18.85
0.5	1.5	9.74
0.5	3.0	26.3
0.5	6.0	97.8
1.0	–	40.35
1.0	6.0	285
2.0	–	147
2.0	6.0	641

Table 3: Corrosion of Armco® iron in HNO_3 with and without HCl at room temperature [23]

Corrosion of steel in pure hydrochloric acid (5.8 mol/l) at a rate of 0.8 g/m^2 h is increased to 2.1 g/m^2 h by additions of nitric acid (e.g. 40 percent by volume 5.8 mol/l HNO_3), and reduced again when the additions are increased further (e.g. to 0.63 g/m^2 h by 60 percent by volume HNO_3) [24].

According to Table 4, the corrosion of carbon steel at 323 K (50 °C) in concentrated sulfuric acid is intensified by small additions of nitric acid. At a content of 1128 ppm HNO_3, however, passivation of the steel occurs, associated with a sharp decrease in corrosion rates [25].

HNO_3 addition ppm	–	5.6	14	25	56	141	423	1128
Corrosion rate, mm/a (mpy)	0.11 (4.33)	0.23 (9.06)	0.36 (14.17)	0.47 (18.5)	0.66 (25.98)	0.78 (30.71)	1.55 (61.02)	0.034 (1.34)

Table 4: Influence of HNO_3 additions on the corrosion of C steel at 323 K (50 °C) in concentrated H_2SO_4 after 320 h [25]

The corrosion rate of carbon steel at 294 K (21 °C) in 65 % sulfuric acid of 0.89 mm/a (35 mpy) is increased to 10.5 and 14 mm/a (413 and 551 mpy) by additions of 2 and 2.6 % HNO_3. After passing through a maximum of > 30 mm/a (1180 mpy), the corrosion rate drops to about 0.2 mm/a (7.9 mpy) on addition of 8 % HNO_3. Regardless of the concentration of sulfuric acid (62 – 65), corrosion at 311 K (38 °C) reached a rate of < 0.03 mm/a (< 1.18 mpy) after addition of 0.5 % nitric acid after 6 d tests. The corresponding value for 70 % sulfuric acid + 0.5 % HNO_3 was 0.05 mm/a (1.97 mpy) [26].

Figure 2 shows the course of the corrosion rate of carbon steel in sulfuric acid solutions of various concentrations at 310 K (37 °C) as a function of small amounts of nitric acid. The corrosion rate is already < 0.02 mm/a (0.79 mpy) with 0.5 % HNO_3 [27].

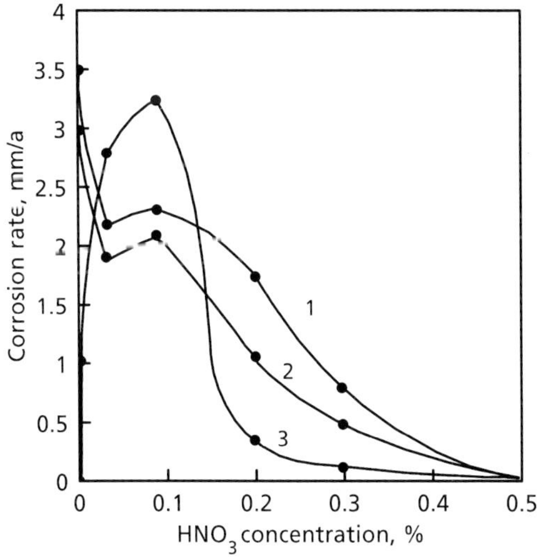

Figure 2: Corrosion rate of C steel in H_2SO_4 with small additions of HNO_3 at 310 K (37 °C) [27] 1) 62.2; 2) 64.9 and 3) 70.7 % H_2SO_4

At concentrations which are not too high (< 100 g/l HNO_3), silver nitrate causes chemical passivation of iron in nitric acid, as shown in Figure 3, whereby the material consumption rate does not exceed 0.1 g/m^2 h. In contrast, the corrosion of iron in 90 g/l $AgNO_3$ solution greatly increases with a growing concentration of nitric acid [28]. About 5, 15 and 40 g/l silver nitrate are needed in a solution containing 30, 75 and 150 g/l HNO_3 to inhibit the corrosion of iron to < 0.1 g/m^2 h.

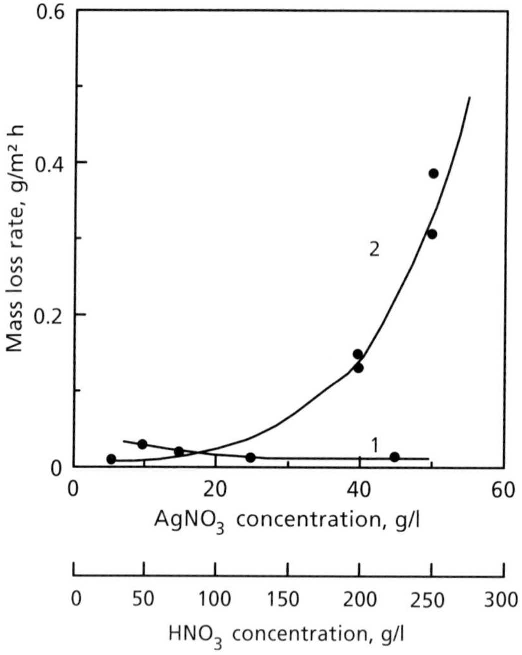

Figure 3: Dependence of the material consumption rate of passivated iron on the concentration of HNO_3 and $AgNO_3$ [28]
1) chemical passivation in 75 g/l HNO_3 at varying $AgNO_3$ concentrations
2) chemical passivation in 90 g/l $AgNO_3$ at varying HNO_3 concentrations

An inhibitor solution of 5 – 10 % nitric acid + 15 g/l urea + 2 g/l thiourea is used to clean plants made of carbon steel and allowed to act at 293 or 313 K (20 or 40 °C) for 5 h. While the rate of dissolution is 460 or 895 g/m^2 h respectively without an inhibitor, it is reduced to 1.5 and 7.1 g/m^2 h in the presence of the inhibitor mixture (inhibition efficiency > 99 %) [29].

Unalloyed cast iron

Mixtures of 45 % H_2SO_4 + 50 % HNO_3 + 5 % H_2O attack cast iron at higher temperatures [30]. Silicon-containing cast iron, high-silicon cast iron, is superior to unalloyed cast iron in its corrosion behavior in nitric acid [31].

Cast iron tanks made of pearlitic gray cast iron with a lamellar graphite structure are attacked during regeneration of waste sulfuric acid (about 8 to 10 mm/a (314 to 394 mpy)). On the other hand, if 5 % nitric acid is added as a passivator at the start of the test, corrosion of the cast iron is reduced to about 1.5 mm/a (59 mpy) [32].

High-alloy cast iron, high-silicon cast iron

The material consumption rate of high-silicon cast iron containing 14.5 % Si in 5 to 65 % nitric acid between 293 and 373 K (20 and 100 °C) is < 0.1 g/m^2 h [33]. As the silicon content increases, the corrosion behavior of cast iron in nitric acid is increasingly improved [34].

According to Table 5, cast iron with a low silicon content (Fe-(3.4 – 3.8)C-(1.9 – 3.1)Si-(0.6 – 0.9)Mn) is severely attacked in nitric acid at 298 K (25 °C) [35].

Cast iron	*HNO_3 concentration, %*		
	1	*5*	*10*
	Material consumption rate, g/m^2 h		
Nodular graphite	42	285	490
Lamellar graphite	70	365	671

Table 5: Corrosion of nodular graphite cast iron containing 2.5 – 3.1 % Si and cast iron with lamellar graphite containing 1.9 – 2.4 % Si in HNO_3 at 298 K (25 °C) [35]

The corrosion rate of the gray cast iron ChS-13 with a carbon and silicon content of 0.9 and < 9 % respectively in 40 % nitric acid was > 1 mm/a (> 40 mpy), and that with an Si content of 12.3 % was 0.025 mm/a (0.98 mpy) (tempered at 1773 K (1500 °C) for 30 min before the test). Corresponding tests in a solution of 69 to 71 % sulfuric acid + 25 to 27 % nitric acid at 363 K (90 °C) resulted in a corrosion rate of 0.01 to 0.02 mm/a (0.39 to 0.79 mpy). At the same time, the improved flexural strength was 250 – 280 N/mm^2 and the hardness 100 N/mm^2 [36] A high-silicon cast iron containing not less than 14 % silicon should be chosen for mixtures of nitric acid and sulfuric acid at the boiling point [34].

Iron-silicon alloys (Tantiron® N) containing 14.25 to 15.25 % Si are absolutely resistant to nitric acid. However, where particularly severe corrosive attack is feared, e.g. in mixtures of HNO_3 and H_2SO_4 at higher temperatures, Tantiron® E with a silicon content of 16 to 18 % is recommended [37].

Cast iron containing 15 % silicon has an adequate resistance to nitric acid containing potassium chloride at room temperature, with a corrosion rate of about 0.15 mm/a (5.91 mpy) [38].

Alloyed cast iron (Fe-3.3C-3.2Si-0.3Mn-0.3P-0.06S) is attacked by a solution of 0.3 % $FeCl_3$ + 0.1 % HCl + 0.1 % HNO_3 at room temperature with a material consumption rate of about 6.0 and 0.59 g/m^2 h after 10 and 90 d respectively. The rate decreases further after more prolonged exposure giving, for example, about 0.23 g/m^2 h after 800 d [35].

Structural steels with up to 12 % chromium

The ferritic chromium steel ((%) Fe-0.1C-9.0Cr-1Mo-0.47Mn-0.44Si-0.017P-0.017S) shows only mild general corrosion without preferential local dissolution after exposure to boiling 65 % nitric acid for 24 h. A test carried out on a weld point of the steel showed no intercrystalline corrosion [39].

The austenitic chromium-nickel steels predominantly used in the nitric acid industry can be replaced by the steel 1.4512 (SAE 409 (%) Fe-<0.08C-(10.5 – 12.5)Cr-<1 Mn-<1.0 Ti-<1.0Si-0.2Al-<0.045P-<0.030S) for moderately hot, (sub)-azeotropic nitric acid. In azeotropic nitric acid free from oxidizing agents, the steel corrodes at a material consumption rate of < 0.01 g/m^2 h due to the formation of protective surface layers of aluminium and chromium oxides [40].

Addition of 20 to 400 ppm chloride to 5, 23 and 56 % nitric acid has no influence on the corrosion rate of chromium steel ((%) Fe-(0.16 – 0.24)C-12Cr-0.6Mn-0.6Ni-0.6Si-<0.035P-<0.030S), which is about 0.79, 0.26 and 0.13 mm/a (31.1, 10.24 and 5.12 mpy) respectively at room temperature [41].

While hardened chromium steels ((%) Fe-(0.22 – 0.29)C-(4.1 – 14)Cr – (0.12 – 0.36)Ni – (0.25 – 0.5)Mn – (0.07 – 0.48)Si) with a chromium content of 4 to 11 % corrode in boiling 65 % nitric acid at a rate of about 9 g/m^2 h, this figure drops to < 0.015 g/m^2 h at 14 % chromium [42].

Ferritic chromium steels with more than 12 % chromium

All chromium steels with a chromium content of ≥ 14 % are resistant to all concentrations of nitric acid at room temperature [43]. Chromium steels of the types Fe-0.015(C + N)-26Cr-1Mo and Fe-0.025(C + N)-29Cr-4Mo (Ti- or Nb-stabilized) as well as Fe-0.025(C + N)-29Cr-4Mo-2Ni (Ti- or Nb-stabilized) are absolutely resistant in boiling 65 % nitric acid [44].

A chromium steel with 30 % chromium shows a material consumption rate of < 0.1 g/m^2 h in 5 to 65 % nitric acid in the temperature range 293 to 373 K (20 to 100 °C). At 373 K (100 °C), the corrosion rate is > 0.5 g/m^2 h in > 30 % nitric acid [33]. Addition of molybdenum leads to no increase in the resistance.

Table 6 lists some test results on the corrosion behavior of E-Brite® 26-1 (26Cr-1Mo) in nitric acid as the temperature rises [45].

Corrosion rates of the chromium-molybdenum steel E-Brite® 26-1 ((%) Fe-0.001C-26.0Cr-1.0Mo-0.25Si-0.12(Ni + Cu)-0.01Mn-0.01P-0.01S-0.01N) in nitric acid can also be seen from Table 7. If hydrogen fluoride is present at the same time, the corrosion rate of the steel is increased considerably, rising, for example, to about 2.5 mm/a (98.43 mpy) in 2 mol/l HNO_3 + 0.5 mol/l HF [46]. E-Brite® 26-1 is suitable for coolers and condensers for nitric acid. The resistance of E-Brite® 26-1 to nitric acid is confirmed again and again [47].

HNO_3 concentration %	Temperature K (°C)	Corrosion rate mm/a (mpy)
10	394 (121)	0.012 (0.47)
	422 (149)	0.093 (3.7)
	436 (163)	0.059 (2.32)
	477 (204)	21.1 (831)
30	379 (106)	0.011 (0.43)
	422 (149)	0.151 (5.94)
	452 (179)	0.152 (5.98)
60	366 (93)	0.024 (0.94)
	393 (120)	0.097 (3.8)
	422 (149)	3.52 (139)
65	394 (121)	1.61 (63.4)
	422 (149)	18.9 (744)
70	366 (93)	0.12 (4.7)
	394 (121)	0.15 (5.9)
90	364 (91)	1.09 (42.9)

Table 6: Corrosion rate of E-Brite® 26-1 in HNO_3 at various temperatures [45]

HNO_3 concentration %	State steam/liquid	Temperature K (°C)	Corrosion rate mm/a (mpy)
10	steam	433 (160)	0.081 (3.19)
20			0.122 (4.8)
30			0.325 (12.8)
65	liquid	boiling	0.10 – 0.13 (3.94 – 5.12)
75			0.356 (14.02)
85			0.737 (29.02)
70	liquid	299 (26)	< 0.025 (< 0.98)
90	liquid		< 0.025 (< 0.98)
95	liquid, fuming		< 0.025 (< 0.98)
90	liquid	333 (60)	0.25 (9.84)

Table 7: Corrosion of E-Brite® 26-1 in various concentrations of HNO_3 [46]

The corrosion rate of the steel 07Ch13AG20 ((%) Fe-0.038C-14.3Cr-19.5Mn-0.7Ni-0.02Ti-0.35Si-0.12N) in 50 % nitric acid at 333 K (60 °C) is about 0.1 mm/a (3.94 mpy) after 300 h, regardless of the exposure time [48].

The corrosion behavior of the nickel-free or low-nickel chromium-manganese steel AS-43 ((%) Fe-0.05C-18.2Cr-16.3Mn-0.36Si-0.4Nb-0.005B-0.05Ce-0.008S-0.014P-0.83-Ni) and, for comparison, that of the austenitic chromium-nickel steel Ch18N10T (SAE 321, 1.4541) in 300 h tests in hot nitric acid is shown in Table 8. While the material consumption rates of the steels AS-43 and Ch18N10T in boiling 45 % nitric acid are 0.60 and 0.22 g/m^2 h after 100 h, the rates are 0.50 and 0.40 g/m^2 h after 500 h. After heat treatment between 873 and 1173 K (600 and 900 °C), chromium-manganese steel is not susceptible to intercrystalline corrosion [49].

Steel	*HNO_3 concentration %*	*Temperature, K (°C)*			
		333 (60)	*363 (90)*	*373 (100)*	*boiling point*
		Material consumption rate, g/m^2 h			
AS-43	45	0.012	0.070	0.20	0.55
	65	0.028	0.180	0.55	7.0
Ch18N10T	45	0.045	0.014	0.062	0.33
	65	0.008	0.050	0.30	1.4

Table 8: Corrosion behavior of the steels AS-43 and Ch18N10T (SAE 321, 1.4541) in hot HNO_3 [49]

The test developed by Huey with boiling 65 % nitric acid indicates intercrystalline corrosion damage due to chromium carbide precipitates and sigma phase formation in Mo-alloyed and Ti- or Ta/Nb-stabilized steels [50].

The Huey test is generally used to evaluate the resistance of stainless austenitic steels to exposure to nitric acid (65 to 67 %). The specimens are exposed for 5 periods of 48 h each, the nitric acid being renewed and the weight loss being determined after each period. The steels on which the weight loss does not rise noticeably over the entire test duration are described as resistant.

The "Superferrite" X 1 CrMoNb 30 2 (Japanese steel) with the composition (%) Fe-<0.004C-0.14Si-<0.02Mn-0.014P-0.012S-30.52Cr-1.75Mo-0.17Ni-0.009N-<0.02Cu-0.12Nb was investigated, inter alia, in [51] on this basis, but with an extended test duration of a total of 50 periods and with measurement of the depth of intergranular attack. The material showed increasing material consumption rates and an increase in the depth of intercrystalline corrosion both in the delivery state and after additional annealing (1123 K (850 °C)/20 min/air). The material consumption rates reached 0.12 to 0.15 g/m^2 h after a test duration of 100 d. The material is not suitable for long-term exposure in hot nitric acid [51]. The "Superferrite" X1CrNiMoNb28-4-2 (25-4-4, UNS S32803, 1.4575) is also unsuitable.

The chromium steels Fe-35Cr and CrMo26-1 corrode in boiling 65 % nitric acid with a corrosion rate of 0.15 and 0.20 mm/a (5.91 and 7.87 mpy), respectively [52]. No intercrystalline corrosion was found.

Vacuum-smelted, highly pure ferritic 25Cr steel ((%) Fe-(0.005 – 0.007)C-(24 – 25)Cr-(0.42 – 0.63)Mn-(0.04 – 0.35)Si-(0.014 – 0.015)S-(0.024 – 0.028)P-(0.005 – 0.010)N) with a low carbon content is superior to normal ferritic steels in its corrosion resistance. To keep the susceptibility to intercrystalline corrosion in boiling 65 % nitric acid as low as possible, as well as the general corrosion, stabilization annealing at 1053 K (780 °C) for about 1 h with subsequent quenching in water is necessary. The sensitivity to corrosion increases above 1073 K (800 °C) and as a result of longer annealing times [53].

The high-chromium ferritic steel with a low carbon and nitrogen content 005Ch25B ((%) Fe-0.005C-0.007N-25Cr; Nb-stabilized), heat treated at 1573 K (1300 °C) for 30 min, corroded in boiling 65 % nitric acid with a rate of 0.21 mm/a (8.27 mpy) (5 cycles of 48 h each). The corrosion resistance, especially towards intercrystalline corrosion, is improved at a ratio of Nb/(C + N) 20 [54]. Addition of 0.6 to 4.7 % molybdenum alters the corrosion behavior in 65 % nitric acid giving corrosion rates of about 8 mm/a (315 mpy) for 0.6 % Mo, 0.4 mm/a (15.75 mpy) for 2.3 % Mo and 3.6 mm/a (142 mpy) for 4.7 % Mo.

Nitrogen-alloyed chromium-manganese steel ((%) Fe-0.04C-0.455N-19.2Cr-10.6Mn-0.6Si) corrodes in boiling 65 % nitric acid the slowest in the quenched state at 0.49 g/m^2 h, compared with steel tempered at 873 and 1173 K (600 and 900 °C). The steel tempered at 873 K (600 °C) for 20 or 2 h corrodes at a rate of 3.8 and 0.66 g/m^2 h, respectively [55]. The susceptibility to intercrystalline corrosion caused in particular during longer tempering times at 873 K (600 °C) is reduced by higher tempering temperatures.

The chromium-manganese steel Ch18AG14 with 18 % chromium, 14 % manganese and 0.5 % nitrogen behaves similarly in boiling 65 % nitric acid. The lowest material consumption rate here is 1.2 g/m^2 h [56, 57].

The resistance of chromium-manganese steels of the type ((%) Fe-<0.1C-(13 – 21)Cr-(7 – 17)Mn-(0.03 – 0.88)N) to intercrystalline corrosion in boiling 65 % nitric acid is increased by nickel. Although the steel containing (%) Fe-18.1Cr-7.4Mn-0.3N is resistant to intercrystalline corrosion in the absence of nickel, it is attacked at a rate of about 0.7 mm/a (27.6 mpy) [58]. Replacement of the chromium by manganese makes the steel highly susceptible to intercrystalline corrosion.

The corrosion behavior of nitrogen-chromium-manganese steels ((%) Fe-0.03C-0.57N-11.0Mn-21.3Cr) with a ferrite content of 5 % in boiling 65 % nitric acid depends greatly on the temperature and duration of annealing. After an annealing time of 20 to 60 min at 873 K (600 °C), a material consumption rate of about 0.7 g/m^2 h results after 24 h, but after annealing for 120 and 480 min, values of about 3.2 and 10.7 g/m^2 h respectively occur. Annealing for 2 h at 1273 K (1000 °C) led to a material consumption rate of 0.6 g/m^2 h [59].

20CrMo steels with 3 or 5 % molybdenum and 1 to 5 % nickel show average material consumption rates of about 0.2 g/m^2 h in the Huey test. At the same time, however, mild to moderate intercrystalline attack is found, this occurring preferentially in the region of weld seams. In contrast, all 28CrMo steels with 2 % molybdenum and 2 to 4 % nickel are completely resistant to intercrystalline attack with an average material consumption rate of 0.1 g/m^2 h [60].

The titanium-stabilized chromium-molybdenum steel ELI 18-2CrMo(Ti) ((%) Fe-0.02C-17.5Cr-2.1Mo-0.23Mn-0.26Ti-<0.02(P + S)-<0.8 (Ni + Cu)-0.60Si) gives a corrosion rate between 2.6 and 5.5 mm/a (102 and 217 mpy) in the welded state after exposure to boiling 65 % nitric acid for 3 periods of 48 h each. Heat treatment at 1113 K (840 °C) for 6 h reduces the figure to about 1.8 mm/a (71 mpy) [61].

The material consumption rate of pipe specimens of 18Cr2Mo steel ((%) Fe-0.018C-17.5Cr-2.52Mo-0.67Ti-0.4Sr-0.04Mn-0.009P-0.006S), produced from powder and cast ingots, in boiling 65 % nitric acid is 0.67 and 0.41 g/m^2 h, respectively [62].

Molybdenum-free chromium steel ((%) Fe-(0.003 – 0.006)C-17.3Cr-0.003N) corrodes in boiling 65 % nitric acid, after quenching from 1323 K (1050 °C), with a material consumption rate of 0.657 g/m^2 h after 250 h. The same steel containing 3 % molybdenum gave a figure of 1.05 g/m^2 h under the same test conditions. If the two steels are tempered at 973 K (700 °C) for 15 h, the corrosion of the Mo-free steel is reduced to 0.42 g/m^2 h and that of the Mo-containing steel increased to 9.4 g/m^2 h [63].

The chromium steel Fe-<0.006C-17.3Cr, solution-annealed at 1323 K (1050 °C) for 0.5 h and then quenched in water, corroded in boiling 65 % nitric acid at a rate of 0.65 g/m^2 h in the Huey test. After additional tempering at 973 K (700 °C) for 1 h, the material consumption rate dropped to 0.44 g/m^2 h. An alloying addition of 3.08 % molybdenum to the Cr steel led to an increase in the first case to 1.2 and in the second case to about 6.0 g/m^2 h [64].

The corrosion of the chromium steels SAE 430 ((%) Fe-0.07C-17Cr-0.45Mn-0.45Si-0.04P-0.01S-0.025N) and ASTM XM-27 (E-Brite® 26-1: (%) Fe-0.002C-26Cr-1Mo-0.10Ni-0.05Mn-0.25Si-0.01P-0.01S-0.10Nb-0.01N) and of the CrMoNi steel 29-4-2 ((%) Fe-0.005C-29Cr-4Mo-2Ni-0.05Mn-0.10Si-0.015P-0.010S-0.013N) in boiling 65 % nitric acid (Huey test, ASTM A 262-C) is shown in Table 9 [65].

Steel	*Cr-Content %*	*Corrosion rate mm/a (mpy)*
SAE 430	17	0.5 – 0.9 (19.7 – 35.4)
ASTM XM-27	26	0.11 (4.3)
Steel 29-4-2	29	0.06 (2.4)

Table 9: Corrosion rates of some Cr steels in boiling 65 % HNO_3 [65]

Some data on the influence of heat treatment on the corrosion of the steel SAE 430 (1.4016 (%) Fe-0.079C-15.85Cr-0.63Ni-0.42Mn-0.05Cu-0.39Si-0.025N-0.015S-0.017P) and SAE 446 (cf. 1.4762 (%) Fe-0.098C-24.96Cr-0.38Ni-0.94Mn-0.05Cu-0.24Si-0.21N-0.013S-0.10P) in boiling 65 % nitric acid are shown in Table 10 [66].

Steel	*Heat treatment temperature, K (°C)*				
	920 (647)	*1140 (867)*	*1255 (982)*	*1475 (1202)*	*1530 (1257)*
	Material consumption rate, g/m² h				
SAE 430	0.54	1.50	1.10	4.58	3.75
SAE 446	0.13	0.12	0.12	0.63	0.60

Table 10: Influence of heat treatment for 1 h on the corrosion of SAE 430 and 446 in boiling 65 % HNO_3 after 240 h [66]

The chromium steels W 4027 ((%) Fe-0.14C-14.2Cr-1.45Mn-0.21Ni-0.64Si) and W 4059 ((%) Fe-0.19C-16.9Cr-1.19Mn-1.43Ni-0.67Si) are resistant in 1 to 67 % nitric acid after 12 h at room temperature; the corrosion rates being 0.035 and 0.017 in 1 % HNO_3, 0.0017 and < 0.001 mm/a (0.067 and < 0.039 mpy) in 10 % HNO_3, < 0.001 (< 0.039 mpy) for both in 25 % HNO_3 and 0.0015 and 0.00011 mm/a (0.06 and 0.004 mpy) in 67 % HNO_3 [67].

A non-welded specimen of the steel 0Ch25T (Russian grade; (%) Fe-<0.01C-25Cr, Ti-stabilized) showed a material consumption rate of 2.7 g/m^2 h after 220 h at 453 K (180 °C) in 70 % nitric acid [68]. The niobium-stabilized CrMo steel 18 2 is used in preference in the processing of nitric acid because of its resistance. The same steel, but stabilized with titanium, is cheaper and has almost the same resistance [69].

The chromium-manganese steel 08Ch18G8N2T ((%) Fe-0.08C-18Cr-8Mn-2Ni; Ti-stabilized), which is resistant to sulfuric acid, is also said to be suitable for tank wagons used for transportation of dilute nitric acid and ammonium nitrate [70].

According to Table 11, the influence of the crystal structure on the corrosion of the alloy Fe-47Cr in nitric acid solutions at 338 K (65 °C) is considerable [71]. The sigma phase of the alloy thus corrodes faster than the ferrite phase by a factor of 42.

Medium	*Temperature K (°C)*	*Ferrite phase*	*Sigma phase*
		Material consumption rate, g/m² h	
65 % HNO_3	boiling	0.085	3.59
10 % HNO_3 + 3 % HF	338 (65)	0	0.25

Table 11: Influence of the crystal structure on the corrosion of Fe-47Cr in HNO_3 solutions [71]

To prevent intercrystalline corrosion of chromium-molybdenum steel ((%) Fe-0.012C-26.13Cr-1.1Mo-0.014N) in hot 10 % nitric acid + 3 % hydrofluoric acid at 343 K (70 °C), 0.26 to 0.30 % titanium was alloyed to the steel. A similar effect was also achieved by an addition of 0.33 – 0.44 % niobium [72].

Steels of the type (%) Fe-(0.023 – 0.032)C-(0.012 – 0.023)N-(0.23 – 0.48)Ti-18Cr-2Mo were exposed, in accordance with ASTM-262, Practice D, to a solution of 10 % HNO_3 + 3 % HF at 343 K (70 °C) for (2 + 2) h and the influence of the sheet thickness and the degree of stabilization Ti/(C + N) on the extent of corrosion was investigated

(Table 12). While there is no relationship between general corrosion and intercrystalline corrosion, one exists between the degree of stabilization and the tendency to undergo intercrystalline corrosion [73].

Content, %				Sheet thickness, mm					
				4		2		1	
C	N	Ti	Ti/(C + N)	mm/d	µm	mm/d	µm	mm/d	µm
0.026	0.015	0.23	5.6	0.237	550	0.127	530	0.127	400
0.032	0.019	0.33	6.5	0.247	320	0.060	100	0.050	50
0.026	0.018	0.41	9.3	0.120	80	0.127	50	0.137	70
0.027	0.018	0.45	10.0	0.100	35	0.187	< 20	0.160	< 20
0.023	0.012	0.48	13.7	0.100	< 20	0.143	30	0.170	40

Table 12: Corrosion rate (mm/d) of sheets of the CrMoTi steel 18 2 with various thicknesses and different Ti, N and C contents, and the maximum depth of intercrystalline attack in µm [73]

The corrosion rate of the chromium-manganese steel AS-43 ((%) Fe-0.05C-18.4Cr-16.5Mn-1.60Ni-0.31Nb-0.26N-0.01Ce-0.004B) in hot nitric acid-hydrogen fluoride solutions passes through a minimum of 0.057 and 0.26 mm/a (2.24 and 10.24 mpy) at about 5 mol/l HNO_3 + 0.01 or 0.1 mol/l HF respectively. The corresponding values for 1 and 12 mol/l nitric acid are higher, 0.18 or 1.8 respectively and 0.14 or 1.23 mm/a (5.5 or 48.4 mpy) respectively [74].

The chromium-manganese steels 07Ch17G17DAMB ((%) Fe-0.06C-17.6Cr-15.2Mn-0.43Mo-0.3Nb-0.005B-0.38N) and 07Ch17G15NAB ((%) Fe-0.05C-18.4Cr-16.5Mn-1.6Ni-0.01Ce-0.005B-0.32N) corrode in 0.5 mol/l nitric acid at 293 K (20 °C) at a rate of 0.009 mm/a (0.35 mpy), and at rates of 0.35 and 0.17 mm/a (13.78 and 6.69 mpy) respectively with addition of 0.01 mol/l hydrogen fluoride. These values increase at 373 K (100 °C) to 0.034 and 0.029 mm/a (1.34 and 1.14 mpy) respectively without hydrogen fluoride, and to 0.46 and 0.33 mm/a (18.11 and 12.99 mpy) respectively with 0.01 mol/l hydrogen fluoride [75]. In the case of chromium-manganese steels, good resistance in solutions containing nitric acid can be achieved with additions of HF or HF + HCl only after addition of molybdenum.

Table 13 shows the corrosion behavior of the two chromium-manganese steels 07Ch17G17DAMB and 06Ch17G15NAB ((%) Fe-0.05C-18.36Cr-16.5Mn-1.6Ni-0.31Nb-0.12Si-0.01Ce-0.017P-0.014S) in nitric acid with and without hydrofluoric acid [76]. According to this table, both steels are unsuitable in boiling nitric acid.

The influence of small additions of hydrogen fluoride (0.01 and 0.1 mol/l) and 5 g/l chloride on the corrosion of three chromium-manganese steels in 10 mol/l nitric acid at 293 and 323 K (20 and 50 °C) can be seen from Table 14. According to this, small additions of chloride improve the resistance of the steels in nitric acid containing HF [77].

Steel	Medium mol/l	Test duration h	Material consumption rate, g/m^2 h	
			alloy	weld sample
07Ch17G17DAMB	10 HNO_3	25	0.37	0.36
		200	–	0.52
	10 HNO_3 + 0.01 HF	25	0.68	2.01
		200	–	1.46
06Ch17G15NAB	10 HNO_3	25	0.36	0.43
		200	–	0.54
	10 HNO_3 + 0.01 HF	25	1.37	2.06
		200	–	1.53

Table 13: Average material consumption rates of two CrMn steels as alloy and weld specimens in nitric acid solutions at 373 K (100 °C) [76]

Steel	Additions, mol/l	Corrosion rate, mm/a (mpy)	
		293 K (20 °C)	323 K (50 °C)
06Ch17G15NAB	none	0.009 (0.35)	0.025 (0.98)
	0.01 HF	0.074 (2.91)	0.40 (15.75)
	0.1 HF	0.29 (11.42)	1.52 (59.84)
	0.01 HF + 5 g/l Cl^-	0.037 (1.46)	0.14 (5.51)
	5 g/l Cl^-	0.019 (0.75)	0.079 (3.11)
07Ch17G17DAMB	none	0.009 (0.35)	0.014 (0.55)
	0.01 HF	0.029 (1.14)	0.22 (8.66)
	0.1 HF	0.30 (11.81)	1.49 (58.66)
	0.01 HF + 5 g/l Cl^-	0.027 (1.06)	0.14 (5.51)
	5 g/l Cl^-	0.016 (0.63)	0.024 (0.94)
08Ch18G8N2T	none	0.020 (0.79)	0.034 (1.34)
	0.01 HF	0.18 (7.09)	0.058 (2.28)
	0.1 HF	0.16 (6.3)	7.15 (281.5)
	0.01 HF + 5 g/l Cl^-	0.050 (1.97)	0.19 (7.48)
	5 g/l Cl^-	0.020 (0.79)	0.058 (2.28)

Table 14: Corrosion rate of CrMn steels in 10 mol/l HNO_3 with additions of HF and Cl^- at 293 and 323 K (20 and 50 °C) [77]

The corrosion behavior of the chromium steels 10Ch13 (Fe-0.10C-13Cr, cf. 1.4006, SAE 410) and 10Ch17 (Fe-0.10C-17Cr, cf. 1.4571, SAE 316 Ti) from powdered and sintered material and as rolled steel specimens in solutions containing HNO_3 is summarized in Table 15 [78]. The corrosion is increased drastically if sulfuric acid is present at the same time.

Medium %	Temperature K (°C)	Test duration h	Sintered		Rolled	
			10Ch13	10Ch17	10Ch13	10Ch17
			Corrosion rate, mm/a (mpy)			
20 HNO_3	293 (20)	–	0.005 (0.2)	–	–	–
	323 (50)	72	0.005 (0.2)	0.002 (0.1)	0.1 (3.9)	0.1 (3.9)
20 HNO_3 + 60 H_2SO_4	323 (50)	72	0.40 (15.8)	0.37 (14.6)	–	0.21 (8.3)
4.7 HNO_3 + 87.7 H_2SO_4	323 (50)	47	0.18 (7.1)	0.04 (1.6)	0.13 (5.1)	0.13 (5.1)

Table 15: Corrosion of powdered and sintered material and rolled specimens (7.5 % pores) of two chromium steels in solutions containing HNO_3 [78]
10Ch13 (cf. 1.4006, SAE 410); 10Ch17 (cf. 1.4571, SAE 316 Ti)

According to Table 16, nickel-free chromium-manganese steel ((%) Fe-0.08C-17.7Cr-17.55Mn-0.015P-0.019S-0.53N) is not as corrosion-resistant in HNO_3 containing solutions at 382 K (109 °C) as the CrNi steel 18 8 (SAE 302 SS, cf. 1.4310).

Medium, mol/l	CrMn steel	SAE 302 SS
	Corrosion rate, mm/a (mpy)	
6 HNO_3	0.15 (5.9)	0.032 (1.3)
6 HNO_3 + 0.1 $HSO_3 \times NH_2$	0.60 (23.6)	0.050 (2.0)
6 HNO_3 + 0.3 $HSO_3 \times NH_2$	0.84 (33.1)	0.077 (3.0)
6 HNO_3 + 0.1 H_2SO_4	0.20 (7.9)	0.049 (1.9)
6 HNO_3 + 0.3 H_2SO_4	0.53 (20.9)	0.070 (2.8)

Table 16: Corrosion of the Ni-free CrMn steel and the CrNi steel 18 8 (SAE 302 SS) in solutions containing HNO_3 at 382 K (109 °C) after 240 h [79]

The chromium-manganese steels 08Ch18G8N2T ((%) Fe-0.08C-18Cr-8Mn-2Ni; Ti-stabilized), 10Ch14G14N4T ((%) Fe-0.10C-14Cr-14Mn-4Ni; Ti-stabilized) and 12Ch13G18D ((%) Fe-0.12C-13Cr-18Mn-Cu) corrode in a hot solution of 40.4 g/l HNO_3 + 18.2 g/l oxalic acid at 373 K (100 °C) with corrosion rates of 0.027, 0.021 and 0.023 mm/a (1.06, 0.83 and 0.91 mpy) respectively [80].

The material consumption rate of the Cr steel 1Ch13 (Fe-0.1C-13Cr, cf. 1.4006, SAE 410), which is about 7000 g/m^2 h in a boiling solution of 8 mol/l nitric acid containing 10 g/l NaF + 20 g/l $Al(NO_3)_3$ + 0.01 g/l K_2CrO_4, can be reduced again from about 6000 to about 5 g/m^2 h by addition of 4 g/l chromate [81].

A solution of 20 percent by volume nitric acid + 22 g/l sodium bichromate is used at about 320 K (47 °C) for passivation of chromium steel containing 12 to 14 % chromium [82].

To detect carbides finely distributed in the structure of the ultra high-strength steel X41CrMoV5-1 (1.7783, SAE 610, UNS T20811), the steel is etched with methanolic 2 % nitric acid at room temperature for about 5 s. A solution of 90 ml methanol + 10 ml HNO_3 is used at room temperature for 12 s for deep etching for scanning electron microscopy photographs [83].

Ferritic-austenitic steels with more than 12 % chromium

The ferritic-austenitic chromium steel VLX 562® (1.4462, UNS S31803(%) Fe-0.03C-(21–23)Cr-(2.5–3.5)Mo-(4.5–6.5)Ni-2Mn-1Si-0.03P-0.02S-(0.08–0.20)N) with twice the tensile strength of the chromium-nickel steel SAE 316 L and a very good resistance to stress corrosion cracking corrodes in boiling 65 % nitric acid at a rate of < 0.6 mm/a (< 23 mpy) [84].

The ferrite phase present to the extent of 31 to 40 % in the ferritic-austenitic chromium-nickel steel Ch22N5 (Russian grade; (%) Fe-0.07C-21.54Cr-5.73Ni) is more severely attacked in 0.1 mol/l nitric acid at 303 K (30 °C) than the austenite with 19.5 % chromium, in spite of its higher chromium content. Heat treatment at 1323 or 1523 K (1050 or 1250 °C) produced no improvement in the corrosion behavior [85].

The ferritic-austenitic steel X2CrMnNiMoN26-5-4 (cf. 1.4467, A 905, (%) Fe-<0.04C-25.5Cr-5.8Mn-3.7Ni-2.3Mo-0.37N) has, in the quenched state, a structure with in each case 50 % ferrite and austenite and a 0.2 % offset yield strength of at least 590 N/mm^2. It is therefore superior in strength to the previous steels and has an equally good corrosion resistance (about 0.1 g/m^2 h in the Hucy test). Its weldability and toughness are also good [86].

The corrosion behavior of three martensitic chromium-nickel steels 13-4 (C) (cf. 1.4313, UNS S41500, (%) Fe-0.082C-11.6Cr-4.2Ni-0.54Mn-0.31Mo-0.26Si-0.009P-0.012S-0.022N), 13-4 (B) ((%) Fe-0.032C-13.2Cr-3.9Ni-0.66Mn-0.36Mo-0.66Si-0.008P-0.002S-0.026N) and 13-4-1 ((%) Fe-0.043C-12.7Cr-3.9Ni-1.5Mo-0.68Mn-0.39Si-0.009P-0.013S-0.030N) in 5 % nitric acid at room temperature is shown in Figure 4. The higher strength of the cast steels achieved by suitable heat treatment also resulted in a lower material consumption rate [87]. In addition to uniform attack, intercrystalline corrosion was found along segregations in more highly tempered states.

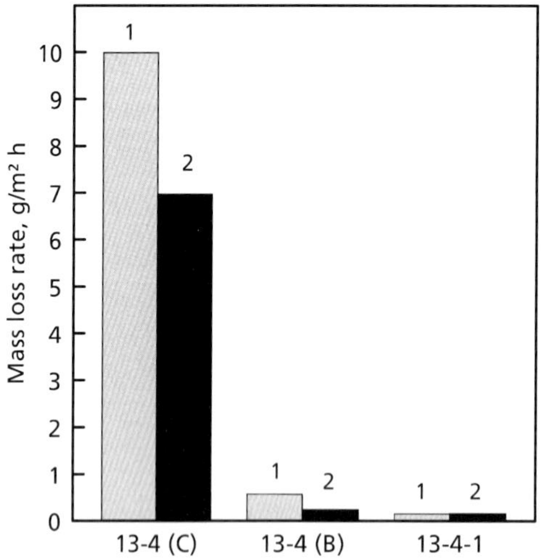

Figure 4: Material consumption rate of the CrNi steels 13-4 (C), 13-4 (B) and 13-4-1 in 5 % HNO_3 at room temperature [87]
1) low strength; 2) higher strength

The corrosion resistance of the heat-affected zone can be increased considerably by increasing the rate of cooling of the weld seam on the ferritic-austenitic steel 0Ch21N5T to 450 K/s. The resulting corrosion rate in boiling 65 % nitric acid corresponds to that of the unaffected steel in Table 17 [88]. Titanium-free, low-carbon steel (≤ 0.04 % C) is more resistant to corrosion.

*Composition of the steel, %**							*Corrosion rate mm/a (mpy)*
C	*Cr*	*Ni*	*Si*	*Mn*	*Ti*	*Ti/C*	
0.04	20.0	5.56	0.25	0.49	–	0	0.24 (9.5)
0.08	20.95	5.8	0.56	0.7	0.2	2.5	0.58 (22.8)
0.05	20.23	5.83	0.33	0.57	0.25	5	0.32 (12.6)
0.06	22.02	6.51	0.42	0.54	0.58	9.2	0.26 (10.2)

* balance Fe

Table 17: Corrosion rates of some steels of the type 0Ch21N5T in boiling 65 % HNO_3, test duration 40 h [88]

The CrNi steel 18 3 is resistant at 373 K (100 °C) in up to 40 % HNO_3 with a corrosion rate of < 0.1 mm/a (< 4 mpy) [33].

Austenitic chromium-nickel steels

The steel SAE 304 L ((%) Fe-0.03C-18.5Cr-10.5Ni-<0.5Mo-0.6Si-0.030P-0.030S) is generally used in nitric acid plants. By reducing the contents of Si, P and S ((%) Fe-0.20C-<0.1Si-18.5Cr-11Ni-<0.1Mo-0.015P-0.010S), the corrosion behavior of the steel SAE 304 L (HNO_3 grade) at higher concentrations of nitric acid can be improved, as shown in Figure 5. With increasing demands of corrosion resistance in nitric acid, either a modified steel SAE 310 ((%) Fe-0.020C-24.5Cr-20.5Ni-<0.3Mo-<0.3Si-0.020P-0.015S) or the steel 2RE 69 (1.4466, SAE 310 mod., (%) Fe-0.20C-25Cr-22Ni-2.1Mo-<0.4Si-0.020P-0.015S-0.12N) is suitable [89]. The paper on corrosion problems in stainless steel heat exchangers in a nitric acid plant should also be mentioned [90].

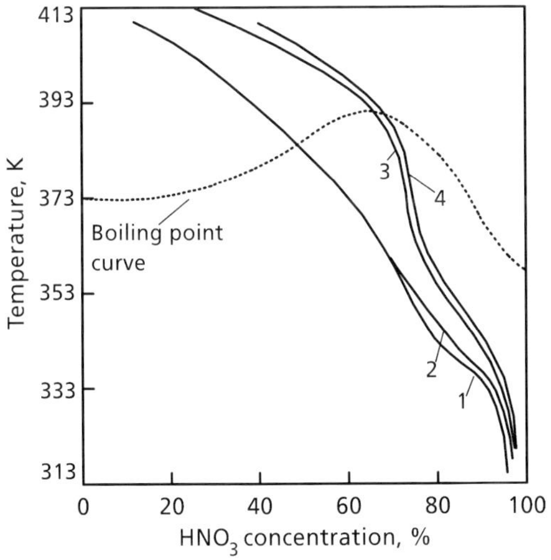

Figure 5: Isocorrosion curves (0.1 mm/a (3.94 mpy)) of the steels (1) SAE 304 L, (2) 304 L (HNO_3 grade), (3) SAE 310 (HNO_3 grade) and (4) Sandvik® 2RE-69 in nitric acid [89]

The steel SAE 304 is unsuitable for the housing and worm wheel shafts of nitric acid pumps, since the pump corrodes and the shaft seizes up. The corrosion products consist mainly of complex salts and aluminium compounds [91].

In the production of highly concentrated nitric acid by the HOKO process by reaction of liquid nitric oxide with oxygen and water under an increased pressure of 5.0 MPa, the CrNiSi steel 18 15 2, which is resistant to nitric acid up to the boiling point, is used for the storage tank (5,000 m^3 capacity) and the heat exchanger jackets and pumps [92].

The chromium-nickel steel 12Ch18N10T (1.4541, SAE 321) is used for the production of nitric acid and ammonium nitrate. In the presence of 56 % nitric acid, it can be replaced by the steel 10Ch14G14N4T ((%) Fe-0.10C-(13–15)Cr-(3.5–4.5)Ni-(13–15)Mn-0.65Ti), but only below 320 K (47 °C) [93].

The use of CrNi 18 8 steels is recommended for a nitric acid plant for oxidation of ammonia and further processing to 60 % acid [94].

The steel Fe-<0.01C-18Cr-15Ni-4Si was developed to keep intercrystalline and general corrosion low when using steels for factories processing nitric acid. However, since completely austenitic steels with a high silicon and low carbon content tend to crack in the weld region, the additional material must be alloyed so that 4 to 10 % δ-ferrite is present in the welding material [95, 96].

Doping a CrNi steel 18 9 with boron not only achieves an increase in the deformation resistance and tensile strength, but also increases the resistance to nitric acid [97].

Even high-alloy chromium-nickel steels, such as, for example, FeCr21Ni32TiAl, can be sensitive to intercrystalline corrosion in boiling concentrated nitric acid and corrode at a material consumption rate of >> 1 g/m^2 h. Cold-rolled specimens of these steels with 15–46 % deformation and subsequently heat-treated at 1,173 K (900 °C) showed a reduction in the material consumption rate to values of < 1 g/m^2 h [98].

Chromium-nickel-titanium steel (Fe-0.1C-18Cr-8Ni-1Ti) is recommended as a substitute for the steels SAE 304 and for the absorption column in a nitric acid plant [99].

The titanium-stabilized steel Fe18Cr10NiTi used for the absorption column (47 – 55 % HNO_3, 323 K (50 °C) and 70 N/mm^2) in a nitric acid plant showed intercrystalline corrosion after 5 years of operation, especially in the region of the weld joints, where TiC had formed. During repair, steel wires made of 01Ch19N9 were used for the welding. Nowadays, the more resistant steel 03Ch18N11 (1.4306, SAE 304 L) would be preferred [100].

A 5000 m^3 capacity, fixed-cover tank for storage of highly concentrated nitric acid was produced, inter alia, from the special alloy X1CrNiSi18-15-4 (cf. 1.4361, A 336) with silicon and showed no noticeable corrosion phenomena after 4 years of use [101].

The shaft cover on a chemical pump made of the steel X1CrNiSi18-15-4 for the delivery of concentrated nitric acid showed signs of local corrosion after 2 years. This was primarily due to contact with glass fiber-reinforced bellows made of polytetrafluoroethylene (PTFE), since the protective layer which formed was constantly damaged [102].

Figure 6 shows the dependence of the corrosion rate of some chromium-nickel steels in boiling nitric acid on their concentration. While the material consumption rate of the steel X1CrNiSi18-15-4 with about 4 % silicon passes through a maximum at about 1 g/m^2 h in 50 % HNO_3, decreases again and reaches about 0.05 g/m^2 h in 98 % HNO_3, those of the steels 1.4306 (X2CrNi19-11, SAE 304 L) and 1.4465 (X1CrNiMoN25-25-2, SAE 310 MoLN) increase [103].

The austenitic chromium-nickel cast steel Noricid® 9.4306 (GX3CrNiSiN20-13-5), which is resistant to nitric acid even at higher temperatures, is a suitable material for centrifugal pumps for solutions containing nitric acid [104].

If it is not possible to eliminate the internal mechanical stresses generated by weld joints in the steels 12Ch18N10T (cf. 1.4541, SAE 321) and 03Ch18N11 (SAE 304 L)

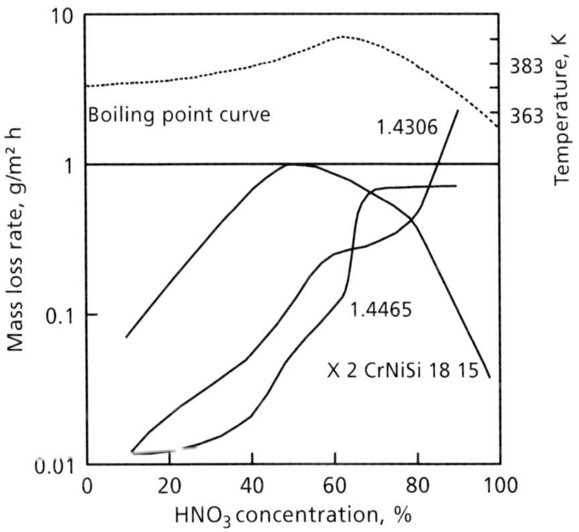

Figure 6: Dependence of the material consumption rate of the steels X 2 CrNiSi 18 15, 1.4306 and 1.4465 on the concentration of boiling HNO_3, including the boiling point curve [103]

during construction of the plant, intercrystalline corrosion occurs during production of dilute nitric acid. The steel 03Ch18N11 is still said to be better in its behavior [105].

Nitrogen tetroxide can be used as a heat transfer agent in nuclear energy plants, since it hardly attacks stainless steels, such as Ch18N10T, at 5.0 MPa and operating temperatures of about 870 K (597 °C). The material consumption rate in N_2O_4 under 5.0 MPa at 773 K (500 °C) for example, is thus 0.0002 g/m^2 h. Nitric acid formed due to moisture leads to a noticeable increase in the corrosive attack [106].

Contact with lead causes an even greater increase in the material consumption rate of the steel 1Ch18N10T (cf. 1.4541, SAE 321) in 4 % nitric acid at 368 K (95 °C). The rate was 0.005 without contact and 55.6 g/m^2 h with contact. Contact with nickel, which is readily polarizable, in contrast to lead, caused an increase to only 0.14 g/m^2 h [107].

Stainless steels, such as SAE 304 L, are preferentially attacked by hot nitric acid in the crevices which form for example during welding or lining, or between the steel and the plastic sealing discs. While the steel SAE 304 L corrodes at a rate of 0.05 and 0.07 mm/a (1.97 and 2.76 mpy) after 100 and 500 h respectively in 15 % nitric acid at 363 K (90 °C), rates of 0.27 and 0.54 mm/a (10.63 and 21.26 mpy) respectively occur after 500 h in the crevice at the weld and between the steel and the sealing disc. Corrosion tests were performed on the steels SAE 304 L, and a CrNi steel 18 10 in nitric acid to investigate the corrosion mechanism in the crevice. For more details see [108].

Regardless of the nickel content of between 2 and 5 %, the corrosion rate of chromium-manganese-nickel steels of low nickel content (e.g. (%) Fe-0.036C-13.5Cr-20.4Mn-(2 – 5)Ni-0.085N) at room temperature in 20 and 60 % nitric acid is about 0.002 mm/a (0.08 mpy). At 363 K (90 °C), the corresponding values are about 0.03 (20 % HNO_3) and 0.68 mm/a (1.18 and 26.77 mpy) (60 % HNO_3) [109].

The steel of the type SAE 310 ((%) Fe-0.020C-25.6Cr-19.41Ni-0.40Mo-0.060N) shows a corrosion rate in 20 and 40 % nitric acid at 422 K (149 °C) of about 0.075 and 0.20 mm/a (2.95 and 7.87 mpy) respectively. Corresponding values of 0.089 and 0.21 mm/a (3.5 and 8.27 mpy) were determined at 394 K (121 °C) in 60 and 80 % nitric acid respectively [110].

By reducing the dew point in a hydrogen-containing atmosphere to 233 to 223 K (–40 to –50 °C), passivating treatment of the steels 12Ch18N10T (1.4541, SAE 321) and 12Ch17G9AN4 ((%) Fe-0.12C-16.8Cr-4.1Ni-9.5Mn-0.27Si-0.02P-0.01S-0.22N) is not necessary, as comparative corrosion tests in boiling 30 % nitric acid have shown. The material consumption rates were 0.070 and 0.098 g/m^2 h respectively [111].

The oxide film which forms on the steel SUS 304 (cf. 1.4301, SAE 304, Fe-0.06C-18.25Cr-8.42Ni-0.93Mn-0.54Si) within 30 min in hot 30 % nitric acid at 333 K (60 °C) consists of a mixture of CrO_3 (or $Cr_2O_7^{2-}$), Cr_2O_3, Fe_2O_3 and a small amount of CrOOH and γ-FeOOH [112].

The relationships between cold working, precipitation properties and corrosion resistance in boiling 5 mol/l nitric acid were investigated on specimens taken from a pipe made of the steel X1CrNiSi18-15-4 (cf. 1.4361, A 336, (%) Fe-0.009C-4.24Si-1.33Mn-0.008P-0.001S-17.32Cr-14Ni-0.01Mo-0.13-Nb-0.03Ti-0.012N). Cold-worked (4 to 20 %) steel subsequently heat-treated at 973 K (700 °C) showed higher material consumption rates than the unworked material. This is based on precipitates ($M_{23}C_6$, Cr-rich carbide and $Cr_5Ni_3FeSi_2$) produced at the grain boundaries by deformation and subsequent annealing, the amount of which increases with the degree of deformation. The specimens were annealed beforehand in air at 1393 K (1120 °C)/10 min/water to establish a recrystallized, precipitation-free starting structure. The material consumption rates per unit area are shown as a function of the treatment state and test duration in Table 18 [113].
As a result of galvanic contact of the unworked and deformed specimens with the same treatment state, the material consumption rate of the deformed specimen was 2 to 3 times higher than that of the unworked specimen. Deformation of the steel and stress-induced precipitates impair the corrosion resistance in nitric acid [113].

Treatment state	*Surface* cm^2	*Test duration* h	*Material consumption rate* g/m^2 h
Non-deformed, non-sensitized	16.1 (16.1)*	48 (72)*	0.08 (0.10)*
Deformed (20 %), non-sensitized	12.1 (12.3)*	72 (72)*	0.12 (0.28)*
Non-deformed, sensitized	16.2 (16.4)*	48 (72)*	0.08 (0.09)*
Deformed (20 %), sensitized	11.6 (11.6)*	72 (72)*	0.08 (0.17)*

*values of specimens in galvanic contact

Table 18: Material consumption rates of specimens of the steel X 2 CrNiSi 18 15 with different treatment states in 5 mol/l boiling HNO_3 [113]

Table 19 shows that as the polarization resistance decreases, the material consumption rate of steel with a low carbon content ((%) Fe-0.02C-23.8Cr-20.67Ni-1.10Mn-0.02Mo-0.60Si-0.015P-0.014S) in boiling nitric acid increases [114]. After calibration, the polarization resistance can be used for rapid determination of the corrosion.

HNO_3 %	Temperature K (°C)	Material consumption rate $g/m^2\,h$	Polarization resistance Ω/cm^2
35	382 (109)	0.028	12 000
45	386 (113)	0.049	7 390
55	390 (117)	0.100	5 210
65	393 (120)	0.270	2 578

Table 19: Corrosion rate and polarization resistance R_p of steel in boiling nitric acid [114]

The steels 03Ch18N11 (1.4306, SAE 304 L) and 12Ch18N10T (1.4541, SAE 321) can be used in up to 40 % boiling nitric acid, since the corrosion rates are about 0.12 and 0.10 mm/a (4.7 and 3.9 mpy), but 0.35 and 0.78 mm/a (13.8 and 30.7 mpy) respectively at 65 % [115].
On the other hand, in contrast to the steel 12Ch18N10T, 03Ch18N11 shows no crevice corrosion between room temperature and the boiling point in nitric acid (> 50 %). The corrosive attack on the steel and also on its welded joint is uniform [116].

Alternating current impedance with radioindicator measurements were used for short-term determination (over a few days) of the corrosion rate of the steel Fe-18Cr-13Ni-1Nb giving values of 0.00005 to 0.001 mm/a (0.002 to 0.039 mpy) in 7.8 mol/l nitric acid at room temperature [117].

The resistance of the chromium-nickel steel NAR-SN-5 (Jap. type) with 27 % chromium and 7 to 10 % nickel to stress corrosion cracking in nitric acid of moderate concentration can be improved by nitrogen contents of 0.09 to 0.13 %, which cause an increase in the ferritic phase of 40 to 65 %. Small additions of molybdenum, copper or niobium led to a deterioration in the corrosion behavior [118].

According to Table 20, tests under evaporation conditions with heat transfer in the vaporizer in a titanium tank with heating steam coils made of the steels Ch18N10T and El-943 (0Ch23N28M3D3T: (%) Fe-<0.01C-23Cr-28Ni-3Mo-3Cu; Ti-stabilized) with 55 % nitric acid resulted in material consumption rates on the steels due to the acid which were up to 3 times higher in comparison with normal boiling tests when the temperature of the heating steam is between 403 and 418 K (130 and 145 °C) [119]. According to these tests, CrNiMo steel and not CrNi steel should be used for heating coils in the vaporizer.

After 60 d tests in hot 56 % nitric acid at 333 and 391 K (60 and 118 °C) (boiling point), the nitrogen-alloyed chromium-nickel steel X5CrNiN19-9 (1.4315) corrodes at a rate of 0.005 and 0.42 mm/a (0.197 and 16.5 mpy) respectively, which is comparable to that of the steels X5CrNi18-10 (1.4301, SAE 304) and X8CrNiTi18-10

Steel	Pressure kP	Temperature K (°C)	Wall temperature* K	Material consumption rate, g/m^2 h		
				on the pipe	on the wall	in the solution at 388 K (115 °C)
Ch18N10T	274.8	403 (130)	400	0.45	0.33	0.16
	421.7	418 (145)	409	0.59	0.58	0.16
El-943	274.8	403 (130)	400	0.10	0.13	0.08
	421.7	418 (145)	409	0.21	0.25	0.08

*calculated

Table 20: Corrosion rate of the steels Ch18N10T and EI-943 after 25 h in 55 % HNO_3 under various conditions [119]

(1.4878, SUS 321) of higher nickel content of 0.007 and 0.01 and 0.36 and 0.64 mm/a (0.28 and 0.39 and 14.2 and 25.2 mpy) respectively [120].

Welded joints of the steels 12Ch18N10T (cf. 1.4541, SAE 321), 08Ch18N10T (cf. 1.4541, SAE 321), 08Ch22N6T and 08Ch18G8N2T as well as 03Ch18N11 (cf. 1.4306, SAE 304 L) are attacked only insignificantly in nitric acid solutions containing dicarboxylic acids in the manufacture of adipic acid. After 11 000 operating hours the material consumption rate, for example, of the 03Ch18N11 steel still-head from the concentration column containing 57 % nitric acid at 343 K (70 °C) was 0.0024 and that of the absorption column containing 60 % nitric acid at 313 K (40 °C) was 0.0007 g/m^2 h [121]. The corrosion of the steel 03Ch18N11 also remained uniform, with material consumption rates between 0.0050 and 0.0071 g/m^2 h, in the column where the nitric oxides are driven off and the other steels mentioned showed intercrystalline corrosion and knife-line corrosion.

The ferritic phase occurring to the extent of about 10 % has no influence on the corrosion resistance of the stainless steels Ch18N10T, 000Ch18N11 (cf. 1.4306, SAE 304 L) and 02Ch19N9 in boiling 65 % nitric acid. According to potentiostatically recorded polarization curves, this behavior can be expected at potentials both in the passive region and in the passive/transpassive transition region [122].

Steels without a particular heat treatment but good strength, toughness and weldability as well as corrosion resistance usually contain 4 – 6 % Mn, 6 – 9.5 % Ni, 20 – 21.5 % Cr, 0.25 – 0.35 % N and possibly up to 2.5 % Mo and 1 % Nb. According to the Huey test, a steel of this type (completely austenitic and non-magnetic) corrodes at a rate of 0.15 mm/a (5.91 mpy) (tempered) and 0.17 mm/a (6.69 mpy) (sensitized) [123].

The steel X2CrNiSi18-15 (cf. 1.4361, A 336) with 2 % silicon has a material consumption rate in boiling concentrated nitric acid of 2 g/m^2 h and with 6 % silicon of 0.03 g/m^2 h [124].

Intercrystalline corrosion on weld seams caused by nitric acid can be prevented if 6 to 8 % silicon is added to steels of Fe ≤ 0.02C-(5.8 – 8.5)Cr-(21 – 25)Ni [125].

The resistance of the austenitic CrNiSi steel 17 14 4 to boiling 65 % nitric acid can be further improved by additions of niobium, zirconium or tantalum. However, nitrogen and titanium additions cause a deterioration. With its good mechanical properties and weldability, the steel (%) Fe-≥0.05C-17Cr-14Ni-4Si-0.8Nb exhibits an excellent resistance to concentrated nitric acid [126].

The corrosion rate of the CrNiSi steel 17 14 4 in boiling 65 % nitric acid only becomes independent of the heat treatment temperature between 870 and 1220 K (597 and 947 °C) if about 1 % niobium, zirconium or tantalum is alloyed with the steel [127]. The material consumption rate is between 0.4 and 0.6 g/m^2 h.

The Nb- and Ta-containing steel (%) Fe-0.05C-18.18Cr-9.06Ni-1.51Mn-0.7(Nb + Ta)-0.55Si-0.027P-0.006S-0.018-N-0.0031B is said to corrode in boiling 65 % nitric acid with a material consumption rate of only 0.09 g/m^2 h [128].

Titanium-stabilized CrNi steel ((%) Fe-0.09C-18.1Cr-13.4Ni-1.17Mn-0.37Mo-0.51Ti-0.017N), which precipitates TiC at the grain boundaries after sensitization at 1070 – 1170 K (797 – 897 °C), is resistant to intercrystalline corrosion, especially at low temperatures, but attacked at the $M_{23}C_6$ and TiC precipitates in the Huey test (boiling 65 % HNO_3) [129].

The austenitic chromium-nickel steels Fe-0.08(C + N)-18Cr-8Ni (Ti-stabilized) and Fe-0.08(C + N)-18Cr-10Ni-2Mo (Ti-stabilized) are resistant in boiling 65 % nitric acid according to [44]. This conclusion is limited, however, by the material consumption rates, reported elsewhere, of the steels X8CrNiTi18-10 (1.4878, SAE 321 H) and X5CrNi18-10 (1.4301, SAE 304) in boiling 56 % nitric acid of 0.58 and 0.32 g/m^2 h [130]. When alloying elements dissolved in the acid become concentrated, the corrosion can be increased further by a factor of 2 to 3 [131, 132].

The corrosion resistance of the steel 03Ch18N11 (1.4306, SAE 304 L) depends greatly on the Si and C content, especially in the sensitized state. According to Table 21, the presence of more than 0.3 % silicon has an adverse effect on the resistance at carbon contents of about 0.03 %. A reduction in the carbon content leads to a decrease in the corrosion rate of the steel [133].

Alloy content, %		***Corrosion rate, mm/a (mpy)***		
Si	C	1348 K (1075 °C)	1393 K (1120 °C)	1423 K (1150 °C)
0.17	0.030	0.27 (10.6)	0.24 (9.5)	0.24 (9.6)
0.49	0.030	6.52 (256)	0.96 (37.8)	0.63 (24.8)
0.78	0.030	16.34 (643)	4.65 (183)	1.44 (57.0)
0.28	0.020	0.25 (9.8)	0.29 (11.4)	0.24 (9.6)
0.48	0.023	0.23 (9.1)	0.20 (7.9)	0.21 (8.3)
0.75	0.012	0.18 (7.1)	0.17 (6.7)	0.17 (6.7)

Table 21: Corrosion rates on test specimens from industrial smelting of the steel 03Ch18N11 (SAE 304 L) with various Si and C contents in boiling 65 % HNO_3 after prior annealing at 1348 to 1423 K (1075 to 1150 °C)/water quenching and subsequent tempering at 923 K (650 °C) [133]

The steel 03Ch18N10 with 0.020 to 0.028 % C is improved in its austenitic stability by addition of 0.017 to 0.172 % N, the deformability remaining constant. At the same time, the susceptibility to intercrystalline corrosion in boiling 65 % nitric acid is reduced [134].

In corrosion tests with solution-annealed and sensitized CrNi 18 8 steels, the steel with the lower carbon content ((%) Fe-0.035C-0.37Si-0.79Mn-16.65Cr-9.80Ni-0.005P-0.024S) also had the lower material consumption rates of 0.50 and 1.22 g/m^2 h, in comparison with the corresponding values of 0.80 and 5.5 g/m^2 h for the steel with the higher carbon content ((%) Fe-0.08C-0.60Si-1.40Mn-18.10Cr-9.20Ni-1.43Mo-0.015P-0.028S), after exposure for 3 times 47 h to boiling 65 % nitric acid. Steels of comparable composition but additionally doped with 0.66 or 0.78 % niobium gave higher corrosion rates [135].

The material consumption rate of the steel 12Ch18N10T (SAE 321) in boiling 65 % nitric acid could be reduced from 1.5 to about 1 g/m^2 h by addition of 0.01 % cerium or vanadium [136].

Chemical stabilization of the steel SAE 321 SS ((%) Fe-0.066C-18.2Cr-9.0Ni-1.1Mn-0.48Ti-0.47Si-0.001S-0.012P) can be lost by incorrect heat treatment. Undesirable sensitization both during processing and use of the steel in question, and therefore intercrystalline corrosion in boiling nitric acid, could be prevented by heat treatment at 1173 K (900 °C) for 2 h [137]. These circumstances also apply to other steels.

The steel SAE 308 (cf. 1.4303; (%) Fe-0.040C-20.95Cr-9.82Ni-1.76Mn-0.41Si-0.008S-0.016P) has an austenitic form after heat treatment up to 1473 K (1200 °C) for 1 h and a duplex form (ferritic + austenitic structure) above 1473 K (1200 °C). Table 22 shows the corrosion behavior after quenching in water and various forms of aging [138]. The lower the aging temperature, the more the two steels corroded. The material consumption rate of the steel SAE 308 SS (completely austenitic) in boiling 65 % nitric acid of 0.121 g/m^2 h was accelerated to 81.6 g/m^2 h by addition of 4 g/l sodium bichromate ($Na_2Cr_2O_7$) (test duration 144 h) [138].

1473 K (1200 °C)/ Water aging	***Material consumption rate g/m^2 h***	***1573 K (1300 °C)/ Water aging***	***Material consumption rate g/m^2 h***
–	0.177	–	0.341
1 h at 973 K (700 °C)	0.802	1 h at 973 K (700 °C)	0.465
100 h at 973 K (700 °C)	2.88	100 h at 973 K (700 °C)	2.25
5 h at 873 K (600 °C)	0.358	2 h at 873 K (600 °C)	1.87
96 h 873 K (600 °C)	> 10	96 h at 873 K (600 °C)	3.14
–	–	5 h at 873 K (600 °C)	4.92
–	–	300 h at 823 K (550 °C)	1.54

Table 22: Influence of aging on the corrosion of SAE 308 in boiling 65 % HNO_3 after heat treatment at 1473 and 1573 K (1200 and 1300 °C) with subsequent quenching in water [138]

Some material consumption rates for the steel (%) Fe-0.019C-18.60Cr-1.05Ni-1.01Mn-0.03Mo-0.49Si-0.017P-0.007S-0.022N in boiling 65 % nitric acid (Huey test) are shown in Table 53 as a further example of the strong influence of the tempering temperature over a period of 300 h on the corrosion of chromium-nickel steel. As the tempering time increases, the sensitivity to corrosion increases significantly, especially at temperatures below 800 K (527 °C). Above 1250 K (977 °C), the influence of the tempering time is no longer significant [139].

Tempering temperature, K (°C)	673 (400)	773 (500)	823 (550)	873 (600)	923 (650)	973 (700)	1073 (800)	1123 (850)
Material consumption rate, g/m^2 h	0.1	1.05	12	20	1.5	0.5	0.15	0.13

Table 23: Corrosion rate of steel in the Huey test after solution annealing at 1323 K (1050 °C) for 15 min and after 300 h of tempering at various temperatures [140]

The corrosion behavior of stabilized steels moreover depends on the precipitation of carbonitride, since this is severely attacked by boiling nitric acid in the Huey test.

In agreement with this, after correct heat treatment (i.e. a sufficiently high tempering temperature), the steel Remanit® 4306 (X2CrNi19-11, 1.4306, SAE 304 L; (%) Fe-0.019C-19.09Cr-10.06Ni-1.59Mn-0.03Mo-0.003Al-0.16Si-0.018P-0.004S-0.0005B-0.042N with 5.03 % ferrite) showed, after annealing at 1393 K (1120 °C) for 10 min and quenching in water, material consumption rates of 0.10 and 0.11 g/m^2 h after 12 and 15 periods respectively of 48 h each in the Huey test. Subsequent annealing at 973 K (700 °C) for 30 min caused an increase in the material consumption rate to 0.26 g/m^2 h [141].

Remanit® 4335-So ((%) Fe-0.012C-25.46Cr–19.93Ni–<0.01Si–1.41Mn–<0.05P–<0.006S-0.140N) heat-treated at 1323 K (1050 °C) for 30 min and quenched in water corrodes in boiling 65 % nitric acid with a material consumption rate of about 0.05 g/m^2 h after 11 and 15 boiling periods of 48 h each. After a final heat treatment in the plant, the rate was 0.50 g/m^2 h [142].

According to the Huey test, SAE 304 can be used for waste tanks in the nuclear reactor industry if the steel is heat treated at 813 K (540 °C) for 1 to 10 h (no longer), since the corrosion rate in boiling 65 % nitric acid does not exceed 0.24 mm/a (9.45 mpy). The steel SAE 304 L with a corrosion rate of 0.15 mm/a (5.91 mpy) can also be used if it is tempered either at 513 or 703 K (240 or 430 °C) for 100 h [143].

The corrosion behavior of the heat-treated steels SAE 304 L ((%) Fe-0.029C-18.37Cr-8.84Ni-1.56Mn-0.52Si-0.032N), SAE 304 ((%) Fe-0.071C-17.52Cr-8.57Ni-1.80Mn-0.68Si-0.075N), LC-20 ((%) Fe-0.022C-21.62Cr-8.67Ni-1.79Mn-0.51Si-0.035N) and MC-20 ((%) Fe-0.06C-22.57Cr-8.55Ni-1.71Mn-0.55Si-0.036N) as cold-rolled specimens in boiling 65 % nitric acid can be seen from Table 24 [144]. The heat treatment temperature has a particular influence on the corrosion rate of the steel SAE 304 L.

Steel	Ferrite content %	Corrosion rate, mm/a (mpy)	
		1200 K (927 °C)	1339 K (1066 °C)
SAE 304 L	–	0.39 (15.35)	0.66 (25.98)
SAE 304	–	3.33 (131.1)	0.31 (12.2)
LC-20	20	0.17 (6.69)	0.17 (6.69)
MC-20	18 – 20	0.23 (9.06)	0.28 (11.02)

Table 24: Corrosion of some CrNi steels containing ferrite in boiling 65 % HNO_3 after heat treatment at 1200 and 1339 K (927 and 1066 °C) for 1 h [144]

According to Figure 7, cold working (here, for example, a decrease in the cylinder height of the steel specimen after deformation) has a varying influence on the corrosion in boiling 65 % nitric acid of the following non-sensitized steels solution-annealed at 1350 K (1077 °C) for 2 h and rapidly quenched: SAE 304 and SAE 304 L as well as SAE 316 ((%) Fe-0.05C-17.27Cr-13.09Ni-2.56Mo-1.76Mn-0.24Cu-0.13Co-0.04V-0.45Si-0.033P-0.015S) [145].

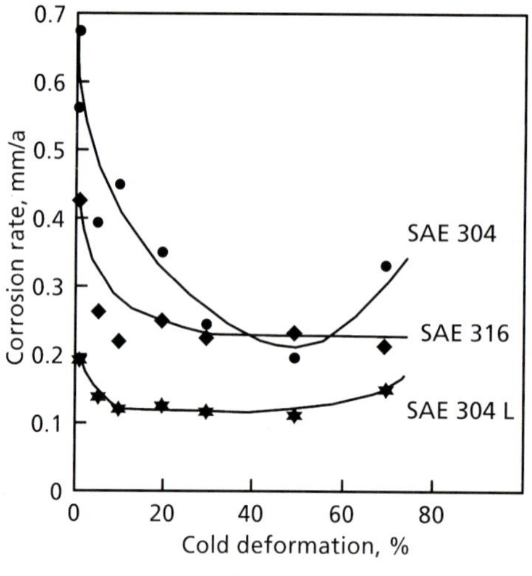

Figure 7: Corrosion of non-sensitized stainless steel in boiling 65 % HNO_3 as a function of the degree of cold working [145]

Compared with a corrosion rate of SAE 304 in boiling 65 % nitric acid of 1.1 mm/a (43.31 mpy) after a cooling rate of 1 K/s (1 °C/s), this drops to about 0.4 mm/a (15.75 mpy) after a cooling rate of 55 K/s (55 °C/s) [146].

According to Table 23, corrosion of the steel SAE 304 in boiling 65 % nitric acid (Huey test) was increased considerably by heat treatment at 923 K (650 °C) of increasing duration, but decreased at 1023 K (750 °C) [147].

Heat treatment		Corrosion rate mm/a (mpy)
Temperature, K (°C)	**Duration, h**	
–	–	0.26 (10.3)
923 (650)	1	0.88 (34.6)
	2	4.2 (165)
	5	8.8 (346)
	50	21.4 (843)
1023 (750)	0.5	1.5 (59.1)
	1	1.3 (51.2)
	2	1.1 (43.3)
	5	1.1 (43.3)

Table 25: Corrosion of SAE 304 in boiling 65 % HNO_3 after various heat treatments [147]

The CrNi steels SAE 304 and SAE 304 L of comparable composition gave corrosion rates which approximately coincide after 240 h in the Huey test – 0.46 mm/a (18.11 mpy) for SAE 304 (in the delivery state) and for SAE 304 L 0.31 mm/a (12.2 mpy) (20 min at 950 K (677 °C)) and 0.61 mm/a (24.02 mpy) (1 h at 950 K (677 °C)) [148].

SAE 304 L (as sheet or pipe), heat treated for 20 min or 1 h at 950 K (677 °C), corroded in boiling 65 % nitric acid at a rate between 0.20 and 0.22 (7.87 and 8.66) or 0.26 mm/a (7.87 mpy) [149].

According to Figure 8, even small amounts of boron reduce the corrosion rate of SAE 304 in boiling 65 % nitric acid. An increase caused by sensitizing heat treatment between 922 and 1033 K (640 and 760 °C) is largely cancelled out by the presence of boron [150]. A content of 4 ppm boron had no influence on the corrosion behavior of a solution-annealed steel. The corrosion rate of this steel after 240 h was 0.065 mm/a (2.56 mpy) with and without boron [150].

Both sulfur (0.03 %) and phosphorus (0.06 %) have no influence on the corrosion behavior of the steel SAE 304 ((%) Fe-0.069C-18.6Cr-9.4Ni-0.01Si-0.003P-0.009S) in boiling 65 % nitric acid (Huey test). The material consumption rates were about 0.27 and 0.10 g/m^2 h, independent of the heat treatment of 100 to 1000 h at 923 and 973 K (650 and 700 °C) [151].

Contact between the steel SAE 304 and platinum, gold or graphite in 65 % nitric acid does not noticeably intensify the corrosion rate either at 293 K (20 °C) (from 0.008 with and without Pt contact to 0.007 with graphite and 0.002 mm/a (0.08 mpy) with Au contact) or 373 K (100 °C), but merely leads to a shift in the potential to more positive values, although these still remain within the passivity range [152].

To prevent the formation of carbides in the steel SUS 304 ((%) Fe-0.04C–18.4Cr–9.0Ni–1.62Mn–0.08Mo–0.02Cu–0.64Si-0.02P-0.007S) by quenching in oil after heat treatment at 1423 K (1150 °C), which causes increased corrosion, cooling in two

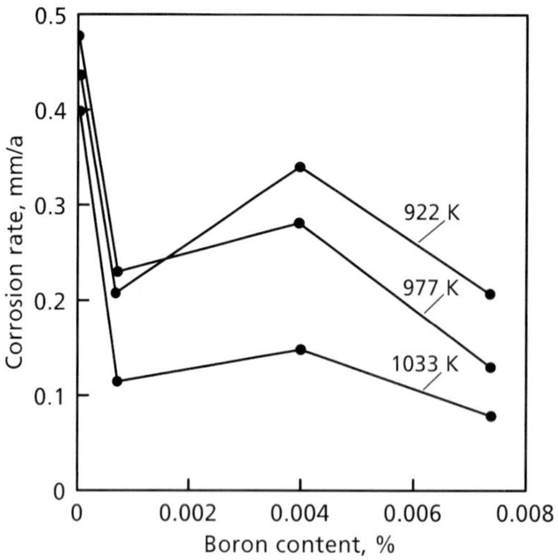

Figure 8: Influence of the boron content in SAE 304 on the corrosion rate of the steel in boiling 65 % HNO_3 as a function of the sensitization temperature [150]

steps – in air down to about 1200 K (927 °C) and then in oil – is recommended. It is possible to reduce corrosion in boiling 65 % nitric acid by a factor of about 3 by this procedure [153].

The stresses caused by welding have the lowest influence on intercrystalline corrosion on the steel 08Ch18N10T (1.4541, SAE 321) in boiling 65 % nitric acid after hardening at 1473 K (1200 °C). Tempering both at 923 and 1023 K (650 and 750 °C) increases the tendency to undergo intercrystalline corrosion. This also has an effect on the material consumption rate (from 8 to 9 up to 65 g/m^2 h). Corresponding tests with the steel 03Ch18N11 (1.4301, SAE 304 L) showed a rise in the material consumption rate from 0.51 to 29 g/m^2 h [154].

In another study, the influence of the dislocation structure on intercrystalline corrosion of the steels 08Ch18N10T (0.07 % C) and 03Ch18N11 (0.03 % C) in the elasto-plastic state was investigated. This state was produced by welding the steel specimens, which had first been quenched from 1473 K (1200 °C) and sensitized at 923 and 1023 K (650 and 750 °C) for 6 h. According to the Huey test (3 × 48 h in boiling 65 % nitric acid), the most severe intercrystalline corrosion occurred on the specimen sensitized at 923 K (650 °C). The site of attack lay in the heat-affected zone about 10–15 mm away from the weld joint [155].

The welded joints of niobium-stabilized chromium-nickel steel ((%) Fe-0.06C-16.8Cr-11.2Ni-1.02Mn-0.02Ti-0.32Si-0.015P-0.008S-0.83Nb) showed a material consumption rate of 0.223 g/m^2 h in boiling 65 % nitric acid without sensitization, while after sensitization at 863 K (590 °C) for 1, 10 and 500 h, the steel corroded at rates of 0.304, 0.593 and 2.45 g/m^2 h, respectively. In contrast to the Ti-stabilized steels, the Nb-stabilized steels are resistant to knife-line corrosion immediately after welding. Only after heating at critical temperatures at which chromium carbides precipitate does the Nb-stabilized steel also become sensitive. However, this can be eliminated by two heat treatments [156].

Spot welding of unstabilized steels ((%) Fe-0.10C-17.5Cr-9.5Ni-1.45Mn-0.5Si-0.05Ti) does not cause susceptibility to intercrystalline corrosion in boiling 65 % nitric acid even with very long welding times. The corrosion rate after 3 × 48 h periods in the Huey test is 0.2 to 0.3 mm/a (7.87 to 11.8 mpy). A potential of about + 1.0 V_{SHE} is established, with chemical resistance [157].

At a very low concentration of the corrosion products (metal nitrates), the corrosion rate of the steel K-299 ((%) Fe-0.04C-17.50Cr-10.12Ni-1.43Mn-0.73Nb-0.69Si) in boiling 65 % nitric acid is 0.20 mm/a (7.87 mpy). As the concentration of the corrosion products increases, so does the corrosion rate, for example, to about 0.78 mm/a (30.7 mpy) at 0.1 g/l [158].

When the steel Sandvik® 2R12 ((%) Fe-<0.020C-<0.10Si-19.5Cr-11Ni-<0.015P-<0.010S) is used, temperatures of 390, 360 and 350 K (117, 87 and 77 °C) should not be exceeded for an acceptable corrosion rate of 0.1 mm/a (3.94 mpy) in 65, 80 and 90 % nitric acid respectively [159].

According to Table 26, the corrosion rate of chromium-nickel steel in boiling 65 % nitric acid is practically unchanged at 0.5 mm/a (19.7 mpy) as the silicon content increases (from 3.8 to 5.9 %), but is reduced in boiling 98 % acid (at the hyperazeotropic point) by a factor of 6 at an Si content of 5.9 % [160].

Chemical composition, %									***Corrosion rate mm/a (mpy)***
Cr	***Ni***	***Mn***	***Mo***	***Si***	***N***	***C***	***S***	***P***	
21.71	15.34	0.88	–	3.78	0.098	0.010	0.011	0.002	0.12 (4.72)
20.70	15.65	0.87	0.21	4.29	0.077	0.009	0.013	0.004	0.08 (3.15)
20.95	15.87	0.92	0.05	4.80	0.132	0.007	0.012	0.003	0.06 (2.36)
20.57	16.71	0.96	–	5.62	0.117	0.006	0.011	0.002	0.04 (1.57)
21.80	17.54	1.07	–	5.87	0.112	0.017	0.013	0.005	0.02 (0.79)

Table 26: Corrosion rate of the CrNiSi steel 20 15 in boiling 98 % HNO_3 as a function of the Si content [160]

Figure 9 shows the corrosion behavior of the steels X 3 CrNi 18 10 (cf. 1.4306, SAE 304 L), X 2 CrNi 25 20 (X1CrNi25 21, 1.4335) and X 2 CrNiSi 18 15 (X1CrNiSi18-15-4, 1.4361, A 336) in boiling nitric acid solutions in the concentration range 67–98 % in three 48 h tests. An excellent resistance (about 0.04 g/m^2 h), even in 98 % acid, is achieved by addition of 4 % silicon to the steel [161]. Welding Si-containing steels requires particular measures.

After 220 h at 453 K (180 °C) in 70 % nitric acid, the steels 08Ch18N10T (1.4541, SAE 321) and 08Ch22N6T are unusable for industrial purposes either as compact alloy or welding material, with corrosion rates of 4.2 and 6.1, 3.9 and 4.5 g/m^2 h respectively. On the other hand, the low-carbon steel 000Ch18N11 (cf. 1.4306, SAE 304 L; Fe-<0.001C-18Cr-11Ni) is suitable for industrial use, as compact alloy or in the welded state, with material consumption rates of 0.4 and 0.53 g/m^2 h respectively [68].

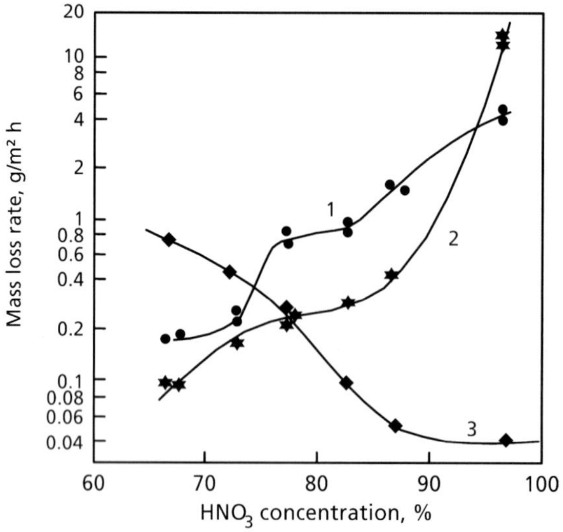

Figure 9: Material consumption rate of the steels 1) X 3 CrNi 18 10, 2) X 2 CrNi 25 20 and 3) X 2 CrNiSi 18 15 4 in 67 to 98 % boiling HNO_3 [161]

The CrNi cast steel 18 8 with a normal Si content of 0.5 to 1 % and a C content slightly above 0.15 % is resistant in up to 80 % nitric acid at 298 (25 °C). Above 80 % HNO_3, an Si content of at least 2.5 to 3 % is necessary. With more than 4 % Si and a C content of between 0.12 and 0.13 % the steel is practically resistant in the entire concentration range of nitric acid. Corrosion rates as a function of the Si and C content can be seen in Figures 10 and 11 [162].

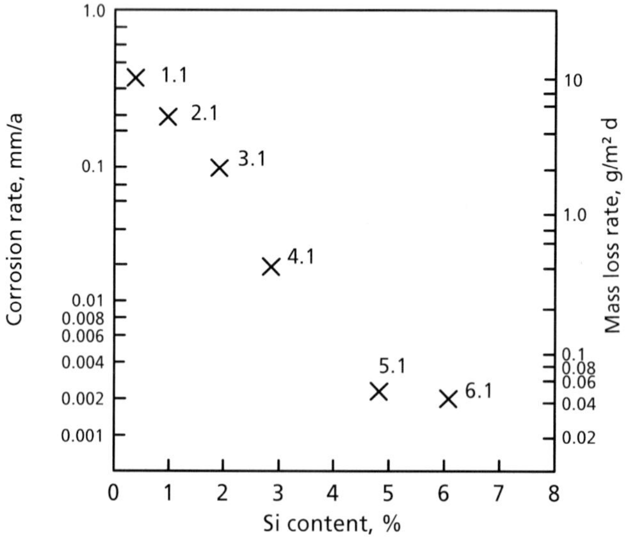

Figure 10: Influence of the Si-content on the corrosion rates of CrNi cast steel 18 8 with about 0.03 % C in 98 % HNO_3 at 298 K (25 °C), test duration 720 h [162]

The silicon-containing steel X 2 CrNiSi 18 15 has found use, above all, for tanks, pipelines, pumps and fittings for hot, highly concentrated nitric acid, since conventional austenitic steels are not suitable in above 70 % HNO_3 [163].

According to Table 27, the corrosion resistance of the steel 000Ch20N20 (Fe-0.03C-18.57Cr-19.40Ni-0.71Mn-0.26Si; austenitic) increases with an increasing silicon content in 24 mol/l nitric acid at 373 K (100 °C), but is reduced in boiling 12 mol/l nitric acid. Above 6 % silicon, an austenitic-ferritic structure is present. If the heat-treated steel is tempered at 1023 to 1173 K (750 to 900 °C), a sigma phase forms at silicon contents above 3 %, leading to susceptibility to intercrystalline corrosion and a reduction in the notched impact strength [164].

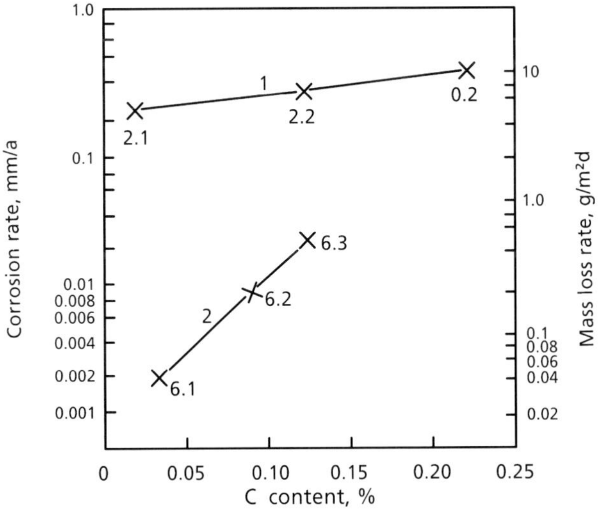

Figure 11: Influence of the C-content on the corrosion rates of the CrNi cast steel 18 8 with Si contents of about 1 and > 6 % in 98 % HNO_3 at 298 K (25 °C), test duration 720 h [162]
1) 1 % Si; 2) 6 % Si

HNO_3 mol/l	*Temperature K (°C)*	*Si-content %*	*Material consumption rate, g/m² h*
24	373 (100)	0.26	3.5
		2.0	2.3
		4.0	1.65
		5.0	0.3
		6.0	0.2
12	boiling point	0.26	0.15
		2.0	0.4
		4.0	0.5
		6.0	0.5

Table 27: Dependence of the corrosion rate of the steel 000Ch20N20S on the Si-content in 24 mol/l HNO_3 at 373 K (100 °C) and 12 mol/l HNO_3 at the boiling point after 100 h [164]

The material consumption rate of stainless steel in concentrated nitric acid (> 95 %) can be greatly reduced (e.g. from 0.26 to < 0.01 g/m^2 h) if ≥ 20 % of the surface of the steel is brought into contact with aluminium. The potential of the partly covered steel is reduced to a value close to that of aluminium in this way [165, 166].

The contributions to the material consumption rate of the metals in the steel X 5 CrNiSi 18 15 in 98 % nitric acid at 323 K (50 °C) correspond quite closely to the proportions present at an anodic polarization of 1.85 and 1.95 V_{SHE} [167].

The corrosion behavior of the special steel X 2 CrNiSi 18 15 (cf. 1.4361, A 336; (%) Fe-(0.014–0.015)C-(17.65–17.92)Cr-(15.0–15.32) Ni-(0.07–0.09)Mo-(0.72–0.79)Mn-(3.90–4.26)Si-(0.007–0.021)S-(0.017–0.021)P) in 98 % industrial nitric acid at 293 K (20 °C) is shown in Table 28 [168]. Iron makes the greatest contribution to the weight loss here.

During manual arc welding on the steel X 2 CrNiSi 18 15 with about 4 % silicon, careful control of the heating of the weld joints is necessary, largely in order to exclude intercrystalline corrosion in the presence of nitric acid [169]. The corrosion rate of the steel in 98 % nitric acid at 323 K (50 °C) rose slightly with the duration of exposure at the start, and then remained virtually constant as the test progressed, after the transpassive region had been reached [170].

Si-content %	Fe	Cr	Ni
	Material consumption rate, g/m^2 h		
3.90	0.0063	0.0015	0.0013
4.26	0.0114	0.0059	0.0028

Table 28: Corrosion of the steel X 2 CrNiSi 18 15 (A 336) with 3.9 and 4.26 % Si in 98 % HNO_3 at 293 K (20 °C) showing the contribution to the weight loss of Fe, Cr and Ni, from 800 h tests [168]

Corrosion studies with the Russian steels 02Ch8N22S6 ((%) Fe-0.02C-8Cr-22Ni-6Si), 02Ch8N22T (Ti-stabilized), 02Ch8N22S6B ((%) Fe-0.02C-8Cr-22Ni-6Si; Nb-stabilized), 02Ch12N10S5, 02Ch12N10S5T, 02Ch12N10S5B ((%) Fe-0.02C-12Cr-10Ni-5Si; Nb-stabilized) and 02Ch12N10S5T, which were rolled at 1370 K (1097 °C) to sheets 12 mm thick and then quenched from 1323 K (1050 °C) in water and finally tempered at 773–1123 K (500–850 °C) for about 5–120 min, also demonstrated the beneficial effect of niobium additions on intercrystalline corrosion in boiling 72 and 98 % nitric acid. The occurrence of a ferritic phase caused a further improvement in the resistance to intercrystalline corrosion [171].

By addition of 2.3 % aluminium to chromium-nickel steel ((%) Fe-0.018C-24.86Cr-19.96Ni-0.12Si), which had been solution-annealed at 1373 K (1100 °C) for about 15 min, the material consumption rate in boiling 98 % nitric acid could be reduced from 5.55 to 0.47 g/m^2 h [172].

According to Figure 12, corrosion of the steel ZI-52 (000Ch20N20S5, Russ. grade: (%) Fe-0.017C-19.2Cr-20.4Ni-5.4Si-0.33Mn) in 23 mol/l nitric acid reacts very sensitively to the temperature and duration of tempering treatment. At 1023 K (750 °C), the corrosion-sensitive sigma phase precipitates. Only after tempering at 1233 K

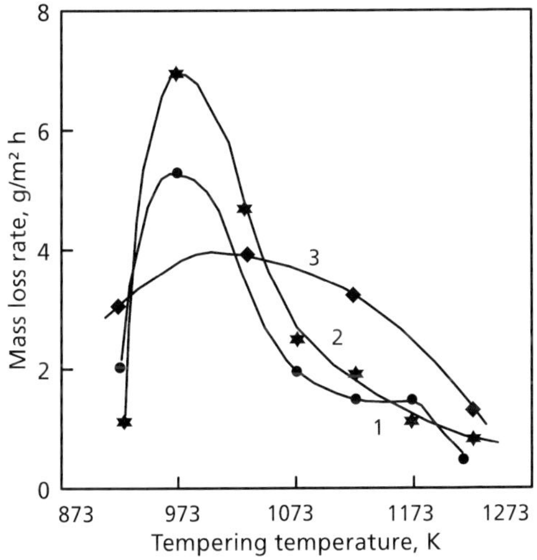

Figure 12: Dependence of the material consumption rate of the steel ZI-52 in 23 mol/l HNO_3 on the temperature and duration of tempering [173]
1) 10 min, 2) 1 h, 3) 100 h

(960 °C) is the susceptibility largely eliminated. This steel should therefore be used only in non-welded form for nitric acid plants [173].

Studies have shown that the sensitivity of chromium-nickel steels to intercrystalline corrosion in nitric acid-chromate solutions increases with increasing grain size, decreasing solution annealing temperature, decreasing carbon content, increasing test temperature, increasing nitric acid and chromate concentration and increasing mechanical stress [174].

The penetration depth of intercrystalline corrosion on the steel SUS 304 L (cf. 1.4306; (%) Fe-0.02C-18.74Cr-11.33Ni-0.98Mn-0.54Si-0.030P-0.008S) in boiling 21 % nitric acid + 4 g/l Cr^{6+}-ions is about 80 μm after 27.8 h; it follows the duration of exposure linearly in the time interval 1 to 100 h [175].

The high-chromium carbides of the type $M_{23}C_6$ which occur in the steels Ch18N10 ((%) Fe-0.08C-18.4Cr-10.2Ni-1.08Mn-0.3Si-0.005P-0.014S-0.005N) and Ch18N14 ((%) Fe-0.035C-18.8Cr-14.6Ni-0.35Mn-0.75Si-0.005P-0.03S-0.004N) after quenching from 1323 K (1050 °C) and subsequent heat treatment of 973 K (700 °C) for 500 and 1000 h are attacked selectively by boiling 27 % nitric acid + 40 g/l Cr^{6+} (as $K_2Cr_2O_7$) [176].

The material consumption rate of the steel 03Ch18N14 (Russ. grade; Fe-0.03C-18Cr-14Ni) in a solution of 27 % nitric acid with 40 g/l Cr^{6+}, with renewal of the solution every 2 hours, was determined in 14 h tests at room temperature as a function of the phosphorus and silicon content. It was 19.1 g/m^2 h with 0.048 % P + 0.16 % Si and 11.6 g/m^2 h with 0.005 % P + 0.75 % Si [177]. Intercrystalline corrosion occurred in the first 6 hours.

The effect of the silicon content in the steel Ch20N20 ((%) Fe-(0.004–0.015)C-(19.4–21.8)Cr-(19.3–20.8)Ni-(0.05–5.40)Si-(0.002–0.1)P) on corrosion in boiling 27 %

nitric acid + 40 g/l Cr^{6+} (as chromate) depends on the phosphorus content of the steel. In 48 h tests, curve (1) in Figure 13 is obtained at a content of 0.002 % P and curve (2) at a content of 0.1 % P [178]. Only above 4 % silicon is the material consumption rate of the steel, especially with a high phosphorus content (0.1 % P), significantly reduced. A relationship exists here between the corrosion rate and the cathodic polarization current density at 1.29 V_{SHE}.
Similar corrosion rate values (as a function of the silicon content) to those in Figure 13, curve ①, are also found with the steel Ch18N11 in boiling 27 % nitric acid + 4 g/l Cr^{6+} [179].

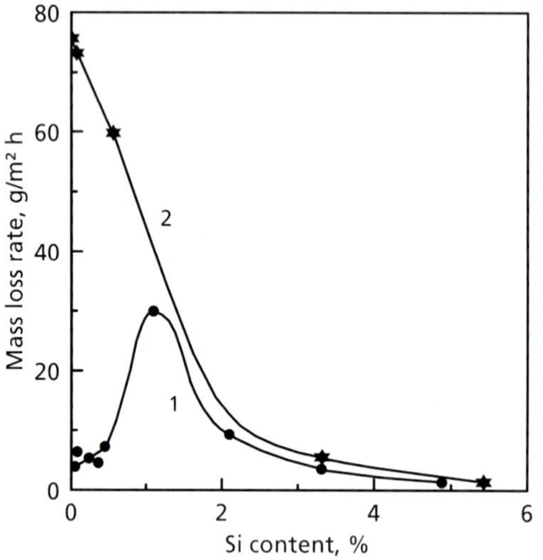

Figure 13: Dependence of the material consumption rate of the steel Ch20N20 in boiling 27 % HNO_3 on the silicon content at ① 0.002 % and ② 0.1 % phosphorus [178]

The dependence of corrosion behavior on the silicon and phosphorus content of the steel 12Ch18N10T (1.4541, SAE 321) in Cr^{6+}-containing nitric acid is also confirmed elsewhere [180].

After 96 h, the material consumption rate of the chromium-nickel steel (%) Fe-(0.006–0.011)C-20Cr-20Ni-0.002P-Si in boiling 27 % nitric acid + 40 g/l Cr^{6+} (as chromate) falls as the silicon content of the steel increases, from 7.6 g/m^2 h at 0.01 % Si to 1.4 g/m^2 h at 4.8 % Si. The corrosion-improving action of silicon recedes as the carbon content increases [181].

Additionally introduced phosphorus (0.002–0.097 % P) causes a marked increase in the intercrystalline corrosion of the steel Ch20N20 ((%) Fe-<0.02C-21Cr-19.5Ni-0.005Mo-0.002Mn-<0.01Cu) in boiling 27 % nitric acid + 40 g/l Cr^{6+} by a factor of about 20. A marked increase was already found with 0.025 % P. In contrast, sulfur has no noticeable influence [182].

The electrochemical corrosion behavior of the steel Ch20N20 in 27 % nitric acid at 313 K (40 °C) and 1.36 V_{SHE} with and without addition of 40 g/l Cr^{6+} is not substantially influenced by the carbon content of the steel (0.004–0.096 % C), regardless of whether the phosphorus content is low (0.002 % P) or high (0.1 % P). In contrast, the phosphorus content increases the susceptibility to intercrystalline corrosion in the transpassivity region. The susceptibility of the phosphorus-containing steel to corrosion also increase as the size of the austenite grain decreases [183]. The material consumption rate was 5.7–6.7 g/m^2 h at a phosphorus content of 0.002 % P and 59 g/m^2 h at 0.09 % P.

To be able to use the steel SAE 304 L (Fe-25Cr-20Ni-Nb) in boiling 8–12 mol/l nitric acid with radioactive material, the concentration of highly oxidizing metal ions, such as, for example, Cr^{6+} and Fe^{3+} of 100–1000 ppm, must be greatly reduced. The occurrence of vapor-liquid phase boundaries should also be prevented in heat exchanger pipes [184].

As a representative of other chromium-nickel steels, Remanit® 4306 (SAE 304 L, X2CrNi19-11: (%) Fe-0.025C-18.35Cr-10.24Ni-0.26Mo-1.54Mn-0.36Si-0.027P-0.003S-0.033B-0.043N; annealed at 1393 K (1120 °C) for 15 min and quenched in water) corrodes in 65 % nitric acid containing 0.5 % Cr^{6+} ions at 313 K (40 °C) with uninterrupted exposure of the specimen at a material consumption rate of 2.75 g/m^2 h after 72 h. If the test solution is renewed periodically every 4 h, the consumption rate is significantly higher at 4.2 g/m^2 h. This difference in corrosion is based on exhaustion of the test solution, which can largely be avoided by a sufficiently large solution volume [185].

To test the usability of the steel UHB 25 L (1.4845, SAE 310 S; (%) Fe-0.016C-24.9Cr-20.8Ni-1.63Mn-0.08Mo-0.03Cu-0.16V-0.45Si-0.012P-0.009S) for reprocessing nuclear fuels, it was investigated in a stringent Huey test (boiling 65 % (13.4 mol/l) nitric acid with and without addition of Cr^{6+} in the form of dissolved CrO_3). Some results are summarized in Table 29 [186].

Material state	***CrO_3 addition g/l***	***Number of test periods***			
		5	***15***	***5***	***15***
		Corrosion rate, mm/a (mpy)		***Depth of local attack, mm***	
Delivery state	–	0.06 – 0.07 (2.36 – 2.76)	0.06 – 0.07 (2.36 – 2.76)	–	0.005 (0.2)
Solution-annealed	–	0.05 – 0.07 (1.97 – 2.76)	0.065 (2.56)	0.002 (0.08)	0.005 0.010 (0.2 – 0.39)
Sensitized for 30 min at 973 K (700 °C)	–	0.119 (4.69)	0.267 (10.51)	0.018 (0.08)	0.064 (2.52)
		0.068 (2.68)	0.120 (4.72)	–	0.048 (1.89)
Solution-annealed	0.050 (1.97)	0.697 (27.44)	–	0.036 (1.42)	–
	0.100 (3.94)	1.616 (63.62)	–	0.075 (2.95)	

Table 29: Corrosion of the steel UHB 25 L (SAE 310 S) in boiling 65 % HNO_3 with and without a CrO_3 addition in 5 to 15 periods of 48 h each [186]

The material consumption rate of the CrNi steel 14 14 also increases in Cr^{6+}-containing boiling 65 % nitric acid as the phosphorus content increases, from about 3 g/m^2 h at 10 ppm P to 83 g/m^2 h at 1000 ppm P. While general corrosion with a material consumption rate of about 3 and < 2 g/m^2 h is found at a silicon content of < 10^3 and > 2 × 10^4 ppm, intercrystalline corrosion occurs between these silicon contents, with a maximum of about 63 g/m^2 h at 6 × 10^3 ppm Si [187].

While additions of 0.005 to 3.4 % molybdenum at a constant silicon and phosphorus content of 0.5 and 0.003 % hardly affect the corrosion rate of the steel (%) Fe-0.05C-17Cr-12.5Ni-(1.60–1.65)Mn-0.012N at all in boiling 65 % HNO_3 + 0.02 g/l Cr^{6+}, increasing phosphorus and silicon contents reduce corrosion of the steel. The material consumption rate is thus reduced from 0.44 to 0.22 g/m^2 h when the Si content rises from 0.5 to 2.08 %, and increased from 0.41 to 0.56 g/m^2 h when the phosphorus content rises from 0.002 to 0.098 % [188].

In nitric acid solutions of ≥ 80 % or lower concentration, but containing Cr^{6+}-ions, it is necessary to use a CrNi cast steel 18 9 or 18 13 with the lowest possible C content (≤ 0.03 % C) and an Si content of about 4 %. If it is not possible to produce a cast steel with such a low C content, steel with C contents of up to 0.12 % C, but only stabilized by Nb can also be used. The specimens of the materials investigated were solution-annealed (1353 K (1080 °C) ± 20 K (20 °C)/2 h/water). In 60 to 70 % HNO_3, the CrNi cast steels 18 9 and 18 13 with a normal Si content and C contents ≤ 0.035 C is adequate. The steel with a higher Ni content showed the better corrosion resistance in the HNO_3 solutions. Valves made of CrNi cast steel 18 9 and 18 13 with Si contents about 4.5 % have proved suitable in concentrated HNO_3 (about 96 %) at 318 K (45 °C) and were found to be several times more resistant than those of cast steel with normal Si contents of around 1 % [189].

An electrochemical method for determination of the susceptibility of stainless steels to intercrystalline corrosion is based on the appearance of an activation branch in the potentiodynamic curve in the potential range − 0.15 to + 0.55 V (against Ag/AgCl) at a measurement rate of 12 V/h. The electrolyte comprises 20 g/l $FeCl_3$ × 6 H_2O with a high redox potential, as well as 5 % nitric acid and 80 mg/l hydrogen chloride [190, 191].

A method for determining the tendency to undergo intercrystalline corrosion on the basis of potential measurements at room temperature under a drop of liquid consisting of 5 % HNO_3 + 20 g/l $FeCl_3$ × 6 H_2O + 70 ml/l HCl (for CrNi steels 18 8) or 240 ml/l HCl (for El-943) at 0.74–0.80 V was tested on the steels Ch18N9T (1.4541, SAE 321) and 06ChN28MDT (Russ. grade El-943; (%) Fe-0.06C-22.8Cr-24.7Ni-0.50Ti-2.77Cu-3.5Mo). The conclusiveness of the results is increased by precise demarcation of the drop, electrical insulation, enlargement of the drop diameter and brief pickling of the steel surface before-hand at room temperature in hydrochloric acid. The method in question is also suitable for in-situ testing of workpieces [192].

In chloride-containing HNO_3 such as occurs, for example, in the fertilizer industry, CrNi steels also undergo intercrystalline corrosion and pitting corrosion, as well as general corrosion. The passive or transpassive state of the steel is eliminated and the surface activated by Cl^- ions.

The occurrence of intercrystalline corrosion and pitting corrosion is determined by the nitric acid and Cl^--ion concentration, the temperature, the flow conditions and the operating time, apart from the material parameters. Figure 14 shows the dependence of the material consumption rates of X6CrNiTi18-10 (1.4541, SAE 321) on the NaCl content and HNO_3 concentration. This figure clearly shows the critical chloride ion concentration at the particular HNO_3 concentration [193].

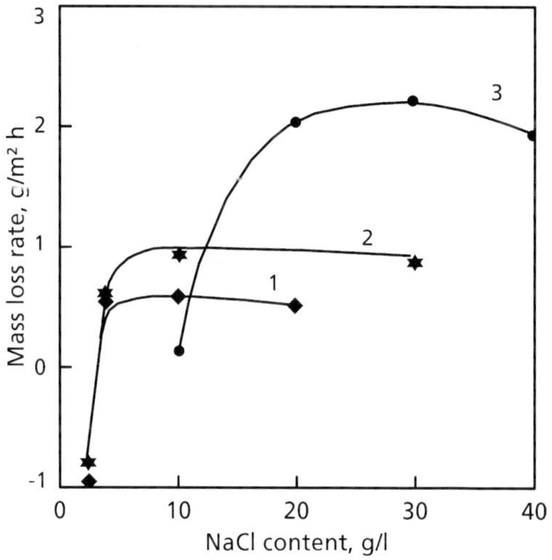

Figure 14: Influence of the NaCl content on the corrosion behavior of X 6 CrNiTi 18 10 in various concentrations of HNO_3 at 300 K (27 °C), test duration 1 h [193]
(1) 1 mol/l; (2) 2 mol/l; (3) 4 mol/l

No pitting corrosion was found on the steel X6CrNiTi18-10 (SAE 321) at nitric acid concentrations between 1.3 and 54.7 % with 14.2 g/l chloride up to 333 K (60 °C), and between 3.1 to 4.9 % HNO_3 with 28.4 g/l chloride [316]. In 45 % HNO_3 containing HCl or Fe(III) chloride, the steels CrTi25, CrNiMoCu28-28, CrMnNiN17-19-4 and CrNiTi18-10 are attacked both by intercrystalline corrosion and pitting [194]. Intercrystalline corrosion on X6CrNiTi18-10 occurred above 323 K (50 °C) in 48.4 % HNO_3 with 3.65 g/l HCl [195].

While the corrosion rate of X5CrNi18-10 (1.4301, SAE 304) in boiling 30 % HNO_3 with 6.8 g/l chloride is 0.03 to 0.33 mm/a (1.18 to 13.0 mpy), in the vapor space it is considerably higher at 1.1 to 3.3 mm/a (43.3 to 130 mpy). This is due to the formation of nitrosyl chloride. At a chloride content of 13.6 g/l, the corrosion rate was > 3.3 mm/a (> 130 mpy) (see Table 30) [196, 197].

The steel 1.4306 (SAE 304 L, nitric acid quality) was investigated as a cooling finger and aftercondenser in a nitric acid condensate with a low chloride content after a test duration of 12 weeks. To record irregularities in the sealing region, the cooling finger was made from pipe rings sealed with PTFE rings. The aftercondenser was produced from bar material. The material consumption rates per unit area of the starting material a) of the pipe rings and b) of the bar material in the delivery state

Steel	HNO_3 %	HCl g/l	Cl^- g/l	Temperature K (°C)	Corrosion rate mm/a (mpy)
X5CrNi18-10 (SAE 304)	30	–	6.8	Boiling	1.1 – 3.3 (43.3 – 130)
	30		13.6	Boiling	> 3.3 (> 130)
	3.7	14.6	–	305 (32)	4.2 (165)
	48.6	167.6	–	305 (32)	84.0 (3307)
CrTi steel, ferritic	38	1.9	–	313 (40)	0.3 – 1.1 (11.8 – 43.3)
	46	3.8	–	313 (40)	0.3 – 1.1 (11.8 – 43.3)
CrNiTi steel, austenitic	38	1.9	–	313 (40)	> 3.3 (> 130)
	46	3.8	–	313 (40)	> 3.3 (> 130)

Table 30: Corrosion behavior of steels in the vapor space above chloride-containing nitric acid solutions [196, 197]

or solution-annealed at 1333 K (1060 °C)/15 min/ water in the Huey test were a) 0.08–0.1 g/m^2 h and b) 0.12 or 0.09 g/m^2 h, respectively. The condensates consisted of 6 mol/l HNO_3 + 45 mg/l Cl^-, 8.5 mol/l HNO_3 + 20 mg/l Cl^- and 10 mol/l HNO_3 + 15 mg/l Cl^- and all contained 1 mg/l F^-. The corrosion medium was renewed weekly. Although the studies showed very low material consumption rates, local roughening, shallow pits and pitting corrosion occurred. The steel 1.4466 (SAE 310 MoLN, X1CrNiMoN25-22-2) is recommended as it is resistant to pitting corrosion and at the same time has a good resistance to sulfuric acid [198].

According to Figure 15, chloride ions have a disastrous influence on the material consumption rate of the steels 04Ch18N10 ((%) Fe-0.013C-17.8Cr-9.9Ni-1.15Mn) and 12Ch18N10T (1.4541, SAE 321(%) Fe-0.09C-17.4Cr-9.4Ni-1.74Mn-0.41Ti; Ti-stabilized) in dilute nitric acid [199]. In order to keep the corrosion rate of the steels below 0.1 mm/a (3.94 mpy) in 1 mol/l nitric acid, the content of chloride ions in the acid should be < 0.1 g/l.

To reduce the corrosive action of chloride-containing 2.3–3.0 mol/l nitric acid, it is treated with potassium permanganate, the chloride largely being removed as chlorine. Subsequent distillation results in 13 % nitric acid with 0.15 g/l chloride ions. The corrosion which occurs during further concentration in a distillation column consisting of chromium-nickel steel was eliminated by further reduction of the chloride content to about 0.02 g/l [200].

The presence of chloride ions in nitric acid reveals complex processes in the low-carbon and Ti-stabilized steels. The cathodic process in chloride-containing nitric acid is thus inhibited only at the carbide inclusions. The kinetics of the reduction of nitric acid are not influenced on the passive surface of the steel 12Ch18N10T (SAE 321) or 04Ch18N10 [201].

Stainless steels are not suitable for plants in which mineral fertilizers are produced with reaction mixtures consisting of nitric acid and potassium chloride (sylvinite). Vitreous carbon, smelting slags and mullite are suitable here [90].

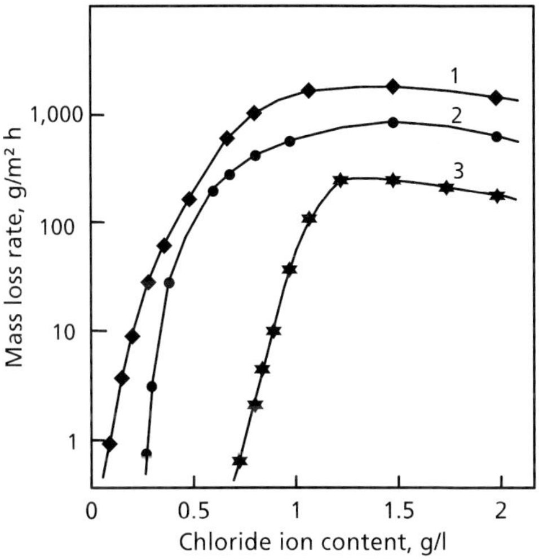

Figure 15: Corrosion of the steels 04Ch18N10 (1, 3) and 12Ch18N10T (2) in boiling 1 mol/l (1, 2) and 0.1 mol/l (3) HNO_3 as a function of the chloride ion content [199]

Table 31 shows the influence of hydrogen fluoride and chloride ions on corrosion of the steels 12Ch18N10T (SAE 321, 1.4541; (%) Fe-0.09C-17.1Cr-10.4Ni-1.28Mn-0.78Ti), 08Ch22N6T ((%) Fe-0.08C-21.7Cr-5.4Ni-1.28Mn-0.78Ti) and 03Ch23N6 ((%) Fe-0.02C-22.7Cr-5.7Ni-1.42Mn) in 10 mol/l nitric acid. The steel 03Ch23N6 is the most resistant for such solutions.

The corrosion rates of the steels 12Ch18N10T, 08Ch22N6T and 03Ch23N6 in 0.5 mol/l nitric acid with and without additions of hydrogen fluoride are summarized in Table 32 [75]. With an addition of 0.1 mol/l HF to the nitric acid, the corrosion rate for the steel 12Ch18N10T of 0.69 mm/a (27.2 mpy) at room temperature is already too high.

Figure 16 shows the corrosion of the steel 12Ch18N10T (SAE 321) in boiling HNO_3/HF solutions [202]. The complex relationship between the acid mixtures and corrosion of the steel can be seen from the isocorrosion curves. According to Figure 17, 0.2 mol/l iron nitrate in the HNO_3/HF solution has an inhibiting influence on corrosion of the steel SAE 304 SS ((%) Fe-0.059C-18.29Cr-9.36Ni-1.0Mn-0.48Si-0.029P-0.002S) at 223 K (-50 °C) [203]. According to ESCA studies, the oxide layer formed on the chromium-nickel steel in nitric acid consists of an accumulation of chromium oxide alongside iron oxides on the outside with a layer of SiO_2 of varying thickness [204].

Attack on the steel 12Ch18N10T (SAE 321) at 353 K (80 °C) in 12 mol/l nitric acid + 0.12 mol/l hydrogen fluoride, which initially proceeds at a material consumption rate of about 30 g/m^2 h, drops to about 3.5 g/m^2 h after 300 h [205].

Steel	Addition, mol/l	293 K (20 °C)	323 K (50 °C)
		Corrosion rate, mm/a (mpy)	
12Ch18N10T	–	0.018 (0.71)	0.020 (0.79)
	0.01 HF	0.12 (4.72)	0.48 (18.9)
	0.1 HF	0.77 (30.31)	5.08 (200)
	0.01 HF + 5 g/l Cl^-	0.052 (2.05)	0.16 (6.3)
	5 g/l Cl^-	0.014 (0.55)	0.036 (1.42)
08Ch22N6T	–	0.008 (0.31)	0.022 (0.87)
	0.01 HF	0.067 (2.64)	0.26 (10.2)
	0.1 HF	0.47 (18.5)	1.29 (51)
	0.01 HF + 5 g/l Cl^-	0.027 (1.06)	0.11 (4.33)
	5 g/l Cl^-	0.013 (0.51)	0.037 (1.46)
03Ch23N6	–	0.010 (0.39)	0.016 (0.63)
	0.01 HF	0.023 (0.91)	0.16 (6.3)
	0.1 HF	0.093 (3.66)	0.68 (26.8)
	0.01 HF + 5 g/l Cl^-	0.033 (1.3)	0.087 (3.43)
	5 g/l Cl^-	0.011 (0.43)	0.018 (0.71)

Table 31: Influence of HF and Cl^- on the corrosion rate of some CrNi steels in 10 mol/l HNO_3 at 293 and 323 K (20 and 50 °C) [77]

Steel	293 K (20 °C)			323 K (50 °C)			373 K (100 °C)		
	0	0.01 HF	0.1 HF	0	0.01 HF	0.1 HF	0	0.01 HF	0.1 HF
	Corrosion rate, mm/a (mpy)								
12Ch18N10T	0.014 (0.55)	0.040 (1.57)	0.61 (24.02)	0.020 (0.79)	0.043 (1.69)	0.82 (32.3)	0.031 (1.22)	0.087 (3.43)	1.95 (76.8)
08Ch22N6T	0.014 (0.55)	0.016 (0.63)	0.17 (6.69)	0.014 (0.55)	0.019 (0.75)	0.28 (11.0)	0.034 (1.34)	0.045 (1.77)	0.64 (25.2)
03Ch23N6	0.012 (0.47)	0.016 (0.63)	0.05 (1.97)	0.017 (0.67)	0.026 (1.02)	0.29 (11.4)	0.033 (1.3)	0.037 (1.46)	0.87 (34.3)

Table 32: Corrosion of CrNi steels in 0.5 mol/l HNO_3 with and without additions (mol/l) of HF between 293 and 373 K (20 and 373 °C) [75]

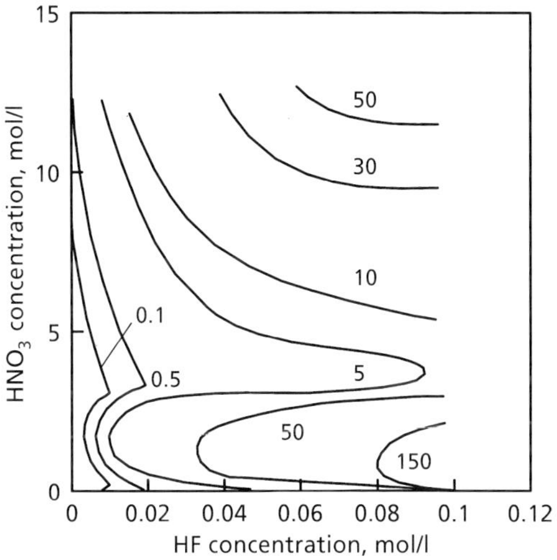

Figure 16: Isocorrosion curves (g/m^2 h) of the steel 12Ch18N10T (SAE 321) in boiling HNO_3/HF solutions [202]

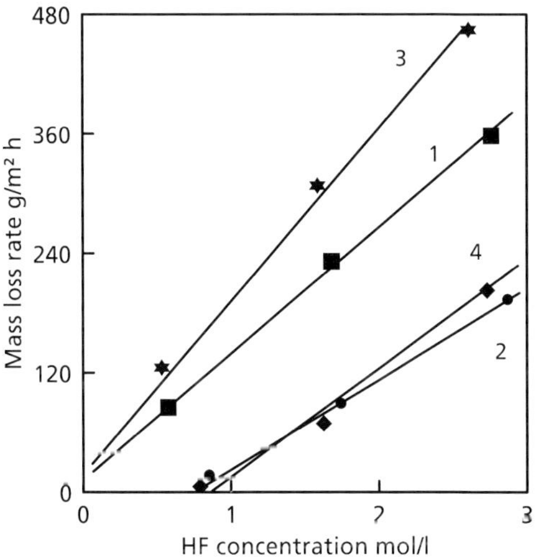

Figure 17: Influence of the HF concentration on the rate of dissolution of the steel SAE 304 SS in HNO_3/HF solutions with and without iron nitrate at 223 K (–50 °C) [203]

1) 3.5 mol/l HNO_3
2) 3.5 mol/l HNO_3 + 0.2 mol/l $Fe(NO_3)_3$
3) 0.8 mol/l HNO_3
4) 0.8 mol/l HNO_3 + 0.2 mol/l $Fe(NO_3)_3$

In a solution of (14 – 16) % HNO_3 + 3.8 % HF, the steel 12Ch18N10T (SAE 321) corrodes between 328 and 333 K (55 and 60 °C) with a material consumption rate of 3.83 g/m^2 h (4.3 mm/a) which is too high [206].

Thorium-containing fuel elements require nitric acid containing hydrogen fluoride for processing. While Inconel® 690 (2.4642) is recommended for the reaction vessel at about 400 K (127 °C), SAE 304 L can also be used up to 368 K (95 °C) [207].

Solutions of 10% HNO_3 + 39% HF at 343 K (70 °C) and 5% HNO_3 + 1% $FeCl_3$ are used to investigate the susceptibility of weld seams of CrNiTi 18 9 0.5 steels to intercrystalline corrosion [208].

According to a previous report, stainless steel proved suitable for evaporation of the primary solution of HNO_3 + HF + $Al(NO_3)_3$ + plutonium nitrate, as well as for the heating pipes [209].

The corrosion problems which arise during processing of nuclear fuels (plutonium) with 12 mol/l nitric acid + 0.44 mol/l hydrogen fluoride + 0.48 mol/l aluminium nitrate (for the purpose of complexing with fluorides), especially during evaporation and recovery of the nitric acid, cannot be solved by CrNi steels, but more favorably by special nickel-chromium-alloys, such as, for example, Ni-34Cr-1.0Ti-3.7Si-0.6Mn [210].

In addition to this, the material consumption rates of the steels 12Ch18N10T (SAE 321, 1.4541) and 03Ch23N6 in boiling 1 mol/l nitric acid with 1 g/l hydrogen fluoride + 2.7 g/l aluminium nitrate are 3.2 and 0.55 g/m^2 h [211].

In nitric acid solutions containing additions, such as, for example, sodium nitrate, secondary reactions have an influence on the corrosion process and the redox potential of steels [212].

Isocorrosion curves (1 and 0.5 mm/a (39.4 and 19.7 mpy)) for various CrNi and CrNiMo steels in nitric acid solutions containing sodium fluoride show that the steels of lower nickel content are more resistant than the chromium-nickel steel 18 10 [213].

According to Table 33, corrosion of the three chromium-nickel steels, which differ only in their carbon content, in hydrazine-containing nitric acid at 388 K (115 °C) is increased drastically by sodium fluoride. In its absence the influence of 2 g/l hydrazine can be ignored [214].

Figure 18 shows the dependence of the corrosion of the steel 12Ch18N10T (SAE 321) in 4 mol/l nitric acid + 10 g/l $K_2Cr_2O_7$ + NaF on the temperature and the sodium fluoride concentration. The material consumption rates of the steel at the boiling point are given in Table 34 [215]. The optimum passivating action of the fluoride ions lies between 0.03 and 0.05 mol/l.

Pitting and knife-line corrosion occurs on welded joints of chromium-nickel steels in a solution of 500 g/l nitric acid + 600 g/l sodium nitrate + 50 g/l copper nitrate + 5 g/l sodium fluoride (Table 35) [216].

Corrosion of the steel 12Ch18N10T (SAE 321) in boiling 6 mol/l nitric acid is inhibited by fluorides if oxidizing agents, such as, for example, 0.2 mol/l CrO_3 or ammonium vanadate (NH_4VO_3), are present at the same time. At a constant acid concentration, the corrosion rate reaches a minimum at a fluoride ion concentration of between 0.05 and 0.15%, the position of the minimum depending on the composition and nature of the oxidizing agent and the structure of the particular alloy [217].

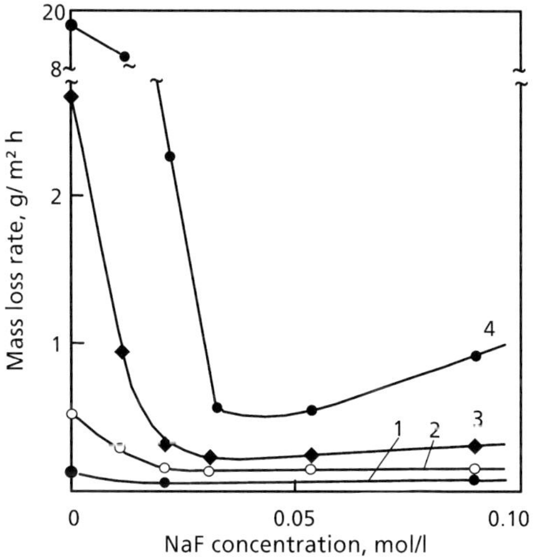

Figure 18: Material consumption rate of the steel 12Ch18N10T (SAE 321) in 4 mol/l HNO_3 + 10 g/l $K_2Cr_2O_7$ + NaF as a function of the NaF concentration at (1) 323 K (50 °C), (2) 348 K (75 °C), (3) 363 K (90 °C) and (4) the boiling point [215]

Concentration, g/l			***Material consumption rate, g/m² h***		
HNO₃	***NaF***	***N₂H₄***	***12Ch18N10T*** *	***08Ch18N10T*** *	***04Ch18N10T*** *
75	0	2	0.11 ± 0.04	0.16 ± 0.04	0.2 ± 0.1
75	0.05	2	0.7 ± 0.1	2.4 ± 0.1	5.1 ± 0.6
75	0.5	2	11 ± 1	16 ± 1	15 ± 1
75	5	2	96 ± 2	183 ± 8	131 ± 11
75	2	0	1.0 ± 0.1	1.2 ± 0.4	2.2 ± 0.7
75	2	2	47 ± 1	46 ± 5	68 ± 2
75	2	10	53 ± 2	66 ± 7	56 ± 3
100	2	2	51 ± 1	95 ± 7	81 ± 5
500	2	2	5.8 ± 1	3.8 ± 0.8	3.6 ± 0.8

Table 33: Corrosion rate of some chromium-nickel steels in HNO_3 with NaF and hydrazine [214]
* cf. SAE 321 with different carbon content

$K_2Cr_2O_7$ mol/l	NaF, mol/l					
	0	0.01	0.02	0.03	0.05	0.10
	Material consumption rate, g/m² h					
0.017	10.5	2.0	0.4	0.4	0.5	0.7
0.034	16.5	8.0	2.7	0.7	0.8	1.2
0.068	33.0	18.2	9.7	2.0	1.2	1.7

Table 34: Dependence of the corrosion of the steel 12Ch18N10T in boiling 6 mol/l HNO_3 + $K_2Cr_2O_7$ on the concentration of NaF [215]

Steel	Chemical composition, %						Corrosion rate mm/a (mpy)	Type of corrosion
	C	Mn	Si	Cr	Ni	Others		
00Ch18N10	0.015	1.7	0.7	17.3	10.4	–	1.2 (47.2)	pitting corrosion
0Ch18N10T	0.06	1.5	0.8	18.0	9.8	0.6 Ti	0.98 (38.6)	pitting corrosion
0Ch18N12B	0.05	0.9	0.3	17.7	12.1	0.7 Nb	0.79 (31.1)	pitting corrosion
1Ch18N9T	0.11	1.3	0.8	17.2	7.9	0.7 Ti	1.13 (44.5)	knife-line corrosion and pitting corrosion
Ch28N18	0.16	1.7	1.1	22.6	18.0	0.4 Ti	2.90 (114)	intercrystalline and pitting corrosion

Table 35: Corrosion rate of welded joints of austenitic steels in a solution containing 500 g/l HNO_3 + 600 g/l $NaNO_3$ + 50 g/l $Cu(NO_3)_2$ + 5 g/l NaF at room temperature [216]

Nevertheless, the material consumption rate of the steel in 0.5 mol/l nitric acid can be increased considerably (130 to 160 g/m^2 h) by between 0.1 and about 0.35 mol/l fluoride ions (KF, NaF, ammonium fluoride and hydrogen fluoride). As the fluoride ion concentration increases further, the material consumption rate then decreases to 30 to 40 g/m^2 h. Corrosion in the presence of fluoride ions proceeds differently in dilute nitric acid to that in concentrated acid, in which the material consumption rate is < 0.1 g/m^2 h [218].

According to the temperature/concentration graph of corrosion of the steel Sandvik® 2RE10 ((%) Fe-<0.02C-24.5Cr-20.5Ni; corresponds to SAE 310 L) in hot nitric acid, its resistance can be described as good. In hot 65 % nitric acid at 393 K (120 °C), in 80 % HNO_3 at 353 K (80 °C) and in 90 % HNO_3 at 325 K (52 °C), the corrosion rate is about 0.1 mm/a (3.94 mpy). This steel is also used for reaction tanks in the explosives industry for the production of nitroglycerine in a solution of 55 % nitric acid and 45 % sulfuric acid, and in the production of TNT (trinitrotoluene) in a solution of 25 % HNO_3 + 75 % H_2SO_4, where temperatures between 350 and 365 K (77 and 92 °C) occur during the course of the reaction [219].

Grinding grooves caused intercrystalline corrosion on a storage tank made of SAE 321 (1.4541) in an anhydrous mixed acid solution containing 90 % HNO_3 + 10 % H_2SO_4. The same result is also observed on coarse grinding grooves on the steel Fe-18Cr-15Ni-4Si, although if polished adequately, this is resistant to highly concentrated nitric acid [220].

According to Figure 19, the corrosion rate in 10 % sulfuric acid is reduced from 1.2 or 10 mm/a (47.2 or 394 mpy), depending on the type of steel, to about 0.1 mm/a (3.94 mpy) by addition of 1 to 3.5 % HNO_3. For comparison, two chromium-nickel-molybdenum steels are shown alongside the two chromium-nickel steels 08Ch22N6T and 12Ch18N10T (SAE 321). Above an addition of 8 % HNO_3, the corrosion rate is the same for all four steels at about 0.001 mm/a (0.04 mpy) [221].

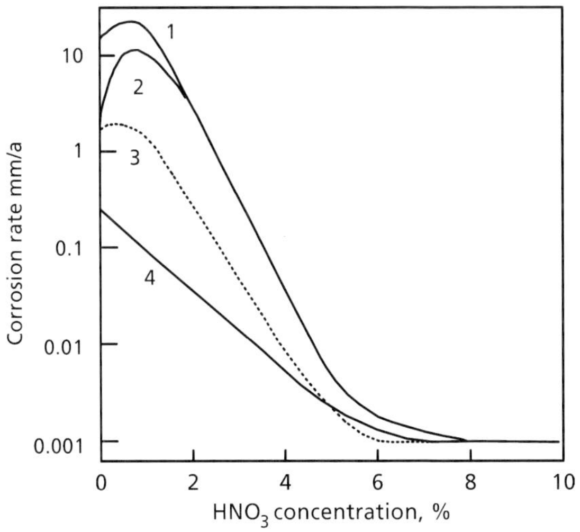

Figure 19: Dependence of the corrosion of some steels in HNO_3-containing 10 % H_2SO_4 on the HNO_3 content at room temperature [221]
(1) 08Ch22N6T
(2) 12Ch18N10T (SAE 321)
(3) 08Ch21N6M2T
(4) 06ChN28MDT

If nitrous gases are washed out with a mixture of sulfuric acid with 1 to 5 % nitric acid instead of with 10 to 45 % sulfuric acid in the production of sulfuric acid, corrosion of the steels used can be greatly reduced. The concentration of nitric acid required is lower, the higher the sulfuric acid concentration. The increased resistance occurs in particular at a somewhat elevated temperature (328 K (55 °C)) [221]. Some test values are summarized in Table 36 (06ChN28MDT: (%) Fe-0.06C-22.4Cr-27.9Ni-2.6Mo-2.9Cu-0.76Ti-0.30Mn-0.69Si and 08Ch21N6M2T (%) Fe-0.08C-21Cr-6Ni-2Mo, Ti-stabilized).

In agreement with the results in Table 36, the steels 08Ch22N6T and 12Ch18N10T (SAE 321, 1.4541) corroded in 50 % sulfuric acid at 333 K (60 °C) with an addition of 0.1 % HNO_3 at an expected rate of 0.03 and 0.01 mm/a (1.18 and 0.39 mpy). This rate is also approximately maintained at 1 % HNO_3 [222].

Steel	Temperature K (°C)	10 % H_2SO_4 + 1 % HNO_3	10 % H_2SO_4 + 5 % HNO_3	30 % H_2SO_4 + 1 % HNO_3	30 % H_2SO_4 + 5 % HNO_3	50 % H_2SO_4 + 1 % HNO_3	50 % H_2SO_4 + 5 % HNO_3
		Corrosion rate, mm/a (mpy)					
08Ch22N6T	295 (22)	32 (1260)	0.003 (0.12)	0.004 (0.16)	0.003 (0.12)	0.004 (0.16)	0.001 (0.04)
	328 (55)	0.3 (11.8)	0.009 (0.35)	0.01 (0.39)	0.02 (0.79)	0.009 (0.35)	0.02 (0.79)
12Ch18N10T	295 (22)	2.1 (82.7)	0.004 (0.16)	0.002 (0.08)	0.001 (0.04)	0.002 (0.08)	0.001 (0.04)
	328 (55)	0.3 (11.8)	0.009 (0.35)	0.01 (0.39)	0.01 (0.39)	0.008 (0.31)	0.02 (0.79)
06ChN28MDT	295 (22)	0.2 (7.87)	0.003 (0.12)	0.001 (0.04)	0.001 (0.04)	0.004 (0.16)	0.001 (0.04)
	328 (55)	0.01 (0.39)	0.005 (0.20)	0.006 (0.24)	0.006 (0.24)	0.003 (0.12)	0.002 (0.08)
08Ch21N6M2T	295 (22	3.0 (118)	0.002 (0.08)	0.002 (0.08)	0.001 (0.04)	0.004 (0.16)	0.001 (0.04)
	328 (55)	0.03 (1.18)	0.009 (0.35)	0.01 (0.39)	0.008 (0.31)	0.002 (0.08)	0.008 (0.31)

Table 36: Corrosion rate of some steels in H_2SO_4/HNO_3 mixtures at 295 and 328 K (22 and 55 °C) [221]

The steel SAE 304, which corrodes in pure sulfuric acid at room temperature at a material consumption rate of about 70 g/m^2 h, gives values of < 0.1 g/m^2 h in a mixture of 90 % H_2SO_4 + 10 % nitric acid [223]. It also behaves similarly in a mixture of 90 % 1.5 mol/l H_2SO_4 + 6 % HCl + 4 % HNO_3.

The corrosion behavior of the steel 12Ch18N10T in 10 % sulfuric acid with nitric acid additions at 333 K (60 °C) after 25 and 100 h tests is summarized in Table 37 [224]. Here also, corrosion of the steel in sulfuric acid is reduced by additions of nitric acid.

HNO_3 content mol/l	With mixing	Without mixing
	Material consumption rate, g/m^2 h	
–	2.83	8.51
		2.39*
0.003	3.62	9.6
0.100	0.16	0.22
0.150	0.11	0.19
0.300	0.02	0.02
		0.001*

Table 37: Influence of HNO_3 on the corrosion of 12Ch18N10T (SAE 321, 1.4541) in 10 % H_2SO_4 at 333 K (60 °C) after 25 and 100* h [224]

According to Table 38, contact between the steel Ch18N10T and graphite or platinum in HNO_3 or sulfuric acid/nitric acid mixtures at 363 K (90 °C) causes an increase in the material consumption rate [225]. Additions of potassium bichromate (1 – 10 g/l) in turn increase the corrosion of the steel in the H_2SO_4/HNO_3 mixtures.

The material consumption rate could be reduced from 18.6 to 1.07 g/m^2 h by an addition of 0.3 mol/l nitric acid to a solution of 10 % sulfuric acid + 1.2 % fluoride ions at 333 K (60 °C) [224].

In 5 % sulfuric acid with additions of 8 % NaCl + 8 % Na_2SO_3 + 2 % HNO_3, the steels 12Ch18N10T and 08Ch22N6T corrode with corrosion rates of 37 and 41 mm/a (1457 and 1614 mpy) at 313 K (40 °C) [226].

The corrosion rate of the steel 12Ch18N10T (SAE 321, 1.4541) in a solution of 36.6 % H_2SO_4 + 23.3 % HNO_3 + 39.1 % SO_3 after 200 h tests is about 0.1 mm/a (3.94 mpy) [227]. In a solution of 28 % H_2SO_4 + 12 % HNO_3 + 60 % SO_3, the corrosion rate was only 0.06 mm/a (2.36 mpy). At 403 K (130 °C), the corrosion of the steel in the first and second solution achieves rates of 0.26 and 2.1 mm/a (10.2 and 82.7 mpy) after 250 h tests [227].

According to Table 39, the steel Ch18N10T, which is resistant in 2.5 mol/l nitric acid at room temperature, corrodes with a high material consumption rate after addition of 0.22 mol/l H_2SO_3, but this decreases again after addition of 0.82 mol/l H_2SO_3 at potentials of + 0.34 and + 0.03 V_{SHE} respectively [228].

Acid mixture, % $H_2SO_4/HNO_3/H_2O$	*Contact material*	S_1/S_2*	*Potential* V_{SHE}	*Material consumption rate* g/m^2 h
0/57/43	–	–	1.160	0.05
	Pt	5	1.290	1.07
	graphite	1	1.280	1.05
10/51/39	–	–	1.100	0.28
	Pt	0.17	1.285	1.4
	graphite	1.0	1.300	2.1
40/33/27	Pt	0.17	1.355	3.0
	graphite	0.38	1.395	5.0
70/16/14	Pt	0.17	1.405	3.1
	graphite	1.0	1.546	3.8

* S_1/S_2 = contact area/total area

Table 38: Corrosion rate and potential of the steel Ch18N10T in H_2SO_4/HNO_3 mixtures at 363 K (90 °C) with and without graphite or Pt contact [225]

Concentration, mol/l		Potential V_{SHE}	Current Density mA/cm^2	Material Consumption Rate $g/m^2\,h$
HNO_3	H_2SO_3			
2.5	0.22	–0.11	310	275
	0.38	+0.07	1.0	5
	0.82	+0.34	0.2	0.02
5.0	0.22	–0.11	300	270
	0.38	–0.09	80	90
	0.82	+0.03	2.0	10

Table 39: Corrosion of the steel Ch18N10T at room temperature in HNO_3 with additions of H_2SO_3 and different potentials [228]

The steels 03Ch18N11 (SAE 304 L, 1.4306), 03ChN28MDT, 03Ch21N21M4GB and 08Ch22N6T can be recommended in the fertilizer industry for solutions of superphosphoric acid + nitric acid up to 333 K (60 °C) (the test solution consisted of 100 g $H_4P_2O_7$ + 246 g 47 % HNO_3) [229].

HNO_3 concentration %	Addition mol/l	Corrosion rate mm/a (mpy)	
		353 K (80 °C)	383 K (110 °C)
15	–	0.002 (0.08)	0.017 (0.67)
30	–	0.004 (0.16)	0.038 (1.5)
15	0.001 oxalic acid	0.003 (0.12)	0.022 (0.87)
15	0.1 oxalic acid	0.007 (0.28)	0.023 (0.91)
30	0.001 oxalic acid	0.006 (0.24)	0.044 (1.73)
30	0.1 oxalic acid	0.009 (0.35)	0.052 (2.05)
15	0.1 glutaric acid	0.001 (0.04)	0.013 (0.51)
15	0.1 glutaric acid	0.001 (0.04)	0.010 (0.39)
30	0.01 glutaric acid	0.003 (0.12)	0.032 (1.26)
30	0.1 glutaric acid	0.002 (0.08)	0.024 (0.94)
15	0.01 adipic acid	0.003 (0.12)	0.007 (0.28)
15	0.1 adipic acid	0.002 (0.08)	0.005 (0.2)
30	0.01 adipic acid	0.006 (0.24)	0.033 (1.3)
30	0.1 adipic acid	0.003 (0.12)	0.026 (1.02)
15	0.001 azelaic acid	0.001 (0.04)	0.011 (0.43)
15	0.015 azelaic acid	0.001 (0.04)	0.015 (0.59)
30	0.001 azelaic acid	0.003 (0.12)	0.024 (0.94)
30	0.015 azelaic acid	0.004 (0.16)	0.026 (1.02)

Table 40: Corrosion of the steel Ch18N10T in a mixture of HNO_3 and dicarboxylic acid after 200 h [230]

The steels 12Ch18N10T and 08Ch22N6T corrode in a hot solution of 40.4 g/l HNO_3 + 18.2 g/l oxalic acid at 373 K (100 °C) with a corrosion rate of 0.007 mm/a (0.28 mpy) [80].

According to the corrosion rates in Table 40, the steel Ch18N10T can be used as a material for the reactor for the production of saturated dicarboxylic acids by oxidation of kerogen by means of nitric acid [230].

The steel SAE 304 was treated with 25.7 g/l U (as $UO_2(NO_3)_2 \times 6H_2O$) + 81 g/l Cd ($Cd(NO_3)_2 \times 4H_2O$) + 2.5 g/l Te ($K_2TeO_4 \times 2H_2O$) + 4.1 g/l K ($KNO_3$) + 3.8 g/l Sr ($Sr(NO_3)_2$) + 16.0 g/l Zr ($ZrO(NO_3)_2 \times 2H_2O$) + 19.4 g/l Mo ($K_2MoO_4 \times H_2O$) + 6.1 g/l Fe ($Fe(NO_3)_3 \times 9H_2O$) + 47.2 g/l lanthanides ($Ce(OH)_4$) + 6.9 g/l Ba ($Ba(NO_3)_2$) in boiling 0.8 mol/l nitric acid, activated with neutrons and left for a maximum of 90 d for the purpose of simulating a solution containing radioactive waste. The corrosion rate on this steel was between 0.13 and 0.7 mm/a (5.12 and 27.6 mpy) [227]. These results are approximately the same as those from the tests carried out on SAE 304 L in boiling 0.9 mol/l nitric acid under comparable conditions without radioactive substances [231, 232].

Figure 20 provides further information on the influence of uranium (UO_2^{2+}) in nitric acid/fluoride solutions from 333 K (60 °C) up to the boiling point on the material consumption rate of 12Ch18N10T (SAE 321, 1.4541). According to Figure 20 b), the corrosion drops sharply as the ratio of metal and fluoride ions increases [233].

The susceptibility of the steels 08Ch18N10 (SAE 304, 1.4301) and 06ChN40B ((%) Fe-0.055C-17.01Cr-39.04Ni-1.99Mn-0.50Nb-0.60Si-0.013S-0.022P) to intercrystalline corrosion is intensified by irradiation with neutrons and subsequent heat treatment. Damage caused by the Huey test has less effect, the higher the resistance of the steel in the non-irradiated state [234].

If the boiling nitric acid contains 400 g/l uranyl nitrate, corrosion rates on the steel in the delivery state of 0.044 mm/a (1.73 mpy) and on the sensitized steel of

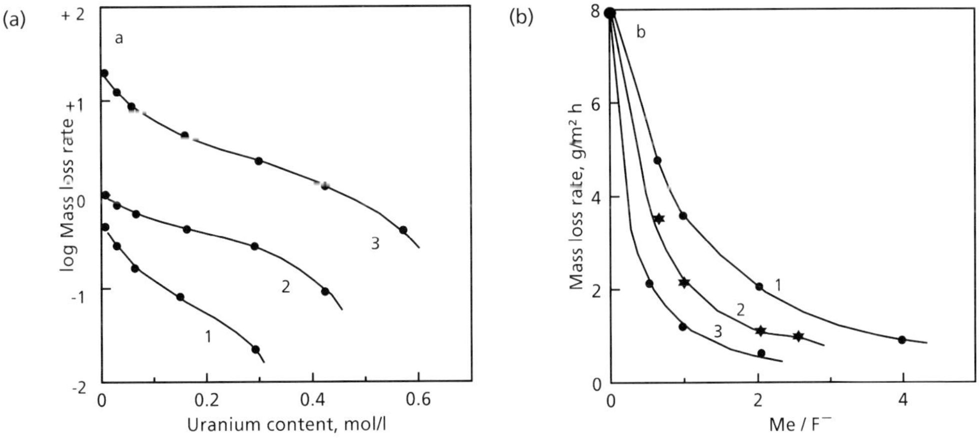

Figure 20: Influence of the uranium content on the corrosion behavior of the steel 12Ch18N10T in HNO_3 + 0.05 mol/l F^- at 353 K (80 °C) (Figure a: (1) 0.5, (2) 5 and (3) 10 mol/l HNO_3) and influence of the Me/F^--ratio (Figure b: (1) U, (2) Zr and (3) Al) on the corrosion in 5 mol/l HNO_3 + 0.05 mol/l F^- [233]

0.06 mm/a (2.36 mpy) result after 240 h tests; i.e. the presence of uranium ions causes only a slight increase in the corrosion rates. When boiling 5.5 mol/l nitric acid was used, the corresponding corrosion rates were only 0.006 and 0.008 mm/a (0.31 mpy). Mild intercrystalline attack was found on the sensitized specimens [235].

For preparation of pure silver nitrate for the photographic industry, the chromium-nickel steel 08Ch22N6T ((%) Fe-0.08C-22Cr-6Ni; Ti-stabilized) is recommended for the reaction chamber where the salting-out of $AgNO_3$ (2.2 kg/l in 70% HNO_3) takes place [236].

According to Table 41, corrosion of the steel Ch18N10T in 5 mol/l nitric acid is greatly increased by addition of cerium nitrate and ozone. When 0.02 mol/l silver nitrate was added instead of cerium nitrate, the material consumption rate was 10 to 13 g/m^2 h [237].

Temperature K (°C)	*$Ce(NO_3)_2$ mol/l*	*Ozone %**	*Material consumption rate g/m^2 h*
293 (20)	–	–	0.004
	0.02	2 – 6	5.05
368 (95)	–	–	0.015
	0.02	2 – 6	11.5

* percent by volume

Table 41: Influence of cerium ions on corrosion in ozonized 5 mol/l HNO_3 [237]

According to Table 42, laboratory tests in nitric acid with and without additions of ammonium nitrate show that the steel 12Ch18N10T (cf. 1.4541) is a suitable material for the production of ammonium nitrate by the one-stage process [238]. The steel 12Ch18N9T behaves similarly.

To avoid additions of hydrofluoric acid to nitric acid for cleaning pipelines made of the steel 0Ch18N10T (1.4541, SAE 321), a solution of 75 ml/l nitric acid (d = 1.32) + 25 ml/l hydrogen superoxide (30%) is proposed. This produces good results after a period of 20 min at 333 K (60 °C) [239].

Addition of thiourea ($CS(NH_2)_2$) accelerated corrosion of the steel 12Ch18N10T in 0.5 mol/l nitric acid between 293 and 333 K (20 and 60 °C) [240].

An increase in the phosphorus content from 0.001 to 0.11 or 0.12% in the CrNi steels 04Ch18N10, 04Ch18N27 and 04Ch18N40 increases the material consumption rate in boiling 65% nitric acid only insignificantly, i.e. from 0.20 to 0.25, from 0.12 to 0.17 and from 0.10 to 0.14 g/m^2 h, respectively [241]. The phosphorus content has a considerably greater influence on the material consumption rate of the three steels in 5 mol/l HNO_3 + 22 g/l $K_2Cr_2O_7$ (4 to 24 g/m^2 h, 4.5 to 37 g/m^2 h and 5.5 to 50 g/m^2 h, respectively).

If fluoride-containing nitric acid is used for etching CrNi steels (e.g. in the case of the steel 12Ch18N10T in 1.5 mol/l nitric acid containing 0.1 mol/l sodium fluoride at various polarization potentials), the corrosion rate above 370 K (97 °C) with heat transfer remains practically constant, while without heat transfer it increases sharply as the temperature rises. The corrosion rate of the steel in the passive range (+ 0.7 V_{SHE}) at temperatures up to 373 K (100 °C) is only slightly below 1.2 mm/a (47.2 mpy), regardless of the heat transfer [242].

State of steel	*HNO_3 %*	*Temperature K (°C)*	*NH_4NO_3 %*	*Material consumption rate g/m^2 h*
Not welded	47	Boiling	0	0.00088
			10	0.00092
			20	0.00278
Welded	57	Boiling	0	0.00128
			10	0.00148
			20	0.00248
Not welded	50	373 (100)	0	0.00045*
			5	0.00036*
			20	0.00021*
Welded	58	373 (100)	0	0.00040
			20	0.00030

*test duration 100 h

Table 42: Corrosion of the steel 12Ch18N10T (SAE 321, 1.4541) in HNO_3 with and without NH_4NO_3 in laboratory tests at various temperatures, test duration 200 h [238]

Nitric acid/hydrofluoric acid solutions (10:1) at 303 K (30 °C) and suitable potentials are used for rapid removal of the oxide layer on the steel SAE 304. Maximum etching with a reflection of 85 % was achieved after only 3 min between – 0.3 and – 0.5 V_{SMSE}, while a reflection value of only about 5 % was obtained without a polarization potential under otherwise identical conditions, even after 5 min [243].

To avoid crevice corrosion on chromium-nickel and chromium-nickel-molybdenum steels, combined treatment of 10 min in 10 % sulfuric acid at 333 K (60 °C) + 30 min in 25 % nitric acid at 313 K (40 °C) is recommended, instead of surface treatment in the usual nitric acid/hydrogen fluoride solution (65 % HNO_3 + 1 percent by volume HF), for workplace-friendly reasons (avoidance of hydrofluoric acid) [244].

A hot solution of 10 % nitric acid + 1 % hydrogen fluoride at 368 K (95 °C) and a treatment time of about 6 min has also proved appropriate for removal of the oxide layer on rolled and heat-treated SAE 304. Replacement of HF by hydrogen chloride caused pitting corrosion during the pickling operation [245].

Treatment for 1–3 h in a solution of 20 % HNO_3 + 2 % H_2O_2 at 343 K (70 °C) is also proposed for removal of oxide layers on the steels 08Ch17T (cf. 1.4510) and 12Ch18N10T (1.4541, SAE 321) [246].

It was demonstrated by electrochemical etching for 30 s in concentrated nitric acid at 1 V that additions promote intercrystalline corrosion of the steels Ch18N10T (1.4510, SAE 439) and Ch18N40T (1.4541, SAE 321) in that they migrate to the grain boundaries under mechanical stress [247].

Weld seams of type X 12 CrNi 18 8 (1.4300) steels with a high degree of scaling are freed completely from the scale layer, to give a silky matt surface, after dipping in a 30 % aqueous solution of 1.3 mol/l nitric acid + 0.85 mol/l hydrofluoric acid + 1.0 mol/l sulfuric acid + 0.001 % fluorosurfactant, based on the total amount of the pickling solution, at 333 K (60 °C) for 5 min [248].

Tanks made of SAE 304 containing 0.5 boron (e.g. (%) Fe-0.068C-17.98Cr-8.04Ni-1.39Mn-0.02Al-0.65Si-0.024P-0.022S-0.53B) are used for storage of the spent fuel from nuclear reactors. The boride precipitates which form are more resistant than the steel matrix to etching solutions of 10 ml HNO_3 + 20 ml HCl + 30 ml water [249].

Pure nitric acid is not suitable for etching stainless steels. Hence the steel Ch18N10T is not etched in non-agitated 12 % acid even after 240 min. At an acid flow rate of 0.4 m/s, etching occurs after 20 min, with a deposit of pickling sludge [250].

A solution of 30 % HNO_3 + 40 % HCl + 8 % H_2SO_4 + 12 % water is recommended for removing the scale formed on the steel SUS-310 Nb ((%) Fe- < 0.1C-25Cr-20Ni-1Nb) during rolling or after heat treatment. This solution also generates an excellent surface, as well as removing the oxide layer [251].

The rate of etching of a stainless steel at 353 K (80 °C) in an etching solution of (parts by volume) 1 conc. HNO_3 + 1 conc. HCl + 3 water was about 4.8 mm/h [252].

The most favorable concentration range for chemical etching of 12Ch18N10T (SAE 321) is shown in triangular diagrams with HNO_3, H_2SO_4 and HCl [253].

A pickling solution of 8 percent by volume each of concentrated nitric acid + hydrochloric acid + sulfuric acid + water as the remainder is used, inter alia, at room temperature for removal of the annealing colors after welding of stainless steels [254]. Passivating after-etching in dilute nitric acid is advisable in order to obtain the thin protective chromium oxide layer.

A mixed solution of (parts by volume) 3 HCl + 2 HNO_3 + 2 CH_3COOH is also used for etching chromium-nickel steels [255].

A pickling solution containing 12–15 percent by volume nitric acid (d = 1.38) + 1–3 % hydrofluoric acid (d = 1.17) + 82–87 % water with 0.1 g/l wetting agent and barium sulfate concentrate as a carrier has proved to be the most favorable pickling paste for chromium-nickel steels [256].

Steels of the type SAE 304 corrode at a material consumption rate of 0.15 or 0.13 g/m^2 h in boiling 65 % nitric acid after mechanical or electrolytic polishing [257]. In the case of solution-annealed material, the mechanically polished specimens are attacked more severely than those polished electrolytically. Sensitized material also behaves in a similar way, the attack also being intercrystalline.

The etching time for stainless steels is shortened from 240 to 20 min by flowing 12 mol/l nitric acid (0.4 m/s). Furthermore, no rhythmic variations in potential, which are found in stationary acid, occur [258].

To protect the CrNi steel 18 8 (V2A, cf. SAE 321) during removal of deposits by nitric acid, 0.12 mol/l thiourea is added to the 10 % nitric acid [259].

At certain proportions of nitric and hydrochloric acid, nitrosyl chloride is formed. This attacks the steel SAE 304 severely and with vigorous evolution of heat. Corrosion can be inhibited by additions of thiourea [260, 261].

The corrosion rate of the steel X8CrNiTi18-10 (1.4878, SAE 321 H) in 96 % sulfuric acid at 393 K (120 °C) is reduced to 0.1 mm/a (3.94 mpy) by addition of 0.03 – 0.04 % N_2O_3 as a result of passivation [262].

The material consumption rates of the steel 12Ch18N10T in solutions of 0.5 and 10 mol/l nitric acid containing 0.05 mol/l sodium fluoride at 353 K (80 °C) of about 0.6 and 20 g/m^2 h, respectively are reduced by additions of 0.3 and 0.6 mol/l UO_2^{2+} to 0.03 and 0.7 g/m^2 h, respectively. The corrosion decreases even further as a result of aluminium ions in the nitric acid [263].

Austenitic chromium-nickel-molybdenum steels

According to Figure 6, the material consumption rate of the steel X1CrNiMoN25-25-2 (SAE 310 MoLN, 1.4465) of 0.01 g/m^2 h in boiling 10 % nitric acid increases constantly as the acid concentration increases and reaches a constant value of about 0.8 g/m^2 h at 65 % and above [103].

Three isocorrosion curves of the niobium-stabilized steel 000Ch21N21M4B (Russ. type; (%) Fe-0.03C-(20–22)Cr-(20–21)Ni-(3.4–3.7)Mo-<0.6Mn-<0.6Si-<0.03P-<0.02S-(0.45–0.8)Nb) are shown in the temperature-concentration graph in Figure 21 [264].

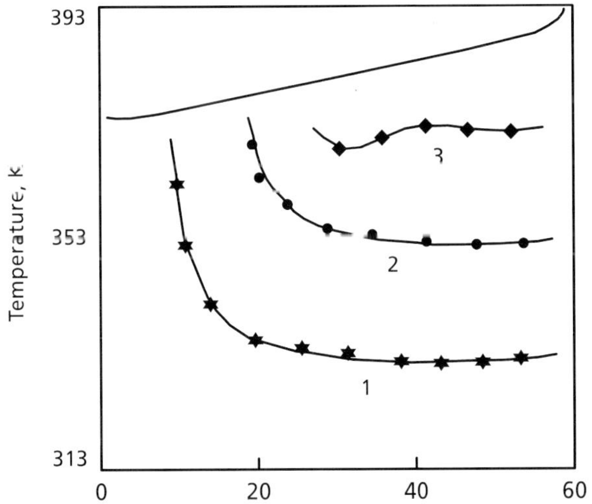

Figure 21: Isocorrosion curves of the steel 000Ch21N21M4B in a temperature/concentration graph at (1) 0.01, (2) 0.02 and (3) 0.3 g/m^2 h [264]

The solution-annealed and water-quenched steels GX5CrNiMo19-11-2 (UNS J92900, 1.4408; (%) Fe-(0.03–0.07)C-19.1Cr-(9.5–9.7)Ni-(2.08–2.35)Mo-(0.03–0.04)Nb-(0.59–0.68)Mn-(0.38–0.66)Si with a δ-ferrite phase of 8.5–11.5 %) corrode in 5 % nitric acid with a corrosion rate of < 0.01 mm/a (< 0.39 mpy) at both 293 and 343 K (20 and 70 °C) [265]. Comparable CrNi steels with a low molybdenum content of 0.14 to 0.18 % of the type GX5CrNi19-10 (UNS J92600, 1.4308) show a similarly good corrosion resistance under the same test conditions.

As can be seen from Table 43, the susceptibility of the steel EI-943 to intercrystalline corrosion (0Ch23N28M3D3T) in nitric acid is determined by the Ti/C ratio [266]. At Ti/C values above 15, i.e. at a very low carbon content, no intercrystalline corrosion occurs.

Chemical composition, %						*Ratio Ti/C*	*Observations*
C	*Cr*	*Ni*	*Ti*	*Cu*	*Mo*		
0.05	23.42	20.2	0.76	2.88	2.56	15.2	–
0.06	22.87	27.17	0.5	2.77	2.5	8.3	intercrystalline corrosion
0.038	23.71	27.13	0.77	2.69	2.77	20.0	–
0.03	23.01	27.55	0.71	2.7	2.88	23.6	–
0.04	23.4	27.0	0.33	–	3.45	7.5	intercrystalline corrosion

Table 43: Chemical composition and intercrystalline corrosion of the steel EI-943 (after tempering at 973 K (700 °C)) in 5 % HNO_3 [266]

The corrosion-related weight loss of the steel SAE 316 L, which was produced by powder metallurgy, depends greatly on the rate of cooling after heat treatment. After 24 h tests at room temperature in 10 % nitric acid, the weight loss was about 0.01 % at a nitrogen content of < 0.001 % and about 2.5 % with 0.8 % nitrogen [267].

The chromium-nickel-molybdenum steels 06ChN28MDT (Russ. type; (%) Fe-0.06C-25Cr-28Ni-(1–3)Mo-(2–3)Cu; Ti-stabilized) and 08Ch21N6M2T ((%) Fe-0.08C-21Cr-6Ni-2Mo; Ti-stabilized) corrode in 10 and 15 % nitric acid containing 1 % H_2SiF_6 at room temperature at rates of about 0.1 and 0.2, and 0.3 and 0.7 mm/a (3.93 and 7.87, and 11.8 and 27.6 mpy) respectively after 48 h tests [268].

MnS impurities, which are the starting points for pitting corrosion in other media, are eliminated during passivation of the surface of the steel SAE 316 ((%) Fe-0.047C-17.8Cr-10.0Ni-1.72Mo-2.06Mn-0.22Cu-0.88Al-0.067W-0.040Ti-0.96Si-0.067P-0.017S) for 30 min in 20 % nitric acid at 323 K (50 °C) [269].

After a 144 d test in boiling 4.75 mol/l nitric acid, the material consumption rate of Incoloy® 825 was 0.0146 g/m^2 h [270].

The chromium-nickel-molybdenum steel FMN (1.4462, UNS S39209; (%) Fe-0.05C-25Cr-5Ni-2Mo-0.8Si-0.2N) corrodes in boiling 30 % nitric acid at a material consumption rate of 0.21 g/m^2 h [271].

Other material consumption rates for the low-carbon steels Fe21Cr24Ni3Mo ((%) Fe-0.03C-21.26Cr-24.31Ni-3.30Mo-0.62Mn-1.25Si-0.011P-0.010S) and Fe20Cr26Ni4-Mo1Cu ((%) Fe-0.03C-20.41Cr-25.52Ni-4.50Mo-0.67Mn-1.1Cu-0.15Ti-1.11Si-0.017P-0.020S) in boiling nitric acid can be seen from Table 44 [114]. The molybdenum-free steel (%) Fe-0.02C-23.8Cr-20.67Ni-1.10Mn-0.60Si-0.015P-0.014S shows a better corrosion behavior, with a corrosion rate of 0.028 g/m^2 h in 35 % HNO_3 at 382 K (109 °C) and 0.270 g/m^2 h in 65 % HNO_3 at 393 K (120 °C). This is also confirmed generally elsewhere [140].

HNO_3 %	Temperature K (°C)	Fe21Cr24Ni3Mo	Fe20Cr26Ni4Mo1Cu
		Material consumption rate, g/m^2 h	
35	382 (109)	0.130	0.125
45	386 (113)	0.310	0.170
55	390 (117)	0.610	0.460
65	393 (120)	0.850	0.780

Table 44: Material consumption rates of two steels in HNO_3 at various concentrations and temperatures [114]

Cermets of stainless steel and UO_2 (steel/UO_2) from spent fuel elements are placed in a basket made of perforated niobium sheets with end holders made of fluorinated plastic in a titanium tank (as the cathode) and dissolved, for uranium processing, in 6.5 mol/l nitric acid at a rate of 0.66 g A/h and a current density of 1.59 A/cm^2. The potential difference between the titanium tank and the fuel element operating as the anode (cermet of steel/UO_2) is 20 V. Less than 0.05 % uranium remains in undissolved form in the total undissolved reside of 1 – 3 % [272].

The CrNiMo steels Sanicro® 28 ((%) Fe-0.011C-26.6Cr-31.1Ni-3.43Mo-1.0Cu-0.039N) and Carpenter® 7-Mo (cf. SAE 329; (%) Fe-0.054C-27.0Cr-4.2Ni-1.44Mo) shown in Table 45 corrode in nitric acid, especially at relatively high temperatures and concentrations, faster than the CrMo steel E-Brite® ((%) Fe-0.002C-26.1Cr-0.1Ni-1.00Mo-0.1Nb-0.010N) [45]. However, corrosion of E-Brite® proceeds faster than that of Carpenter® 7-Mo or Sanicro® 28 in 60 % nitric acid at 422 K (149 °C) and in 70 % nitric acid at 366 K (93 °C).

Tests under evaporation conditions with heat transfer in the evaporator from a titanium tank with heating steam coils made of steel containing 55 % nitric acid at 403 K (130 °C) showed that, according to Table 20, the chromium-nickel-molybdenum-copper steel EI-943 ((%) Fe-<0.01C-23Cr-28Ni-3Mo-3Cu) is superior in its corrosion behavior to normal chromium-nickel steel [119].

However, the greatest advantage of the molybdenum-containing steels is their good resistance to intercrystalline corrosion in nitric acid, such as, for example, the Swedish ELC steel Sandvik® 2RE10 (UNS S31002, 1.4335) with about 4.5 % molybdenum [273].

HNO_3 %	Temperature K (°C)	Sanicro® 28	Carpenter® 7-Mo	E-Brite®
		Corrosion rate, mm/a (mpy)		
40	366 (93)	0.020 (0.79)	–	–
	384 (111)	0.223 (8.78)	0.067 (2.64)	0.021 (0.83)
	408 (135)	–	0.176 (6.93)	0.146 (5.75)
	422 (149)	–	0.750 (29.5)	0.226 (8.9)
60	366 (93)	0.026 (1.02)	–	0.024 (0.94)
	393 (120)	0.409 (16.1)	0.166 (6.54)	0.097 (3.82)
	422 (149)	–	2.87 (113)	3.52 (139)
70	366 (93)	0.049 (1.93)	–	0.119 (4.69)
	394* (121)	0.642 (25.3)	0.251 (9.88)	0.146 (5.75)

*boiling point

Table 45: Corrosion of some steels in nitric acid [45]

Figure 22 shows the dependence of the corrosion rate of the steel X 2 CrNiMo 18 12 ((%) Fe-0.03C-17.5Cr-12.4Ni-2.83Mo-0.97Mn-0.66Si-0.016N-0.012P-0.012S) in boiling 50 and 65 % nitric acid on the immersion time. While it has reached a stationary value of about 0.2 mm/a (7.87 mpy) after 200 h in 50 % acid, in 65 % acid it is still rising considerably even after 250 h. After 100 % cold-working, corrosion of the steel after 250 h tests in boiling 65 % nitric acid drops from 1.35 to 0.62 mm/a (53.2 to 24.4 mpy), and in 50 % acid from 0.24 to 0.12 mm/a (9.45 to 4.72 mpy) [274].

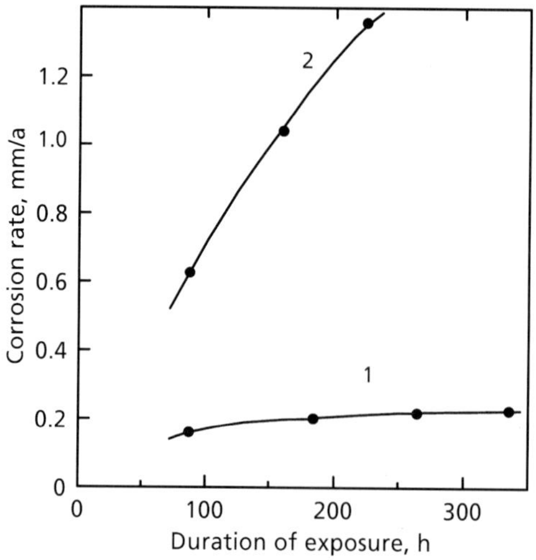

Figure 22: Corrosion rate of the solution-annealed steel X 2 CrNiMo 18 12 in (1) boiling 50 % and (2) 65 % HNO_3 as a function of the exposure time [274]

Pipe specimens of SAE 316 L ((%) Fe-0.07C-16.5Cr-11.8Ni-2.45Mo-0.6Mn-0.6Sr-0.002P-0.009S), produced from powder or cast ingots, corrode in boiling 65 % nitric acid at material consumption rates of 0.41 and 0.37 g/m^2 h respectively [62].

On the basis of studies in the Huey test, CrNiMo steels of the type (%) Fe-(0.034–0.043)S-(12.7–16.4)Cr-(3.9–5.9)Ni-(1.1–1.6)Mo-(0.28–0.039)Si-(0.030–0.037) N-(0.009–0.022)P-(0.007–0.020)S are not recommended for use in nitric acid [275, 276].

In agreement with the above values, steels of the type SAE 316 L heat-treated at 950 K (677 °C) for 20 min showed a corrosion rate of between 0.64 and 0.71 mm/a (25.2 and 28.0 mpy) in the Huey test [149]. The steel SAE 316 gave a comparable corrosion behavior.

The corrosion-increasing effect of molybdenum in boiling 65 % nitric acid is also found in the Huey test (5 × 48 h in boiling 65 % HNO_3) on the steel Fe-0.1C-18Cr-16Ni-Mo, which corrodes with a material consumption rate of 0.0871 g/m^2 h without molybdenum and 0.0879 and 0.120 g/m^2 h with 2.5 and 5 % molybdenum, respectively [277]. According to the Huey test, the steel SAE 316 L is attacked by boiling nitric acid, especially in the region of the σ-phase [278].

Even expensive high-alloy steels, such as, for example, Incoloy® 825 ((%) Fe-0.03C-20.8Cr-42.0Ni-3.0Mo-1.0Ti-1.74Cu) and Incoloy® 901 ((%) Fe-0.03C-13.1Cr-42.2Ni-6.2Mo-2.5Ti) are attacked with corrosion rates of 0.2 and 3.4 mm/a (7.87 and 134 mpy) in boiling 65 % nitric acid, since the chromium content is too low, especially in the latter, and the molybdenum content too high [279].

Figure 23 shows the tempering time and temperature of a CrNiMo steel ((%) Fe-0.020C-17.35Cr-13.70Ni-2.69Mo-1.67Mn-0.53Si-0.018P-0.008S) in the form of an iso-corrosion curve at 0.25 and 0.35 g/m^2 h, determined in the Huey test after solution-annealing at 1573 K (1300 °C) for 15 min [140].

Table 46 shows the dependence of the material consumption rate of some specimens of the steel 1.4435 (SAE 316 L) in boiling 65 % nitric acid (Huey test) on the heat treatment. The relatively large differences in corrosion which occur on the basis of uncontrollable changes within the tolerance range are remarkable [280].

The austenitic CrNiMo steel VEW A-963 with 6.3 % molybdenum and a low carbon content ((%) Fe-<0.03C-17.0Cr-16.0Ni 6.3Mo-0.15N) corrodes in the Huey test (5 × 48 h in boiling 65 % HNO_3) with a material consumption rate of < 0.20 g/m^2 h [281]. Comparable values are obtained with the quenched and sensitized steel 05Ch16N15M3 (Russian type; Fe-0.05C-16Cr-15Ni-3Mo) with material consumption rates of 0.16 and 0.32 g/m^2 h at polarization potentials of 1.2 and 1.1 V respectively [282]. The intercrystalline corrosion which is simultaneously observed results from the precipitation of the carbides $(Cr, Fe, Mo)_{23}C_6$ at the grain boundaries. The carbide $Cr_{15.5}Fe_{6.1}Mo_{1.4}C_6$ corrodes at 1.2 and 1.1 V with material consumption rates of 16 and 5 g/m^2 h, respectively.

Welded pipes of CrNiMo steel ((%) Fe-<0.1C-20.25Cr-24.5Ni-6.25Mo-1.5Mn-0.50Si) available under the name AL-6X for cooling plants are also resistant to 65 % nitric acid [283].

Figure 24 shows the change with respect to time in the corrosion rate of the steel 05Ch16N15M3 (Fe-0.05C-16Cr-15Ni-3Mo) and the carbides $Cr_{16.9}Fe_{6.1}C_6$ and $Cr_{15.5}Fe_{6.1}C_6$ which precipitate at the grain boundaries during heat treatment in boiling 65 % nitric acid at a polarization potential of 1.2 V_{SHE} [284]. Here as well, corrosion of the carbides is greater than that of the steel by more than a power of ten.

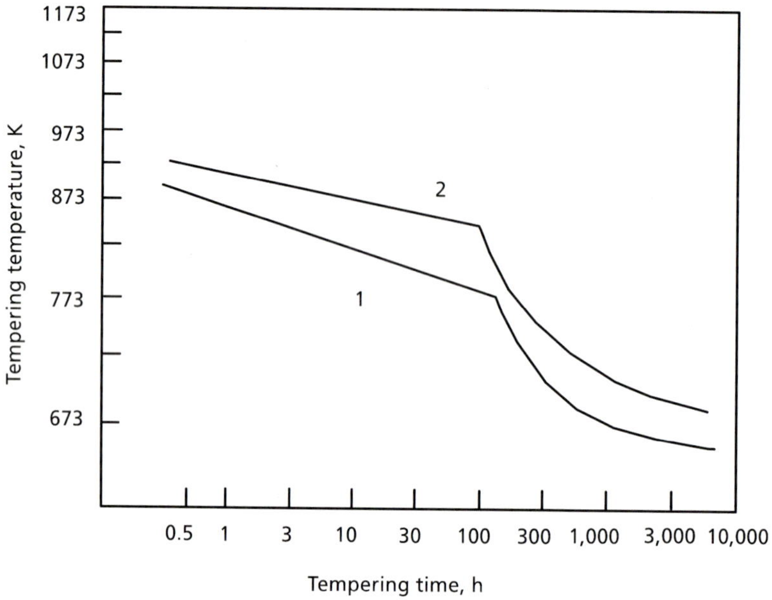

Figure 23: Isocorrosion curves for (1) 0.25 and (2) 0.35 g/m^2 h in the Huey test on the CrNiMo steel after solution-annealing at 1573 K (1300 °C) for 15 minutes [140]

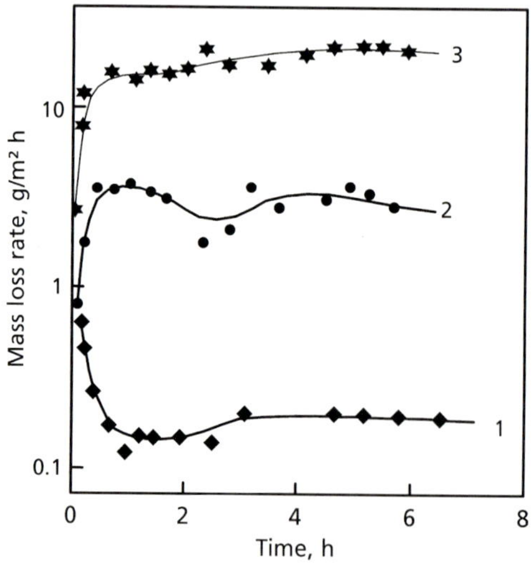

Figure 24: Change with respect to time in the material consumption rate (1) of the steel 05Ch16N15M3 and (2) of the carbides $Cr_{16.9}Fe_{6.1}C_6$ and (3) $Cr_{15.5}Fe_{6.1}Mo_{1.4}C_6$ in boiling 65 % HNO_3 at 1.2 V_{SHE} [284]

Chemical composition, %						Material consumption rate, g/m² h			
C	Si	Mn	Cr	Mo	Ni	a	b	c	d
0.024	0.64	0.86	17.40	2.81	14.47	0.29	1.39	3.8	7.3
0.020	0.70	0.92	17.99	2.75	14.53	0.26	4.59	13.7	18.9
0.024	0.72	0.89	17.67	3.02	14.77	0.24	8.1	51.2	33.3
0.015	0.65	0.88	17.68	2.82	14.59	0.33	0.31	0.39	1.5

a: quenched
b: quenched + heat treatment at 1073 K (800 °C) for 3 min
c: quenched + heat treatment at 1073 K (800 °C) for 15 min and
d: quenched + heat treatment at 1073 K (800 °C) for 30 min

Table 46: Composition of the steel 1.4435 (SAE 316 L) and corrosion behavior in the Huey test after various heat treatments at 1073 K (800 °C) [280]

No.	Chemical composition, %								Ferrite content %
	C	Si	Mn	S	Cr	Ni	Mo	N	
1	0.029	0.34	0.70	0.016	15.02	14.02	3.12	–	0
2	0.045	–	4.10	–	16.98	15.68	2.13	–	0
3	0.020	0.34	0.14	0.033	17.52	11.22	2.05	0.16	0.2
4	0.03	0.25	1.24	0.017	18.06	13.40	2.96	–	2.3
5	0.025	0.30	1.89	0.020	18.17	14.81	2.61	–	0
6	0.01	0.36	1.60	0.014	18.98	16.48	3.80	–	0

Table 47: Chemical composition of the steels shown in Table 48 [285]

The influence of the 5 test periods of 48 h each in the Huey test on the material consumption rates of some chromium-nickel-molybdenum steels with the composition shown in Table 47 can be seen from Table 48. On some steels, the corrosion rate rises considerably with the number of weighings [285]. According to these tables, steels 5 and 6 show the greatest resistance in boiling 65 % nitric acid.

These results are confirmed by earlier studies on the low-carbon austenitic steels 000Ch16N(13–16)M(2–4) (identical to SAE 316), where the material consumption rates of the quenched (hardened) and welded specimens in the Huey test were 0.20–0.33 and 0.25–0.58 g/m^2 h. Two ferritic-austenitic steels 000Ch21N10M2 ((%) Fe-0.020C-19.8Cr-10.5Ni-2.1Mo) and 000Ch21N6M2 ((%) Fe-0.036C-21.1Cr-6.5Ni-2.4Mo) have also been shown in Table 49 for comparison [286].

No.	Test period					Mean
	1	2	3	4	5	
	Material consumption rate, g/m² h					
1	0.703	0.889	1.040	–	–	0.887*
2	0.426	0.334	0.339	0.385	0.416	0.382
3	0.233	0.227	0.297	0.378	0.215	0.278
4	0.260	0.240	0.252	0.551	0.833	0.428*
5	0.212	0.127	0.130	0.228	0.126	0.143
6	0.391	0.207	0.272	0.387	0.215	0.278

* corrosion rate increasing with the number of periods

Table 48: Material consumption rates of the steels shown in Table 47 in boiling 65 % HNO_3 after 5 test periods of 48 h each [285]

Steel	Specimen 1[1)]	Specimen 2[2)]	Susceptibility to intercrystalline corrosion (IC)
	Material consumption rate, g/m² h		
000Ch16N13M2	0.22 – 0.25	0.25 – 0.27	no or weak IC
000Ch16N13M3	0.26	0.35	none, only in HZ*
000Ch16N16M2	0.20	0.25	none, only weak IC in HZ
000Ch16N16M3	0.28	0.36	none, only weak IC in HZ
000Ch16N16M4	0.37	0.77	IC in HZ
000Ch18N13M2	0.25	0.19	none
000Ch18N16M2	0.17	0.25	none
000Ch18N16M4	0.34	0.82	IC in HZ and on the surface
000Ch21N10M2	0.28	0.22	none
000Ch21N6M2	0.43	0.48	none

HZ – heat-affected zone
[1)] specimen 1 quenched
[2)] specimen 2 welded

Table 49: Corrosion behavior of some CrNiMo steels in boiling 65 % HNO_3 in the Huey test [286]

CrNiMo steels with a ferritic-austenitic structure are used for special applications in the presence of nitric acid. The material consumption rate of the steel X2CrNiMoN22-5-3 (UNS S39209, 1.4462; (%) Fe-0.020C-22.47Cr-5.21Ni-3.08Mo-1.82Mn-0.43Si-0.165N-0.031P-0.006S) in the Huey test (3 periods of 48 h in boiling 65 %

nitric acid) is about 50 to 70 g/m^2 h after an annealing time of between 1 and 5 h at 973 K (700 °C), and about 0.6 to 0.8 g/m^2 h at 1073 K (800 °C). In the delivery state, the steel corroded at a material consumption rate of 0.45 g/m^2 h. According to Figure 25, relatively high annealing temperatures are more favorable for the corrosion behavior in nitric acid. The rate of corrosion of the steel is independent of the annealing time at 1423 K (1150 °C) [287].

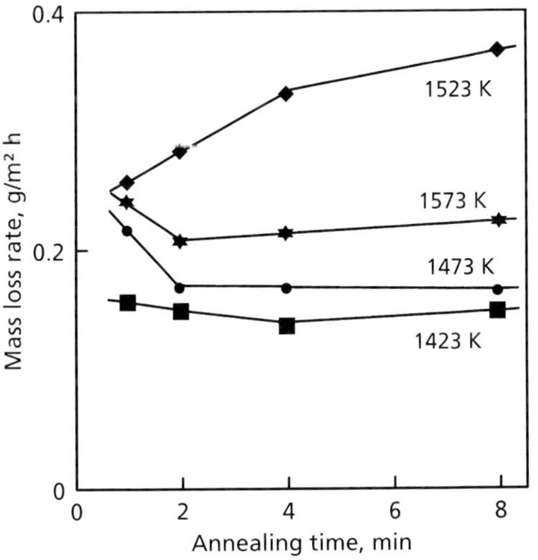

Figure 25: Material consumption rates of the steel X2CrNiMoN22-5-3 (UNS S39209) in the Huey test after short annealing times at high temperatures [287]

The decisive influence of the correct heat treatment on the corrosion of stainless steels can be seen from Table 50 with the example of the niobium-containing steel ITM-43 ((%) Fe-0.074C-24.8Cr-3.93Ni-4.05Mo-0.51Nb-0.12Si-0.013P-0.002S-0.0117N) [288].

Heat treatment temperature K (°C)	Test cycles	Heat treatment duration			
		5 min	20 min	1 h	5 h
		Material consumption rate g/m^2 h			
923 (650)	3	25.0	–	–	59.1
1023 (750)	3	33.5	31.0	31.0	29.0
1073 (800)	5	18.8	23.1	15.6	2.11
1123 (850)	5	10.01	0.36	0.35	0.24
1173 (900)	5	0.20	0.20	0.26	0.15

Table 50: Corrosion of the steel ITM-43 in the Huey test after 3 and 5 cycles (48 h each) and various heat treatment durations [288]

If the heat treatment temperature is below 1100 K (827 °C), even a period of 5 h is too short to achieve a sufficiently low material consumption rate.

The influence of correct heat treatment on the corrosion of the steels AF-22 (UNS S39209, 1.4462; (%) Fe-0.028C-21.8Cr-5.00Ni-3.12Mo-1.63Mn-0.45Si-0.031P-0.012S-0.113N) and AF-22 + Mo ((%) Fe-0.027C-21.5Cr-7.21Ni-4.88Mo-1.67Mn-0.49Cu-0.29Si-0.021P-0.009S-0.140N) in boiling 65 % nitric acid can be seen in Figures 26 and 27. According to Figure 26, the corrosion rate of the steel AF-22 reaches the lowest values after heat treatment times of 20 min or 30 h between 573 and 673 K (300 and 400 °C) and above 1100 K (827 °C) respectively, and the highest value at about 900 K (627 °C).

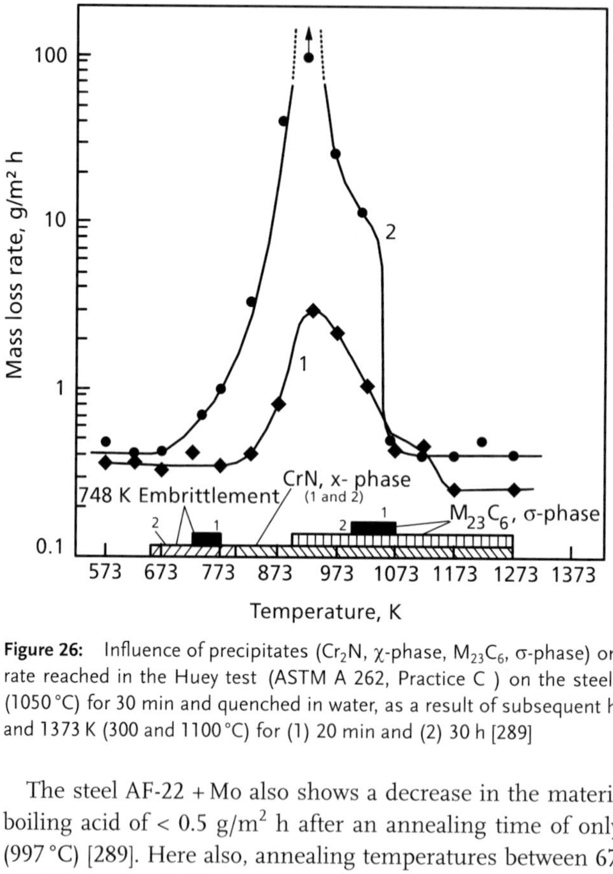

Figure 26: Influence of precipitates (Cr_2N, χ-phase, $M_{23}C_6$, σ-phase) on the material consumption rate reached in the Huey test (ASTM A 262, Practice C) on the steel AF-22, annealed at 1323 K (1050 °C) for 30 min and quenched in water, as a result of subsequent heat treatment between 573 and 1373 K (300 and 1100 °C) for (1) 20 min and (2) 30 h [289]

The steel AF-22 + Mo also shows a decrease in the material consumption rate in boiling acid of < 0.5 g/m^2 h after an annealing time of only 10 min above 1270 K (997 °C) [289]. Here also, annealing temperatures between 670 and 1070 K (397 and 797 °C) are unsuitable.

As a further example, Figure 28 shows the marked influence of the heat treatment (temperature and time) on the corrosion rate of the steel SAE 316 L in boiling 65 % nitric acid after exposure for 4 × 48 h periods. In all cases, the attack was intercrystalline. The corrosive attack on a steel specimen which had been heat-treated at 948 K (675 °C) for longer than 2 h was particularly intense [290].

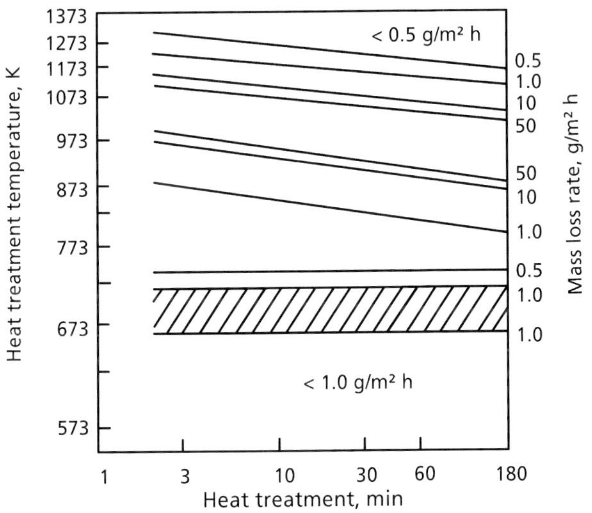

Figure 27: Influence of the duration and temperature of heat treatment after prior annealing at 1373 K (1100 °C) for 20 min and quenching in water on the corrosion of the steel AF-22 + Mo in the Huey test (ASTM A 262, Practice C) [289]

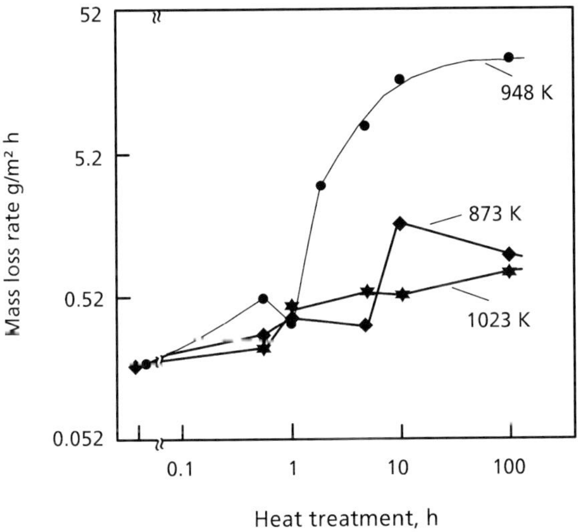

Figure 28: Material consumption rate of the steel SAE 316 L in boiling 65 % HNO_3 as a function of the temperature and duration of heat treatment [290]

To improve the corrosion resistance of the steel SAE 316 L ((%) Fe-0.02C-17.37Cr-15.45Ni-2.45Mo-1.73Mn-0.029Al-0.35Si-0.024P-0.014S-0.015N) in nitric acid, the occurrence of the ferrite and σ-phase should be prevented during heat treatment. A low carbon content is also necessary to avoid carbide precipitates. Similarly, the silicon, phosphorus and sulfur content should be kept low to prevent precipitates at the grain boundaries (intercrystalline corrosion) [291].

The influence of delayed cooling of the steel X2CrNiMoN17-13-5 (SAE 317 LMN, 1.4439) after solution-annealing at 1343 K (1070 °C) for 15 min on the corrosion rate in the Huey test after 96 h (in this case only 2 × 48 h) is informative in this connection. Figure 29 shows the influence of delayed cooling in the maximum precipitation range on corrosion [292].

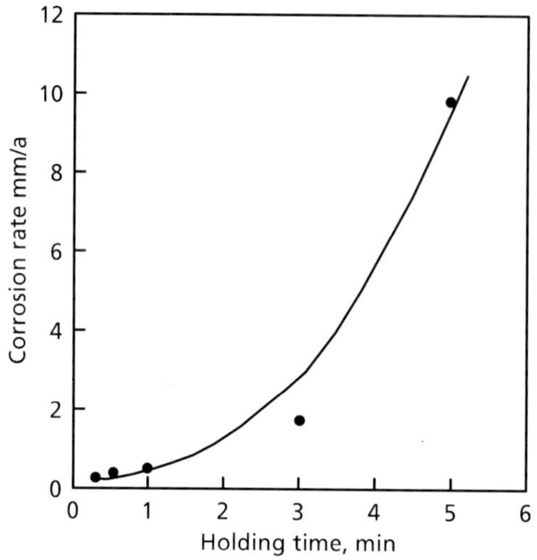

Figure 29: Influence of the holding time at 1073 K (800 °C) with subsequent quenching after solution-annealing of 15 min at 1343 K (1070 °C) on the corrosion rate of the steel X2CrNiMoN17-13-5 (SAE 317 LMN) in the Huey test after 96 h [292].

Figure 30 also shows the marked influence of the annealing temperature and time on the corrosion of chromium-nickel-molybdenum steels in boiling 65 % nitric acid (Huey test). Below the solution-annealing temperature, the influence of precipitates becomes particularly critical as a result of the various periods of time at the susceptibility maximum of the steel Remanit® 4462 (UNS S39209, X2CrNiMoN22-5-3; (%) Fe-0.025C-22.42Cr-5.30Ni-3.06Mo-1.83Mn-0.40Si-0.022P-0.004S-0.110N) at 970 K (697 °C) [142]. The tests on the steel Remanit® 4362 (UNS S32304, X2CrNiN23-4, (%) Fe-0.010C-23.56Cr–3.97Ni–<0.01Mo–1.15Mn–0.15Si–<0.005P-<0.003S-0.110N) performed for comparison show a considerably lower susceptibility to corrosion in the Huey test as a result of the annealing temperature and time.

While addition of up to 2 % niobium to an untreated and solution-annealed CrNiMoCu steel ((%) Fe-0.09C-20.0Cr-29.78Ni-2.31Mo-3.20Cu-0.91Mn-0.64Si-0.95N-0.018P-0.0 21S) caused no change in the corrosion rate of 0.18 mm/a (7.09 mpy) in boiling 65 % nitric acid, corrosion of sensitized, and of solution-annealed + stabilized + sensitized steel, which corroded more severely (1.25 and 0.75 mm/a (49.2 and 29.5 mpy)), was reduced to 0.13 mm/a (5.12 mpy) by addition of 1.4 % Nb [293]. This value should not be exceeded.

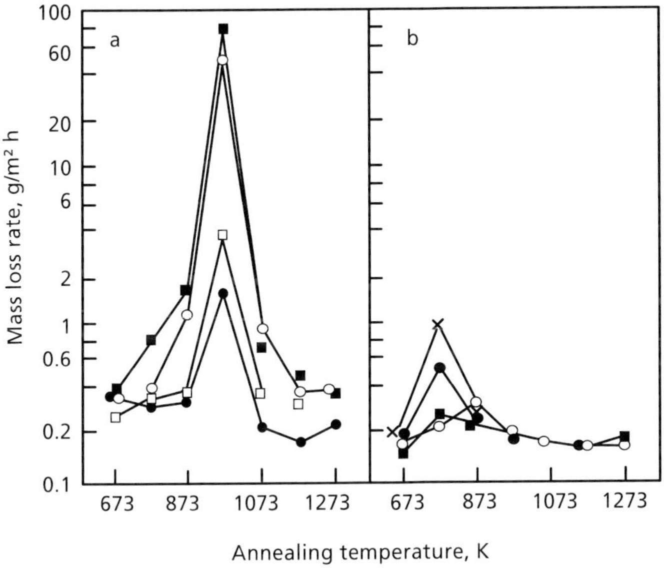

Figure 30: Influence of the annealing temperature and time on the material consumption rate in the Huey test on a) Remanit® 4462 (UNS S39209) and b) Remanit® 4362 (UNS S32304) [142]
a): ● = 5 min; □ = 15 min; ○ = 1 h; ■ = 10 h
b): ○ = 1 h; ■ = 10 h; ● = 30 h; x = 100 h

After 240 h tests in boiling 65 % nitric acid, a rapidly quenched steel SAE 316 ((%) Fe-0.05C-17.27Cr-13.09Ni-2.56Mo-0.24Cu-1.76Mn-0.13Co-0.04V-0.45Si-0.033P-0.015S) corroded with a corrosion rate of only 0.5 mm/a (19.7 mpy), and after only 30 % cold-working reached a rate of 0.23 mm/a (9.06 mpy), which remained constant with further cold-working [145].

Figure 31 shows particularly clearly the change with respect to time in the material consumption rate of the steel SAE 316 LN (1.4429, X2CrNiMoN17-13-3) in boiling 65 % nitric acid in the Huey test due to the degree of deformation, or the number of deformation stages. After passing through a corrosion maximum of > 12 g/m^2 h at a degree of deformation of about 40 %, the material consumption rate drops to about 0.3 g/m^2 h at about 50 % deformation [294]. It reaches a minimum of < 0.1 g/m^2 h after a degree of deformation of only 4 %.

Studies of damage to acid pumps made of Alloy-20 (Fe-(0.05–0.07)C-(19.5–20.5)Cr-(29–30)Ni-2.5Mo-3.2Cu-(0.5–1.0)Si) showed that the behavior under erosion corrosion conditions cannot be predicted by the Huey test in the laboratory [295].

The corrosion behavior and chemical composition of some Remanit steels in hot 86 % nitric acid at 333 K (60 °C) after 3 × 24 h tests can be seen from Table 51 [296]. The corresponding values in the Huey test (5 × 48 h in boiling 65 % nitric acid) are higher and are: 0.45, 0.40, 0.35 and 0.30 g/m^2 h. The steels Remanit® 4438 (SAE 317 L) and Remanit® 4565 (UNS S34565) are accordingly the most corrosion-resistant.

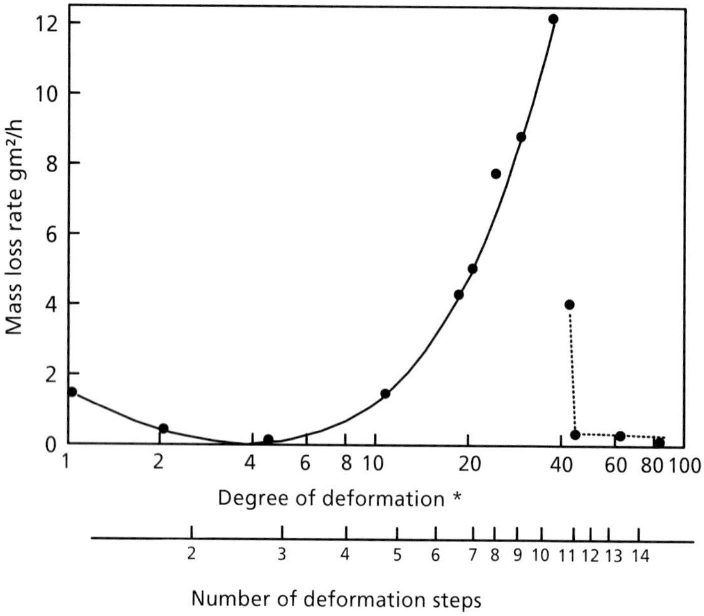

Figure 31: Influence of the degree of deformation on the material consumption rate of SAE 316 LN after 5 × 48 h periods in boiling 65 % HNO_3 [294].
* starting cross-section / deformed cross-section

Remanit® Steel	***No.****	***Chemical composition, %***								***Material consumption rate, g/m² h***
		C	***Si***	***Mn***	***Cr***	***Mo***	***Ni***	***N***	***Nb***	
4462	1	0.028	0.49	1.71	22.20	3.06	5.39	0.12	–	0.125
4429	2	0.022	0.30	1.31	17.26	2.68	13.36	0.15	–	0.091
4438	3	0.022	0.15	1.82	18.32	3.34	14.77	0.14	–	0.032
4565	4	0.020	0.62	5.79	23.14	3.27	16.53	0.39	0.22	0.032

* 1: 2205, UNS S39209, X2CrNiMoN22-5-3,
2: SAE 316 LN, X2CrNiMoN17-13-3,
3: SAE 317 L, X2CrNiMoN18-15-4,
4: UNS S34565X3CrNiMnMoNbN25-18-5-4

Table 51: Chemical composition and material consumption rates of the Remanit® steels investigated [296]

A valve made of the steel Alloy-20 CN-7M ((%) Fe-0.1C-20Cr-30Ni-2.3Mo-3.2Cu-0.9Mn) and also a pump of the same material were subjected to corrosion in 99 % nitric acid. The Huey test gave a corrosion rate of only 0.36 mm/a (14.2 mpy) on the base steel, while the casting surface corroded at a rate of 0.78 mm/a (30.7 mpy) [297]. Intercrystalline corrosion was found on the propeller of the acid pump.

The structure of the corrosion layer on the steels SAE 316 Ti (1.4571, X6CrNiMoTi17-12-2) and SAE 310 MoLN (1.4465, X1CrNiMoN25-25-2) after exposure to concentrated nitric acid for 3 years was investigated by Auger and X-ray fluorescence measurements [204]. Similar studies were also carried out on the oxidized surface layers on the austenitic-ferritic cast material Noridur® 1.4593 (GX3CrNiMoCuN24-6-2-3; (%) Fe-0.036C-24.73Cr-7.01Ni-2.46Mo-3.21Cu-1.11Mn-1.14Si-0.144N-0.021P-0.010S) [298].

The austenitic steel Sanicro® 28 (UNS N08028, 1.4563; (%) Fe-0.020C-27Cr-31Ni-3.5Mo-1.0Cu) has been developed for very severe corrosion conditions, such as, for example, for highly concentrated, boiling nitric acid [299].

The high-alloy steels which have already been mentioned several times, such as, for example, 03ChN28MDT and 03Ch21N21M4GB, are recommended for installations in fertilizer factories where highly concentrated nitric acid solutions occur [300]. The corrosion rate under the conditions which arise here is ≤ 0.1 mm/a (3.94 mpy).

Remanit® 4465 (1.4465, X1CrNiMoN25-25-2) and Remanit® 4575 (1.4575, 25-4-4, UNS S44635) still have an acceptable resistance to a hot solution of 10% HNO_3 + 2% HF at 343 K (70 °C) with material consumption rates of 0.6 and 0.2 g/m^2 h respectively [301]. This is confirmed for Remanit® 4575 under the same conditions with a material consumption rate of 0.18 g/m^2 h [302].

In a solution of 10% nitric acid + 3% hydrofluoric acid at 323 K (50 °C), the steel EI-943 (06ChN28MDT) corrodes between + 0.3 and + 0.6 V_{SHE} with a corrosion rate of < 1 mm/a (39.4 mpy). This figure is < 0.1 mm/a (3.94 mpy) with an addition of 1% hydrofluoric acid at + 0.5 V_{SHE} [303].

Weld joints of the steel 06ChN28MDT corrode in 14 to 16% nitric acid containing 3.8% hydrogen fluoride between 328 and 333 K (55 and 60 °C) at a rate of 4.5 to 4.7 mm/a (177 to 185 mpy) after 336 h tests. The Nb-stabilized steel 04Ch21N21M4B has a higher corrosion resistance with a corrosion rate of 1.1 mm/a (43.3 mpy), but is unusable in this medium [206].

Even the high-molybdenum steel SAE 317 LMN (1.4439, X2CrNiMoN17-13-5) is noticeably attacked by nitric acid containing hydrogen fluoride. Figure 32 shows the dependence of the corrosion of welded specimens of the steel in 20% nitric acid at 303 K (30 °C) on the hydrofluoric acid content after 70 h tests. On the other hand, the material consumption rate of the steel X 5 CrNiMo 18 10 also plotted in Figure 32 increases greatly as the concentration of HF increases [304].

The resistance of steels of the type ChN28MDT ((%) Fe-(0.03–0.046)C-(22.2–23.5)Cr-(26.55–27.88)Ni-(2.55–3.06)Mo-(2.68–3.38)Cu-(0.54–0.76)Ti-(0.15–0.30)Mn-(0.39–0.69)Si-(0.021–0.43)P-(0.008–0.017)S) to nitric acid containing hydrogen fluoride depends on the carbide precipitates caused by heat treatment and on the occurrence of the σ-phase, which increases corrosion. The risk can certainly be reduced by reducing the carbon content, but can still not be adequately eliminated. It is advisable for the steels to be additionally stabilized with niobium or zirconium [305].

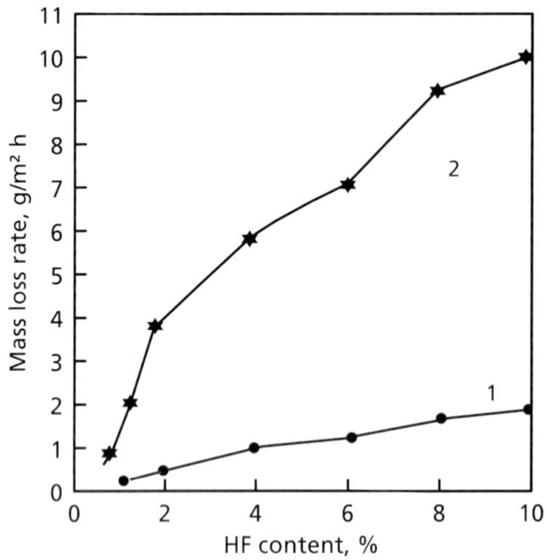

Figure 32: Material consumption rate on welded specimens of the steels (1) SAE 317 LMN (1.4439) and (2) SAE 316 (1.4401, X5CrNiMo17-12-2) in HF-containing 20 % HNO_3 at 303 K (30 °C) as a function of the HF concentration, test duration 70 h [304]

The influence of a carbon content of 0.06 and 0.03 %, shown in Figure 33, on corrosion of the steels 03ChN28MDT (Russ. types; (%) Fe-(0.03–0.06)C-22.3Cr-27Ni-2.3Mo-3.0Cu–(0.55–0.85)Ti–0.40Si–(0.006–0.010)N–0.010S-(0.018–0.033)P) in hot 10 mol/l nitric acid + 0.1 mol/l hydrogen fluoride at 363 K (90 °C) is thus to be expected. According to Figure 33, movement of the acid, expressed as the rotational velocity of the specimen, also has a corrosion-increasing influence [306].

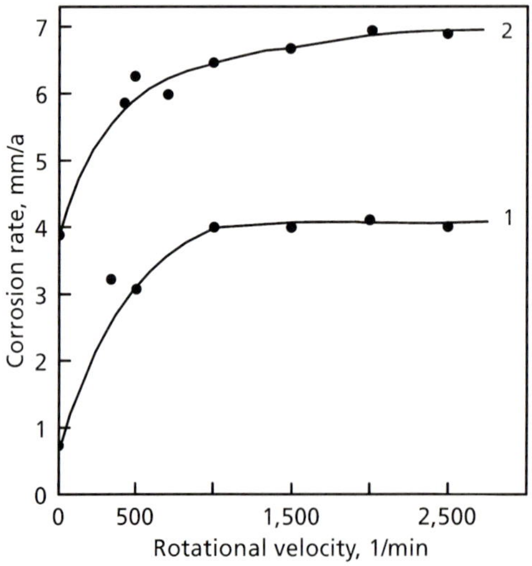

Figure 33: Influence of the rotational velocity and carbon content on the corrosion rate of the steels (1) 03ChN28MDT with 0.03 % C and (2) 06ChN28MDT with 0.06 % C in 10 mol/l HNO_3 + 0.1 mol/l HF at 363 K (90 °C) [306]

According to ESCA analyses of the surface, the chromium content in the oxide layer on the steel SAE 316 increases in the sequence of treatments in dry air, in nitric acid + hydrofluoric acid and by passivation in 30 % nitric acid, with the potential simultaneously rising from + 0.4 to + 1.05 V_{SCE} [307, 308].

According to Table 52, corrosion of the steel 03Ch21N21M4B in nitric acid with and without hydrogen fluoride is increased both by additions of niobium and hydrofluoric acid [309]. The corrosion rate of the steel 06ChN28MDT in 40 % nitric acid containing 2 % hydrofluoric acid at 368 K (95 °C) is also increased from 11.4 to 38.5 mm/a (449 to 1515 mpy) by titanium additions of 0.43 to 0.86 %.

Nb Content %	**Medium**	**Temperature** K (°C)	**Duration** h	**Corrosion rate** mm/a (mpy)
–	65 % HNO_3	boiling	240	0.13 (5.12)
0.45				0.22 (8.66)
0.90				0.32 (12.60)
–	40 % HNO_3 + 2 % HF	368 (95 °C)	48	14.2 (559)
0.45				15.9 (625)
0.90				27.6 (1087)

Table 52: Influence of the niobium content on the corrosion resistance of the steel 03Ch21N21M4B in nitric acid with and without HF [309]

According to [213] the steel 08Ch21N6M2T has a superior resistance to the molybdenum-free chromium-nickel steels in nitric acid containing fluoride (8 to 12 mol/l HNO_3 + 0.02 to 0.12 mol/l NaF) between 323 and 373 K (50 and 100 °C).

To simulate the influence of flowing acid on the corrosion rate of the steels 03ChN28MDT and 06ChN28MDT, rotating steel specimens were exposed to attack by 10 mol/l nitric acid + 0.1 mol/l hydrogen fluoride at 363 K (90 °C) at an increasing rotational velocity. While the corrosion rates in the stationary state were 0.6 and 3.8 mm/a (23.6 and 150 mpy), they rose to 3.4 and 6, and 4.0 and 6.7 mm/a (134 and 236, 157 and 264 mpy) at 500 and 2500 rpm respectively [310].

Carpenter® 20Cb-3 (cf. 2.4660; (%) Fe-0.036C-19.76Cr-33.70Ni-2.25Mo-3.14Cu-0.79Nb-0.23Mn-0.38Si-0.020P-0.04S), which has been annealed at 1477 K (1204 °C) for 1 h and quenched in water, corrodes in the Huey test (in boiling 65 % HNO_3) at a rate of 0.18 mm/a (7.09 mpy). Intercrystalline corrosion was observed neither in the Huey test nor in the test with 10 % HNO_3 + 3 % HF at 343 K (70 °C) (in accordance with ASTM A 262-D).

The corrosion rates of the steels 08Ch21N6M2T ((%) Fe-0.05C-20.6Cr-5.6Ni-2.4Mo-0.52Mn-0.25Ti) and 08Ch18G8N2M2T ((%) Fe-0.08C-18.2Cr-3.42Ni-8.9Mn-2.32Mo-0.22Ti) in 0.5 and 10 mol/l nitric acid with and without additions of 0.01 and 0.1 mol/l hydrofluoric acid or 5 g/l chloride at various temperatures are summarized in Table 53 [75, 77].

Steel	Addition mol/l	0.5 mol/l HNO_3				10 mol/l HNO_3	
		293 K (20 °C)	323 K (50 °C)	373 K (100 °C)	boiling point	293 K (20 °C)	323 K (50 °C)
		Corrosion rate, mm/a (mpy)					
08Ch21N6M2T	–	0.011 (0.43)	0.020 (0.79)	0.025 (0.98)	0.041 (1.61)	0.008 (0.31)	0.030 (1.18)
	0.01 HF	0.010 (0.39)	0.018 (0.71)	0.046 (1.81)	0.067 (2.64)	0.060 (2.36)	0.39 (15.35)
	0.1 HF	0.20 (7.87)	0.062 (2.44)	0.41 (16.14)	0.53 (20.87)	0.42 (16.54)	1.18 (46.46)
	0.01 HF + 5 g/l Cl^-	0.022 (0.87)	0.025 (0.98)	0.062 (2.44)	0.040 (1.57)	0.038 (1.5)	0.14 (5.51)
	5 g/l Cl^-	0.023 (0.91)	0.012 (0.47)	0.018 (0.71)	0.048 (1.89)	0.016 (0.63)	0.042 (1.65)
08Ch18G8N2M2T	–	0.016 (0.63)	0.016 (0.63)	0.030 (1.18)	0.053 (2.09)	0.013 (0.51)	0.019 (0.75)
	0.01 HF	0.036 (1.42)	0.025 (0.98)	0.075 (2.95)	0.096 (3.78)	0.080 (3.15)	0.31 (12.20)
	0.1 HF	0.73 (28.74)	0.90 (35.43)	2.11 (83.07)	11 (433)	0.38 (14.96)	1.42 (55.91)
	0.01 HF + 5 g/l Cl^-	0.22 (8.66)	0.47 (18.5)	0.19 (7.48)	5.69 (224)	0.036 (1.42)	0.16 (6.3)
	5 g/l Cl^-	0.028 (1.10)	0.019 (0.75)	0.018 (0.71)	0.058 (2.28)	0.025 (0.98)	0.044 (1.73)

Table 53: Corrosion of two CrNiMo steels in 0.5 and 10 mol/l HNO_3 with and without 0.01 or 0.1 mol/l HF or Cl^- ions at various temperatures [75, 77]

The welding material Ch17N18M2T ((%) Fe-0.09C-16.9Cr-12.3Ni-1.9Mo-1.4Mn-0.6Si) corrodes in fluoride-containing nitric acid (500 g/l HNO_3 + 600 g/l $NaNO_3$ + 50 g/l ($Fe(NO_3)_3$ + $Cu(NO_3)_2$) + 5 g/l NaF) at a material consumption rate of 1.97 g/m^2 h (about 2.4 mm/a (95 mpy)), knife-line corrosion and pitting corrosion also occurring [311].

The steels SAE 316, Carpenter® 20 Cb-3 and Hastelloy® C corrode in 25 % sulfuric acid containing 4 % nitric acid at 343 K (70 °C) at rates of 0.008, 0.005 and 0.08 mm/a (0.31, 0.20 and 3.15 mpy) respectively [312]. In acid mixtures containing 5.8 mol/l (HCl + HNO_3), the material consumption rate of SAE 316 is about 2 powers of ten higher, at about 10 g/m^2 h, at a mixing ratio of 50 percent by volume compared with that in mixtures with < 30 percent by volume HNO_3 [313]. The material consumption rate at about 80 percent by volume nitric acid is < 0.1 g/m^2 h.

The steel 08Ch21N6M2T corroded in the two mixtures of 36.6 % H_2SO_4 + 23.3 % HNO_3 + 39.1 % SO_3 and 28 % H_2SO_4 + 12 % HNO_3 + 60 % SO_3 with material consumption rates of 0.61 g/m^2 h (0.69 mm/a) and 1.30 g/m^2 h (1.46 mm/a) after 250 h tests [227].

The corrosion rate of the steels 10Ch17N13M2T (SAE 316 Ti, 1.4571) and 08Ch21N6M2T in 50 % sulfuric acid at 333 K (60 °C) is reduced from 171 and 361 mm/a to 0.02 mm/a (6,733 and 14,213 mpy to 0.79 mpy) by additions of ≥ 0.050 % nitric acid. A further increase in the nitric acid addition to 0.5 and 1 % caused no further reduction in the corrosion rate [222].

An austenitic steel of relatively high nickel and molybdenum content, Sandvik® 2RK65 (SAE 904 L, 1.4539; (%) Fe-0.020C-19.5Cr-25.0Ni-4.5Mo-1.5Cu), is used in the explosives industry for the reaction chamber for the production of TNT (trinitrotoluene) in an acid mixture of 25 % nitric acid and 75 % sulfuric acid which heats up during the course of the reaction [213].

The corrosion rate of the steels 06ChN28MDT, 03Ch21N21M4GB and 10Ch17N13M2T (SAE 316 Ti) increases with nitric acid-containing sulfuric acid (0.03 – 0.045 % HNO_3 in 75.3 to 75.6 % H_2SO_4) from < 0.05 at 333 K (60 °C) to 0.4, 0.45 and 0.85 mm/a (15.7, 17.7 and 33.5 mpy) respectively at 393 K (120 °C) [314].

According to Figure 34, the corrosion rate of the steels X 8 CrNiMoTi 18 11 (cf. 1.4541, SAE 321) and X5NiCrMoCuTi20-18 (1.4506) in hot 96 % sulfuric acid at 393 K (120 °C) is reduced from 1.2 and 0.6 mm/a (47.2 and 23.6 mpy) respectively to about 0.15 mm/a (5.91 mpy) by small amounts of nitric acid, expressed as N_2O_3, e.g. with 0.03 % N_2O_3. At the same time, instead of the crater-like local attack, general corrosion occurs. For the steels to be used in plants for further processing of concentrated sulfuric acid at 390 K (117 °C) (Müller-Kühne process), the acid must have the necessary concentration of 0.03 % N_2O_3 [262].

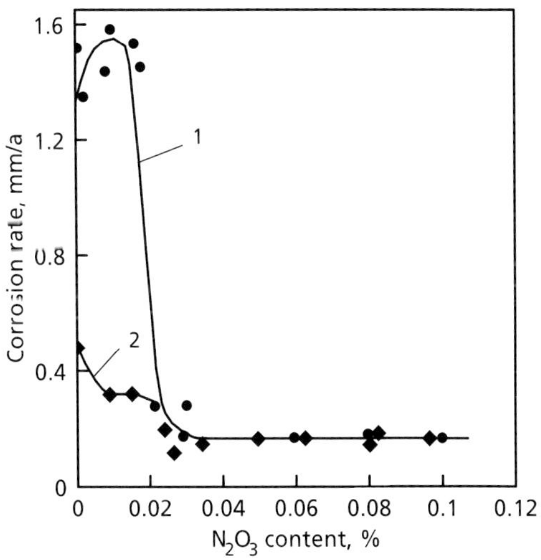

Figure 34: Dependence of the corrosion rate of austenitic steels in 96 % H_2SO_4 on the HNO_3 content (as N_2O_3) at 393 K (120 °C) [262]
(1) X 8 CrNiMoTi 18 11 (cf. 1.4541, SAE 321)
(2) X5NiCrMoCuTi20-18 (1.4506)

For surface pretreatment of implant steels used for replacement joints based on chromium-nickel-molybdenum steels, such as, for example, URX 2 CrNiMoN 18 ((18.12)) 12(0), the steels can be pickled either for 30 min at 293 K (20 °C) with 20 or 65 % nitric acid or for 30 min at 333 K (60 °C) with a solution of 20 % HNO_3 + 5 % H_2SO_4 + 4 % NaF. No increase in the pitting corrosion potential occurs with the latter solution, and the resistance to pitting corrosion is retained [315].

Table 54 shows the corrosion of the CrNiMo steels 10Ch17N13M3T (SAE 316 Ti, 1.4571; (%) Fe-0.10C-17Cr-13Ni-3Mo; Ti-stabilized) and 06ChN28MDT ((%) Fe-0.06C-(22–25)Cr-(26–29)Ni-(2.5–3.5)Mo-(2.5–3.5)Cu; Ti-stabilized) in hot nitric acid with various amounts of sodium fluoride and hydrazine (N_2H_4). If in each case 2 g/l sodium fluoride and hydrazine are simultaneously present, the material consumption rate of both steels increases by a factor of about 20, regardless of the concentration of nitric acid (75–100 g/l) [214].

Concentration			*10Ch17N13M3T*	*06ChN28MDT*
HNO_3	*NaF*	*N_2H_4*	*Material consumption rate*	
	g/l		*g/m² h*	
75	0	2	0.2 ± 0.1	0.08 ± 0.05
75	0.05	2	0.17 ± 0.13	0.2 ± 0.1
75	0.5	2	0.41 ± 0.16	0.16 ± 0.07
75	5.0	2	0.70 ± 0.1	0.11 ± 0.04
75	2	2	4.8 ± 0.5	0.3 ± 0.1
75	2	10	5.2 ± 0.3	0.2 ± 0.1
100	2	2	4.7 ± 0.7	0.27 ± 0.14
500	2	2	4.7 ± 1.2	1.7 ± 0.4

Table 54: Material consumption rate of CrNiMo steels in nitric acid with sodium fluoride and hydrazine at 368 K (95 °C) [283]

The steel 0Ch23N28M3D3T ((%) Fe-<0.01C-23Cr-28Ni-3Mo-3Cu; Ti-stabilized) corrodes in 5 % sulfuric acid containing 1.2 % HCl + 2 % HNO_3 at 333 K (60 °C) with a corrosion rate of 0.3 mm/a (11.8 mpy), which corresponds to a current density of about 0.03 mA/cm^2 [316].

The steels summarized in Table 55 were developed based on a given corrosion problem for the purpose of discovering a steel resistant to a solution of 25 % HNO_3 + 10 % HCl + 65 % water at 303 and 333 K (60 °C). The additional cobalt content in some steels was added to improve the forming and mechanical properties. At a low silicon content, the corrosion resistance increases as the nickel content decreases.

Comparable tests on steels with the composition (%) Fe-0.020C-(24–29)Cr-24Ni-4Mo resulted in corrosion rates which were already higher by more than 2 powers of ten at 303 K (30 °C) [317].

Steel composition, %	*Average material consumption rate g/m² h*
Fe-0.052C-0.33Si-1.54Mn-28.56Cr-9.03Ni-3.08Mo-0.40N	0.07
Fe-0.048C-0.37Si-1.40Mn-27.25Cr-8.94Ni-3.32Mo-0.042N	0.02
Fe-0.023C-0.43Si-1.52Mn-27.97Cr-5.27Ni-5.16Mo-3.29Co	0.08
Fe-0.018C-0.35Si-0.84Mn-27.19Cr-5.14Ni-3.62Mo	0.02
Fe-0.028C-0.44Si-1.62Mn-28.36Cr-2.98Ni-4.88Mo-3.29Co	0.03
Fe-0.026C-0.47Si-1.54Mn-28.75Cr-0.15Ni-4.36Mo-3.27Co	0.01

Table 55: Composition and average weight loss of specimens in 25 % HNO_3 + 10 % HCl + 65 % H_2O at 333 K (60 °C) [317]

The appearance of passivity on the steel Ch23N28M3D3T ((%) Fe-<0.1C-23Cr-28Ni-3Mo-3Cu; Ti-stabilized) is made difficult by the simultaneous presence of potassium chloride in nitric acid, so that a high degree of dissolution already occurs at room temperature [38].

The steel 08Ch21N6M2T corroded at 373 K (100 °C) in a solution of 40.4 g/l HNO_3 + 18.2 g/l oxalic acid at a corrosion rate of 0.004 mm/a (0.16 mpy) [80].

The corrosion rate of three steels used in a reactor for the synthesis of 3-chloro-2-hydroxypropanoic acid (RP) is shown in Table 56 [318].

Reaction medium, %	*Steel*	*Corrosion rate mm/a (mpy)*
Liquid phase		
* 45 – 46 HNO_3 + 20 – 21 epichlorohydrin + 28 – 29 H_2O + 5.3 – 5.4 HCHO	08Ch22N6M2T 08Ch17N13M2T	1.7 – 3.45 (67 – 136) 0.85 – 1.24 (33 – 49)
** 29 – 30 RP + 9 – 12 $H_2C_2O_4$ + 55 – 56 H_2O	06ChN28MDT	0.29 – 0.78 (11 – 31)

* starting composition
** reaction products

Table 56: Corrosion rate of steels in HNO_3-containing process solution after 1270 h tests at 330 K (57 °C) [318]

While phosphoric acid, even in high concentrations, does not influence the corrosion of the steels 06ChN28MDT and 08Ch21N6M2T in 15 % nitric acid, according to Table 57 addition of only 1 % H_2SiF_6 causes a considerable increase in the corrosion rate [268].

The steels already mentioned in Table 57 are recommended for the reactor containing nitric and phosphoric acid as well as small amounts of hydrogen fluoride in the production of Nitrophoska. Figure 35 shows the corrosion behavior of the steels in various production solutions of HNO_3 + H_3PO_4 + H_2SiF_6 + HF [319].

Composition of solution, %	06ChN28MDT	08Ch21N6M2T
	Corrosion rate, mm/a (mpy)	
15 HNO_3	0.007 (0.28)	0.001 (0.04)
15 HNO_3 + 15 H_3PO_4	0.013 (0.51)	0.001 (0.04)
15 HNO_3 + 15 H_3PO_4 + 1 H_2SiF_6	0.32 (12.6)	0.24 (9.45)
15 HNO_3 + 1 H_2SiF_6	0.32 (12.6)	0.62 (24.4)
15 HNO_3 + 1 H_2SiF_6 + 0.1 HF	0.54 (21.3)	0.72 (28.3)

Table 57: Corrosion of the steels 06ChN28MDT and 08Ch21N6M2T in 15 % HNO_3 with and without additions of H_3PO_4, HF and H_2SiF_6 at room temperature [268]

For the steel proposed for urea production to pass the Huey test, it must have a carbon content of < 0.03 % (for the welded regions). To obtain a ferrite-free structure at chromium contents of 17 to 18 % together with 2.6 to 3.0 % molybdenum, the nickel content should be not less than 15 %. The acceptable corrosion value in the Huey test is given as 0.54 g/m^2 h. As has already been pointed out several times, attention should also be paid to the correct heat treatment [320].

From their successful corrosion resistance in the Huey test, the steels UHB-724L (SAE 316 L, 1.4435; (%) Fe-<0.03C-17.4Cr-14.4Ni-2.6Mo) and UHB-725LN (SAE 310 MoLN, 1.4466; (%) Fe-0.020C-25.0Cr-22.0Ni-2.1Mo; nitrogen-containing) can be recommended for urea plants (specific corrosion rates for the urea region: 0.60 mm/a (23.62 mpy) for UHB-724L and 0.27 mm/a (10.6 mpy) for UHB-725LN) [321].

If > 2 % molybdenum is added to chromium-nickel steel, the nitrogen content should be increased, this preventing the formation of intermetallic compounds as well as improving the structure. If the processing conditions and heat treatment of the steel are carefully observed, the resistance to nitric acid, which can also contain halide ions, is increased by the higher molybdenum content [322].

To evaluate the corrosion resistance of steels of the type SAE 316 L in urea synthesis solutions at 468 K (195 °C), the corrosion behavior of the steel in the Huey test (48 h in boiling 65 % HNO_3) was determined for comparison. The corresponding corrosion rates are compared in Table 58 [323]. According to these, steels with corrosion rates in the Huey test of between 0.27 and 0.65 mm/a (10.6 and 25.6 mpy) can be recommended as materials for a urea synthesis plant.

The corrosion which occurs on the steels Ch21N6M2T and 0Ch23N28M3D3T under the conditions which prevail during production of methacrylic acid is shown in Table 59, using the following artificially prepared reaction mixtures (solution 1: 4 mol/l N_2O_4 + 1 mol/l HNO_3 + isobutylene, solution 2: 4 mol/l N_2O_4 + 1 mol/l HNO_3 + 0.5 mol/l isobutylene and solution 3: 4 mol/l N_2O_4 + 1 mol/l HNO_3) [324]. The normal CrNi steel 18 8 (cf. 1.4541, SAE 321) has also been shown for comparison. Relatively severe pitting corrosion was observed on the steels during continuous operation, with the exception of 0Ch23N28M3D3T.

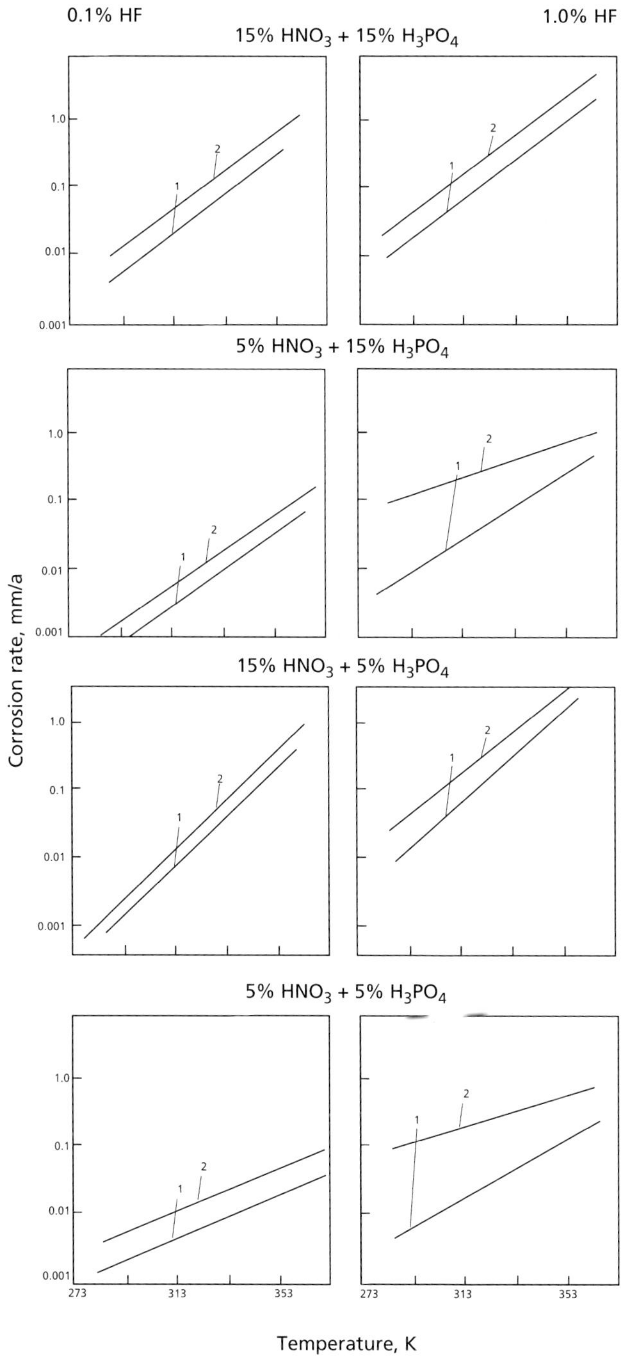

Figure 35: Temperature-dependence of the corrosion of (1) 06ChN28MDT and (2) 08Ch21N6M2T in (5–15) % HNO_3 + 1 % H_2SiF_6 + (0.1–1.0) % HF + (5–15) % H_3PO_4 [319]

Steel composition, %	*Huey test*	*Urea solution*
	Corrosion rate, mm/a (mpy)	
Fe-0.019C-17.01Cr-12.04Ni-2.51Mo-0.93Mn-0.50Si	0.274 (10.79)	0.086 (3.39)
Fe-0.021C-17.24Cr-12.32Ni-2.11Mo-0.83Mn-0.45Si	0.365 (14.37)	0.080 (3.15)
Fe-0.010C-17.49Cr-12.75Ni-2.25Mo-1.73Mn-0.66Si	0.602 (23.70)	0.062 (2.44)
Fe-0.013C-17.12Cr-12.16Ni-2.09Mo-1.28Mn-0.67Si	0.274 (10.79)	0.079 (3.11)
Fe-0.014C-17.28Cr-13.12Ni-2.26Mo-1.41Mn-0.52Si	0.274 (10.79)	0.073 (2.87)
Fe-0.021C-16.40Cr-12.68Ni-2.31Mo-0.99Mn-0.53Si	0.657 (25.87)	0.099 (3.9)
Fe-0.014C-17.58Cr-12.48Ni-2.40Mo-1.00Mn-0.72Si	0.402 (15.83)	0.063 (2.48)

Table 58: Comparison of the corrosion rates of the steel SAE 316 L in a urea synthesis solution at 468 K (195 °C) and in the Huey test as a function of the composition [323]

Steel	*Solution 1*	*Solution 2*	*Solution 3*
	Material consumption rate g/m² h		
Ch21N6M2T	0.00065	0.0011	0.0007
0Ch23N28M3D3T	–	0.00007	0.00082
Ch18N10T	0.0002	0.00047	0.00133

Table 59: Corrosion of steels in 3 artificially prepared reaction solutions under the operating conditions in methacrylic acid production [324]

The sliding gates made of the steel (%) Fe-<0.1C-18Cr-10Ni-3Mo in sliding gate housings are also resistant to nitric acid at low temperatures [325].

The ferritic-austenitic steel Sandvik® 3RE60 (UNS S31500, 1.4417; (%) Fe-0.030C-18.5Cr-4.9Ni-2.7Mo-1.5Mn-1.7Si-<0.03P-<0.03S) which was used for coolers in petroleum production in the North Sea and was in contact with gaseous nitric acid at 373 K (100 °C) at the intake and 313 K (40 °C) at the discharge for 2.5 years showed stress corrosion cracking [326].

Austenitic chromium-nickel steels with special alloying additions

The complex and costly CrNiMoCu steels X1NiCrMoCu31-27-4 (UNS N08028, 1.4563) often bring no improvement under the influence of nitric acid compared with CrNi steels with 18 to 20 % chromium, 10 to 14 % nickel, a low carbon content (< 0.05 % C) and the correct heat treatment, and in some cases they even worsen the corrosion resistance. Figure 36 shows the temperature dependence of corrosion of this steel in azeotropic nitric acid. The corrosion rate k in mm/a measured at 373 K (100 °C) for the range 20 to 75 % HNO_3 can be described by the equation

$$k = 1.95 \times 10^{-4} + 4.0 \times 10^{-6}\,[HNO_3]$$

where $[HNO_3]$ denotes the concentration of nitric acid in % [327].

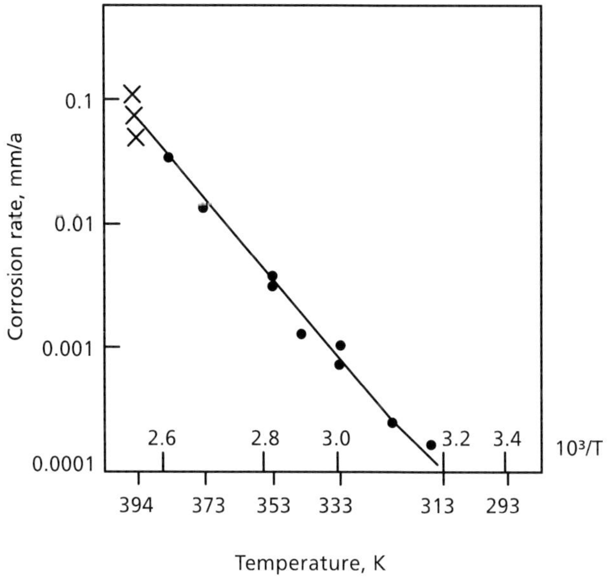

Figure 36: Temperature dependence of the corrosion of X 1 NiCrMoCu 31 27 in azeotropic nitric acid [327]
(x = values from the Huey test)

Table 60 summarizes steels which are particularly suitable for use in the presence of nitric acid. Of the steels listed, SAE 304 L (1.4306) gives the best results in the cost and behavior comparison if high purity (low C content, for example) and good homogeneity of composition are ensured [327].

Ferralium® 255, a CrNiMoCu steel 25 5 3 2, is also resistant to stress corrosion cracking in nitric acid at high concentrations and temperatures [328].

According to Figure 37, the Si-containing steel 1815-LCSi (UNS S30600, 1.4361) of very low carbon content ((%) Fe-0.006C-18.3Cr-15.1Ni-1.5Mn-4.1Si-0.005S-0.010P-0.010N) has an adequate resistance in boiling nitric acid between 28 and 95 %, apart from the range 40 to 90 %. Hot-rolled steel specimens heat-treated at 1423 K (1150 °C) for 10 min, quenched in water and annealed again at 1423 K (1150 °C) for 30 min after welding corroded in boiling 95 % nitric acid with a corrosion rate of 0.11 to 0.12 mm/a (4.33 to 4.72 mpy) after 1296 h [329].

DIN-Mat. No.	Composition, %				Huey test[1)]	azeotr. HNO_3[2)]	Relative cost[3)]	Area of application
	Cr	Mo	Ni	others	Material consumption rate g/m^2 h			
1.4306 (SAE 304 L)	19	–	11	–	0.20 – 0.50	–	0.7	–
1.4306S	20	–	12	4)	0.07 – 0.09	0.09	1.0[5)]	–
1.4335 (UNS S31002)	25	–	21	–	0.05 – 0.07	0.07	1.36[5)]	–
1.4547 (UNS S31254)	24	–	20	Nb	0.07 – 0.11	–	–	–
1.4435 (UNS S31603)	17.5	2.6	14	–	0.13 – 0.30	–	1.15	6)
1.4466 (SAE 310 MoLN)	25	2.3	22	N	0.04 – 0.08	0.09	1.27	6)
1.4563 (UNS N08028)	27	3.5	31	Cu	0.07 – 0.08	0.08	1.72	7)
2.4858 (UNS N08825)	21	3	40	Cu, Ti	0.15 – 0.20	0.25	1.9	7)

1) determined on seamless pipes, 5 test periods of 48 h each
2) 50 test periods of 48 h each in boiling, azeotropic HNO_3 with continuous removal of the corrosion products (distillation process)
3) for hot-rolled sheets 2 – 3 mm thick; acceptance about 3 t, position: March 1985 in the FRG
4) with low C, Si and Mo contents and somewhat increased Cr and Ni contents
5) ESR material
6) urea industry
7) for mixed acids

Table 60: Steels for use in nitric acid of approximately azeotropic composition with their material consumption rate, relative cost and main area of application [327]

Steels of the type 10Ch14G14N4T ((%) Fe-<0.1C-(13 – 15)Cr-(3.5 – 4.5)Ni-(13 – 15)Mn-0.65Ti) can also be used for production of nitric acid and ammonium nitrate in the presence of 56 % nitric acid below 320 K (47 °C) [93]. The corrosion rate of CrMnNi steels with 20 % manganese and (2 – 5) % nickel at room temperature and 363 K (90 °C) in 60 % nitric acid is about 0.002 and 0.68 mm/a (0.08 and 26.8 mpy) respectively, regardless of the nickel content of between 2 and 5 % [109].

Alloying up to 2 % germanium with 25Cr6Ni duplex steel for the purpose of increasing the ferrite content has no influence on corrosion in boiling 65 % nitric acid [330].

While addition of 0.042 % cerium to the steel 02Ch18N11 (quenched in water after heating for 30 min and tempering at 923 K (650 °C) for 1 h) causes no noticeable increase in corrosion rates in the Huey test (5 × 48 h in boiling 65 % HNO_3),

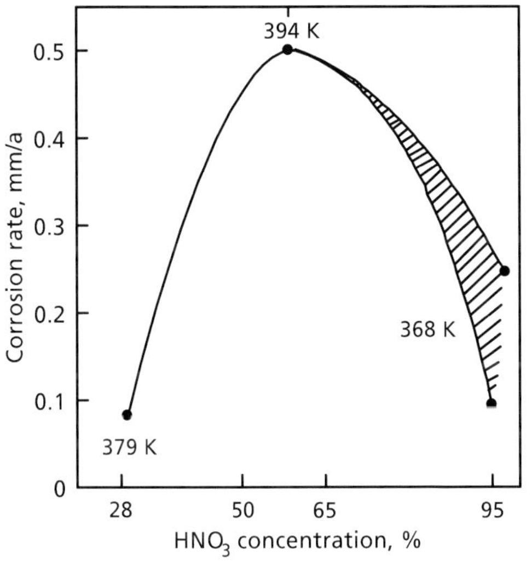

Figure 37: Corrosion rate of the steel 1815-LCSi (UNS S30600) in boiling nitric acid (28 to 95 %) [329]

the effect of boron is considerable even at 0.0015 % (e.g. an increase from 0.26 to 5.38 mm/a (10.2 to 211 mpy), or from 0.23 to 11.6 mm/a (9.06 to 457 mpy) after tempering at 1353 and 1423 K (1150 °C) respectively) [331].

A protective SiO_2 layer is formed from the high silicon content in the steel. It should be mentioned that the corrosion rate of the steels SAE 304 L and 310 L is lower up to about 70 % HNO_3. Only above 80 % HNO_3 is the steel 1815-LCSi (UNS S30600) superior to the other two in respect of corrosion behavior [332].

The material consumption rate of the steel KV-80 ((%) Fe-0.018C-17.7Cr-14.3Ni-0.19Cu-1.22Mn-0.05Si-0.006P-0.011S) and KV-81 ((%) Fe-0.016C-24.7Cr-19.8Ni-1.31Mn-0.17Cu-0.10Si-0.006P-0.016S) in boiling 98 % nitric acid could be reduced from 7 and 5.5 g/m^2 h respectively to 0.07 and 0.31 g/m^2 h respectively by addition of 4 % aluminium. While corrosion of the steel KV-81 is not improved sufficiently, a considerable improvement comparable to that on the steel KV-80 of 0.16 g/m^2 h is achieved on the steel KV-82 ((%) Fe-0.022C-18.4Cr-9.4Ni-0.25Mo-0.22Cu-1.18Mn-0.06Si-0.006P 0.016S) [166].

Welds produced with welding wires of the CrNiSi steels 02Ch17N(10 – 18)S6 ((%) Fe-0.02C-(4 – 6.5)Si-(0.43 – 0.52)Mn-(16.3 – 18.0)Cr-(10.5 – 18.2)Ni-(0.005 – 0.008)S-(0.012 – 0.014)P) and 02Ch8N(22 – 60)S6 ((%) Fe-0.02C-(5.5 – 6.5)Si-(0.004 – 0.005)Mn-(7.4 – 9.0)Cr-(21.1 – 61.0) Ni-(0.006 – 0.016)S-(0.013 – 0.014)P) show only a low corrosion rate, which does not exceed 0.1 mm/a (3.94 mpy), in boiling 98 % nitric acid in the entire alloy range after 1000 h tests [333]. In practical tests in apparatuses for dissolution of Zr-plated uranium, Carpenter® 20 (cf. 2.4660), Durco® D-10, Durimet® 20 (CN-7M) and SAE 309 Nb (cf. 1.4828; (%) Fe-0.20-C(22 – 24)Cr-(12 – 15)Ni-2Nb) showed corrosion rates of 0.38 – 0.75 mm/a (15.0 – 29.5 mpy) during two batches (a year is calculated with 180 batches) in (mol/l) 10 nitric acid

+4 hydrogen fluoride. The stainless steel SAE 309 Nb had a higher corrosion resistance than Carpenter® 20. The corrosion rates were about 2 and 3.1 mm/a (79 and 122 mpy) at 313 and 353 K (40 and 80 °C) respectively. No galvanic action occurs when the steel is brought into contact with zirconium in the acid mixture [334]. Lining the steel tank with Teflon® (PTFE) had no effect, since the coverings are not sufficiently impermeable to acid.

The weld seam of the chromium-nickel steel 1Ch18N9T ((%) Fe-0.1C-19.0Cr-9.9Ni-0.68Mn-1.18V-0.76Nb-0.22Ti-1.24Si) shows pitting corrosion and intercrystalline corrosion at 303 K (30 °C) in a solution of 10 % HCl + 5 % HNO_3 [335].

The CrNiTi steel X6CrNiTi18-10 (SAE 321, 1.4541; (%) Fe-0.1C-18.0Cr-10.0Ni-(5 x C)Ti) corrodes in 0.1, 0.5 and 1.0 mol/l nitric acid at 353 K (80 °C) with a material consumption rate of 0.03, 0.07 and 0.09 g/m^2 h, respectively, and if 0.05 mol/l ammonium cerium(IV)nitrate is also present in the acid, with rates of 0.16, 6.5 and 44 g/m^2 h [336].

The material consumption rate of chromium-nickel steels with additions of 4 % silicon and niobium in boiling 56 % nitric acid +0.2 % ammonium bichromate ($(NH_4)_2Cr_2O_7$) is shown in Table 61 [337]. The corrosion-reducing effect of silicon on the high-nickel CrNi steel which has already been mentioned above is also confirmed here.

Steel	*Composition of steel, %*						*Material consumption rate g/m^2 h*	
	C	*Cr*	*Ni*	*Si*	*Mn*	*Nb*	*(a)*	*(b)*
ChN40SB	0.040	18.8	39.4	4.3	0.06	0.63	0.25	0.30
ChN40S	0.031	20.0	38.9	4.2	0.05	0.13	0.27	0.32
Ch14N40SB	0.034	14.4	38.9	4.0	0.05	0.63	0.48	0.45
ChN40B	0.032	18.2	40.4	0.08	0.05	0.49	2.60	10.3

Table 61: Corrosion rate of steels after (a) quenching and (b) quenching and tempering at 923 K (650 °C) for 2 h in 56 % HNO_3 + 0.2 % $(NH_4)_2Cr_3O_7$ [337]

Solutions of 150 to 250 g/l HNO_3 + 75 to 100 g/l NH_4 HF have proved suitable for descaling the CrNiTiAl steel 36NChTJu (EI-702; (%) Fe-0.03C-12.1Cr-35.95Ni-3.03Ti-1.13Al-1.03Mn-0.03Si) after dispersion annealing and quenching from 1263 K (990 °C), excellent, silver-looking surfaces being produced. The scale is removed within 3 min. The steel itself is attacked in such a pickling solution at a rate of about 1 g/m^2 min, which corresponds to one tenth of the rate of dissolution of the scale [338].

Special iron-based alloys

Iron of high silicon content (10 to 18 % Si) is also resistant to nitric acid and can be cast into complicated shapes. The standard grade (Tantiron® N), which complies with the British Standard BS 1591/49, has a silicon content of 14.3 to 15.3 %, good physical properties and an excellent resistance to nitric acid. Tantiron® E with 16 to 18 % Si is recommended under severe corrosion conditions if mechanical properties are of secondary importance [339, 340].

The manganese-aluminium steels ((%) Fe-(0.1–1)C-(20–32.5)Mn-(7–10)Al) are not resistant in nitric acid at room temperature. Only the steel Fe-0.95C-23.3Mn-7.4Al, which is not resistant in nitric acid below 50 %, can be used in industry at higher concentrations. The material consumption rate of the steel at room temperature, e.g. in 65 % nitric acid, is thus about 0.21 g/m^2 h [341]. The corrosion potential here is – 0.59 V_{SCE}.

Figure 38 shows the corrosion of the chromium-manganese steel AS-43 (Russ. type; (%) Fe-0.05C-16.5Mn-18.4Cr-1.60Ni-0.31Nb-0.01Ce-0.26N-0.004B) in nitric acid containing 0.1 and 0.01 mol/l hydrofluoric acid at 293 and 313 K (20 and 40 °C). The corrosion rate minimum is at 313 K (40 °C) at about 0.3 mm/a (11.8 mpy) in 5 mol/l HNO_3 + 0.1 mol/l HF [74].

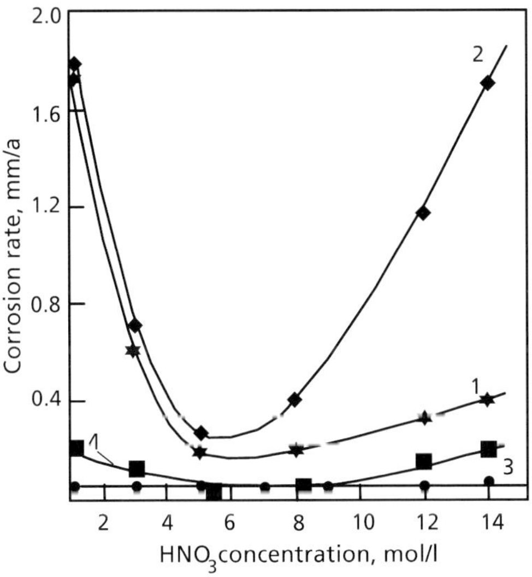

Figure 38: Dependence of the corrosion rate of the steel AS-43 on the HNO_3 concentration in HNO_3 containing 0.1 mol/l HF at (1) 293 K (20 °C) and (2) 313 K (40 °C) and in HNO_3 containing 0.01 mol/l HF at (3) 293 K (20 °C) and (4) 313 K (40 °C) [74]

Sintered stainless steels with an excellent absorption capacity for neutrons ((%) Fe-≤0.15C-(0.5–4)Si-≤0.6Mn-(7–50)Cr-≤25Ni-≤5Mo-≤5Cu-≤5Ti with and without ≤4Nb, mixed with 1 to 55 percent by volume metal boride) corrode in boiling 65 % nitric acid at a material consumption rate of 38 g/m^2 h [342].

Current density/potential curves using the iron alloy (atomic percent) Fe-40Ni-16P-6B in 15 % nitric acid show a higher corrosion rate for the crystalline alloy than for the amorphous alloy. However, at a polarization potential of about – 0.2 V_{SHE}, a corrosion current density of 0.2 mA/cm^2 occurs on both, i.e. the corrosion rate is too high for industrial use [343].

Iron-nickel alloys with 10 to 30 % nickel are susceptible to corrosion in nitric oxide (NO) under 0.0027 MPa between 498 and 548 K (225 and 275 °C), iron oxide coatings being formed. The corrosion decreases as the nickel content increases [344].

For chemical surface treatment of the iron-nickel-cobalt alloy Kovar® (UNS K94610, 1.3981, Fe-29Ni-18Co), it is pickled in a solution of 2 H_2SO_4:1 HCl:2 H_2O (percent by volume) at 343 K (70 °C) for 2 min and then polished in a solution of 30 HNO_3 : 1 HCl : 70 CH_3COOH (percent by volume) at the same temperature for 7 seconds. A layer of 2.5 μm is corroded here during the pickling operation, with a higher loss of nickel than cobalt. During the polishing operation, a 7.8 μm layer dissolves, but the ratio of the dissolved metals corresponds to that of the alloy [345].

An aqueous solution of 70 to 130 g tartaric acid + 80 to 120 ml hydrochloric acid + 30 to 70 ml nitric acid + 0.5 to 2 g/l of a non-ionic wetting agent is also proposed for polishing Kovar® [346].

For analyzing the structure of high-speed steels ((%) Fe-(0.59–1.10)C-(3.85–4.17)Cr-(0.25–0.32)Mn-(0.34–9.21)Mo-(1.1–1.9)V-(6.2–17.6)W-(0–17)Co-(0.23–0.30) Si), the best results in respect of ease of differentiation between austenite grain boundaries and the basic structure were achieved by pickling in a solution of 10 ml conc. HCl + 3 ml conc. HNO_3 + 80 ml methyl alcohol [347].

Bibliography

[1] Komp, M. E.; Mathay, W. L.
Steels for the pulp and paper industry
Mater. Performance *16* (1977) 6, p. 22

[2] Vakulenko, L. I.; Kozlovskaya, N. A.; Shedenko, L. I.
The effect of sulfamic acid on the corrosion and electrochemical behavior of iron in nitric acid solutions
Sov. Progress Chem. *48* (1982) 9, p. 62

[3] Pakhomov, V. S.
Nitric acid as a corrosion medium (in Russian)
Tr. Mosk. Inst. Khim. Mashinostr. *67* (1975) p. 102

[4] Gajduchok, V. M.; Kasperskij, G. A.
Effect of plastic working on the oxidation of steel (in Russian)
Probl. Treniya Iznashivaniya (1972) 2, p. 118

[5] Gavrilenko, A. G.
Dissolution of iron in nitric acid (in Russian)
Voronezh. Thekhnol. Inst., Voronezh, USSR, Deposited Doc. (1974), VINITI 1527-74, 8 p

[6] Krutikov, P. G.; Nemirov, N. V.; Papurin, N. M.
State of the surface of the steel 16GS after various chemical actions (in Russian)
Zashch. Met. *19* (1983) 3, p. 455

[7] Hasegawa, M.; Osawa, M.
Anomalous corrosion of hydrogen-containing ferritic steels in aqueous acid solution
Corrosion *39* (1983) 4, p. 115

[8] Cihal, V.; Kubelka, J.
Corrosion cracks on steel in nitrate solutions (in Czech)
Strojirenstvi *13* (1963) 11, p. 837

[9] Makwana, S. C.; Patel, N. K.; Vora, J. C.
Corrosion of 3S aluminium in the mixture of acid solutions
J. Indian Chem. Soc. *51* (1974) p. 1051

[10] Krotil, B.
Oceli pro difúzní chromování (in Czech)
Koroze Ochr. Mater. *18* (1974) 4, p. 53

[11] Abd El Haleem, S. M.; Khedr, M. G. A.; Killa, H. M.
Corrosion behaviour of metals in HNO_3. I. Contribution to the study of the dissolution of steel in HNO_3
Br. Corros. J. *16* (1981) 1, p. 42

[12] El-Basiouny, M. S.; Elnagdi, M. H.; Ismail, A. R.
Einfluß von Benzolsulfonsäure-Derivaten auf die Auflösungsgeschwindigkeit von Eisen in Salpetersäure. Grenzen der Anwendbarkeit thermometrischer Verfahren zum Messen von Korrosionsgeschwindigkeiten
(Effect of benzenesulfonic acid derivatives on the dissolution rate of iron in nitric acid. Limits of applying thermometric methods for measurement of corrosion rates) (in German)
Metalloberfläche *35* (1981) 12, p. 482

[13] Zavrazhina, V. I.; Artemenko, A. I.
The effect of the potassium salts of some hydroxamic acids on the corrosion of steel in aqueous solutions (in Russian)
Sb. Tr. Belgor. Tekhnol. Inst. Stroit. Mater. *22* (1976) p. 67

[14] Ponomarenko, V. I.; Novachek, L. A.; Fedorov, Yu. V.
The effect of N-decyl and N-dodecylquinoline iodide on the acid corrosion of carbon steels (in Russian)
Vopr. Khim. i. Khim. Tekhnol. *71* (1983) p. 14

[15] Sayed, S. M.; Gouda, V. K.
Korrosionsinhibierung von Weichstahl in verschiedenen Säurelösungen durch Petrolsulfonate
(Corrosion inhibition of mild steel in various acid solutions using petroleum sulfonate) (in German)
Metalloberfläche *32* (1978) 7, p. 298

[16] Uzlyuk, M. V.; Fedorov, Yu. V.
Effect of pyridine thiocyanates on the corrosion of iron in nitric acid (in Russian)
Okislit. Vosstanov. i Adsorbts. Protsessy na Poverkhn. Tverd. Metallov, Izhevsk (1980) 2, p. 29

[17] Fedorov, Yu. V.; Uzlyuk, M. V.
Effect of mixture of indole and ammonium thiocyanate on the corrosion of steel in acid solutions
Sov. Progress Chem. *37* (1971) 1, p. 75

[18] Uzlyuk, M. V.; Fedorov, Yu. B.; Zelenin, B. M.
Effect of mixtures of sodium sulfide with indole and thiourea on steel corrosion in nitric acid (in Russian)
Zashch. Met. *10* (1974) 4, p. 482

[19] Brown, M. H.
Behavior of austenitic stainless steels in evaluation tests for the detection of susceptibility to intergranular corrosion
Corrosion *30* (1974) 1, p. 1

[20] Sathianandhan, B.; Balakrishnan, K.; Subramanyan, N.
Triazoles as inhibitors of corrosion of mild steel in acids
Br. Corros. J. *5* (1970) 4, p. 270

[21] Demyanets, A. A.; Ezau, Ya. Ya.
The effect of chloride ions on the corrosion and electrochemical behavior of iron and carbon steels in nitric acid (in Russian)
Khim. i Khim. Tekhnol., Minsk *15* (1980) p. 21

[22] Demyanets, A. A.; Ezau, Ya. Ya.
Effect of chloride ions on the anodic behavior of carbon steels in nitric acid (in Russian)
Khim. i Khim. Tekhnol., Minsk *18* (1983) p. 3

[23] Vdovenko, I. D.; Vakulenko, L. I.; Kozlovskaya, N. A.
Effect of catapin A on corrosive and electrochemical behavior of iron in hydrochloric and nitrogen acid mixtures (in Russian)
Ukr. Khim. Zh. *46* (1980) 4, p. 360

[24] Dhirendra; Pandey, G. N.; Sanyal, B.
Comparative corrosion of iron and iron alloys by vapours of single and mixed acids
Corros. Prev. Control *29* (1982) 1, p. 10

[25] Andersen, T. N.; Vanorden, N.; Schlitt, W. J.
Effects of nitrogen oxides, sulfur dioxide, and ferric ions on the corrosion of mild steel in concentrated sulfuric acid
Metall. Trans. A *11A* (1980) Aug., p. 1421

[26] Miller, R. F.; Rhodes, P. R.
Nitric acid a corrosion inhibitor for carbon steel in sulfuric acid
Mater. Protection and Performance *9* (1970) 10, p. 33

[27] McDowell jr., D. W.
Handling mixed nitric and sulfuric acid
Chem. Eng. *81* (1974) 23, p. 133

[28] Donchenko, M. I.; Sribnaya, O. G.; Zheleznyak, Yu. Yu.
Retarding iron corrosion in nitric acid by means of organic and inorganic additives (in Russian)
Zashch. Met. *17* (1981) 2, p. 156

[29] Apostolache, S.
Influence of temperature on corrosion inhibitors in the system Fe-HNO_3 (in Rumanian)
Revista de Chimie *30* (1979) 7, p. 693

[30] Tsejtlin, Kh. L.; Sorokin, Yu. I.; Ryzhkova, Zh. S.
Corrosion of equipment and problems of accident prevention (in Russian)
Khim. Prom. (1977) 3, p. 179

[31] Vaccari, J. A.
Cast irons
Mater. Eng. *79* (1974) 6, p. 52

[32] Salminkeit, V.; von Plessen, H.; Vollmüller, H.
Die Korrosion der Gußeisenkessel beim Pauling-Verfahren
(Corrosion of cast iron boiler in the Pauling process) (in German)
Chem.-Ing.-Tech. *53* (1981) p. 822

[33] Dilthey, U.; Wanke, R.
Hochleistungs-Schweißplattierverfahren für den Chemie-Apparatebau
(Highly efficient plating in chemical plant construction) (in German)
Chem.-Ing.-Tech. *46* (1974) 11, p. 467

[34] Saldanha, B. J.; Streicher, M. A.
Effect of silicon on the corrosion resistance of iron in sulfuric acid
Mater. Perform. *25* (1986) 1, p. 37

[35] Telmanova, O. N.; Karyazin, P. P.; Shtanko, V. M.
Behavior of pig iron with lamellar and globular-shaped graphite in dilute acids (in Russian)
Zashch. Met. *16* (1980) 1, p. 57

[36] Sukhodolskaya, E. A.; Odarchenko, V. V.
Improvement in the properties of corrosion-resistant cast iron (in Russian)
Liteinoe Proizvod. (1980) 10, p. 8

[37] Ford, E.
Korrosionsbeständige Metalle
(Corrosion-resistant metals) (in German)
Metall *28* (1974) 5, p. 459

[38] Voroshilov, I. P.; Shilov, V. R.; Voroshilova, E. P.; Shapovalova, L. P.
Corrosion behavior of certain structural materials in a mixture of nitric acid and potassium chloride (in Russian)
Zashch. Met. *11* (1975) 5, p. 602

[39] Poulson, B.
The sensitization of ferritic steels containing less than 12 % Cr
Corros. Sci. *18* (1978) p. 371

[40] Horn, E.-M.; Storp, S.; Holm, R.
Korrosion und Deckschichtenaufbau von X 5 CrTi 12, Werkstoff-Nr. 1.4512, bei Beanspruchung in Salpetersäure
(Corrosion resistance and formation of protective surface films on X 5 CrTi 12, material no. 1.4512, in nitric acids)
(in German)
Werkst. Korros. *37* (1986) p. 69

[41] Cepero, A. E.; Guedes, J.
Influencia de los iones cloruros en la corrosividad del acido nitrico
(Influence of chloride ions on the corrosiveness of nitric acid) (in Spanish)
R. Tecnológica *13* (1975) 4, p. 80

[42] Truman, J. E.
Corrosion resistance of 13 % chromium steels as influenced by tempering treatments
Br. Corros. J. *11* (1976) 2, p. 92

[43] Komp, M. E.; Mathay, W. L.
Steels for the pulp and paper industry
Mater. Performance *16* (1977) 6, p. 22

[44] Tomashov, N. D.
Corrosion resistant alloys and the prospects for their development (in Russian)
Zashch. Met. *17* (1981) 1, p. 16

[45] Johnson, M. J.; Kearns, J. R.; Deverell, H. E.
The corrosion of the new ferritic stainless steels in nitric acid
Corrosion '84, International Corrosion Forum Devoted Exclusively to the Protection and Performance of Materials, New Orleans, April 1984 (Proc. Conf.), Paper No. 144

[46] Knoth, J.
High purity ferritic Cr-Mo stainless steel – Five years' successful fight against corrosion in the process industry
Werkst. Korros. *28* (1977) p. 409

[47] Vaccari, J. A.
New ferritic stainless steels beat stress corrosion, ease fabrication
Mater. Eng. *82* (1975) 7, p. 24

[48] Glazkova, S. A.; Karasyuk, T. N.; Bukanova, G. S.; Zheltova, G. A.; Lejbzon, V. M.; Kalinin, B. P.
Corrosion resistance of weld joints of the steel 08Ch13AG20
(in Russian)
Khim. Neft. Mashinostr. (1985) 2, p. 28

[49] Sotnichenko, A. L.; Agapov, G. N.; Yarkovoj, V. S.
Corrosion resistance of the nickel-free chromium manganese steel AS-43 in nitric acid (in Russian)
Khim. Neft. Mashinostr. (1974) 10, p. 23

[50] Mertins, K.
Metallographic investigations as a supplement to short-time corrosion tests on high alloy chromium and chromium-nickel steels
Prakt. Metallogr. *10* (1973) 2, p. 75

[51] Horn, E.-M.; Schoeller, K.
Temperatur- und Konzentrationsabhängigkeit der Korrosion nichtrostender austenitischer und ferritischer Werkstoffe in 20- bis 75%iger Salpetersäure
(Corrosion resistance of austenitic and ferritic stainless alloys in 20 to 75% nitric acid as a function of temperature and concentration) (in German)
Werkst. Korros. *41* (1990) 3, p. 97

[52] Demo, J. J.
Weldable and corrosion-resistant ferritic stainless steels
Metallurg. Trans. 5 (1974) Nov., p. 2253

[53] Tokareva, T. B.; Ershova, N. I.; Zubchenko, A. S.
Intercrystalline corrosion of low-carbon, high-chromium ferritic steels (in Russian)
Metalloved. Term. Obrab. Met. (1974) 12, p. 33

[54] Tokareva, T. B.; Kolyada, A. A.; Smolin, V. V.; Medvedev, E. A.
Corrosion mechanical properties of high-chromium content ferritic steels produced by vacuum melting (in Russian)
Zashch. Met. *13* (1977) 5, p. 529

[55] Dzhambazova, L.; Rashev, Ts.; Zlateva, G.
Corrosion of nitrogenized CrMn steel in nitric acid (in Russian)
Zashch. Met. *14* (1978) 4, p. 465

[56] Dzhambazova, L.; Zlateva, G.; Kamenova, Ts.
Effect of the characteristics of the nitride phase on the corrosion behavior of nitrogen chromium-manganese steels (in Bulgarian)
Tekh. Mis'l *15* (1978) 4, p. 93

[57] Dzhambazova, L. D.; Zlateva, G.; Kamenova, Ts.
Effect of nitride phase on the corrosion behavior of chrome manganese steels alloyed with nitrogen (in Russian)
Zashch. Met. *15* (1979) 2, p. 202

[58] Dimov, I.; Rashev, T.; Dzhambazova, L.; Andreev, Ch.
Corrosion of austenitic chromium-manganese steels with increased nitrogen content (in Bulgarian)
Materialoznanie Tekhnol. (1976) 2, p. 20

[59] Kowatschewa, R.; Djambazowa, L.
Precipitation and corrosion processes in ferrite containing nitrogen-chromium-manganese steels after isothermal annealing
Prakt. Metallogr. *17* (1980) p. 560

[60] Kiesheyer, H.; Lennartz, G.; Brandis, H.
Korrosionsverhalten hochchromhaltiger, ferritischer, chemisch beständiger Stähle (Corrosion behaviour of high-chromium ferritic stainless steels) (in German)
Werkst. Korros. *27* (1976) p. 416

[61] Fortunati, S.; Tamba, A.
Mechanical and corrosion properties of welded ELI 18-2 CrMo(Ti) steel pipes
Arch. Eisenhüttenwes. *53* (1982) 3, p. 105

[62] Åslund, C.; Gemmel, G.; Andersson, T.
Stranggepreßte Rohre auf pulvermetallurgischer Basis
(Extruded tubes based on powder metallurgy) (in German)
Bänder Bleche Rohre *22* (1981) 9, p. 223

[63] Charbonnier, J. C.; Lena, M.; Thomas, B. J.
Influence du molybdène sur la corrosion intergranulaire d'un acier pur à 17 % de chrome
(Effect of molydenum on the intergranular corrosion of a pure 17 % chromium steel) (in French)
Rev. Métall. *76* (1979) p. 469

[64] Charbonnier, J. C.; Lena, M.; Thomas, B. J.
The influence of molybdenum on the intergranular corrosion of a pure ferritic steel with 17 % chromium
Corros. Sci. *19* (1979) p. 23

[65] Moller, G. E.; Franson, I. A.; Nichol, T. J.
Experience with ferritic stainless steel in petroleum refinery heat exchangers
Mater. Performance *20* (1981) 4, p. 41

[66] Streicher, M. A.
The role of carbon, nitrogen, and heat treatment in the dissolution of iron-chromium alloys in acids
Corrosion *29* (1973) 9, p. 337

[67] Schmidt, L.
(The corrosion resistance of chromium steels) (in Rumanian)
Revista de Chimie *37* (1986) 1, p. 68

[68] Sedenko, A. M
Corrosion stability of stainless steels and titanium alloys in nitric acid (in Russian)
Ukrain. Khim. Zh. *51* (1985) 11, p. 1226

[69] Miska, K. H.
Ferritic stainless steels
Mater. Eng. *85* (1977) 4, p. 69

[70] Zhuravleva, L. V.; Kasinskaya, L. K.; Nosivets, L. A.
Corrosion resistance of steel 08Ch18G8N2T in sulfuric acid (in Russian)
Khim. Neft. Mashinostr. (1979) 12, p. 21

[71] Steigerwald, R. F.
The effects of metallic second phases in stainless steels
Corrosion *33* (1977) 9, p. 338

[72] Dundas, H. J.; Bond, A. P.
Niobium and titanium requirements for stabilization of ferritic stainless steels
Intergranular Corrosion of Stainless Alloys, Toronto (Canada), 1977 (Proc. Conf.)
American Society for Testing and Materials (1978), p. 154

[73] Troselius, L.; Andersson, I.; Andersson, T.; Bernhardsson, S. O.; Degerbeck, J.; Henrikson, S.; Karlsson, A.
Corrosion resistance of type 18Cr2MoTi stainless steel
Br. Corros. J. *10* (1975) 4, p. 174

[74] Lomovtsev, V. I.; Konstantinova, E. V.
Corrosion of austenitic chromium-manganese steel AS-43 in nitrofluoride media (in Russian)
Zashch. Met. *14* (1978) 60, p. 702

[75] Konstantinova, E. V.; Lomovtsev, V. M.
Corrosion behavior of economically alloyed steels in dilute nitric acid solutions with activating additives (in Russian)
Zashch. Met. *18* (1982) 1, p. 82

[76] Konstantinova, E. V.; Ryabova, N. I.; Lomovtsev, V. I.
Corrosion resistance of welded joints in low-nickel and nickel-free steels
Weld. Prod. (USSR) *29* (1982) 2, S. 8

[77] Konstantinova, E. V.; Lomovtsev, V. I.
Corrosion of stainless steels containing little or no nickel in concentrated nitric acid with hydrofluoric acid or chloride ions (in Russian)
Zashch. Met. *19* (1983) 6, p. 933

[78] Napara-Volgina, S. G.
Properties of materials based on alloy powders prepared by diffusion saturation from point sources, and areas of their use (in Russian)
Poroshk. Metall. *24* (1984) 11; p. 1

[79] Tomashov, N. D.; Chernova, G. P.; Rutten, M. Ya.
Investigation of the electrochemical behavior of stainless steels in solutions of nitric acid with additions of iron chloride and hydrochloric acid (in Russian)
Zashch. Met. *18* (1982) 6, p. 850

[80] Migaj, L. L.; Malchevskij, E. G.
Stability of sparingly alloyed stainless steels in rare metal industrial media (in Russian)
Zashch. Met. *16* (1980) 2, p. 143

[81] Razygraev, V. P.; Mirolubov, E. N.
Influence of Cr(VI) on the kinetics of anodic and cathodic processes during corrosion of stainless steels and alloys in nitric acid solutions (in Russian)
Zashch. Met. 9 (1973) 1, p. 44

[82] Brown, R. S.
The three-way trade off in stainless-steel selection
Mater. Eng. *96* (1982) 5, p. 58

[83] Fleer, R.; Rickel, J.; Draugelates, U.
Metallographic detection of carbides in the steel X 41
CrMoV 5 1 after different austenizing processes
Prakt. Metallogr. *16* (1979) p. 105

[84] Anonymous
Vallourec® VLX 562 high strength austenitic-ferritic alloys with superior resistance to stress corrosion cracking
Hydrocarbon Processing *65* (1986) 11, p. 39

[85] Khokhlova, P. M.; Levin, I. A.
Selective structural corrosion of the ferritic austenitic chromium-nickel steel Ch22N5 (in Russian)
Zashch. Met. *10* (1974) 6, p. 674

[86] Koren, M.; Hochörtler, G.
Eigenschaften des ferritisch-austenitischen Stahles X 3 CrMnNiMoN 25 6 4
(Properties of the ferritic-austenitic steel X 3 CrMnNiMoN 25 6 4) (in German)
Stahl Eisen *102* (1982) 10, p. 509

[87] Süry, P.; Brezina, P.
Kurzzeitprüfmethoden zur Untersuchung des Einflusses von Wärmebehandlung und chemischer Zusammensetzung auf die Korrosionsbeständigkeit martensitischer Chrom-Nickel-(Molybdän)-Stähle mit tiefem Kohlenstoff-Gehalt
(Short-duration test for evaluating the influence of heat-treatment and chemical composition on the corrosion resistance of low carbon martensitic chromium nickel (molybdenum) steels) (in German)
Werkst. Korros. *30* (1979) 5. p. 341

[88] Melkumov, S. B.
Effect of a welding thermal cycle on the corrosion resistance welding zone of the ferritic-austenitic steel 0Ch21N5T in a 65 % solution of HNO_3 (in Russian)
Zashch. Met. 9 (1973) 2, p. 185

[89] Blom, U.; Kvarnback, B.
The importance of high purity in stainless steels for nitric acid service – Experience from plant service
Mater. Performance *14* (1975) 7, p. 43

[90] Voroshilov, I. P.; Shilov, V. R.; Voroshilova, E. P.; Shapovalova, L. P.
Corrosion behavior of certain structural materials in a mixture of nitric acid and potassium chloride (in Russian)
Zashch. Met. *11* (1975) 5, p. 602

[91] van Maaren, P. W.
Possible applications of optical methods in failure
Prakt. Metallogr. *18* (1981) p. 494

[92] Anonymous
Lagertank aus austenitischem CrNiSi-Stahl
(Storage tank made of austenitic CrNiSi steel) (in German)
Nickel-Ber. *38* (1973) p. 6

[93] Vazhenin, S. F.; Remashevskaya, Z. V.; Bojko, A. Z.; Korotenko, N. D.
Possible use of nickel-free and economically alloyed, low-nickel stainless steels in the production of nitric acid and ammonium nitrate (in Russian)
Vopr. Khim. i Khim. Tekhnol., Kharkov *67* (1982) p. 99

[94] Bingham jr., E. C.
Compact design pays off at new nitric acid plant
Chem. Eng. *73* (1966) 11, p. 116

[95] Kügler, A.
Die Wahl der Edelstähle bei aggressiven Medien
(The selection of high-grade steels in aggressive media) (in German)
VDI Z. *119* (1977) 8, p. 411

[96] Donat, H.; Schäfer, K.
Stand und Entwicklungstendenzen des Schweißens von korrosionsbeständigen Stählen
(Current state and trends in development in the welding of corrosion-resistant steels) (in German)
Schweissen und Schneiden *27* (1975) 9, p. 343

[97] Kato, T.; Fujikura, M.; Ichikawa, J.
Some properties of boron-containing 18Cr9Ni stainless steel for nuclear engineering
Denki Seiko *49* (1978) p. 108

[98] Stefec, R.; Protiva, K.
Improvement of the corrosion resistance of type FeCr21Ni-32TiAl steel by a cold work and anneal combination treatment
(in Czech)
Kovove Mater. *20* (1977) p. 225

[99] Pavlović Milovanovi, J.
Corrosion of stainless steels in nitric acid production plants (Serbo-Croatian)
Hemijska Industrija, Beograd *27* (1973) p. 203

[100] Malakhova, E. K.; Kuzyukov, A. N.
Plant experience with equipment in chemicals that cause intercrystalline corrosion and corrosion cracking
(in Russian)
Tr. Vses. Konstr. Inst. Khim. Mashinostr *78* (1977) p. 110

[101] Horn, E.-M.; Kügler, A.
Entwicklung, Eigenschaften, Verarbeitung und Einsatz des hochsiliciumhaltigen austenitischen Stahls X 2 CrNiSi 18 15
(Development, properties, processing and applications of high-silicon steel grade X 2 CrNiSi 18 15) (in German)
Z. Werkstofftech. *8* (1977) p. 41

[102] Gramberg, U.; Günther, T.; Palla, H.
Erfahrungen beim Einsatz der Rasterelektronenmikroskopie im Rahmen der Untersuchung von Schadensfällen unter korrosiven Bedingungen
(Experience gained with the scanning electron microscopy during the evalution of case histories with corrosive conditions)
(in German)
Werkst. Korros. *26* (1975) 6, p. 461

[103] Gräfen, H.
Die Bedeutung der Schadensanalyse für Werkstoffentwicklung, Konstruktion und Fertigung im Chemieapparatebau
(The importance of case histories with respect to materials, development, design and construction of chemical plant)
(in German)
Werkst. Korros. *26* (1975) 9, p. 675

[104] Schmitz, D.
Werkstoffe für Kreiselpumpen
(Materials for centrifugal pumps)
(in German)
3R International *22* (1983) p. 276

[105] Kuzyukov, A. N.; Zajtseva, L. V.; Khanzadeev, I. V.
Effect of stress on intergranular corrosion of austenitic steels in the production of weak nitric acid (in Russian)
Zashch. Met. *18* (1982) 3, p. 413

[106] Sukhotin, A. M.; Lantratova, N. Ya.; Trubnikov, V. P.; Atroshenko, E. I.
The corrosion resistance of structural materials in N_2O_4 and coolant technology
(in Russian)
Atomnaya Energiya *36* (1974) p. 496

[107] Gorelik, G. N.; Nikonova, E. A.
Influence of contact with lead on the behavior of stainless steels and titanium
(in Russian)
Zashch. Met. 6 (1970) 4, p. 416

[108] Harrison, J. M.; Shaw, R. D.; Worthington, S. E.; Thomas, J. G. N.; Andon, R. J. L.; Pemberton, R. C.
Localised corrosion of stainless steels in nitric acid
UK Corrosion, Wembley, Middlesex (GB) (Proc. Conf.) *2* (1984) p. 180

[109] Storchaj, E. I.; Kuznetsova, M. V.; Elchinova, L. N.; Ezhov, N. V.
Corrosion Electrochemical behavior of steel 03Ch13AG19 and 07Ch13N4AG20 (in Russian)
Khim. Neft. Mashinostr. (1984) 3, p. 37

[110] Anonymous
Rapid anodization of aluminium (alloys) in aqueous solution of organic acids
Jap. Pat. 1066 243 and 1066 244 (Fuji Sash Ltd.; Dec. 5, 1974)

[111] Shapovalov, E. T.; Kazakova, G. V.; Andrushova, N. V.
An investigation of the processes of pickling and passivation in a non-contact electrochemical pickling bath (in Russian)
Stal USSR (1983) 1, p. 39

[112] Tokunaga, K.; Sakitani, K.
Thin oxide film on stainless steel in 30 % nitric acid solution (in Japanese)
Corros. Eng. (Boshoku Gijutsu) *32* (1983) 4, p. 221

[113] Herbsleb, G.; Jäkel, U.; Schwaab, P.
Einfluß von Verformung und spannungsinduzierten Ausscheidungen auf die Korrosionsbeständigkeit von siliciumlegiertem nichtrostendem Stahl X 2 CrNiSi 18 15 in Salpetersäure
(Effect of deformation and stress-induced precipitations on the corrosion resistance of silicon alloyed stainless steel X 2 CrNiSi 18 15 in nitric acid) (in German)
Werkst. Korros. *41* (1990) 4, p. 170

[114] Eremias, B.; Prazak, M.
Polarization resistance measurements of high alloyed austenitic steels with low carbon content in boiling solutions of concentrated nitric acid
Corrosion *35* (1979) 5, p. 216

[115] Maslov, V. A.; Semenova, L. A.
Evaluation of the corrosion resistance of welded joints in titanium alloys in nitric acid solutions
Weld. Prod. (USSR) (1982) 6, p. 29

[116] Maslov, V. A.; Semenova, A. A.
Corrosion resistance of the steel 03Ch18N11 and its weld joints in nitric acid solutions (in Russian)
Svar. Proizvod. (1980) 11, p. 29

[117] Williams, D. E.; Asher, J.
Measurement of low corrosion rates: Comparison of a.c. impedance and thin layer activation methods
Corros. Sci. *24* (1984) 3, p. 185

[118] Kobayashi, M.; Yoshida, T.; Aoki, M.; Ikeda, N.; Takahashi, M.
Development of nitric acid corrosion resistant duplex stainless steel NAR-SN-5
Nippon Stainless Tech. Rep. *17* (1982) 23

[119] Zilberman, B. Ya.; Kotlyar, N. Z.; Sakulin, S. V.
Effect of the temperature of the heat-transfer surface of an evaporation apparatus on the corrosion rate (in Russian)
Zashch. Met. *11* (1975) 3, p. 354

[120] Naumann, G.
Korrosionsverhalten des austenitischen Chrom-Nickel-Stahles X 5 CrNiN 19.7 in Salpetersäure bei erhöhten Temperaturen
(Corrosion behaviour of the austenitic chromium nickel steel X 5 CrNiN 19.7 in nitric acid at elevated temperatures) (in German)
Chem. Tech. *30* (1978) 2, p. 94

[121] Kachanov, V. A.; Nikitin, D. G:; Klyushnikova, L. A.; Kabashnyj, A. I.; Ponomarenko, V. I.
Corrosion of welded joints in derivatives of adipic acid (in Russian)
Zashch. Met. *17* (1981) 6, p. 739

[122] Volikova, I. G.; Shapiro, M. B.; Subbotina, L. A
Effect of ferrite on the corrosion resistance of the chromium-nickel steel Ch18N10T (in Russian)
Zashch. Met. *8* (1972) 5, p. 555

[123] Denhard jr., E. E.
Austenitic stainless steel combining strength and resistance to intergranular corrosion
US Pat. 3 645 725 (Febr. 1972)

[124] Hochörtler, G.; Horn, E.-M.
Austenitic stainless steel with approximately 5.3 % silicon
8th Internat. Congress Metallic Corrosion, Mainz, 1981
(Proc. Conf.) Vol. *2*, p. 1447

[125] Zhadan, T. A. et al.
Corrosion resistant steel alloy used for weld seams USSR Pat. 377 404 (April 17, 1973); (C.A. *79* (1973) 9559s)

[126] Kobayashi, M.; Fujiyama, S.; Araya, Y.; Wada, S.; Sunayama, Y.
Study on concentrated nitric acid-resistant stainless steel
Nippon Sutenreso Giho (1976) 12, p. 1

[127] Kabayashi, M.; Miki, M.; Ohkubo, K.
Nitric acid resistant stainless steels (in Japanese)
Bull. Jpn. Inst. Met. *22* (1983) p. 320

[128] Ozawa, R.
Grain boundary corrosion-resistant austenitic stainless steels
Jap. Pat. 78 19 915 (February 23, 1978)

[129] Box, S. M.; Wilson, F. G.
Effect of carbide morphology and composition on the intergranular corrosion of titanium stabilized austenitic stainless steels
J. Iron Steel Inst. *210* (1972) p. 71

[130] Zitter, H.
Prüfung geschweißter und ungeschweißter austenitischer Chrom-Nickel-Stähle auf interkristalline Korrosion
(Testing of welded and non-welded austenitic chromium-nickel steels for intercrystalline corrosion) (in German)
Arch. Eisenhüttenwes. *28* (1957) 7, p. 401

[131] Naumann, G.
Korrosionsverhalten des austenitischen Chrom-Nickel-Stahles X 5 CrNiN 19.7 in Salpetersäure bei erhöhten Temperaturen
(Corrosion behaviour of the austenitic chromium nickel steel X 5 CrNiN 19.7) (in German)
Chem. Tech. *30* (1978) 2, p. 94

[132] Smallwood, R. E.
Heat exchanger tubing reliability
Mater. Performance *16* (1977) 2, p. 27

[133] Lozovatskaya, L. P.; Levin, I. A.; Burtseva, I. K.; Goldshstejn, Ya. E.; Shmatko, M. N.; Piskunova, A. I.
Increasing the resistance of the steel 03Ch18N11 against ICC by means of correcting its chemical composition (in Russian)
Zashch. Met. *20* (1984) 3, p. 411

[134] Eremenko, A. S.; Shchesno, L. P.; Taraban, A. I.; Severina, L. S.
Effect of alloying with nitrogen on the properties of low-carbon stainless steels
Sov. Mater. Sci. *11* (1975) p. 672

[135] Staronka, A.; Holtzer, M.
Korrosionsbeständigkeit von CrNi-Stahlguß (Typ 18/9 bzw. 18/13) mit erhöhtem Siliciumgehalt in konzentrierten Salpetersäurelösungen
(Corrosion resistance of chromium nickel cast steel (types 18/9 and 18/13) with increased silicon content in concentrated nitric acid solutions) (in German)
Werkst. Korros. *38* (1987) p. 431

[136] Lazebnov, P. P.; Savonov, Yu. N.; Aleksandrov, A. G.
Influence of titanium nitride, cerium and vanadium on the corrosion resistance of welded chromium-nickel metals (in Russian)
Avtom. Svarka (1981) 8, p. 69

[137] Anonymous
The effect of heat treatments in the prevention of intergranular corrosion failures of AISI 321 stainless steel
Mater. Performance *22* (1983) 9, p. 22

[138] Devine, T. M.; Drummond, B. J.
Use of accelerated intergranular corrosion tests and pitting corrosion tests to detect sensitization and susceptibility to intergranular stress corrosion cracking in high temperature water of duplex 308 stainless steel
Corrosion *37* (1981) 2, p. 104

[139] Herbsleb, G.; Schüller, H.-J.; Schwaab, P.
Ausscheidungs- und Korrosionsverhalten unstabilisierter und stabilisierter 18/10 Chrom-Nickel-Stähle nach kurzzeitigem sensibilisierendem Glühen
(Precipitation and corrosion behavior of unstabilized and stabilized 18 10 CrNi steels after short term sensibilizing annealing) (in German)
Werkst. Korros. *27* (1976) p. 560

[140] Herbsleb, G.; Westerfeld, K.-J.
Der Einfluß von Stickstoff auf die korrosionschemischen Eigenschaften lösungsgeglühter und angelassener austenitischer 18/10 Chrom-Nickel- und 18/12 Chrom-Nickel-Molybdän-Stähle – II. Interkristalline Korrosion in Kupfersulfat-Schwefelsäure-Lösungen und in siedender 65 %iger Salpetersäure
(Influence of nitrogen on the corrosion behavior of solution treated and annealed austenitic 18/10 chromium-nickel and 18/12 chromium-nickel-molybdenum steels – II. Intercrystalline corrosion in copper sulfate sulfuric acid solution and in boiling 65 % nitric acid) (in German)
Werkst. Korros. *27* (1976) p. 404

[141] Kiesheyer, H.
Korrosionsverhalten von betrieblich erzeugtem chemisch beständigen Stahl X 2 CrNi 19 11 (Remanit® 4306) in Salpetersäure
(Corrosion behavior of the works-produced, chemically resistant steel X 2 CrNi 19 11 (Remanit® 4306) in nitric acid) (in German)
Thyssen Edelst. Tech. Ber. *14* (1988) 1, p. 35

[142] Chlibec, G.; Gümpel, P.; Ladwein, T.
Einsatzmöglichkeiten verschiedener Werkstoffe in salpetersäurehaltigen Medien
(Application possibilities of various materials in nitric acid-containing media) (in German)
Thyssen Edelst. Tech. Ber. *14* (1988) 1, p. 39

[143] Slate, S. C.; Maness, R. F.
Corrosion experience in nuclear waste processing at Battelle-Northwest
Mater. Performance *17* (1978) 6, p. 13

[144] Bodine jr., G. C.; Sump, C. H.
Development of wrought austenitic-ferritic stress corrosion resistant alloys with less than 40 % ferrite for pipe and tube
Mater. Performance *16* (1977) 6, p. 13

[145] Hahin, C.; Stoss, R. M.; Nelson, B. H.; Reucroft, P. J.
Effect of cold work on the corrosion resistance of nonsensitized austenitic stainless steels in nitric acid
Corrosion *32* (1976) 6, p. 229

[146] Hahin, C.
Effect of cooling rate and subsequent plastic deformation on the corrosion rate of AISI 304 in boiling nitric acid
Corrosion *38* (1982) 2, p. 116

[147] Chung, P.; Szklarska-Smialowska, S.
The effect of heat treatment on the degree of sensitization of Type 304 stainless steel
Corrosion *37* (1981) 1, p. 39

[148] Brown, M. H.
Behavior of austenitic stainless steels in evaluation tests for the detection of susceptibility to intergranular corrosion
Corrosion *30* (1974) 1, p. 1

[149] Brown, M. H.
Behavior of austenitic stainless steels in evaluation tests for the detection of susceptibility to intergranular corrosion
Corrosion *30* (1974) 1, p. 1

[150] Robinson, F. P. A.; Scurr, W. G.
The effect of boron on the corrosion resistance of austenitic stainless steels
Corrosion *33* (1977) 11, p. 408

[151] Briant, C. L.
The effects of sulfur and phosphorus on the intergranular corrosion of 304 stainless steel
Corrosion *36* (1980) 9, p. 497

[152] Casarini, G.; Colonna, C.; Songa, T.
Comportamento elettrochimico dell'acciaio inossidabile AISI 304 accoppiato con Pt, Au e grafite in soluzioni concentrate di HNO_3 (65 % in peso)
(Electrochemical behavior of the stainless steel AISI 304 in contact with Pt, Au and graphite in concentrated nitric acid (65 percent by weight)) (in Italian)
Metall. Ital. *62* (1970) p. 183

[153] Ishigami, I.; Tsunasawa, E.; Yamanaka, K.
Carburization during vacuum oil-quenching of SUS 304 and a proposal of prevention of it (in Japanese)
J. Jpn. Inst. Met. *46* (1982) 12, p. 1168

[154] Kuzyukov, A. N.
Effect of welding seam tensions on the intercrystalline corrosion of 08Ch18N10T and 03Ch18N11 steels (in Russian)
Zashch. Met. *14* (1978) 4, p. 430

[155] Kuzyukov, A. N.; Levchenko, V. A.
Influence of substructure on intergranular corrosion of 08Ch18N10T and 03Ch18N11 steels
Sov. Mater. Sci. *16* (1980) p. 329

[156] Chihal, V.; Lehká, N.; Malík, J. K.
Probleme der interkristallinen und der Messerlinienkorrosion von Schweißverbindungen der mit Niob stabilisierten korrosionsbeständigen Stähle
(Intercrystalline and knife corrosion of welded joints of chromium-nickel niobium-stabilized corrosion resisting steels) (in German)
Metalloberfläche *26* (1972) 2, p. 453

[157] Dorn, L.; Anik, S.; Günaltan, A.
Untersuchungen zur Beständigkeit widerstandspunktgeschweißter Verbindungen aus X 12 CrNi 18 8 gegenüber interkristalliner Korrosion
(Investigations relating to the intergranular corrosion resistance of resistance spot welded joints in X 12 CrNi 18 8) (in German)
Schweissen Schneiden *32* (1980) 1, p. 5

[158] Vrabely, E.
Effect of the corrosion product on the intergranular corrosion tendency of austenitic corrosion-resistant steel in nitric acid tests (in Hungarian)
Banyasz. Kohasz. Lapok Kohasz. *107* (1974) p. 507

[159] Kvärnbach, B.
Sandvik® 2R12 AISI 304 L steel for nitric acid
Chem. Age India *25* (1974) 1, p. 49

[160] Chronister, D. J.; Spence, T. C.
Influence of higher silicon levels on the corrosion resistance of modified CF-type cast stainless steels
Corrosion in Sulfuric Acid, Boston (USA), March 1985 (Proc. Conf.), p. 75
NACE, Houston, Tx., 1985

[161] Bäumel, A.
Korrosion in der Wärmeeinflußzone geschweißter chemisch beständiger Stähle und Legierungen und ihre Verhütung
(Corrosion in the heat-affected zone of welds in chemically resistant steels and alloys, and respective preventive measures) (in German)
Werkst. Korros. *26* (1975) 6, p. 433

[162] Holtzer, M.
Effect of carbon and silicon on the structure and corrosion resistance of 17Cr-8Ni cast steel in concentrated nitric acid solutions
Werkst. Korros. *41* (1990) 1, p. 25

[163] Drehsen, H.; Strassburg, F. W.
Das Schweißverhalten neuer austenitischer Chrom-Nickel-Stähle und Nickel-Chrom-Eisen-Legierungen
(The welding behavior of new austenitic chromium-nickel steels and nickel-chromium-iron alloys) (in German)
Praktiker *27* (1975) 8, p. 146

[164] Babakov, A. A.; Novokschenova, S. M.; Levin, F. L.; Zhadan, T. A.; Sharonova, T. N.
Silicon as an alloying element in 000Ch20N20 steel (in Russian)
Zashch. Met. *10* (1974) 5, p. 552

[165] Ohkubo, M.; Takekawa, T.; Miki, M.
Corrosion prevention for stainless steel
Jap. Pat. 73 47 453 (Sumitomo Chemical Co., Ltd.; July 5, 1973)

[166] Miki, M.
Studies on cathodic protection of stainless steel by contact of aluminum metal and aluminum bearing stainless steel in strong concentrated nitric acid solution (in Japanese)
Corros. Eng. (Boshoku Gijutsu) *32* (1983) 12, p. 701

[167] Vehlow, J.
Eine Methode zur kontinuierlichen und elementspezifischen Messung des korrosiven Abtrags von metallischen Werkstoffen
(A method of continuous and element specific measurement of corrosive abrasion in metallic materials) (in German)
Z. Werkstofftech. *12* (1981) 9, p. 324

[168] Horn, E.-M.; Kilian, R.; Schoeller, K.
Deckschichtbildung und Korrosionsverhalten des hochsiliciumlegierten austenitischen Sonderstahles X 2 CrNiSi 18 15 in Salpetersäuren
(The formation of surface films and the corrosion resistance of the silicon containing austenitic steel X 2 CrNiSi 18 15 in nitric acid) (in German)
Z. Werstofftech. *13* (1982) 8, p. 274

[169] Horn, E.-M.; Kügler, A.
Development, properties, processing and applications of high-silicon steel grade X 2 CrNiSi 18 15
Z. Werkstofftech. *8* (1977) p. 362

[170] Vehlow, J.
Kontinuierliche Verfolgung der Korrosion des Stahls X 2 CrNiSi 18 15 in 98 %iger Salpetersäure
(Continuous measurement of corrosion on steel X 2 CrNiSi 18 15 in 98 % nitric acid) (in German)
Z. Werkstofftech. *13* (1982) p. 286

[171] Lipodaev, V. N.; Yushenko, K. A.; Shulskij, V. Yu. et al.
Effect of stabilizers and the ferrite phase on the corrosion resistance and ductility of welded joints of austenitic steels with a high silicon content (in Russian)
Avtom. Svarka (1986) 6, p. 9

[172] Ito, N.; Kobayashi, M.; Okubo, K.; Miki, M.
Stainless steel for use in concentrated nitric acid
Jap. Pat. 77 156 122 (December 26, 1977)

[173] Zhadan, T. A.; Babakov, A. A.; Sharonova, T. N.; Vasilieva, N. M.
Inclination of steel 000Ch20N20S5 (ZI-52) to intercrystalline corrosion (in Russian)
Zashch. Met. 9 (1973) 1, p. 42

[174] Herbsleb, G.
Werkstoff- und Korrosionsprobleme in wäßrigen Lösungen bei hohen Temperaturen und Drücken
(Material and corrosion problems in aqueous solutions at high temperatures and pressures) (in German)
Werkst. Korros. *27* (1976) 3, p. 145

[175] Kumada, M.
Intergranular corrosion and fatigue strength of nonsensitized austenitic stainless steels (in Japanese)
J. Jpn. Inst. Met. *44* (1980) 5, p. 562

[176] Yudina, N. S.; Shapovalov, E. T.
Preferential dissolution of high-chrome carbides of the chomium-nickel steels (in Russian)
Zashch. Met. *18* (1982) 5, p. 686

[177] Yudina, N. S.; Shapovalov, E. T.; Bortsov, A. N.
Intercrystalline corrosion of austenitic chromium-nickel steel in a strongly oxidizing medium
Prot. Met. (U.S.S.R.) *20* (1984) 1, p. 58

[178] Kasparova, O. V.; Bogolyubski, S. D.; Kolotyrkin, Ya. M.; Milman, V. M.; Lubnin, E. N.; Shapovalov, E. T.; Yudina, N. S.
Role of silicon in the intergranular corrosion of phosphorus steel Ch20N20 (in Russian)
Zashch. Met. *18* (1982) 3, p. 336

[179] Lozovatskaya, L. P.
Effect of silicon on the stability of grain boundaries in chromium-nickel steels in strong oxidizing media (in Russian)
Zashch. Met. *19* (1983) 6, p. 923

[180] Kolotyrkin, Ya. M.
The development of the corrosion theory its successes and tasks (in Russian)
Zashch. Met. *16* (1980) 6, p. 660

[181] Kasparova, O. V.; Bogolyubski, S. D.; Kolotyrkin, Ya. M.; Milman, V. M.
Enhancing the stability of austenitic stainless steels to intergranular corrosion in strongly oxidizing media by regulating the composition of impurities (in Russian)
Zashch. Met. *20* (1984) 6, p. 844

[182] Kasparova, O. V.; Bogolyubskij, S. D.; Kolotyrkin, Ya. M.; Ulyanin, E. A.; Vasyukov, A. B.; Yudina, N. S.; Kostromina, S. V.
Investigation of the effect of additions of phosphorus, sulphur and molybdenum on the corrosion-electrochemical behavior of the steel Ch20N20 in acid media in the over-passivation area (in Russian)
Zashch. Met. *15* (1979) 2, p. 147

[183] Kasparova, O. V.; Bogolyubski, S. D.; Kolotyrkin, Ya. M.; Milman, V. M.; Mekhryusheva, L. I.; Smakhtin, L. A.; Yudina, N. S.
Effect of carbon and phosphorus on the intergranular corrosion of tempered steel Ch20N20 in nitric acid solutions (in Russian)
Zashch. Met. *18* (1982) 1, p. 18

[184] Yamamoto, K.; Koumi, N.
Corrosion of 25Cr-20Ni-Nb used in nuclear plant handling nitric acid
Internat. Conf. on World Nuclear Energy Accomplishments and Perspectives, Illinois (USA), 1980 (Proc. Conf.), p. 231

[185] Arlt, N.; Gümpel, P.; Michel, E.
Korrosionsverhalten von X 2 CrNi 19 11 (Remanit® 4306) in Salpetersäure mit Zusatz von sechswertigem Chrom (Corrosion behavior of X 2 CrNi 19 11 (Remanit® 4306) in nitric acid with the addition of hexavalent chromium) (in German)
Thyssen Edelst. Tech. Ber. *14* (1988) 1, p. 26

[186] Kraft, R.; Leistikow, S.; Pott. E.
Untersuchungen zur Korrosion des austenitischen 25Cr20Ni-Stahls UHB 25 L in Salpetersäure und salpetersauren Lösungen oxidierender Metallionen (Investigation of the corrosion of the austenitic steel 25Cr20Ni UHB 25 L in nitric acid and its solutions containing metal ions) (in German)
Kernforschungszentrum Karlsruhe, ISSN 0303-4003, 1983, 40 p.

[187] Sinigaglia, D.; Re, G.
Intergranular corrosion (in Italian)
Metall. Ital. *67* (1975) 11, p. 660

[188] Kowaka, M.; Yamanaka, K.
The effect of alloying elements and testing methods on intergranular fracture in chloride stress corrosion cracking of austenitic stainless steels (in Japanese)
Corros. Eng. (Boshoku Gijutsu) *32* (1983) 9, p. 449

[189] Staronka, A.; Holtzer, M.
Korrosionsbeständigkeit von CrNi-Stahlguß (Typ 18/9 bzw. 18/13) mit erhöhtem Siliciumgehalt in konzentrierten Salpetersäurelösungen
(Corrosion resistance of chromium nickel cast steel (types 18/9 and 18/13) with increased silicon content in concentrated nitric acid solutions) (in German)
Werkst. Korros. *38* (1987) 8, p. 431

[190] Volikova, I. G.; Petrova, T. V.; Lebedev, B. V.
Quick electrochemical method for determining the inclination of low-carbon chromium-nickel steels to intercrystalline corrosion (in Russian)
Zashch. Met. *15* (1979) 4, p. 502

[191] Chernova, G. P.; Rutten, M. Ya; Tomashov, N. D.
Accelerated electrochemical method for determining the susceptibility of stainless steels to intercrystalline corrosion (in Russian)
Korroz. Zashch. (1979) 8, p. 3

[192] Tomashov, N. D.; Chernova, G. P.; Rutten, M. Ya.; Radetskaya, G. K.; Yakovleva, L. F.
Testing the tendency of stainless steels to undergo intercrystalline corrosion when exposed to drops of electrolyte (in Russian)
Korroz. Zashch. *12* (1979) 10, p. 3

[193] Mirolyubov, E. N.; Kazakov, A. P.; Kurtepov, M. M.
The influence of chlorides on the corrosion resistance of stainless steels in nitric acid in: Corrosion of Metals and Alloys (Izdatel'stvo Metallurgiya), Ed. Tomashow, N. D.; Miralyubov, E. N., Part 2 (1965) p. 101

[194] Kuzyukov, A. N.; Zaytseva, L. V.; Khanzadeev, I. V.
Effect of stress on intergranular corrosion of austenite steels in the production of weak nitric acid (in Russian)
Zashch. Met. *18* (1982) 3, p. 413

[195] Glukhova, A. I.
Effect of temperature and the nature of corrosive destruction on stainless steel in acid solutions (in Russian)
Zh. Prikl. Khim. *33* (1960) 8, p. 1853

[196] Gupta, S.; Vajpeyi, Dhirendra; Pandey, G. N.
Determination of corrosion of stainless steel (AISI 304) by mixed vapours of HCl and HNO_3
Corros. Prev. Control *33* (1986) 4, p. 47

[197] Anonymous
Lexikon der Korrosion, Abschnitt "Salpetersäure" (Corrosion lexicon, section "Nitric acid") (in German)
Mannesmann-Werke *2* (1970) p. 150

[198] Horn, E.; Schoeller, K.
Korrosion nichtrostender austenitischer Stähle in (kondensierender) chloridhaltiger Salpetersäure
(Corrosion of austenitic stainless steels in (condensing) nitric acid containing chlorides) (in German)
Werkst. Korros. *42* (1991) 11, p. 569

[199] Razygraev, V. P.; Lebedeva, M. V.; Egorikhin, E. V.; Ponomareva, E. Yu.; Lobanova, L. P.
Safe ion concentration of Ch18N10 type steel in boiling solutions of dilute nitric acid (in Russian)
Zashch. Met. *16* (1980) 5, p. 586

[200] Lang, G. P.; Weidman, S. W.; Armstrong, W. P.; Martin, G. L.; Bradford, W. G.
Removal of chloride from nitric acid by oxidation with permanganate
U.S. At. Energy Comm. MCW 1436 (1958) 28

[201] Razygraev, V. P.; Yegorikhin, E. V.
Self-passivation of Ch18N10 steel in boiling, dilute nitrogen chloride solutions (in Russian)
Zashch. Met. *15* (1979) 1, p. 67

[202] Razygrayev, V. P.; Lobanova, L. P.; Lebedeva, M. V.
Electrochemical behavior of carbide inclusions during the corrosion of Ch18N10 type steel in boiling nitro-fluoride solutions (in Russian)
Zashch. Met. *16* (1980) 6, p. 728

[203] Covino jr., B. S.; Scalera, J. V.; Driscoll, T. J.; Carter, J. P.
Dissolution behavior of 304 stainless steel in HNO_3/HF mixtures
Metall. Trans A *17A* (1986) Jan., p. 137

[204] Diekmann, H.; Gräfen, H.; Holm, R.; Horn, E.-M.; Storp, S.
ESCA – Untersuchungen zum Aufbau von Oxidschichten auf austenitischen Stählen nach Salpetersäurebeanspruchung
(ESCA – Investigations of the passive films formed on austenitic stainless steels in nitric acid) (in German)
Z. Werkstofftech. 9 (1978) 2, p. 37

[205] Konstantinova, E. V.; Muravyev, L. L.
Time factor in corrosion research (in Russian)
Zashch. Met. *12* (1976) 5, p. 599

[206] Kakhovskij, N. I.; Lipodaev, V. N.; Deryugin, B. G.
Corrosion resistance of high-alloyed steels and their weld joints (in Russian)
Avtom. Svarka (1980) 2, p. 72

[207] Ondrejcin, R. S.; McLaughlin, B. D.
Corrosion of high NiCr-alloys and type 304 L stainless steel in HNO_3-HF
Du Pont de Nemours & Co., DP-1550, April 1980, 46

[208] Henthorne, M.
Corrosion testing of weldments
Corrosion *30* (1974) 2, p. 39

[209] Low, A. J.
Plutonium recovery from scrap
Chem. Eng. *74* (1967) 11, p. 132

[210] Mah, R.; Terada, K.; Cash, D. L.
Corrosion of distillation equipment by HNO_3-HF solutions during HNO_3 recovery
Mater. Performance *14* (1975) 11, p. 28

[211] Konstantinova, E. V.; Mikhlynina, T. A.; Feldgandler, E. G.
Comparative investigation of corrosion of the sheet and bar rolled steels 12Ch18N10T, 08Ch18N10T and 03Ch23N6 in boiling nitric acid (in Russian)
Zashch. Met. *20* (1984) 4, p. 622

[212] Razygraev, V. P.; Lebedeva, M. V.
Influence of some secondary reactions on the redox-potential and corrosion process in nitric acid solutions (in Russian)
Zashch. Met. *19* (1983) 5, p. 771

[213] Lebedev, A. N.; Blinova, V. A.
The use of low nickel content stainless steel in nitrofluorine solutions (in Russian)
Zashch. Met. *13* (1977) 6, p. 710

[214] Kopeliovich, D. Kh.; Saprykina, N. P.; Glagolenko, Yu. V.
Corrosion of certain materials in nitro-fluoride media containing hydrazine (in Russian)
Zashch. Met. *18* (1982) 2, p. 232

[215] Kopeliovich, D. H.; Anashkin, R. D.; Ivlev, V. I.
Inhibition of intercrystalline corrosion on the steel 12Ch18N10T by fluorides in nitric acid solutions containing other oxidizers (in Russian)
Zashch. Met. *14* (1978) 4, p. 457

[216] Zamiryakin, L. K.; Chikunov, V. K.
Corrosion stability of welded joints of austenitic steels in nitric acid solutions (in Russian)
Zashch. Met. 5 (1969) 4, p. 422

[217] Razygraev, V. P.; Lebedeva, M. V.
Inhibition of corrosion and cathodic reactions on stainless steels in nitric acid media by fluoride ions (in Russian)
Zashch. Met. *18* (1980) 2, p. 227

[218] Lukin, B. V.; Kurtepov, M. M.; Kazarin, V. I.
Electrochemical corrosion behavior of stainless steel in nitrogen fluoride solutions (in Russian)
Zashch. Met. *15* (1979) 4, p. 445

[219] Edström, J. O.; Carlén, J. C.; Kampinge, S.
Nouveaux máteriaux de construction pour le génie chimique
(New construction materials for chemical engineering) (in French)
La Métallurgie *101* (1969) 8/9, p. 391

[220] Risch, K.; Althen, W.
Gezieltes Beizen von Apparaten aus chemisch beständigen Stählen
(Controlled pickling of apparatus made of chemical-resistant steels) (in German)
Z. Werkstofftech. *12* (1981) 1, p. 23

[221] Vasileva, V. A.; Dzutsev, V. T.; Moskvichev, I. F.; Ulyanin, E. A.; Feldgandler, E. G.; Lobova, M. V.
Nitric acid effect on steels corrosion resistance in diluted sulphuric-nitric acid mixtures (in Russian)
Khim. Prom. (1982) p. 165

[222] Fajngold, L. L.; Shatova, T. Yu.
The corrosion and electrochemical behavior of stainless steels in sulfuric acid (in Russian)
Zashch. Met. *17* (1981) 3, p. 312

[223] Vajpeyi, M.; Gupta, S.; Dhirendra; Pandey, G. N.
Corrosion of stainless steel (AISI 304) in H_2SO_4 contaminated with HCl and HNO_3
Corros. Prev. Control *32* (1985) 5, p. 102

[224] Glejzer, M. M.; Tsejtlin, Kh. L.; Sorokin, Yu. N.; Isaenko, G. I.; Babitskaya, S. M.
The effect of organic and inorganic oxidizers on the corrosion of the steel 12Ch18N10T and VT1-0 titanium in sulfuric acid (in Russian)
Zashch. Met. *13* (1977) 6, p. 684

[225] Chak, R. O.; Turkovskaya, A. V.
Effect of redox potential and contact with precious metals on behavior of the steel Ch18N10T in mixtures of sulfuric and nitric acids (in Russian)
Zashch. Met. *9* (1973) 6, p. 705

[226] Sorokin, Yu. I.; Zeitlin, Kh. L.; Gleizer, M. M.; Potapova, K. V.; Isayenko, G. I.
Effect of certain salts on the corrosion of metals in diluted sulfuric acid (in Russian)
Zashch. Met. *20* (1984) 5, p. 780

[227] Sidorenko, N. R.; Chekhovskii, A. V.; Prokhorov, V. M.; Kuznetsov, V. N.; Makartsev, V. V.
Study of corrosion resistance of structural materials in sulfur-nitrogen mixtures with high free sulfur trioxide content (in Russian)
Khim. Mashinostr. *12* (1980) p. 160

[228] Oknin, I. V.; Gabrielyan, S. G.
Corrosion of the steel Ch18N10T in a solution of nitric acid with sulphite additions (in Russian)
Zashch. Met. *6* (1970) 3, p. 274

[229] Shiganova, L. N.; Pakhomova, N. M.; Obnosova, V. A.
Corrosion resistance of some steels in superphosphoric acid (in Russian)
Khim. Prom., Ser.: Azotn. Prom-st (1979) 5, p. 68

[230] Merendi, Yu. Ya.; Metsik, R. E.
Behavior of stainless steels in models of the production process of saturated dicarbonic acids by the destructive oxidation of nitric acid kerogene (in Russian)
Zashch. Met. *19* (1983) 3, p. 444

[231] Clark, A. T.
Processing of power reactor fuels
AEC Report DP-740 (1962) Jan.–April

[232] Maness, R. F.
Effects of Cr(VI) and Fe(III) on purex plant corrosion
AEC Report HW-72076 (1962) Jan.

[233] Kurtepov, M. M.
Effect of uranium on the corrosion of stainless steel in nitric fluorine solutions (in Russian)
Zashch. Met. *21* (1985) 1, p. 104

[234] Votinov, S. N.; Kazennov, Yu. I.; Bogoyavlenskij, V. L.; Belokopytov, V. S.; Krylov, E. A.; Klestova, L. M.; Reviznikov, L. I.
The effect of reactor irradiation on the tendency of austenitic steels to intercrystalline corrosion (in Russian)
Atomnaya Energiya *41* (1976) 6, p. 405

[235] Kraft, R.; Leistikow, S.; Pott. E.
Untersuchungen zur Korrosion des austenitischen 25Cr20Ni-Stahls UHB 25 L in Salpetersäure und salpetersauren Lösungen oxidierender Metallionen (Investigation of the corrosion of the austenitic steel 25Cr20Ni UHB 25 L in nitric acid and its solutions containing metal ions) (in German)
Kernforschungszentrum Karlsruhe, ISSN 0303-4003, 1983, 40

[236] Babchuk, G. M. et al.
Selection of a material for equipment in the production of chemically pure silver nitrate for use in the ciné industry (in Russian)
Vses. Nauchno-Issled. Inst. Khim. Reakt. Osob. Chist. Khim. Veshchestv. *44* (1978) p. 166

[237] Zhuravlev, V. K.; Kurtepov, M. M.; Bochkareva, E. F.
Effect of cerium and silver ions on the corrosion and electrochemical behavior of stainless steel in ozonized nitric acid solutions (in Russian)
Zashch. Met. *10* (1975) 3, p. 294

[238] Kilman, Ya. I.; Zaichko, N. D.
Structural materials selection in ammonium nitrate production by single-stage non-steaming method and production safety questions (in Russian)
Khim. Prom. (1983) 1, p. 28

[239] Pavlova, F. S.; Kuznetsova, V. N.; Gerasimov, V. V.
Perfection of the cleaning technology of pipe lines made from the steel 0Ch18N10T (in Russian)
Zashch. Met. *5* (1969) 4, p. 451

[240] Kopeliovich, D. Kh.; Ivlev, V. I.
Corrosion of structural materials in nitrate solutions of thiourea (in Russian)
Zashch. Met. *22* (1986) p. 121

[241] Gulyaev, V. M.; Chulkova, V. M.
Effect of phosphorus on the corrosion properties of stainless steel (in Russian)
Zashch. Met. *12* (1976) 3, p. 287

[242] Razygraev, V. P.; Ponomareva, E. Yu.
Behavior of the steel 12Ch18N10T in nitric acid solutions containing fluoride during heat emission (in Russian)
Zashch. Met. *14* (1978) 4, p. 461

[243] Azzerri, N.; Tamba, A.
Potentiostatic pickling: a new technique for improving stainless steel processing
J. Appl. Electrochem. 6 (1976) p. 347

[244] Sydberger, T.
Influence of the surface state on the initiation of crevice corrosion on stainless steels
Werkst. Korros. *32* (1981) 3, p. 119

[245] Apparao, B. V.
Pickling treatment of hot rolled stainless steel (304) problem of pitting corrosion in local industry
J. Electrochem. Soc. India *30* (1981) 3, p. 213

[246] Kudryavtseva, E. F.; Dolinkin, V. N.
Chemical methods for removing high-temperature oxide films (in Russian)
Zashch. Met. *13* (1977) 5, p. 567

[247] Shuvalov, V. A.; Emel'yantseva, Z. I.; Gerasimov, V. V.
Relationship of the character of distribution of dislocations in austenitic steels to corrosion under stress
Sov. Mater. Sci. *12* (1976) 3, p. 313

[248] Schmidt, M.; Kowall, D.; Schöler, W.
Mittel zum chemischen Entzundern, Beizen und Glänzen von Edelstahl (Agent used for chemical descaling, pickling and shining) (in German)
Galvanotechnik *72* (1981) p. 664

[249] Loria, E. A.; Isaacs, H. S.
Type 304 stainless steel with 0.5 % boron for storage of spent nuclear fuel
J. Met. *32* (1980) 12; p. 10

[250] Devyatkina, T. S.; Linkin, Ya. N.; Korosteleva, T. K.
Effect of the circulation of nitric acid solution on the etching kinetics of stainless steels (in Russian)
Zashch. Met. *14* (1978) 1, p. 92

[251] Konda, K.
Descaling of stainless steels
Japan Pat. 77 44 743 (April 8, 1977)

[252] Anonymous
Formteilätzen
(Etching of molded parts) (in German)
Oberfläche Surf. *19* (1978) p. 36

[253] Lipkin, Ya. N.; Bershadskaya, T. M.; Albrut, V. M.; Pugacheva, L. V.; Chepurova, L. G.
Special features of chemical polishing of steels (in Russian)
Zashch. Met. *20* (1984) 6, p. 908

[254] Killing, R.
Beseitigung der Anlauffarben beim Schweißen nichtrostender Stähle (Removal of annealing colours from the welding of stainless steels) (in German)
Praktiker *24* (1972) 11, p. 224

[255] Searson, P. C.; Latanision, R. M.
A comparison of the general and localized corrosion resistance of conventional and rapidly solidified AISI 303 stainless steel
Corrosion *42* (1986) 3, p. 161

[256] Steinhäuser, S.; Graubner, E.; Goldschmidt, V.
Oberflächenbehandlung im Schweißnahtbereich nichtrostender und säurebeständiger Stähle mit Beizpasten (Surface treatment in the weld seam region of stainless and acid-resistant steels with pickling pastes) (in German)
Schweisstechnik (Berlin) *20* (1970) 6, p. 266

[257] Buttinelli, D.; Capotorto, C.; Memmi, M.
Corrosion behavior of austenitic chromium-nickel steels in boiling nitric acid (Huey test). III. Effect of polishing the surface on the corrosion rate (in Italian)
Ann. Chimica *63* (1973) p. 181

[258] Devyatkina, T. S.; Lipkin, Ya. N.; Korosteleva, T. K.
Effect of circulation of nitric acid solution on the kinetics of etching of stainless steels (in Russian)
Zashch. Met. *14* (1978) 1, p. 92

[259] Cojocaru, V.
Inhibitors for fighting corrosion during removal of deposits by chemical treatment of steels (in Rumanian)
Revista de Chimie *25* (1974) 11, p. 908

[260] Gupta, S.; Kumar, Y.; Sanyal, B.; Pandey, G. N.
The exothermic reaction of stainless steel (AISI 304) in mixtures of HCl and HNO_3 and the inhibition of its corrosion
Corros. Prev. Control *30* (1983) 1, p. 11

[261] Gupta, S.; Vajpeyi, M.; Dhirendra, N. I.; Pandey, G. N.
Determination of corrosion of stainless steel (AISI 304) by mixed vapours of HCl and HNO_3
Corros. Prev. Control *33* (1986) 2, p. 47

[262] Niendorf, K.; Peuker, R.
Zum Einfluß des Nitrosegehaltes konzentrierter Schwefelsäure auf das Korrosionsverhalten austenitischer Chrom-Nickel-Stähle bei hohen Temperaturen (The influence of the nitrogen content of sulfuric acid on the corrosion behavior of austenitic chromium-nickel steels at elevated temperatures) (in German)
Chem. Tech. *28* (1976) 4, p. 236

[263] Kurtepov, M. M.
Effect of uranium on the corrosion of stainless steel in nitric fluoride solutions (in Russian)
Zashch. Met. *21* (1985) 1, p. 104

[264] Babakov, A. A.; Posysaeva, L. I.; Petrovskaya, V. A.; Sidorkina, Yu. S.
New, high-alloyed anti-corrosion steel 000Ch21N21M4B (in Russian)
Zashch. Met. *7* (1972) 2, p. 99

[265] Süry, P.
Korrosionseigenschaften gegossener Chrom-Nickel- und Chrom-Nickel-Molybdänstähle in aggressiven Medien (Corrosion properties of cast chromium-nickel and chromium-nickel-molybdenum steels in aggressive media) (in German)
Chemie-Technik *4* (1975) 7, p. 241

[266] Tomashov, N. D.; Chernova, G. P.; Rutten, M. Ya.
Determination of the susceptibility of stainless steels to intercrystalline corrosion (in Russian)
Korroz. Zashch. (1978) 4, p. 3

[267] Nayar, H. S.; German, R. M.; Johnson, W. R.
Effect of sintering on the corrosion resistance of 316L stainless steel
Ind. Heat. *48* (1981) 12, p. 23

[268] Ogneva, V. K.; Kiselev, V. D.; Chernova, T. V.
Stainless steels corrosion resistance in media of nitric acid decomposition of apatite (in Russian)
Khim. Prom. (1980) p. 480

[269] Barbosa, M. A.
The pitting resistance of AISI 316 stainless steel passivated in diluted nitric acid
Corros. Sci. *23* (1983) 12, p. 1293

[270] Vogg, H.; Braun, H.; Löffel, R.; Lubecki, A.; Merz, A.; Schmitz, J.; Schneider, J.; Vehlow, J.
Anwendung der Radionuklidtechnik in Chemie und Verfahrenstechnik
(Use of radionuclide techniques in chemistry and chemical engineering) (in German)
J. Radioanal. Chem. *32* (1976) p. 495

[271] Anonymous
FMN duplex steel for chemical process plant
Anticorros. Methods Mater. *26* (1979) 10, p. 10

[272] Lakey, L. T.; Kerr, W. B.
Pilot plant development of an electrolytic dissolver for stainless steel alloy nuclear fuels
Ind. Eng. Chem. Proc. Des. Develop. 6 (1967) 2, p. 174

[273] Edström, J. O.; Carlén, J. C.; Kämpinge, S.
Anforderungen an Stähle für die chemische Industrie
(Properties required for steels in the chemical industry) (in German)
Werkst. Korros. *21* (1970) 10, p. 812

[274] Süry, P.
Untersuchungen zum Einfluss der Kaltverformung auf die Korrosionseigenschaften des rostfreien Stahles X 2 CrNiMo 18 12
(Studies on the effect of cold deformation on corrosion properties of the stainless steel X 2 CrNiMo 18 12) (in German)
Material u. Technik *8* (1980) 4; p. 163

[275] Süry, P.; Brezina, P.
Kurzzeitprüfmethoden zur Untersuchung des Einflusses von Wärmebehandlung und chemischer Zusammensetzung auf die Korrosionsbeständigkeit martensitischer Chrom-Nickel-(Molybdän)-Stähle mit tiefem Kohlenstoffgehalt
(Short-duration test for evaluating the influence of heat-treatment and chemical composition on the corrosion resistance of low carbon martensitic chromium nickel (molybdenum) steels) (in German)
Werkst. Korros. *30* (1979) 5; p. 341

[276] Shapiro, M. B.; Gorlenko, A. P.; Adugina, N. A.; Shadrukhina, E. I.
A new higher strength corrosion-resistant steel with nitrogen (in Russian)
Khim. Neft. Mashinostr. (1977) 9, p. 26

[277] Charbonnier, J.-C.
Influence du molybdène sur la résistance des aciers inoxydables à différents types de corrosion, en présence de solutions aqueuses minérales ou de milieux organiques (étude bibliographique)
(Effect of molybdenum on the resistance of stainless steels to different types of corrosion in aqueous mineral solutions or organic media. Bibliographic study) (in French)
Métaux-Corros.-Ind. (1975) Nr. 598, p. 201

[278] Leymonie, C.
Étude de l'influence de la structure sur la résistance à la corrosion des soudures d'acier 18-10 au Mo à bas carbone (316 L)
(Study of the influence of the structure of welds of the steel 18-10 (316 L) with Mo and low in carbon on the corrosion resistance) (in French)
Rev. Metall. (1971) p. 289

[279] Scarberry, R. C.; Graver, D. L.; Stephens, C. D.
Alloying for corrosion control. Properties and benefits of alloy materials
Mater. Protection *6* (1967) 6, p. 54

[280] Weingerl, H.; Straube, H.; Blöch, R.
Über die Auswirkung von Seigerungen auf das Korrosionsverhalten austenitischer Cr-Ni-Mo-Stähle
(Contribution to the effect of segregation on the corrosion resistance of austenitic Cr-Ni-Mo-steels) (in German)
Werkst. Korros. *27* (1976) 2, p. 69

[281] Kohl, H.; Hochörtler, G.; Kriszt, K.; Koren, M.
Sonderstähle für den Apparatebau mit sehr guter Korrosionsbeständigkeit und erhöhter Festigkeit
(Special steels with superior corrosion resistance and strength for chemical equipment manufacture) (in German)
Werkst. Korros. *34* (1983) 1, p. 1

[282] Plaskeev, A. V.; Knyazheva, V. M.; Dergach, T. A.; Dembrovskij, M. A.
The corrosion behavior of CrNiMo steels in nitric acid (in Russian)
Zashch. Met. *14* (1978) 4, p. 393

[283] Anonymous
New stainless tubing for salt water service
Mater. Eng. *84* (1976) 3, p. 20

[284] Knyazheva, V. M.; Chigal, V.; Kolotyrkin, Ya. M.
Role of surplus phases in the corrosion resistance of stainless steels (in Russian)
Zashch. Met. *11* (1975) 5, p. 531

[285] Vyklick, M.; Brenner, O.; Hamouz, E.; Podhradsk, M.
Langzeit-Korrosionsuntersuchungen an austenitischen Stählen im Hochdruckteil einer Harnstoffanlage
(Longterm corrosion field tests with austenitic steels in the high pressure parts of a urea plant) (in German)
Werkst. Korros. *32* (1981) 2, p. 85

[286] Feldgandler, E. G.; Kareva, E. N.; Brusentsova, V. M.
Corrosion resistance of low-carbon stainless steels under urea synthesis conditions (in Russian)
Zashch. Met. *6* (1970) 5, p. 540

[287] Wehner, H.; Speckhardt, H.
Zum Ausscheidungs- und Korrosionsverhalten eines ferritisch-austenitischen Chrom-Nickel-Molybdän-Stahles nach kurzzeitiger Glühbehandlung unter besonderer Berücksichtigung des Schweißens
(Structural changes and corrosion behaviour of a ferritic-austenitic chromium-nickel-molybdenum steel after short time heat treatments with special consideration of welding) (in German)
Z. Werkstofftech. *10* (1979) 9, p. 317

[288] Rondelli, G.; Vicentini, B.; Sinigaglia, D.
Investigation into precipitation phenomena following heat treatment of ELI ferritic stainless steels and their influence on mechanical and corrosion properties
Microstructural Sci. *12* (1985) p. 73

[289] Herbsleb, G.; Schwaab, P.
Precipitation of intermetallic compounds, nitrides and carbides in AF 22 duplex steel and their influence on corrosion behavior in acids
Mannesmann Forschungsber. (1983) No. 957, 26 p

[290] Devine, T. M.; Briant, C. L.; Drummond, B. J.
Mechanism of intergranular corrosion of 316 L stainless steel in oxidizing acids
Scr. Metall. *14* (1980) p. 1175

[291] Hooper, R. A. E.; Honess, C. V.
Corrosion of type 316 L in nitric acid
Stainless Steel, London, Sept. 1977 (Proc. Conf.)
Climax Molybdenum Co., New York (1978), p. 247

[292] Stanz, A.; Schäfer, K.
Einfluß einer Wärmenachbehandlung auf die mechanischen Eigenschaften und das Korrosionsverhalten nichtrostender Stähle
(Influence of an ultimate heat treatment on the mechanical properties and the corrosion behavior of stainless steels) (in German)
Werkst. Korros. *27* (1976) p. 701

[293] Minick, G. A.; Wilfley, A. R.; Olson, D. L.
Effect of columbium additions in austenitic stainless steel castings to inhibit intergranular corrosion
Mater. Performance *14* (1975) 9, p. 41

[294] Böhm, K.; Frohberg, M. G.
Korrosionsverhalten austenitischer Stähle bei der Prüfung nach Huey in Abhängigkeit vom Verformungsgrad
(Corrosion behaviour of austenitic steels in the Huey test as a function of the degree of deformation) (in German)
Arch. Eisenhüttenwes. *44* (1973) 7, p. 553

[295] Klodt, D. T.; Minick, G. A.
Acid pump impeller corrosion
Mater. Protection and Performance *12* (1973) 6, p. 28

[296] Grundmann, R.; Gümpel, P.; Michel, E.
Betrachtungen über die Einsatzmöglichkeiten eines ferritisch-austenitischen und eines hochfesten austenitischen Stahles im Chemikalientankerbau
(Considerations regarding the application possibilities of a ferritic austenitic, high-strength steel in chemical tanker construction) (in German)
Thyssen Edelst. Tech. Ber. *14* (1988) 1, p. 49

[297] Herrnstein, W. H.; Cangi, J. W.; Fontana, M. G.
Effect of carbon pickup on the serviceability of stainless steel alloy castings
Mater. Performance *14* (1975) 10, p. 21

[298] Holm. R.; Horn, E.-M.; Storp, S.
Oxidische Deckschichten auf dem austenitisch-ferritischen Gußwerkstoff G-X 3 CrNiMoCuN 24 6
(Oxide surface layers on the austenitic-ferritic cast iron material G-X 3 CrNiMoCuN 24 6) (in German)
VDI Z. *124* (1982) 23/24, p. 917

[299] Hummel, O. H.
Werkstoffe im Behälter- und Apparatebau
(Materials in container and apparatus construction) (in German)
Chem. Tech. *11* (1982) 8, p. 930

[300] Sidorkina, Yu. S.; Larionova, R. F.; Bekoeva, G. P.; Khakhlova, N. V.
Corrosion resistance of high-alloyed steels and alloys based on iron and nickel in reaction media used in fertilizer production (in Russian)
Tr. Vses. Nauchno-Issled. Konstrukt. Inst. Khim. Mashinostr. (1977) No. 78, p. 61

[301] Anonymous
Korrosionsbeständigkeit von Superferrit (Remanit® 4575) – 6. Korrosionsbeständigkeit
(Corrosion resistance of superferrite (Remanit® 4575) – 6. Corrosion resistance) (in German)
Thyssen Edelst. Tech. Ber. *5* (1979) 1, p. 48

[302] Anonymous
Korrosionsbeständigkeit von Superferrit (Remanit® 4575). 6. Korrosionsbeständigkeit
(Corrosion resistance of superferrit (Remanit® 4575). 6. Corrosion resistance) (in German)
Thyssen Edelst. Tech. Ber. *5* (1979) 1, p. 48

[303] Bershadskaya, T. M.; Makarov, V. A; Lipkin, Ya. N.; Artamonova, N. M.; Albrut, V. M.
Analysis of corrosion conditions in the operation of pickling baths made of EI-943 alloy (in Russian)
Zashch. Met. *17* (1981) 5, p. 565

[304] Bäumel, A.; Horn, E. M.; Siebers, G.
Entwicklung, Verarbeitung und Einsatz des stickstofflegierten, hochmolybdänhaltigen Stahles X 3 CrNiMoN 17 13 5
(Development, processing and use of nitrogen alloyed high molybdenum steel X 3 CrNiMoN 17 13 5) (in German)
Werkst. Korros. *23* (1972) 11, p. 973

[305] Sidorkina, Yu. S.; Zhadan, T. A.; Khakhlova, N. V.; Shapiro, M. B.; Shetkov, A. Ya.; Larinova, R. F.
Effect of carbon and of stabilizing elements on the corrosion stability of the ChN28MDT alloy (in Russian)
Zashch. Met. *18* (1982) 1, p. 71

[306] Konstantinova, E. V.; Chechetina, N.A.; Lomovtsev, V. I.
Construction materials for hot nitrofluoride solutions (in Russian)
Zashch. Met. *16* (1980) 1, p. 66

[307] Storp. S.; Holm, R.
ESCA investigation of the oxide layers on some Cr containing alloys
Surf. Sci. *68* (1977) p. 10

[308] Asami, K.; Hashimoto, K.
An X-ray photo-electron spectroscopic study of surface treatments of stainless steels
Corros. Sci. *19* (1979) p. 1017

[309] Sidorkina, Yu. S.; Khakhlova, N. V.; Bekoyeva, G. P.; Sergeyeva, G. V.
Corrosion of the steels 03Ch21N21M4B and 06ChN28MDT in nitric acid containing fluoride ions (in Russian)
Zashch. Met. *13* (1977) 1, p. 39

[310] Arslanov, V. V.; Gevorkyan, O. M.; Ogarev, V. A.
Strength and stability of adhesion compounds formed between anodized magnesium and polymers
Colloid J. of the USA, New York; (Kolloidny Zh. *43* (1981) 5, p. 952)

[311] Zamiryakin, L. K.; Chikunov, V. K.
Corrosion stability of welded joints made of austenitic steels in nitric acid solutions (in Russian)
Zashch. Met. *5* (1969) 4, p. 422

[312] Liening, E. L.
Practical applications of electrochemical techniques to plant localized corrosion problems
Mater. Performance *19* (1980) 2, p. 35

[313] Dhirendra, Pandey, G. N.; Sanyal; B.
Comparative corrosion of unalloyed and alloyed steels in single and mixed mineral acids
Corros. Prev. Control *28* (1981) 1, p. 19

[314] Poluboyartseva, L. A.; Rejfer, A. A.
Steel and alloy corrosion resistance in production sulfuric acid (in Russian)
Khim. Prom. (1980) p. 481

[315] Werner, H.; Erkel, K.-P.
Der Einfluß der Oberflächenvorbehandlung auf die Lochfraßkorrosionsbeständigkeit des austenitischen Implantatstahles UR X 2 CrNiMoN 18.12(0)
(The effect of surface treatment on the pitting corrosion resistance of the austenitic implant steel UR X 2 CrNiMoN 18 12(0)) (in German)
Korrosion *16* (1985) 3, p. 131

[316] Shulgat, G. A.; Pakhomov, V. S.; Klinov, I. Ya.
Local fracture of steel 12Ch18N10T in nitric acid containing chlorides (in Russian)
Khim. Neft. Mash. (1976) 9, p. 21

[317] Buchner, K. H.
Versuch zur Optimierung von Werkstoffeigenschaften durch gezielte Variation der Legierungszusätze und Anwendung der Multiregressionsanalyse
(An attempt to optimize material by deliberate variation of alloying additions and application of multi-regression analysis) (in German)
Z. Metallkd. *65* (1974) 11, p. 667

[318] Ponomareva, O. V.; Khramushina, M. I.; Etlis, V. S.
Corrosion resistance of steel and alloys and the effect of some metal cations on the synthesis of 3-chloro-2-hydroxypropanoic acid (in Russian)
Khim. Prom. (1983) 10, p. 600

[319] Ogneva, V. K.; Sidlin, Z. A.; Stroev, V. S.; Dobrolyubov, V. V.; Chernova, T. V.
Economically alloyed corrosion-resisting steels used in Nitrophoska® production (in Russian)
Khim. Prom. (1982) p. 605

[320] Weingerl, H.; Kahler, E.; Kriszt, K.; Kühnelt, G.; Plessing, R.
Auswirkung der chemischen Zusammensetzung und der Erschmelzung auf die Eigenschaften von Stählen für Harnstoffreaktoren
(Effects of the chemical composition and smelting on the properties of steels for urea-reactors) (in German)
Berg Hüttenmann. Monatsh. *122* (1977) 9, p. 388

[321] Nordin, S.
Rostfreie Spezialstähle für die chemische Industrie. Teil I: Korrosionseigenschaften und Anwendungsbereiche
(Stainless special steels in the chemical industry. Part I: Corrosion properties and fields of use) (in German)
Chem.-Anl. + Verfahren (1979) 5, p. 67

[322] Anonymous
Influence of molybdenum on austenitic chromium-nickel steels (in Bulgarian)
Metallurgiya (Sofia) *39* (1984) 12, p. 12

[323] Matsumoto, K.; Shinohara, T.; Kasamatsu, A.
Effects of process conditions on corrosion of 316 L in urea reactor and evaluation of corrosion resistance (in Japanese)
Corros. Eng. (Boshoku Gijutsu) *33* (1984) 11, p. 643

[324] Ponomareva, O. V.; Shvarts, G. L.; Sycheva, V. N.
Corrosion stability of stainless steel and titanium under conditions of methacrylic acid production (in Russian)
Khim. Prom. (1972) p. 668

[325] Kemplay, J., Schieber
Chem. Process Eng. *47* (1966) 7, p. 53

[326] Eden, A.
Recent advances in the use of stainless steel for offshore and seawater applications
Anticorros. Methods Mater. *26* (1979) 11, p. 7

[327] Horn, E.-M.; Kohl, H.
Werkstoffe für die Salpetersäure-Industrie
Materials for the nitric acid industry (in German)
Werkst. Korros. *37* (1986) 2, p. 57

[328] Product Information (High Technology Materials Division, Kokomo)
Testing will prove Ferralium® alloy 255 outperforms ordinary stainless steels
Chem. Eng. Progr. *77* (1981) 3, p. 29

[329] Kirchheiner, R. R.; Hofmann, F.; Hoffmann, T.; Rudolph, G.
A silicon-alloyed stainless steel for highly oxidizing conditions
Mater. Performance *26* (1987) 1, p. 49

[330] Imai, H.; Fukumoto, I.; Masuko, N.
The influence of germanium on the corrosion resistance of 25Cr-6Ni duplex steel (in Japanese)
Corros. Eng. (Boshoku Gijutsu) *34* (1985) p. 339

[331] Lozovatskaya, L. P.; Shmatko, M. N.; Goldshtejn, Ya. N.; Piskunova, A. E.
Effect of micro-alloying with boron and cerium on the corrosion stability and hot ductility of the steel 02Ch18N11 (in Russian)
Zashch. Met. *21* (1985) 1, p. 98

[332] Kirchheiner, R. R.; Heubner, U.; Hofmann, F.
Increasing the life-time of nitric acid equipment using improved stainless steels and nickel alloys
Corrosion 88, St. Louis-Missouri, 1988, Nr. 318, 22 p.

[333] Lipodaev, V. N.; Yushchenko, K. A.; Skulskij, V. Yu.; Tikhonovskaya, L. D.; Dzykovich, I. Ya.
Effect of nickel on the structure and properties of corrosion-resistant welds with high silicon content (in Russian)
Avtom. Svarka (1985) 9, p. 9

[334] Shuler, W. E.
Corrosion by fluoride solutions
U.S. At. Energy Comm. DP-348 (1959) 15

[335] Chen, N. G.; Bocharov, V. A.; Fursov, P. F.; Shust, T. F.; Dektyareva, V. K.; Borozdina, R. R.; Yudina, S. M.
Delaying the corrosion of weld seams in carbon and stainless steels in acid solutions (in Russian)
Zashch. Met. *1* (1965) p. 726

[336] Wiedemann, K. H.
Korrosionsuntersuchungen zur Entwicklung von Dekontaminationslösungen für Nuklearanlagen. Teil 1: Das Korrosionsverhalten verschiedener Werkstoffe der Kerntechnik in sauren und alkalischen Redoxlösungen
(Investigation of corrosion in the development of decontamination solutions for nuclear systems. Part I: The corrosion behavior of several nuclear materials in acid and alkaline redox solutions) (in German)
Werkst. Korros. *39* (1988) 5, p. 185

[337] Grishin, A. M.; Kondrashin, Yu. V.; Sentyurev, V. P.
Effect of silicon on the corrosion properties of ChN40B alloy (in Russian)
Zashch. Met. *16* (1980) 6, p. 718

[338] Ivanov, E. S.; Makovetski, R. D.
Investigation of the etching of 36NChTYu alloy after thermal treatment (in Russian)
Zashch. Met. *17* (1981) 2, p. 195

[339] Ford, E.
Silizium macht Eisen säurefest
(Silicon makes iron acid-resistant) (in German)
Technica *21* (1962) 16, p. 1381

[340] Ford, E.
(Tantiron®) (in German)
Oberflächentechnik *50* (1973), Metall-Journal 2/73, p. A 3

[341] Cavallini, M.; Felli, F.; Fratesi, R.; Veniali, F.
Aqueous solution corrosion behaviour of "poor man" high manganese-aluminum steels
Werkst. Korros. *33* (1982) 5; p. 281

[342] Anonymous
Sintered stainless steels with excellent neutron absorbability
Japan Pat. 81 23 256 (Daido Steel Co., Ltd.; March 5, 1981)

[343] Melnik, P. I.; Shumilov, V. N.; Rimskaya, E. A.; Sokolovskij, M. F.
Electrochemical behavior of alloys based on nickel and iron (in Russian)
Fiz. Khim. Mekh. Mater. *20* (1984) 2, p. 33

[344] Roque, R.
Mecanismo de la oxidación de las aleaciones Fe-Ni (de 5 a 30 %)
(Oxidation mechanism of FeNi alloys with 5 to 30 % Ni) (in Spanish)
Revista Cenic *11* (1980) 1–2, p. 79

[345] Musina, A. S.; Lange, A. A.; Bukhman, S. P.; Semashko, T. S.
Change in the surface state of the alloys Permalloy® and Kovar® after chemical treatment (in Russian)
Izv. Akad. Nauk Kaz. SSR. Ser. Khim. (1984) 3, p. 12

[346] Nakamura, M.; Ito, H.
Polishing solution for iron-nickel-cobalt alloy
Japan Pat. 78 47 335 (April 27, 1978)

[347] Leitner, S.; Köstler, H. J.
Investigations into the application of etching reagents for austenite grain boundaries in tempered high-speed steels
Prakt. Metallogr. *15* (1978) p. 66

Phosphoric Acid

Unalloyed steels and cast steel

Unalloyed steels and cast steel are unsuitable for handling all concentrations of phosphoric acid [1–25].

The corrosion rates are always recorded as being above 1 mm/a (39.4 mpy) [2, 5, 8, 23, 25] rates above 1.27 mm/a (50.0 mpy) being mentioned for non-aerated and aerated acid over the entire concentration range [8].

The corrosion rate for steel 10 (cf. SAE 1010, 1.0301, C10) in 73 % phosphoric acid at 353 K (80 °C) is stated as being 707 mm/a (27,836 mpy) [13]. At higher temperatures, the corrosion rates in concentrated acid are even greater.

A summary of corrosion rates in phosphoric acid and phosphoric acid containing solutions of varying concentration and temperature is contained in Table 1 (after laboratory studies lasting 3 hours on steel specimens with 0.054 % C, 0.02 % Si, 0.021 % S, 0.013 % P and 0.094 % Mn, balance Fe, in analytically pure phosphoric acid solutions) and 2, in comparison with high-alloy steels and titanium.

Table 1 illustrates the sometimes considerably higher corrosion rates of unalloyed steel than shown by the ratings of more than 1 mm/a (39.4 mpy) in Table 2.

H_3PO_4 mol/l	*Temperature K (°C)*	*Corrosion rate mm/a (mpy)*
0.33	306 (33)	5.57 (219)
0.66	306 (33)	6.42 (253)
1.00	306 (33)	9.8 (386)
1.33	306 (33)	12.4 (488)
1.67	306 (33)	17.7 (697)
1.83	298 (25)	15.3 (602)
1.83	306 (33)	28.9 (1,138)
1.83	313 (40)	34.4 (1,354)
1.83	318 (45)	49.0 (1,929)
1.83	323 (50)	73.7 (2,902)
1.83	328 (55)	96.4 (3,795)
1.83	333 (60)	151 (5,945)
2.00	306 (33)	19.3 (760)
2.33	306 (33)	20.4 (803)
2.00	306 (33)	23.7 (933)
3.33	306 (33)	25.0 (984)

Table 1: Corrosion rates of unalloyed steel (0.054 % C, 0.02 % Si, 0.021 % S, 0.013 % P and 0.094 % Mn, balance Fe) in analytically pure phosphoric acid, test duration 3 h [25]

Medium/concentration		Temperature K (°C)	Chemical composition of the steels[2)], %												
			Cr	–	13	17	18	18	17	18	17	20	25	Ti	Ti-Pd
			Ni	–	–	–	–	9	12	14	15	25	5		
			Mo	–	–	–	2	–	2.5	3.5	4.5	4.5	1.5		
			Cu	–	–	–	–	–	–	–	–	1.5	–		
Phosphoric acid, chemically pure	1 %	293 (20)		2	0	0	0	0	0	0	0	0	0	0	0
H_3PO_4	1 %	373 (100) = b.p.		2	2	1	0	0	0	0	0	0	0	0	0
	1 %	413 (140)		2	1	1	0	0	0	0	0	0	0	0	0
	3 %	373 (100) = b.p.		2	2	0	0	0	0	0	0	0	0	1	0
	5 %	293–333 (20–60)		2	0	0	0	0	0	0	0	0	0	0	0
	5 %	358 (85)		2	0	0	0	0	0	0	0	0	0	1	0
	5 %	373 (100) = b.p.		2	2	1	0	0	0	0	0	0	0	2	0
	10 %	313 (40)		2	2	0	0	0	0	0	0	0	0	0	0
	10 %	333 (60)		2	2	0	0	0	0	0	0	0	0	1	0
	10 %	353 (80)		2	2	0	0	0	0	0	0	0	0	2	1
	10 %	374 (101) = b.p.		2	2	2	0	0	0	0	0	0	0	2	2
	20 %	308 (35)		2	2	2	0	0	0	0	0	0	0	0	0
	20 %	333 (60)		2	2	2	0	0	0	0	0	0	0	1	1
	20 %	375 (102) = b.p.		2	2	2	0	0	0	0	0	0	0	2	2
	30 %	293–308 (20–35)		2	2	2	0	0	0	0	0	0	0	0	0
	30 %	333 (60)		2	2	2	0	0	0	0	0	0	0	1	1
	30 %	373 (100)		2	2	2	1	1	0	0	0	0	0	2	2

Table 2: Corrosion rates (for the rating see footnote[1)]) of steels and titanium in phosphoric acid and phosphoric acid solutions of varying compositions [23]

Table 2: Continued

Medium/concentration		*Temperature K (°C)*	*Chemical composition of the steels*[2)], %												
			Cr	**–**	***13***	***17***	***18***	***18***	***17***	***18***	***17***	***20***	***25***	***Ti***	***Ti-Pd***
			Ni	**–**	**–**	**–**	**–**	***9***	***12***	***14***	***15***	***25***	***5***		
			Mo	**–**	**–**	**–**	***2***	**–**	***2.5***	***3.5***	***4.5***	***4.5***	***1.5***		
			Cu	**–**	**–**	**–**	**–**	**–**	**–**	**–**	**–**	***1.5***	**–**		
H_3PO_4	40 %	308 (35)		2	2	2	0	0	0	0	0	0	0	1	0
	40 %	323 (50)		2	2	2	0	0	0	0	0	0	0	2	1
	40 %	373 (100)		2	2	2	1	1	0	0	0	0	0	2	2
	40 %	379 (106) = b.p.		2	2	2	2	2	1	1	0	1	0	2	2
	50 %	293 (20)		2	2	2	0	0	0	0	0	0	0	0	0
	50 %	308 (35)		2	2	2	0	0	0	0	0	0	0	1	0
	50 %	323 (50)		2	2	2	0	0	0	0	0	0	0	2	1
	50 %	358 (85)		2	2	2	1	0	0	0	0	0	0	2	2
	50 %	373 (100)		2	2	2	2	1	1	1	0	0	0	2	2
	50 %	383 (110) = b.p.		2	2	2	2	2	2	1	1	1	0	2	2
	60 %	293 (20)		2	2	2	0	0	0	0	0	0	0	1	0
	60 %	308 (35)		2	2	2	0	0	0	0	0	0	0	1	1
	60 %	373 (100)		2	2	2	2	2	1	1	1	0	0	2	2
	60 %	389 (116) = b.p.		2	2	2	2	2	2	2	2	1	1	2	2

Table 2: Corrosion rates (for the rating see footnote[1)]) of steels and titanium in phosphoric acid and phosphoric acid solutions of varying compositions [23]

Table 2: Continued

Medium/concentration		Temperature K (°C)	Chemical composition of the steels[2)], %												
			Cr	–	13	17	18	18	17	18	17	20	25	Ti	Ti-Pd
			Ni	–	–	–	–	9	12	14	15	25	5		
			Mo	–	–	–	2	–	2.5	3.5	4.5	4.5	1.5		
			Cu	–	–	–	–	–	–	–	–	1.5	–		
H_3PO_4	70 %	308 (35)		2	2	2	0	0	0	0	0	0	0	2	1
	70 %	363 (90)		2	2	2	2	2	1	0	0	0	0	2	2
	70 %	399 (126) = b.p.		2	2	2	2	2	2	2	2	1	1	2	2
	80 %	293 (20)		2	2	2	0	0	0	0	0	0	0	2	1
	80 %	308 (35)		2	2	2	0	0	0	0	0	0	0	2	2
	80 %	353 (80)		2	2	2	2	1	0	0	0	0	0	2	2
	80 %	373 (100)		2	2	2	2	2	1	1	1	1	0	2	2
	80 %	419 (146) = b.p.		2	2	2	2	2	2	2	2	2	2	2	2
	85 %	293 (20)		2	2	2	0	0	0	0	0	0	0	2	1
	85 %	323 (50)		2	2	2	1	0	0	0	0	0	0	2	2
	85 %	368 (95)		2	2	2	2	2	1	1	1	0	0	2	2
	85 %	429 (156) = b.p.		2	2	2	2	2	2	2	2	2	2	2	2

Table 2: Corrosion rates (for the rating see footnote[1)]) of steels and titanium in phosphoric acid and phosphoric acid solutions of varying compositions [23]

Table 2: Continued

Medium/concentration				*Temperature K (°C)*	*Chemical composition of the steels²⁾, %*												
					Cr	**–**	**13**	**17**	**18**	**18**	**17**	**18**	**17**	**20**	**25**	***Ti***	***Ti-Pd***
					Ni	**–**	**–**	**–**	**–**	**9**	**12**	**14**	**15**	**25**	**5**		
					Mo	**–**	**–**	**–**	**2**	**–**	**2.5**	**3.5**	**4.5**	**4.5**	**1.5**		
					Cu	**–**	**–**	**–**	**–**	**–**	**–**	**–**	**–**	**1.5**	**–**		
Phosphoric acid, industrial																	
H_3PO_4	$+ SO_4^{2-}$	$+ Cl^-$	$+ F^-$														
40 %	1.8 %	0.03 %	1.3 %	298 (25)		2	2	2		1	0	0	0	0	0	2	1
				348 (75)		2	2	2		2	1	0	0	0	1	2	2
40 %	1.8 %	0.06 %	1.3 %	298 (25)		2	2	2		1	0	0	0	0	0	2	1
				348 (75)		2	2	2		2	1	0	0	0	1	2	2
42 %	1.8 %	0.02 %	1.5 %	293 (20)		2	2	2		2	0	0	0	0	0	2	1
				348 (75)		2	2	2		2	1	0	0	0	2	2	2
75 %	3 %	0.03 %	0.5 %	323 (50)		2	2	2		2	0	0	0	0	0	2	2
76.5 %	3 %	0.03 %	0.5 %	323 (50)		2	2	2		2	0	0	0	0	0	2	2
76.5 %	4.1 %	0.005 %	0.1 %	343 (70)		2	2	2		2	0	0	0	0	0	2	2
76.5 %	4.1 %	0.02 %	0.1 %	343 (70)		2	2	2		2	0	0	0	0	0	2	2
76.5 %	4.1 %	0.035 %	0.1 %	343 (70)		2	2	2		2	2		0	0	2	2	2
76.5 %	4.1 %	0.075 %	0.01 %	343 (70)		2	2	2		2	2		1	1	2	2	2

Table 2: Corrosion rates (for the rating see footnote[1]) of steels and titanium in phosphoric acid and phosphoric acid solutions of varying compositions [23]

Table 2: Continued

Medium/concentration				*Temperature K (°C)*	*Chemical composition of the steels[2], %*												
					Cr	–	*13*	*17*	*18*	*18*	*17*	*18*	*17*	*20*	*25*	*Ti*	*Ti-Pd*
					Ni	–	–	–	–	*9*	*12*	*14*	*15*	*25*	*5*		
					Mo	–	–	–	*2*	–	*2.5*	*3.5*	*4.5*	*4.5*	*1.5*		
					Cu	–	–	–	–	–	–	–	–	*1.5*	–		
Superphosphoric acid																	
103 %	1 %	0	0.3 %	393 (120)							0	0	0	0			
				423 (150)							1	1	1	1			
Phosphoric acid + ammonium nitrate																	
H_3PO_4		$+ NH_4NO_3$															
10 %		30 %		353 (80)		2	1	0	0	0	0	0	0	0	0	2	1
Phosphoric acid + ammonium nitrate + ammonium sulfate + nitric acid																	
H_3PO_4	$+ NH_4NO_3$	$+ (NH_4)_2SO_4$	$+ HNO_3$														
4 %	15 %	9 %	9 %	373 (100)						0	0	0	0	0	0	0	0
Phosphoric acid + ammonium sulfate + sulfuric acid																	
	H_3PO_4	$+ (NH_4)_2SO_4$	$+ H_2SO_4$														
	10 %	25 %	1.5 %	363 (90)							0	0	0	0	0		
	15 %	25 %	1 %	373 (100)						2	0	0	0	0			
	15 %	25 %	3 %	373 (100)						2	0	0	0	0			
	15 %	30 %	3 %	363 (90)						2	0	0	0	0			

Table 2: Corrosion rates (for the rating see footnote[1]) of steels and titanium in phosphoric acid and phosphoric acid solutions of varying compositions [23]

Table 2: Continued

Medium/concentration			Temperature K (°C)	Chemical composition of the steels[2], %												
				Cr	–	13	17	18	18	17	18	17	20	25	Ti	Ti-Pd
				Ni	–	–	–	–	9	12	14	15	25	5		
				Mo	–	–	–	2	–	2.5	3.5	4.5	4.5	1.5		
				Cu	–	–	–	–	–	–	–	–	1.5	–		
15 %	20 %	20 %	293 (20)						2	0	0	0	0			
16 %	9 %	1 %	353 (80)						0	0	0	0	0	0		
Phosphoric acid + calcium sulphate + sulfuric acid																
H_3PO_4	+ $CaSO_4$	+ H_2SO_4														
4 %	50 %	2 %	323 (50)		2	2	2		1	0	0	0	0	0	1	0
22 %	traces	1 %	343 (70)		2	2	1		1	0	0	0	0	0	2	1
Phosphoric acid + chromic acid + sulfuric acid																
H_3PO_4	+ CrO_3	+ H_2SO_4														
57 %	9 %	14 %	353 (80)							2	2	2				
Phosphoric acid + chromic acid																
H_3PO_4	+ CrO_3		293 (20)						0	0	0	0	0	0	0	0
80 %	10 %		293 (20)						0	0	0	0	0	0	0	0
80 %	10 %		333 (60)						2	2	2	2	1	2	0	0

Table 2: Corrosion rates (for the rating see footnote[1]) of steels and titanium in phosphoric acid and phosphoric acid solutions of varying compositions [23]

Table 2: Continued

Medium/concentration			*Temperature K (°C)*	*Chemical composition of the steels*[2], %												
				Cr	–	13	17	18	18	17	18	17	20	25	Ti	Ti-Pd
				Ni	–	–	–	–	9	12	14	15	25	5		
				Mo	–	–	–	2	–	2.5	3.5	4.5	4.5	1.5		
				Cu	–	–	–	–	–	–	–	–	1.5	–		
Phosphoric acid + fluorosilicic acid + sulfuric acid																
H_3PO_4	+ H_2SiF_6	+ H_2SO_4														
30 %	1 %	3 %	343 (70)		2	2	2	2	2	0	0	0	0	1	2	2
55 %	1 %	1 %	363 (90)		2	2	2	2	2	1	1	1	1	2	2	2
Phosphoric acid + hydrofluoric acid																
	H_3PO_4	+ HF														
	1.5 %	1 %	323 (50)		2	2	2	2	2	2	2	1	1	2	2	2
	41.4 %	0.5 %	293 (20)		2	2	1	1	1	0	0	0	0	0	1	1
	41.4 %	0.5 %	323 (50)		2	2	2	2	2	1	1	0	0	1	1	1
	41.4 %	0.5 %	333 (60)		2	2	2	2	2	1	1	1	0	1	2	1
	41.4 %	1 %	293 (20)		2	2	2	2	2	1	1	0	0	1	1	1
	41.4 %	1 %	313 (40)		2	2	2	2	2	1	1	1	0	1	2	1
	41.4 %	1 %	333 (60)		2	2	2	2	2	2	1	1	1	2	2	2
	41.4 %	2.5 %	293–313 (20–40)		2	2	2	2	2	1	1	1	0	1	2	1
	41.4 %	2.5 %	333 (60)		2	2	2	2	2	2	1	1	1	2	2	2
	76 %	0.5 %	293 (20)		2	2	1	1	1	0	0	0	0	0	2	2

Table 2: Corrosion rates (for the rating see footnote[1]) of steels and titanium in phosphoric acid and phosphoric acid solutions of varying compositions [23]

Table 2: Continued

Medium/concentration			**Temperature K (°C)**	**Chemical composition of the steels[2), %**												
				Cr	**–**	**13**	**17**	**18**	**18**	**17**	**18**	**17**	**20**	**25**	***Ti***	***Ti-Pd***
				Ni	**–**	**–**	**–**	**–**	**9**	**12**	**14**	**15**	**25**	**5**		
				Mo	**–**	**–**	**–**	**2**	**–**	**2.5**	**3.5**	**4.5**	**4.5**	**1.5**		
				Cu	**–**	**–**	**–**	**–**	**–**	**–**	**–**	**–**	**1.5**	**–**		
	76 %	0.5 %	313 (40)		2	2	2	1	1	0	0	0	0	0	2	2
	76 %	0.5 %	333 (60)		2	2	2	2	2	1	0	0	0	1	2	2
	76 %	0.5 %	353 (80)		2	2	2	2	2	1	1	1	1	1	2	2
	76 %	1 %	293 (20)		2	2	2	1	1	0	0	0	0	0	2	2
	76 %	1 %	313 (40)		2	2	2	2	2	0	0	0	0	1	2	2
	76 %	1 %	333 (60)		2	2	2	2	2	1	1	1	0	1	2	2
	75 %	1 %	353 (80)		2	2	2	2	2	1	1	1	1	2	2	2
	75 %	2 %	293 (20)		2	2	2	2	2	0	0	0	0	1	2	2
	75 %	2.5 %	293 (20)		2	2	2	2	2	1	0	0	0	1	2	2
	80 %	1 %	393 (120)		2	2	2	2	2	2	2	2	2	2	2	2
Phosphoric acid + hydrofluoric acid + nitric acid																
H_3PO_4	+ HF	+ HNO_3														
18 %	0.2 %	28 %	338 (65)		2	2	2	2	2	1	1	1	1	1	1	1
19.3 %	0.1 %	31.2 %	338 (65)		2	2	2	2	2	0	0	0	0	1	1	1

Table 2: Corrosion rates (for the rating see footnote[1)) of steels and titanium in phosphoric acid and phosphoric acid solutions of varying compositions [23]

Table 2: Continued

Medium/concentration				*Temperature K (°C)*	*Chemical composition of the steels*[2], %												
					Cr	–	13	17	18	18	17	18	17	20	25	Ti	Ti-Pd
					Ni	–	–	–	–	9	12	14	15	25	5		
					Mo	–	–	–	2	–	2.5	3.5	4.5	4.5	1.5		
					Cu	–	–	–	–	–	–	–	–	1.5	–		
	19.3 %	0.1 %	31.2 %	363 (90)		2	2	2	2	2	1	1	1	1	1	2	2
Phosphoric acid + hydrofluoric acid + nitric acid + sulfuric acid																	
H_3PO_4	+ HF	+ HNO_3	+ H_2SO_4														
7.9 %	0.1 %	12 %	25.8 %	363 (90)		2	2	2	2	2	1	1	1	1	1	2	2
Phosphoric acid + hydrogen halides																	
H_3PO_4	+ HF	+ HCl	+ HBr														
30 %	0	0	0	323 (50)						0	0						
30 %	7 ppm	0	0	323 (50)						0	0						
30 %	0	12 ppm	0	323 (50)						0	0						
30 %	0	0	27 ppm	323 (50)						0	0						
30 %	0	0	0	353 (80)						0	0						
30 %	7 ppm	0	0	353 (80)						1	1						
30 %	0	12 ppm	0	353 (80)						0	0						
30 %	0	0	27 ppm	353 (80)						0	0						
30 %	0	0	0	373 (100)						0	1						

Table 2: Corrosion rates (for the rating see footnote[1]) of steels and titanium in phosphoric acid and phosphoric acid solutions of varying compositions [23]

Table 2: Continued

Medium/concentration				*Temperature K (°C)*	*Chemical composition of the steels[2], %*												
					Cr	–	13	17	18	18	17	18	17	20	25	Ti	Ti-Pd
					Ni	–	–	–	–	9	12	14	15	25	5		
					Mo	–	–	–	2	–	2.5	3.5	4.5	4.5	1.5		
					Cu	–	–	–	–	–	–	–	–	1.5	–		
30 %	7 ppm	0	0	373 (100)						2	2						
30 %	0	12 ppm	0	373 (100)						0	1						
30 %	0	0	27 ppm	373 (100)						0	1						
50 %	0	0	0	323 (50)						0	0						
50 %	6 ppm	0	0	323 (50)						0	0						
50 %	0	11 ppm	0	323 (50)						0	0						
50 %	0	0	24 ppm	323 (50)						0	0						
50 %	0	0	0	353 (50)						0	0						
50 %	6 ppm	0	0	353 (80)						2	1						
50 %	0	11 ppm	0	353 (80)						0	0						
50 %	0	0	24 ppm	353 (80)						0	0						
50 %	0	0	0	373 (100)						0	2						
50 %	6 ppm	0	0	373 (100)						2	2						
50 %	0	11 ppm	0	373 (100)						2	1						

Table 2: Corrosion rates (for the rating see footnote[1]) of steels and titanium in phosphoric acid and phosphoric acid solutions of varying compositions [23]

Table 2: Continued

Medium/concentration				Temperature K (°C)	Chemical composition of the steels[2], %												
					Cr	–	13	17	18	18	17	18	17	20	25	Ti	Ti-Pd
					Ni	–	–	–	–	9	12	14	15	25	5		
					Mo	–	–	–	2	–	2.5	3.5	4.5	4.5	1.5		
					Cu	–	–	–	–	–	–	–	–	1.5	–		
50 %	0	0	24 ppm	373 (100)						0	1						
70 %	0	0	0	323 (50)						0	0						
70 %	5 ppm	0	0	323 (50)						1	0						
70 %	0	10 ppm	0	323 (50)						2	1						
70 %	0	0	21 ppm	323 (50)						0	0						
70 %	0	0	0	353 (80)						1	0						
70 %	5 ppm	0	0	353 (80)						2	1						
70 %	0	10 ppm	0	353 (80)						2	1						
70 %	0	0	21 ppm	353 (80)						1	1						
70 %	0	0	0	373 (100)						2	2						
70 %	5 ppm	0	0	373 (100)						2	2						
70 %	0	10 ppm	0	373 (100)						2	2						
70 %	0	0	21 ppm	373 (100)						2	2						
85 %	0	0	0	323 (50)						0	0						

Table 2: Corrosion rates (for the rating see footnote[1]) of steels and titanium in phosphoric acid and phosphoric acid solutions of varying compositions [23]

Table 2: Continued

Medium/concentration				*Temperature K (°C)*	*Chemical composition of the steels[2], %*												
					Cr	–	**13**	**17**	**18**	**18**	**17**	**18**	**17**	**20**	**25**	***Ti***	***Ti-Pd***
					Ni	–	–	–	–	**9**	**12**	**14**	**15**	**25**	**5**		
					Mo	–	–	–	**2**	–	**2.5**	**3.5**	**4.5**	**4.5**	**1.5**		
					Cu	–	–	–	–	–	–	–	–	**1.5**	–		
85 %	5 ppm	0	0	323 (50)						1	0						
85 %	0	9 ppm	0	323 (50)						2	0						
85 %	0	0	19 ppm	323 (50)						0	0						
85 %	0	0	0	353 (80)						0	1						
85 %	5 ppm	0	0	353 (80)						2	1						
85 %	0	9 ppm	0	353 (80)						2	1						
85 %	0	0	19 ppm	353 (80)						1	0						
85 %	0	0	0	373 (100)						2	1						
85 %	5 ppm	0	0	373 (100)						2	2						
85 %	0	9 ppm	0	373 (100)						2	2						
85 %	0	0	19 ppm	373 (100)						2	0						

Table 2: Corrosion rates (for the rating see footnote[1]) of steels and titanium in phosphoric acid and phosphoric acid solutions of varying compositions [23]

Table 2: Continued

Medium/concentration			*Temperature K (°C)*	*Chemical composition of the steels[2)], %*												
				Cr	–	13	17	18	18	17	18	17	20	25	Ti	Ti-Pd
				Ni	–	–	–	–	9	12	14	15	25	5		
				Mo	–	–	–	2	–	2.5	3.5	4.5	4.5	1.5		
				Cu	–	–	–	–	–	–	–	–	1.5	–		
Phosphoric acid + nitric acid + sulfuric acid																
H_3PO_4	+ HNO_3	+ H_2SO_4														
43 %	2 %	45 %	373 (100)							1	1	1				
43 %	2.2 %	45 %	378 (105)							1	1	1	1			
64 %	1.9 %	23 %	363 (90)						0	0	0	0	0			
66 %	7.2 %	10.5 %	373 (100)							1	1	1	1			
78 %	3 %	18 %	368 (95)							1	1	1	1			
Phosphoric acid + sodium chloride																
	H_3PO_4	+ NaCl														
	76 %	6.2 ppm	313 (40)					2		0						
	76 %	6.2 ppm	353 (80))					2		1						
	76 %	312 ppm	313 (40)					2		0						
	76 %	312 ppm	353 (80)					2		1						

Table 2: Corrosion rates (for the rating see footnote[1)]) of steels and titanium in phosphoric acid and phosphoric acid solutions of varying compositions [23]

Table 2: Continued

Medium/concentration				*Temperature K (°C)*	*Chemical composition of the steels*[2], %												
					Cr	–	13	17	18	18	17	18	17	20	25	Ti	Ti-Pd
					Ni	–	–	–	–	9	12	14	15	25	5		
					Mo	–	–	–	2	–	2.5	3.5	4.5	4.5	1.5		
					Cu	–	–	–	–	–	–	–	–	1.5	–		
Phosphoric acid + sulfuric acid																	
		H_3PO_4	$+H_2SO_4$														
		40 %	2 %	b.p.		2	2	2	2	2	2	2	2	2	2	2	2
		41.4 %	2 %	353 (80)		2	2	2	2	2	0	0	0	0	1	2	2
		41.4 %	3.5 %	353 (80)		2	2	2	2	2	1	0	0	0	1	2	2
		43 %	47 %	343 (70)		2	2	2	2	2	1	1	1	1	2	2	2
		53 %	15 %	333 (60)		2	2	2	2	2	2	2	2	1	2	2	2
		76 %	3.5 %	353 (80)		2	2	2	2	2	1	0	0	0	1	2	2
Phosphoric acid, 75 % + sodium salts fluorides, chlorides, sulfates plus iron(III)oxide)																	
F^-	$+Cl^-$	$+SO_4^{2-}$	$+Fe_2O_3$														
0 %	0.05 %	5.0 %	1.5 %	313 (40)							0	0					
0.4 %	0.002 %	1.4 %	0.3 %	313 (40)							1	0					
0.4 %	0.002 %	1.4 %	1.7 %	313 (40)							1	1					
0.4 %	0.002 %	5.6 %	0.3 %	313 (40)							1	0					

Table 2: Corrosion rates (for the rating see footnote[1]) of steels and titanium in phosphoric acid and phosphoric acid solutions of varying compositions [23]

Table 2: Continued

Medium/concentration				Temperature K (°C)	Chemical composition of the steels[2)], %												
					Cr	–	13	17	18	18	17	18	17	20	25	Ti	Ti-Pd
					Ni	–	–	–	–	9	12	14	15	25	5		
					Mo	–	–	–	2	–	2.5	3.5	4.5	4.5	1.5		
					Cu	–	–	–	–	–	–	–	–	1.5	–		
0.4 %	0.058 %	5.6 %	0.3 %	313 (40)							1	1					
0.4 %	0.058 %	5.6 %	1.7 %	313 (40)							2	1					
0.5 %	0 %	0 %	0 %	313 (40)							0						
0.5 %	0.01 %	0 %	0.5 %	313 (40)							1	1					
0.5 %	0 %	2.0 %	0.5 %	313 (40)							1	0					
0.5 %	0 %	2.0 %	1.5 %	313 (40)							1	1					
0.5 %	0.01 %	2.0 %	0 %	313 (40)							2	0					
0.5 %	0.01 %	2.0 %	0.5 %	313 (40)							1	1					
0.5 %	0.05 %	2.0 %	0 %	313 (40)							1	0					
0.5 %	0.05 %	2.0 %	0.5 %	313 (40)							2	1					
0.5 %	0.05 %	2.0 %	1.5 %	313 (40)							1	1					
0.5 %	0 %	5.0 %	0.5 %	313 (40)							1	0					
0.5 %	0.01 %	5.0 %	0 %	313 (40)							2	0					
0.5 %	0.01 %	5.0 %	0.5 %	313 (40)							1	1					
0.5 %	0.01 %	5.0 %	1.5 %	313 (40)							2	1					

Table 2: Corrosion rates (for the rating see footnote[1)]) of steels and titanium in phosphoric acid and phosphoric acid solutions of varying compositions [23]

Table 2: Continued

Medium/concentration				*Temperature K (°C)*	*Chemical composition of the steels[2], %*												
					Cr	**–**	**13**	**17**	**18**	**18**	**17**	**18**	**17**	**20**	**25**	***Ti***	***Ti-Pd***
					Ni	**–**	**–**	**–**	**–**	**9**	**12**	**14**	**15**	**25**	**5**		
					Mo	**–**	**–**	**–**	**2**	**–**	**2.5**	**3.5**	**4.5**	**4.5**	**1.5**		
					Cu	**–**	**–**	**–**	**–**	**–**	**–**	**–**	**–**	**1.5**	**–**		
0.5 %	0.05 %	5.0 %	0 %	313 (40)							2	0					
0.5 %	0.05 %	5.0 %	0.5 %	313 (40)							2	1					
1.0 %	0 %	0 %	0 %	313 (40)							1						
1.0 %	0.01 %	0 %	0.5 %	313 (40)							2	1					
1.0 %	0.01 %	0 %	1.5 %	313 (40)							1	1					
1.0 %	0.01 %	2.0 %	0 %	313 (40)							2	0					
1.0 %	0.01 %	2.0 %	0.5 %	313 (40)							2	1					
1.0 %	0.05 %	2.0 %	0 %	313 (40)							1	0					
1.0 %	0.05 %	2.0 %	0.5 %	313 (40)							2	1					
1.0 %	0 %	5.0 %	0.5 %	313 (40)							1	1					
1.0 %	0.01 %	5.0 %	0 %	313 (40)							2	0					
1.0 %	0.01 %	5.0 %	0.5 %	313 (40)							2	1					
1.0 %	0.05 %	5.0 %	0 %	313 (40)							2	0					
1.0 %	0.05 %	5.0 %	0.5 %	313 (40)							2	1					
1.1 %	0.002 %	1.4 %	1.7 %	313 (40)							1	1					

Table 2: Corrosion rates (for the rating see footnote[1]) of steels and titanium in phosphoric acid and phosphoric acid solutions of varying compositions [23]

Table 2: Continued

Medium/concentration				Temperature K (°C)	Chemical composition of the steels[2], %												
					Cr	–	13	17	18	18	17	18	17	20	25	Ti	Ti-Pd
					Ni	–	–	–	–	9	12	14	15	25	5		
					Mo	–	–	–	2	–	2.5	3.5	4.5	4.5	1.5		
					Cu	–	–	–	–	–	–	–	–	1.5	–		
1.1 %	0.002 %	5.6 %	0.3 %	313 (40)							2	1					
1.1 %	0.002 %	5.6 %	1.7 %	313 (40)							1	1					
1.1 %	0.058 %	5.6 %	0.3 %	313 (40)							2	1					
Phosphoric acid, 75 % with 5 % Na_2SO_4, 1 % NaF, 0.05 % NaCl and 1.5 % Fe_2O_3 + aluminium oxide + silicon dioxide + magnesium oxide + calcium oxide																	
Al_2O_3	+ SiO_2	+ MgO	+ CaO														
0 %	0 %	0 %	0 %	313 (40)							2	1					
0 %	0 %	0 %	0.5 %	313 (40)							0	0					
0.5 %	0.2 %	0 %	0 %	313 (40)							2						
0.5 %	0.2 %	0 %	0.25 %	313 (40)							1						
0.5 %	0.2 %	0.25 %	0.5 %	313 (40)							0						
0.5 %	0.2 %	0.5 %	0 %	313 (40)							1						
0.5 %	0.2 %	0.5 %	0.5 %	313 (40)							0						
1.0 %	0.2 %	0 %	0 %	313 (40)							1	1					
1.0 %	0.2 %	0 %	0.25 %	313 (40)							1	0					
1.0 %	0.2 %	0 %	0.5 %	313 (40)							0	0					
1.0 %	0.2 %	0.25 %	0 %	313 (40)							1	0					

Table 2: Corrosion rates (for the rating see footnote[1]) of steels and titanium in phosphoric acid and phosphoric acid solutions of varying compositions [23]

Table 2: Continued

Medium/concentration				*Temperature K (°C)*	*Chemical composition of the steels[2)], %*												
					Cr	–	13	17	18	18	17	18	17	20	25	Ti	Ti-Pd
					Ni	–	–	–	–	9	12	14	15	25	5		
					Mo	–	–	–	2	–	2.5	3.5	4.5	4.5	1.5		
					Cu	–	–	–	–	–	–	–	–	1.5	–		
1.0 %	0.2 %	0.25 %	0.25 %	313 (40)							0	0					
1.0 %	0.2 %	0.5 %	0 %	313 (40)							1	0					
1.0 %	0.2 %	0.5 %	0.5 %	313 (40)							0	0					
Phosphoric acid, 75 % with 5 % Na_2SO_4, 1 % NaF and 0.05 % NaCl + iron(III)oxide + aluminium oxide + silicon dioxide																	
	Fe_2O_3	+ Al_2O_3	+ SiO_2														
	0.5 %	0 %	0 %	313 (40)							2	1					
	0.5 %	0 %	0.4 %	313 (40)							1	1					
	0.5 %	0.5 %	0.4 %	313 (40)							1	0					
	0.5 %	1.0 %	0 %	313 (40)							1	1					
	0.5 %	1.0 %	0.2 %	313 (40)							1	0					
	0.5 %	1.0 %	0.4 %	313 (40)							0	0					
	1.5 %	0 %	0 %	313 (40)							2	1					
	1.5 %	0.5 %	0.4 %	313 (40)							1	0					
	1.5 %	1.0 %	0 %	313 (40)							1	1					
	1.5 %	1.0 %	0.4 %	313 (40)							1	0					

Table 2: Corrosion rates (for the rating see footnote[1)]) of steels and titanium in phosphoric acid and phosphoric acid solutions of varying compositions [23]

Table 2: Continued

Medium/concentration	Temperature K (°C)	Chemical composition of the steels[2)], %												
		Cr	–	13	17	18	18	17	18	17	20	25	Ti	Ti-Pd
		Ni	–	–	–	–	9	12	14	15	25	5		
		Mo	–	–	–	2	–	2.5	3.5	4.5	4.5	1.5		
		Cu	–	–	–	–	–	–	–	–	1.5	–		
Phosphoric anhydride, phosphorus pentoxide, P_2O_5														
	dry	293						0	0	0	0	0	0	0
	moist	293						1	0	0	0	0	0	1

1) 0 ≙ < 0.1 mm/a (3.94 mpy)
   1 ≙ 0.1–1.0 mm/a (3.94–39.4 mpy)
   2 ≙ > 1.0 mm/a (39.4 mpy)

2)
13Cr	1.4000 (SAE 403, X6Cr13)
17Cr	1.4016 (SAE 430, X6Cr17)
18Cr2Mo	1.4521 (SAE 443/444, X2CrMoTi18-2)
25Cr5Ni1.5Mo	1.4460 (SAE 329, X3CrNiMoN27-5-2)
18CriNi	1.4301 (SAE 304, X5CrNi18-10)
	1.4541 (SAE 321, X6CrNiTi18-10)
	1.4550 (SAE 347/348, X6CrNiNb18-10)
	1.4306 (SAE 304 L, X2CrNi19-11)
	1.4311 (SAE 304 LN, X2CrNiN18-10)
17Cr12Ni2.5Mo	1.4401 (SAE 316, X5CrNiMo17-12-2)
	1.4404 (SAE 316 L, X2CrNiMo17-12-2)
	1.4571 (SAE 316 Ti, X6CrNiMoTi17-12-2)
	1.4436 (SAE 316, X3CrNiMo17-13-3)
	1.4435 (SAE 316 L, X2CrNiMo18-14-3)
	1.4429 (SAE 316 LN, X2CrNiMoN17-13-3)
18Cr14Ni3.5Mo	1.4438 (SAE 317 L, X2CrNiMo18-15-4)
20Cr25Ni4.5Mo1.5Cu	1.4539 (SAE 904 L, X1NiCrMoCu25-20-5)

Table 2: Corrosion rates (for the rating see footnote[1)]) of steels and titanium in phosphoric acid and phosphoric acid solutions of varying compositions [23]

Unalloyed steel is attacked at a rate of 0.1 or 0.46 mm/a (3.94 or 18.1 mpy) by 3.34 % H_3PO_4 at room temperature, without further details of the acid [26]. The corrosive attack at 353 K (80 °C) increases as the concentration of the phosphoric acid increases as follows:
1 %, 6.6 mm/a (260 mpy); 2.5 %, 14.7 mm/a (579 mpy); 5 %, 28.1 mm/a (1,106 mpy); 10 %, 48 mm/a (1,890 mpy); 20 %, 100 mm/a (3,937 mpy); 35 %, 205 mm/a (8,071 mpy); 50 %, 381 mm/a (15,000 mpy); 65 %, 364 mm/a (14,331 mpy); 80 %, 196 mm/a (7,717 mpy), and in 89 % acid, likewise as before at 353 K (80 °C) 25 mm/a (984 mpy) [2].

The corrosion behavior of unalloyed steel in phosphoric acid solutions can be improved, for example, by diffusion boronization. Such boride coatings reduce the corrosion rate in 0.3 mol/l H_3PO_4 at 293 K (20 °C) from 9.5 mm/a (374 mpy) to 0.1 mm/a (3.94 mpy), and at 333 K (60 °C) from about 20 mm/a (787 mpy) to 0.7 mm/a (27.6 mpy) [20]. The resistance is, of course, guaranteed only as long as the coating is present.

As well as by electrochemical determination of the corrosion rates in phosphoric acid [27], corrosion can also be monitored electrochemically [28], and cathodic [29] or anodic protection [30] can take place or be carried out. The latter is possible only if passivity is possible or passivity is established under the corresponding conditions.

Phosphoric acid is used as a pickling acid for unalloyed steels together with the corresponding pickling inhibitors [31–33].

The inhibitors cause detachment of the scale or oxide coatings and very greatly reduce attack on the underlying material. The fluorosilicic acid (H_2SiF_6) which remains in a small amount in the crude phosphoric acid during production acts as such an inhibitor. Phosphoric acid as a pickling acid has the advantage that it can simultaneously be used for degreasing, and phosphate residues do not affect subsequent enamelling [32].

In 10 % H_3PO_4 as a pickling acid for unalloyed steel, the inhibition efficiency by DBSO (dibenzyl sulfoxide) and EMBI (ethoxylated mercaptobenzimidazole) at 328 K (55 °C) and 358 K (85 °C) is in each case 99 %. The two inhibitors are known as the commercial products Preventol® C1-5 and 6 [33].

The inhibition efficiency in 10 % H_3PO_4 at 303 K (30 °C) and at a concentration of 10 mmol/l is 69 % for triazole, 70 % for aminotriazole, 89 % for benzotriazole and 83 % for thiourea [34]. In 20 % H_3PO_4 at 293 K (20 °C) the corrosion rate for Armco® iron (0.002 C, 0.027 Mn, 0.0022 S, 0.008 P, 0.004 Si, 0.018 Cr, 0.025 Ni, 0.027 Ce, 0.046 Al, balance Fe) is 14.4 mm/a (567 mpy). By addition of 0.001 mol/l triphenylbenzyl-phosphorus bromide, the corrosion rate drops to 0.8 mm/a (31.5 mpy), and at 0.004 mol/l to 0.6 mm/a (23.6 mpy) [21].

In H_3PO_4 at a concentration of 0.03 mol/l, corrosion rates of 19.1 mm/a (752 mpy) are determined electrochemically at 303 K (30 °C). They are reduced to 4.5 mm/a (177 mpy) by addition of 0.002 mol/l thiourea as an inhibitor, to 2.8 mm/a (110 mpy) at 0.005 mol/l, to 2.3 mm/a (90.6 mpy) at 0.01 mol/l, and to 2 mm/a (78.7 mpy) at 0.05 mol/l [35].

For further information on the resistance of unalloyed steel, see also under Section Unalloyed cast iron and Tables 13–15 and 31, and Figures 21 and 22.

Unalloyed cast iron

Like unalloyed steel, cast iron is attacked by phosphoric acid [2, 5, 6, 8, 12–15, 18, 26, 36, 37]. The corrosion rates exceed 1.27 mm/a (50 mpy) in aerated and non-aerated acid at room temperature at all concentrations [8].

The corrosion rate at room temperature in 3.34 % phosphoric acid is 0.6 mm/a (23.6 mpy) [26]. In spite of this higher corrosion rate, cast iron is said to have better behavior in phosphoric acid than steel (0.1–0.5 mm/a (3.94–19.7 mpy)) [26].

The corrosion behavior of cast iron is comparable to that of unalloyed steel (see Section Unalloyed steels and cast steel).

Like unalloyed steel, cast iron is also the anode in combination with Ms 60 (CuZn40, CW509L) in phosphoric acid as the corrosion medium. It corrodes faster than brass. Table 3 contains the corresponding corrosion rates [36, 38]. As can be seen from this table, the increase in corrosion in combination with brass is considerable.

H_3PO_4 concentration mol/l	*Cast iron*	*Cast iron in contact with Ms 60*
	Corrosion rate, mm/a (mpy)	
0.32	5.4 (213)	11.3 (445)
0.63	13.8 (543)	20.0 (787)
0.93	19.2 (756)	25.0 (984)

Table 3: Corrosion behavior of cast iron (2.96 % C, 2.05 % Si, 1.01 % P, 0.11 % S, 0.41 % Mn, balance Fe) in phosphoric acid and in a conductive combination with Ms 60 (CuZn40, CW509L) at 298 K (25 °C) test duration 6 d [36, 38]

According to [38], corrosion of cast iron is reduced after relatively severe initial corrosion in phosphoric acid as a consequence of the formation of a deposit. After 14 days, the cast iron achieved only 10 % of its initial corrosion rate in phosphoric acid. The studies were carried out in 500 ml acid at 298 K (25 °C). The weight loss measurements were made daily. The same acid was used. It was renewed only in cases where the removal of material was too great. These cases were not identified in the work. Additions of other acids, such as sulfuric acid and hydrochloric acid, increase the corrosion rates [36, 38].

The effect of inhibitors on the corrosion behavior of cast iron in phosphoric acid at 298 K (25 °C) can be seen from Table 4.

Phosphoric acid is the only mineral acid which is capable of forming a surface film on iron and cast iron even in dilute acid solution. In comparison with hydrochloric acid or sulfuric acid, it is considerably more weakly dissociated. Because of this weak dissociation of phosphoric acid, it is even possible, by addition of Na_2HPO_4 alone, to suppress the dissociation of phosphoric acid still further, to reduce the dissolving capacity of the acid and to improve the film formation. These processes are represented in a graph for 0.63 mol/l H_3PO_4 according to Table 4 in Figure 1.

H_3PO_4 mol/l	Addition	Inhibition efficiency, %					
		After	1 d	3 d	5 d	7 d	14 d
0.32	+ 2 % gelatine		51	47	45	– 2	– 2
0.32	+ 5 % Na_2HPO_4		40	51	52	23	23
0.32	+ 2 % gelatine + 5 % Na_2HPO_4		51	63	69	40	52
0.63	+ 2 % gelatine		41	28	21	4	– 3
0.63	+ 2 % Na_2HPO_4		12	23	25	11	8
0.63	+ 2 % gelatine + 5 % Na_2HPO_4		53	52	55	41	30
0.93	+ 2 % gelatine		31	28	33	3	1
0.93	+ 5 % Na_2HPO_4		4	16	25	7	8
0.93	+ 2 % gelatine + 5 % Na_2HPO_4		51	40	41	22	16

Table 4: Effect of inhibitors on the corrosion resistance of cast iron (see Table 3) in phosphoric acid solutions at 298 K (25 °C) [36, 38]

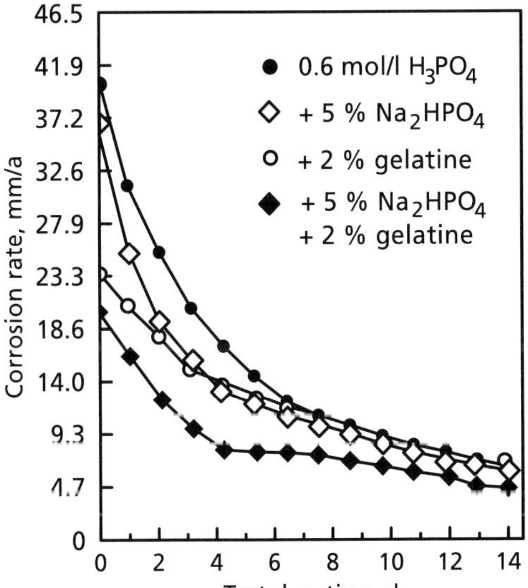

Figure 1: Influence of additions on the corrosion behavior of cast iron in 0.6 mol/l H_3PO_4 at 298 K (25 °C) (see Table 4) [36]
* calculated from g/m^2 d

With increasing time the inhibition is reduced and the corrosion increases as demonstrated in Table 4. A decrease in the inhibition efficiency as the action time increases is likewise to be recorded by addition of other acids, such as sulfuric acid and phosphoric acid [38]. In spite of the relatively short test durations and the rapid decrease in the corrosion rates (see Figure 1), the corrosion rates are too high for industrial use.

About the same corrosion behavior in contact with brass is to be observed when inhibitors are present. $SnCl_2$ (5 %) and 1 % of size (no further information) have been used as inhibitors. Here also, cast iron corroded more severely than brass. The inhibitors mainly inhibited the electrochemical action of the local cells, and had hardly any influence on the cast iron-brass contact corrosion [36]. For further information on the resistance of unalloyed cast iron under the action of phosphoric acid, see Table 31.

The resistance of cast iron in phosphoric acid is influenced by alloying elements in the same way as that of rolled steels. For information on the increasing resistance, see under Sections High-alloy cast iron, high-silicon cast iron to Special iron-based alloys. In the case of these materials, the equivalent cast materials are similar in their resistance.

High-alloy cast iron, high-silicon cast iron

High-silicon cast iron in particular shows a good corrosion resistance under exposure to phosphoric acid, and especially in pure phosphoric acid [2, 5, 6, 12, 14, 15, 39–44].

It should be remembered that precisely phosphoric acids originating from the wet process, which contain hydrogen fluoride and fluorosilicic acid, are very aggressive toward high-silicon cast iron containing about 15 % Si, and are already so at room temperature.

Figures 2 and 3 give an overview of the position of the isocorrosion curves as a function of the concentration and temperature of the acid. The graph in Figure 2 was plotted in pure acid with access of air on the surface of the test solution, and Figure 3 was plotted under the same conditions but with aeration [14, 40].

The Si alloy Superchlor® SD51 can be used with relatively low corrosion rates under very corrosive conditions under which steels reach corrosion rates of more than 0.5 mm/a (19.7 mpy). SD51, according to ASTM specification A 518, is composed of 4–5 % Cr, 14.2–14.75 % Si, 0.7–1.1 % C, max. 1.5 % Mn, balance Fe [44]. It is approximately comparable to Sicro® 5 (G-X 90 SiCr 15 5). This alloy is melted with 0.9 % C, 15 % Si, 0.5 % Mn, 0.05 % P, 0.04 % S, 5 % Cr, 1 % Cu, balance Fe [45].

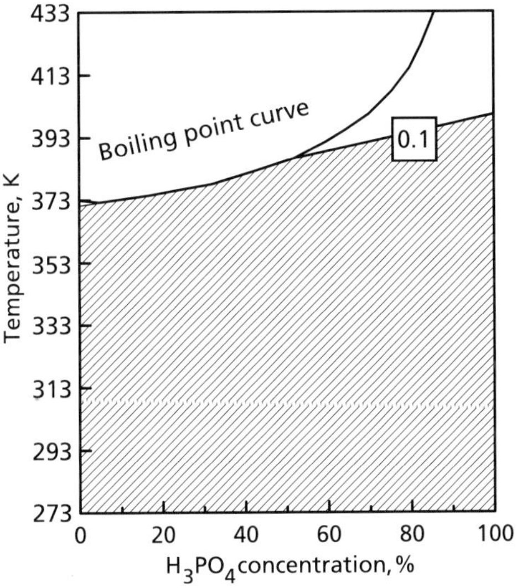

Figure 2: Isocorrosion curve (0.1 mm/a (3.94 mpy)) for high-silicon cast iron (G-X 70 Si 15) in pure phosphoric acid, access of air to the surface of the medium [40]

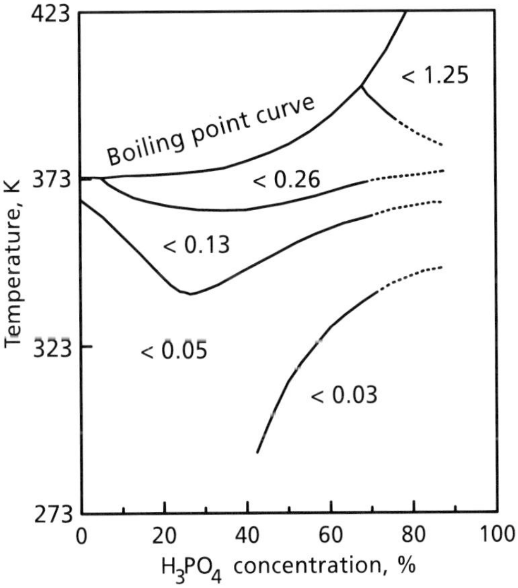

Figure 3: Isocorrosion curves (mm/a) for high-silicon cast iron (G-X 70 Si 15) in pure aerated H_3PO_4 [14]

< 0.03 mm/a (< 1.18 mpy)   < 0.13 mm/a (< 5.12 mpy)   < 1.25 mm/a (< 49.2 mpy)

< 0.05 mm/a (< 1.97 mpy)   < 0.26 mm/a (< 10.2 mpy)

The range of use for this alloy with Cr and Si is shown in Figure 4.

The very good resistance of high-silicon cast iron in comparison with other materials and the ranges of use are shown in Figure 4. The isocorrosion curves indicate the regions for a maximum corrosion rate of 0.5 mm/a (19.7 mpy) in pure acid. Phosphoric acid becomes more aggressive as a result of contamination, in particular as a result of fluorine content when high-Si cast iron is used. In such cases, the use ranges of the materials in Figure 4 shift by one level, and even more in very heavily contaminated acids. This means that the cast alloy 1.4408 (UNS J92900, GX5CrNiMo19-11-2) cannot be used. This may be the case, for example, at a high chloride content, which may trigger off pitting corrosion.

Table 6 contains a summary of corrosion rates for high-alloy cast materials from various sources (see also Figure 4 and Table 5).

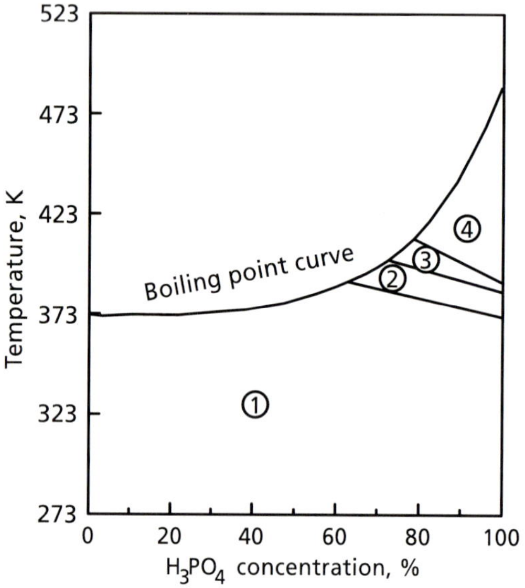

Figure 4: Isocorrosion curves (0.5 mm/a (19.7 mpy)) and ranges of use of high-alloy cast materials in pure phosphoric acid (with contamination and increasing aggressiveness, the range of use of the particular material falls by at least one level) [44]
① CF-8M (cf. 1.4408, UNS J92900, A 744)
② CD-4MCu (cf. A 744, 1.4463)
③ CN-7M (cf. 1.4500, UNS N08007)
④ SD51 (cf. A 518)

Table 5 shows a summary of corrosion rates for high-silicon cast iron in phosphoric acid under various conditions.

Figures 5–10 contain further isocorrosion diagrams of high-alloy cast alloys. For the resistance of these materials, see under Sections Ferritic chromium steels with more than 12% chromium (for Figures 5, 6), Austenitic CrNi steels (for Figure 7), Austenitic chromium-nickel steels with special alloying additions (for Figure 8). For further corrosion rates for high-alloy cast materials under the action of phosphoric acid, see also Tables 31 and 41.

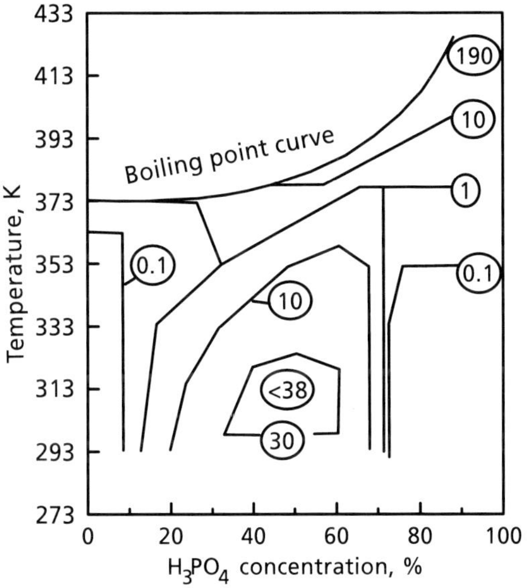

Figure 5: Isocorrosion curves (mm/a) in pure phosphoric acid for G-X 20 Cr 14 (1.4027) [14, 40]

0.1 mm/a (3.94 mpy)	10 mm/a (394 mpy)	38 mm/a (1,496 mpy)
1.0 mm/a (39.4 mpy)	30 mm/a (1,181 mpy)	190 mm/a (7,480 mpy)

Alloy content		*H_3PO_4 concentration*	*Tempera-ture*	*Corrosion rate mm/a (mpy)*	*Comparison material*
Si	*Mo*	*%*	*K (°C)*		
%					
13.9	–	77	373 (100)	< 1.0 (< 39.4)	G-X 70 Si 15
18.2	–	77	373 (100)	0.07 (2.76)	G-X 70 Si 15
14.5	–	1 to 85	343 (70)	0.05 (1.97)	G-X 70 Si 15
14.5	3	1 to 85	343 (70)	0.05 (1.97)	G-X 70 SiMo 15 3
14.5		1 to 80	373 (100)	< 0.5 (< 19.7)	G-X 70 Si 15
14.5	3	1 to 80	373 (100)	< 0.5 (< 19.7)	G-X 70 SiMo 15 3
15 to 16	–	all conc.	up to b.p.	0.1 (3.94)	G-X 70 Si 15
16	–	conc. (80 to 85)	473 (200)	0.1 to 0.3 (3.94 to 11.8)	G-X 70 Si 15
16	–	50	370 (97)	< 0.1 (< 3.94)	G-X 70 Si 15

Table 5: Summary of corrosion rates for high-silicon cast materials in phosphoric acid. (The sometimes quite different and high corrosion rates are due to impurities, in particular HF) [2, 14]

Table 5: Continued

Alloy content		H_3PO_4 concentration	Tempera-ture	Corrosion rate mm/a (mpy)	Comparison material
Si	*Mo*	%	K (°C)		
%					
16	–	dilute crude acid	370 (97)	3.0 (118)	G-X 70 Si 15
14.5 to 16	–	80	293 (20)	< 0.1 (< 3.94)	G-X 70 Si 15
14.5 to 16	–	80	333 (60)	< 0.1 (< 3.94)	G-X 70 Si 15
14.5 to 16	–	80	b.p.	> 10 (> 394)	G-X 70 Si 15
16 to 18	–	10	b.p.	< 0.1 (< 3.94)	G-X 70 Si 15
16 to 18	–	80	333 (60)	< 0.1 (< 3.94)	G-X 70 Si 15
16 to 18	–	80	b.p.	3 to 10 (118 to 394)	G-X 70 Si 15
16.8	2.6	17	313 (40)	4.9 (193)	G-X 70 SiMo 15 3
16.8	2.6	25	313 (40)	4.7 (185)	G-X 70 SiMo 15 3
16.8	2.6	42	313 (40)	0.7 to 3.7 (27.6 to 146)	G-X 70 SiMo 15 3
16.8	2.6	62	313 (40)	0.7 to 11.4 (27.6 to 449)	G-X 70 SiMo 15 3
16.8	2.6	75	313 (40)	0.65 (25.6)	G-X 70 SiMo 15 3
16.8	2.6	75	333 (60)	3.5 (138)	G-X 70 SiMo 15 3

Table 5: Summary of corrosion rates for high-silicon cast materials in phosphoric acid. (The sometimes quite different and high corrosion rates are due to impurities, in particular HF) [2, 14]

Material		Alloying elements %	H_3PO_4 concentration %	Temperature K (°C)	Corrosion rate mm/a (mpy)
CF-8 M; 0[1]	cf. 1.4408 GX5CrNiMo19-11-2 UNS J92900	16.7 Cr, 15 Ni, 2.5 Mo	40	boiling	0.040 (1.58)
			85	boiling	0.868 (34.2)
CF-8 M; 15[1]		20.2 Cr, 9.6 Ni, 2.4 Mo	40	boiling	0.019 (0.75)
			85	boiling	0.053 (2.09)
CF-8 M; 38[1]		21.9 Cr, 9.5 Ni, 2.65 Mo	40	boiling	0.006 (0.24)
			85	boiling	0.055 (2.17)
Ni-Resist® 1	0.6655, A 436, EN-GJLA-XNiCu Cr15-6-2	14 Ni, 6 Cr, 5 Cu	50	RT	0.13[2] (5.12)
			15	303 (30)	1.3[2] (51.2)
Ni-Resist® D-2C	0.7670, A 439, EN-GJSA-XNi22	21-24 Ni, 2.5 Si	85, aerated	303 (30)	5.4[2] (213)
Ni-Resist® 2	0.6660; A 436 (Type 2), GGL-NiCr 20 2	18-22 Ni, 1-2.5 Cr	85, aerated	303 (30)	2.2[2] (86.6)
NiCuCr-cast steel	approx. 0.6656, A 436 (Type 1B)	13-16 Ni, 6-8 Cu, 1.5-4.5 Cr	10	293 (20)	0.2 (7.87)
			10	343 (70)	0.7 (27.6)
NiCrMo-cast steel	approx. 1.4500, CN-7M	24-26 Ni, 19-21 Cr, 2.5-3.5 Mo	25	353 (80)	< 0.13 (< 5.12)
CF-16F	1.4312, GX10CrNi18-8	18 Cr, 8 Ni	crude acid	RT	not resistant
GX70Cr Mo29-2	1.4136	27-29 Cr, 2-2.5 Mo	45	boiling	< 0.13 (< 5.12)
Durimet® 20	approx. 1.4500	29 Ni, 20 Cr, 3 Cu, 2 Mo	5 to 80	373 (100)	< 0.05 (< 1.97)
			10 to 85	398 (125)	< 0.5 (< 19.7)
			5 to 70	343 (70)	< 0.05 (< 1.97)

1) These numbers signify the ferrite content in the structure in % by volume
2) flow rate: 5 m/min
3) impure acid with HF and H_2SiF_4

Table 6: Summary of corrosion rates of high-alloy cast materials in phosphoric acid solutions under various conditions [14]

Table 6: Continued

Material		*Alloying elements %*	*H_3PO_4 concentration %*	*Temperature K (°C)*	*Corrosion rate mm/a (mpy)*
Worthite®	approx. 1.4585, GX7CrNiMoCu Nb18-18	24 Ni, 20 Cr, 3 Mo, 1.8 Cu	up to 80	393 (120)	< 0.05 (< 1.97)
			40	623 (350)	not resistant
			up to 60$^{3)}$	333 (60)	< 0.05 (< 1.97)
			up to 60$^{3)}$	373 (100)	< 0.5 (< 19.7)
Langalloy® 20 V	1.4500, CN-7M	29 Ni, 20 Cr, 4 Mo, 4 Cu	up to 35	boiling	resistant
Langalloy® 3 V	1.4408, UNS J92900	18 Cr, 10 Ni, 3 Mo	50	353 (80)	0.402 (15.8)
			89	353 (80)	0.153 (6.0)
Langalloy® 40 V	1.4340, UNS J92615, GX40CrNi27-4	25 Cr, 5 Ni + Cu + Mo	50	353 (80)	0.015 (0.59)

1) These numbers signify the ferrite content in the structure in % by volume
2) flow rate: 5 m/min
3) impure acid with HF and H_2SiF_4

Table 6: Summary of corrosion rates of high-alloy cast materials in phosphoric acid solutions under various conditions [14]

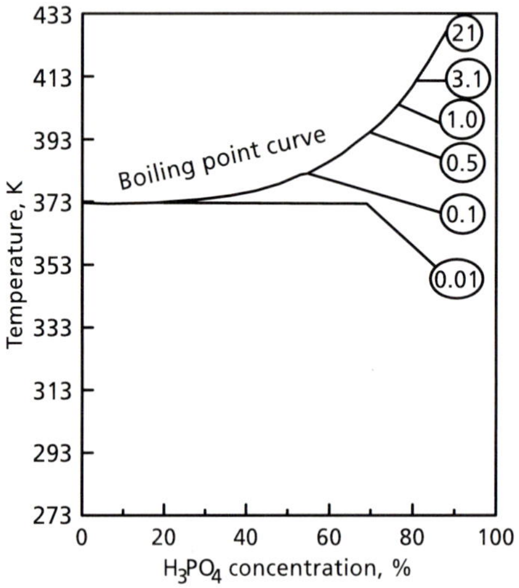

Figure 6: Isocorrosion curves (mm/a) in pure phosphoric acid for 1.4085 (GX70Cr29) [14, 40]

0.01 mm/a (0.39 mpy) 0.5 mm/a (19.7 mpy) 3.1 mm/a (122 mpy)
0.1 mm/a (3.94 mpy) 1.0 mm/a (39.4 mpy) 21 mm/a (827 mpy)

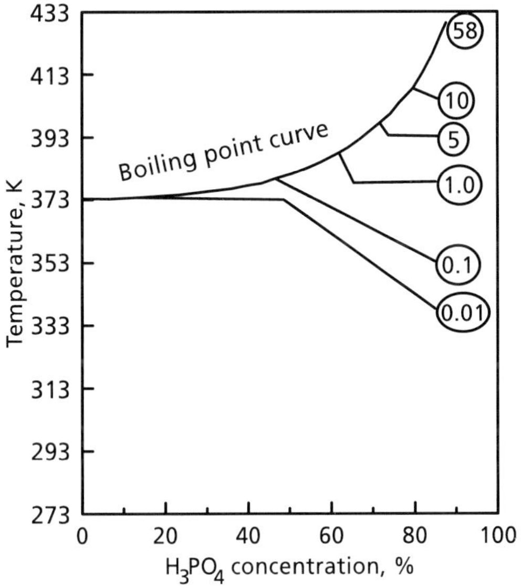

Figure 7: Isocorrosion curves (mm/a) in pure phosphoric acid for 1.4312 (CF-16F, GX10CrNi18-8) [14]

0.01 mm/a (0.39 mpy)	1.0 mm/a (39.4 mpy)	10 mm/a (394 mpy)
0.1 mm/a (3.94 mpy)	5.0 mm/a (197 mpy)	58 mm/a (2,283 mpy)

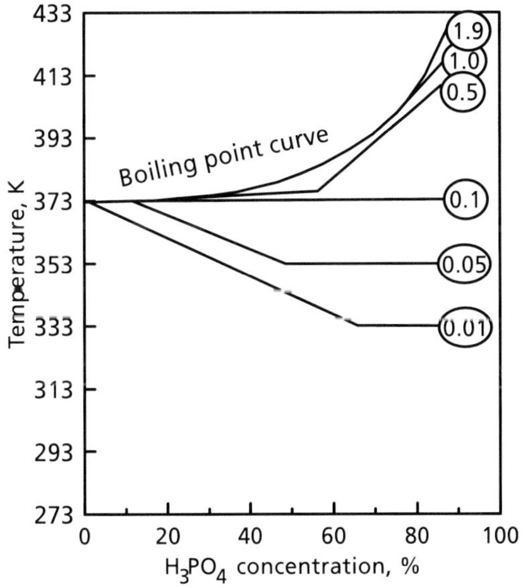

Figure 8: Isocorrosion curves (mm/a) in pure phosphoric acid for 1.4500 (CN-7M, GX7NiCrMo-CuNb25-20) [14]

0.01 mm/a (0.39 mpy)	0.1 mm/a (3.94 mpy)	1.0 mm/a (39.4 mpy)
0.05 mm/a (1.97 mpy)	0.5 mm/a (19.7 mpy)	1.9 mm/a (74.8 mpy)

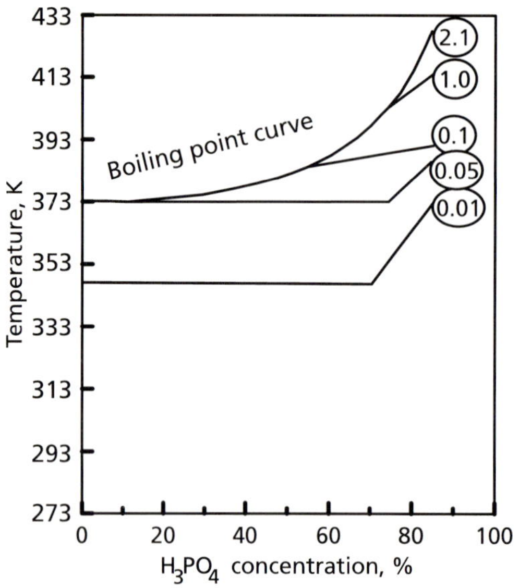

Figure 9: Isocorrosion curves (mm/a) in pure phosphoric acid for 2.4883 (Hastelloy® C alloy, UNS N10002) [14]

0.01 mm/a (0.39 mpy)	0.1 mm/a (3.94 mpy)	2.1 mm/a (82.7 mpy)
0.05 mm/a (1.97 mpy)	1.0 mm/a (39.4 mpy)	

Structural steels with up to 12 % chromium

The resistance of these steels depends on the chromium content. Compared with sulfuric acid, phosphoric acid is the less aggressive acid. As the chromium content increases, the resistance of these steels to phosphoric acid also increases.

In principle, steels with up to 12 % chromium cannot be used in phosphoric acid [5, 8]. They do not have a substantially better resistance in phosphoric acid than unalloyed or low-alloy steels. Steels with up to 12 % chromium, and not only these, have a corrosion resistance which depends on the heat treatment. The best resistance, if resistance can be referred to at all in phosphoric acid, is to be expected from steels having a largely homogeneous structure. This means that steels which have been given a heat treatment for the purpose of increasing the mechanical properties are even less resistant. They tend to suffer more from local corrosion than those without precipitates.

The corrosion rates of steels with about 12 % Cr in phosphoric acid of all concentrations at about room temperature are reported as being above 1.27 mm/a (50.0 mpy), regardless of whether the acid is aerated or non-aerated [8].

The increasing resistance of these steels as the Cr content increases and the dependence of this on the heat treatment is described in [28, 46–48].

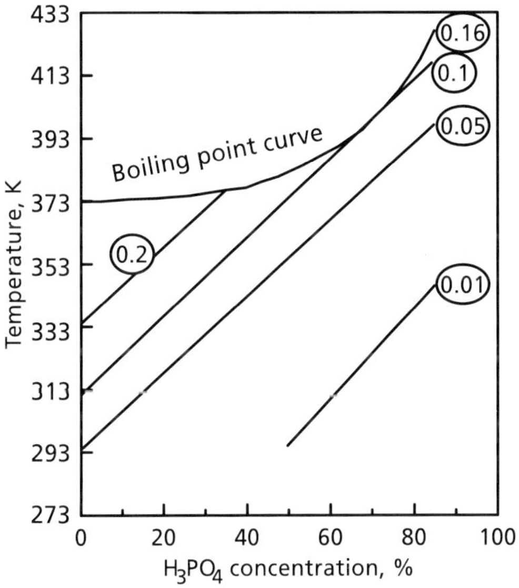

Figure 10: Isocorrosion curves (mm/a) in pure phosphoric acid for 2.4882 (Hastelloy® B alloy, UNS N10001) [14]

0.01 mm/a (0.39 mpy)	0.1 mm/a (3.94 mpy)
0.05 mm/a (1.97 mpy)	0.16 mm/a (6.30 mpy)

Although determined over a test duration of only 24 h, Figure 11 shows a drastic reduction in the corrosion rates of hardened chromium steels in 35 % H_3PO_4 at 293 K (20 °C) as the chromium content increases, and in particular above 12 % [47].

Figure 12 shows a better overview of the corrosion behavior of chromium steels in phosphoric acid over the entire concentration range. This graph shows that industrially usable corrosion rates are to be expected only in very dilute phosphoric acid, or at relatively high Cr contents [48]. In this context, see also under Section Ferritic chromium steels with more than 12 % chromium.

Figure 13 shows a summarizing graph of the corrosion resistance in pure phosphoric acid (less aggressive than contaminated). This figure shows the isocorrosion diagram for the steel SAE 409 (cf. 1.4512, X2CrTi12; 1.4720, X7CrTi12). This material of relatively high C content and with 10.5–12.5 % Cr is at the upper limit of the Cr content in this section. This steel can be used in phosphoric acid in ranges of up to 0.13 mm/a (5.12 mpy). Nevertheless, there are restrictions at about 20 % and 80 % acid.

On the other hand, this graph contradicts [8], according to which steels with 12 % Cr have corrosion rates above 1.27 mm/a (50.0 mpy) in the entire concentration range at room temperature. From this reference, no conclusions can be drawn on the range below 10 % H_3PO_4. This steel is resistant in 80 % acid (non-aerated) at 339 K (66 °C) [8]. See also Figures 5, 6.

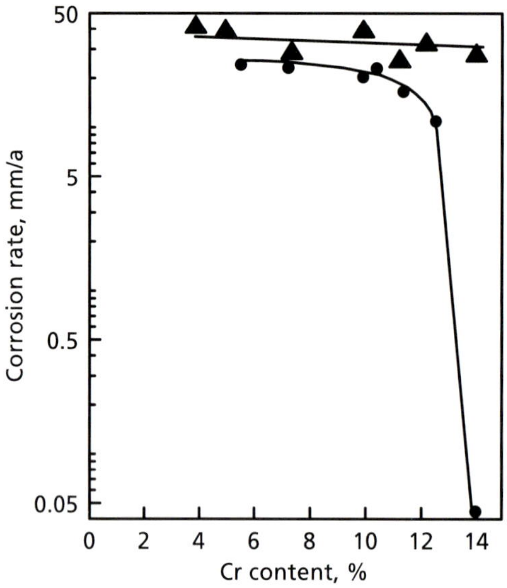

Figure 11: Corrosion rates of hardened Cr steels in 35 % H_3PO_4 at 293 K (20 °C, ●) as a function of the chromium content, test duration 24 h [47]
▲ 2.5 % H_2SO_4 at 293 K (20 °C) for comparison

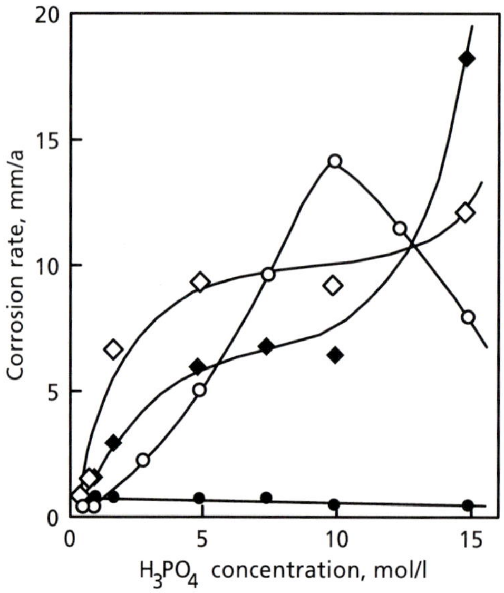

Figure 12: Influence of the Cr content of steels on the corrosion behavior in phosphoric acid of varying concentration at 293 K (20 °C) test duration 112 h [48]
◆ 7.5 % Cr, ◇ 10 % Cr, ○ 12 % Cr, ● 29 % Cr

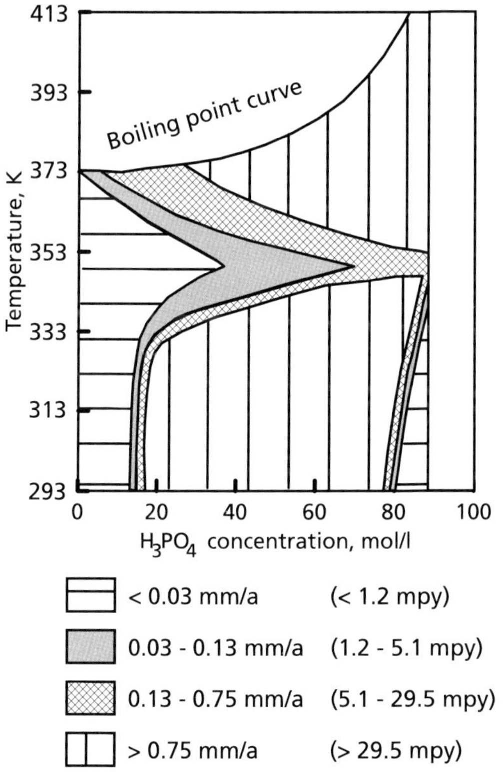

Figure 13: Isocorrosion diagram for SAE 409 (cf. 1.4512, X2CrTi12; 1.4720, X7CrTi12) in pure phosphoric acid [49]

From these comments, it can be concluded that steels with up to 12 % Cr should not be used in phosphoric acid. Exceptions would be short-term use, very low concentrations or prior corrosion studies under the concrete conditions.

From this group of materials, the material CA6NM (cf. UNS J91540, 1.4313) is the most suitable for production as a grinding body material for comminution of phosphate rock under the operating conditions of a plant in Florida (see Tables 13–15 and Figures 21 and 22) [18]. Nevertheless, these materials cannot be referred to as ferritic steels. The classification was made according to the chromium content (Table 13).

Ferritic chromium steels with more than 12 % chromium

As can be seen from Figures 11 and 12, an increasing resistance toward phosphoric acid is to be expected as the Cr content of the steels increases.

According to [8], corrosion rates of more than 1.27 mm/a (50.0 mpy) are reported for 17 % Cr steels, as well as for those with 12 % Cr, in the entire concentration range at room temperature. The 0–10 % acid range is excluded.

The high resistance of "corrosion resistant steels" to chemicals is associated with the formation of a very thin passive film on the surface under the corrosion conditions. This passive film, which forms spontaneously, is a film of metal oxide only a few layers of atoms thick. Increasing chromium contents promote the formation of this passive film and consequently lead to an increasing resistance of these steels.

In non-oxidizing acids, such as in phosphoric acid, there is the danger that no stable passive film can form. In this case, the steel corrodes in the active state with severe surface removal. Low pH values and high temperatures counteract passivation. As already mentioned, the resistance of these steels is increased by, in addition to chromium, high Ni and Mo contents, and in particular cases also Cu contents. Such steels achieve low passivation current densities and, if they do not remain passive, may remain industrially usable in the active state due to sufficiently low corrosion rates. The passive state can be promoted by oxidizing agents in the medium, such as oxygen and Fe^{3+} and Cu^{2+} ions. Under these conditions, corrosion is severely limited. On the other hand, if such additives do not lead to passivity, they may intensify corrosion in the active state. Chloride and fluoride make passivation in non-oxidizing media more difficult. Fluoride increases the corrosion rate of high-alloy steels in the passive state. The presence of chloride can cause pitting corrosion, crevice corrosion or stress corrosion cracking [50].

Molybdenum in particular helps to increase the stability of the passive layer, and already does so at a 13 % Cr content. It is an important alloying constituent which in particular increases the resistance of the materials to pitting corrosion [50, 51].

The increase in corrosion resistance of steels under exposure to phosphoric acid by the measures discussed can be seen from Figure 14, with the aid of the isocorrosion curves at 0.1 mm/a (3.94 mpy). The use range in phosphoric acid rises in the sequence 18 % Cr steel with 2 % Mo ①, CrNi steel 18 9 ②, CrNiMo steel 18 12 3 ③ and CrNiMoN steel 17 13 5 ④ [50].

This illustration of the sequence of corrosion resistance of high-alloy steels is particularly important since the range of steels above 13 % Cr is widely spread, on the one hand by the chromium content of up to 30 % or more, and on the other hand by the addition of molybdenum or other alloying elements. Preliminary selection of Cr steels under the action of phosphoric acid should thus be facilitated.

Corrosion studies in 10–80 % phosphoric acid at 353 K (80 °C) (deaerated by flushing with nitrogen, specimens were activated) showed the following results:

- The corrosion resistance of a 17 % Cr steel with 3 % Mo is greater than that of the same steel with only 1 % Mo.
- Copper contents and silicon contents (2 % Cu, 1 % Si) increase the corrosion resistance of 17 % Cr steel with 3 % Mo at all concentrations.
- Simultaneous addition of all improving alloying elements does not lead to an additional improvement in the corrosion resistance of Cr steels [51].

A large number of isocorrosion diagrams with sometimes also different illustrations of the corrosion resistance are founded on this. Figures 5 and 6 show isocorrosion curves for Cr steels with about 13 % Cr and 30 % Cr. With 17 % Cr, the steel

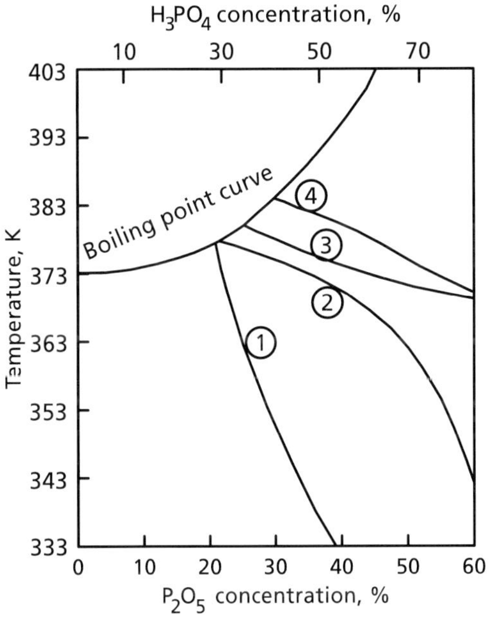

Figure 14: Influence of alloying elements on the increase in corrosion resistance of high-alloy steels in pure phosphoric acid with the aid of the isocorrosion curves (0.1 mm/a (3.94 mpy)) [50, 52]

① 18 Cr, 2.5 Mo, Ti	approx. 1.4521, SAE 444
② 18 Cr, 9 Ni	approx. 1.4306, SAE 304 L
③ 18 Cr, 12 Ni, 3 Mo	approx. 1.4436, SAE 316
④ 17 Cr, 13 Ni, 5 Mo, N	approx. 1.4439, SAE 317 LMN

1.4016 (SAE 430, X6Cr17) is between these two steels in its resistance. Figure 15 shows the isocorrosion diagram.

If the three figures are compared with one another, a significant increase in resistance can be seen as the Cr content increases (see also Figure 67). The steel with 17 % Cr shows the low resistance mentioned in [8] in phosphoric acid when used industrially (in this case, corrosion rates of not more than 0.1 mm/a (3.94 mpy) count as resistance).

While Figure 15 shows a less good to poor resistance in the concentration range between about 5 % and 65 % H_3PO_4, according to Figure 16 [49], 17 % Cr steel can be used, with a corrosion rate of about or less than 0.1 mm/a (3.94 mpy).

According to [40] and [49], the two steels correspond to one another in their chemical analysis. Pure phosphoric acid was used for the graph in Figure 15. Air had free access to the surface of the test solution [40].

Table 2 gives an overview of the resistance of high-alloy steels. It shows a significant increase in resistance under exposure to phosphoric acid when Mo is added to 17 % Cr steels. Given such serious differences, their use should be doubted, or certainty should be achieved by in-house studies under concrete conditions.

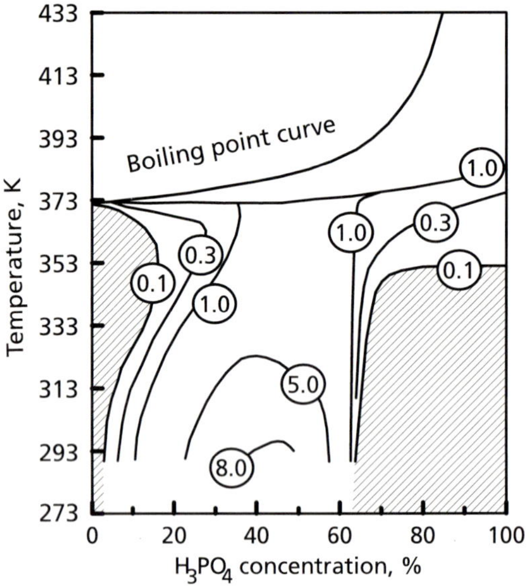

Figure 15: Isocorrosion diagram (mm/a (mpy)) for chromium steel (0.22 C, 17 Cr, traces of Ni), approximately 1.4016 (SAE 430, X6Cr17), in pure phosphoric acid (air has access to the test solution, see also Figure 16) [40]

0.1 mm/a (3.94 mpy) 1.0 mm/a (39.4 mpy) 8.0 mm/a (315 mpy)
0.3 mm/a (11.8 mpy) 5.0 mm/a (197 mpy)

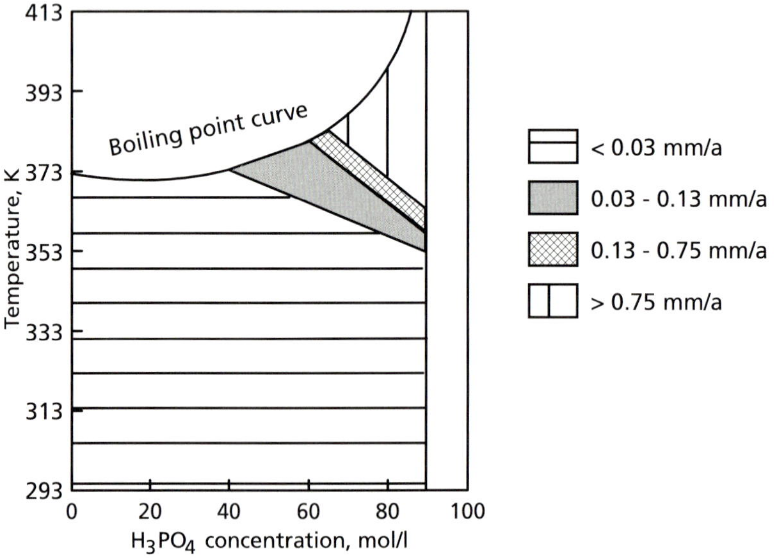

Figure 16: Resistance ranges for Cr steel, Type SAE 430 (1.4016, X6Cr17) in pure phosphoric acid (see also Figure 15) [49]

Medium	Concen-tration	Tempera-ture K (°C)	1.4000 1.4002 1.4005 1.4006 1.4021 1.4024 1.4034 1.4104 1.4116 1.4117 1.4125 1.4512	1.4016 1.4057 1.4112 1.4120 1.4122 1.4305 1.4510 1.4511	1.4113 1.4542 1.4568	1.4301 1.4303 1.4306 1.4310 1.4311 1.4541 1.4550	1.4521 1.4522	1.4401 1.4404 1.4406 1.4429 1.4435 1.4436 1.4460 1.4571 1.4580	1.4465• 1.4505^+
Section			**A**	**A**	**A/B**	**C**	**A**	**B**	**D/E**
Phosphoric acid, chem. pure H_3PO_4	1 %	293 (20)	–	0	0	0	0	0	
		boiling	1	1	0	0	0	0	
	10 %	293 (20)	2	1	0	0	0	0	
		boiling	2	2	0	0	0	0	
	45 %	293 (20)	2	2	1	0	1	0	
		boiling	3	2	2	2	–	1	1^+/0•
	60 %	293 (20)	2	2	1	0	1	0	
		boiling	3	3	2	2	–	1	1^+•
	70 %	293 (20)	2	2	1	0	1	0	
		boiling	3	3	2	2	2	2	1^+•
	80 %	293 (20)	2	2	1	1	1	0	
		boiling	3	3	3	3	–	2	1^+
	conc.	293 (20)	2	2	1	1	1	0	
		boiling	3	3	3	3		3	
Phosphoric anhydride = phosphoric pentoxide P_2O_5	dry or moist	293 (20)	–	–	1	1	1	0	

0 ≙ < 0.11 mm/a (< 4.33 mpy); 1 ≙ 0.11–1.1 mm/a (4.33–43.3 mpy); 2 ≙ 1.1–11 mm/a (43.3–433 mpy); 3 ≙ > 11.0 mm/a (> 433 mpy)

A: see section Ferritic chromium steels with more than 12 % chromium
B: see section Ferritic-austenitic steels with more than 12 % chromium
C: see section Austenitic CrNi steels
D: see section Austenitic CrNiMo(N) steels
E: see section Austenitic chromium-nickel steels with special alloying additions

Table 7: Overview of corrosion rates (for the ratings, see footnote) of high-alloy materials in pure phosphoric acid (see also Table 2). For the corresponding material designations, see Table 8 [55] (similar data in [53, 54])

Tables 2 and 7 contain a tabular overview of corrosion rates (see also Figure 27). Columns 1 and 2 (Table 7) contain mainly 13 and 17 % Cr steels with a very low resistance in phosphoric acid. The steel 1.4113 (SAE 434, X6CrMo17-1 with 0.5 to 1.5 % Mo) already shows a better resistance at low temperatures because of the Mo content, and this also becomes particularly noticeable at an increase in concentration and temperature. It is comparable to 1.4542 (SAE 630, Section Ferritic-austenitic steels with more than 12 % chromium) and 1.4568 (SAE 631, Section Austenitic CrNi steels) in its range of use. Higher-alloy Mo containing Cr steels (column 5) 1.4521 (SAE 444, X2CrMoTi18-2) and 1.4522 (SAE 443, X2CrMoNb18-2) with 1.8-2.3 % Mo show a visibly better resistance according to this table [55]. About the same use ranges are mentioned in [56] for cast materials.

DIN-Mat. No.	***DIN-designation***	***Comparable SAE types***[1)]	***UNS No.***
1.4000	X6Cr13	403, 410[1)]	cf. S40300
1.4002	X6CrAl13	405	cf. S40500
1.4005	X12CrS13	416	cf. S41600
1.4006	X12Cr13	410[1)]	cf. S41000
1.4016	X6Cr17	430	cf. S43000
1.4021	X20Cr13	420[1)]	cf. S42000
1.4024	X15Cr13	UNS J91201	cf. J91201
1.4034	X46Cr13	–	–
1.4057	X17CrNi16-2	431	cf. S43100
1.4104	X14CrMoS17	430 F	cf. S43020
1.4112	X90CrMoV18	440 B	cf. S44003
1.4113	X6CrMo17-1	434	cf. S43400
1.4116	X50CrMoV15	–	–
1.4117	X38CrMoV15	–	–
1.4120	X20CrMo13	–	–
1.4122	X39CrMo17-1	–	–
1.4125	X105CrMo17	440 C	cf. S44004
1.4301	X5CrNi18-10	304	cf. S30400
1.4303	X4CrNi18-12	305 L	cf. S30800
1.4305	X8CrNiS18-9	303	cf. S30300

1) SAE grades may correspond to several DIN materials

Table 8: Overview of the materials used in Table 7 and their resistance in phosphoric acid [55]

Table 8: Continued

DIN-Mat. No.	*DIN-designation*	*Comparable SAE types*[1)]	*UNS No.*
1.4306	X2CrNi19-11	304 L	cf. S30403
1.4310	X10CrNi18-8	302	cf. S30100
1.4311	X2CrNiN18-10	304 LN	cf. S30453
1.4401	X5CrNiMo17-12-2	316	cf. S31600
1.4404	X2CrNiMo17-12-2	316 L	cf. S31603
1.4406	X2CrNiMoN17-11-2	316 LN[1)]	cf. S31653
1.4429	X2CrNiMoN17-13-3	SAE 316 LN[1)]	cf. S31653
1.4435	X2CrNiMo18-14-3	SAE 316 L	cf. S31603
1.4436	X3CrNiMo17-13-3	SAE 316	cf. S31600
1.4438	X2CrNiMo18-15-4	317 L	cf. S31703
1.4439	X2CrNiMoN17-13-5	317 LMN	cf. S31703
1.4460	X3CrNiMoN27-5-2	329	cf. S31260
1.4465	X1CrNiMoN25-25-2	310 MoLN	cf. N08310
1.4505	X4NiCrMoCuNb20-18-2	–	–
1.4510	X3CrTi17	430 Ti	cf. S43035
1.4511	X3CrNb17	(430 Cb)	–
1.4512	X2CrTi12	409	cf. S40900
1.4521	X2CrMoTi18-2	443, 444	cf. S44400
1.4522	X2CrMoNb18-2	443	cf. S44300
1.4541	X6CrNiTi18-10	321	cf. S32100
1.4542	X5CrNiCuNb16-4	630	cf. S17400
1.4550	X6CrNiNb18-10	348	cf. S34700
1.4568	X7CrNiAl17-7	631	cf. S17700
1.4571	X6CrNiMoTi17-12-2	316 Ti	cf. S31635
1.4580	X6CrNiMoNb17-12-2	316 Cb	cf. S31640

1) SAE grades may correspond to several DIN materials

Table 8: Overview of the materials used in Table 7 and their resistance in phosphoric acid [55]

In boiling 50 % phosphoric acid, the CrMo steel 18 2 with a corrosion rate of 0.12 mm/a (4.72 mpy) is superior to steel 1.4401 (SAE 316) with 0.2 mm/a (7.87 mpy), and even very superior to steel 1.4301 (SAE 304) with more than 10 mm/a (394 mpy), test duration only 24 h [57, 58].

With a sulfur content of 0.3 %, this steel is used as machining steel. As a result of its Mo content, it has a low susceptibility to pitting corrosion and stress corrosion cracking [59].

Machining steels are given increased contents of sulfur to improve machinability. For further details in this context, see under Section Austenitic CrNi steels. Since manganese sulfides are less corrosion resistant than titanium sulfides, the effect of titanium on the corrosion resistance of machining steels compared with standard steels was also investigated by an appropriate titanium addition. Table 9 contains a summary of corrosion rates.

It can be seen from this table that sulfur containing ferritic chromium-molybdenum steel has a lower corrosion resistance than standard steel 1.4113 (SAE 434). In contrast, the sulfur and titanium containing steel has a significantly better corrosion resistance than the standard material. Table 9a contains the alloy composition.

The machinability of these machining steels is only slightly impaired by this alloying measure. This steel shows no susceptibility to pitting corrosion even in up to 4 % NaCl solution in the pH range from 0 to 7, while this corrosion is already observed at low NaCl concentrations with standard steel, and more vigorously with sulfur containing steel [60]. As already mentioned at the beginning, the corrosion resistance of chromium steels increases with an increasing alloying content. This is particularly the case with steel 1.4575 (25-4-4, Cronifer 2803 Mo) under exposure to phosphoric acid [60–63]. See also Table 7.

According to Table 10, it shows the best resistance in 60 % boiling phosphoric acid, and under these conditions (only short-term testing) is even superior to the higher-alloy steels. The same good resistance of 1.4575 is shown in 74.5 % phosphoric acid (54 % P_2O_5), as can be seen from Figure 17 [63]. In this acid with 3 % SO_3, 2 % HF, 1 % Fe^{3+} and 0.1 % Cl^-, the steel is even superior to the Ni alloy 2.4811 (NiCr 20 Mo 15†) at 363 K (90 °C) with a corrosion rate of only about 0.1 mm/a (3.94 mpy). This composition corresponds to industrial phosphoric acid [63].

In addition to this generally good corrosion resistance, the high-alloy ferritic chromium steels are distinguished by a good resistance to pitting corrosion, in particular as a result of their molybdenum content. As well as having metallurgical reasons, chlorine ions are almost exclusively the reason for the occurrence of local types of corrosion, such as pitting corrosion, crevice corrosion and transcrystalline stress corrosion cracking , in commercially available high-alloy steels exposed to chemicals. Because the areas acting as anodes are very small in relation to the cathodic passive surface, rapid destruction of the material (for example leakages) occurs at such points. Comparisons in respect of the pitting corrosion behavior of the various materials can be made from their pitting corrosion potentials.

H_3PO_4 concentration %	1.4301[1]			1.4436[1]			1.4113[1]		
	–	+ S	+ S + Ti	–	+ S	+ S + Ti	–	+ S	+ S + Ti
	Corrosion rates, mm/a (mpy)								
10	0.02 (0.79)	95.0 (3,740)	0	0	0	0	0	1.34 (52.8)	0
20	41.0 (1,614)	–	81.0 (3,189)	0	0	0	0.02 (0.79)	17.0 (669)	0.02 (0.79)
30	72.2 (2,843)	–	149 (5,866)	0	0	0	–	–	–
40	136(5,354)	–	270 (10,630)	0.49 (19.3)	0.11 (4.33)	0.07 (2.76)	0.62 (24.4)	–	0
50	–	–	–	0.78 (30.7)	0.32 (12.6)	0.21 (8.27)	0.44 (17.3)	–	0.25 (9.84)
60	–	–	–	1.08 (42.5)	1.55 (61.0)	0.51 (20.1)	–	–	–
70	–	–	–	2.35 (92.5)	9.76 (384)	4.30 (169)	0.92 (36.2)	–	0.73 (28.7)
85	–	–	–	–	–	–	34.7 (1,366)	1072 (42,205)	20.0 (787)

1) for the chemical composition, see Table 9a

Table 9: Corrosion behavior of automatic steels (with S or S + Ti) in comparison with standard grades in boiling phosphoric acid [60]

DIN-Mat. No.[1)]	Chemical composition, %									Section
	C	Si	Mn	P	S	Cr	Mo	Ni	Ti	
1.4113	0.070	0.42	1.26	0.026	0.008	16.98	0.83	0.10	< 0.01	A
1.4113 + S	0.066	0.40	1.32	0.010	0.240	17.24	0.84	0.01	< 0.01	A
1.4113 + S + Ti	0.063	0.42	1.40	0.007	0.280	17.23	0.84	0.06	1.31	A
1.4301	0.029	0.31	1.97	0.007	0.014	18.39	0.02	9.56	< 0.01	C
1.4301 + S	0.030	0.40	1.82	0.009	0.264	18.88	< 0.01	10.33	< 0.01	C
1.4301 + S + Ti	0.026	0.50	1.82	0.005	0.226	17.67	< 0.01	10.40	1.30	C
1.4301 + S	0.029	0.45	1.68	0.006	0.250	17.53	0.05	10.90	< 0.01	C
1.4436	0.016	0.43	1.68	0.009	0.017	17.83	2.79	14.16	< 0.01	D
1.4436 + S	0.038	0.43	1.74	0.007	0.230	17.84	2.78	14.53	< 0.01	D
1.4436 + S + Ti	0.018	0.44	1.88	0.010	0.235	17.65	2.81	14.02	1.37	D

1) 1.4113 (SAE 434); 1.4301 (SAE 304); 1.4436 (SAE 316)

A: see section Ferritic chromium steels with more than 12 % chromium
C: see section Austenitic CrNi steels
D: see section Austenitic CrNiMo(N) steels

Table 9a: Chemical composition of automatic steels (standard material plus S and Ti) for corrosion studies according to Table 9 [60]

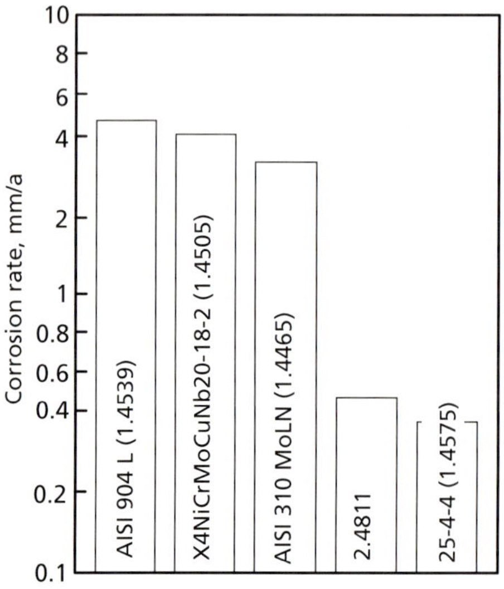

Figure 17: Corrosion behavior of high-alloy chemical plant materials in comparison with 1.4575 (UNS S44635) in industrial phosphoric acid (54 % P_2O_5, 3 % SO_3, 2 % HF, 1 % Fe^{3+}, 0.1 % Cl^-) at 363 K (90 °C) (the corrosion rates were calculated from g/m^2 h) [63]

H_3PO_4 concentration %	Temperature K (°C)	Material designation							Literature
		1.4016 (cf. SAE 430)	1.4113 (cf. SAE 434)	1.4521/22 (cf. SAE 444/ SAE 443)	1.4575 (cf. UNS S44635)	1.4460 (SAE 329)	1.4404 (SAE 316 L)	1.4577 (X3CrNi-MoTi25-25)	
		Corrosion rate, mm/a (mpy)							
50	b.p.			0.01 (0.39)			0.35 (13.8)	0.01 (0.39)	[62]
60	b.p.	> 10 (> 394)	1–10 (39.4–394)		< 0.1 (< 3.94)	0.1–1 (3.94–39.4)			[61]
60	b.p.			0.13 (5.12)			0.88 (34.6)	1.2 (47.3)	[62]
70	b.p.			0.51 (20.1)			3.5 (138)	2.1 (82.7)	[62]

b.p.: boiling point

Table 10: Corrosion behavior of Cr steels in comparison with higher-alloy steels in boiling H_3PO_4 after short-term studies over 24 h [61, 62]

In neutral, air-saturated chloride solutions, passivatable materials assume a rest potential of about +400 mV (based on the hydrogen electrode). Figure 18 contains such a graph in 3 % aerated NaCl solution for various temperatures. Resistance to pitting corrosion exists in this or any other test medium if the pitting corrosion potential of the material under investigation is more positive than its rest potential. From Figure 18, it can be seen that steel 1.4575 (25-4-4, Cronifer 2803 Mo) remains free from pitting corrosion in this solution up to the boiling point. All other steels are susceptible to pitting corrosion at increasing temperature when the +400 mV limit is exceeded.

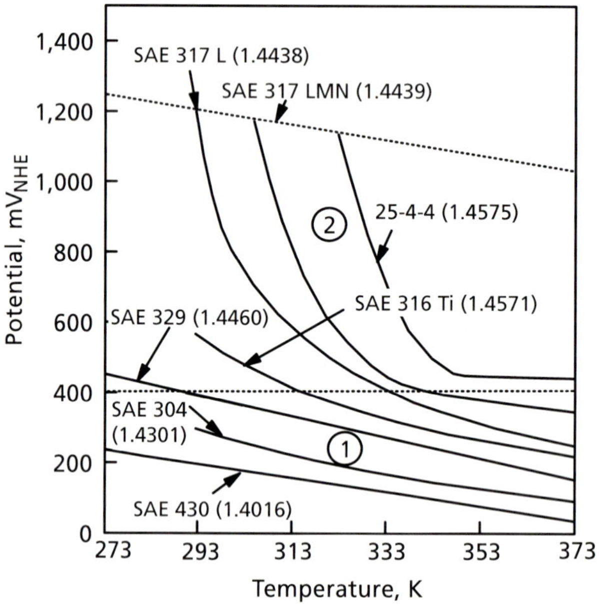

Figure 18: Pitting corrosion potentials of high-alloy steels in 3 % aerated NaCl solution [61, 63]
① region of susceptibility to pitting corrosion
② region of resistance to pitting corrosion

In other media, including those containing phosphoric acid, this threshold will be at a different point than +400 mV even if other comparison electrodes are used. It is important that the pitting corrosion potential must lie outside the passive range of the steel used, and in fact must be more positive than this, at least above the rest potential. In this context, see also [64–66].

This also explains in some cases the good resistance of 1.4575 in Figure 17 in Cl^- containing phosphoric acid. The pitting resistance equivalent (PRE) of the materials is a measure of the increase in resistance to pitting corrosion

PRE = 1 x % Cr + 3.3 x % Mo + 30 x % N

This is an empirical equation based on experimental results (see Table 42).

It can be deduced from this equation that Cr and in particular Mo are the critical alloying elements for improving the resistance to pitting and crevice corrosion in the material. It must be added that an adequate Ni content is the basis of this. Nitrogen also has a very favorable effect on the increase in resistance to pitting corrosion according to this equation. However, this applies only in Mo containing steels. In this context, see also Section Ferritic-austenitic steels with more than 12% chromium.

The so called superferrites with an extremely high chromium content show very good abrasion corrosion behavior, in comparison with other high-alloy materials, in the production of wet phosphoric acid in suspensions containing sulfuric acid. Table 11 contains a summary of two suspensions, and Table 12 contains the materials investigated. Of the suspensions investigated, "I" in Table 11 proved to be the more aggressive for all the alloys used, including under plant conditions. Figures 19 and 20 contain the material consumption rates as a function of the stirring velocity and temperature. Under the more adverse corrosive conditions of solution "I", the superferrite with 31% Cr (No. 1, Table 12) shows a material consumption rate of about 13 g/m^2 d at 353 K (80 °C) (test duration 8 h). This approximately corresponds to a corrosion rate of 0.61 mm/a (24.0 mpy) [67].

Content %	**Suspension I**		**Suspension II**	
	Filtrate	**Solid**	**Filtrate**	**Solid**
P_2O_5	28.9	1.27	29.2	0.75
CaO	0.21	32.5	0.20	31.9
SO_3	2.51	43.4	2.85	43.4
MgO	0.72	0.29	0.70	0.49
H_2SO_4	3.07	–	3.49	–
CO	0.023	0.43	0.021	0.39
SiO_2	0.32	1.29	0.43	1.19
FeO_3	0.15	0.02	0.15	0.01
Al_2O_3	0.10	0.00	0.09	0.00
F^-	1.27	2.58	1.25	2.10
Cl^- (ppm)	460	85.0	452	98.0
Solids content, %	28.4		38.6	
Density, g/cm^3	1.476		1.590	

Table 11: Chemical composition of phosphate suspensions for production of phosphoric acid by the wet process [67]

Serial No.	Material	Designation/ number	Section	Chemical composition, %[1)]						
				C	Cr	Ni	Mo	Mn	Cu	Si
1	Fonte au chrome (F-Cr)	–	A	1	31	1.85	1.95	0.68	–	2.81
2	Uranus® B6 (UB6)	SAE 904 L 1.4539	E	0.02	20	26.8	4	1.28	2	0.2
3	Uranus® 50 (U50)	UNS S32404	B	0.03	20.3	8.50	2.5	1	1.58	0.6
4	Hastelloy® G (HG)	UNS N06007 NiCr22Mo6Cu 2.4618		0.03	23	43.4	6.9	1.3	2.13	
5	SAE 316 L	1.4404 1.4435	D	0.03	18	12	2.5	1.5	–	–

1) balance Fe

A: see section Ferritic chromium steels with more than 12 % chromium
B: see section Ferritic-austenitic steels with more than 12 % chromium
D: see section Austenitic CrNiMo(N) steels
E: see section Austenitic chromium-nickel steels with special alloying additions

Table 12: Chemical composition of materials which were investigated in crude phosphate suspensions containing sulfuric acid for production of phosphoric acid by the wet process, and their abrasion corrosion resistance (see Table 11) [67]

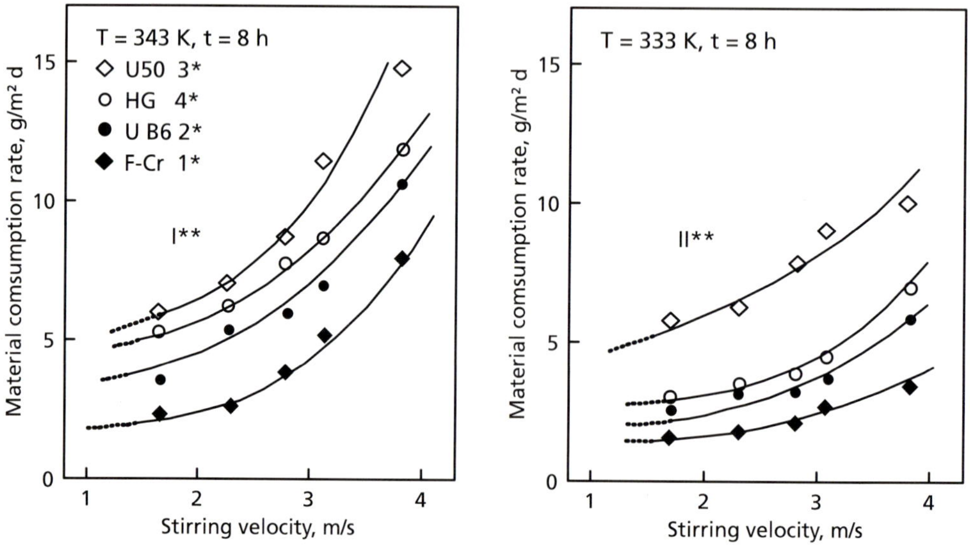

Figure 19: Influence of the stirring velocity on the material consumption rate of high-alloy materials (Table 12) in phosphate suspensions containing sulfuric acid (Table 11) in production of wet phosphoric acid [67]
* see Table 12
** see Table 11

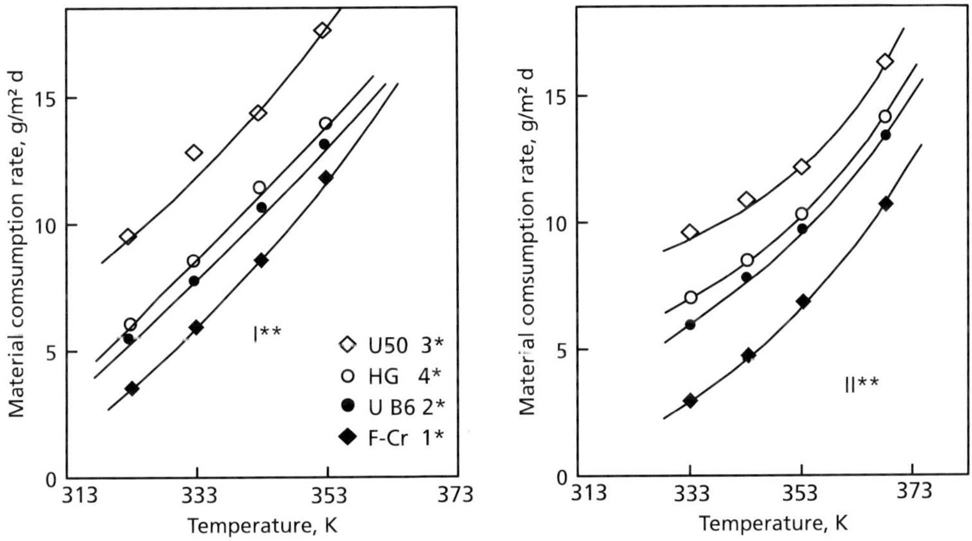

Figure 20: Influence of temperature on the material consumption rate of high-alloy materials (Table 12) in phosphate suspensions containing sulfuric acid (Table 11) in the production of wet phosphoric acid, stirring velocity 3.92 m/s, test duration 8 h [67]
(At a density of 7.85, 1 g/m^2 d approximately corresponds to 0.05 mm/a (1.97 mpy))
* see Table 12
** see Table 11

Extensive laboratory studies with a number of metallic materials as grinding bodies for comminution of phosphate rock in phosphoric acid waste water, corresponding to the operating conditions of a plant in Florida for production of phosphoric acid, are described in [18]. Table 13 contains the materials used, Table 14 the composition of the phosphoric acid waste water, and Tables 15 and 16 the test results. Figure 21 contains the associated graph.

Trade name	*DIN-Mat. No.*	*Material designation*	*Section*	*Chemical composition, %*								*HB*1)
				C	*Cr*	*Fe*	*Mn*	*Mo*	*Ni*	*Si*	*Others*	
Hastelloy® C-276	2.4819	NiMo16Cr15W		0.005	15.94	5.61	0.41	15.67	56.82	0.04	3.58 W, 1.82 Co, 0.11 V	183
Incoloy® 825	2.4858	NiCr21Mo		0.03	23.20	27.70	0.39	2.76	43.10	0.11	1.69 Cu, 0.91 Ti	187
Cronifer® 1713 LCN	1.4439	X2CrNiMoN17-13-5	D	0.016	17.50	63.51	1.72	3.87	12.80	0.03	0.42 Cu, 0.13 N	187
Illium® P	–	–	E	0.20	26.90	57.63	0.53	2.18	8.58	0.76	3.22 Cu	269
SAE 316	1.4436	X3CrNiMo17-13-3	D	0.10	17.30	66.10	1.74	2.24	12.30	0.26	n.k.	212
Ni-Resist® 1	0.6655	EN-JL3011	G	2.93	3.72	70.82	1.11	0.018	14.5	1.75	5.15 Cu	167
NP 599			A	0.09	16.40	71.55	8.71	0.21	2.16	0.41	0.31 Cu, 0.16 N	217
CA-6NM	1.4313	X3CrNiMo13-4	G	0.03	11.80	83.45	0.03	0.61	3.92	0.16	n.k.	359
Cr-Mo cast (white)	approx. 0.9640	EN-JN3029	A	2.46	13.20	82.35	0.51	1.40	0.35	0.19	n.k.	606
NiHard® 4			G	3.47	9.09	78.78	1.04	0.20	5.66	1.76	n.k.	601
NiHard® 1	0.9625	EN-JN2039	G	3.41	3.35	87.56	0.74	0.025	4.20	0.72	n.k.	555
C steel			F	1.00	0.46	96.89	1.02	0.13	0.17	0.40	0.18 Cu	450

1) Brinell hardness
n.k. not known

A: see section Ferritic chromium steels with more than 12 % chromium
D: see section Austenitic CrNiMo(N) steels
E: see section Austenitic chromium-nickel steels with special alloying additions
F: see section Unalloyed steels and cast steel
G: see section Structural steels with up to 12 % chromium

Table 13: Summary of materials used as grinding bodies (cylinders 50 mm ∅, 50 mm long) for comminution of phosphate rock in a tumbling mill (60 cm ∅) [18]

Contents	Intake pH 1.6	After 1 h grinding, filtered pH 3.0
		g/l
P_2O_5	17.5	16.2
Al	0.082	0.009
Ca	0.950	1.780
Cl	0.110	0.095
F	8.290	0.560
Fe	0.230	0.030
K	0.240	0.140
Mg	0.270	1.310
Na	1.870	1.650
Si	2.450	0.510
SO_4^{2-}	5.100	3.150

Table 14: Analysis of phosphoric acid waste water from a plant in Florida [18]

Serial No.	Material	A		B		C	
		mm/a (mpy)	Rank	mm/a (mpy)	Rank	mm/a (mpy)	Rank
1	Hastelloy® C-276	14.2 (559)	1	7.9 (311)	2	< 0.03 (< 1.18)	1
2	Incoloy® 825	23.7 (933)	6	14.9 (587)	11	< 0.03 (< 1.18)	1
3	Cronifer® 1713 LCN	20.2 (795)	4	11.4 (449)	7	0.03 (1.18)	3
4	Illium® P	20.1 (791)	3	11.2 (441)	6	0.03 (1.18)	3
5	SAE 316	27.8 (1,094)	7	14.8 (583)	10	0.03 (1.18)	3
6	Ni-Resist®	37.3 (1,469)	9	24.6 (969)	12	4.70 (185)	8
7	NP 599	14.6 (575)	2	6.8 (268)	1	0.03 (1.18)	3
8	CA6NM	22.4 (882)	5	8.3 (327)	3	0.60 (23.6)	7
9	CrMo-cast	62.6 (2,465)	12	10.8 (425)	5	18.0 (709)	11
10	NiHard® 4	49.1 (1,933)	10	12.3 (484)	9	15.0 (591)	9
11	NiHard® 1	55.4 (2,181)	11	10.4 (409)	4	15.4 (606)	10
12	C steel	36.1 (1,421)	8	11.6 (457)	8	19.7 (776)	12

A: erosion rate on grinding of phosphate rock in a 60 cm tumbler mill after a grinding time of 1 h in phosphoric acid waste water (see Table 14)
B: erosion rate on grinding of phosphate rock in pure water (probably drinking water) after 1 h grinding in a tumbler mill as A
C: corrosion rate in non-agitated phosphoric acid waste water (mean value according to Table 16)

Table 15: Corrosion rates of grinding body materials (see Table 13) under conditions A-C at 310 K (37 °C) (plant conditions in Florida) – for a graphical presentation, see Figure 21 [18]

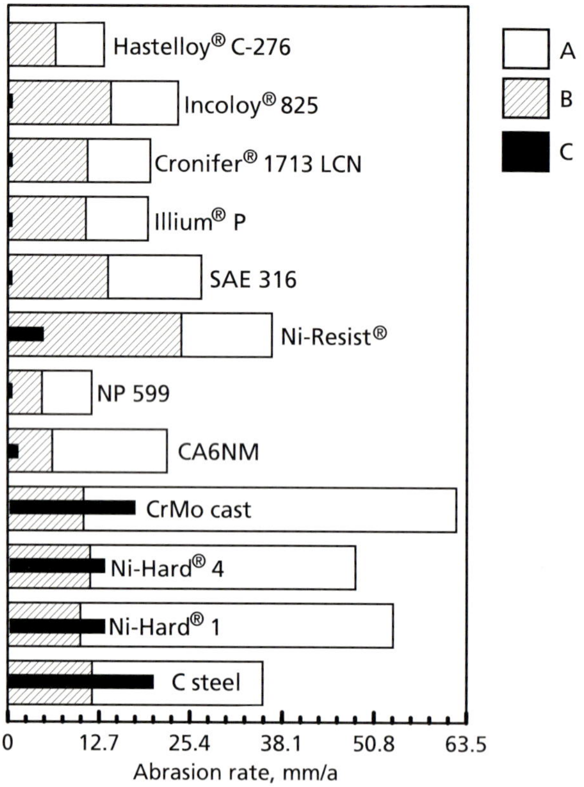

Figure 21: Overview of the wear rates to be expected on the materials listed (Table 13) during communication of phosphate rock at 310 K (37 °C) (see Table 15) [18]

Of the materials described under Section Ferritic chromium steels with more than 12 % chromium, steel NP 599 shows the best erosion corrosion behavior, and is to be regarded as equivalent to the Ni alloy Hastelloy® C-276. The grinding cycles extended over a period of only 1 h. Figure 22 contains an overview of the corrosion rates over several grinding cycles. It can also be seen from this that steel NP 599 can be used as a grinding body material [18]. Only the results of the studies with the 60 cm tumbler mill have been taken from this work; results with the 12 cm tumbler mill cannot be concluded from them. The results with the larger mill will rather simulate the trend in operating plants [18].

Depending on the plant conditions (in respect of costs), a preliminary selection of materials may be made under similar conditions.

Short-term studies over 100 h in wet phosphoric acid confirm a very good to adequate corrosion resistance, which is comparable to or corresponds to that of steels under Section Austenitic chromium-nickel steels with special alloying additions, for the high-chromium steel with Mo (90Ch28MFTAL). Table 17 contains a summary of corrosion rates [68]. According to this table, the steel can be regarded as resistant up to 363 K (90 °C) in wet phosphoric acid (30–36 % P_2O_5, 1.8 % F, 2.2 % SO_3).

Material	Phosphoric acid pH 1.6	Waste water[1)] pH 3.0	Average corrosion rate mm/a (mpy)
	Corrosion rate, mm/a (mpy)		
Hastelloy® C-276	0.03 (1.18)	0.03 (1.18)	0.03 (1.18)
Incoloy® 825	0.03 (1.18)	0.03 (1.18)	0.03 (1.18)
Cronifer® 1713 LCN	0.05 (1.97)	0.03 (1.18)	0.03 (1.18)
Illium® P	0.05 (1.97)	0.03 (1.18)	0.03 (1.18)
SAE 316	0.03 (1.18)	0.03 (1.18)	0.03 (1.18)
Ni-Resist®	7.37 (290)	1.96 (77.2)	4.70 (185)
NP 599	0.05 (1.97)	0.03 (1.18)	0.03 (1.18)
CA-6NM	1.19 (46.9)	0.03 (1.18)	0.60 (23.6)
CrMo-cast	21.4 (842)	14.7 (579)	18.0 (709)
NiHard® 4	20.8 (819)	9.14 (360)	15.0 (591)
NiHard® 1	19.6 (772)	11.2 (441)	15.4 (606)
C-steel	22.4 (882)	17.0 (669)	19.7 (776)

1) see Table 14

Table 16: Corrosion rates of grinding body test materials (see Table 13) in phosphoric acid waste water at on average 310 K (37 °C) (plant conditions in Florida), test duration 1 h [18]

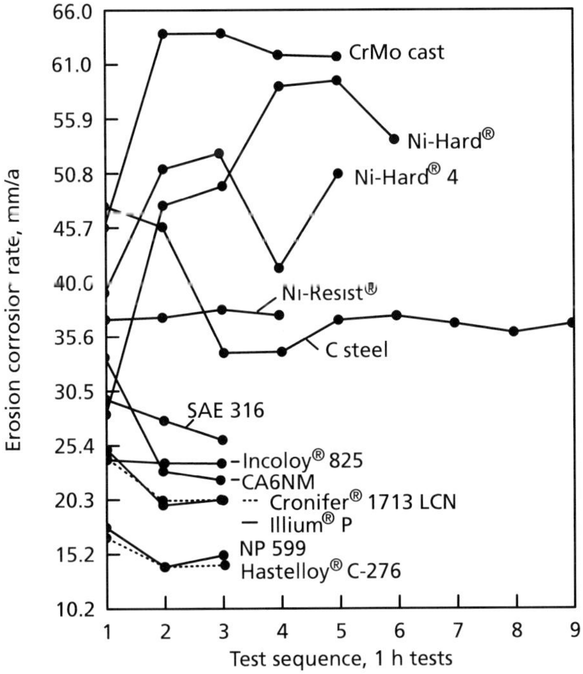

Figure 22: Erosion corrosion rate of grinding materials (see Table 13) as a function of the test sequence under conditions A (see Table 15) at 310 K (37 °C) [18]

Material		*Temperature, K (°C)*				*Section*
		343 (70)	*353 (80)*	*363 (90)*	*373 (100)*	
		Corrosion rate, mm/a (mpy)				
90Ch28MFTAL[1)]						A
% Mo	1.53	0.03 (1.18)	0.05 (1.97)	0.07 (2.76)	0.22 (8.66)	
	2.50	0.06 (2.36)	0.06 (2.36)	0.11 (4.33)	0.22 (8.66)	
	3.50	0.06 (2.36)	0.06 (2.36)	0.13 (5.12)	0.43 (16.9)	
06ChN28MDT[2)]		0.02 (0.79)	0.05 (1.97)	0.26 (10.2)	0.34 (13.4)	E

1) 0.9 C, 28 Cr, 1.53–3.5 Mo, V, Ti
2) 0.06 C, 22–55 Cr, 26–29 Ni, 2.5–3.0 Mo, 2.5–3.5 Cu

A: see section Ferritic chromium steels with more than 12 % chromium
E: see section Austenitic chromium-nickel steels with special alloying additions

Table 17: Corrosion behavior of CrMo steel and CrNiMoCu steel in wet phosphoric acid (30–36 % P_2O_5, 1.8 % F, 2.2 % SO_3) as a function of the temperature, test duration 100 h [68]

With a corrosion rate of 0.13 mm/a (5.12 mpy), even the 3.5 % Mo steel can be used according to this summary, and is therefore superior to the austenitic steel 06ChN28MDT, with a rate of 0.26 mm/a (10.2 mpy). The studies on steel 90Ch28MFTAL were carried out with various Mo and Cr contents. The result of these studies showed that these steels have an optimum resistance at 28 % Cr and 2.5 % Mo (see Figure 67). Studies have been carried out with such a steel in media of varying phosphoric acid content, and the results are summarized in Table 18. According to this table, the CrMo steel 28 2.5 has a wide field of use, even up to relatively high temperatures [68].

Corrosion medium %	*Temperature K (°C)*	*Test duration h*	*Material*	
			06ChN28MDT[1)]	*90Ch28MFTAL[2)]*
			Corrosion rate, mm/a (mpy)	
Wet phosphoric acid ($29.5\ P_2O_5$; 1.8 F)	343 (70)	100	0.010 (0.39)	0.014 (0.55)
	353 (80)	100	0.022 (0.87)	0.056 (2.2)
Wet phosphoric acid ($37.5\ P_2O_5$; 1.8 F)	363 (90)	100	0.063 (2.48)	0.119 (4.69)
	373 (100)	100	0.160 (6.30)	0.242 (9.53)

1) 0.06 C, 22–25 Cr, 26–29 Ni, 2.5–3.0 Mo, 2.5–3.5 Cu
2) 0.9 C, 28 Cr, 2.5 Mo, V, Ti

Table 18: Corrosion rates for two high-alloy steels in phosphoric acid and media containing phosphoric acid (see also Table 17) [68]

Table 18: Continued

Corrosion medium %	*Temperature* K (°C)	*Test duration* h	*Material*	
			06ChN28MDT[1)]	*90Ch28MFTAL*[2)]
			Corrosion rate, mm/a (mpy)	
Wet phosphoric acid from Karatau crude phosphate (22 P_2O_5; 1.28 F)	363 (90)	100	0.018 (0.71)	0.024 (0.94)
Wet phosphoric acid from Moroccan crude phosphate (27.5 P_2O_5; 2.28 F)	353 (80)	20	0.113 (4.45)	0.243 (9.57)
Wet phosphoric acid (47.8 P_2O_5; 1.35 F)	375 (102)	2,000	0.100 (3.94)	0.400 (15.8)
Wet phosphoric acid (evaporated) (54 P_2O_5; 0.6 F)	353 (80)	100	0.034 (1.34)	0.076 (2.99)
	363 (90)	100	0.106 (4.17)	0.140 (5.51)
	373 (100)	100	0.210 (8.27)	0.380 (15.0)
	393 (120)	100	0.350 (13.8)	0.960 (37.8)
Superphosphoric acid (45 P_2O_5)	353 (80)	100	0.008 (0.31)	0.019 (0.75)
	373 (100)	100	0.066 (2.6)	0.091 (3.58)
	393 (120)	100	0.081 (3.19)	0.124 (4.88)
	413 (140)	100	0.217 (8.54)	0.416 (16.4)
Superphosphoric acid (71.1 P_2O_5)	353 (80)	100	0.00	0.00
	373 (100)	100	0.010 (0.39)	0.013 (0.51)
	393 (120)	100	0.070 (2.76)	0.050 (1.97)
	413 (140)	100	0.130 (5.12)	0.196 (7.72)
Complete fertilizer suspension	378 (105)	2,000	0.01 (0.39)	0.01 (0.39)
Aqueous complex fertilizer	353 (80)	100	0.003 (0.12)	0.004 (0.16)

1) 0.06 C, 22–25 Cr, 26–29 Ni, 2.5–3.0 Mo, 2.5–3.5 Cu
2) 0.9 C, 28 Cr, 2.5 Mo, V, Ti

Table 18: Corrosion rates for two high-alloy steels in phosphoric acid and media containing phosphoric acid (see also Table 17) [68]

In contrast, the Mo-free steel SAE 446 (1.4762, X10CrAl24) [43] – in [42] it was redesignated as X18CrN28 (1.4749) – showed corrosion rates of more than 15 mm/a (591 mpy) in an evaporator of a pilot plant under operating conditions in chloride-containing crude phosphoric acid at temperatures between 355 and 366 K (82 and 93 °C) (see Figure 64 and Table 41).

Ferritic-austenitic steels with more than 12 % chromium

Similarly to the ferritic steels under Section Ferritic chromium steels with more than 12 % chromium, a varying resistance of the ferritic-austenitic steels is to be expected, depending on their composition, higher-alloy grades here also proving to have a better resistance to the action of chemicals.

Table 19 contains a summary of commercially available ferritic-austenitic steels (duplex steels).

Material designation/trade name/ DIN-Mat. No.		*Chemical composition, %*[1)]							
		C	*Si*	*Mn*	*Cr*	*Mo*	*Ni*	*Cu*	*N*
3RE60, 1.4417	GX2CrNiMoN25-7-3	0.03	2.0	2.0	18.5	2.7	4.9	–	–
Uranus® 50	UNS S32404	0.06	1.0	2.0	21.0	2.5	7.5	1.5	
SAF 2205 , 1.4462	X2CrNiMoN22-5-3	0.03	1.0	2.0	22.0	3.0	5.5	–	0.74
Ferralium® 255	cf. 1.4507	0.02	1.4	1.4	26.0	3.3	5.5	3.0	0.20
Ferralium® 255-3SC		0.05			25.0	3.0	6.0	2.5	0.18
Ferralium® 255-3SF		0.04			25.0	3.0	6.0	1.8	0.18
SAE 329, 1.4460	X3CrNiMoN27-5-2	0.10	1.0	2.0	27.0	2.0	4.5	–	–
Fermanel®	WV49	0.06			27.0	3.1	8.5	1.0	0.23
1.4530	X1CrNiMoAlTi12-9	0.03			25.0	3.5	5	1.5	
A 905, 1.4467	X2CrMnNiMoN26-5-4	0.04	0.35	5.8	25.5	2.3	3.7	–	0.37
Noridur® 9.4460	GX3CrNiMoCuN24-6-2-3, cf. 1.4593	0.03	1.5	1.5	24.0	2.5	6	3	0.15

1) balance Fe

Table 19: Overview of the composition of ferritic-austenitic steels (duplex steels), see also Table 24

The increasing improvement in their processability makes these steels interesting for chemical apparatus construction, because they have

- a yield strength which is about twice as high as that of the austenitic steels,
- a high resistance to erosion corrosion and corrosion fatigue,
- a good resistance to corrosion, pitting corrosion and crevice corrosion,

- a high resistance to stress corrosion cracking in media containing chloride and hydrogen sulfide and
- a good weldability.

They have a very good to adequate resistance toward phosphoric acid [55, 56, 69–74].

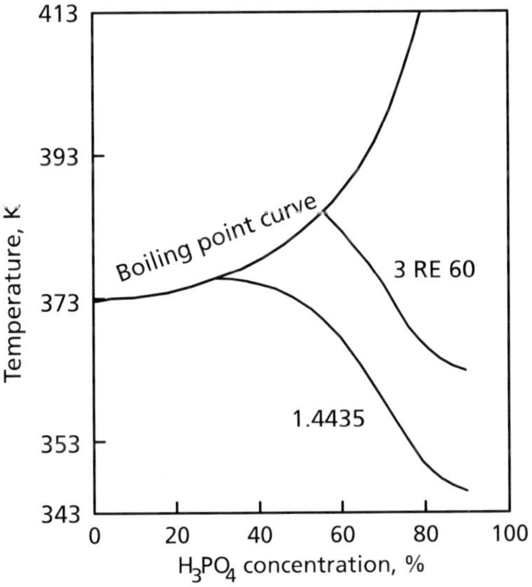

Figure 23: Isocorrosion curves (0.1 mm/a (3.94 mpy)) for the duplex steel 3RE60 (Table 19) in comparison with 1.4435 (SAE 316 L) [75]

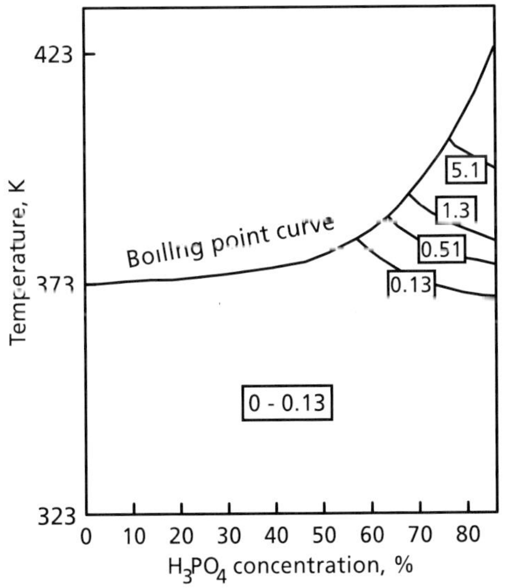

Figure 24: Isocorrosion curves and corrosion range (mm/a) for Ferralium® 255 (Table 19) in pure phosphoric acid (see also Figure 4) [76]

0–0.13 mm/a (0–5.12 mpy)	0.51 mm/a (20.1 mpy)	5.1 mm/a (201 mpy)
0.13 mm/a (5.12 mpy)	1.3 mm/a (51.2 mpy)	

Tables 2 and 7 contain a summary of corrosion rates (1.4542), mainly in pure acid. Figures 23–26 contain isocorrosion curves for individual duplex steels listed in Table 19, also for pure acid [23, 75–78]

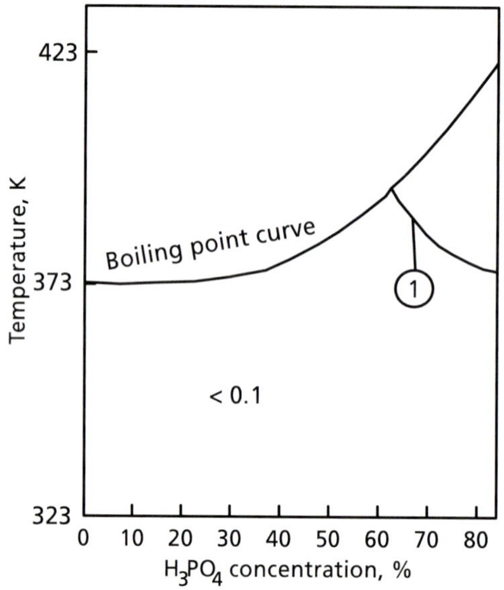

Figure 25: Isocorrosion curve ① (0.1 mm/a (3.94 mpy)) for Fermanel® (Table 19) in pure phosphoric acid [78]

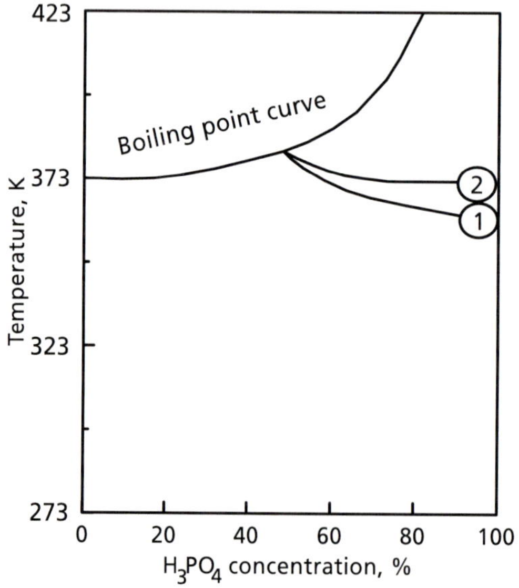

Figure 26: Isocorrosion curves (0.3 mm/a (11.8 mpy)) for the duplex steel A 905 ① (Table 19) in comparison with 1.4571 (SAE 316 Ti) ② in pure phosphoric acid [77]

As can be seen from Figure 27, the resistance of the duplex steel with 25 % Cr, 5 % Ni and 1.5 % Mo is significantly better than that of the ferritic CrMo steel 18 2 (see also Figure 67). It is also better than that of the austenitic CrNi steels 18 9 and CrNiMo steels 17 12 2.5, and according to this graph, exceeds even higher-alloy steels. See also Figures 5 and 23.

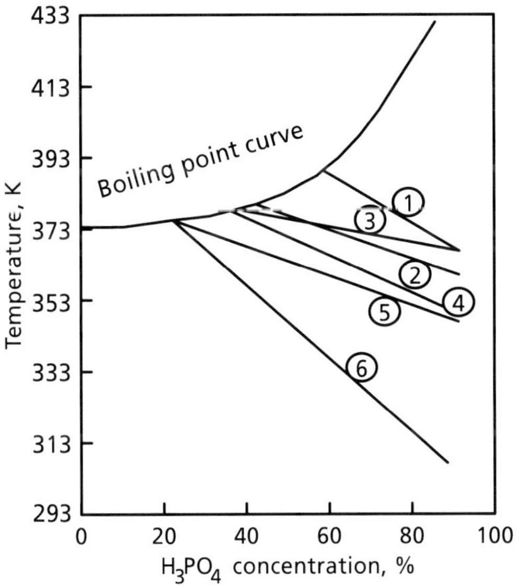

Figure 27: Isocorrosion diagram for Cr steels and CrNi steels (ferritic, ferritic-austenitic and austenitic) with isocorrosion curves at 0.1 mm/a (3.94 mpy), corresponding to Table 2 [23]

① CrNiMo 25 5 1,5 ③ CrNiMo 20 25 4,5 1,5 ⑤ CrNi 18 9
② CrNiMo 17 15 4,5 ④ CrNiMo 17 12 2,5 ⑥ CrMo 18 2

If the corrosion behavior of duplex steels in pure phosphoric acid is compared, the steels Ferralium® 255 and Fermanel® (see Table 19) are the most suitable. Both steels contain relatively large amounts of chromium and molybdenum. A recommendation for the use of high-alloy materials in industrial phosphoric acid can be seen from Figure 28. According to this figure, the duplex steel 1.4417 (3RE60, UNS S31500) and the austenitic steel 1.4435 (SAE 316 L), are the least resistant. However, 1.4417 is an inexpensive variant for use in industrial phosphoric acid over the entire concentration range at low temperatures. The use of duplex steels, remembering their materials-specific peculiarities, as a material for chemical tankers cannot be excluded, because of their higher strength compared with austenitic steels and relatively good corrosion resistance [79]. See later for Remanit® 4462 (1.4462, UNS S31803).

Ferralium® 255 is significantly superior to the CrNiMo steels 18 10 3 in resistance to phosphoric acid in all concentrations up to 353 K (80 °C). In boiling acid, it is suitable up to a concentration of 80 %. Because of its very good resistance in contaminated phosphoric acid (fluorine compounds, chlorides and sulfuric acid), this material is particularly suitable for pumps and valves and other components sub-

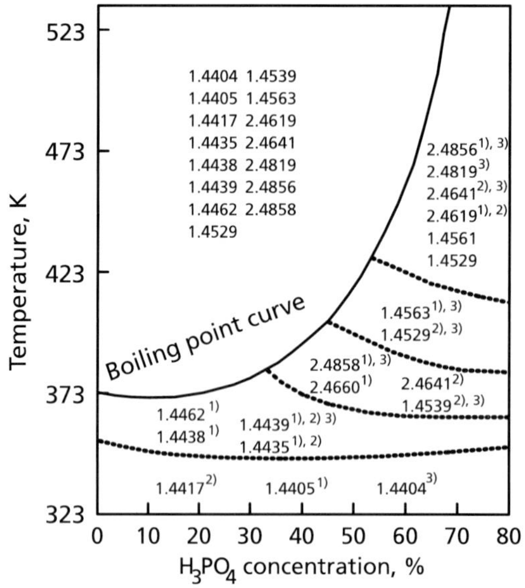

Figure 28: Use ranges of steels and nickel alloys (see Table 40) in industrial phosphoric acid [19, 80, 81]

DIN-Mat. No.		DIN-Mat. No.	
1.4404	cf. SAE 316 L	1.4561	X1CrNiMoTi18-13-2
1.4405	GX4CrNiMo16-5-1	1.4563	UNS N08028
1.4417	UNS S31500	2.4619	UNS N06985
1.4435	cf. SAE 316 L	2.4641	UNS N08042
1.4438	cf. SAE 317 L	2.4660	UNS N08020
1.4439	cf. SAE 317 LMN	2.4819	UNS N10276
1.4462	2205	2.4856	UNS N26625
1.4529	UNS N08926	2.4858	UNS N08825
1.4539	SAE 904 L		

1) [19] no information on corrosion rates of the isocorrosion curves, no information on the acid (probably industrial acid)

2) [79] Industrial phosphoric acid, no information on corrosion rates of the isocorrosion curves

3) [80] Industrial acid, isocorrosion curves 0.5 mm/a (19.7 mpy)

jected to high stresses during production of phosphoric acid by the wet process due to its excellent wear resistance and erosion resistance. After many years of experience, Ferralium® 255 is superior to conventional materials in phosphoric acid production by the wet process, especially for conveying and mixing hot abrasive mashes. This duplex steel is resistant up to 343 K (70 °C) in a mixture of 50 % H_3PO_4 and 50 % HNO_3 [82].

Ferralium® 255 has a good resistance towards intercrystalline corrosion and carbide precipitations at the grain boundaries in the heat-affected zone of weld seams, so that heat treatment after welding is not necessary [82]. This does not exclude the possibility of this being necessary in specific cases.

The cast materials Ferralium® 255-3SC [83] and Ferralium® 255-3SF [84] are distinguished by at least the same resistance under exposure to phosphoric acid.

Extensive electrochemical studies in crude phosphoric acid (66–75 % H_3PO_4) with additions of chloride and fluoride ions, as are usual in practice, on the duplex steel X 2 CrNiMoCuN 25 5 (0.027 C, 0.44 Si, 0.47 Mn, 0.011 P, 0.016 S, 24.73 Cr, 3.24 Mo, 5.03 Ni, 0.08 V, 1.78 Cu, 0.16 N_2, balance Fe) at temperatures up to 353 K (80 °C) have demonstrated that this steel experiences a significant increase in its corrosion rate at higher temperatures in the active range by heat treatment after welding. The ferritic structural constituent is preferentially attacked here. Specimens investigated in the passive range, on the other hand, showed no difference in the corrosion behavior of the weld seam, the heat-affected zone and the base material. The steel shows no macroscopically detectable corrosive attack in the passive range at all temperatures (up to 353 K (80 °C)) in crude phosphoric acid which is free from fluoride ions but contains chloride ions. The corrosive attack in the active range is changed only slightly by addition of fluoride ions to the acid. In the active range, a significant increase in corrosion is to be recorded [85, 86].

Table 20 contains a summary of corrosion rates for Ferralium® 255.

In boiling 60 % phosphoric acid, the duplex steel 1.4460 (SAE 329, X3CrNiMoN27-5-2) is attacked at a rate of 0.1–1.0 mm/a (3.94–39.4 mpy) [61].

In laboratory studies on steels SAF® 2205 (1.4462, X2CrNiMoN22-5-3), 1.4435 (SAE 316 L, X2CrNiMo18-14-3) and 1.4438 (SAE 317 L, X2CrNiMo18-15-4), the duplex steel mentioned first showed the better resistance in comparison with the two austenitic steels. In 70 % H_3PO_4 at 383 K (110 °C) it had a corrosion rate of only 0.22 mm/a (8.66 mpy), 1.4435 had a rate of 3.9 mm/a (154 mpy) and 1.4438 a rate of 2.4 mm/a (94.5 mpy). In 80 % acid, the corrosion rate at 373 K (100 °C) was 0.35 mm/a (13.8 mpy), and for 1.4435 it was 1.3 mm/a (51.2 mpy) [87].

Phosphoric acids produced by the wet process always contained corrosive impurities. In such an acid containing 54 % P_2O_5 (74.5 % H_3PO_4), 0.06 % HCl, 1.1 % HF, 1.1 % H_2SO_4, 0.27 % Fe_2O_3, 0.17 % Al_2O_3, 0.17 % SiO_2, 0.20 % CaO and 0.70 % MgO at 333 K (60) the corrosion rate for SAF® 2205 (1.4462) was only 0.08 mm/a (3.15 mpy), while that for 1.4539 (SAE 904 L) was 1.2 mm/a (47.2 mpy), and that for 1.4436 (SAE 316) was more than 5 mm/a (197 mpy) [88]. The corrosion rate is also more favorable in a less concentrated acid, as can be seen from Table 21.

H_3PO_4 concentration %	Temperature K (°C)	Ferralium® 255[1)]	Ferralium® 255-3SC[1)]	Ferralium® 255-3SF[1)]
		Corrosion rate, mm/a (mpy)		
all	353 (80)	< 0.15 (< 5.91)		
all	373 (100)		< 0.15 (< 5.91)	< 0.15 (< 5.91)
0–40	boiling	< 0.15 (< 5.91)		
0–60	boiling		< 0.15 (< 5.91)	< 0.15 (< 5.91)
30	boiling	0.01 (0.39)		
40–80	boiling	< 0.5 (< 19.7)		
60–70	393 (120)		< 0.5 (< 19.7)	< 0.5 (< 19.7)
85	339 (66)	< 0.01 (< 0.39)		
88	boiling	0.5–1.5[2)] (19.7–59.1)	≥ 0.5[2)] (≥ 19.7)	≥ 0.5 (≥ 19.7)
34.5–48.3[3)]	353 (80)	< 0.15[2)] (< 5.91)	< 0.15[2)] (< 5.91)	< 0.15[2)] (< 5.91)
62.1–75.9[3)]	353–363 (80–90)	< 0.15[2)] (< 5.91)	< 0.15[2)] (< 5.91)	< 0.15[2)] (< 5.91)

1) see Table 19
2) plant studies necessary
3) wet process, fertilizer production

Table 20: Corrosion rates to be expected for the duplex steel Ferralium® 255 in phosphoric acid [82, 89–92]

Material designation DIN-Mat. No.		Solution A[1)] 333 K (60 °C)	Solution B[1)] 325 K (52 °C)	Section
		Corrosion rate, mm/a (mpy)		
X3CrNiMo17-13-3 1.4436	SAE 316	> 5 (> 197)	> 5 (> 197)	AD
X1NiCrMoCu25-20-5 1.4539	SAE 904 L	1.2 (47.2)	0.16 (6.30)	AE
X2CrNiMoN22-5-3 1.4462	UNS S39209 2205	0.08 (3.15)	0.10 (3.94)	AB

1) chemical composition, %

	P_2O_5	HCl	HF	H_2SO_4	Fe_2O_3	Al_2O_3	SiO_2	CaO	MgO
A:	54	0.06	1.1	4.1	0.27	0.17	0.10	0.20	0.70
B:	27.5	0.34	1.3	1.72	0.4	0.01	0.3	0.02	–

AB: see section Ferritic-austenitic steels with more than 12 % chromium
AD: see section Austenitic CrNiMo(N) steels
AE: see section Austenitic chromium-nickel steels with special alloying additions

Table 21: Corrosion behavior of steels in contaminated phosphoric acid [88]

Table 22 summarizes the corrosion behavior of high-alloy steels in extraction acid containing 35.73 % P_2O_5 (49.31 % H_3PO_4), 1.25 % SO_3 and 1.88 % F, with varying additions of HNO_3. This table shows an increase in the corrosion rate as the content of nitric acid increases. The corrosion rates hardly differ from those of higher-alloy steels, taking account of the very low test duration of only 100 h. According to these laboratory studies, duplex steels prove to be particularly suitable for use in contaminated phosphoric acid [93].

Table 23 shows a further summary of corrosion rates of the steels in Table 22. This summary of corrosion rates shows the influence of H_2SO_4 and HF [94]. It can be seen from this that there is practically no difference between the austenitic steels and the duplex steels in the two phosphoric acid solutions, including in the 41.4 % H_3PO_4 solution with up to 5 % H_2SO_4. Additions of hydrofluoric acid and of sulfuric acid and hydrofluoric acid worsen the corrosion resistance of duplex steels under these conditions such that they can no longer be used [94].

Material	***HNO_3 additive %***	***Corrosion rate, mm/a (mpy)***		***Section***
		293 K (20 °C)	***353 K (80 °C)***	
06ChN28MDT[1)]	0	0.002 (0.08)	0.067 (2.64)	E
	1.0	0.002 (0.08)	0.075 (2.95)	
	2.5	0.002 (0.08)	0.101 (3.98)	
	5.0	0.005 (0.20)	0.120 (3.98)	
03Ch21N21M4GB[2)]	0	0.003 (0.12)	0.072 (2.83)	D
	1.0	0.003 (0.12)	0.083 (3.27)	
	2.5	0.004 (0.16)	0.113 (4.45)	
	5.0	0.006 (0.24)	0.128 (5.04)	
08Ch22N6T[3)]	0	0.004 (0.16)	0.080 (3.15)	B
	1.0	0.004 (0.16)	0.108 (4.25)	
	2.5	0.006 (0.24)	0.123 (4.84)	
	5.0	0.007 (0.28)	0.157 (6.18)	
08Ch21N6M2T[4)]	0	0.003 (0.12)	0.078 (3.07)	B
	1.0	0.003 (0.12)	0.095 (3.74)	
	2.5	0.005 (0.20)	0.128 (5.04)	
	5.0	0.006 (0.24)	0.162 (6.38)	

	% Cr	% Ni	% Mo	% Cu
1)	22–25	26–29	2.5–3.5	2.5–3.5
2)	20–22	20–22	3.4–3.7	–
3)	21–23	5.3–6.3	–	–
4)	20–22	5.5–6.5	–	–

B: see section Ferritic-austenitic steels with more than 12 % chromium
D: see section Austenitic CrNiMo(N) steels
E: see section Austenitic chromium-nickel steels with special alloying additions

Table 22: Laboratory studies on the corrosion behavior of high-alloy steels in phosphoric acid (35.73 % P_2O_5 (49.31 % H_3PO_4), 1.25 % SO_3, 1.88 % F) with varying additions of HNO_3, test duration 100 h [93]

Additive	*06ChN28MDT*[1)]*(see E)*	*03Ch21N21M4GB*[1)] *(see D)*	*08Ch21N6M2T*[1)] *(see B)*	*08Ch22N6T*[1)] *(see B)*
	Corrosion rate, mm/a (mpy)			
no additive	0.004 (0.16)	0.006 (0.24)	0.004 (0.16)	0.004 (0.16)
1 % H_2SO_4	0.003 (0.12)	0.005 (0.20)	0.004 (0.16)	0.008 (0.31)
5 % H_2SO_4	0.003 (0.12)	0.001 (0.04)	0.003 (0.12)	0.003 (0.12)
0.5 % HF	0.077 (3.03)	7.5–11.4 (295–449)	0.96 (37.8)	3.34 (132)
1.5 % HF	0.89 (35.0)	0.19–0.75 (7.5–29.5)	3.05 (120)	7.41 (292)
1 % H_2SO_4 + 0.5 % HF	0.14 (5.51)	1.61 (63.4)	2.54 (100)	6.41 (252)
1 % H_2SO_4 + 1.5 % HF	0.87 (34.3)	0.90 (35.4)	5.07 (200)	12.6 (496)
5 % H_2SO_4 + 0.5 % HF	0.20 (7.87)	0.36 (14.2)	7.81 (307)	21.2 (835)
5 % H_2SO_4 + 1.5 % HF	1.57 (61.8)	1.77 (69.7)	21.5 (846)	48.6 (1,913)
production acid[2)]	0.04 (1.57)	0.02 (0.79)	0.05 (1.97)	0.01 (0.39)

1) see Table 22
2) 28.6 % P_2O_5 (39.5 % H_3PO_4), 2.26 % SO_3, 1.82 % F

B: see section Ferritic-austenitic steels with more than 12 % chromium
D: see section Austenitic CrNiMo(N) steels
E: see section Austenitic chromium-nickel steels with special alloying additions

Table 23: Corrosion of high-alloy steels in phosphoric acid (41.4 % H_3PO_4) with various additions of H_2SO_4 and HF and in production acid at 343 K (70 °C) test duration 100 h [94]

The composition of crude acid after the wet process lies within the following ranges: 46–56 % P_2O_5, 0.8–5.0 % H_2SO_4, 0.1–1.3 % F, 0.2–1.7 % Fe_2O_3, 0.3–1.2 % Al_2O_3, about 1.7 % MgO, about 0.5 % CaO and about 0.05 % Cl [95].

Corrosion studies in such a crude phosphoric acid solution at 323 K (50 °C) over a period of 2–3 weeks on high-alloy steels gave the corrosion rates which can be seen in Figure 29. The values were calculated from material consumption rates (g/m^2 h).

According to these studies, there is clearly a trend toward the duplex steels with 25 % Cr, 4–7 % Ni and 2–3 % Mo. They are also superior to higher-alloy steels in a crude phosphoric acid containing 49 % P_2O_5, 5.0 % H_2SO_4, 0.5 % F^-, 0.05 % Cl^- and 1.0 % Fe^{3+} [96].

The corrosion rates of these steels, at 0.1 mm/a (3.94 mpy), are significantly below those of the other steels shown in Figure 29 [95], such as the austenitic CrNiMo steels and the NiMoCr alloy, the corrosion rate of which is about 0.1 mm/a (3.94 mpy). These studies relate to selection of materials for transportation tanks for crude phosphoric acid [95].

As a result of these studies, the alloying constituents Cr, Mo and Cu proved to be the most effective for increasing corrosion resistance in this crude phosphoric acid, the elements Mo and Cr in particular increasing the passivity of the steels. The study results in Figures 30 for Mo and 31 for Cr confirm this fact.

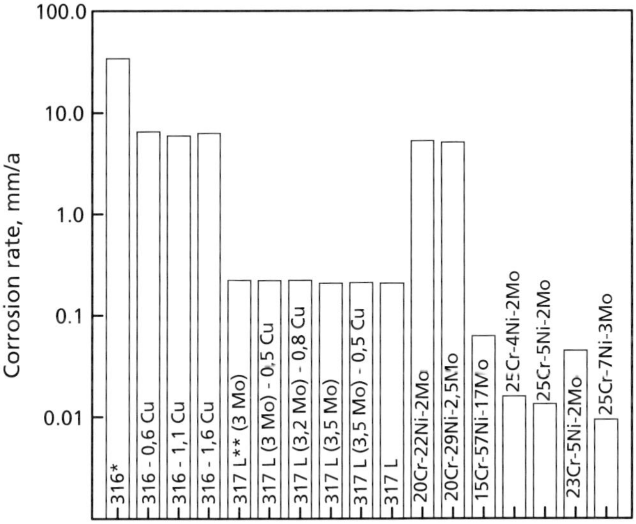

Figure 29: Corrosion behavior of high-alloy materials in crude phosphoric acid (49 % P_2O_5, 5.0 % H_2SO_4, 0.5 % F^-, 0.05 % Cl^- and 1.0 % Fe^{3+}) at 323 K (50 °C) test duration 2–3 weeks [95]
* 316 ≙ 1.4401/1.4436 (X5CrNiMo17-12-2/X3CrNiMo17-13-3)
** 317 L ≙ 1.4438 (X2CrNiMo18-15-4)

As can be seen from Figure 30, a significant improvement in corrosion resistance is detectable in the range from 1 to 2 % Mo in duplex steel, and above 2 % to 3 % it is significantly lower. It should be remembered here that the steel is already resistant in this crude acid in the range from 1 to 2 % Mo with a corrosion rate of 0.1 mm/a (3.94 mpy). Figure 31 shows the noticeable increase in corrosion resistance when the Cr content increases from 23 % to 25 %.

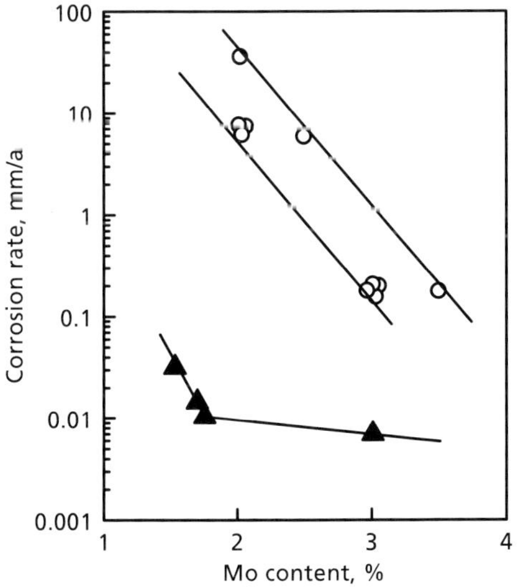

Figure 30: Influence of the molybdenum content on the corrosion behavior of duplex steels and austenitic steels (see Figure 29) in crude phosphoric acid (49 % P_2O_3, 5.0 % H_2SO_4, 0.5 % F^-, 0.05 % Cl^-, 1.0 % Fe^{3+}), test duration 2–3 weeks [95]
○ austenitic steels ▲ duplex steels

The hardly noticeable influence of the P_2O_5 concentration in this crude acid on the corrosion behavior of duplex steels is demonstrated in Figure 32, while austenitic steel reacts sensitively to such a change in concentration.

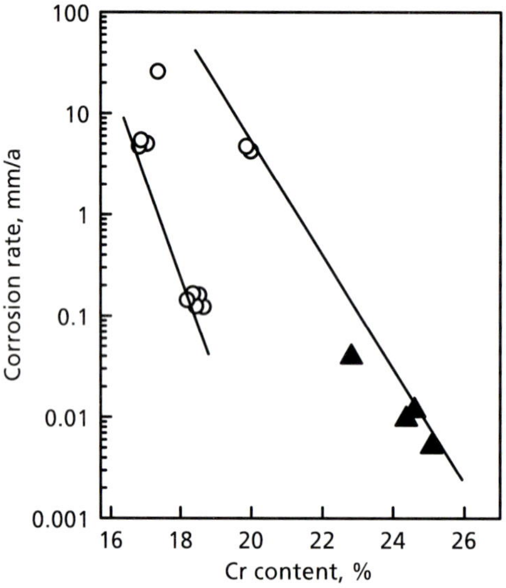

Figure 31: Influence of the chromium content on the corrosion behavior of duplex steels and austenitic steels (see Figure 29) in crude phosphoric acid (49 % P_2O_5, 5.0 % H_2SO_4, 0.5 % F^-, 0.05 % Cl^- and 1.0 % Fe^{3+}), test duration 2–3 weeks [95]
○ austenitic steels ▲ duplex steels

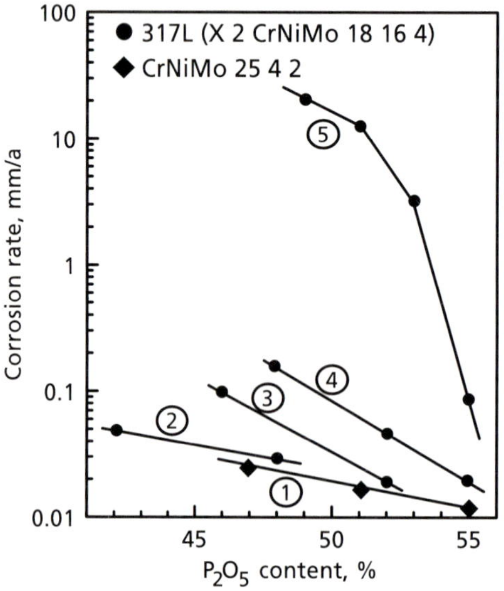

%	H_2SO_4	F^-	Fe^{3+}	Cl^-
①	4.0	0.9	0.5	0.05
②	4.8	0.2	0.9	0.05
③	6.0	0.43	1.3	0.05
④	3.6	0.43	0.9	0.05
⑤	3.0	0.9	0.9	0.05

Figure 32: Influence of the P_2O_5 content of crude phosphoric acid on the corrosion behavior of duplex and austenitic steel at 323 K (50 °C) test duration 2–3 weeks (cf. Figure 33) [95]

A marked increase in corrosion rates is to be expected in crude phosphoric acid at 323 K (50 °C) above an H_2SO_4 content of 4 % (Figure 33).

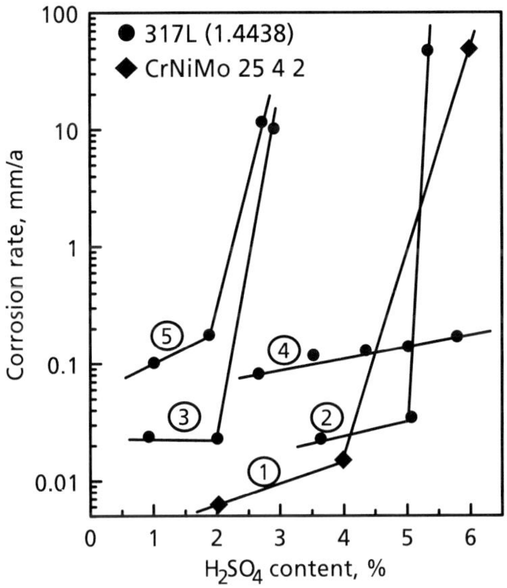

%	P_2O_5	F^-	Fe^{3+}	Cl^-
①	51	0.9	0.5	0.05
②	51	0.2	0.9	0.05
③	48	0.2	0.5	0.05
④	47	0.5	1.3	0.05
⑤	51	0.5	0.5	0.05

Figure 33: Influence of the H_2SO_4 content of crude phosphoric acid on the corrosion behavior of the steels mentioned at 323 K (50 °C) test duration 2–3 weeks (for the duplex steel, see also Figure 32) [95]

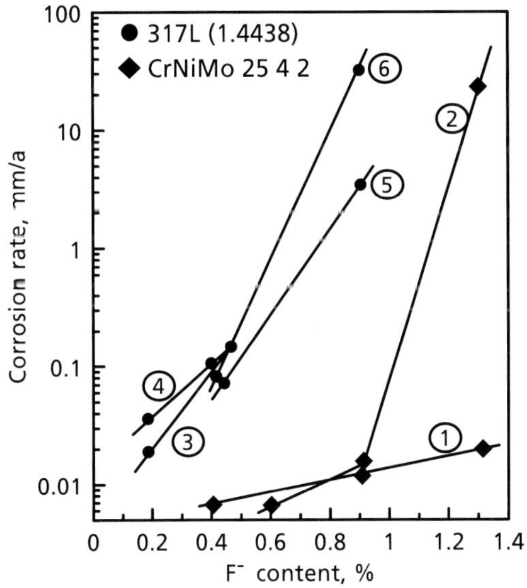

%	P_2O_5	H_2SO_4	Fe^{3+}	Cl^-
①	51	4.0	0.9	0.05
②	51	4.0	0.5	0.05
③	51	1.9	0.5	0.05
④	46	6.0	1.3	0.05
⑤	49	1.8	0.9	0.05
⑥	49	2.7	0.9	0.05

Figure 34: Influence of the F^- ion content on the corrosion behavior of duplex steel and 1.4438 in crude phosphoric acid at 323 K (50 °C) test duration 2–3 weeks [95]

F^- contents also have a less corrosive effect on duplex steel than on the austenitic steel SAE 317 L (Figure 34).

As the Fe^{3+} ions in the crude acid increase, a noticeable decrease is detectable in the corrosion rate for the duplex steel with 25 % Cr, 4 % Ni and 2 % Mo. According to Figure 35, this is the case in particular above 0.5 % Fe^{3+}.

Finally, Figure 36 shows the temperature dependence of the corrosion rates for the duplex steel CrNiMo 25 4 2, in which at about 348 K (75 °C) the practical resistance in the crude acid of 0.1 mm/a (3.94 mpy) is exceeded.

In Figure 37, as a result of the extensive studies, the resistance ranges of the duplex steel are shown as a function of the crude phosphoric acid composition at 323 K (50 °C) and a Cl^- concentration of not more than 0.05 % [95].

Figures 29–37 show the entire spectrum of corrosion influences of impurities in crude phosphoric acid and of temperature and the improvement in the resistance of the material as a function of its alloying elements by the example of ferritic-austenitic steel (in this context, see also Tables 31–34, 37–39).

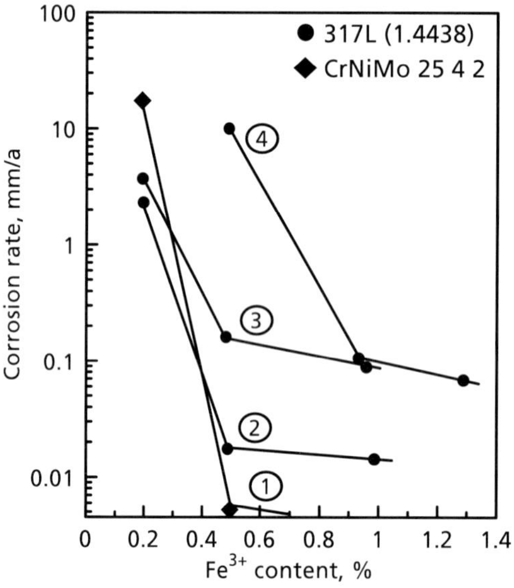

%	P_2O_5	H_2SO_4	Fe^{3+}	Cl^-
①	51	2.0	0.9	0.05
②	52	1.9	0.2	0.05
③	51	1.8	0.4	0.05
④	48	2.7	0.4	0.05

Figure 35: Influence of the Fe^{3+} ion content in crude phosphoric acid at 323 K (50 °C) on the corrosion behavior of duplex steel CrNiMo 25 4 2 and SAE 317 L (1.4438), test duration 2–3 weeks [95]

Under erosive conditions, such as exist during wet decomposition of phosphates, the duplex steel Uranus® 50 is inferior to the other steels investigated, especially the ferritic steel (see Figures 19 and 20 and Tables 11 and 12) [67].

Electrochemical studies on abrasion corrosion are contained in [97, 98]. Addition of molybdenum to steel, in contrast to tungsten and copper, causes better protection against corrosion and wear [98].

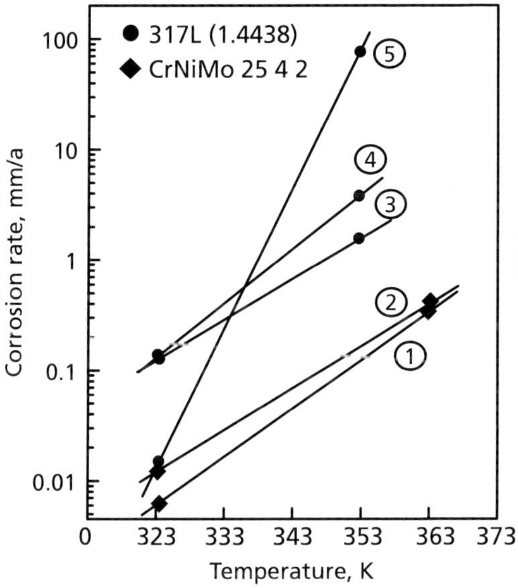

%	P_2O_5	H_2SO_4	Fe^{3+}	F^-	Cl^-
①	49	2.0	0.9	0.6	0.05
②	49	4.0	0.9	1.0	0.05
③	46	6.0	0.43	1.3	0.05
④	48	5.3	0.43	0.9	0.05
⑤	48	4.0	0.2	1.3	0.05

Figure 36: Influence of the temperature of crude phosphoric acid on the corrosion behavior of the duplex steel CrNiMo 25 4 2 and SAE 317 L (1.4438), test duration 2–3 weeks [95]

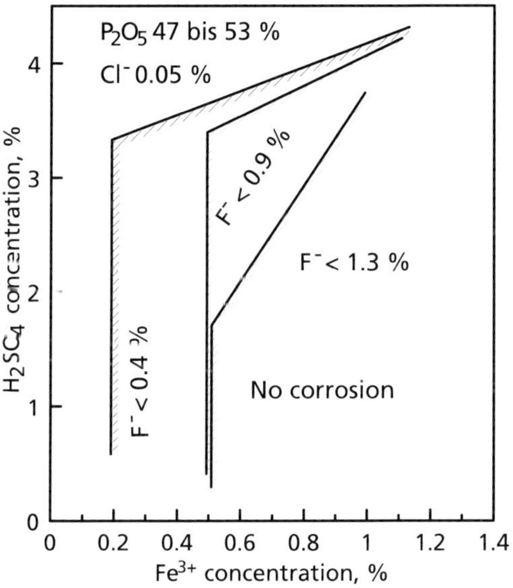

Figure 37: Ranges of corrosion resistance (on the right, below the curves) for the steel CrNiMo 25 4 2 in crude phosphoric acid at 323 K (50 °C) as a function of the H_2SO_4, F^- and Fe^{3+} content (see also Figure 48) [95]

Pitting corrosion

Pitting corrosion and the associated crevice corrosion are the main causes of failure of high-alloy materials. These two types of corrosion are caused almost exclusively by Cl^- ions. Another consequence of these types of corrosion is the occurrence of stress corrosion cracking.

The empirically determined action sum gives or can give a measure of the resistance of high-alloy materials in comparison with one another (in this context, see also under Section Ferritic chromium steels with more than 12% chromium).

In addition to the particular importance of chromium for improving the resistance of steels to pitting corrosion, molybdenum is included with a factor of 3.3. The content of nitrogen is rated extremely highly, and in fact with a factor of 30 (see Table 42). It should be pointed out here that this factor is set too high in some cases; different opinions on its level still exist.

Nitrogen in CrNiMo steels significantly improves their yield strength, tensile strength and hardness, and therefore goes a long way to compensating the reduction caused by reduction in the C content. Nitrogen has an improving influence on the erosion resistance of these steels, it stabilizes the austenitic structure, and it delays possible precipitations [80].

Especially its last mentioned property, the austenitic stabilization, may be the reason for its high rating in the pitting resistance equivalent (PRE)

$$\text{PRE} = 1 \times \%\,\text{Cr} + 3.3 \times \%\,\text{Mo} + 30 \times \%\,\text{N}.$$

In high Mo-containing steels with 6% Mo, the factor of 30 seems to be set too high [72]. It is currently assumed to be between 15 and 30, and is also rated as variable in this range for different steels, depending on their Mo content.

For the user of such steels, it is important to know that nitrogen has a critical influence on the improvement in resistance to pitting corrosion. Its improving action on the resistance of such steels to pitting corrosion is certainly due rather to its austenite-stabilizing action, which delays precipitation of Cr and Mo from the mixed crystal and upgrades their content in the equation.

If the use ranges of high-alloy steels and nickel alloys in contaminated phosphoric acid are compared, their resistance increases with an increasing Cr and Mo content (see Figure 28). The pitting resistance equivalent (PRE) derived for chloride-containing acid media consequently can also be used as a selection criterion for contaminated phosphoric acid. At an adequate Ni content, it guarantees increasing resistance in this [80] (see also Section Unalloyed steels and cast steel).

It should be added that pitting corrosion and crevice corrosion can occur on high-alloy steels only in the passive state, in the region of their corrosion resistance. This passive state is interrupted locally by chlorides, and in particular generally by a considerably higher local current density, with consequent damage to the material. In the active state, on the other hand, the steels corrode more severely, but uniformly.

A measurement parameter for the determination of the resistance to pitting corrosion in the particular medium is the determination of the pitting corrosion potential. If this is greater than the free corrosion potential for the steel under the given conditions (more positive), no pitting corrosion is to be expected.

Chloride ions in phosphoric acid solutions generally intensify corrosion. In industrial phosphoric acid (30% P_2O_5 before concentration, 50% thereafter, 1–3% H_2SO_4, HF + H_2SiF_6 0.5 to 1.5%, chlorides 0.1% + gypsum), the steel Uranus® 50 shows corrosion rates at 333 K (60 °C) of about 0.05 mm/a (1.97 mpy), and 0.2 mm/a (7.87 mpy) at a Cl^- content of 1%, and 0.1 mm/a (3.94 mpy) at 373 K (100 °C) without Cl^- and 0.5 mm/a (19.7 mpy) with addition of 1% Cl^- (Figure 49) [99].

Since no further information is available in this reference, industrial phosphoric acid with 30% P_2O_5 (41.4% H_3PO_4) is to be assumed.

By comparison with the CrNiMo steel 25 4 2 (Figures 29–37), corresponding resistance graphs for the duplex steel Remanit® 4462 (2205, 1.4462, X2CrNiMoN22-5-3) with the corresponding influencing factors are shown in comparison with austenitic steels in Figures 38–40.

With a pitting resistance equivalent (PRE) of more than 35, the duplex steel 1.4462 shows an expected good corrosion behavior in phosphoric acid containing Cl^- ions. Even at high Cl^- concentrations, it is superior to the austenitic standard steels 1.4429 (SAE 316 LN) and 1.4438 (SAE 317 L) (Figure 38 and 39).

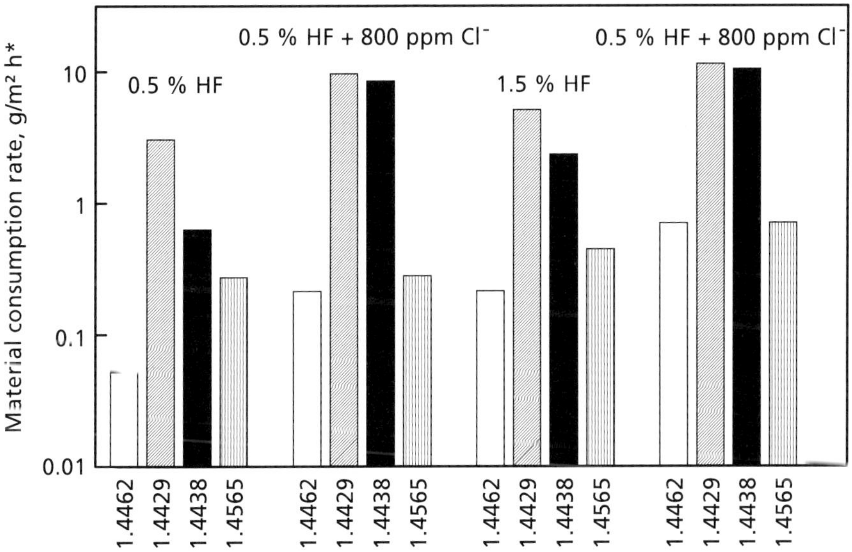

Figure 38: Influence of the content of impurities on the material consumption rates of CrNiMo steels when tested in crude phosphoric acid (52% P_2O_5 + 4% H_2SO_4 + 1% Fe^{3+}), test conditions: 373 K (100 °C) test duration 24 h [81]

1.4462 ≙ 2205

1.4429 ≙ SAE 316 LN

1.4438 ≙ SAE 317 L

1.4565 ≙ UNS S34565

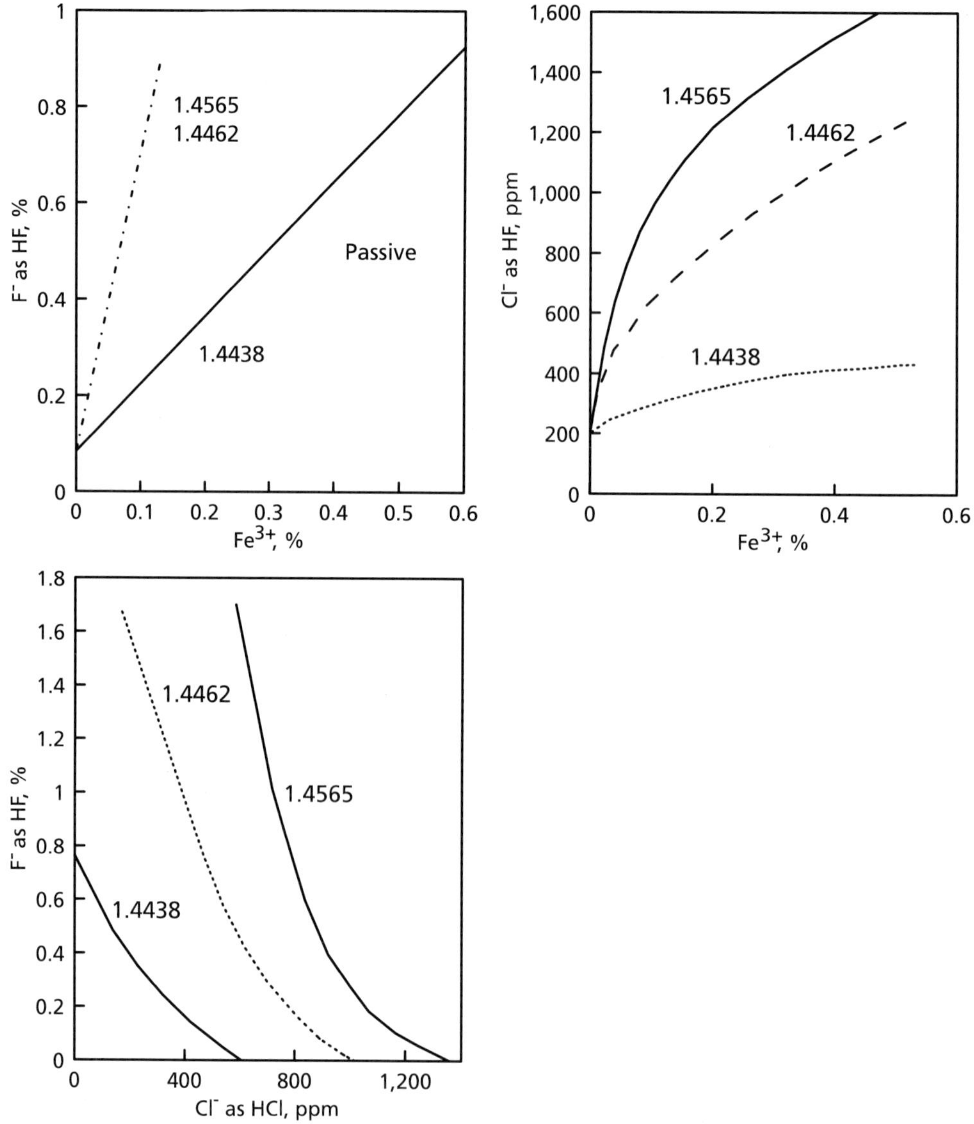

Figure 39: Dependence of the self-passivation behavior of high-alloy steels (see Figure 40) in 71.7% H_3PO_4 (52% P_2O_5) at 323 K (50 °C) on the impurities thereof (above the curves, the corrosion rate is significantly above 0.1 mm/a (3.94 mpy)) – results by potential measurements on zinc-activated specimens – [81]
* at a density of 7.85, corresponds to 1 g/m^2 h ≙ 1.1 mm/a (43.3 mpy)
1.4462 ≙ 2205
1.4438 ≙ SAE 317 L
1.4565 ≙ UNS S34565

In superphosphoric acid (Figure 40), the steels (Figure 38) are in the stable passive state, which can be seen from the very low corrosion rates. The duplex steel is also more resistant than the standard materials under these conditions.

Table 24 contains a summary of other high-alloy steels and alloys which are of interest when handling crude phosphoric acid.

Material designation		***DIN-Mat. No.***	***Average chemical composition***[1] ***%***							***Structure***	***Section***
			C	***Cr***	***Ni***	***Mo***	***Cu***	***Fe***	***N***		
UR45N	2205	1.4462	0.02	22	5.7	2.8	–	balance	0.12	ferr.-aust.	B
UR 47 N		(cf. 1.4410)	0.02	25	6.5	3.0	0.2	balance	0.17	ferr.-aust.	B
UR 50[2]		–	≤ 0.03	22	6.5	2.8	1.5	balance	–	ferr.-aust.	B
UR 52 N		(cf. 1.4507)	0.02	25	6.5	3.0	1.5	balance	0.17	ferr.-aust.	B
Ferralium®[2]		–	0.02	26	5.5	3.3	3.0	balance	0.20	ferr.-aust.	B
SAE 316 L		1.4404	≤ 0.03	17	12	2.2	–	balance	–	austenitic	D
		1.4435									
SAE 316 LN		1.4406	≤ 0.03	18	12	3.0	–	balance	0.15	austenitic	D
		1.4429									
SAE 317 L		1.4438	≤ 0.03	18	14	3.0	–	balance	–	austenitic	D
SAE 317 LN		(cf. 1.4442)	≤ 0.03	18	13	3.5	–	balance	0.13	austenitic	D
170 HE		1.4439	≤ 0.03	18	14	4.5	–	balance	0.14	austenitic	D
UR B 6	SAE 904 L	1.4539	≤ 0.02	20	25	4.4	1.5	balance	< 0.15	austenitic	E
UR B 28	UNS N08028	1.4563[3]	≤ 0.02	27	31	3.5	1.0	balance		austenitic	E
UR SB 8		–	0.01	25	25	4.8	1.5	balance	0.20	austenitic	E
UR 625	UNS N06625	2.4856	≤ 0.05	22	balance	9		4		austenitic	
HB 2		2.4615	≤ 0.02	< 1	balance	28		2		austenitic	
		2.4617									

1) data from the literature, differences are therefore possible
2) see Table 19
3) see Table 42

B: see section Ferritic-austenitic steels with more than 12% chromium
D: see section Austenitic CrNiMo(N) steels
E: see section Austenitic chromium-nickel steels with special alloying additions

Table 24: Average composition of high-alloy materials for use in phosphoric acid solutions (see Figures 41 and 42), [72, 100]

The application ranges of these materials when handling the usual phosphoric acid are shown in Figures 41 and 42. These summaries also show the sometimes very good suitability of the ferritic-austenitic steels when handling contaminated phosphoric acid (see also the pitting resistance equivalent (PRE)) [72].

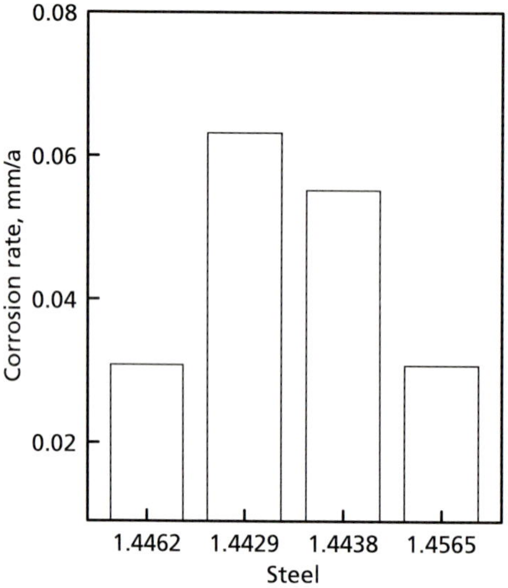

Figure 40: Corrosion rates for high-alloy steels in superphosphoric acid (70 % P_2O_5 + 3 % H_2SO_4 + 1 % Fe^{3+} + 0.6 % HF) at 373 K (100 °C) test duration 24 h [81]
1.4462 ≙ 2205
1.4429 ≙ SAE 316 LN
1.4438 ≙ SAE 317 L
1.4565 ≙ UNS S34565

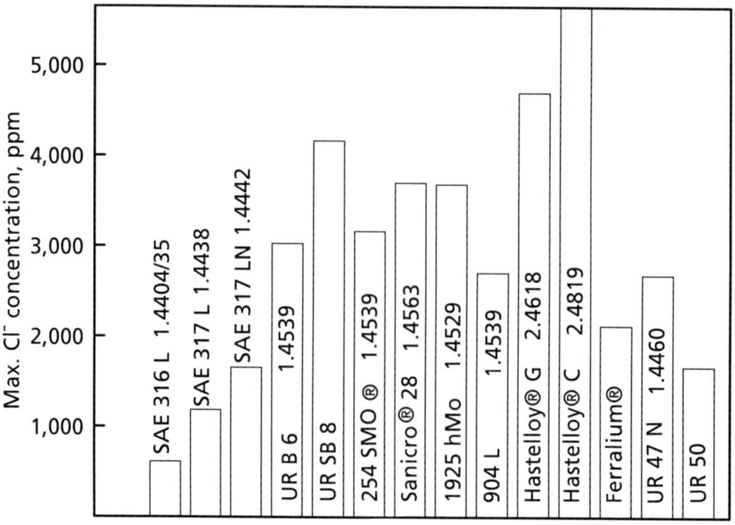

Figure 41: Application ranges (upper limit of the Cl^- content) for high-alloy materials in boiling 41.4 % H_3PO_4 (30 % P_2O_5) in decomposition of natural phosphate [72, 100]

Stress corrosion cracking

In media of the foodstuffs industry (0.8 % NaCl + 0.2 % H_3PO_4, pH 1.5–2, T = 414 K (141 °C)), U specimens of Ferralium® 255 (Table 19) displayed cracking. The test duration was 30 days.

In a boiling solution (374 K (101 °C) of 4 % NaCl and 1 % H_3PO_4, Ferralium® 255 showed no cracking. Cracking occurred in the same solution in an autoclave at 414 K (141 °C) (test duration 30 days, U specimens) [73].

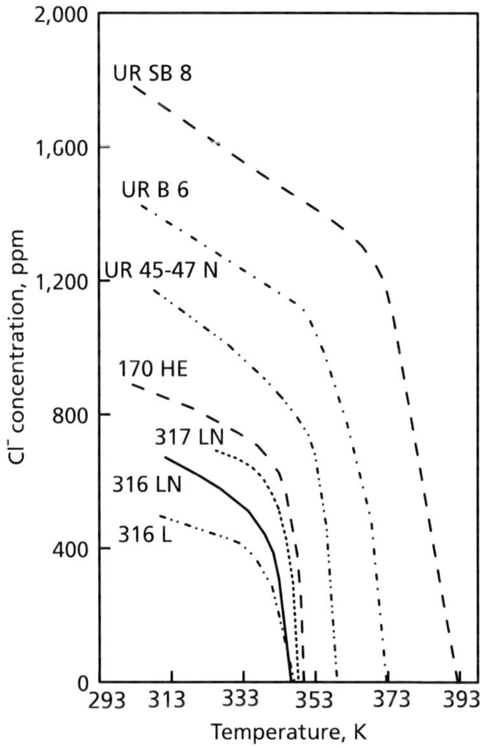

Figure 42: Application ranges (below the curves in the left-hand region) for high alloy materials (see Table 24) in 74.5 % H_3PO_4 (54 % P_2O_5, ≤ 4 % H_2SO_4, ≤ 1 % F^-, < 0.2 % HF) as a function of their temperature and Cl^- concentration [72, 100]

Austenitic CrNi steels

It was stated under Sections Ferritic chromium steels with more than 12 % chromium and Ferritic-austenitic steels with more than 12 % chromium that the most effective alloying constituents in ferrous materials for improving corrosion resistance toward phosphoric acid are chromium and molybdenum. Both alloying elements, moreover, are responsible for a good resistance of the materials to pitting corrosion, as can be seen from the pitting resistance equivalent (PRE) (in this context see Section Ferritic-austenitic steels with more than 12 % chromium).

These facts indicate that the resistance of molybdenum-free CrNi steels in phosphoric acid will sometimes be below that of the high-alloy ferrites or of duplex steels, as has already been found in the two preceding sections.

Steels 301 and 302 (cf. 1.4310, X10CrNi18-8) thus cannot be used either in pure phosphoric acid with 45–55 % P_2O_5 or in one containing 20–30 % P_2O_5, and cannot be used at all in crude acids of this type. Steel 304 (cf. 1.4301, X10CrNi18-8) has a somewhat better resistance. It also cannot be used in the crude acid mentioned, but can be used in pure acid with 20–30 % P_2O_5 (27.6–41.4 % H_3PO_4) at room temperature. Nevertheless, failure of the steel must be expected [6].

The steel 1.4301 can be used in pure phosphoric acid – and only in such – in the entire concentration range up to about 353 K (80 °C) and in pure boiling phosphoric acid of up to about 50 % strength (36.23 % P_2O_5) [80].

Under these conditions, corrosion rates (at the extremes) of up to 0.1 mm/a (3.94 mpy) or slightly above can be expected. In these steels in particular, small amounts of impurities in the phosphoric acid have an effect on the resistance. For possible use of these steels, a test under the conditions to be expected is absolutely advisable.

In industrial phosphoric acid in particular, CrNi steels tend to assume unstable states. Passivation and activation proceed slowly in it, which is why it is difficult to determine the corrosion rate by measuring the weight losses. Large differences may be obtained if the corrosion studies are carried out without and with artificial activation (contact with zinc dust). An extended corrosion resistance is to be expected in Mo-containing steels. See under Section Austenitic CrNiMo(N) steels.

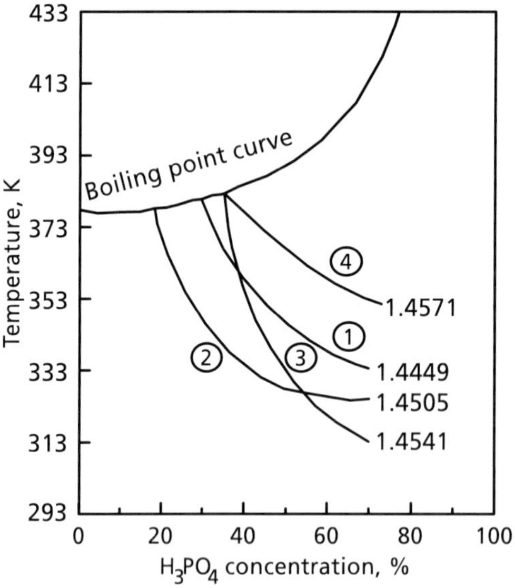

Figure 43: Corrosion behavior of CrNi steels in pure phosphoric acid, isocorrosion curves at 0.1 mm/a (3.94 mpy) [102]
① 1.4449, (SAE 317, X3CrNiMo18-12-3)
② 1.4505, (X4NiCrMoCuNb20-18-2)
③ 1.4541, (SAE 321, X6CrNiTi18-10)
④ 1.4571, (SAE 316 Ti, X6CrNiMoTi17-12-2)

The increase in resistance can be seen from Figures 7, 13, 27, 28.

The isocorrosion diagram for a CrNi steel with 0.1 % C, 18 % Cr and 8 % Ni [40] is similar to Figure 7, and for this reason it has not been shown as an extra figure. The same also applies to the steel SAE 304 (1.4301) in pure acid [101].

Figures 43 and 44 contain graphs of the corrosion behavior of the usual standard steels in pure and industrial phosphoric acid [102].

The designations of the phosphoric acid in this literature reference are confused; industrial phosphoric acid is considerably more aggressive than pure acid.

As can be seen from the two graphs, steel 1.4541 (SAE 321) is considerably more resistant in pure acid than in industrial acid. In industrial acid, it is far behind Mo-containing steels in its resistance (Figure 44).

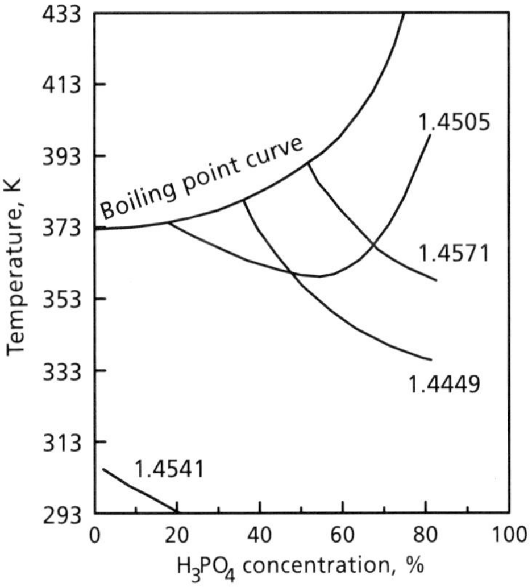

Figure 44: Corrosion behavior of CrNi steels in industrial phosphoric acid, isocorrosion curves at 0.1 mm/a (3.94 mpy) [102]

Tables 2 and 7 contain a summary of corrosion rates to be expected for CrNi steels in phosphoric acid. These are the materials under Section Austenitic CrNi steels in Table 2 and 7 and the materials with the numbers 1.4568 (SAE 631) and 1.4305 (SAE 303) in Table 7. The resistance of the last two steels mentioned is lower than that of those listed under Austenitic CrNi steels.

According to [8], the corrosion rates for steel 304 (1.4301, X5CrNi18-10) in aerated and non-aerated phosphoric acid in the range from 10 to 80 % and up to 339 K (66 °C) are reported as being not more than 0.05 mm/a (1.97 mpy). In up to 20 % phosphoric acid, the temperature can be exceeded somewhat. At higher concentrations, the corrosion sometimes increases rapidly. In this case, the acid is pure acid.

In 85 % acid at 339 K (66 °C) the corrosion rate is 0.03 mm/a (1.18 mpy) [103].

The problems of short-term studies in phosphoric acid and how difficult it is to determine corrosion rates have been referred to above. Such studies were carried out in [104] in phosphoric acid in the concentration range from 5 to 87 % at temperatures up to the boiling point. The test durations up to 373 K (100 °C) were 7–9 h, and above this temperature only 4–6 h. Apart from the steel 08Ch18N10T (cf. SAE 321, 1.4541), the other materials cannot be encoded into DIN materials. The corresponding sections and the chemical compositions are stated for these, so that they can be compared with corresponding DIN materials. The test results are summarized in Table 25. When the corrosion rates of steel 08Ch1810NT under the test conditions are compared with the practical resistance limit of 0.1 mm/a (3.94 mpy), as expected the use of this material is considerably limited, as shown in Table 25. If chloride-free acid is a prerequisite of these studies, the resistance in chloride-containing acid falls even further, as confirmed by the short-term studies in [24].

In this context, see also Figure 18 and under Section Ferritic chromium steels with more than 12 % chromium, (pitting resistance equivalent (PRE)). A lack of Mo content in the steels is the reason for the low resistance to pitting corrosion (see also under Section Austenitic CrNiMo(N) steels).

Electrochemical studies in an aqueous solution of 40 % H_3PO_4 and 4 % H_2SO_4 with addition of 3,000 ppm Cl + 24 g/l SiC at 333 K (60 °C) on the CrNi steel 18 10 with 1.5 and 4 % added Mo showed a significant decrease in the corrosion rate, the jump from 0 to 1.5 % Mo in particular being very large. The corrosion current dropped by a factor of 100 by addition of 1.5 % Mo, and by a factor of 60 by addition of between 1.5 % Mo and 4 % Mo [98].

The effect of halogens on the corrosion behavior of SAE 304 and 316 is summarized in Table 26. In principle, the attack on the Mo-free steels is greater than on the Mo-containing steels, although these can be referred to as having resistance (corrosion rates not more than 0.1 mm/a (3.94 mpy)). In 30 and 50 % H_3PO_4, the effect decreases in the sequence $F^- > Cl^- > Br^-$, while in 70 and 85 % H_3PO_4 the sequence is $Cl^- > F^- > Br^-$ [105]. See also [106–108].

In 75 % H_3PO_4 at 302–311 K (29–38 °C) the corrosion rate rose from 0.035 mm/a (1.38 mpy) at a chloride content of 8 ppm to 0.28 mm/a (11.0 mpy) at 15 ppm. An increase to 23 ppm already led to corrosion rates of 13 mm/a (512 mpy). During this already vigorous corrosion, hydrogen was evolved, and of course also at higher chloride contents [42, 43]. The test duration was 23 h.

During industrial use, an explosion due to evolution of hydrogen (a consequence of too vigorous corrosion) cannot be ruled out.

Machining steels

Machining steels are alloyed with sulfur to improve machinability. The manganese-rich sulfides distributed in the steel are the reason for short-breaking swarfs, low cutting forces, higher permitted cutting speeds and higher service lives of the tools.

Addition of sulfur has an adverse effect on the corrosion resistance of the machining steels. It is reduced to the extent that these steels can only be used in slightly corrosive media. Manganese sulfides are already attacked in weakly acid solutions. In this case, pits and pores containing decomposition products are formed on the steel surface. These are difficult to repassivate and are the starting points for further corrosion.

H_3PO_4 %	Temp. K (°C)	08Ch18N10T[1)] C	09Ch16N15M3B[2)] D	03Ch16N15M3[3)] D	08Ch17N15M3B[4)] D	06ChN28MDT[5)] E	ChN77TJuR[6)]
		Corrosion rate, mm/a (mpy)					
5	313 (40)	0.06 (2.36)	0.06 (2.36)	0.16 (6.3)	0.24 (9.45)	0.04 (1.57)	0.07 (2.76)
	333 (60)	0.24 (9.45)	0.04 (1.57)	0.26 (10.2)	0.27 (10.6)	0.13 (5.12)	0.09 (3.54)
	353 (80)	0.13 (5.12)	0.21 (8.27)	0.09 (3.54)	0.27 (10.6)	0.12 (4.72)	0.16 (6.3)
	363 (90)	0.12 (4.72)	0.41 (16.1)	0.15 (5.91)	0.23 (9.06)	0.06 (2.36)	0.21 (8.27)
	boiling	0.16 (6.3)	0.28 (11.0)	0.30 (11.8)	0.76 (29.9)	0.36 (14.2)	0.24 (9.45)
10	313 (40)	0.08 (3.15)	0.09 (3.54)	0.20 (7.87)	0.14 (5.51)	0.35 (13.8)	0.01 (0.39)
	333 (60)	0.24 (9.45)	0.14 (5.51)	0.18 (7.09)	0.27 (10.6)	0.16 (6.3)	0.13 (5.12)
	353 (80)	0.12 (4.72)	0.26 (10.3)	0.27 (10.6)	0.11 (4.33)	0.07 (2.76)	0.20 (7.87)
	363 (90)	0.13 (5.12)	0.13 (5.12)	0.28 (11.0)	0.26 (10.2)	0.08 (3.15)	0.19 (7.48)
	boiling	0.37 (14.6)	0.58 (22.8)	0.39 (15.4)	0.84 (33.1)	0.59 (23.2)	0.36 (14.2)
20	313 (40)	0.12 (4.72)	0.02 (0.79)	0.13 (5.12)	0.08 (3.15)	0.11 (4.33)	0.04 (1.57)
	333 (60)	0.24 (9.45)	0.13 (5.12)	0.18 (7.09)	0.16 (6.3)	0.09 (3.54)	0.17 (6.69)
	353 (80)	0.09 (3.54)	0.13 (5.12)	0.18 (7.09)	0.27 (10.6)	0.09 (3.54)	0.29 (11.4)
	363 (90)	0.10 (3.94)	0.14 (5.51)	0.09 (3.54)	0.40 (15.6)	0.33 (13.0)	0.33 (13.0)
	boiling	0.90 (35.4)	0.47 (18.5)	0.45 (17.7)	1.51 (59.5)	1.47 (57.9)	0.48 (18.9)

Table 25: Corrosion rates of steels and a Ni alloy in phosphoric acid (no information) – test duration 7–9 h up to 373 K (100 °C) above 373 K (100 °C) 4–6 h (for short-term studies, see Section Austenitic CrNi steels) [104]

Table 25: Continued

H_3PO_4 %	Temp. K (°C)	08Ch18N10T[1)] C	09Ch16N15M3B[2)] D	03Ch16N15M3[3)] D	08Ch17N15M3B[4)] D	06ChN28MDT[5)] E	ChN77TJuR[6)]
		Corrosion rate, mm/a (mpy)					
30	313 (40)	0.08 (3.15)	0.04 (1.57)	0.18 (7.09)	0.03 (1.18)	0.08 (3.15)	0.05 (1.97)
	333 (60)	0.12 (4.72)	0.05 (1.97)	0.06 (2.36)	0.07 (2.86)	0.08 (3.15)	0.03 (1.18)
	353 (80)	0.13 (5.12)	0.09 (3.54)	0.09 (3.54)	0.09 (3.54)	0.16 (6.30)	0.12 (4.72)
	363 (90)	0.19 (7.48)	0.11 (4.33)	0.14 (5.51)	0.09 (3.54)	0.22 (8.66)	0.32 (12.6)
	373 (100)	0.16 (6.30)	0.28 (11.0)	0.16 (6.30)	0.87 (34.3)	0.18 (7.09)	0.40 (15.8)
	boiling	0.45 (17.7)	0.89 (35.0)	0.43 (16.9)	1.90 (74.8)	0.71 (27.9)	0.57 (22.4)
40	313 (40)	0.08 (3.15)	0.11 (4.33)	0.08 (3.15)	0.16 (6.3)	0.05 (1.97)	0.06 (2.36)
	333 (60)	0.26 (10.2)	0.09 (3.54)	0.07 (2.86)	0.19 (7.48)	0.05 (1.97)	0.07 (2.76)
	353 (80)	0.32 (12.6)	0.13 (5.12)	0.09 (3.54)	0.27 (10.6)	0.06 (2.36)	0.18 (7.09)
	363 (90)	0.62 (24.4)	0.25 (9.84)	0.11 (4.33)	0.10 (3.94)	0.18 (7.09)	0.26 (10.2)
	373 (100)	0.92 (36.2)	0.21 (8.27)	0.51 (20.1)	0.11 (4.33)	0.19 (7.48)	0.48 (18.9)
	boiling	0.48 (18.9)	0.91 (35.8)	1.05 (41.3)	0.91 (35.8)	0.51 (20.1)	0.51 (20.1)

Table 25: Corrosion rates of steels and a Ni alloy in phosphoric acid (no information) – test duration 7–9 h up to 373 K (100 °C) above 373 K (100 °C) 4–6 h (for short-term studies, see Section Austenitic CrNi steels) [104]

Table 25: Continued

H_3PO_4 %	Temp. K (°C)	08Ch18N10T[1] C	09Ch16N15M3B[2] D	03Ch16N15M3[3] D	08Ch17N15M3B[4] D	06ChN28MDT[5] E	ChN77TJuR[6]
		Corrosion rate, mm/a (mpy)					
50	313 (40)	0.09 (3.54)	0.39 (15.4)	0.81 (31.9)	0.40 (15.8)	0.03 (1.18)	0.06 (2.36)
	333 (60)	0.25 (9.84)	0.11 (4.33)	0.09 (3.54)	0.05 (1.97)	0.16 (6.30)	0.08 (3.15)
	353 (80)	0.24 (9.45)	0.13 (5.12)	0.18 (7.09)	0.04 (1.57)	0.20 (7.87)	0.23 (9.06)
	363 (90)	0.84 (33.1)	0.13 (5.12)	0.11 (4.33)	0.07 (2.76)	0.20 (7.87)	0.19 (7.48)
	373 (100)	0.12 (4.72)	1.02 (40.2)	0.13 (5.12)	0.10 (3.94)	0.98 (38.6)	0.26 (10.2)
	boiling	0.23 (9.06)	1.08 (42.5)	1.35 (53.2)	2.08 (81.9)	1.68 (66.1)	0.29 (11.4)
60	313 (40)	0.09 (3.54)	0.08 (3.15)	0.07 (2.76)	0.13 (5.12)	0.11 (4.33)	0.07 (2.76)
	333 (60)	0.10 (3.94)	0.09 (3.54)	0.14 (5.51)	0.27 (10.6)	0.23 (9.06)	0.07 (2.76)
	353 (80)	0.44 (17.3)	0.11 (4.33)	0.19 (7.48)	0.10 (3.94)	0.45 (17.7)	0.09 (3.54)
	363 (90)	0.38 (15.0)	0.13 (5.12)	0.18 (7.09)	0.08 (3.15)	0.10 (3.94)	0.96 (37.8)
	373 (100)	0.32 (12.6)	0.42 (16.5)	0.49 (19.3)	0.08 (3.15)	0.53 (20.9)	0.63 (24.8)
	393 (120)	1.40 (55.1)	1.12 (44.1)	3.02 (119)	1.05 (41.3)	1.06 (41.7)	0.66 (26.0)
	boiling	1.65 (65.0)	19.8 (779)	7.78 (306)	3.98 (157)	2.93 (115)	0.58 (22.8)

Table 25: Corrosion rates of steels and a Ni alloy in phosphoric acid (no information) – test duration 7–9 h up to 373 K (100 °C) above 373 K (100 °C) 4–6 h (for short-term studies, see Section Austenitic CrNi steels) [104]

Table 25: Continued

H_3PO_4 %	Temp. K (°C)	08Ch18N10T[1)] C	09Ch16N15M3B[2)] D	03Ch16N15M3[3)] D	08Ch17N15M3B[4)] D	06ChN28MDT[5)] E	ChN77TJuR[6)]
		Corrosion rate, mm/a (mpy)					
70	313 (40)	0.04 (1.57)	0.10 (3.94)	0.09 (3.54)	0.18 (7.09)	0.13 (5.12)	0.08 (3.15)
	333 (60)	0.09 (3.54)	0.11 (4.33)	0.27 (10.6)	0.07 (2.76)	0.13 (5.12)	0.07 (2.76)
	353 (80)	0.12 (4.72)	0.13 (5.12)	0.31 (12.2)	0.06 (3.26)	0.09 (3.54)	0.10 (3.94)
	363 (90)	0.29 (11.4)	–	0.36 (14.2)	0.14 (5.51)	0.25 (9.84)	–
	373 (100)	0.24 (9.45)	0.80 (31.5)	1.08 (42.5)	1.70 (66.9)	0.91 (35.8)	0.68 (26.8)
	393 (120)	2.04 (80.3)	3.65 (144)	3.96 (156)	0.05 (1.97)	1.03 (40.6)	1.08 (42.5)
	boiling	10.5 (413)	26.6 (1,047)	9.04 (356)	14.5 (571)	3.35 (132)	1.12 (44.1)
80	313 (40)	0.09 (3.54)	0.10 (3.94)	0.27 (10.6)	0.09 (3.54)	0.06 (2.36)	0.08 (3.15)
	333 (60)	0.13 (5.12)	0.11 (4.33)	0.70 (27.6)	0.10 (3.94)	0.06 (2.36)	0.08 (3.15)
	353 (80)	0.09 (3.54)	0.13 (5.12)	0.26 (10.2)	0.54 (21.3)	0.16 (6.30)	0.09 (3.54)
	363 (90)	0.23 (9.06)	0.25 (9.84)	0.73 (28.7)	0.14 (5.51)	0.26 (10.2)	0.15 (5.91)
	373 (100)	0.49 (19.3)	0.56 (22.1)	4.75 (187)	1.20 (47.2)	0.71 (27.9)	0.69 (27.2)
	393 (120)	6.20 (244)	8.14 (329)	9.28 (365)	4.27 (168)	1.58 (62.2)	1.93 (76.0)
	413 (140)	65.7 (2,587)	69.5 (2,736)	15.4 (606)	19.2 (756)	4.8 (189)	3.01 (119)
	boiling	301 (11,850)	106 (4,173)	27.4 (1,078)	27.5 (1,083)	5.79 (228)	15.8 (622)

Table 25: Corrosion rates of steels and a Ni alloy in phosphoric acid (no information) – test duration 7–9 h up to 373 K (100 °C) above 373 K (100 °C) 4–6 h (for short-term studies, see Section Austenitic CrNi steels) [104]

Table 25: Continued

H_3PO_4 %	Temp. K (°C)	08Ch18N10T[1)] C	09Ch16N15M3B[2)] D	03Ch16N15M3[3)] D	08Ch17N15M3B[4)] D	06ChN28MDT[5)] E	ChN77TJuR[6)]
		Corrosion rate, mm/a (mpy)					
87	313 (40)	0.06 (2.36)	0.11 (4.33)	0.06 (2.36)	0.04 (1.57)	0.15 (5.91)	0.09 (3.54)
	333 (60)	0.04 (1.57)	0.12 (4.72)	0.09 (3.54)	0.06 (2.36)	0.16 (6.30)	–
	353 (80)	0.09 (3.54)	0.12 (4.72)	0.10 (3.94)	0.20 (7.87)	–	0.12 (4.72)
	373 (100)	0.37 (14.6)	0.53 (20.9)	0.22 (8.66)	1.67 (65.6)	0.66 (26.0)	1.03 (40.6)
	393 (120)	7.40 (291)	10.7 (421)	5.91 (233)	5.27 (207)	0.53 (20.9)	2.80 (110)
	413 (140)	–	24.3 (957)	27.2 (1,071)	43.0 (1,693)	4.23 (167)	10.9 (429)
	433 (160)	264 (10,394)	110 (4,330)	53.6 (2,110)	64.2 (2,528)	6.7 (264)	100 (3,937)
	473 (200)	676 (26,614)	119 (4,685)	83.1 (3,272)	110 (4,330)	21.2 (835)	251 (9,882)

1) ≡ 1.4541 (SAE 321, X6CrNiTi18-10)
2) ≤ 0.09 C, 15-17 Cr, 14-16 Ni, 2.5-3.0 Mo, Ti balance Fe
3) ≤ 0.03 C, 15-17 Cr, 14-16 Ni, 2.5-3.0 Mo, Ti balance Fe
4) ≤ 0.08 C, 16-18 Cr, 14-16 Ni, 3.0-4.0 Mo, Ti balance Fe
5) ≤ 0.06 C, 22-25 Cr, 26-29 Ni, 2.5-3.0 Mo, 2.5-3.5 Cu, Ti, balance Fe
6) ≤ 0.07 C, 19-22 Cr, 2.4-2.8 Ti, balance Ni

C: see section Austenitic CrNi steels
D: see section Austenitic CrNiMo(N) steels
E: see section Austenitic chromium-nickel steels with special alloying additions

Table 25: Corrosion rates of steels and a Ni alloy in phosphoric acid (no information) – test duration 7–9 h up to 373 K (100 °C) above 373 K (100 °C) 4–6 h (for short-term studies, see Section Austenitic CrNi steels) [104]

The corrosion resistance of machining steels can be improved by alloying with titanium, which leads to titanium sulfides instead of manganese sulfides, in accordance with the sulfur content. In comparison with manganese sulfides, titanium sulfides are resistant even in strongly acid media.

Table 9 contains a summary of corrosion rates in various concentrations of phosphoric acid at the boiling point.

As can be seen from the steels investigated, the Mo-free CrNi steel 18 10 has a very good resistance in 10% H_3PO_4, while the sulfur-containing steel can no longer be used with corrosion rates of 95 mm/a (3,740 mpy). In contrast, the S and Ti-containing steel has a slightly improved corrosion behavior. As well as having a better corrosion resistance, the resistance of these steels to pitting corrosion is increased by addition of S and Ti. They do not achieve the values of sulfur-containing grades in respect of machinability. The chemical composition of the steels is contained in Table 9a [60].

As can be furthermore seen from this table, a significantly improved corrosion resistance can be achieved even in these steels by addition of Mo (see Section Austenitic CrNiMo(N) steels). Corrosion studies on U-specimens (133 × 13 × 3 mm strips of sheet metal curved over a 25 mm ø) of SAE 304 (1.4301) and SAE 304 L (1.4306) showed crevice corrosion and stress corrosion cracking under the following conditions: 0.2% H_3PO_4 + 0.08% NaCl at 414 K (141 °C) and 1% H_3PO_4 + 4% NaCl at 414 K (141 °C) [109].

The reason for this local destruction is the presence of chlorides.

The corrosion rate for SAE 304 and 304 L in 1% H_3PO_4 + 4% NaCl was only 0.15 mm/a (5.91 mpy). In the solution containing 0.2% H_3PO_4, destruction occurred after 10 days, although corrosion rates of 0.14 mm/a (5.51 mpy) for steel SAE 304 and 0.03 mm/a (1.18 mpy) for SAE 304 L in principle demonstrate that these are resistant and, respectively, very resistant.

The corrosion behavior of CrNi steels under operating conditions for the production of wet phosphoric acid (Figure 64) can be seen from Tables 31, 35, 38 and 41.

Austenitic CrNiMo(N) steels

The austenitic chromium-nickel-molybdenum steels, and to a lesser extent the chromium-nickel steels under Section Austenitic CrNi steels, are chiefly used for the manufacture of tanks for phosphoric acid.

A prerequisite for this is their suitability for the concentration in question, the temperature and the purity of the acid.

Chromium-nickel steels and chromium-nickel-molybdenum steels tend to assume unstable states in phosphoric acid, and especially in industrial acid. In these, both passivation and activation proceed slowly [102].

Material	***Halogens***	***Tempera-ture K (°C)***	***H_3PO_4 concentration, %***			
			30	***50***	***70***	***85***
			Corrosion rate, mm/a (mpy)			
SAE 304 (1.4301, X5CrNi18-10)	–	323 (50)	0.023 (0.91)	0.005 (0.2)	0.014 (0.55)	0.023 (0.91)
	F	323 (50)	0.023 (0.91)	0.005 (0.2)	1.00 (39.4)	0.280 (11.0)
	Cl	323 (50)	0.023 (0.91)	0.005 (0.2)	3.14 (124)	24.0 (945)
	Br	323 (50)	0.023 (0.91)	0.005 (0.2)	0.070 (2.76)	0.046 (1.81)
	–	353 (80)	0.023 (0.91)	0.009 (0.35)	0.540 (21.3)	0.070 (2.76)
	F	353 (80)	0.302 (11.9)	1.16 (45.7)	1.023 (40.3)	99.0 (3,898)
	Cl	353 (80)	0.046 (1.81)	0.009 (0.35)	14.0 (551)	145 (5,708)
	Br	353 (80)	0.023 (0.91)	0.009 (0.35)	0.511 (20.1)	0.600 (23.6)
	–	373 (100)	0.093 (3.66)	0.023 (0.91)	2.79 (110)	254 (10,000)
	F	373 (100)	1.75 (68.9)	6.28 (247)	4.51 (178)	238 (9,370)
	Cl	373 (100)	0.070 (2.76)	15.0 (591)	54.4 (2,142)	400 (15,748)
	Br	373 (100)	0.070 (2.76)	0.023 (0.91)	1.77 (69.7)	1.26 (49,600)

Table 26: Influence of halogens on the corrosion behavior of the CrNi steel SAE 304 and the CrNiMo steel SAE 316 in phosphoric acid (analytically pure) with and without their addition (40 mmol/l) [105]

Table 26: Continued

Material	*Halogens*	*Temperature K (°C)*	*H_3PO_4 concentration, %*			
			30	*50*	*70*	*85*
			Corrosion rate, mm/a (mpy)			
SAE 316 (cf. 1.4401/1.4436, X5CrNiMo17-12-2/ X3CrNiMo17-13-3)	–	323 (50)	0.005 (0.2)	0.005 (0.2)	0.009 (0.35)	0.009 (0.35)
	F	323 (50)	0.005 (0.2)	0.005 (0.2)	0.023 (0.91)	0.023 (0.91)
	Cl	323 (50)	0.005 (0.2)	0.005 (0.2)	0.116 (4.57)	0.023 (0.91)
	Br	323 (50)	0.005 (0.2)	0.005 (0.2)	0.023 (0.91)	0.014 (0.55)
	–	353 (80)	0.009 (0.35)	0.023 (0.91)	0.093 (3.66)	0.600 (23.6)
	F	353 (80)	0.139 (5.47)	0.420 (16.5)	0.370 (14.6)	0.190 (7.48)
	Cl	353 (80)	0.023 (0.91)	0.023 (0.91)	0.560 (22.1)	0.370 (14.6)
	Br	353 (80)	0.023 (0.91)	0.023 (0.91)	0.116 (4.57)	0.023 (0.91)
	–	373 (100)	0.400 (15.8)	0.070 (2.76)	1.813 (71.4)	0.450 (17.7)
	F	373 (100)	1.140 (44.9)	1.420 (55.9)	1.930 (76.0)	3.240 (128)
	Cl	373 (100)	0.139 (5.47)	0.210 (8.27)	2.650 (104)	1.170 (46.1)
	Br	373 (100)	0.116 (4.57)	0.139 (5.47)	1.540 (60.6)	0.046 (1.81)

Table 26: Influence of halogens on the corrosion behavior of the CrNi steel SAE 304 and the CrNiMo steel SAE 316 in phosphoric acid (analytically pure) with and without their addition (40 mmol/l) [105]

The range of chromium-nickel-molybdenum steels on the market is very diverse. The resistance of these steels to phosphoric acid increases as the chromium content increases, and increases significantly with the molybdenum content [5, 6, 10, 11, 51, 52, 75, 80, 82, 89–91, 102, 106, 110–119].

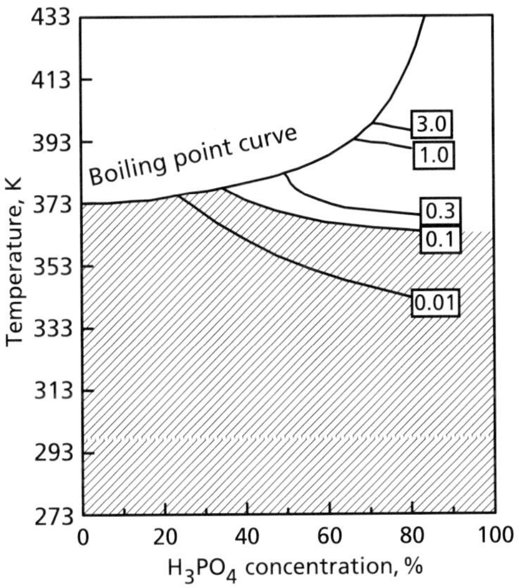

Figure 45: Isocorrosion curves (mm/a) for CrNiMo steel (0.05 % C, 18 % Cr, 10 % Ni, 2 % Mo – approximately corresponds to SAE 316, 1.4401, X5CrNiMo17-12-2) in pure phosphoric acid [40, 77]

0.01 mm/a (0.39 mpy)	0.3 mm/a (11.8 mpy)	3.0 mm/a (118 mpy)
0.1 mm/a (3.94 mpy)	1.0 mm/a (39.4 mpy)	

An overview of the corrosion behavior of CrNiMo steels of increasing Mo content is given in Figure 45 for a steel with 2 % Mo and Figure 46 for a steel with 5 % Mo. Grades which are currently commercially available have been assigned to the material compositions stated. The studies to plot these isocorrosion diagrams were carried out in chemically pure phosphoric acid without special aeration. Air had access to the surface of the medium [40].

According to these diagrams, steel 1.4401 can be used in all concentrations of pure phosphoric acid – and only in this acid – up to 353 K (80 °C) and in only up to about 50 % boiling pure phosphoric acid if corrosion rates of 0.1 mm/a (3.94 mpy) may not be exceeded [40, 80]. The use of pure acid is also the reason why the difference in the corrosion resistance of the two steels is not so marked (0.1 mm/a (3.94 mpy) curve). This difference only becomes more clearly visible if the phosphoric acid contains impurities. As such, Cl^- and F^- ions and sulfuric acid stimulate corrosiveness, while Fe_2O_3, Al_2O_3, SiO_2, CaO and MgO inhibit corrosion [120]. See also under Austenitic CrNi steels.

Studies on the steel X2CrNiMo18-14-3 (SAE 316 L, 1.4435) which is often used in chemical apparatus construction, in respect of modification of its corrosion resistance by cold working, led to the following results:

Uniform corrosion in boiling 50 % phosphoric acid is not adversely influenced.

The susceptibility to stable pitting corrosion in chloride-containing solutions at elevated temperature (threshold potential of stable pitting corrosion) is not adversely influenced by cold working at 20–100 %, although repassivatable pitting corrosion is promoted by cold working. (Such damage scarcely occurs by this type of pitting corrosion).

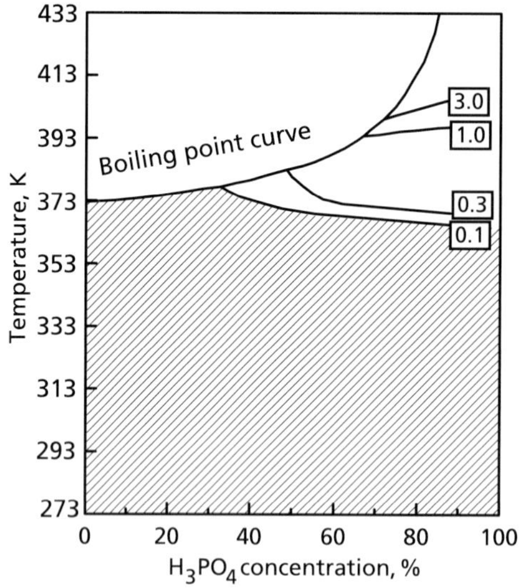

Figure 46: Isocorrosion curves (mm/a) for CrNiMo steel (0.05 C, 17 % Cr, 13 % Ni, 5 % Mo – approximately corresponds to SAE 317 LMN, 1.4439; X2CrNiMoN17-13-5) in pure phosphoric acid [40]

0.1 mm/a (3.9 mpy)	1.0 mm/a (39.4 mpy)
0.3 mm/a (11.8 mpy)	3.0 mm/a (118 mpy)

The very low sensitivity toward intercrystalline corrosion resulting from the very low carbon content of this steel is also not increased by cold working [121].

The difference in corrosion resistance of Mo-containing and Mo-free CrNi steels can be seen by comparing Figures 7 and 45. The varying behavior in pure and industrial acid can be seen from Figures 43 and 44. Figure 44 in particular (contaminated industrial phosphoric acid) shows, as already indicated above, the better resistance of the Mo-containing steels 1.4449 (SAE 317), 1.4505 and 1.4571 (SAE 316 Ti). According to this graph, steel 1.4571 is practically resistant in boiling industrial phosphoric acid up to concentrations of 55 %, the corrosion rate according to this graph being not more than 0.1 mm/a (3.94 mpy). At higher concentrations up to 80 %, the 0.1 mm/a (3.94 mpy) limit curve drops relatively sharply [102].

Table 2 contains a summary of corrosion rates in phosphoric acid and its mixtures as the molybdenum content increases. This also shows that molybdenum-containing steels are more suitable in phosphoric acid [23].

Table 7 contains a summary of corrosion rates of commercially available steels [48]. The ferritic-austenitic steel 1.4460 (SAE 329, Table 8), with a comparatively good resistance, is classified under Section Ferritic-austenitic steels with more than 12 % chromium. The copper-containing steel 1.4505 (Table 8) is distinguished by a higher resistance (see Figures 63 and 64 and Austenitic chromium-nickel steels with special alloying additions) [55].

For additional information, Table 25 contains corrosion rates for steels from the former Soviet Union [17].

The corrosive action of phosphoric acid on "stainless steels" decreases at concentrations above 80 % to 95 %. This dependence of the corrosion rate as the concentration increases in this range is shown by the example of steel 1.4401 (SAE 316, X5CrNiMo17-12-2) in Figure 47 [42, 43].

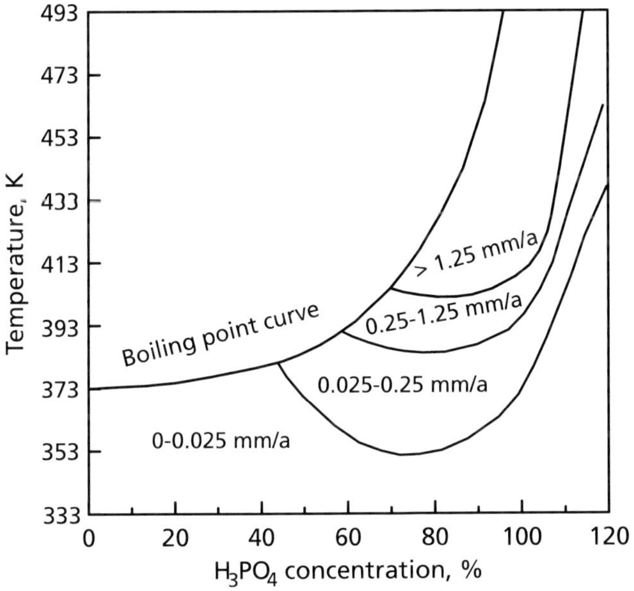

Figure 47: Corrosion behavior of steel 1.4401 (SAE 316, X5CrNiMo17-12-2) in electrothermally produced phosphoric acid, weakly agitated medium [6, 42, 43]

0–0.025 mm/a (0–0.98 mpy)	0.25–1.25 mm/a (9.84–49.2 mpy)
0.025–0.25 mm/a (0.98–9.84 mpy)	> 1.25 mm/a (> 49.2 mpy)

Electrochemical methods are rapid methods which are now being used more and more frequently for predicting corrosion behavior under given conditions [27, 98, 122].

A preliminary selection of materials for the particular conditions can be made according to Figures 4, 14, 27, 28. The resistance of high-alloy materials in phosphoric acid is determined by their impurities, and in particular by chlorides [106]. In these, Mo-containing steels prove to be advantageous [24].

Figures 17, 29, 38 and 40 show the corrosion rates to be expected for different materials, in comparison with CrNiMo steels, in phosphoric acid containing the usual impurities. The more highly Mo-alloyed steels prove to be the better materials here, as shown by the example of 1.4438 (SAE 317 L) in Figure 39. Its more adverse corrosion behavior under the conditions in Figure 38 and the better resistance of superferrite according to Figure 17 can also be explained with the aid of the pitting resistance equivalent (PRE). This is an empirically determined standard for the increase in resistance of materials, primarily toward pitting corrosion and crevice corrosion (in this context, see also Sections Ferritic chromium steels with more than 12 % chromium – Austenitic CrNi steels and Figures 30 and 36 for steel 317 L (1.4438)).

The usability of this steel in crude phosphoric acid is shown in Figure 48 [95]. In comparison with Figure 37, it is less resistant than the superferrite CrNiMo 25 4 2 (Section Ferritic chromium steels with more than 12 % chromium). For further corrosion rates of CrNiMo steels in the usual phosphoric acid with appropriate impurities see Tables 21–23, 26.

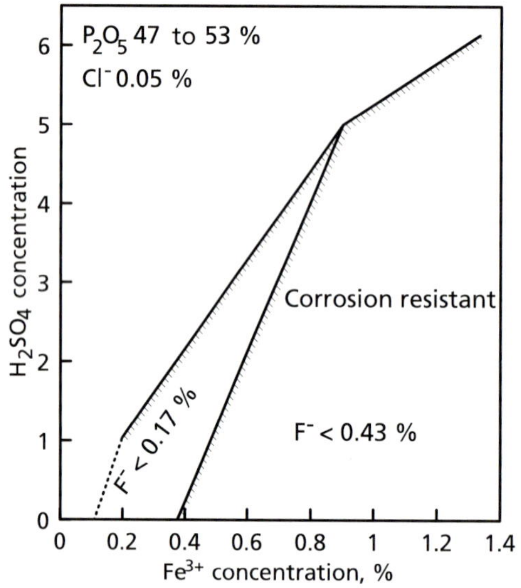

Figure 48: Ranges of corrosion resistance (on the right under the curves) for steel SAE 317 L (1.4438, X2CrNiMo18-15-4) in crude phosphoric acid at 323 K (50 °C) as a function of its H_2SO_4, F^- and Fe^{3+} content (see also Figure 37) [95]

Additions of H_2SO_4 and HF have a highly corrosive action (Table 23) [94]. Under such conditions, the use of CrNiMo steels is already doubtful, with corrosion rates of equal to or greater than 1 mm/a (39.4 mpy).

Phosphoric acid production

For the production of phosphoric acid by the thermal process (IG process here) all apparatus components which come into contact with P_2O_5-containing waste gas and the acid tower are or can be produced from steel 1.4404 (SAE 316 L, X2CrNiMo17-12-2). The acid tower is resistant to concentrated phosphoric acid under these conditions up to 373 K (100 °C) [1].

In 70 % H_3PO_4 at 383 K (110 °C) corrosion rates of 3.9 mm/a (154 mpy) are reported for steel 1.4435 (SAE 316 L, X2CrNiMo18-14-3) and 2.4 mm/a (94.5 mpy) for 1.4438 (SAE 317 L, X2CrNiMo18-15-4). In 80 % H_3PO_4 at 373 K (100 °C) 1.4435 experiences corrosion of "only" 1.3 mm/a (51.2 mpy) [87]. The ferritic-austenitic steel 1.4462 (2205, X2CrNiMoN22-5-3) is more suitable for use under these conditions, with corrosion rates of 0.22 (8.7) and, respectively, 0.35 mm/a (13.8 mpy) (see also Section Ferritic-austenitic steels with more than 12 % chromium) [87].

As already mentioned above, the corrosion of high-alloy steels is intensified considerably by fluorine – and in particular by chloride ions – while iron ions and copper ions (see Figure 52) inhibit it [19].

This fact is illustrated by the example of industrial phosphoric acid in Figure 49 [99].

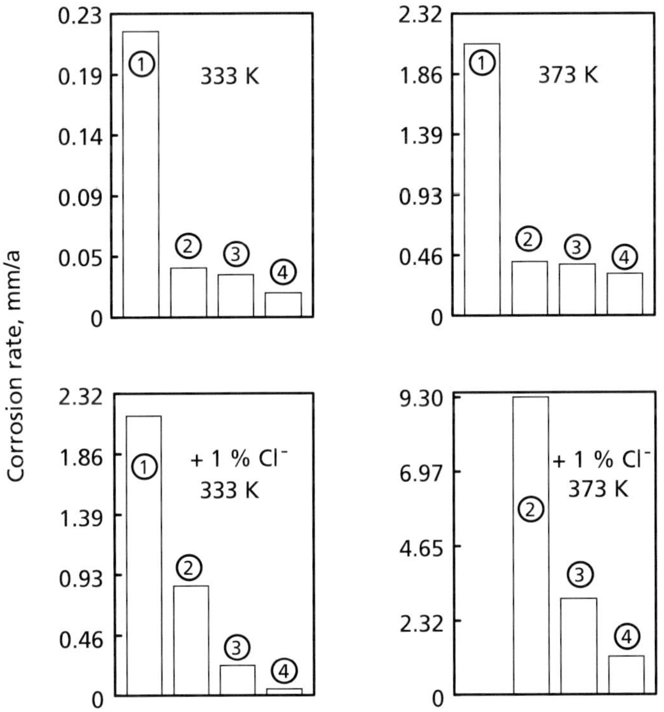

Figure 49: Influence of chloride ions on the corrosion behavior of high-alloy steels in industrial phosphoric acid (see under Section Ferritic-austenitic steels with more than 12 % chromium) at 333 and 373 K (60–100 °C) [99]
① 1.4404 (SAE 316 L, X2CrNiMo17-12-2), Austenitic CrNiMo(N) steels
② 1.4435 (SAE 316 L, X2CrNiMo18-14-3), Austenitic CrNiMo(N) steels
③ Uranus® 50 (UNS S32404), A 19
④ Uranus® B6 (1.4539, X1NiCrMoCu25-20-5), Austenitic chromium-nickel steels with special alloying additions

This graph clearly shows the increase in corrosion resistance of CrNiMo steels of increasing Mo content in industrial phosphoric acid (30 % P_2O_5 before, 50 % P_2O_5 after concentration, 1–3 % H_2SO_4, 0.5–1.5 % HF + H_2SiF_6, 0.1 % chlorides and gypsum). This section includes steels ① (1.4404) and ② (1.4435). Steel 1.4435 is usable for a prolonged period of time at 333 K (60 °C) and its use is still possible at 373 K (100 °C).

The two steels can no longer be used at even higher chloride contents. Under such conditions, a higher-alloy steel of even higher Mo content, like 1.4539 (SAE 904 L,

X1NiCrMoCu25-20-5), should be used (see under Section Austenitic chromium-nickel steels with special alloying additions).

An expected exception is the duplex steel Uranus® 50 (Section Ferritic-austenitic steels with more than 12 % chromium). An explanation for this is given by the pitting resistance equivalent (PRE), which is a measure for the resistance to chloride (resistance to pitting corrosion and crevice corrosion are also included here).

The high-alloy steel 1.4539 (Uranus® B6) has been introduced and proved itself as a material for the production of phosphoric acid in an ever more aggressive acid [99].

As an example of the increase in corrosiveness of phosphoric acid, Figure 50 shows the corrosion rates for steel 1.4401 (SAE 316) as a function of its HF and H_2SO_4 concentration, for "wet acid" at 361 K (88 °C) [42, 43]. According to this figure, steel 1.4401 can be used for only a short time – if at all – under these conditions, with corrosion rates of 1 mm/a (39.4 mpy).

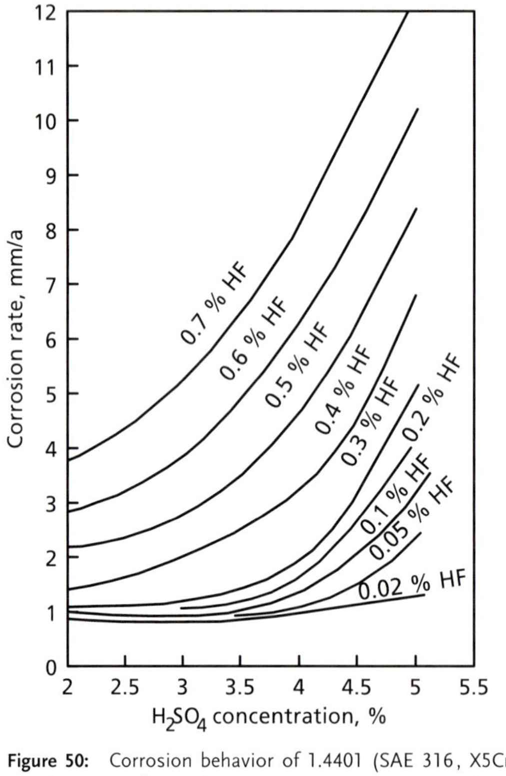

Figure 50: Corrosion behavior of 1.4401 (SAE 316, X5CrNiMo17-12-2) in 35 % strength H_3PO_4 (25 % P_2O_5), with 1.5 % F as H_2SiF_6, 0.02–0.7 % HF and 2–5 % H_2SO_4 [42, 43]

Figure 51 shows an overview of the use limits (corrosion rates of the isocorrosion curves 0.3 mm/a (11.8 mpy)) of metallic materials at 373 K (100 °C) during production of wet acid (Florida acid) [19].

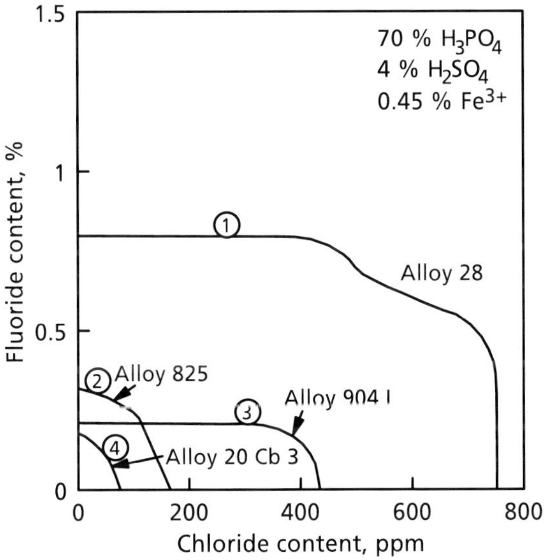

Figure 51: Application ranges (isocorrosion curves 0.3 mm/a (11.8 mpy)) for high-alloy materials in synthetic phosphoric acid ("Florida acid") at 373 K (100 °C) as a function of the fluoride and chloride content [19]
① Alloy 28 (1.4563, X1NiCrMoCu31-27-4)
② Alloy 825 (2.4858, NiCr21Mo)
③ Alloy 904 L (1.4539, X1NiCrMoCu25-20-5)
④ Carpenter® 20 Cb 3 (2.4660, NiCr20CuMo)

The composition of the Florida acid is reported as being as follows:

46.7 % P_2O_5, 2.9 % H_2SO_4 1.1 % F, 1.4 % Fe_2O_3, 0.7 % Al_2O_3, 0.4 % MgO, traces of CaO, no Cl [95].

Figure 51 shows that under these conditions CrNiMo standard steels cannot be used, since corrosion rates of 0.3 mm/a (11.8 mpy) are already above the resistance limit of 0.1 mm/a (3.94 mpy). The resistance-promoting effect of copper in CrNiMo steels is dealt with in Section Austenitic chromium-nickel steels with special alloying additions (see also, inter alia, Figure 51).

Copper(II)-ions can already be present in the medium, or can be introduced into this at the start of corrosion of copper-containing materials. The decrease in corrosion rate in 85 % H_3PO_4 as a function of its copper content is shown by the example of steel 1.4404 (SAE 316 L, X2CrNiMo17-12-2) (Figure 52). This shows that a copper content of about 0.2 % in the steel reduces the corrosion rate from more than 40 mm/a (1,575 mpy) to less than 15 mm/a (590 mpy). Although no resistance results, the trend is detectable [42, 43].

A quite comprehensive summary of possible uses of, inter alia, steel 316 in phosphoric acid solutions in comparison with Ni alloys is contained in [123]. It should be mentioned that the range of this steel in the summary mentioned is quite limited, and corresponds to the stated limits (see above).

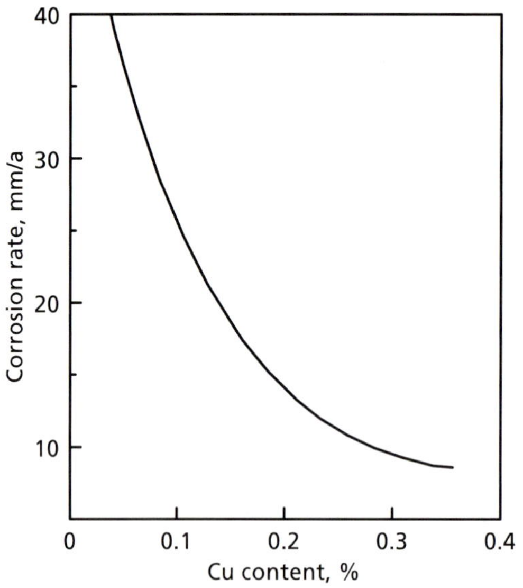

Figure 52: Influence of the copper content in steel 1.4404 (SAE 316 L) on its corrosion behavior in boiling 85 % H_3PO_4 [6, 42, 43]

Stress corrosion cracking

Stress corrosion cracking is to be expected in chloride-containing phosphoric acid solutions. On steel SAE 316 L (1.4404/1.4435) it occurred in a solution of 0.2 % H_3PO_4 and 0.8 % NaCl at 414 K (141 °C) and a pH of 1.5–2.5, and in a solution of 1 % H_3PO_4 and 4 % NaCl at 374 K (101 °C) [73, 109].

Machining steels

Sulfur is alloyed to automatic steels to improve machinability (see also Sections Ferritic chromium steels with more than 12 % chromium and Austenitic CrNi steels).

The corrosion resistance of such steels is improved noticeably by addition of Mo and Ti in comparison with steels containing only S. This fact can also be seen from Table 9. Table 9a contains the composition of the steels (alloyed amounts of the elements mentioned) [60].

Of the steels investigated, Mo-containing automatic steel, as expected, shows the best resistance.

Grinding body materials

Figures 21 and 22 show results of studies on various grinding body materials during decomposition of phosphates for production of wet acid under the conditions in Florida (for the composition of the acid, see also above under this section). According to these graphs, steel 316 shows an adequate resistance, in comparison with the other materials [18].

Erosion corrosion

The requirements of grinding body materials as described above already fall under this category. Erosion corrosion studies according to [67] in phosphate suspensions (Table 11) on a number of materials (Table 12) do not mention steel 316 in Figures 19 and 20. Its resistance is only adequate under such conditions.

Steel SAE 316 L cannot be used under the erosion corrosion conditions of production of wet acid in chloride-containing acid. The corrosion rates are always above 1 mm/a (39.4 mpy), and in fact sometimes considerably above this figure [112, 114].

Storage and transportation of phosphoric acid

When transporting wet acid, it is absolutely essential to ensure that it is not too warm when the corresponding tanks are filled, since hot acid is more corrosive than cold. It is important to know the precise composition of the crude acid, so that its corrosiveness can be estimated.

Crude phosphoric acid contains sludge, the majority forming when the crude acid cools in tanks, or generally during cooling. Sludge can consequently also form during transportation. Under such conditions, the corrosion problems which arise can scarcely be overlooked. Pitting corrosion can develop under deposits. Recirculation (5–6 h per day) [80] can prevent formation of a solid layer of sediment. Seawater should not be used for cleaning, since its chlorides are highly corrosion-promoting (see above under this section). Transportation tanks should be cleaned quickly, since dried sludge can be removed only with difficulty. Such deposits crack and form an ideal base for crevice corrosion. Alkaline cleaning agents cause hardening of the sludge. These facts are of particular importance for transportation by ship [80, 115–118].

Steel 317 L is used as a plating material in tanker construction for transportation of superphosphoric acid [124]. See also Figures 41 and 42, in which the use ranges for steels 316 L and 317 LN for transportation of superphosphoric acid are outlined [72]. The CrNiMo steel 18 8 has become the standard material for transportation of phosphoric acid, a minimum content of 2.5 %, and often even 2.7 % Mo being prescribed [115]. Special steels for transportation of chemicals by sea are the grades D24LN (cf. SAE 316 LN, 1.4429) with 2.7 % Mo, E24LN with 2.9 % Mo, and otherwise the same composition as D24LN, B734LN (approximately SAE 317 LN, 1.4442) with 3.3 % Mo, C734LN with 3.5 % Mo and C34LN (cf. SAE 317 LMN, 1.4439) with 4.5 % Mo. These steels contain between 0.12 and 0.15 % nitrogen to guarantee the maximum possible yield strength, which is noticeably above that of the nitrogen-free grades [115].

Ships in particular intended for transportation of phosphoric acid are made of steel 734 L (cf. SAE 317 L, 1.4438). With corrosion rates of not more than 0.1 mm/a (3.94 mpy) in wet phosphoric acid at 333 K (60 °C) it is resistant to a great extent.

Table 27 summarizes the corresponding corrosion rates as a function of the impurities in the wet acid [117].

It can be seen from this table that a CrNiMo standard steel 18 8 could not be used in this context.

Impurity, %			Corrosion rate, mm/a (mpy)	
SO_4^{2-}	F^-	Cl^-	CrNiMo 18 8	734 L 1)
4.0	1.00	0.05	0.07 (2.8)	0.03 (1.2)
6.0	1.0	0.05	11.0 (433)	0.04 (1.6)
4.0	1.25	0.05	10.0 (394)	0.04 (1.6)
4.0	1.5	0.05	8.0 (315)	0.16 (6.3)
4.0	1.5	0.10	10.0 (394)	0.11 (4.3)

1) cf. 1.4438 with 3.3 % Mo

Table 27: Corrosion rates for CrNiMo steels in wet phosphoric acid (49 % P_2O_5, 0.89 % Fe^{3+}, 0.37 % Al^{3+}, 0.17 % Mg^{2+}) at 333 K (60 °C) [117]

The materials chiefly used for chemical tankers are the steels Avesta® 832 SKR-4 (SAE 316 LN, 1.4429) and Avesta® 832 SNR-4 (SAE 317 LN, 1.4442). A molybdenum-containing steel is always necessary for transportation of phosphoric acid, and the use of 317 LN has become more frequent in recent times [116, 80].

A typical concentration of wet acid is 53 % P_2O_5 (73 % H_3PO_4). This acid of course contains quite different contents of impurities, which have a marked influence on its corrosiveness, depending on the nature of the various crude phosphates. Results from studies in more than 250 different phosphoric acid compositions based on 53 % P_2O_5 are shown below in Figures 53–60 according to [116]. From these cases, it can be decided in a specific case whether an existing tanker or tank can be used for transportation of the crude acid in question. A change in the acid composition can even become possible. The results originate from electrochemical studies with the aid of the corrosion potential. To obtain extensively realistic results, the specimens were activated before the measurements (see above on establishing the active-passive equilibrium state). The test duration was 20 days, corresponding to the transportation time by ship, and 308 and 323 K (35 and 50 °C) were chosen as test temperatures. Table 28 contains a summary of the test acids. The curves in Figures 53–60 correspond to isocorrosion curves at a corrosion rate of 0.1 mm/a (3.94 mpy) [116]. This is designated as the passivity threshold in the figures. It should be pointed out here that a residual current flows in the passive state. As a result, corrosion is measurable. It depends on the corrosion conditions, and in the case of the high-alloy steels is as a rule less than 0.1 mm/a (3.94 mpy) and below, so that according to the definition these can be referred to as being resistant. With anodic protection (see under – Anodic protection –), this passive current must be low enough for the corrosion rate to lie within the limits mentioned.

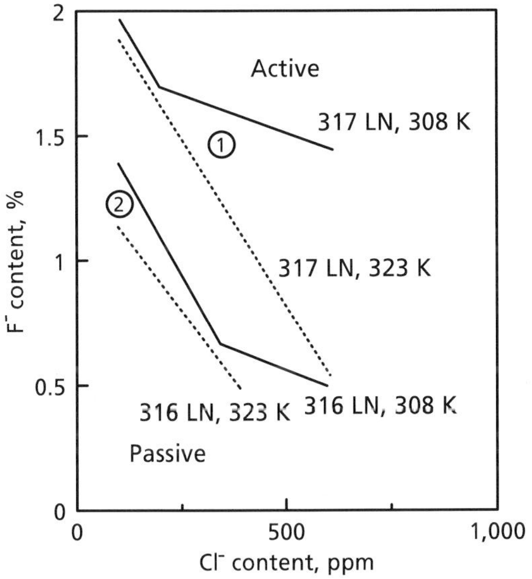

Figure 53: Passivity threshold (0.1 mm/a (3.94 mpy)) for 316 LN and 317 LN in phosphoric acid "A" (see Table 28) at 308 and 323 K (35 and 50 °C) as a function of the F^- and Cl^- content [116]
① SAE 317 LN (1.4442, X2CrNiMoN18-15-4)
② SAE 316 LN (1.4429, X2CrNiMoN17-13-3)

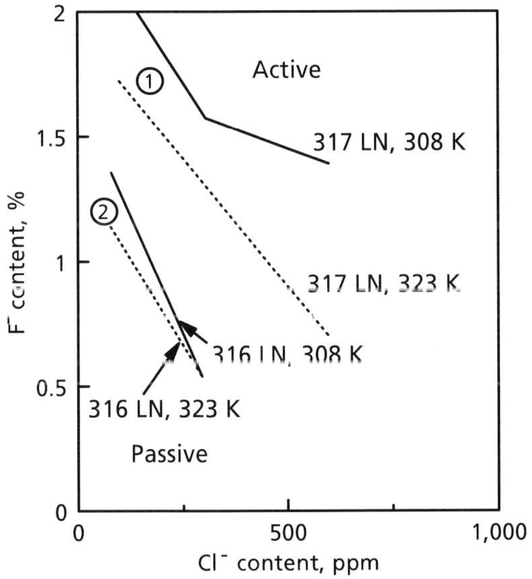

Figure 54: Passivity threshold (0.1 mm/a (3.94 mpy)) for 316 LN and 317 LN in phosphoric acid "B" (see Table 28) at 308 and 323 K (35 and 50 °C) as a function of the F^- and Cl^- content [116]
① SAE 317 LN (1.4442, X2CrNiMoN18-15-4)
② SAE 316 LN (1.4429, X2CrNiMoN17-13-3)

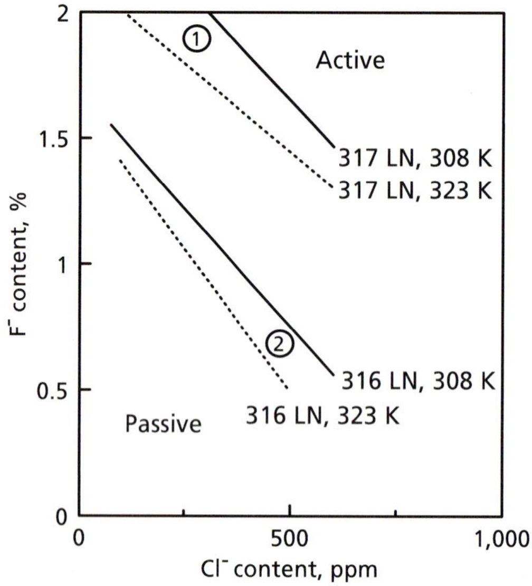

Figure 55: Passivity threshold (0.1 mm/a (3.94 mpy)) of 316 LN and 317 LN in phosphoric acid "C" (see Table 28) at 308 and 323 K (35 and 50 °C) [116]
① SAE 317 LN (1.4442, X2CrNiMoN18-15-4)
② SAE 316 LN (1.4429, X2CrNiMoN17-13-3)

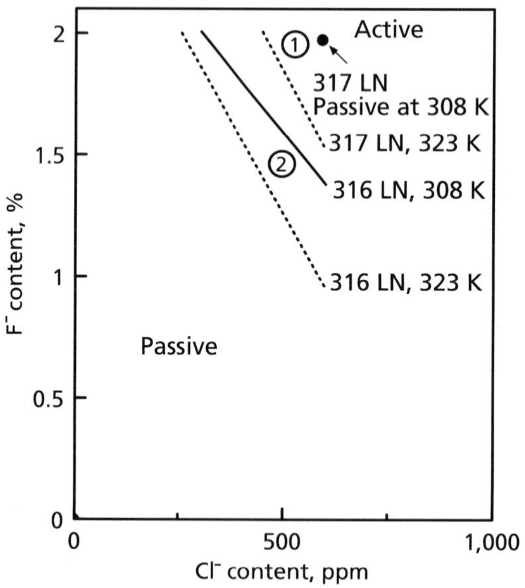

Figure 56: Passivity threshold (0.1 mm/a (3.94 mpy)) for 316 LN and 317 LN in phosphoric acid "D" (see Table 28) at 308 and 323 K (35 and 50 °C) [116]
① SAE 317 LN (1.4442, X2CrNiMoN18-15-4)
② SAE 316 LN (1.4429, X2CrNiMoN17-13-3)

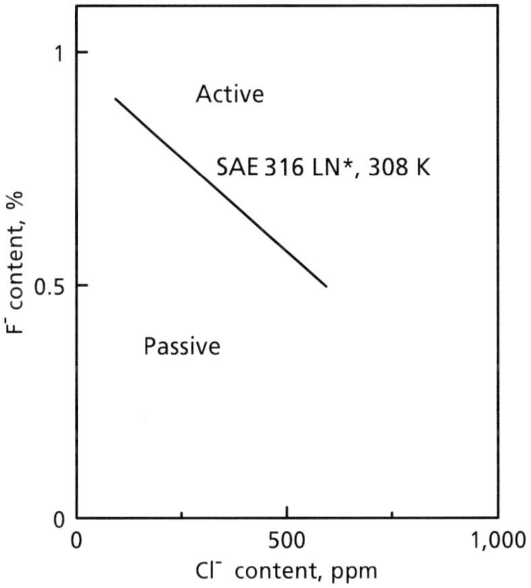

Figure 57: Passivity threshold (0.1 mm/a (3.94 mpy)) for 316 LN in phosphoric acid "E" (see Table 28) at 308 K (35 °C) [116]
* SAE 316 LN (1.4429, X2CrNiMoN17-13-3)

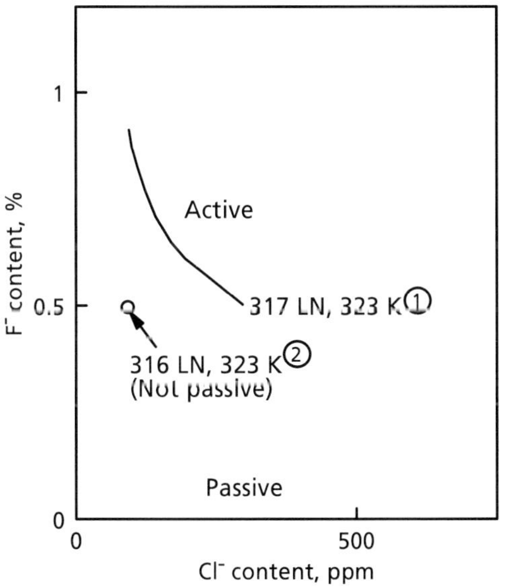

Figure 58: Passivity threshold (0.1 mm/a (3.94 mpy)) for 317 LN and 316 LN in phosphoric acid "F" (see Table 28) at 323 K (50 °C) as a function of the F^- and Cl^- content [116]
① SAE 317 LN (1.4442, X2CrNiMoN18-15-4)
② SAE 316 LN (1.4429, X2CrNiMoN17-13-3)

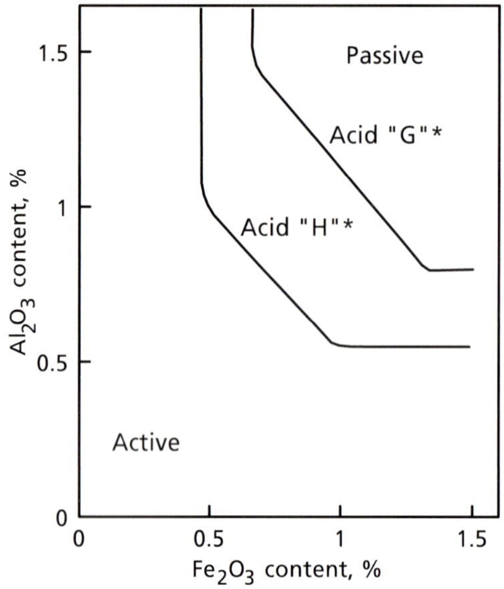

Figure 59: Passivity threshold (0.1 mm/a (3.94 mpy)) of SAE 316 LN (1.4429, X2CrNiMoN17-13-3) in phosphoric acid "G" and "H" (*see Table 28) at 323 K (50 °C) as a function of the Al_2O_3 and Fe_2O_3 content [116]

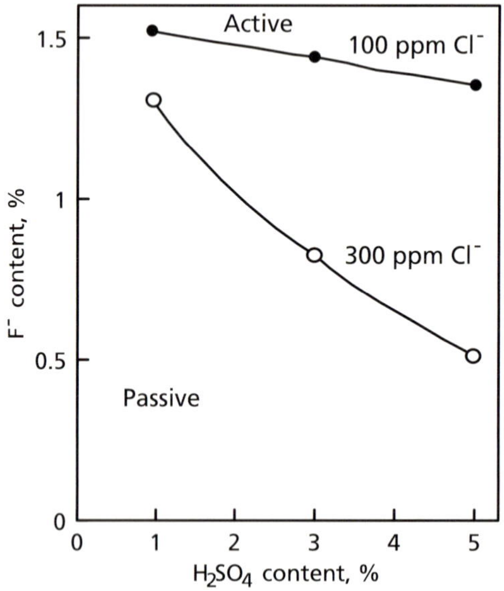

Figure 60: Influence of H_2SO_4 on the passivity (0.1 mm/a (3.94 mpy)) of SAE 316 LN (1.4429, X2CrNiMoN17-13-3) in phosphoric acid "I" (see Table 28) corresponding to Figures 53 and 54 at 308 K (35 °C) [116]

Figures 53–58 show the severely corrosion-promoting influence of chlorine and fluorine and, as expected, the better resistance of the steel 1.4442 of higher Mo content. Figure 59 shows the inhibiting action of Al_2O_3 and Fe_2O_3, which also depends on the acid. Finally, the corrosion-intensifying action of phosphoric acid, which rises further as the chlorine concentration increases, as has already been explained above, and is of great importance for use in practice for transportation of crude acid, can be seen in Figure 60.

Even more vigorous corrosion may occur in suspensions of the phosphoric acid or in deposits from this under the layers. In an agitated medium, which should be aimed for, this risk of corrosion is largely excluded [116].

Inhibition

The inhibitor action of Al_2O_3 and Fe_2O_3 in wet acid of similar composition to that in [116] is referred to in [118] by the same authors. See also Figure 59.

Phosphoric acid[1)]	***Cl ppm***	***Content of impurities, %***							***Temperature K (°C)***	***Figure***
		F	***H_2SO_4***	***Fe_2O_3***	***Al_2O_3***	***SiO_2***	***CaO***	***MgO***		
A	variable	variable	3.0	0.9	0.6	0.2	0.4	0.7	308, 323 (35, 50)	53, 60
B	variable	variable	5.0	0.9	0.6	0.2	0.4	0.7	308, 323 (35, 50)	54, 60
C	variable	variable	5.0	1.5	0.6	0.2	0.4	0.7	308, 323 (35, 50)	55
D	variable	variable	5.0	1.5	1.1	0.2	0.4	0.7	308, 323 (35, 50)	55, 56
E	variable	variable	3.0	0.5	0.4	0.2	0.4	0.7	308 (35)	57
F	variable	variable	3.0	0.3	0.2	0.2	0.4	0.7	323 (50)	58
G	300	0.7	5.0	variable	variable		0.4	0.7	323 (50)	59
H	150	1.1	3.0	0.3	0.2	0.2	0.4	0.7	323 (50)	59
I	100		1.0	0.9	0.6	0.2	0.4	0.7	308 (35)	60

1) The basis of the phosphoric acids studied is a P_2O_5 content of 53 % (typical for transportation by ship)

Table 28: Overview of the composition of wet phosphoric acids such as may be encountered in transportation by sea – basis for corrosion studies, the results of which are shown in the figures illustrated – [116]

Both additions or contents in the phosphoric acid have the same inhibiting effect. The action of the sum (Fe_2O_3 + Al_2O_3) as a content or addition in wet acid is shown in Figure 61 using an example of such an acid with 53 % P_2O_5.

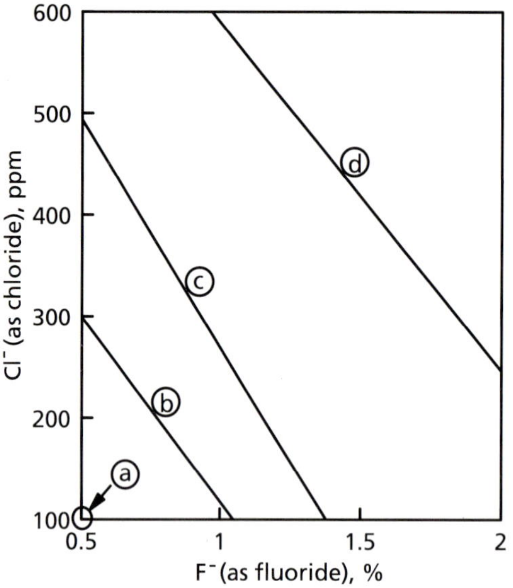

Figure 61: Influence of the inhibitor content on the passivity threshold (0.1 mm/a (3.94 mpy)) in wet acids with about 53 % P_2O_5 for steel SAE 316 LN (1.4429) [118]
ⓐ $Fe_2O_3 + Al_2O_3 = 0.5\%$
ⓑ $Fe_2O_3 + Al_2O_3 = 1.5\%$
ⓒ $Fe_2O_3 + Al_2O_3 = 2.1\%$
ⓓ $Fe_2O_3 + Al_2O_3 = 2.6\%$

This graph demonstrates that its addition in the range from 0.5 to 2.6 % $Fe_2O_3 + Al_2O_3$ has the effect of a substantial increase in the use range of steel 1.4429 in the above-mentioned phosphoric acid, in spite of exposure to fluoride and chloride. Below 0.5 %, the exposure to fluoride and chloride should only be low [118].

In this reference [118], usability diagrams are given for Avesta® 832 SKR-4 (SAE 316 LN) and Avesta® 832 SNR-4 (SAE 317 LN) (1.4429 and 1.4442 respectively) in some common acids (53–55 % P_2O_5) with various impurities approximately in the form of Figure 59, these containing the MgO or SiO_2 content as further parameters. Above the passivity curve, the particular steel studied is resistant. Since it is doubtful whether a comparison acid is to be found in a concrete case, such graphs have not been adopted. Preliminary selection and determination of the aggressiveness of the phosphoric acid is already possible from Figures 53–60. The best check for determination of the passive range would be plotting of current/potential curves.

The expenditure on determination of these curves is low in comparison with the freight and the secondary costs. On the other hand, such documentation may be available on site. A trend toward more highly resistant materials currently exists.

Anodic protection

As already mentioned elsewhere, CrNi steels and CrNiMo steels can be used, as here in phosphoric acid, only if they are passive in the medium. Passive means that the corresponding material is in the passive region of the current/potential curve. Under these conditions, corrosion proceeds at a low residual current and uniformly, with corrosion rates of far less than 0.1 mm/a (3.94 mpy). These conditions are disturbed in phosphoric acid containing fluoride and/or chloride. Corrosion proceeds more vigorously, and especially as a result of Cl^- ions, as has already been seen from the Figures (38, 39, 41, 49, 53, 54, 57–61). In special cases with only local corrosion, pitting corrosion may occur and proceeds far more vigorously than uniform corrosion due to local corrosion current peaks. The current/potential curves also provide information on the position of the free corrosion potential (corrosion without polarization) and the position of the pitting corrosion potential (as a rule determined with polarization).

Steel (CrNi) can be provided with anodic protection by applying a foreign potential at the level to be determined. It is brought into a corrosion region in which it corrodes at corrosion rates (described under transportation by ship with the passivity threshold of 0.1 mm/a (3.94 mpy)) of far less than 0.1 mm/a (3.94 mpy) under a low residual current, that is to say it is practically resistant. The applicability of the method depends on a large number of factors, including the size and shape of the component to be protected [125, 126].

In a very corrosive phosphoric acid with (%) 54 P_2O_5, 4 SO_4^{2-}, 1.05 F^-, 0.0617 Cl^- (= 617 ppm), 0.27 Fe_2O_3, 0.17 Al_2O_3, 0.1 SiO_2, 0.2 CaO, 0.7 MgO, corrosion studies on steel SAE 316 L (1.4435, X2CrNiMo18-14-3) at room temperature showed corrosion rates of more than 4 mm/a (157 mpy), but considerably less than 0.1 mm/a (3.94 mpy) with anodic protection [125, 126].

Anodic protection is used in practice on tanks of 1.4435 in fertilizer production, for example a tank (13 m ∅ and 10 m high) [125, 126].

Phosphoric acid as a cleaning agent in the foodstuffs industry

Residues in CrNi steel plants in the foodstuffs industry consist mainly of proteins, fats, carbohydrates and salts. Alkaline, neutral and acid cleaning agents are used for cleaning the plants to remove these substances, according to their solubility. Inorganic salts are thus best removed with acid cleaning agents. In the ideal case, these should not attack the material and they must be very easy to remove.

Nitric acid is used as a cleaning agent, the acid concentration being 2 %. Additions of phosphoric acid for cleaning cause a change in the adsorption layers and a reduction in the corrosion rates, which is of particular importance here. Figures 62 and 63 show the results of studies on steel 1.4401 (SAE 316, X5CrNiMo17-12-2) in such nitric acid/phosphoric acid mixtures [127].

In the concentration range of 2 % acid which is important in practice, improvements in the corrosion behavior of the cleaning solution were found as the H_3PO_4 content increased, as shown in Figure 62 at 363 K (90 °C) and confirmed by studies at 343 and 353 K (70 and 80 °C) [127]. Addition of 4 % H_3PO_4 to HNO_3 proved to be the most favorable value. Studies at 413 K (140 °C) (Figure 63) also showed a lower corrosion rate as the addition of H_3PO_4 to the solution increased. In contrast to other studies, however, a very low corrosion rate for nitric acid was also found in low concentrations, which cannot be explained [127].

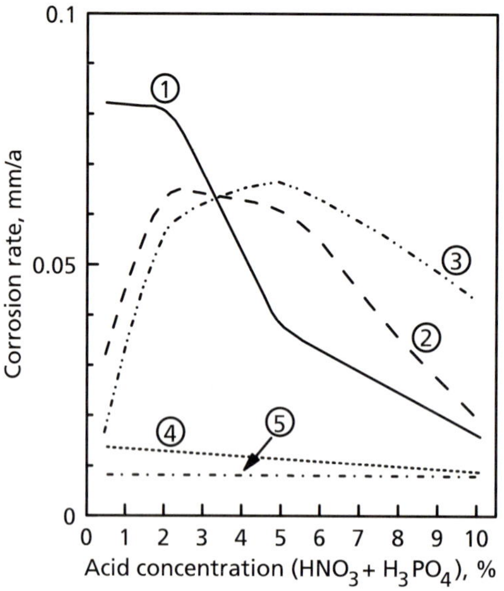

	HNO_3, %	H_3PO_4, %
①	100	–
②	98	2
③	97	3
④	96	4
⑤	95	5

Figure 62: Corrosion rates for steel 1.4401 (SAE 316, X5CrNiMo17-12-2) in acid cleaning solutions ① to ⑤ in the foodstuffs industry (test duration 24 h, 363 K (90 °C) the corrosion rates were calculated from g/m^2 h) [127]

However, Figure 63 also shows that addition of phosphoric acid to acid cleaning solutions considerably reduces the corrosion rates. These corrosion rates, of course, cannot be compared with those at otherwise customary exposure to corrosion, since the cleaning effect is of prime importance here.

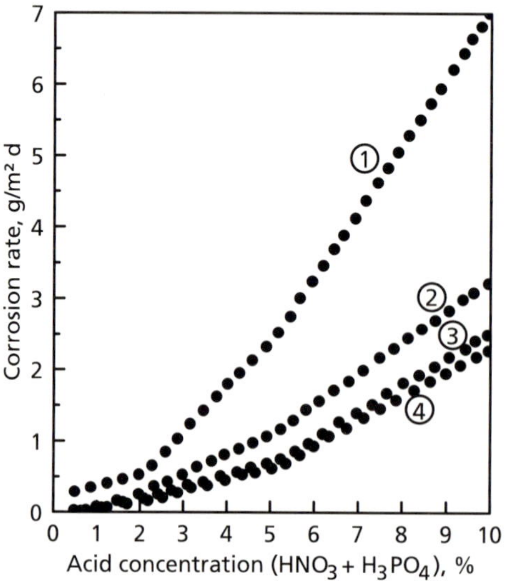

	HNO_3, %	H_3PO_4, %
①	100	–
②	98	2
③	95	5
④	90	10

Figure 63: Corrosion rates for steel 1.4401 (SAE 316, X5CrNiMo17-12-2) in acid cleaning solutions ① to ④ in the foodstuffs industry (test duration 24h, at 413 K (140 °C)) [127]

Corrosion behavior of CrNiMo steels under phosphoric acid production conditions in practice

This chapter is of particular interest in respect of handling phosphoric acid and is a useful transition to the steels under Section Austenitic chromium-nickel steels with special alloying additions. A detailed assessment of the effect of impurities in phosphoric acid has been presented in this chapter. In contaminated phosphoric acid, the use of CrNi steels is only made possible by the addition of molybdenum.

A flow chart for the phosphoric acid production by the wet decomposition process up to highly concentrated phosphoric acid is shown in Figure 64.

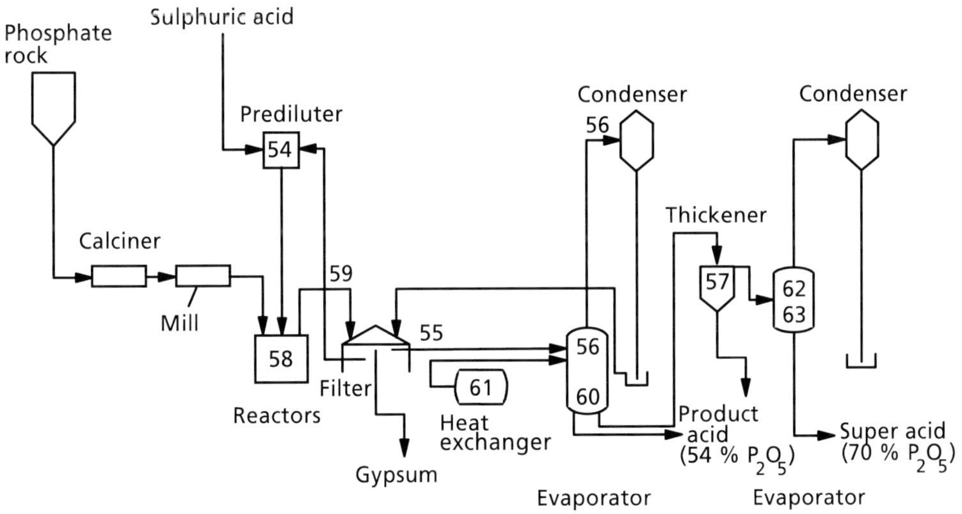

Figure 64: Flow chart for the production of phosphoric acid by the wet decomposition process – numbers designate the Tables where materials and corrosion rates are listed for these components – [42, 43, 80]

According to [80], the possible process temperatures can be classified into two temperature ranges:

- Process stages which proceed in the decomposition reactor, in the filter plant and in the pipelines and as a rule scarcely exceed temperatures of 333 K (60 °C) and the transportation of the phosphoric acid, during which the temperature likewise does not exceed 333 K (60 °C)
- Temperatures above 333 K (60 °C) including the boiling temperature which prevail in heat exchangers and in evaporators.

Impurity, %			Corrosion rate, mm/a (mpy)
SO_4^{2-}	F^-	Cl^-	
4.0	0	0	< 0.005 (< 0.2)
0	1.0	0	0.030 (1.2)
0	0	0.02	< 0.005 (< 0.2)
4.0	1.0	0	0.073 (2.9)
4.0	0	0.02	5.4 (213)
0	1.0	0.02	4.4 (173)
4.0	1.0	0.02	9.9 (390)

Table 29: Influence of sulfates, fluorides and chlorides on the corrosion resistance of steel SAE 317 L with only 3.5 % Mo (similar to 1.4438) in 70 % phosphoric acid at 313 K (40 °C) [80]

It is appropriate and of practical interest to consider the two use ranges separately in respect of the behavior of materials. In the temperature range up to 333 K (60 °C) corrosion studies in 70 % H_3PO_4 on steel 317 L (1.4438 with only 3.5 % Mo) at 313 K (40 °C) already showed a marked dependence on the impurities present in the phosphoric acid, as can be seen from Table 29 (in this context, see also Tables 21 and 26). Under these conditions, low additions of 200 ppm Cl^- or 4 % SO_4^{2-} have practically no effect on the corrosion of the steel. The corrosion increases if the fluoride content is 1 % F^-, regardless of whether or not 4 % SO_4^{2-} is added. However, the steel still

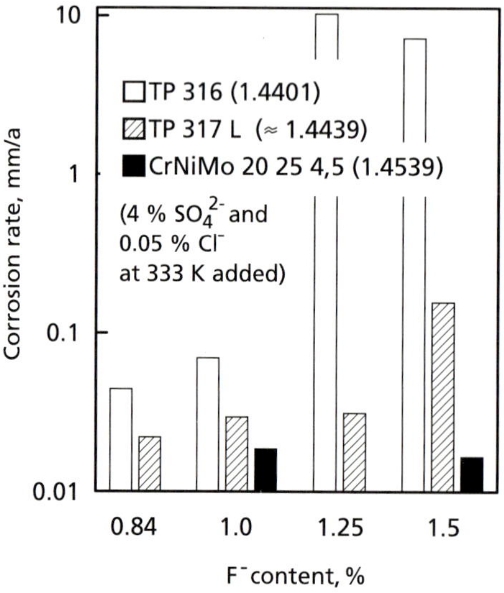

Figure 65: Influence of the fluoride concentration on the corrosion resistance of high-alloy steels in 70 % industrial phosphoric acid (see also Figures 17, 28–40) [80]

has a very good resistance, with corrosion rates significantly below 0.1 mm/a (3.94 mpy). If chloride ions are present (see also above under pitting corrosion) in combination with sulfates and/or fluorides, the corrosion increases even on this steel with a relatively high Mo alloying content (3.5 % Mo). This steel can then no longer be used for such plants, with corrosion rates above 4.4 mm/a (173 mpy). Its use is possible if inhibiting constituents are present in the acid, and these must be present in a sufficient ratio or sufficient concentration. They include Al^{3+} and Mg^{2+} ions, which form complexes with fluorides (Al^{3+} more strongly than Mg^{2+}) and thus noticeably reduce the corrosion of high-alloy steels. If these are absent or are present only in an inadequate concentration, a higher-alloy material must be used, as can be seen from Figure 65.

As a result of corrosion studies, it was found that if the molar ratio of F^-: (Al^{3+} + Mg^{2+}) significantly exceeds 3:1, stainless steels are activated in 70 % H_3PO_4 at 333 K (60 °C) thus leading to corrosion rates of more than 1 mm/a (39.4 mpy).

Material[1)]	*Trade name (DIN-Mat. No.)*	*Section*	*Corrosion rate*[1)] *mm/a (mpy)*	*Pitting depth mm*
G-X 6 NiCrCu Mo 29 20 4	Aloyco® 20	E	0.038 (1.50)	
NiCr21Mo	Incoloy® 825, 2.4858		0.069 (2.72)	0.13
NiCrMoCu 56 23 6 6	Illium® G		0.076 (2.99)	
NiCr 19 Nb 5 Mo	Inconel® 718, 2.4667		0.081 (3.19)	0.13
NiCr13Mo6Ti3	Incoloy® alloy 901 , 2.4662		0.155 (6.10)	0.13
NiCrMoCu 68 21 5 3	Illium® R		0.175; 0.360 (6.89; 14.2)	
NiCu 30 Al	Monel® K-500, 2.4375		0.790 (31.1)	0.08
NiCu30 Fe	Monel® 400, 2.4360		0.970 (38.2)	
NiMo 30†	Hastelloy® B, 2.4810		1.98 (77.9)	
X3CrNiMo18 12 3	1.4449, SAE 317	D	> 3.80[2)] (> 150)	
X5CrNiMo17-12-2	1.4401, SAE 316	D	> 4.50[2)] (> 177)	

† old designation
1) thick layers had formed on all specimens
2) stress corrosion cracking in the region of the markings

D: see section Austenitic CrNiMo(N) steels
E: see section Austenitic chromium-nickel steels with special alloying additions

Table 30: Corrosion behavior of high-alloy steels and nickel alloys (see also Table 40) in a wet decomposition plant (prediluter – see Figure 64) under operating conditions (on the base of the prediluter, test duration 42 d, temperature 355–383 K (82–110 °C) on average 366 K (93 °C) moderate aeration, agitation by convection, 28 % H_3PO_4 + 21 % H_2SO_4 + 1 – 1.5 % F^-, probably as H_2SiF_6). Concentrated sulfuric acid was diluted continuously by phosphoric acid from the process [42, 43, 80]

Material designation (trade name)		Section	Test 1[1)]	Test 2[1)]
			Corrosion rate, mm/a (mpy)	
High-purity lead	PB990R, PB985R 2.3010†, 2.3020†		0[2)]	
Aloyco® 20	GX6NiCrCuMo 29 20 4	E		0.005 (0.20)
Worthite®	approx. 1.4585, GX7CrNiMo-CuNb18-18	E		0.008 (0.32)
Carpenter® 20 Cb	cf. UNS N08020, 2.4660	E		0.008 (0.32)
SAE 329	cf. 1.4460, X3CrNiMoN27-5-2	B		0.008[3)] (0.32)
X3CrNiMo18-12-3	1.4449, SAE 317	D	0.023 (0.91)	0.010 (0.39)
SAE 316	cf. 1.4401, X5CrNiMo17-12-2	D	0.041 (1.61)	
CF-3M	cf. 1.4404, SAE 316 L	D		0.013 (0.51)
Hastelloy® F alloy				0.015 (0.59)
INCOLOY® alloy 825	2.4858, NiCr21Mo			0.038 (1.50)
SAE 304	1.4301, X5CrNi18-10	C	0.170[4)] (6.70)	
INCONEL® alloy 600	2.4816, NiCr15Fe		> 3.60[5)] (> 142)	
Nickel 200	2.4060, Ni 99.6		> 4.20[6)] (> 165)	
Monel® 400	2.4360, NiCu 30 Fe		> 4.30[6)] (> 169)	> 2.00[6)] (> 78.7)
HASTELLOY® B alloy	cf. NiMo 30†			> 2.40[6)] (> 94.5)
Unalloyed steel		F	> 8.60[6)] (> 339)	

Table 31: Corrosion rates on steels, Ni alloys (see also Table 40) and cast iron under the operating conditions of wet phosphoric acid production (see Figure 64) [42, 43]

Table 31: Continued

Material designation (trade name)		*Section*	*Test 1*[1)]	*Test 2*[1)]
			Corrosion rate, mm/a (mpy)	
Ni-Resist® 1	0.6655, EN-GJLA-XNiCuCr15-6-2	E	> 15.0[5)] (> 591)	
Cast iron		G	> 22.5[6)] (> 886)	

† old designation
1) Test conditions:

	Test 1	Test 2
Medium:	33 % H_3PO_4 (≙ 24 % P_2O_5)	42 % H_3PO_4 (≙ 30.5 % P_2O_5)
	occasionally a little H_2SO_4	2 % H_2SO_4
	3 % H_2SiF_6	H_2SiF_6 (amount unknown)
	traces of HF	HF (amount unknown)
	$CaSO_4$ x 2 H_2O (amount unknown) suspended	2 % $CaSO_4$ x 2 H_2O suspended
Test site:	concentrator after the filter unit	tank for circulation acid
Duration:	33 days	72 days
Temperature:	average:	350 K (77 °C)
	highest: 355 K (82 °C)	highest: 336 K (63 °C)
	lowest: 344 K (71 °C)	lowest: 330 K (57 °C)
Agitation:	5 m/s	57...115 l/min
Aeration:	no information	none

2) but increasing trend
3) values of one specimen; part of the second specimen broke off, probably as a result of intercrystalline corrosion
4) pitting corrosion: max depth 0.33 mm
5) value of one specimen; the second specimen was completely destroyed
6) specimen completely destroyed
B: see section Ferritic-austenitic steels with more than 12 % chromium
C: see section Austenitic CrNi steels
D: see section Austenitic CrNiMo(N) steels
E: see section Austenitic chromium-nickel steels with special alloying additions
F: see section Unalloyed steel and cast steel
G: see section Unalloyed cast iron

Table 31: Corrosion rates on steels, Ni alloys (see also Table 40) and cast iron under the operating conditions of wet phosphoric acid production (see Figure 64) [42, 43]

Steel 1.4571 (SAE 316 Ti , X6CrNiMoTi17-12-2) cannot be used in a decomposition sludge (33.5 % P_2O_5, 3.0 % SO_4^{2-}, 2.89 % F^-, 20 ppm Cl^-, 3.6 % Si, 1.17 % Fe^{3+} and 1 % Al^{3+}) at a temperature of 301 K (28 °C). For transportation of phosphoric acid by ship, steel 1.4439 (SAE 317 LMN) is the safest choice of material for transportation of wet acid [80], and it is also increasingly used as the aggressiveness of the acid increases.

The use of CrNiMo standard steels in the prediluters of a wet decomposition plant for the production of phosphoric acid (Figure 64) was not to be expected, with corrosion rates of more than 3.8 mm/a (150 mpy) (Table 30) and the occurrence of stress corrosion cracking. The average temperature was also significantly above 333 K (60 °C).

According to Figure 64, temperatures of about 333 K (60 °C) also occur in the combination of filter plant and evaporator, as can be seen from Table 31 (test 1). Under operating conditions at about 333 K (60 °C) in crude acid with 42 % H_3PO_4 + 2 % H_2SO_4, corrosion rates of less than 0.1 mm/a (3.94 mpy) resulted on the materials to be classified here under Section Austenitic CrNiMo(N) steels [42, 43].

CrNiMo standard steels also show a good resistance under the conditions of test 1 of Table 32 at the top of the evaporator in the crude acid mist at temperatures of up to 333 K (60 °C). Here also, the corrosion rates are below 0.1 mm/a (3.94 mpy) [42, 43].

Vapor which arises during evaporation contains suspended substances and a mist of liquid consisting of phosphoric, sulfuric, hydrofluoric and hexafluorosilicic acid. The concentrations of the acids under these conditions are low, but highly corrosive. According to Figure 64, similar conditions are to be encountered in the condenser downstream of the evaporator. In the concentrator at 323 to 338 K (50 to 65 °C) under the conditions according to Table 33 in 54 % H_3PO_4, steel 1.4401 (SAE 316) showed corrosion rates of about 0.5 mm/a (19.7 mpy) [42, 43].

Material	*DIN-Mat. No., SAE (trade name)*	*Section*	*Test 1[1)]*	*Test 2[1)]*	*Test 3[1)]*
			Corrosion rate, mm/a (mpy)		
NiCrMoCu 56 23 6 6	(Illium® G)			0.074 (2.9)	0.240 (9.5)
NiCrMoCu 68 21 5 3	(Illium® G)			0.079 (3.1)	
NiCr 22 Mo				0.081 (3.2)	0.147 (5.8)
NiCr21Mo	2.4858 (INCOLOY® alloy 825)		0.008 (0.3)	0.107 (4.2)	0.025; 0.117 (1.0; 4.6)
NiCrFe 55 36	(Corronel® 230)			0.107 (4.2)	
NiCr20CuMo	2.4660 (Carpenter® 20 Cb 3)	E	0.010 (0.39)	0.127 (5.0)	0.018 (0.7)
	(Aloyco® 20)	E		0.165 (6.5)	
NiMo 30†	2.4810		0.360 (14.2)	0.168[2)] (6.6)	

Table 32: Corrosion rates of steels and nickel alloys (see also Table 40) under the operating conditions of wet phosphoric acid production (see Figure 64) [42, 43]

Table 32: Continued

Material	***DIN-Mat. No., SAE (trade name)***	***Section***	***Test 1[1)]***	***Test 2[1)]***	***Test 3[1)]***
			Corrosion rate, mm/a (mpy)		
GX7CrNiMo-CuNb18-18	cf. 1.4585 (Worthite®)	E		0.206 (8.1)	
X3CrNiMoN27-5-2	1.4460, SAE 329	B		0.211 (8.3)	0.008 (0.3)
X3CrNiMo18-12-3	1.4449, SAE 317	D	0.012 (0.5)	0.260 (10.2)	0.074; 0.380 (2.9; 15.0)
X5CrNiMo17-12-2	1.4401, SAE 316	D	0.074 (2.9)	0.270 (10.6)	
X5CrNiMo17-12-2 (sensibilized)		D		0.690 (27.2)	
X2CrNiMo17-12-2	1.4404, SAE 316 L	D			0.530 (20.9)
NiCr15Fe	2.4816 (INCONEL® alloy 600)		0.380 (15.0)		
NiCu 30 Fe	2.4360 (Monel® 400)		0.760 (30.0)		

† old designation
1) Test conditions:

	Test 1	Test 2	Test 3
Medium:	vapors and mist of H_3PO_4, H_2SiF_6, HF and H_2SO_4 (amounts unknown) from the wet process	vapors and mist of H_3PO_4, H_2SiF_6, HF and H_2SO_4 (amounts unknown) from the wet process	15 % H_3PO_4 (≙ 11 % P_2O_5), 20 % H_2SiF_6, 1 % H_2SO_4
Test site:	vapor dome of the evaporator	vapor line downstream of the vapor dome	in the liquid of the condenser for the vapor from the evaporator
Duration:	41.5 days	59 days	16 days
Temperature:	298–338 K (25–65 °C) (average: 323 K (50 °C))	333–344 K (60–71 °C) (average: 338 K (65 °C))	348–358 K (75–85 °C) (average: 352 K (79 °C))
Agitation:	1.2 m/s	vigorous	slight
Aeration:	none	none	high

2) crevice corrosion, depth 0.1 mm

B: see section Ferritic-austenitic steels with more than 12 % chromium
D: see section Austenitic CrNiMo(N) steels
E: see section Austenitic chromium-nickel steels with special alloying additions

Table 32: Corrosion rates of steels and nickel alloys (see also Table 40) under the operating conditions of wet phosphoric acid production (see Figure 64) [42, 43]

Material	*DIN-Mat. No. (trade name)*	*Section*	*Corrosion rate*[1) mm/a (mpy)]
High-purity lead	PB990R, PB985R 2.3010†, 2.3020†		0.010 (0.39)
NiCr21Mo	2.4858 (INCOLOY® alloy 825)		0.013 (0.51)
NiCr20CuMo	2.4660 (Carpenter® 20 Cb 3)	E	0.023 (0.91)
X5CrNiMo17-12-2	1.4401, SAE 316	D	0.500 (19.7)

1) Test conditions:
Medium: 53.8 % H_3PO_4, (≙ 39 % P_2O_5), 2.05 ... 2.15 % H_2SiF_6, 1.5 ... 2.5 % H_2SO_4, 2 % $CaSO_4$
Test site: specimens in the concentrator 45 cm below the surface of the liquid
Duration: 51 days
Temperature: 323–338 K (50–65 °C) average temperature: 328 K (55 °C)
Agitation: slight
Aeration: None
D: see section Austenitic CrNiMo(N) steels
E: see section Austenitic chromium-nickel steels with special alloying additions

Table 33: Corrosion rates of steels and nickel alloys (see also Table 40) and also lead under operating conditions in a concentrator for wet phosphoric acid production (see Figure 64) [42, 43]

Material	*DIN-Mat. No., SAE (trade name)*	*Section*	*Corrosion rate*[1)2)] *mm/a (mpy)*
NiCr22Mo9Nb	2.4856		0.018 (0.71)
NiCr22Mo6Cu	2.4618 (Hastelloy® G alloy)		0.028 (1.1)
NiCrMoCu 55 28 8 6†	(Illium® 98)		0.069 (2.72)
Lead, containing antimony			0.081 (3.19)
NiCr20CuMo	2.4660 (Carpenter® 20 Cb 3)	F	0.165 (6.5)
GX7CrNiMoCuNb18-18	cf. 1.4585 (Worthite®)	E	0.173 (6.81)
NiCr21Mo	2.4858, UNS N08825		0.188 (7.4)
GX7NiCrMoCuNb25-20	1.4500 (Durimet® 20)	E	0.295 (11.6)
CW-6M	UNS N30107 (Chlorimet® 3)		0.310 (12.2)

Table 34: Corrosion rates of materials (see also Table 40) in the decomposition tank for the crude phosphates in wet phosphoric acid production (see Figure 64) under operating conditions [42, 43]

Table 34: Continued

Material	*DIN-Mat. No., SAE (trade name)*	*Section*	*Corrosion rate*[1)2)] *mm/a (mpy)*
NiCrMoCu 56 23 6 6†	(Illium® G)		0.380 (15.0)
X3CrNiMo18-12-3	1.4449, SAE 317	D	0.560 (22.1)
G-X 8 CrNiMoCu 28 8†	(Illium® P)	B	0.690 (27.2)
X2CrNiMo17-12-2	1.4404, SAE 316 L	D	1.04 (41.0)
X5CrNiMo17-12-2 (sensibilized)	1.4401, SAE 316	D	1.30 (51.2)
GX5CrNiMo19-11-2	1.4408, CF-8M	D	> 5.30 (> 210)
NiMo 30 (%, 61 Ni, 28 Mo, 5 Fe)	2.4810, UNS N10001		> 1.70[3)] (> 66)
NiMo 30 (%, 66 Ni, 31 Mo, 2 Fe)			> 7.90[3)] (> 310)
NiCrMoCu 68 21 5 3†	(Illium® R)		> 2.30[3)] (> 90)

† old designation

1) Test conditions:
Medium: 39 % H_3PO_4, (≙ 28 % P_2O_5) from the wet process, 2 % H_2SO_4, traces of H_2SiF_6 and HF (about 1.2 % total content of fluorides), 20 % $CaSO_4$, 2 H_2O, suspended
Test site: specimens in the sludge at the bottom of the reaction vessel
Duration: 96 days
Temperature: 350–357 K (77–84 °C) average temperature: 355 K (82 °C)
Agitation: vigorous
Aeration: moderate
2) no pitting corrosion occurred
3) specimens completely destroyed
B: see section Ferritic-austenitic steels with more than 12 % chromium
D: see section Austenitic CrNiMo(N) steels
E: see section Austenitic chromium-nickel steels with special alloying additions
F: see section Special iron based alloys

Table 34: Corrosion rates of materials (see also Table 40) in the decomposition tank for the crude phosphates in wet phosphoric acid production (see Figure 64) under operating conditions [12, 13]

Material	*DIN-Mat. No.*	*Section*	*Corrosion rate*[1) 2)] *mm/a (mpy)*
X12CrNi22-12	1.4829, UNS S30980	C	0.015 (0.59)
X3CrNiMo18-12-3	1.4449, SAE 317	C	0.040 (1.57)
X5CrNiMo17-12-2	1.4401, SAE 316	C	0.110 (4.33)
X5CrNi18-10	1.4301, SAE 304	C	0.340 (13.4)
X10NiCrAlTi32-20	1.4876, UNS N08800	F	0.370 (14.6)
High-purity lead	PB990R, PB985R 2.3010†, 2.3020†		0.435 (17.1)
NiCu 30 Fe	2.4360, UNS N04400		1.15 (45.3)
NiCr15Fe	2.4816, UNS N06600	C	> 1.85[3)] (> 72.8)

† old designation

1) Test conditions:
Medium: 30 % solids, 39 % H_3PO_4, (≙ 29 % P_2O_5), 1.4 % H_2SO_4, 2 % fluorides, probably H_2SiF_6
Test site: specimens in the washing plant between two reaction vessels
Duration: 35 days
Temperature: 344–364 K (71–91 °C) average temperature: 355 K (82 °C)
Agitation: 1.5 m/s
Aeration: high
2) no pitting corrosion occurred
3) specimen completely destroyed

C: see section Austenitic CrNi steels
F: see section Special iron based alloys

Table 35: Corrosion rates of various materials (see also Table 40) in a phosphoric acid/calcium sulfate sludge in wet phosphoric acid production (see Figure 64) under operating conditions [42, 43]

During production of wet phosphoric acid, temperatures above 333 K (60 °C) generally arise. Under such conditions, materials are exposed to considerable stress as a result of intensified corrosion. As expected, the CrNiMo standard steels (see above, described under Section Austenitic CrNiMo(N) steels) cannot be used here, or can be used only under very favorable conditions, as can be seen from the exceptions in Tables 30–39. Such exceptions include, for example, the conditions in Table 35 in the phosphoric acid/calcium sulfate sludge, under which corrosion rates of 0.04 mm/a (1.57 mpy) have been determined for steel X3CrNiMo17-13-3 (1.4436, SAE 316) and 0.11 mm/a (4.33 mpy) for steel X5CrNiMo17-12-2 (1.4401, SAE 316) [42, 43].

Phosphoric acid can also be produced by means of hydrochloric acid by wet decomposition. Table 41 summarizes test results on evaporation of phosphoric acid containing sulfuric acid and hydrochloric acid. The materials listed can no longer be used under these conditions, and specifically also the high-alloy CrNiMo steels are unusable. Only the Ni alloy NiMo16Cr can be used to a limited extent, compared with the other materials, with a corrosion rate of 0.28 mm/a (11.0 mpy).

Material designation		*Section*	*Corrosion rate[1) 2)] mm/a (mpy)*
NiCr20CuMo	2.4660, Carpenter® 20 Cb 3	E	0.120 (4.72)
NiCr21Mo	UNS N08825, 2.4858		0.160 (6.30)
X3CrNiMo17-13-3	1.4436, cf. SAE 316	D	0.260 (10.2)
NiCu 30 Fe	Monel alloy 400, 2.4360		0.640 (25.2)
X5CrNiMo17-12-2	1.4401, cf. SAE 316	D	1.10 (43.3)
NiCr15Fe	UNS N06600, 2.4816		> 33.0[3)] (> 1,300)
NiMo 30†	UNS N10001, 2.4810		> 38.0[3)] (> 1,500)

† old designation

1) Test conditions:
Medium: 53 % H_3PO_4, (≙ 38 % P_2O_5), 1–2 % H_2SO_4, 1.2–1.5 % fluorides
Test site: specimens were in the liquid at the bottom of the evaporator
Duration: 42 days
Temperature: 394 K (121 °C)
2) no pitting corrosion occurred; after 12 days, all the specimens were covered with a coating
3) specimen completely destroyed

D: see section Austenitic CrNiMo(N) steels
E: see section Austenitic chromium-nickel steels with special alloying additions

Table 36: Corrosion rates for nickel-containing materials (see also Table 40) in an evaporator for wet phosphoric acid production (see Figure 64) under operating conditions [42, 43]

Phosphoric acid which has been produced by the electrothermal process is less contaminated. The use limits for CrNiMo steels are thus widened [42, 43]. In summary, Tables 30–39 and 41 as had already been stated several times under this section, show that in addition to the impurities, the operating conditions also play a decisive role in the use of CrNiMo steels in phosphoric acid. Even under conditions where their use seems possible, corrosion studies are advisable in all cases.

Material designation		*Section*	*Corrosion rate[1)] mm/a (mpy)*
High-purity lead	(PB990R, PB985R 2.3010†, 2.3020†)		0.033 (1.30)
NiCrMoCu 56 23 6 6†	Illium® G		0.220 (8.66)
X3CrNiMoN27-5-2	1.4460	B	0.250 (9.84)
NiCr21Mo	UNS N08825, 2.4858		0.390 (15.4)
NiCr20CuMo	2.4660, Carpenter® 20 Cb 3	E	0.510 (20.1)
X3CrNiMo17-13-3	1.4436, cf. SAE 316	D	0.710 (27.9)

Table 37: Corrosion rates for various materials (see also Table 40) in a heat exchanger in wet phosphoric acid production (see Figure 64) under operating conditions [42, 43]

Table 37: Continued

Material designation		*Section*	*Corrosion rate*[1] mm/a (mpy)
X5CrNiMo17-12-2	1.4401, cf. SAE 316	D	8.90 (350)
X5CrNiMo17-12-2, sensibilized	1.4401, cf. SAE 316	D	> 53.0[2] (> 2082)

† old designation

1) Test conditions:

Medium: 65 % H_3PO_4, (≙ 47 % P_2O_5), 3–4 % H_2SO_4, traces of H_2SiF_6 and HF, 5–7 % solids, suspended
Test site: specimens were in the heat exchanger with forced circulation connected to an evaporator
Duration: 4.1 days
Temperature: average temperature: 355 K (82 °C)
Agitation: vigorous
Aeration: moderate

2) specimen completely destroyed

B: see section Ferritic-austenitic steels with more than 12 % chromium
D: see section Austenitic CrNiMo(N) steels
E: see section Austenitic chromium-nickel steels with special alloying additions

Table 37: Corrosion rates for various materials (see also Table 40) in a heat exchanger in wet phosphoric acid production (see Figure 64) under operating conditions [42, 43]

Material designation		*Section*	*Test 1*[1]	*Test 2*[1]	*Test 3*[1]	*Test 4*[1]
			Corrosion rate, mm/a (mpy)			
NiCrMoCu 56 23 6 6†	Illium® G		0.520 (20.5)	0.520 (20.5)	0.140 (5.51)	0.180 (7.09)
NiCrMoCu 58 21 5 3†	Illium® R		0.530 (20.9)	0.530 (20.9)		
NiCr21Mo	2.4858, UNS N08825		1.17 (46.1)	0.510 (20.1)	0.230 (9.06)	0.200 (7.87)
NiCr20CuMo	2.4660, UNS N08020	E	1.22 (48.0)	0.570 (22.4)	0.260 (10.2)	0.180[3] (7.09)
G-X 6 NiCrCuMo 29 20 4	Aloyco® 20	E	1.24 (48.8)	0.560 (22.1)		
X3CrNiMoN27-5-2	1.4460, SAE 329	B	1.50 (59.1)	0.790[4] (31.1)		
X 5 NiCrMoCu 24 20 3 2	Worthite®	E	1.70 (66.9)	0.710 (27.9)		
X3CrNiMo18-12-3	1.4449, SAE 317	D	1.80 (70.9)	0.760 (29.9)	0.460 (18.1)	0.230[3] (9.06)
X2CrNiMo17-12-2	1.4404, SAE 316 L	D			0.530 (20.9)	0.360[6] (14.2)

Table 38: Corrosion rates of chromium-nickel steels and nickel alloys (see also Table 40) in an evaporator of a pilot plant under the operating conditions of wet phosphoric acid production (see Figure 64) [42, 43]

Table 38: Continued

Material designation		*Section*	*Test 1*[1)]	*Test 2*[1)]	*Test 3*[1)]	*Test 4*[1)]
			Corrosion rate, mm/a (mpy)			
X5CrNiMo17-12-2	1.4401, SAE 316	D	2.00 (78.7)	0.940[5)] (37.0)		
X5CrNiMo17-12-2		D			0.530 (20.9)	0.360[6)] (14.2)
X5CrNiMo17-12-2, sensibilized		D	2.10 (82.7)	0.990[7)] (39.0)	0.530 (20.9)	0.460 (18.1)
NiCu 30 Fe	2.4360, UNS N04400				0.690 (27.2)	0.990 (39.0)
X5CrNi18-10	1.4301, SAE 304	C			0.970 (38.2)	0.740 (29.1)

† old designation

1) Test conditions:

Medium: intake: 70 % H_3PO_4 (≙ 52 % H_2P_5) (obtained from crude phosphate from Florida); liquid phase in the evaporator:
about 97 % H_3PO_4 (≙ 70 % P_2O_5)

	Test 1	Test 2	Test 3	Test 4
	also: 5 % H_2SO_4, 0.5 % HF, other impurities	H_2O vapor (main constituent), SiF_4, HF (traces), mist of H_3PO_4 and H_2SiF_6	$CaSO_4 \cdot 2\ H_2O$, Fe and Al phosphates, H_2SiF_6, HF	H_2 vapor (main constituent), SiF_4 (traces), mist of H_3PO_4 and H_2SiF_6
Test site:	immersed in liquid	in the vapor dome	immersed in the acid tank of the vapor dome	in the vapor dome
Duration:	total: 38.5 days, 3.8 of these in 80–86 % H_3PO_4 (≙ 58–66 % P_2O_5)	total: 38.5 days, 3.8 of these in 80–86 % H_3PO_4 (≙ 58–66 % P_2O_5)	total: 76 days, 55 of these cooled from 450 K (177 °C) to 302 K (29 °C) 1 day at 373 K (100 °C) in water	total: 76 days, 55 of these cooled from 450 K (177 °C) to 302 K (29 °C) 1 day at 373 K (100 °C) in water
Temperature:	average: 461 K (188 °C) highest: 477 K (204 °C) lowest: 433 K (160 °C)	average: 461 K (188 °C) highest: 477 K (204 °C) lowest: 433 K (160 °C)	average: 499 K (226 °C) highest: 461 K (188 °C) lowest: 302 K (29 °C)	average: 499 K (226 °C) highest: 461 K (188 °C) lowest: 302 K (20 °C)
Agitation:	vigorous	vigorous	unknown	vigorous
Aeration:	none	none	none	none

2) crevice corrosion: max. depth 0.30 mm
3) pitting corrosion: max. depth 0.13 mm
4) circular corrosion region on the spacer of the specimen holder, probably caused by residual liquid: max. depth 0.15 mm
5) circular corrosion region on the spacer of the specimen holder, probably caused by residual liquid: max. depth 0.075 mm
6) pitting corrosion: max. depth 0.30 mm
7) circular corrosion region on the spacer of the specimen holder, probably caused by residual liquid: max. depth 0.13 mm
8) crevice corrosion: max. depth 0.10 mm
9) crevice corrosion: max. depth 0.15 mm

B: see section Ferritic-austenitic steels with more than 12 % chromium
C: see section Austenitic CrNi steels
D: see section Austenitic CrNiMo(N) steels
E: see section Austenitic chromium-nickel steels with special alloying additions

Table 38: Corrosion rates of chromium-nickel steels and nickel alloys (see also Table 40) in an evaporator of a pilot plant under the operating conditions of wet phosphoric acid production (see Figure 64) [42, 43]

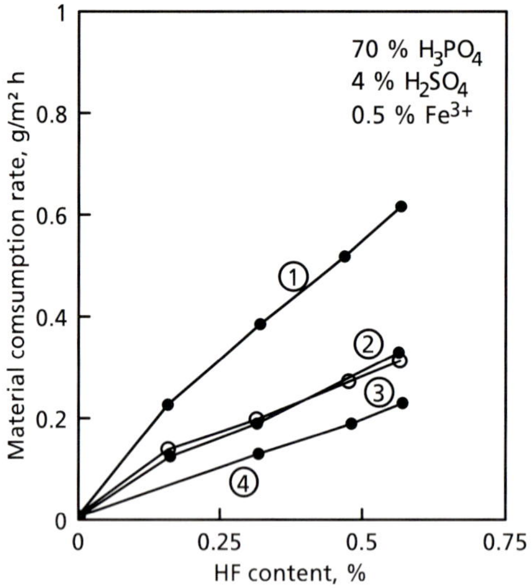

Figure 66: Influence of the hydrochloric acid concentration in chloride-free industrial phosphoric acid on the corrosion behavior of high-alloy steels (Table 42) at 373 K (100 °C) test duration 12 h [128]
① X1NiCrMoCuN25-20-7 (1.4529, UNS N08925)
② X1CrNiMoN25-25-2 (1.4465, SAE 310 MoLN)
③ Remanit® 4565S + Cu (cf. 1.4565, UNS S34565)
④ X1NiCrMoCu31-27-4 (1.4563, UNS N08028)

New processes and also increasing contamination of the raw materials require continual development of new materials. In addition to being distinguished by an increased general resistance, these are also distinguished by a high resistance to pitting corrosion and crevice corrosion, which is "measurable" from their action sum.

Corrosion studies were carried out on the high-alloy steels summarized in Table 42 under conditions which correspond to the very corrosive circumstances during concentration of industrial phosphoric acid. The test results are summarized in Figures 66–68.

Figure 66 shows the corrosion behavior in chloride-free industrial phosphoric acid as a function of the fluoride concentration (as hydrogen fluoride). As expected, the material consumption rates increase as the fluoride concentration in the acid increases. The almost copper-free steel 1.4465 (Remanit® 4465, SAE 310 MoLN) showed a relatively good corrosion behavior under these conditions in comparison with copper-containing high-alloy steels, but is inferior to steel 1.4563 (UNS N08028, see under Section Austenitic chromium-nickel steels with special alloying additions).

The material consumption rates nevertheless suggest that the steels were in the passive state, since higher corrosion rates are to be expected in the active state [128].

The two graphs 84 and 85 show an increasing resistance of the steels investigated as the Cr content increases. Accordingly, chromium is evidently the decisive alloying element (see also under Section Ferritic chromium steels with more than 12 % chromium) for counteracting the corrosive action of fluoride in industrial phosphoric acid [128].

Material		**Section**	**Test 1[1)]**	**Test 2[1)]**	**Test 3[1)]**	**Test 4[1)]**
			Corrosion rate, mm/a (mpy)			
NiCrMoCu 56 23 6 6†	Illium® G		0.086 (3.39)	0.180 (7.09)	0.180 (7.09)	0.074 (2.91)
NiCrMoCu 58 21 5 3†	Illium® R		0.099 (3.9)	0.180 (7.09)	0.190[4)] (7.48)	0.079 (3.11)
NiCr20CuMo	2.4660, UNS N08020	E	0.119 (4.69)	0.240 (9.45)	0.220[5)] (8.66)	0.127 (5.0)
NiCr21Mo	2.4858, UNS N08825		0.107 (4.21)	0.210 (8.27)	0.220 (8.66)	0.107 (4.21)
X3CrNiMoN27-5-2	1.4460, SAE 329	B	0.210 (8.27)	0.380 (15.0)	0.270 (10.6)	0.210 (8.27)
G-X 6 NiCrCuMo 29 20 4	Aloyco® 20	E	0.117 (4.61)	0.230 (9.06)	0.300 (11.8)	0.165 (6.5)
X 5 NiCrMoCu 24 20 3 2	Worthite®	E	0.147 (5.79)	0.320 (12.6)	0.300[5)] (11.8)	0.210 (8.27)
X3CrNiMo18-12-3	1.4449, SAE 317	D	0.210 (8.27)	0.430 (16.9)	0.390[5)] (15.4)	0.260[5)] (10.2)
X5CrNiMo17-12-2	1.4401, SAE 316	D	0.360 (14.2)	0.580 (22.8)	0.460 (18.1)	0.280 (11.0)
X5CrNiMo17-12-2, sensibilized		D	0.360 (14.2)	0.560 (22.0)	1.14 (44.9)	0.690 (27.2)

† old designation

1) Test conditions:

Medium: intake: 72 % H_3PO_4 (≙ 52 % P_2O_5) (obtained from crude phosphate from Western USA); liquid phase in the evaporator: about 97 % H_3PO_4 (≙ 70 % P_2O_5)

	Test 1	Test 2	Test 3	Test 4
	also: 5 % H_2SO_4, 0.5 % HF	H_2O vapor (main constituent), SiF_4, HF (traces), mist of H_3PO_4 and H_2SiF_6	other constituents unknown	H_2O vapor (main constituent), SiF_4, HF (traces), mist of H_3PO_4 and H_2SiF_6
Test site:	pilot plant, specimens immersed in liquid	pilot plant, specimens in the vapor phase	production plant, specimens immersed in the liquid of the circulation pipeline	production plant, specimens in the vapor phase
Duration:	25 days	25 days	41.5 days	41.5 days
Temperature:	average: 466 K (193 °C) highest: 478 K (205 °C) lowest: 450 K (177 °C)	average: 466 K (193 °C) highest: 478 K (205 °C) lowest: 450 K (177 °C)	average: 461 K (188 °C) highest: 475 K (202 °C) lowest: 453 K (180 °C)	average: 322 K (49 °C) highest: 338 K (65 °C) lowest: 300 K (27 °C)
Agitation:	vigorous	vigorous	vigorous	vigorous
Aeration:	none	none	none	none

2) crevice corrosion: max. depth 0.18 mm
3) crevice corrosion: max. depth 0.13 mm
4) crevice corrosion: max. depth 0.10 mm
5) crevice corrosion: max. depth 0.08 mm
6) stress corrosion cracking in the region of the hammered numbers
7) crevice corrosion in the spacer region: max. depth 0.10 mm

B: see section Ferritic-austenitic steels with more than 12 % chromium
D: see section Austenitic CrNiMo(N) steels
E: see section Austenitic chromium-nickel steels with special alloying additions

Table 39: Corrosion rates of CrNiMo steels and nickel alloys (see also Table 40) in the evaporator of a pilot plant for production of highly concentrated phosphoric acid from wet acid (see Figure 64) under operating conditions [42, 43]

Designation or commercial name	Corresponds (approx.) to German material	DIN-Mat. No.	Chemical composition, %[1)]								Other
			Ni	Fe	Cr	Mo	Cu	C	Si	Mn	
Chromium-nickel steels (rolled- and forged steels)											
	X 12 CrNi 18 8[†]	1.4300	9	70	18	–	–	max. 0.15	max. 1.0	max. 2.0	–
SAE 304	X5CrNi18-10	1.4301	9.5	70	18	–	–	max. 0.08	max. 1.0	max. 2.0	–
SAE 304 L	X2CrNi19-11	1.4306	10	69	18	–	–	max. 0.03	max. 1.0	max. 2.0	–
SAE 309	X12CrNi22-12	1.4828	13.5	61	23	–	–	max. 0.20	max. 1.0	max. 2.0	–
SAE 310	X12CrNi25-20	1.4841	20	52	25	–	–	max. 0.25	max. 1.0	max. 2.0	–
SAE 316	X5CrNiMo17-12-2	1.4401	13	65	17	2.5	–	max. 0.08	max. 1.0	max. 2.0	–
SAE 316 L	X2CrNiMo17-12-2	1.4404	13	65	17	2.5	–	max. 0.03	max. 1.0	max. 2.0	–
SAE 317	X3CrNiMo18-12-3	1.4449	14	61	19	3.5	–	max. 0.08	max. 1.0	max. 2.0	–
SAE 317 L	X2CrNiMo18-15-4	1.4438	13	63	18	3.5	–	max. 0.025	max. 0.5	max. 1.75	–
SAE 321	X6CrNiTi18-10	1.4541	10	69	18	–	–	max. 0.08	max. 1.0	max. 2.0	Ti ≥ 5 X C
SAE 329	X3CrNiMoN27-5-2	1.4460	4.5	67	27	1.5	–	max. 0.10	max. 1.0	max. 1.0	–
SAE 347	X6CrNiNb18-10	1.4550	11	68	18	–	–	max. 0.08	max. 1.0	max. 2.0	Nb-Ta ≥ 10 X C
Chromium-nickel steels											
ACI-Type CD4MCu			5.4	62	26	2.0	3.0	max. 0.04	max. 1.0	max. 1.0	–
ACI CF-3	X2CrNi19-11	1.4306	10	68	19	–	–	max. 0.03	max. 2.0	max. 1.5	–
ACI CF-8	GX5CrNi19-10	1.4308	9.5	68	19.5	–	–	max. 0.08	max. 2.0	max. 1.5	–
ACI CF-3M	X2CrNiMo17-12-2	1.4404	11	65	19	2.5	–	max. 0.03	max. 1.5	max. 1.5	–

Table 40: Chemical composition of the nickel-containing materials used for the corrosion studies and in phosphoric acid plants (some designations out of date) (see Tables 30–39) [42, 43]

Table 40: Continued

Designation or commercial name	Corresponds (approx.) to German material	DIN-Mat. No.	Chemical composition, %[1)]								Other
			Ni	Fe	Cr	Mo	Cu	C	Si	Mn	
ACI CF-8M	GX5CrNiMo19 11-2	1.4408	10.5	65	19.5	2.5	–	max. 0.08	max. 1.5	max. 1.5	–
ACI CG-8M	G-X 5 CrNi 22 10[†]	1.4947	11	64	19.5	3.5	–	max. 0.08	max. 1.5	max. 1.5	–
Illium® P	G-X 8 CrNiMoCu 28 8[2)]		8	58	28	2.5	3				–
Nickel-chromium-molybdenum-copper steels											
Durimet® 20	G-X 6 NiCrCuMo 29 20 5[2)]		29	42	20	2.5	4.5	max. 0.07	1.0	0.7	–
Aloyco® 20	G-X 6 NiCrCuMo 29 20 4[2)]		29	42	20	3	4.0	max. 0.07	max. 1.5	0.7	–
Worthite® stainless	X 5 NiCrMoCu 24 20 3 2[2)]		24	48	20	3.0	1.75	max. 0.07	3.25	0.6	–
Carpenter® 20 Cb	X 5 NiCrMoCuNb 29 20 3 3[2)]		29	44	20	2.5	3.3	max. 0.07	0.6	0.75	Nb (+Ta) 0.6
Carpenter® 20 Cb 3	X 5 NiCrMoCuNb 34 20 3 3[2)]		34	39	20	2.5	3.3	max. 0.07	0.6	0.75	Nb (+Ta) 0.6
ACI CN-7M	GX7NiCrMo-CuNb25-20	1.4500	29	44	20	2.0	3.0	max. 0.07	1.0	1.0	–
Iron-nickel-chromium alloy											
Incoloy® alloy 800	X10NiCrAlTi32-21	1.4876	32	46	21	–	0.40	max. 0.05	0.50	0.75	Ti 0.40; Al 0.40

Table 40: Chemical composition of the nickel-containing materials used for the corrosion studies and in phosphoric acid plants (some designations out of date) (see Tables 30–39) [42, 43]

Table 40: Continued

Designation or commercial name	*Corresponds (approx.) to German material*	*DIN-Mat. No.*	*Chemical composition, %[1)]*								*Other*
			Ni	*Fe*	*Cr*	*Mo*	*Cu*	*C*	*Si*	*Mn*	
Nickel-iron-chromium-molybdenum alloys											
Incoloy® alloy 825	NiCr21Mo	2.4858	41.8	30	21.5	3.0	1.80	0.03	0.35	0.65	Al 0.15; Ti 0.90
Hastelloy® alloy G	NiCr22Mo6Cu	2.4618	45	19.5	22.2	6.5	2.0	0.03	0.35	1.3	W 0.5; Nb + Ta 2.12
Incoloy® alloy 901	NiCr13Mo6Ti3	2.4662	40	38	13	6.0					Ti 2.4
Inconel® alloy 718	NiCr19-Fe19Nb5Mo3	2.4668	52.5	18.5	18.6	3.1	max. 0.10	0.04	0.30	0.20	Nb 5.0; Ti 0.90; Al 0.40
Nickel-chromium-molybdenum alloys											
Illium® 98			55	1	28	8.5	5.5	0.05	0.7	1.25	–
Illium® G			56	6.5	22.5	6.4	6.5	0.20	0.65	1.25	–
Inconel® alloy 625	NiCr22Mo9Nb	2.4856	61	3	22	9.0	0.1	0.05	0.3	0.15	Nb (+ Ta) 4
Illium® R			68	1	21	5	3	0.05	0.7	1.25	–
Hastelloy® alloy C-276	NiMo16Cr15W	2.4819	60	5.5	16	16	–	max. 0.02	max. 0.05	max. 1.0	W 4, Co max. 2.5; V max. 0.35
Chlorimet® 3			60	3	18	18	–	0.07	1	1	–
Nickel-molybdenum alloys											
Hastelloy® B	NiMo 30†	cf. 2.4810, 2.4882	61	5	max. 1	28	–	max. 0.05	–	–	Co max. 2.5; other 3

Table 40: Chemical composition of the nickel-containing materials used for the corrosion studies and in phosphoric acid plants (some designations out of date) (see Tables 30–39) [42, 43]

Table 40: Continued

Designation or commercial name	*Corresponds (approx.) to German material*	*DIN-Mat. No.*	*Chemical composition, %¹⁾*								*Other*
			Ni	*Fe*	*Cr*	*Mo*	*Cu*	*C*	*Si*	*Mn*	
Nickel-alloyed cast iron											
Ni-Resist® 1 (casting material)	EN-GJLA-XNi-CuCr15-6-2	0.6655	15.5	69	2.5	–	6.5	max. 3.0	2.0	1.2	–
Ni-Resist® 2 (casting material)	GGL-NiCr 20 2	0.6660	20	72	2.5	–	–	max. 3.0	2.0	1.0	–
Ni-Resist® 3 (casting material)	GGL-NiCr 30 3	0.6676	30	63	3	–	–	max. 2.6	1.5	0.6	–
Ni-Resist® 4 (casting material)	GGL-NiSiCr 30 5 5	0.6680	30.5	56	5	–	–	max. 2.6	5.5	0.6	–
Ni-Resist® D-2 (casting material)	EN-GJSA-XNiCr20-2	0.7660	20	72	2	–	–	max. 3.0	2.0	1.0	P max. 0.08
Ni-Resist® D-2B (casting material)	GGG-NiCr 20 3	0.7661	20	70.5	3.25	–	–	max. 3.0	2.3	1.1	P max. 0.08
Ni-Resist® D-4 (casting material)	EN-GJSA-XNiSiCr30-5-5	0.7680	30.5	56	5	–	–	max. 2.6	5.5	max. 0.5	P max. 0.08
Nickel-copper alloys											
Monel® alloy 400	NiCu 30 Fe	2.4360	66	1.25	–	–	31.5	0.12	0.15	0.90	–
Monel® alloy K 500	NiCu 30 Al	2.4375	65	1.0	–	–	29.5	0.15	0.15	0.60	Al 2.80; Ti 0.50
Nickel-copper alloy 505	G-NiCu 30Si4	2.4368	64	2.0	–	–	29	0.08	4.0	0.80	–

Table 40: Chemical composition of the nicke -containing materials used for the corrosion studies and in phosphoric acid plants (some designations out of date) (see Tables 30–39) [42, 43]

Table 40: Continued

Designation or commercial name	Corresponds (approx.) to German material	DIN-Mat. No.	Chemical composition, %[1)]								Other
			Ni	Fe	Cr	Mo	Cu	C	Si	Mn	
Copper-nickel alloys											
Copper-nickel CA 715	CuNi30Mn1Fe	CW354H	31	0.55	–	–	67	–	–	max. 1.0	Pb max. 0.05; Zn max. 1.0
Copper-nickel CA 706	CuNi10Fe1Mn	CW352H	10	1.0	–	–	86.5	–	–	max. 1.0	Pb max. 0.05; Zn max. 1.0
Nickel											
Nickel 200	Ni 99,6[3)]	2.4060	99.5	0.15	–	–	0.05	0.06	0.05	0.025	–
Nickel-chromium alloys											
Incoloy® alloy 804			42.5	25.5	29.0	–	0.40	0.06	0.50	0.85	Al 0.25; Ti 0.40
Corronel® alloy 230			55	≤ 5.0	36	–	max. 1.0	max. 0.08	max. 0.6	max. 1.0	Ti max. 1.0; Al max. 0.5
Inconel® alloy 600	NiCr15Fe	2.4816	76.0	7.2	15.8	–	0.10	0.04	0.20	0.20	–
Nickel-chromium-cobalt alloy											
Nimonic® alloy 90	NiCr 20 Co 18 Ti	2.4969	51	5.0	19.5	–	–	0.13	1.5	0.20	Co 18; Ti 2.4; Al 1.4

† old designation
1) mean values, unless stated otherwise
2) not standardized in the Federal Republic of Germany
3) In addition to Ni 99.6 (DIN-Mat. No. 2.4060), the grades Ni 99.2 (DIN-Mat. No. 2.4066) and LC-Ni 99 (DIN-Mat. No. 2.4068) are also used. These materials are standardized in DIN 17 740. LC-Ni 99 is preferred for plants to be welded

Table 40: Chemical composition of the nickel-containing materials used for the corrosion studies and in phosphoric acid plants (some designations out of date) (see Tables 30–39) [42, 43]

Material designation		***Section***	***Corrosion rate mm/a (mpy)***	***Types and depth of corrosion mm***
Worthite®	X 5 NiCrMoCu 24 20 3 2	E	2.95 (116)	
G-NiCu 30Si4	2.6368		3.00 (118)	
Hastelloy® B	2.4810, NiMo 30		4.20 (165)	0.25 2)
Monel® alloy 400	2.4360, NiCu 30 Fe		4.50 (177)	0.08 2)
Ni-Resist® 1	0.6655, EN-GJLA-XNiCuCr15-6-2	E	4.70 (185)	
Ni-Resist® D-4	0.7680, EN-GJSA-XNiSiCr30-5-5	F	5.05 (199)	
Carpenter® 20 Cb	X 5 NiCrMoCuNb 29 20 3 3	E	5.40 (213)	0.41 2)3)
Incoloy® alloy 825	2.4858, NiCr21Mo		> 5.60 (> 220)	0.76 2)3)
Ni-Resist® Type D-2	0.7660, EN-GJSA-XNiCr20-2	C	5.80 (228)	
Inconel® alloy 600	2.4816, NiCr15Fe		> 6.10 (> 240)	0.76 2)
Ni-Resist® Type 2	0.6660, GGL-NiCr 20 2	C	6.40 (252)	
SAE 316	1.4401, X5CrNiMo17-12-2	D	> 6.60 (> 260)	0.48 2)
SAE 317	1.4449,X3CrNiMo18-12-3	D	> 7.10 (> 280)	0.51 2)
SAE 446	X 10 CrAl 24 (SAE 446 (1.4762))	A	> 15 (> 590)	

1) Test conditions:
Medium: usually 58 % H_3PO_4, (≙ 12 % P_2O_5) sometimes rising to 75 % H_3PO_4 (≙ 54 % P_2O_5), 5 % HCl, 3 % H_2SO_4, a little HF, traces of Na_2SiF_6 and K_2SiF_6, $CaSO_4 \cdot 2\,H_2O$ suspended
Test site: specimens were positioned at the top section of a tubular heat exchanger, but were recovered on the bottom of the drain funnel
Duration: 10 days
Temperature: 355–366 K (82–93 °C)
Agitation: vigorous
Aeration: none
2) circular corrosion region on the spacer of the specimen holder, probably caused by residual liquid
3) stress corrosion cracking in the region of the hammered numbers
4) pitting corrosion

A: see section Ferritic chromium steels with more than 12 % chromium
C: see section Austenitic CrNi steels
D: see section Austenitic CrNiMo(N) steels
E: see section Austenitic chromium-nickel steels with special alloying additions
F: see section Special iron based alloys

Table 41: Corrosion rates of high-alloy steels and nickel alloys (see also Table 40) in chloride-containing phosphoric acid in an evaporator from wet decomposition (see Tables 36 and 37) under operating conditions [42, 43]

TEW designation[1)]	Material designation	Abbreviation	Chemical composition, %							Section	PRE[2)]
			C	Cr	Ni	Mn	Mo	Cu	N		
Remanit® 4539	SAE 904 L	X1NiCrMoCu25-20-5	0.019	20.2	24.3	1.56	4.57	1.33	0.077	E	37.4
Remanit® 4465	SAE 310 MoLN	X1CrNiMoN25-25-2	0.014	24.5	24.6	1.37	2.22	0.08	0.12	D	35.4
Remanit® 4563	UNS N08028	X1NiCrMoCu31-27-4	0.010	27.2	31.4	1.53	3.34	1.21	0.12	E	41.8
Remanit® 4529 S		X 2 CrNiMoCuN 20 18 6	0.021	19.9	18.0	0.54	6.28	0.76	0.21	E	46.9
Remanit® 4565 S		X 2 CrNiMnMoN 24 17 6 4	0.019	23.8	18.0	6.14	4.28	0.13	0.43	F	50.8
Remanit® 4565 S + Cu		X 2 CrNiMnMoCuN 24 17 6 4	0.026	25.8	17.6	7.07	4.60	0.63	0.52	F	56.6

1) Thyssen Edelstahlwerke AG
2) PRE (pitting resistance equivalent) ≙ % Cr + 3.3 x % Mo + 30 x % N

D: see section Austenitic CrNiMo(N) steels
E: see section Austenitic chromium-nickel steels with special alloying additions
F: see section Special iron based alloys

Table 42: Chemical composition of high-alloy steels for studies in highly concentrated hot industrial phosphoric acid (see Figures 66–68) [81, 128]

The influence of the alloy composition changes, however, if contamination of the acid by chloride becomes the critical factor. In contrast to fluoride, chloride does not increase the corrosion rate of high-alloy steels in phosphoric acid in the passive state. The action of chloride lies in the activation of these materials. This is a reason for the development of pitting corrosion and crevice corrosion. This resistance is increased as the pitting resistance equivalent (PRE) increases (see, for example, under Sections Ferritic chromium steels with more than 12 % chromium and Austenitic CrNi steels), as can also be deduced from Figure 68. The particular pitting index of the individual steels is also shown in Table 42.

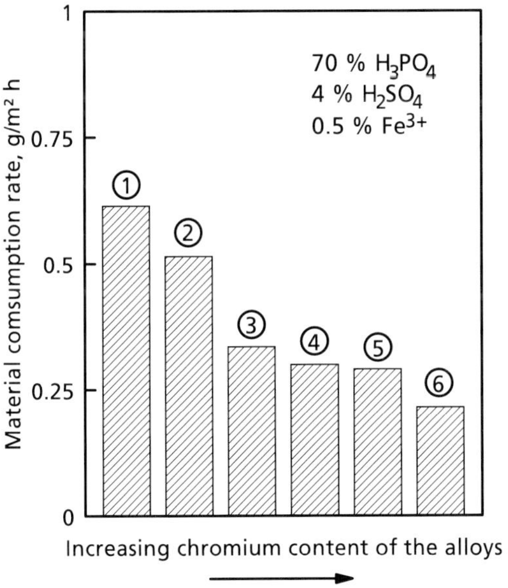

Figure 67: Influence of the chromium content on the material consumption rates of high-alloy steels (Table 42) in industrial phosphoric acid at 373 K (100 °C), test duration 12 h [128]
① Remanit® 4539
② Remanit® 4529 S
③ Remanit® 4565 S
④ Remanit® 4465
⑤ Remanit® 4565 S + Cu
⑥ Remanit® 4563

It is to be assumed that an alloy composition having a high pitting resistance equivalent (PRE) also offers a high resistance towards activation of materials in a chloride-containing acid medium, as with industrial phosphoric acid here [128].

Steel 1.4465 (SAE 310 MoLN, X1CrNiMoN25-25-2), which is dealt with under Section Austenitic CrNiMo(N) steels, has mathematically the lowest pitting resistance equivalent (PRE) [3, 129].

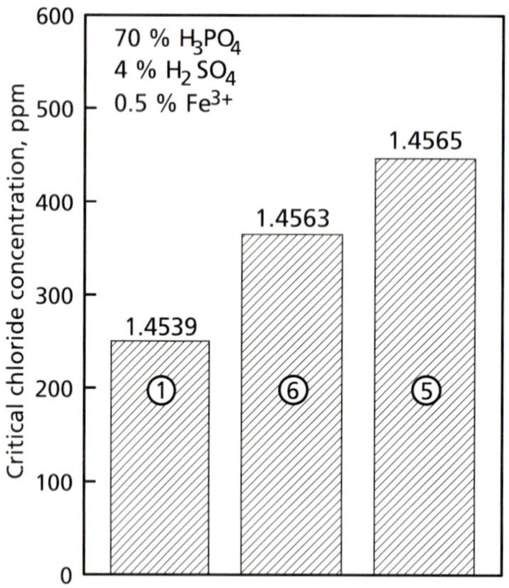

Figure 68: Critical chloride concentration for repassivation of high-alloy steels (see Table 42, Figure 67) in industrial phosphoric acid at 373 K (100 °C) [128]

Austenitic chromium-nickel steels with special alloying additions

Among the austenitic steels, the Mo containing CrNi steels are superior in their corrosion resistance to those without molybdenum (see Sections Austenitic CrNi steels and Austenitic CrNiMo(N) steels).

As is the case under exposure to sulfuric acid, the corrosion resistance of these Mo-containing steels can be increased further by alloying with copper (in this context, compare Figures 8 (with Cu) and 46 (without Cu)). Copper additions to these steels can positively influence the corrosion behavior, especially in non-oxidizing acids [80], and considerably promote resistance under such conditions [128].

Materials which have been introduced and proved suitable for phosphoric acid production by the wet process are the high-alloy steels SAE 904 L (1.4539, X1NiCrMoCu25-20-5) – trade name 2RK65, Uranus® B6 –, UNS N08925 (1.4529, X1NiCrMoCuN25-20-7) and UNS N08028 (1.4563, X1NiCrMoCu31-27-4) – trade name Sanicro® 28 – [39, 51, 70, 71, 77, 75, 79, 130–146].

Steel 1.4563 shows the better resistance in comparison with 1.4539, especially under extremely high exposure to corrosion during phosphoric acid production by the wet process [143].

Figures 28 and 69 show the use ranges of CrNiMoCu steels in comparison with other steels.

Corrosion rates and further corrosion behavior under various conditions in phosphoric acid solutions can be found in Tables 2, 6, 7, 12, 15, 16–18, 21–23, 25, 30–34, 36–39, 41, 42 and in Figures 4, 8, 17, 19–22, 28, 29, 41–44, 49, 51 (summary in Table 45).

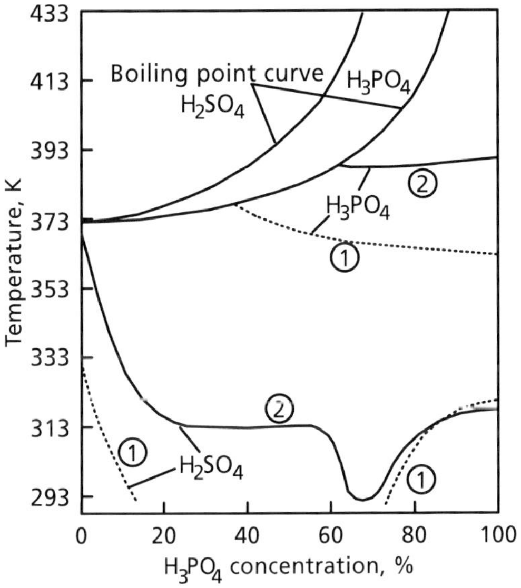

Figure 69: Isocorrosion curves (0.1 mm/a (3.94 mpy)) for CrNiMo and CrNiMoCu steels in chemically pure phosphoric acid in comparison with similar sulfuric acid [111]
① SAE 316 (1.4401, X5CrNiMo17-12-2)
② SAE 904 L (1.4539, X1NiCrMoCu25-20-5)

Copper ions in phosphoric acid and its solutions inhibit the corrosion of stainless steels. Cu(II)-ions are formed at the start of the corrosion of copper-containing alloys. As the concentration of these ions increases, the corrosion rate drops (see Figure 52). Although the corrosion rate drops sharply as the copper content in the steel increases in this example, steel 316 L (1.4404) is still not resistant in boiling 85 % phosphoric acid at a copper content of 0.35 %. Additions of 0.8 % copper to steel 317 L (1.4438) are also not yet sufficient to achieve resistance to crude phosphoric acid at 323 K (50 °C) (Figure 29).

The high corrosion resistance of CrNiMo steels can be achieved only by a higher alloying content of copper. The copper contents are 1.5 % or more. See also Figure 68.

Corrosion studies on a steel with about 25 % Ni, 20 % Cr, 4.5 % Mo, 1.5 % Cu, about 1.5 % Mn and max. 0.015 % C but a varying Si content (0.09–0.75 %) have shown that the resistance of the steel in phosphoric acid increases as the Si content decreases [147]. The steel corresponds to SAE 904 L (1.4539). Corrosion studies on weld joints of steel N08028 (1.4563, X1NiCrMoCu31-27-4) have confirmed that in the case of an additional material of the same type, the weld seam is less resistant to corrosion under aggressive conditions than the base material. Possible pitting corrosion can already be crevice corrosion under such conditions, caused by intercrystalline corrosion at dendrite tips formed as a consequence of segregations. The studies furthermore showed that the purity of the side of the welding metal surface facing the corrosion medium considerably influences corrosion resistance in respect of slag inclusions and the oxide film. Pitting corrosion may occur at such points. They are the reason for the expected destruction of the material.

A significantly over-alloyed weld filler material is to be used with these high-alloy special steels in order to avoid a reduction in the concentration of the alloying elements Cr and Mo, which chiefly determine the corrosion resistance [148] (see also Figure 67) in the micro-range [80, 149].

Overlaps of welding runs in the final layer can be the preferred points of attack for intercrystalline corrosion in the remelting region of the underlying welding run under exposure to severe corrosion. Welding methods which have proved to be more suitable are those using an inert gas and mechanization, because of the better surface quality which can be achieved [80, 149].

The high-alloy copper-containing steel SAE 904 L (1.4539) is very successfully used in the phosphoric acid industry as a material e.g. for stirrers, filters, pipeline systems and preheaters [114, 117].

Table 45 contains a summary of the figures and tables containing corrosion rates for steels SAE 904 L (1.4539), N08925 (1.4529) and N08028 (1.4563) under various conditions.

These steels are a cheaper alternative to the nickel alloys for use for highly aggressive phosphoric acid. Under the conditions according to Table 43, steel 1.4563 even proves to be the better material, compared with 2.4856 (UNS N06625), with a corrosion rate of 0.16 mm/a (6.30 mpy) [80].

Material	*Corrosion rate mm/a (mpy)*	*Literature*	*Section*
Cronifer® 2328 (1.4503)	0.31 (12.2)	[80]	E
Nicrofer® 3127 LC (1.4563)	0.16 (6.3)	[80, 150]	E
Nicrofer® 3127 hMo (1.4563 + 3 % Mo) (1.4562)	0.14 (5.51)	[150]	E
Nicrofer® 4823 hMo (2.4619)	0.15 (5.91)	[80, 150]	
Nicrofer® 6020 hMo (2.4856)	0.25 (9.84)	[80, 150]	
Nicrofer® 6020, mod. (2.4856 + 3 % Cu)	0.20 (7.87)	[80]	

E: see section Austenitic chromium-nickel steels with special alloying additions

Table 43: Corrosion behavior of high-alloy Ni materials in industrial phosphoric acid (72 % H_3PO_4, 4.5 % H_2SO_4, 0.9 % H_2SiF_6, 1.5 % Fe_2O_3, 400 ppm HCl) at 373 K (100 °C)

In phosphoric acid contaminated with a high chloride content, it will be the preferred material at high temperatures, compared with 1.4539. The reason for this is its higher pitting resistance equivalent (PRE) (Table 42) and therefore its higher resistance to pitting corrosion (see also under Section Austenitic CrNiMo(N) steels and Figure 68).

According to Figure 68, the critical chloride concentration in the 70 % H_3PO_4 mentioned therein at 373 K (100 °C) can be about 350 ppm in order to achieve repassivation of steel 1.4563, while it may be 250 ppm for 1.4539 [151].

This also explains the better resistance of 1.4563 (N08028) under the conditions in Figures 70 and 71 in comparison with steel 1.4529 (N08925). In this context, see also Table 44. Under these conditions, there is hardly any difference in resistance, since with the corrosion rate of 0.01 mm/a (0.39 mpy) reported here, 1.4529 is also to be regarded as resistant.

Material		***Corrosion rate, mm/a (mpy)***
Cronifer® 1925 LC	(1.4539)	0.01 (0.39)
Cronifer® 1925 hMo	(1.4529)	0.01 (0.39)
Nicrofer® 3127 LC	(1.4563)	0.00

Table 44: Results of corrosion studies in the laboratory in phosphoric acid (35 % P_2O_5 (48.3 % H_3PO_4), 0.03 % F^-, 0.2 % Cl^-) at 353 K (80 °C) [151]

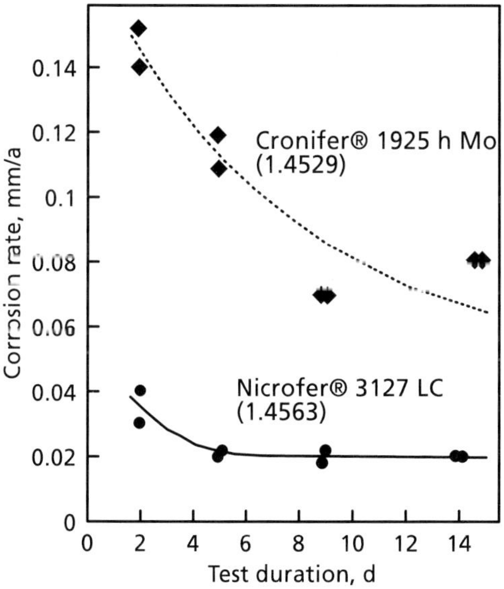

Figure 70: Results of corrosion studies on CrNiMoCu steels in a phosphoric acid pilot plant (33.4–34.6 % P_2O_5 (46–47.8 % H_3PO_4), 1.37–1.51 % F^-, 240–250 ppm Cl^-) at 323 K (50 °C) [151]

Serial No.	*H_3PO_4 %*	*Temperature K (°C)*	*Material*	*Figure No.*[1)]	*Table No.*[2)]	*Literature*
1+	22.4–24.2	310 (37)	(Illium® P)	21, 22	15, 16	[18]
2	38.6	366 (93)	(diverse)		30	[80]
3	39	355 (82)	(diverse)		34	[80]
4+	40	333, 363 (60, 90)	1.4539 (SAE 904 L)	19, 20	11, 12	[67]
5	41–69	333, 373 (60, 100)	1.4539 (SAE 904 L)	49		[99]
6	41.4	353 (80) boiling	1.4539 (SAE 904 L) 1.4539 + N	41		[72]
7	46–48	323 (50)	1.4529 (N08925) 1.4563 (N08028)	70		[151]
8	48.3	353 (80)	1.4529 (N08925) 1.4539 (SAE 904 L) 1.4563 (N08028)		44	[151]
9	53	394 (121)	(diverse)		36	[80]
10	62.1, 98	343, 373 (70, 100)	(Russ. steel)		17, 18	[68]
11	70	333 (60)	1.4539 (SAE 904 L)	65		
12	70	373 (100)	1.4537 2.4660 (20 Cb-3)	51		[19]
13	70	377 (104)	(s. Tab. 66)	66–68		[152]
14	71.8	353 (80)	1.4529 (N08925) 1.4563 (N08028)	71		[151]
15	72	373 (100)	1.4503 1.4563 (N08028) 1.4563 + 3 % Mo		43	[80, 153]
16	72	464 (191)	(diverse)		39	[80]
17	74.5	363 (90)	1.4505 1.4539 (SAE 904 L)	17		[63]
18	74.5	up to 393 (120)	1.4539 (SAE 904 L)	42		[72]
19	all	up to 473 (200)	(diverse)	28		[19, 80]
20	all	up to 473 (200)	(Russ. steel)		25	[17]
21	all	up to 393 (120)	1.4563 (N08028)	72		[154]

+ erosion corrosion
1) see also Figures 4, 8, 27, 43, 69
2) see also Tables 2, 6, 7, 21–23, 30–32, 34–39, 41

Table 45: Overview of figures and tables with corrosion rates of high-alloy CrNiMoCu steels of recent production in phosphoric acid under extreme exposure to corrosion

Under the conditions of Figures 70 and 71, the high-alloy steel 1.4563 showed no corrosion rates above 0.05 mm/a (1.97 mpy), and was resistant to pitting corrosion and crevice corrosion. It was used for the manufacture of a heat exchanger in phosphoric acid preconcentration, and has proved to be suitable for this purpose.

This applies to the concentration range from 26 to 52 % H_3PO_4 with up to 2 % fluorides and 0.2 % chlorides in the temperature range from 323 to 353 K (50 to 80 °C) (see also Figure 41). For welding, the additional material 2.4831 (UNS N06625) was used, and the MIG or TIG process was used as the welding process [151].

Given the very low carbon content of the steel, intercrystalline corrosion is hardly to be expected.

Figure 72 shows the dependence of the corrosion rates of 1.4563 in contaminated phosphoric acid on the temperature [154].

Figure 73 contains the dependence of the corrosion rates of steels 1.4539 (SAE 904 L) and 1.4563 (N08028) on the fluoride content in 70 % H_3PO_4 with 4 % H_2SO_4 and 60 ppm Cl^- at 373 K (100 °C). This graph demonstrates the clear superiority of steel 1.4563 over 1.4539 in H_3PO_4 with F^- contents of up to 1 % [71].

No stress corrosion cracking was observed in phosphoric acid solutions up to 2 % H_3PO_4 with the additions of up to 4 % NaCl, pH about 1.7, and at temperatures of up to 414 K (141 °C) [155]. The test material was the steel Carpenter® 20 Cb 3 (see Table 40).

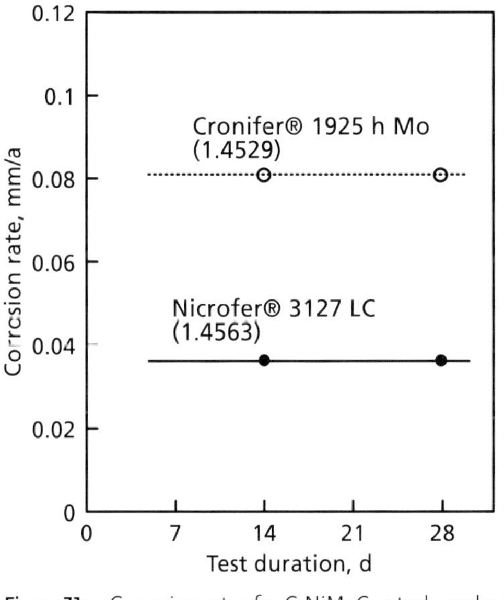

Figure 71: Corrosion rates for CrNiMoCu steels under intensified corrosion conditions in a phosphoric acid plant (52 % P_2O_5 (71.8 % H_3PO_4), 0.23 % F^-, 20–30 ppm Cl^-, 4 % SO_3^{2-}, 2 % solids) at 353 K (80 °C) [151]

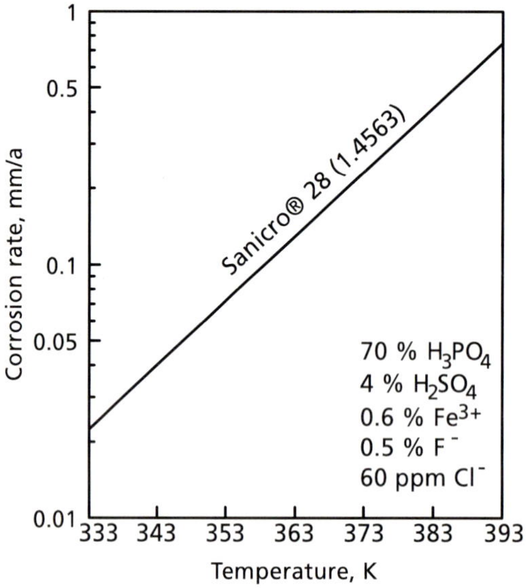

Figure 72: Dependence of the corrosion rate of Sanicro® 28 (N08028) in contaminated 70% H_3PO_4 on the temperature [154]

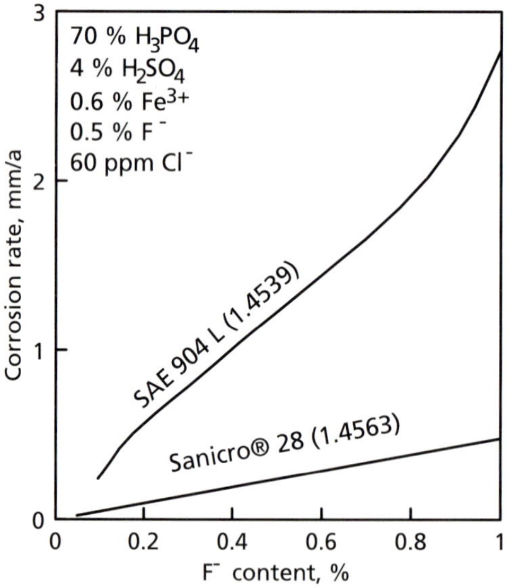

Figure 73: Comparison of the corrosion resistance of 1.4539 (SAE 904 L, X1NiCrMoCu25-20-5) and 1.4563 (N08028, X1NiCrMoCu31-27-4) in contaminated phosphoric acid as a function of the fluoride content at 373 K (100 °C) [71]

According to [153], a new NiCrMoCu steel Nicrofer® 3127 hMo (1.4562) combines the advantages of materials of high chromium content with a molybdenum content of more than 6%. This steel exhibits an outstanding resistance to pitting corrosion and crevice corrosion in aqueous neutral and acid media. It is significantly superior both to the known 6% Mo steels, such as e.g. 1.4529 (N08925), and to the higher-alloy Ni base materials, such as e.g. 2.4619 (UNS N06985). It is a promising material of the future, even under exposure to phosphoric acid, especially since the requirements on materials are constantly increasing due to changes in processes. In comparison with 1.4563, 1.4562 contains 3% more molybdenum [153].

In fact, steel 1.4562 shows a good corrosion behavior during concentration of phosphoric acid similar to that of steel 1.4563, coupled with good machinability, as can be seen from Table 43. Welds effected by the TIG inert gas process (filler material 2.4831 (UNS N06625)) and the manual arc process (filler material 2.4621 (EL-NiCr20Mo9Nb)) have proved to be appropriate, without impairment of the corrosion resistance in comparison with the base material.

As well as in the liquid phase during evaporation above these vapors and mists containing suspended substances, phosphoric, sulfuric, hydrofluoric and hexafluorosilicic acid occur in unknown concentrations. At temperatures in the range from 298 to 358 K (25 to 85 °C) these can be very aggressive. In this case, it is scarcely still possible to use CrNiMo steel. Steel 1.4503 (X3NiCrCuMoTi27-23) shows a corrosion rate of about 0.12 mm/a (4.72 mpy), and the nickel alloy 2.4856 (NiCr22Mo9Nb) a corrosion rate of less than 0.05 mm/a (1.97 mpy) (in this context, see Table 39) [42, 43, 80]. At temperatures of about 473 K (200 °C) the usual stainless steels fail due to excessive surface removal or to crevice corrosion [8].

Erosion corrosion

When handling phosphoric acid, and in particular during its production by the wet process, erosion corrosion is of particular interest. Extremely high corrosion rates occur very often on materials due to abrasion under exposure to corrosion, such as takes place with solids in the form of phosphates.

Figures 18–21 and Tables 15, 16, 34, 41 contain an overview of erosion corrosion studies for materials which could be used under such conditions.

The basic requirement for a high resistance of high-alloy steels to erosion under exposure to phosphoric acid is a molybdenum content [98].

As Figures 19 and 20 demonstrate, the steel Uranus® B 6 (Table 12) shows a quite attractive erosion corrosion behavior (also abrasion corrosion behavior) in phosphate suspensions (Table 11), but is inferior to the high-alloy Cr steel containing only 1.95% Mo (Table 12) [67].

The Uranus® B 6 corresponds to 1.4539 (X1NiCrMoCu25-20-5).

Figures 21 and 22 summarize an overview of the wear rates to be expected under various conditions during comminution of phosphate rock (Table 15). The CrNiMoCu steel Illium® P (Table 13) shows a balanced corrosion behavior in respect of varying corrosive conditions even under erosive conditions [18].

These summaries are quite a good basis for preselecting materials for use as the grinding material under such and similar conditions.

Laboratory studies [113] on steel HV-9A (19 Cr, 25 Ni, 5 Mo, 2 Cu, 0.08 C, balance Fe) at 363 K (90 °C) under the conditions of the premixer in the Nissan process with varying crude phosphates in respect of particle size and chemical composition showed erosion corrosion rates of 0.3–10 mm/a (11.2–394 mpy). In the vapor phase above it, the rates determined were between 0.3 and 1.1 mm/a (11.2–43.3 mpy). The high corrosion rates were found using a phosphate rock with 32.5 % P_2O_5, 1.26 % SO_3, 3.4 % F^- and 900 ppm Cl^-, and a solids content of 2.6 %. Additions of kieselguhr to the aqueous suspension considerably reduced the erosion corrosion. The corrosion rates at an addition of 6 % dropped to 0.13 mm/a (5.12 mpy), while in the vapor phase they rose to 0.31 mm/a (12.2 mpy). At a smaller addition, they are even above this figure, but do not exceed 0.5 mm/a (19.7 mpy) at 2 % kieselguhr addition. The illustrations show that preselection of materials in accordance with Figures 21 and 22 is appropriate for a concrete case. The abrasion corrosion resistance of CrNiMoCu steels proves to be significantly higher than that of the lower-alloy materials. As already pointed out above, the resistance increases as the molybdenum content increases.

Without reporting corrosion rates, it is stated in [80] that in a phosphoric acid with (g/l) 400 H_3PO_4, 30 H_2SO_4, 25 HF and H_2SiF_6, 1 HCl and 630 g of gypsum, 12 g of phosphate and 8 g of Fluosil at 355 K (82 °C) and under a flow rate of more than 1 m/s, the resistance of materials under abrasive conditions increases in the following sequence:

1.4401 (X5CrNiMo17-12-2, cf. SAE 316)
1.4436 (X3CrNiMo17-13-3, cf. SAE 316)
1.4539 (X1NiCrMoCu25-20-5, SAE 904 L)
2.4641 (NiCr21Mo6Cu, UNS N08042)
2.4819 (NiMo16Cr15W, UNS N10276).

Similar conclusions are confirmed by studies in a filter acid with 30.5 % P_2O_5 (42.1 % H_3PO_4), 3.7 % H_2SO_4, 0.75 % F^-, 1000 ppm Cl^-, 0.3 % SiO_2, 0.11 % Al_2O_3 and 0.04 % Fe_2O_3.

The corrosion rates at 353 K (80 °C) deduced from electrochemical measurements decrease significantly, and in particular by a factor of 10 and 50 respectively, under abrasive conditions in the sequence 18 % Cr, 10 % Ni, 2.5 % Mo, 20 % Cr. 25 % Ni, 5 % Mo, nickel alloys (without information on the composition) [80].

Orientation experiments under the particular erosion corrosion conditions are to be recommended in all cases.

Special iron–based alloys

Although other materials are described below under this section, it should be pointed out here that magnetic materials under the name Koerzit® (17–28 % Ni, 12 % Al, 5 % Co, 3–6 % Cu, balance Fe, cf. 1.3728) cannot be used in phosphoric acid [156].

Steels which, in addition to having a high resistance to pitting corrosion and crevice corrosion, are also distinguished by high strength values have been developed for use in chemical process technology. These steels, for example Remanit® 4565 S (Table 42) are characterized by a very high nitrogen content of more than 0.4 %, and by manganese contents of more than 6 % or, in the case of 1.4565, just below 6 % [81, 128].

As Figures 38–40 show, steel 1.4565 (UNS S34565) is distinguished by relatively slowly increasing corrosion rates in contaminated phosphoric acid as the chloride and fluoride content increase [81].

The 0.2 % tensile yield strength of this steel is 420 N/mm^2, and therefore in the region of that of the duplex steel 1.4462 (450 N/mm^2, 2205, X2CrNiMoN22-5-3) [81], which is of particular interest for apparatus dimensions. In comparison, these values are 295 N/mm^2 for 1.4429 (SAE 316 LN, X2CrNiMoN17-13-3) and 195 N/mm^2 for 1.4438 (SAE 317, X2CrNiMo18-15-4) [81].

By the alloying change in the steel Remanit® 4565 S to the very high nitrogen content of 0.43 % and to the reduced Mo content of 4.28 %, the high resistance to pitting corrosion and crevice corrosion, also in the welded state, is largely retained in comparison with the 6 % Mo steels. The addition of copper (0.63 % in Remanit® 4565 S + Cu) is said to intensify the favorable effect of this element on the corrosion resistance of these steels (in this context, see under Section Austenitic chromium-nickel steels with special alloying additions) still further.

Figures 66 and 67 show the results of corrosion studies on these steels. The studies were carried out under conditions which are close to those during concentration of industrial phosphoric acid.

As expected, the material consumption rates in the chloride-free industrial phosphoric acid rise as the fluoride concentration (as hydrochloric acid) increases. Although sometimes considerable material consumption rates exist in Figure 66, it is assumed that the steels were passive, since more severe corrosive attack takes place in the active state (compare Figure 74) [128].

An increase in chromium content in high-alloy steels also leads here to an increase in resistance in chloride-free hot industrial phosphoric acid, as can be seen from Figure 67. The influence of the alloy composition changes when chlorides become decisive as contamination in the phosphoric acid. Chlorides do not increase corrosion in the passive state, like fluorides, but activate material, as can be seen from Figure 74. Vigorous corrosion is to be expected when a critical chloride concentration which prevents repassivation of the material under the given conditions is exceeded (in this context, see also Figure 68). According to Figure 74, activation occurred under practically the same conditions in one test.

Under the conditions mentioned in Figure 68, the steel Remanit® 4565 S + Cu (X 2 CrNiMnMoCuN 24 17 6 4) showed the highest resistance to chloride in industrial phosphoric acid [128]. The high pitting resistance equivalent (PRE) corresponding to the alloy composition (see Table 42) in practice leads to a relatively low passive current density with associated low corrosion rates being established, and thus to a high resistance of the steel in contaminated phosphoric acid.

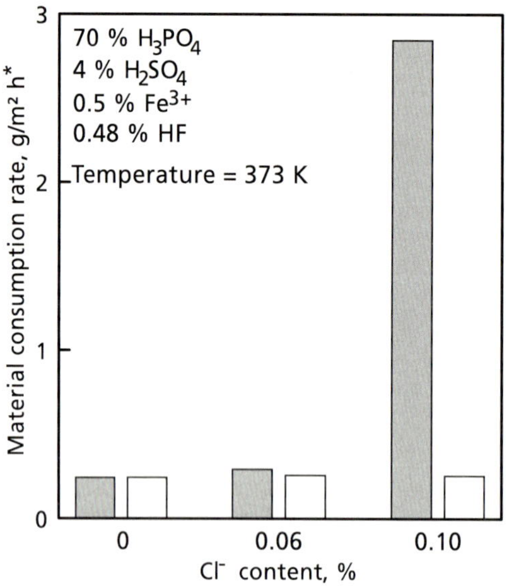

Figure 74: Influence of the chloride content on the corrosion behavior of Remanit® 4565S + Cu in industrial phosphoric acid (results of in each case 2 tests of 72 h test duration) [128]
* at a density of 7.85, 1 g/m^2 h corresponds to 1.1 mm/a (43.3 mpy)

Corrosion studies on the pipe materials of steam generators in pressurized water reactors under a pressure of 17 MPa showed a corrosion rate of 0.12 mm/a (4.72 mpy) on the material 1.4558 (X2NiCrAlTi32-20) in an aqueous solution (flushed to a low oxygen content with nitrogen) of 0.42 mol/l H_3PO_4 at a temperature of 623 K (350 °C) after a test duration of 3,000 h. The pH of this solution was 1.4 at the start of the studies and 4.1 at the end. The nickel alloy Inconel® 600 (2.4816) shows a better resistance under these conditions, with a corrosion rate of 0.09 mm/a (3.54 mpy). No intercrystalline stress corrosion cracking occurred on 1.4558 under these conditions [157].

Amorphous iron alloys cannot be used in 87 % phosphoric acid, with material consumption rates of above 10 g/m^2 h (more than 10 mm/a (394 mpy)), as can be seen in [158].

Bibliography

[1] Hartlapp, G.; Schrödter, K.; Haas, H. et al.
in: Ullmanns Encyklopädie der technischen Chemie
(Ullmann's encyclopedia of industrial chemistry) (in German), 4th ed., vol. 18, p. 301
Verlag Chemie GmbH, Weinheim–Deerfield Beach, Florida–Basel, 1979

[2] DECHEMA–WERKSTOFF-TABELLE "Phosphorsäure"
(DECHEMA–WERKSTOFF-TABELLE "Phosphoric Acid") (in German)
DECHEMA e. V., Frankfurt am Main, December 1966

[3] Harnisch, H.; Heymer, G.; Klose, W.; Schrödter, K.
in: Winnacker–Küchler "Chemische Technologie", Anorganische Technologie
(Chemical technology–inorganic technology I) (in German), 4th ed., vol. 2, p. 204
Carl Hanser Verlag, München–Wien, 1982

[4] Hommel, G.
Handbuch der gefährlichen Güter, Merkblatt 160 "Phosphorsäuren"
(Handbook of dangerous goods, sheet 160 "Phosphoric Acid") (in German)
Springer–Verlag, Berlin–Heidelberg, 1987

[5] Rabald, E.
Corrosion Guide, 2nd revised edition, p. 556
Elsevier Publishing Company, Amsterdam–London–New York, 1968

[6] McDowell, D. W.
Handling phosphoric acids and phosphate fertilizers–II
Chem. Eng. 82 (1975) 18, p. 121

[7] Itou, K.; Suzuki, H.; Iida, T.; Yamada, T.
Corrosion behavior of metals–Copper group and platinum group–in condensed phosphoric acid at higher temperature (in Japanese)
Nagoya Kogyo Daigaku Gakohu 25 (1973) 20, p. 377

[8] Anonymous
Corrosion Data Survey, Metals Section, 6th ed., p. 95
NACE, Houston (Texas/USA), 1985

[9] Anonymous
Bisphenol–Polyester–Harze für die Herstellung von korrosionsbeständigen GfK-Konstruktionen
(Polyester bisphenol resins for the production of corrosion-resistant GRP-constructions) (in German)
Kunststoffberater 24 (1979) 1/2, p. 20

[10] Anonymous
Le resine poliestere bisfenoliche sono ora in grado di soddisfare tutte le necessità per la costruzione di apparecchiature industriali resistenti alla corrosione
(Polyester bisphenol resins are now in a position to meet any requirement in corrosion resistant technical equipment) (in Italian)
Ingegneria chim. 27 (1978) 2, p. 1

[11] Burbridge, J. F.
Corrosion–resistant reinforced polyester for process plant
Anticorros. Methods Mater. 23 (1976) 2, p. 7

[12] Fabian, R. J.; Vaccari, J. A.
How materials stand up to corrosion and chemical attack
Mat. Eng. 73 (1971) 2, p. 36

[13] Gladkij, I. N.; Zajetz, I. L.; Davydenko, N. M.; Demjanenko, V. D.; Serdjukov, B. N.
Effect of diffusional strengthening of pig iron on its stability in concentrated phosphoric acid (in Russian)
Zashch. Met. 13 (1977) 5, p. 581

[14] Katz, W.
Gießen für die Chemie. Phosphor und Phosphorsäure
(Casting in chemistry. Phosphorus and phosphoric acid) (in German), p. 58
VDI–Verlag GmbH, Düsseldorf, 1976

[15] McDowell jr., D. W.
Handling phosphoric acid and phosphate fertilizers – I
Chem. Eng. 82 (1975) 4, p. 119

[16] Zhivotovskaya, G. P.; Chekrygina, L. M.; Zhivotovskij, E. A.; Vasin, Yu. P.
Steel corrosion in aluminium – chrome – phosphate mixture (in Russian)
Zashch. Met. 16 (1980) 3, p. 338

[17] Khvostov, V. P.; Anoshchenko, I. P.
Stainless steel corrosion in phosphoric acid (in Russian)
Zashch. Met. 16 (1980) 3, p. 310

[18] Singleton, D. J.; Blickensderfer, R.
Wear and corrosion of 12 alloys during laboratory milling of phosphate rock in phosphoric acid waste water, p. 16
U.S. Bureau of Mines, Pittsburgh (PA./USA), 1984

[19] Schillmoller, C. M.
Alloy selection in wet process phosphoric acid plants
in: Process Industries Corrosion – The theory and practice, p. 161
NACE, Houston (Texas/USA), 1986

[20] Zorin, A. A.; Kasparova, O. V.; Khokhlov, N. I.
Investigation of corrosion and electrochemical behavior of borized steel (in Russian)
Zashch. Met. 25 (1989) 3, p. 390

[21] Benyaich, A.; Roche, M.; Pagetti, J.; Troquet, M.
Inhibition of Armco® iron corrosion by triphenylbenzylphosphonium bromide in phosphoric acid
Matér. Tech. 76 (1988) 11/12, p. 35

[22] Smith, D. J.; Van der Schijff, O. J.
Corrosion of galvanised steel and carbon steel in deaerated aqueous solutions of industrial fertilizer chemicals
Br. Corros. J. 24 (1989) 3, p. 189

[23] Product Information
Corrosion tables for stainless steels and titanium
Jernkontoret Stockholm (Sweden), 1979, p. 20

[24] Volna, V. F.; Karatsyuba, V. V.; Medyanik, A. A.; Trofimov, Yu. M.
The corrosivity of phosphoric acid (in Russian)
Tr. Leningrad. Proektn. Inst. Osnov. Khim. Prom-st. 27 (1977) p. 97

[25] Mathur, P. B.; Vasudevan, T.
Reaction rate studies for the corrosion of metals in acids – I. Iron in mineral acids
Corrosion 38 (1982) 3, p. 171

[26] Rabald, E.
Werkstoffe, physikalische Eigenschaften und Korrosion
(Materials, physical properties and corrosion) (in German), vol. 1, p. 419
Verlag von Otto Spamer, Leipzig, 1931

[27] Silverman, D. C.; Carrico, J. E.
Electrochemical impedance technique – a practical tool for corrosion prediction
Corrosion 44 (1988) 5, p. 280

[28] Orjéla, G; Boden, P. J.
Monitoring of varying corrosion rate systems: Fe–Cr alloy in H_3PO_4 solutions
Br. Corros. J. 16 (1981) 4, p. 212

[29] Riggs jr., O. L.
Cathodic protection of carbon steel in phosphoric acids
Mater. Performance 9 (1970) 10, p. 21

[30] Sastry, T. P.; Rao, V. V.; Hariveer, A.
Anodic protection of mild steel in phosphoric acid
International Congress on Metallic Corrosion (Proc. Conf.), Toronto (Canada), June 1984, vol. 1, p. 311
National Research Council of Canada, Ottawa (Canada)

[31] Ruff, C. W.
Reinigung und Passivierung von Stahlteilen mittels einer Beize auf Phosphorsäure–Basis
(Cleaning and passivation of steel by a pickle based upon phosphoric acid) (in German)
Galvanotechnik 61 (1970) 12, p. 1019

[32] Anonymous
Das Beizen vor der Direktweißemaillierung
(Pickling before the direct porcelain enamelling) (in German)
GEKT 23 (1972) 5, p. 200

[33] Kuron, D.; Gräfen, H.; Rother, H.-J.
Anwendung von Beizinhibitoren
(Application of pickling inhibitors) (in German)
Werkst. Korros. 37 (1986) 5, p. 223

[34] Santhianandhan, B.; Balakrishnan, K.; Subramanyan, N.
Triazoles as inhibitors of corrosion of mild steel in acids
Br. Corros. J. 5 (1970) 5, p. 270

[35] Metha, G. N.
Tafel intercept and linear polarisation resistance techniques to study acid corrosion of mild steel and effect of thiourea
Bull. Electrochem. 4 (1988) 2, p. 111

[36] Machu, W.; Fouad, M. G.
Über die galvanische Korrosion von Gußeisen und Messing Ms 60 in Säuren bei Ab– und Anwesenheit von Inhibitoren II
(On the galvanic corrosion of cast iron and brass MS 60 in acid during absence and presence of inhibitors II) (in German)
Werkst. Korros. 9 (1958) 11, p. 699

[37] Ried, G.
Übersicht über Werkstoffe für Absperr-Organe, die bei angreifenden Medien zu empfehlen sind
(Survey of materials for shut-off valves recommended for corrosive media) (in German)
Werkst. Korros. 15 (1964) 6, p. 468

[38] Machu, W.; Fouad, M. G.
Über das Verhalten von Gußeisen in Säuren und Säuregemischen bei Ab- und Anwesenheit von Inhibitoren
(On the behavior of cast iron in acids and acid mixtures during absence and presence of inhibitors) (in German)
Werkst. Korros. 9 (1958) 6, p. 369

[39] Product Information
Du Pont Information Service, Corrosion Resistance Charts, April 1975, p. 296
Du Pont de Nemours International S.A., CH-1211 Geneva

[40] Berg, F. F.
Korrosionsschaubilder
(Corrosion Diagrams) (in German), 2nd ed., p. 2
VDI-Verlag GmbH, Düsseldorf, 1969

[41] Swandby, R. K.
Corrosion charts: Guides to materials selection
Chem. Eng. (New York) 69 (1962) November, p. 186

[42] Product Information
Korrosionbeständigkeit nickelhaltiger Werkstoffe gegenüber Phosphorsäure und Phosphaten
(Corrosion resistance of nickel containing materials against phosphoric acid and phosphates) (in German), 1st ed., No. 61, March 1970, p. 1
International Nickel Deutschland GmbH, Düsseldorf

[43] Product Information
Corrosion resistance of nickel-containing alloys in phosphoric acid, p. 1
The International Nickel Company, Inc., New York (USA)

[44] Spence, T. C.; George, G. W.
Material selection for corrosion-resistant centrifugal pumps
World Pumps (1981) 182, p. 531

[45] Product Information
Gußwerkstoffe
(Materials for casting (in German), Technical Information 3, No. 5.50.0042-0383)
Rheinhütte, Wiesbaden

[46] Zvigintsev, N. V.; Berezovskaya, V. V.; Khadyev, M. S.; Rudychev, A. S.
Structure, mechanical properties and cavitation – corrosion resistance of 03Kh 10N5K5M3DTYuS steel (in Russian)
Fiz. Met. Metalloved. 62 (1986) 5, p. 1014

[47] Truman, J. E.
Corrosion resistance of 13 % chromium steels as influenced by tempering treatments
Br. Corros. J. 11 (1976) 2, p. 92

[48] Lewis, G.; Fox, P. G.; Boden, P. J.
The corrosion of Fe-12Cr iron-chromium alloys in o-phosphoric acid
Corros. Sci. 20 (1980) 3, p. 331

[49] Anonymous
Corrosion of stainless steels
Engineering 218 (1978) 11, p. 1207

[50] Arlt, N.; Fleischer, H.-J.; Gebel, W.; Grundmann, R.; Gümpel, P.
Stand und Entwicklungstendenzen auf dem Gebiet der nichtrostenden Stähle
(Status and trends of development in the field of stainless steels) (in German)
Thyssen Edelst. Tech. Ber. 15 (1989) 1, p. 1

[51] Colombier, L.
Molybdän in rost- und säurebeständigen Stählen und Legierungen
(Molybdenum in stainless and acid-resistant steels and alloys) (in German), Information Bulletin, p. 77
Climax Molybdenum GmbH, Düsseldorf

[52] Product Information
Nyby ELI-T-Stähle mit Eigenschaften von besonderem Wert und Gewicht
(Nyby ELI-T-Stähle – high in value and significance) (in German)
Gränges Nyby AB, Torshälla (Sweden)

[53] Product Information
Chemische Beständigkeit nichtrostender Böhler-Edelstähle
(Chemical resistance of stainless special steels from Böhler) (in German), p.12
Böhler Gesellschaft M. B. H., A-8605 Kapfenberg

[54] Product Information
Remy Edelstähle
(Remy special steels) (in German)
Publication for ACHEMA 1991, p. 28
Hagener Gußstahlwerke GmbH, Hagen

[55] Product Information
Chemische Beständigkeit der nichtrostenden Remanit®-Stähle
(Chemical resistance of the stainless Remanit® steels) (in German), December 1985
Thyssen Edelstahlwerke AG, Krefeld

[56] Product Information
Korrosionsbeständiger Edelstahl und Edelstahlguß
(Corrosion resistant special steel and special steel casting) (in German), December 1983
Schmidt + Clemens Edelstahlwerk, Lindlar

[57] Product Information
Ein neuer rostfreier Stahl mit Zukunft 18Cr-2Mo
(A new stainless steel with prospects 18Cr-2Mo) (in German)
Climax Molybdenum GmbH, Düsseldorf

[58] Product Information
An 18Cr-2Mo ferritic stainless steel
Climax Molybdenum GmbH, Düsseldorf

[59] Product Information
Korrosionsbeständiger Automatenstahl 1802
(Corrosion resistant free-machining steel 1802) (in German)
Draht 33 (1982) 4, p. 188

[60] Brandis, H.; Kiesheyer, H.; Lennartz, G.
Einfluß von Titanzusätzen auf die Korrosionsbeständigkeit nichtrostender Automatenstähle
(Influence of titanium additions on the corrosion resistance of stainless free-machining steels) (in German)
Thyssen Edelstahl Tech. Ber. 8 (1982) 2, p. 135

[61] Heimann, W.; Kiesheyer, H.
Ein weichmagnetischer Stahl für höchste korrosionschemische Beanspruchung: X 1 CrNiMoNb 28 4 2 (Remanit® 4575)
(A soft magnetic steel for highest corrosion-chemical stress: X 1 CrNiMoNb 28 4 2 (Remanit® 4575)) (in German)
TEW Tech. Ber. 3 (1977) 1, p. 23

[62] Oppenheim, R.; Lennartz, G.
Eigenschaften eines ferritischen 28-2 Chrom-Molybdän-Stahles in Superferrit-Güte
(Properties of a ferritc 28-2 chromium-molybdenum steel in superferrite quality) (in German)
Chem. Ind. 23 (1971) 10, p. 705

[63] Anonymous
X 1 CrNiMoNb 28 42 (Remanit® 4575), Korrosionsbeständigkeit
(X 1 CrNiMoNb 28 42 (Remanit® 4575), corrosion resistance) (in German)
Thyssen Edelstahl Tech. Ber. 5 (1979) 1, p. 48

[64] Gräfen, H.; v. Baeckmann, W. G.; Föhl, J.; Herbsleb, G.; Huppatz, W.; Kuron, D.; Rother, H.-J.; Rüdinger, K.
Die Praxis des Korrosionsschutzes. Kontakt & Studium
(Corrosion protection in practice; dealings with and study of) (in German), vol. 64
expert verlag, Grafenau, 1981

[65] Anonymous
Korrosion und Korrosionsschutz
(Corrosion and corrosion protection) (in German), 1st ed.,
DIN pocket book 219
Beuth Verlag GmbH, Berlin, 1987

[66] Fischer, W.
Korrosionsschutz durch Information und Normung. Kommentar zum DIN-Taschenbuch 219
(Corrosion protection by information and standardization. Comment to DIN pocket book 219) (in German)
Verlag Irene Kuron, Bonn, 1988

[67] Jallouli, E. M.; Fikrat, M.
Investigation on abrasive corrosion of some stainless steel alloys in industrial phosphoric media
Bull. Cercle Etud. Métaux 15 (1988), p. 18.7.

[68] Kalinichenko, V. A.; Dobrolyubov, V. V.; Grishina, V. A.; Porshneva, A. I.; Kuznetzov, V. I.; Shvetz, V. A.
Corrosion – electrochemical properties of steel Cr28MoV-TiAl in wet process phosphoric acid (in Russian)
Zashch. Met. 24 (1988) 4, p. 621

[69] Kratzer, A.; Heumann, A.; Wittekindt, W.
Tauchmotorpumpen aus korrosions- und verschleißbeständigen Werkstoffen
(Submersible motor pumps made of corrosion and wear-resistant materials) (in German)
Chemie–Technik 13 (1984) 12, p. 56

[70] Product Information
Sandvik nichtrostende Vielzweckstähle für aggressive Medien – SAF 2205, 3RE60, Sanicro® 28, 2RK65
(Sandvik stainless multi-purpose steels for aggressive media – SAF 2205, 3RE60, Sanicro® 28, 2RK65) (in German), March 1985
Sandvik GmbH, Düsseldorf

[71] Katz, W.
ACHEMA '79 – Teil 2: Kurzer Uberblick über Werkstoffe und Werkstoffprobleme
(ACHEMA '79 – Part 2: A short survey of materials and material problems) (in German)
Werkst. Korros. 30 (1979) 9, p. 651

[72] Audouard, J.-P.; Catelin, D.
Quelques cas industriels de corrosion sévères en milieux chlorurés. Solutions proposées
(Some industrial cases of severe corrosion in chlorine containing media. Proposed solutions) (in French)
Matér. Tech. 75 (1987) 7/8, p. 291

[73] Asphahani, A. I.
Corrosion resistance of high performance alloys
Mater. Performance 19 (1980) 12, p. 33

[74] Bock, H. E.
Einfluß von Legierungselementen und Gefüge auf das Korrosionsverhalten nichtrostender Stähle
(The influence of alloy elements and structure on the corrosion behavior of stainless steels) (in German)
VDI Berichte (1986) 600.2, p. 55

[75] Hummel, O. H.
Werkstoffe im Behälter- und Apparatebau
(Materials in vessel construction and apparatus engineering) (in German)
Chemie-Technik 11 (1982) 8, p. 930

[76] Product Information
Haynes® corrosion-resistant alloys. Ferralium® alloy 255, 1986
Cabot Corporation, Kokomo (Ind./USA)

[77] Kohl, H.; Hochörtler, K.; Kriszt, K.; Koren, M.
Sonderstähle für den Apparatebau mit sehr guter Korrosionsbeständigkeit und erhöhter Festigkeit
(Special steels with superior corrosion resistance and strength for chemical equipment manufacture) (in German)
Werkst. Korros. 34 (1983) 1, p. 1

[78] Product Information
Fermanel® alloy, August 1986
Deutsche Langley Alloys GmbH, Frankfurt am Main

[79] Rockel, M. B.
Hochlegierter austenitischer Stahl. Cronifer® 1925 hMo für Anwendungen in der chemischen Industrie
(High-alloyed austenitic steel. Cronifer® 1925 hMo for applications in chemical industry) (in German)
Chemie-Technik 14 (1985) 10, p. 108

[80] Product Information
Korrosionsbeständigkeit nichtrostender Stähle und Nickellegierungen in Phosphorsäure
(Corrosion resistance of stainless steels and nickel alloys in phosphoric acid) (in German), 1988
Mannesmann Edelstahlrohr GmbH, Langenfeld; Vereinigte Deutsche Metallwerke AG, Duisburg

[81] Grundmann, R.; Gümpel, P.; Michel, E.
Betrachtungen über die Einsatzmöglichkeiten eines ferritisch-austenitischen und eines hochfesten austenitischen Stahles im Chemikalientankerbau
(Study on the use of a ferritic-austenitic and a high-tensile austenitic steel for the construction of chemical tank vessels) (in German)
Thyssen Edelstahl Tech. Ber. 14 (1988) 1, p. 49

[82] Product Information
Ferralium® alloy 255. Hochfester korrosionsbeständiger Edelstahl
(Ferralium® alloy 255. High-tensile corrosion resistant special steel) (in German), April 1987
Deutsche Langley Alloys GmbH, Frankfurt am Main

[83] Product Information
Ferralium® alloy 255-3SC. Cast duplex stainless steel, April 1987
Langley Alloys Ltd., Slough (GB)

[84] Product Information
Ferralium® alloy 255-3SF. Wrought duplex stainless steel
Langley Alloys Ltd., Slough (GB)

[85] Schmidtmann, E.; Hesse, H.-D.
Untersuchungen zum Korrosionsverhalten nichtrostender Stähle in Phosphorsäure in Abhängigkeit von dem Chlorid- und Fluoridgehalt und einer Wärmebehandlung beim Schweißen
(Investigations into the corrosion behavior of stainless steels in phosphoric acid in dependence on the chloride and fluoride contents and a heat-treatment during welding) (in German)
Werkst. Korros. 32 (1981) 12, p. 521

[86] Schmidtmann, E.
Der Einfluß von schweißsimulierenden Wärmebehandlungen und der Chlorid- und Fluoridionengehalte auf das Korrosionsverhalten nichtrostender Stähle in Phosphorsäure
(The influence of sweat-simulating heat-treatments and chloride- and fluoride ion contents on the corrosion behavior of stainless steels in phosphoric acid) (in German)
Werkst. Korros. 31 (1980) 8, p. 633

[87] Product Information
Sandvik® SAF 2205 W.-Nr. 1.4462; X 2 CrNiMoN 22 5. Ferritisch-austenitischer nichtrostender Stahl
(Sandvik® SAF 2205 Mat. No. 1.4462; X 2 CrNiMoN 22 5. Ferritic-austenitic stainless steel) (in German), November 1982
Sandvik GmbH, Düsseldorf

[88] Product Information
Avesta. Rost- und hitzebeständige Stähle. Avesta® 2205, W.-Nr. 1.4462
(Avesta. Stainless and heat-resistant steels. Avesta® 2205, Mat. No. 1.4462) (in German)
Avesta GmbH, Düsseldorf

[89] Product Information
Ferralium® alloy 255-3SC. Cast duplex stainless steel, April 1987
Langley Alloys Ltd., Slough (GB)

[90] Product Information
Ferralium® alloy 255–3SF. Wrought duplex stainless steel
Langley Alloys Ltd., Slough (GB)

[91] Product Information
Corrosion Guide. Langley corrosion resistant alloys
Langley Alloys Ltd., Slough (GB)

[92] Davison, R. M.; Redmond, J. D.
Practical guide to using duplex stainless steels
Mater. Performance 29 (1990) 1, p. 57 (Materials Selection & Design)

[93] Dudukina, T. A.; Ogneva, V. K.; Moskvitchev, I. F.; Dobrolyubov, V. V.; Fazylova, S. P.; Ivolgina, N. Yu.
Steels and alloys corrosion stability in extractional phosphoric acid with nitric acid and carbamide admixtures (in Russian)
Khim. Ind., Moskau (1987) 7, p. 420

[94] Kiselev, V. D.; Dobrolyubov, V. V.
Corrosion stability of stainless steels in the systems H_3PO_4 – H_2SO_4-HF (in Russian)
Khim. Ind., Moskau (1979) 3, p. 160

[95] Umemura, F.; Kawamoto, T.
Corrosion of stainless steel in crude phosphoric acid (in Japanese)
Boshoku Gijutsu 31 (1982) 4, p. 275

[96] Product Information
Wiggin Nickellegierungen in der chemischen Verfahrenstechnik
(Wiggin nickel alloys in chemical process engineering) (in German), November 1976, p. 27
Henry Wiggin & Company Limited, Hereford (GB)

[97] Audouard, J.-P.; Vallier, G.
Problems of abrasion corrosion in industrial phosphoric media (in French)
Métaux-Corros.-Ind. (1980) Febr., Nr. 654, p. 42

[98] Guenbour, A.; Faucheu, J.; Ben Bachir, A.; Dabosi, F.; Bui, N.
Electrochemical study of corrosion-abrasion of stainless steels in phosphoric acids
Br. Corros. J. 23 (1988) 4, p. 234

[99] Desolneux, J. P.
Korrosionsprobleme in chlorhaltigen Medien: Lösungsmöglichkeiten durch einige nichtrostende Spezialstähle
(Corrosion problems in chlorid containing media: possible solution by some stainless special steels) (in German)
Werkst. Korros. 28 (1977) 5, p. 325

[100] Charles, J.; Catelin, D.; Dupoiron, F.
Choix de matériaux pour application en milieux extrêmes: aciers inox et alliages en tôles massives ou plaquées
(Materials choice for extreme environment: Stainless steels and alloys in plain or plated sheets) (in French)
Matér. Tech. 75 (1987) 7/8, p. 309

[101] Anonymous
Corrosion of stainless steels
Engineering 218 (1978) 11, p. 1207

[102] Product Information
ABC der Stahlkorrosion
(ABC of steel corrosion) (in German), 2nd ed., 1966, p. 6
Mannesmann AG, Düsseldorf

[103] Anonymous
Verschleißfest und abriebbeständig: LP-Legierungen Tribaloy® für Gußteile und zum Beschichten
(Wear and abrasion resisting: LP alloys Tribaloy® for cast parts and for coating) (in German)
Werkstoffe und ihre Veredlung I (1979) 2, p. 24

[104] Khvostov, V. P.; Anoshchenko, I. P.
Stainless steel corrosion in phosphoric acid (in Russian)
Zashch. Met. 16 (1980) 3, p. 310

[105] Alon, A.; Yahalom, J.; Schorr, M.
Influence of halides in phosphoric acid on the corrosion behavior of stainless steels
Corrosion 31 (1975) 9, p. 315

[106] Nassif, N.
Influence of impurities in phosphoric acid on the corrosion resistance of some commercially produced alloys
Surf. Tech. 26 (1985) 3, p. 189

[107] Badran, M. M.
Pitting corrosion of 304 stainless steel in phosphoric acid solutions
Corros. Prev. Control 34 (1987) 4, p. 97

[108] Saleh, R. M.; Badran, M. M.; El Hosary, A. A.; El Dahan, H. A.
Corrosion inhibition of 304 stainless steel in H_3PO_4-Cl^- solutions by chromium, molybdenum, nitrogen, tungsten, and boron anions
Br. Corros. J. 23 (1988) 2, p. 105

[109] Asphahani, A. I.
Effect of acids on the stress corrosion cracking of stainless materials in dilute chloride solutions
Mater. Performance 19 (1980) 11, p. 9

[110] Kalinichenko, V. A.; Dobrolyubov, V. V.; Grishina, V. A.; Krivonos, V. N.; Abramova, L. V.
Cr-Ni-Mo-foundry steels corrosion and electrochemical behavior (in Russian)
Khim. Ind., Moskau (1988) 4, p. 224

[111] Bock, H. E.
Einfluß von Legierungselementen und Gefüge auf das Korrosionsverhalten nichtrostender Stähle
(The influence of alloy elements and structure on the corrosion behavior of stainless steels) (in German)
VDI Berichte (1986) 600.2, p. 55

[112] Schorr, M.; Weintraub, E.; Andrasi, D.; Finkelstein, N. P.
Erosion-corrosion in wet process phosphoric acid (WPA) production
International Congress on Metallic Corrosion (Proc. Conf.), Mainz, September 1981, vol. 2, p. 1384
DECHEMA e.V., Frankfurt am Main

[113] Verma, K. M.; Ghosh, H.; Pattnaik K. C.; Singh, R. U.
Corrosion – erosion characteristics of rock phosphate in the manufacture of wet process phosphoric acid
Br. Corros. J. 17 (1982) 2, p. 71

[114] Fyfe, D.; Brooks, J. B.
Corrosion and oxidation in sulfuric acid and phosphoric acid manufacture – case histories
Corrosion '79 (Proc. Conf.), Atlanta (Ga./USA), March 1979, p. 120
NACE, Katy (Texas/USA)

[115] Heurling, K.
Stähle für Transport und Lagerung chemischer Produkte
(Steels for transport and storage of chemical products) (in German)
Blech – Rohre – Profile 28 (1981) 10, p. 481

[116] Wallèn, B.; Andersson, I.
Transport of wet-process phosphoric acid. Corrosion studies and experiences, Avesta Stainless Bulletin No. 1, January/March 1980, p. 6
Avesta Jernverks. AB., Avesta (Sweden)

[117] Nordin, S.
Rostfreie Spezialstähle für die chemische Industrie. Teil I: Korrosionseigenschaften und Anwendungsbereiche
(Special stainless steels for the chemical industry. Part I: Corrosion properties and fields of application) (in German)
Chem.-Anlagen Verfahren (1979) 5, p. 67

[118] Product Information
Transport of wet process phosphoric acids containing low amounts of impurities, Avesta Stainless Bulletin No. 1, January/March 1980, p. 19
Avesta Jernverks. AB., Avesta (Sweden)

[119] Guenbour, A.; Faucheu, J.; Ben Bachir, A.
On the mechanism for improved passivation by addition of molybdenum to austenitic stainless steels in o-phosphoric acid
Corrosion 44 (1988) 4, p. 214

[120] Linder, B.
Anodic protection of stainless steel in phosphoric acid containing halide ions
UK Corrosion 1984 (Proc. Conf.), J. Wembley, Middlesex (UK), Nov. 1984, vol. 1 (Papers), p. 73
The Institute of Corrosion, Science and Technology, Birmingham (GB)

[121] Süry, P.
Untersuchungen zum Einfluß der Kaltverformung auf die Korrosionseigenschaften des rostfreien Stahles X 2 CrNi–Mo 18 12
(Studies on the effect of cold deformation on corrosion properties of the stainless steel X 2 CrNiMo 18 12) (in German)
Material und Technik 8 (1980) 4, p. 163

[122] Product Information
Transport of wet process phosphoric acids containing low amounts of impurities, Avesta Stainless Bulletin No. 1, January/March 1980, p. 19
Avesta Jernverks. AB., Avesta (Sweden)

[123] Product Information
Corrosion resistance of Hastelloy® alloys, 1984, p. 3
Haynes International, Inc., Kokomo (Ind./USA)

[124] Anonymous
Three barges for superphosphoric acid use over 2,700 metric tons of stainless-clad steel
Nickel Topics 33 (1980) 3, p. 4

[125] Linder, B.
Anodic protection of stainless steel in phosphoric acid containing halide ions
UK Corrosion 1984 (Proc. Conf.), J. Wembley, Middlesex (UK), Nov. 1984, vol. 1 (Papers), p. 73
The Institute of Corrosion, Science and Technology, Birmingham (GB)

[126] Linder, B.
Anodic protection of stainless steel in phosphoric acid containing halide ions
Industrial Corrosion 5 (1987) 3, p. 12

[127] Puderbach, H.; Grosse–Böwing, W.
Analyse von Adsorptionsschichten auf Edelstahlblechen
(Analysis of adsorption layers on stainless steel sheets) (in German)
Fresenius Z Anal Chem (1984) 319, p. 627

[128] Arlt, N.; Gillesen, C.; Kiesheyer, H.; Michel, E.
Untersuchungen zur Korrosionsbeständigkeit hochlegierter austenitischer nichtrostender Stähle in Säuren
(Investigations on the corrosion resistance of high-alloyed austenitic stainless steels in acids) (in German)
Thyssen Edelstahl Tech. Ber. 17 (1991) 1, p. 29

[129] Weber, H. M.; Hönig, A.
Duratherm®-aushärtbare Federwerkstoffe für vielseitige Anwendungen
(Duratherm®-curable spring materials for versatile applications) (in German)
Metall 30 (1976) 11, p. 1041

[130] Schmidtmann, E.
Der Einfluß von schweißsimulierenden Wärmebehandlungen und der Chlorid- und Fluoridionengehalte auf das Korrosionsverhalten nichtrostender Stähle in Phosphorsäure
(The influence of sweat-simulating heat-treatments and chloride- and fluoride ion contents on the corrosion behavior of stainless steels in phosphoric acid) (in German)
Werkst. Korros. 31 (1980) 8, p. 633

[131] Nordin, S.
Rostfreie Spezialstähle für die chemische Industrie. Teil I: Korrosionseigenschaften und Anwendungsbereiche
(Special stainless steels for the chemical industry. Part I: Corrosion properties and fields of application) (in German)
Chem.–Anlagen Verfahren (1979) 5, p. 67

[132] Product Information
Chemische Beständigkeit nichtrostender Böhler–Edelstähle
(Chemical resistance of stainless special steels from Böhler) (in German), p.12
Böhler Gesellschaft M. B. H., A-8605 Kapfenberg

[133] Product Information
Remy Edelstähle
(Remy special steels) (in German)
Publication for ACHEMA 1991, p. 28
Hagener Gußstahlwerke GmbH, Hagen

[134] Carlén, J.-C.; Kvarnbäck, B.
Erfahrungen mit zwei hochlegierten, hochkorrosionsbeständigen Stählen in der chemischen Industrie
(Experiences with two high alloy, high corrosion resistant steels in chemical industry) (in German)
Werkst. Korros. 25 (1974) 9, p. 653

[135] Bekkers, K. et al.
Schweißen und Schneiden ’83. Korrosionsbeständige Schweißzustände
(Welding and cutting ’83. Corrosion-resistant welding conditions) (in German)
Werkst. Korros. 35 (1984) 2, p. 87

[136] Anonymous
Rostfreie Sonderstähle aus Schweden
(Stainless special steels from Sweden) (in German)
Technica 25 (1981) p. 2353

[137] Nordin, S.
Werkstoffe für den Chemieanlagenbau. NU Stainless 904 L
(Materials for the construction of chemical plants: NU stainless 904 L) (in German)
Blech – Rohre – Profile 28 (1981) 11, p. 54

[138] Product Information
VDM Hochleistungswerkstoffe
(VDM high-power materials) (in German)
Vereinigte Deutsche Metallwerke AG, Werdohl

[139] Product Information
Stahl 904 L
(Steel 904 L) (in German)
Uddeholm-Stahlber. 9 (1971) p. 3

[140] Product Information
Avesta rost- und hitzebeständige Stähle. Avesta® 904 L. W.-Nr. 1.4539
(Avesta rust- and heat-resistant steels. Avesta® 904 L. Mat. No. 1.4539) (in German), ed. 1988
Avesta GmbH, Düsseldorf

[141] Product Information
Avesta rost- und hitzebeständige Stähle. Avesta® 254 SMO
(Avesta corrosion- and heat-resistant steels. Avesta® 254 SMO) (in German), ed. 1989
Avesta GmbH, Düsseldorf

[142] Product Information
Sandvik® 2RK65. Austenitischer nichtrostender Stahl für schwere Korrosionsverhältnisse
(Sandvik® 2RK65. Austenitic stainless steel for heavy corrosion conditions)
(in German), ed. August 1987
Sandvik Steel, Sandviken (Sweden)

[143] Product Information
Sandvik Sanicro® 28. Austenitischer nichtrostender Werkstoff für extreme Korrosionsverhältnisse
(Sandvik Sanicro® 28. Austenitic stainless material for extreme corrosion conditions)
(in German), ed. September 1990
Sandvik Steel, Sandviken (Sweden)

[144] Edström, J. O.; Carlén, J. C.; Kämpinge, S.
Anforderungen an Stähle für die chemische Industrie
(Demands on steels for the chemical industry) (in German)
Werkst. Korros. 27 (1970) 10, p. 812

[145] Rockel, M. B.
Einsatz hochlegierter Stähle und Nickelbasislegierungen im chemischen Anlagenbau
(The application of high-alloy steels and nickel-based alloys in chemical plant construction) (in German)
Werkstoffe und ihre Veredelung 4 (1982) 4/5, p. 153

[146] Charles, J.; Catelin, D.; Dupoiron, F.
Choix de matériaux pour application en milieux extrêmes: aciers inox et alliages en tôles massives ou plaquées
(Materials choice for extreme environment: Stainless steels and alloys in plain or plated sheets) (in French)
Matér. Tech. 75 (1987) 7/8, p. 309

[147] El Safty, M.; Dabosi, F.; Bui, N.
Influence of silicon on the sensitisation and corrosion of the alloy 20Cr-25 Ni
International Congress on Metallic Corrosion (Proc. Conf.), Toronto (Canada), June 1984, vol. 2, p. 597
National Research Council of Canada, Ottawa (Canada)

[148] Kalinichenko, V. A.; Dobrolyubov, V. V.; Grishina, V. A.; Krivonos, V. N.; Abramova, L. V.
Cr-Ni-Mo-foundry steels corrosion and electrochemical behavior (in Russian)
Khim. Ind., Moskau (1988) 4, p. 224

[149] Hoffmann, T.; Renner, M.; Rudolph, G.
Gefüge und Eigenschaften von Schweißgut aus hochlegierten korrosionsbeständigen Nickelwerkstoffen
(Structure and properties of welding deposit of high-alloyed corrosion-resistant nickel materials) (in German)
Schweissen Schneiden 38 (1986) 11, p. 551.

[150] Heubner, U.; Rockel, M.; Wallis, E.
Ein neuer hochlegierter Nickel-Chrom-Molybdän-Stahl für den Chemie-Apparatebau
(A new high-alloyed nickel-chromium-molybdenum steel for the chemical process industry) (in German)
Werkst. Korros. 40 (1989) 7, p. 418

[151] Kirchheiner, R.; Schalk, W.; Müller, H.; Palomino, S. M.
Qualification of Nicrofer® 3127 LC for the preconcentration of phosphoric acid
Werkst. Korros. 40 (1989) 9. p. 545

[152] Arlt, N.; Gillesen, C.; Kiesheyer, H.; Michel, E.
Untersuchungen zur Korrosionsbeständigkeit hochlegierter austenitischer nichtrostender Stähle in Säuren
(Investigations on the corrosion resistance of high-alloyed austenitic stainless steels in acids) (in German)
Thyssen Edelstahl Tech. Ber. 17 (1991) 1, p. 29

[153] Heubner, U.; Rockel, M.; Wallis, E.
Ein neuer hochlegierter Nickel-Chrom-Molybdän-Stahl für den Chemie-Apparatebau
(A new high-alloyed nickel-chromium-molybdenum steel for the chemical process industry) (in German)
Werkst. Korros. 40 (1989) 7, p. 418

[154] Anonymous
Sanicro® 28, a versatile molybdenum stainless steel
Molybdenum Mosaic 5 (1982) 3, p. 6

[155] Asphahani, A. I.
Effect of acids on the stress corrosion cracking of stainless materials in dilute chloride solutions
Mater. Performance 19 (1980) 11, p. 9

[156] Schuchert, H.
Magnetwerkstoffe für die Aufbereitungstechnik
(Magnetic materials for the processing technique) (in German)
Chemie-Technik 4 (1975) 9, p. 321

[157] Effertz, P.-H.; Forchhammer, P.
Aus der Schadenforschung. Korrosionsverhalten von Rohrwerkstoffen für DWR-Dampferzeuger in konditioniertem Sekundärwasser bei 350 °C-Kapselprüfungen
(Corrosion patterns in tube materials for DWR-steam generating plants in conditioned secondary water at 350 °C-capsule-tests) (in German)
Der Maschinenschaden 50 (1977) 5, p. 180

[158] Mitsuhashi, A.; Asami, K.; Kawashima, A.; Hashimoto, K.
The corrosion behavior of amorphous nickel base alloys in a hot concentrated phosphoric acid
Corros. Sci. 27 (1987) 9, p. 957

Sulfuric Acid

Unalloyed steels and cast steel

Sulfuric acid is one of the most important chemicals in the chemical industry. Because of the quantities produced, the use of unalloyed steels and cast steel under exposure to sulfuric acid is of particular interest.

Figure 1 shows the corrosion behavior of unalloyed steel in sulfuric acid over the entire concentration range. According to this graph, unalloyed steel corrodes in concentration ranges other than those just mentioned at corrosion rates too high for its use. This graph applies to steels having carbon contents of up to about 0.4 % and the usual accompanying elements, such as, for example, silicon, manganese, phosphorus and sulfur [1]. The corrosive attack on steels having a higher carbon content than 0.4 % is more severe than that shown in Figure 1, and is milder on steels of lower carbon content. Figure 2 shows the corrosion behavior of steels in sulfuric acid as a function of the carbon content.

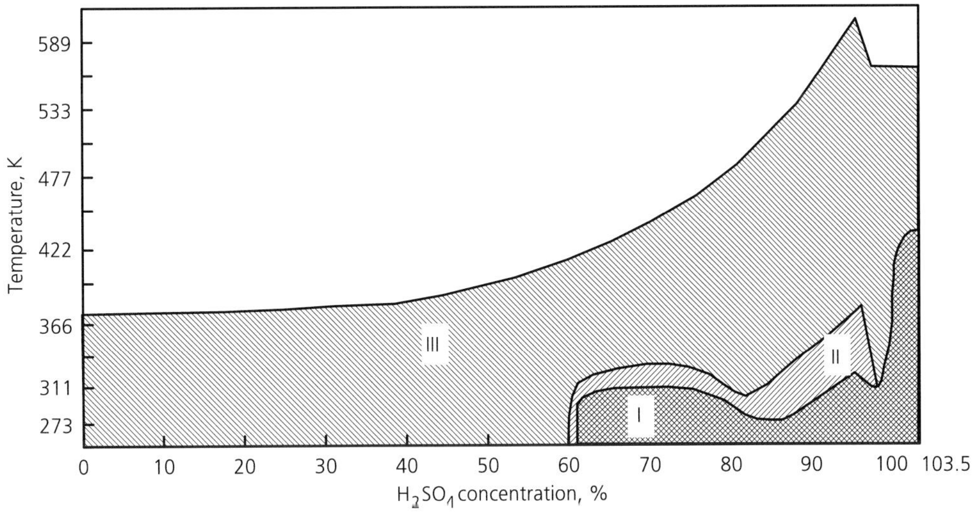

Figure 1: Corrosion rates of unalloyed steel and cast steel containing a carbon content of about 0.4 % and the usual accompanying elements in sulfuric acid solutions (according to G. A. Nelson) [2]
Corrosion rate: I < 0.5 mm/a (< 19.7 mpy)
II 0.5 to 1.25 mm/a (19.7 to 49.2 mpy)
III > 1.25 mm/a (> 49.2 mpy)

A very marked increase in the corrosion rate is associated with the increasing carbon content of the steels. However, at room temperature the corrosion rates at a low carbon content already exceed the limit for possible use. A sharp decrease in corrosion above an acid concentration of 60 % can be seen [3].

Figure 3 shows corrosion rates in static sulfuric acid as a function of the temperature and concentration in the range above 60 %. Acceptable corrosion rates which favor prolonged use of unalloyed cast steel and steel exist only above 95 % H_2SO_4. In

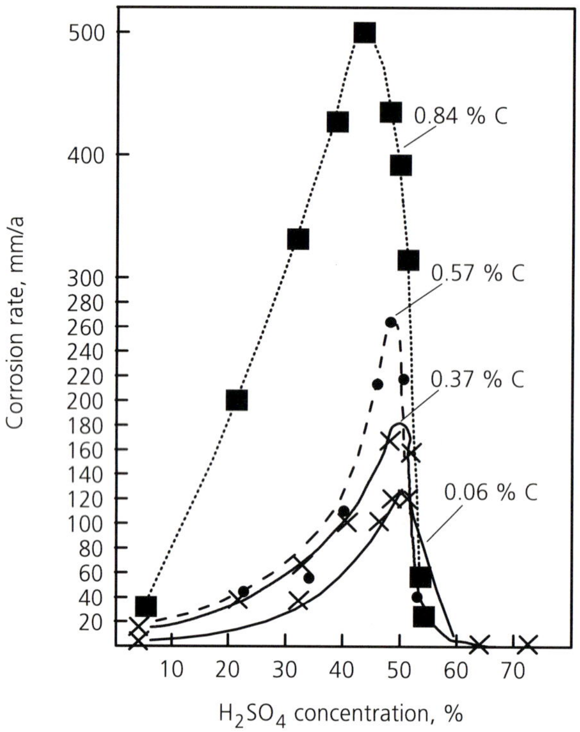

Figure 2: Corrosion of unalloyed steel as a function of the carbon content at 298 K (25 °C) (according to G. H. Damon) [3]

sulfuric acid solutions above a concentration of 60 %, an iron sulfate layer forms on the iron surface. The corrosion rates decrease above 78 % H_2SO_4 as a result of a greatly decreasing solubility of iron sulfate in more highly concentrated sulfuric acid [4]. This decrease in corrosion rates at higher temperatures can also be followed in Figure 3. The corrosion behavior of steel in sulfuric acid solutions is worsened by the presence of chloride ions, since pitting corrosion may occur. At a sulfuric acid concentration of more than 100 %, the corrosion increases drastically in the presence of free SO_3 up to an SO_3 content of 20 %, and then drops again sharply.

Steel is used in concentration ranges of sulfuric acid of between 68 and 100 % at low temperatures for storage and transportation containers. Steel is a preferred material in plants for the contact process at concentrations of between 93 and 98.5 % H_2SO_4. It can be used in oleum only from a concentration of 23 % free SO_3. Chemical enamels are suitable at lower concentrations [5, 6].

The resistance of unalloyed steel in sulfuric acid decreases as the temperature increases (Figure 3), so that in such cases it is used only as a support material with a corresponding acid-resistant masonry lining. Steel can be used in acid lines only up to flow rates of 0.5 m/s [5]. Figure 4 shows the corrosion rate for steel at constant temperature as a function of the sulfuric acid concentration [6, 7]. According to this graph, the corrosion maximum will be at about 83 % H_2SO_4.

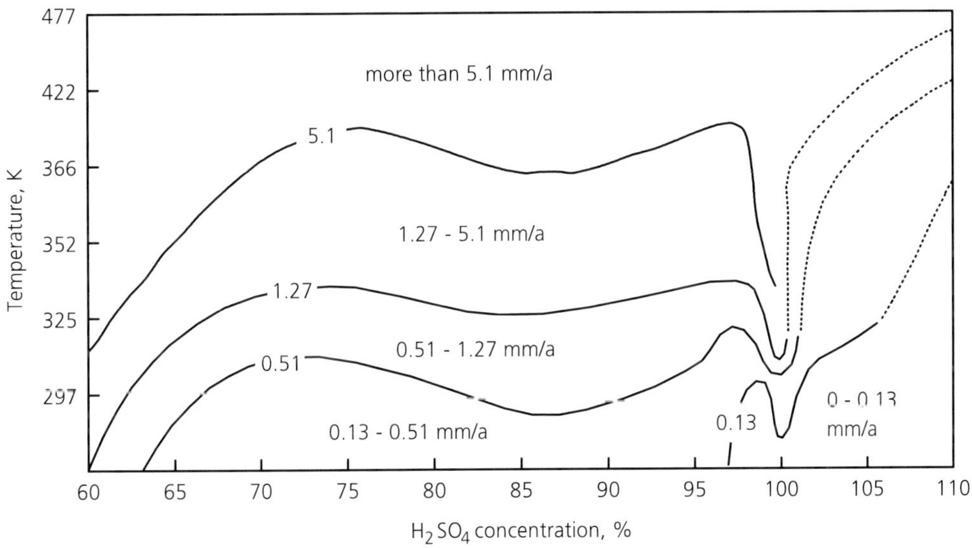

Figure 3: Corrosion rates in mm/a for unalloyed steel and cast steel in static sulfuric acid [4]

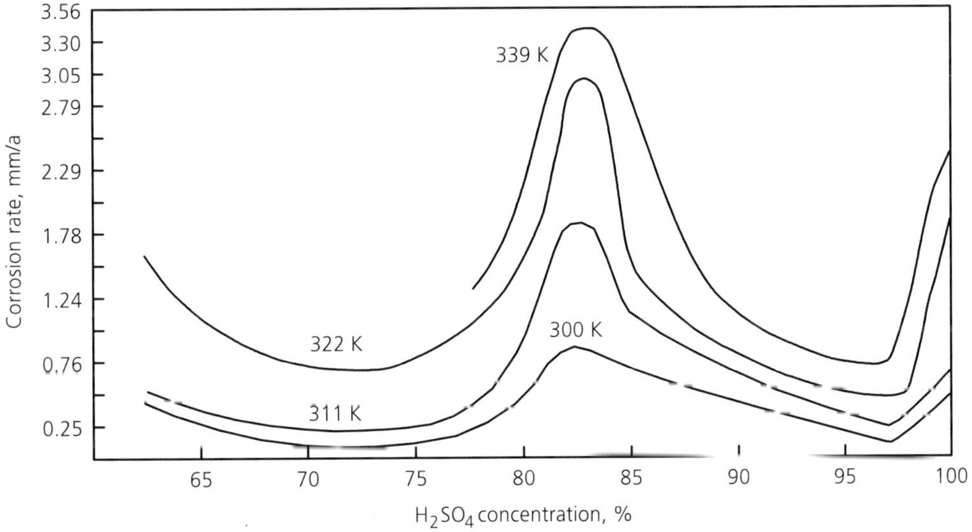

Figure 4: Corrosion of steel in sulfuric acid at constant temperature [6, 7]

Corrosion studies on steel specimens of the composition 0.14 % C, 0.85 % Mn, 0.023 % S and 0.025 % P at 333 K (60 °C) in sulfuric acid gave the corrosion rates plotted in Figure 5. According to this figure, the corrosion maximum is at 83.5 % and the corrosion minimum at 75 % H_2SO_4. As Figure 5 shows, the corrosion maximum corresponds to the maximum solubility of iron sulfate in sulfuric acid and the corrosion minimum to the lowest iron sulfate solubility [8].

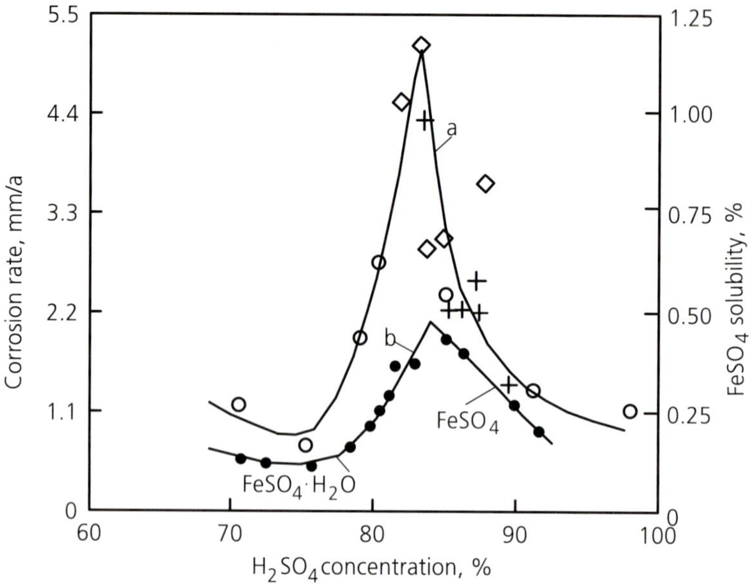

Figure 5: Corrosion rate of a steel (Fe-0.14C-0.85Mn-0.025P-0.023S) in sulfuric acid at 333 K (60 °C) (a) and in comparison with the $FeSO_4$ solubility (b) [8]

As well as the dependence of the corrosive attack on the various acid concentrations and temperatures, the influence of the flow rate of sulfuric acid and its gas content must also be taken into account. Figure 6 shows this influence using the example of a steel containing 0.12 % carbon at room temperature in 0.165 mol/l sulfuric acid. Since no further data are available, saturated solutions are to be assumed. The corrosion rate hardly increases as a function of the flow rate in sulfuric acid containing nitrogen, whereas in air-containing H_2SO_4 a moderate increase, and in oxygen-containing H_2SO_4 a marked increase in the corrosive attack is to be found [9].

This graph is more of theoretical interest, since steel is unsuitable in nitrogen. Figure 7 shows the influence of the flow rate in the area of use of steel in sulfuric acid containing nitrogen. As expected, the curves for the different flow rates run almost parallel. In 100 % sulfuric acid, when the rate increases from 2 m/s to 3.5 m/s there is a disproportionately high corrosion of the material. At higher temperatures, this phenomenon occurs already at lower rates, as shown, for example, in Figure 8. Carbon steel accordingly reacts very sensitively to a flow rate of 2 m/s in 75 to 85 % sulfuric acid at 333 K (60 °C) [10]. According to [5], as already mentioned above, flow rates of up to 0.5 m/s are recommended. The increase in corrosion with the flow rate is due to the surface layer properties which were explained in Figures 3–5.

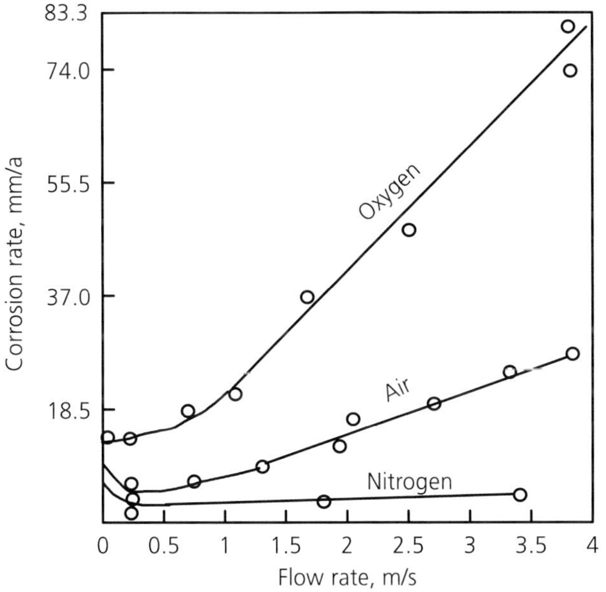

Figure 6: Corrosion of steel containing 0.12 % C in sulfuric acid (0.165 mol/l) at room temperature as a function of the flow rate and the dissolved gas [9]

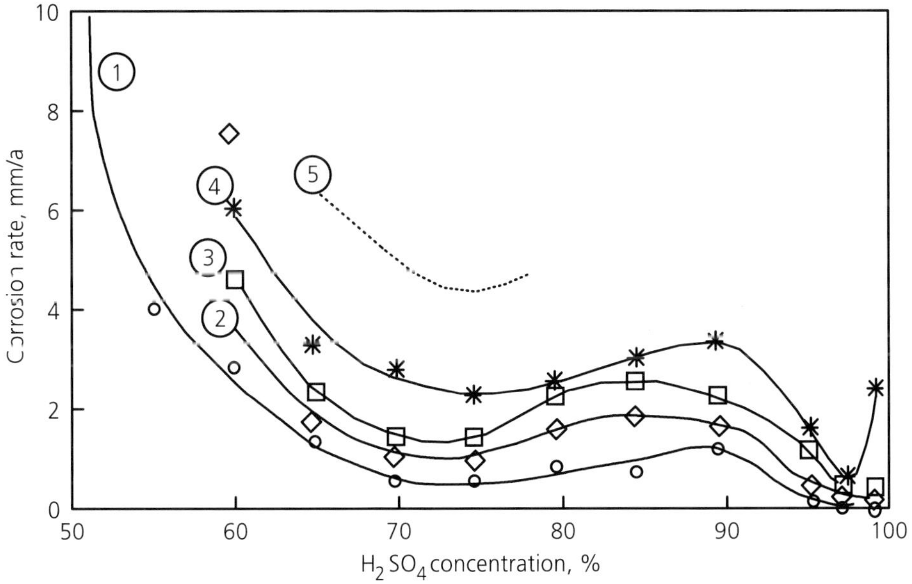

Figure 7: Influence of the sulfuric acid concentration and the flow rate on the corrosion of carbon steel (0.2 % C) in nitrogen-containing acid solutions at 298 K (25 °C) [10]
Flow rate:
① 0 m/s; ② 1 m/s; ③ 2 m/s; ④ 2.5 m/s; ⑤ 5 m/s

The corrosion rates obtained from short-term experiments (1 week) proved to be too high in practice [10]. The reason for this is the high initial corrosion, which changes to lower linear corrosion rates as the test duration or duration of action increases.

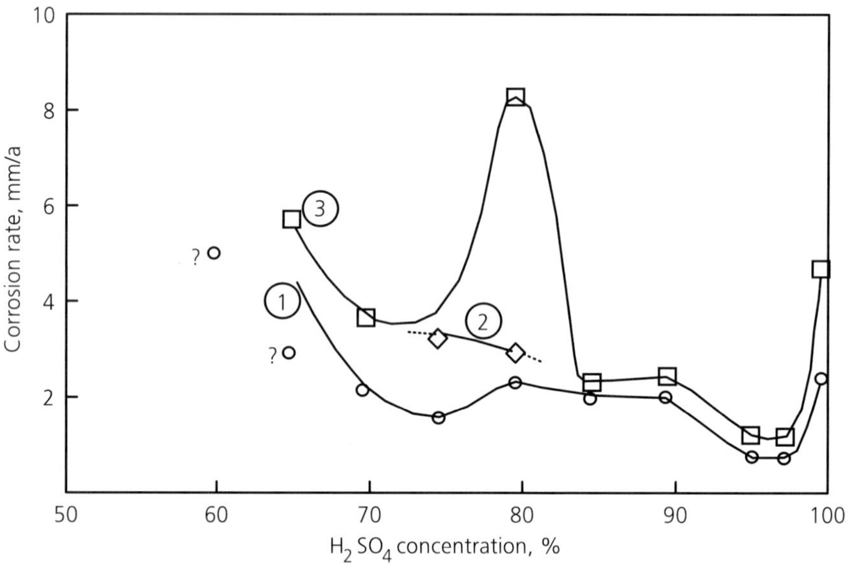

Figure 8: Influence of the acid concentration and the flow rate on the corrosion of carbon steel (0.2 % C) in nitrogen-containing sulfuric acid at 333 K (60 °C) [10]
① 0 m/s
② 1 m/s
③ 2 m/s

The use of unalloyed and low-alloy steels under the action of sulfuric acid is possible and has proved appropriate in practice. However, precisely defined conditions must be observed and control of the corrosive sulfuric acid medium is necessary. Corrosion in the liquid-gaseous phase boundary region, at which particularly severe corrosion occurs under the action of oxygen, is critical [11].

Table 1–3 provide a final comparison of corrosion rates in sulfuric acid under various conditions [12, 13].

They confirm the above comments. Table 3 shows that the carbon content and also the copper content of unalloyed and low-alloy steels has no substantial influence on their corrosion behavior in sulfuric acid, and nor does the flow rate. Table 3, furthermore, shows a good agreement of the corrosion rates determined via weight losses with those obtained by electrochemical measurements [13].

Figure 9 shows a comparison of the corrosion rates of various unalloyed and low-alloy iron materials in sulfuric acid solutions [9].

H_2SO_4 concentration %	Temperature, K (°C)					
	293 (20)		308 (35)		323 (50)	
	O_2	H_2	O_2	H_2	O_2	H_2
	Corrosion rate, mm/a (mpy)					
6.0	8.54 (336)	0.76 (30.0)	7.00 (276)	1.32 (52.0)	11.3 (445)	11.8 (465)
20.0	6.85 (270)	1.90 (74.8)	12.8 (504)	5.60 (220)	19.70 (776)	11.4 (449)
50.0	5.20 (205)	4.60 (181)	20.6 (811)	18.9 (744)	79.0 (3,110)	94.0 (3,701)
96.5	1.58 (62.2)	1.65 (65.0)	2.70 (106)	3.15 (124)	4.67 (184)	4.85 (191)

Table 1: Corrosion rates of steel in various concentrations of sulfuric acid at various temperatures while passing through oxygen or hydrogen [12]

The corrosion behavior of samples of the steel 20 in sulfuric acid solution is changed by the short-term action of a magnetic field.

Table 4 summarizes the corrosion rates. In the concentration range investigated, in which steel cannot be used in sulfuric acid solutions, the magnetic influence leads to a further increase in the corrosion rates, the maximum of which occurs at a magnetic field strength of 39,788.5 A/m [14].

The corrosive attack of dilute sulfuric acid on unalloyed and low-alloy steels is of industrial interest in pickling; for example as pretreatment of surfaces in 9 % sulfuric acid at 348 K (75 °C) before enamelling.

In practice, it has been found that an increasing phosphorus content in the steels reduces their corrosion resistance in dilute sulfuric acid.

Corrosion studies in 9 % sulfuric acid were performed on steels with different phosphorus contents at 348 K (75 °C). Table 5 shows the chemical composition of the steels used.

On the part of the user, lower limits are given for the material consumption rates. The test medium was flushed with nitrogen to remove oxygen. Figure 10 shows the increase in the material consumption per unit area as the phosphorus content increases. All the samples show a continuous increase in weight loss with time at the start of the studies, this increase being linear as the pickling time increases. According to these studies, a constant corrosion rate has still not been reached after 1 h. Furthermore, the studies show a dependence of the material consumption on the copper content such that as the copper content of the steels increases, a decrease in the material consumption rates per unit area is recorded, as can also be seen from Figure 10.

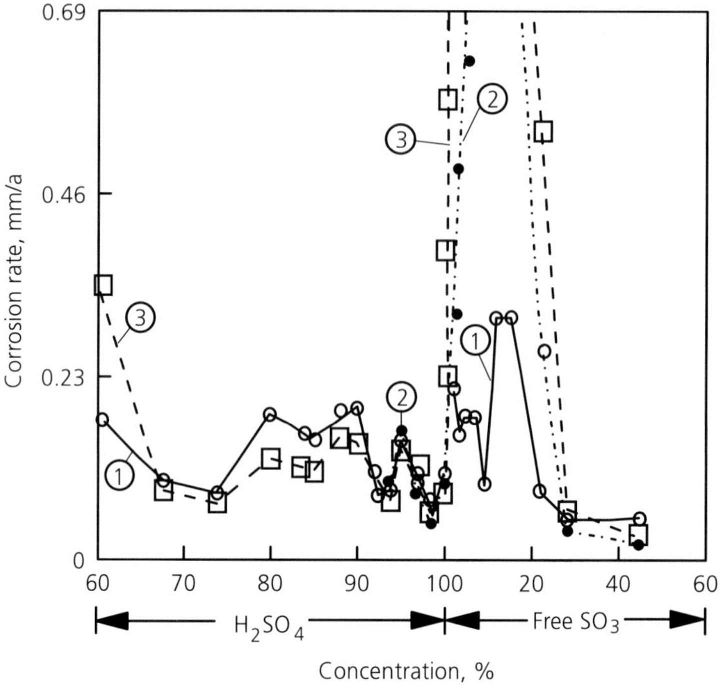

Figure 9: Corrosion of iron materials in static sulfuric acid at room temperature [9]
① Cast iron (3.55 % C)
② Steel (0.155 % C)
③ Wrought iron (0.076 % C)

H_2SO_4 concentration %	Test duration 48 h			Test duration 72 h			Remarks
	Flow rate, m/s						
	0	0.09	0.61	0	0.09	0.61	
	Corrosion rate, mm/a (mpy)						
80.5	0.36 (14.2)		1.09 (42.9)	0.58 (22.8)		0.06 (2.36)	
80.5	0.18 (7.09)		0.30 (11.8)				saturated with air
80.5			2.31 (90.9)				sat. with illuminating gas
89.0	1.47 (57.9)		0.64 (25.2)	0.89 (35.0)		0.00 (0)	
91.9	1.12 (44.1)		4.98 (196)	0.84 (33.1)		3.60 (142)	

Table 2: Corrosion of steel in concentrated sulfuric acid at room temperature as a function of the flow rate (H_2 evolution was observed at the start of every experiment) [12]

Table 2: Continued

H_2SO_4 concentration %	*Test duration 48 h*			*Test duration 72 h*			*Remarks*
	Flow rate, m/s						
	0	*0.09*	*0.61*	*0*	*0.09*	*0.61*	
	Corrosion rate, mm/a (mpy)						
94.8	0.46 (18.1)	0.79 (31.1)	1.98 (78.0)	0.00 (0.0)	0.43 (16.9)	0.84 (33.1)	
97.8	0.53 (20.9)	5.60 (220)	4.60 (181)	0.28 (11.0)	1.12 (44.1)	3.30 (130)	
97.8						26.70 (1,051)	
100.0	5.10 (201)	5.80 (228)	17.50 (689)	0.00 (0.0)	0.00 (0.0)	23.60 (929)	
100.0						76.00 (2,992)	
100.0						21.30 (839)	

Table 2: Corrosion of steel in concentrated sulfuric acid at room temperature as a function of the flow rate (H_2 evolution was observed at the start of every experiment) [12]

Material	*Determined after 24 h via the weight loss*	*Determined after 1 h via electrochem. measurements*
	Corrosion rate, mm/a (mpy)	
SAE 1018 = 1.0401 (C 15)	6.35 (250)	6.02 (237)
Corten® (max. 0.10 C, min. 0.30 Cu)	5.76 (227)	5.87 (231)
SAE 1045 = 1.1191 (Ck 45)	6.91 (272)	6.73 (265)
SAE 1113 = 1.0711 (9 S 20)	6.40 (252)	6.53 (257)

Table 3: Corrosion on rotating steel cylinders (0.9 m/s) in 68 % sulfuric acid at 313 K (40 °C) [13]

Figure 11 shows an overview of the dependence of the corrosion rate in the pickling solution on the pickling time and the phosphorus content of steels containing 0.003 to 0.005 % Cu. According to this graph, there is no linear dependence of the corrosion rate in the region of initial corrosion. This is detectable only after an action time of 30 min, and is pronounced after a pickling time of 60 min. The very high corrosion rates show a significant dependence on the P-content and also differ according to the pickling times. The rapidly destructive action of dilute sulfuric acid can also be deduced from the unacceptably high corrosion rates. For further details on the pickling, see [15].

Magnetic field strength A/m	*H_2SO_4 concentration, %*		
	5	*7*	*17*
		Corrosion rate, mm/a (mpy)	
0	6.5 (256)	7.1 (279)	8.6 (339)
5,968.3	7.2 (283)	7.6 (299)	9.1 (358)
7,957.7	7.8 (307)	8.6 (339)	10.7 (421)
28,873.1	8.4 (331)	9.2 (362)	11.4 (449)
31,830.8	8.8 (346)	9.7 (382)	12.5 (492)
39,788.5	8.9 (350)	10.1 (398)	14.0 (551)
47,746.2	8.3 (327)	9.2 (362)	12.5 (492)
55,703.9	6.9 (272)	8.3 (327)	10.4 (409)

Table 4: Corrosion rates of steel 20 (conversion from g/m^2 h) in flowing sulfuric acid (0.04 m/s) at 293 ± 2 K (20 ± 2 °C) after short-term action (0.24 min) of a magnetic field of various strengths, test duration 70 h [14]

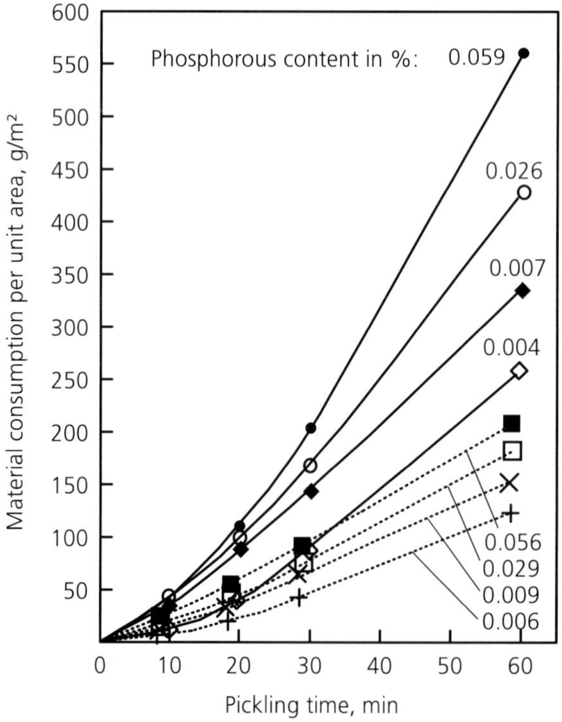

Figure 10: Dependence of the material consumption of various unalloyed steels (see Table 5) on the phosphorus and copper content and on the pickling time in 9 % air-free sulfuric acid at 348 K (75 °C) [15]

——— copper content 0.003 to 0.005 %

- - - - - copper content 0.028 to 0.029 %

As expected, unalloyed steels are destroyed in a pickling solution for cast iron consisting of 5 % H_2SO_4 + 5 % HF + 10 % $FeSO_4$ at 322 K (49 °C). Corrosion rates of 21.3 mm/a (839 mpy) resulted after a test duration of 40 h (Table 62) [16].

Ion implantation provides the possibility in particular cases of modifying the surface of iron by alloying in a manner not possible with melting metallurgy. This is of little importance in general practice. Implantations of Ne^+ and Cu^+ slightly increase the corrosion of iron in 0.5 mol/l H_2SO_4 at room temperature, whereas Au^+ increases it more than 10-fold. Implantation of Pb^+ into the iron surface leads to a marked reduction in corrosion under the conditions mentioned [17, 18].

Chemical composition, %						
C	***Mn***	***P***	***S***	***Cu***	***Cr***	***As***
0.011	< 0.01	0.004	0.004–0.006	0.003–0.005	< 0.002	0.012
0.009		0.007				0.010
0.010		0.026				0.010
0.010		0.059				0.009
0.013	< 0.01	0.004	0.004–0.006	0.028–0.029	< 0.002	0.01
0.009		0.009				
0.008		0.029				
0.007		0.056				

Table 5: Chemical composition of steels of different phosphorus and copper contents for corrosion studies in a pickling solution (9 % sulfuric acid at 348 K (75 °C)) (see Figures 10 and 11) [15]

Studies have shown [19] that a tanning agent containing 1.15 % H_2SO_4 and 0.6 % formic acid is the optimum composition at 293 K (20 °C). Under these conditions, corrosion rates of between 0.21 and 0.76 mm/a (8.27 and 29.9 mpy) occur on steel St 3 (0.14 0.22 % C). The test durations were between 1 and 52 h.

In industry, inorganic acids are often used as reactants or catalysts in organic media and can occur as impurities in solvents.

Figure 12 shows the corrosion behavior of pure iron and the carbon steel 1.0616 (0.83–0.88 C, 0.10–0.30 Si, 0.30–0.70 Mn, ≤ 0.04 P, ≤ 0.04 S; cf. SAE 1086) in the organic solvents ethanol and acetone in comparison with acetic acid and water, all containing 0.05 % sulfuric acid.

The material consumption per unit area of iron is high in organic media, with the exception of acetone, and is comparable to the rate of reaction in the corresponding aqueous solutions. As the temperature increases, the corrosion rate also increases as expected. The corrosion behaviors of iron and steel differ only slightly under these conditions. However, the passivation of pure iron in ethanol containing sulfuric acid is remarkable. The surface of the specimens became golden yellow under these circumstances. Additions of chloride cancelled out the passivity immediately.

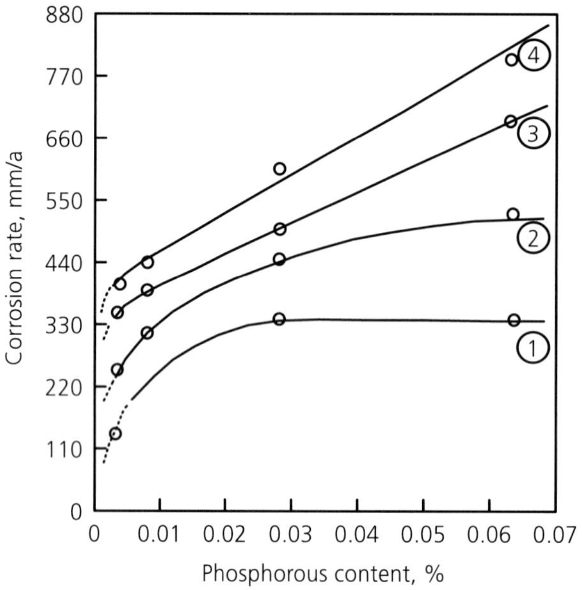

Figure 11: Influence of the phosphorus content on the corrosion rate of steels containing 0.003–0.005 % Cu (Table 5) during pickling in 9 % sulfuric acid at 348 K (75 °C) [15]
Pickling time: ① 10 min; ② 20 min; ③ 30 min; ④ 60 min

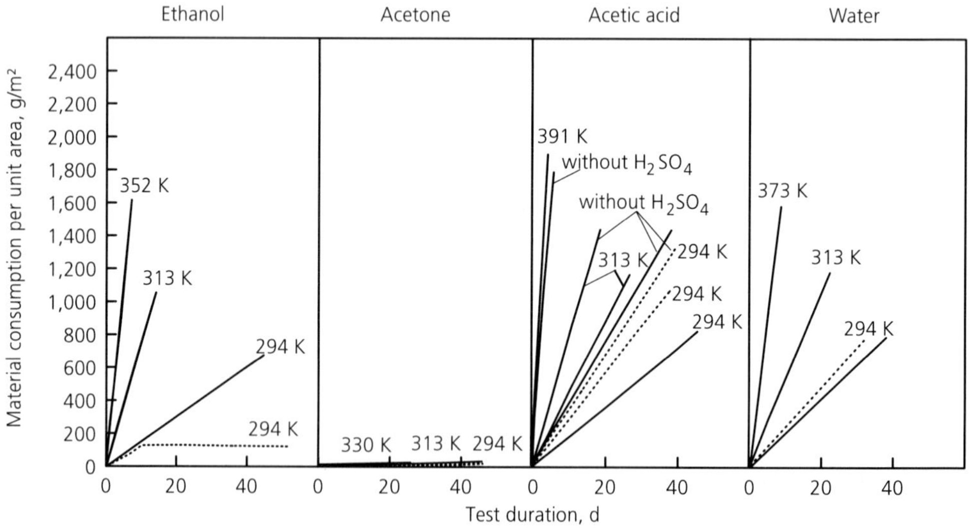

Figure 12: Corrosion of pure iron and C-steel (1.0616, cf. SAE 1086) in solutions containing sulfuric acid (0.05 % H_2SO_4) [20]
——— C-steel
– – – – pure iron

No noticeable corrosion is to be found in the sulfuric acid-acetone system. In both cases, on pure iron and on carbon steel, a thin tarnished layer formed [20].

The previously customary (required) temperatures of the waste gas and boiler water of 453 and 343 K (180 and 70 °C) respectively provided an extensive guarantee against corrosion from flue gas in heating boilers made of unalloyed or low-alloy steel and gray cast iron. The service life of such apparatuses was 20 years or more. The demand for more economical heating boilers led to lower boiler water temperatures. The amount of SO_2 and consequently SO_3 contained in flue gas depends on the amount of sulfur in the fuel, for example oil.

Corrosion studies on specimens of 1.0305 (P235G1TH, old: St 35.8, cf. UNS K01200) and GG 20 in a flue gas from extra-light heating oil with a sulfur content of 0.2 to 0.3 % have shown that corrosion above the dew point of water (about 323 K (50 °C)) is chiefly due to sulfuric acid corrosion (diffusion-controlled transportation of sulfuric acid through the flue gas boundary layer as the rate-determining step of the corrosion rate). The corrosion rates under these conditions are between 0.2 and 0.6 mm/a (7.87 and 23.6 mpy) under continuous burner operation for both materials.

At surface temperatures below the dew point of water, a liquid electrolyte film can be formed which then allows oxygen corrosion to take place. The high residual oxygen content in the flue gases leads to a rapid increase in the corrosion. The corrosion rates under the conditions mentioned are between 5 and 6 mm/a (197 and 236 mpy) at surface temperatures of 308 to 313 K (35 to 40 °C). GG 20 shows a more favorable corrosion behavior under these conditions, with almost uniform corrosion of the surface, whereas P235G1TH has a tendency towards local corrosion (deep pitting and wide pitting). With intermittent burner operation, both materials exhibited the same corrosion behavior [21]. The actual material consumption by corrosion depends on the mode of operation of the plant.

– *Corrosion behavior in sulfuric acid mixtures* –

It is possible to use unalloyed steel instead of, for example, lead in the production of sulfuric acid by the lead chamber process if certain outline conditions are observed. In the Salsas-Sera process, sulfuric acid in the concentration range between 80 and 88 % hardly attacks steel at all if the nitric oxide content is between 7 and 10 %. The reason for this is the formation of a protective surface layer of basic iron sulfate. Figure 13 illustrates the reduction in corrosion as the $FeSO_4$ content increases in 75 % sulfuric acid at 333 K (60 °C) [2, 10]. The graph shows that the particular composition of the sulfuric acid mixture and the condition of the metal surface determine the corrosion behavior of the steel.

Figure 14 shows the corrosion rates to be expected on steel at room temperature in mixtures of sulfuric acid, nitric acid and water. These mixtures are closely related to the chamber acid obtained in the lead chamber process [2]. Table 19 gives a summary of corrosion rates in these mixtures at room temperature. Steel and cast iron are not attacked to a very great degree by acid mixtures of sulfuric and nitric acid containing more than 15 % H_2SO_4 and less than 20 % H_2O at room temperature [22]. The materials are practically resistant in these regions with a maximum corrosion rate of 0.1 mm/a (3.94 mpy).

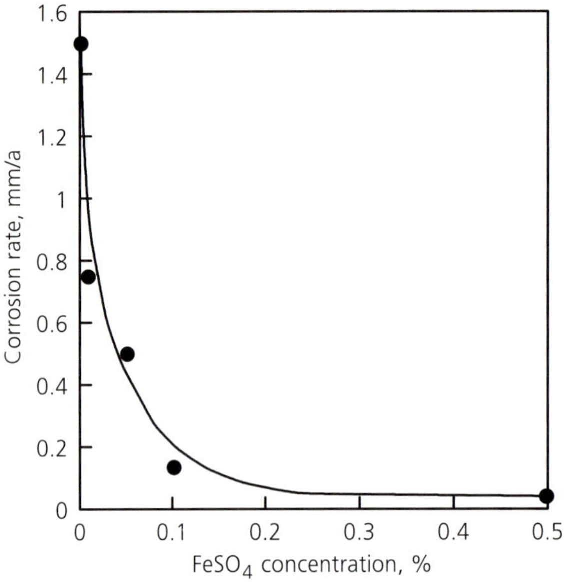

Figure 13: Influence of the $FeSO_4$ concentration on the corrosion rate of carbon steel (0.2 % C) in 75 % sulfuric acid at 333 K (60 °C) [2, 10]

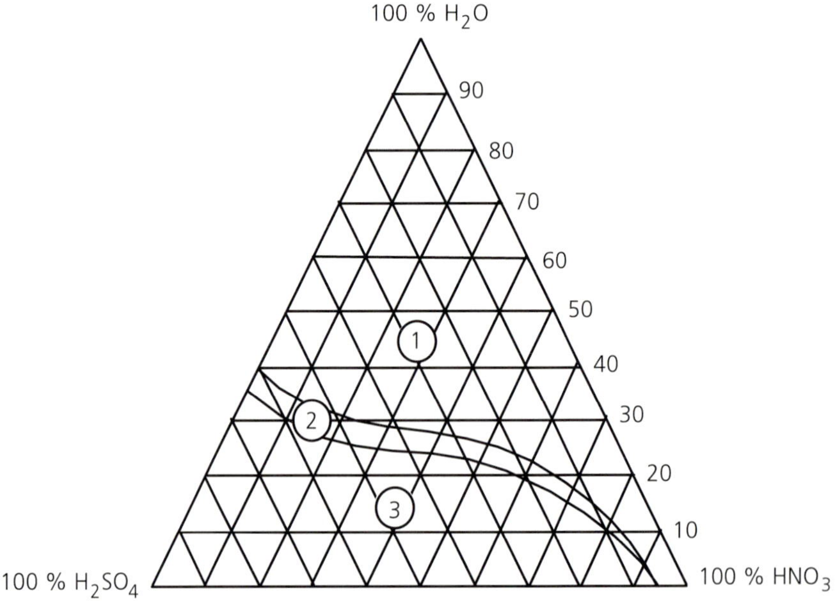

Figure 14: Corrosion areas of unalloyed steel in aqueous solutions of sulfuric and nitric acid at room temperature (according to Nelson) [2]
Corrosion rates:
① ≥ 1.25 mm/a (≥ 49.2 mpy)
② 0.5 to 1.25 mm/a (19.7 to 49.2 mpy)
③ ≤ 0.5 mm/a (≤ 19.7 mpy)

According to [23], corrosion studies in the laboratory and under operating conditions in chamber acid showed good agreement. As already explained above, increased corrosion occurs at the phase boundary. Figure 15 shows the results of laboratory studies on various materials in sulfuric acid in the concentration range 75.3 to 75.6 % H_2SO_4 and 0.035 to 0.045 % HNO_3 at a flow rate between 0.7 and 1.4 m/s. The test durations were 260 h up to 383 K (110 °C) and between 44 and 160 h at 393 to 403 K (120 to 130 °C). This graph shows that, like St 3, the unalloyed steel 3 already experiences corrosion rates in excess of 1 mm/a (39.4 mpy) at 333 K (60 °C) and thus can no longer be used. Studies under conditions resembling those in the practice on specimens in pipes over test durations of between 92 and 1,184 h at flow rates of between 0.7 and 1.4 m/s gave the results summarized in Table 6.

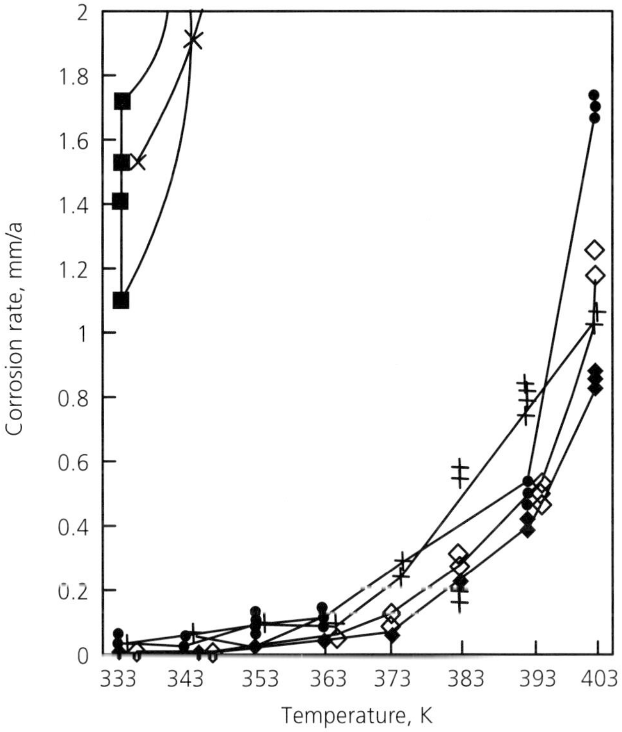

Figure 15: Corrosion of metallic materials in chamber acid (75.3 to 75.6 % H_2SO_4, 0.035 to 0.045 % HNO_3) as a function of the temperature at flow rates between 0.7 and 1.4 m/s [23]
Test conditions: 260 h up to 383 K (110 °C), 44 to 160 h at 393 K (120 °C) and 44 to 160 h at 403 K (130 °C)
■ ST 3
◆ 06ChN28MDT
◇ 03Ch21N21M4B
● 12Ch18N10T (cf. 1.4541, SAE 321)
+ 10Ch17N13M2T (cf. 1.4571, SAE 316 Ti)
× Alloy 4201 (Ti + 33 % Mo)

Here, too, the corrosion rates on unalloyed steel in chamber acids are very high and make any use practically impossible, even at the reduced temperatures downstream of the heat exchanger [23].

In the production of sodium hexafluorosilicate by the sulfate process, a filtrate consisting of an aqueous solution containing up to 1.5 % Na_2SiF_6 and 6 to 8 % H_2SO_4 occurs as a very aggressive medium. It is surpassed in terms of corrosiveness however, by industrial waters containing up to 1 % Na_2SiF_6 and 1 to 2 % H_2SO_4. Table 7 summarizes the corrosion rates of various steels. The steel St 3 (St 37) can no longer be used in either case in the temperature range between 303 and 313 K (30 and 40 °C). The corrosion rates of 2.05 and 8.66 mm/a (80.7 and 341 mpy) respectively are far in excess of the application limit in practice [24].

Gases containing up to 80 % HF, 5 to 10 % H_2SO_4 and 2 to 3 % H_2O are formed during decomposition of fluorspar. The condensates formed from the gases contain 18 to 40 % HF, 35 to 70 % H_2SO_4 and up to 40 % H_2O. Under such conditions, as expected, unalloyed steel is not resistant. In accordance with the corrosion studies, corrosion rates in excess of 5 mm/a (197 mpy) were determined on steel St 3 in apparatuses for prepurification in the liquid phase at 353 K (80 °C) (test duration 1,650 h). The same high weight losses occur in the gas phase between 393 and 403 K (120 and 130 °C) (test duration 1,970 h) [25].

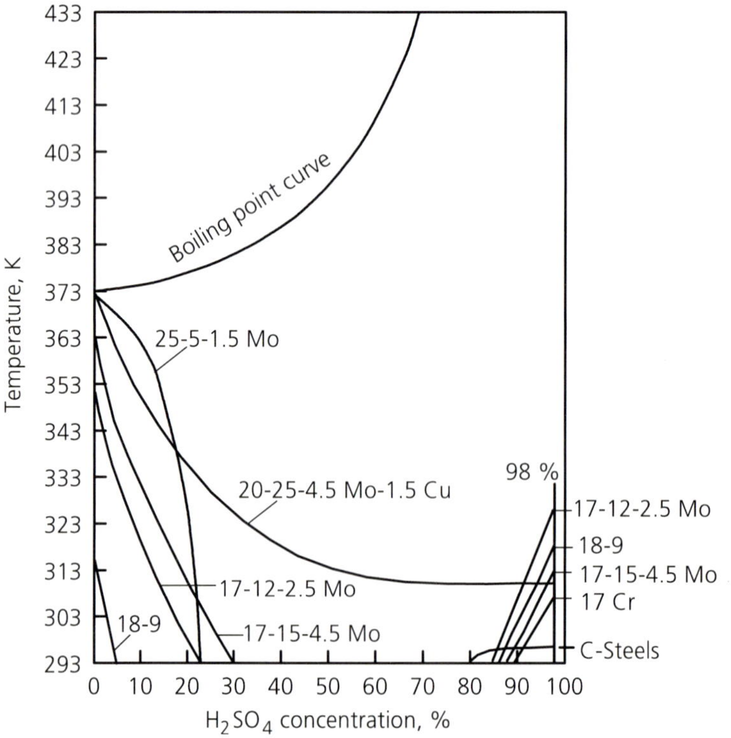

Figure 16: Isocorrosion curves (0.1 mm/a (3.94 mpy)) for various groups of steels, including carbon steels, in chemically pure sulfuric acid [26]

Table 8 gives a summary of the corrosion behavior of steels, divided into groups, in sulfuric acid and sulfuric acid solutions. It allows a comparison of the corrosion resistance of the steels and facilitates the appropriate choice of material. Figure 16 provides an overview of the areas of use of the individual groups of steels. Table 8 and Figure 16 show the small areas of use of unalloyed steel in sulfuric acid and solutions containing sulfuric acid [26].

Material	***Before the cooler***		***After the cooler***	
	A	***B***	***A***	***B***
	Corrosion rate, mm/a (mpy)			
ST 3	6.1 (240)	5.4 (213)	0.65–0.76 (25.6–30.0)	0.99–1.08 (39.0–42.5)
08Ch22N6T	119 (4,685)	0.22 (8.66)	0.004 (0.16)	0.007 (0.28)
12Ch18N10T (cf. 1.4541)	dissolved after 92 h	0.22 (8.66)	0.006 (0.24)	0.007 (0.28)
		0.29 (11.4)	0.005 (0.20)	0.006 (0.24)
			0.005 (0.20)	0.006 (0.24)
10Ch17N13M2T (cf. 1.4571)		0.21 (8.27)	0.008 (0.31)	0.017 (0.67)
06ChN28MDT	0.21–0.28 (8.27–11.0)	0.21 (8.27)		0.011 (0.43)
N70MF	2.12 (83.5)	0.06–0.09 (2.36–3.54)	0.001–0.002 (0.04–0.08)	0.002 (0.08)
	1.92 (75.6)	–	–	–
ChN65MV	3.00–3.14 (118 121)	0.76 (30.0)	0.032–0.038 (1.25 1.50)	–
0Ch20N20M3D3B	0.1–0.13 (3.94–5.12)	–	0.001–0.002 (0.04 0.08)	0.001
08Ch18G8N2T	–	–	0.006–0.007 (0.24–0.28)	0.012–0.017 (0.47–0.67)
Alloy 4201 (Ti + 33 % Mo)	2.28 (98.8)	–	–	–

A 74.2–77.6 % H_2SO_4 + 0.02–0.05 % HNO_3, 365–391 K (92–118 °C) before and 317–356 K (44–83 °C) after the cooler

B 72.6–76.8 % H_2SO_4 + 0.02–0.03 % HNO_3, 344–387 K (71–114 °C) before and 310–339 K (37–66 °C) after the cooler

Table 6: Corrosion behavior of various metallic materials in chamber acid (A and B) according to pilot studies on specimens in pipelines at flow rates between 0.7 and 1.4 m/s over test durations between 92 and 1,184 h [23]

Test location	Medium and conditions		ST 3	08Ch22N6T	12Ch18N10T	10Ch17N13M2T	06ChN28MDT
			Corrosion rate, mm/a (mpy)				
Collecting tank, mother liquor	Na_2SiF_6	1.5 %	2.057 (81.0)	0.727* (28.6)	0.406* (16.0)	0.0045 (0.18)	0.0014 (0.06)
	H_2SO_4	6–8 %					
	303–313 K (30–40 °C)						
Collecting tank, production water	Na_2SiF_6	up to 1 %	8.66 (341)	6.545* (273)	0.569** (22.4)	0.383** (15.1)	0.0049 (0.19)
	H_2SO_4	1–2 %					
	303–313 K (30–40 °C)						

* average value of 4–5 samples
** crevice corrosion with pitting

Table 7: Corrosion on steels in sulfuric acid solutions during production of sodium hexafluorosilicate by the sulfate process (test durations 500 to 525 h) [24]

Concentration, %	Temperature K (°C)	Chemical composition of the steels, %												
		Cr	–	13	17	18	18	17	18	17	20	25	Ti	Ti-Pd
		Ni	–	–	–	–	9	12	14	15	25	5		
		Mo	–	–	–	2	–	2.5	3.5	4.5	4.5	1.5		
		Cu	–	–	–	–	–	–	–	–	1.5	–		
H_2SO_4 (sulfuric acid)														
1	373 (100)		2	2	2	2	2	1	1	1	0	0	1	0
0.5	293 (20)		2	2	1	0	0	0	0	0	0	0	0	0
0.5	323 (50)		2	2	2	2	1	0	0	0	0	0	0	0

0: corrosion rate < 0.1 mm/a (< 3.94 mpy)
1: corrosion rate 0.1–1 mm/a (3.94–39.4 mpy)
2: corrosion rate > 1 mm/a (> 39.4 mpy)
P pitting and crevice corrosion possible
S stress corrosion cracking
BP boiling point

Table 8: Compilation of corrosion rates on steels in sulfuric acid and in sulfuric acid solutions (arranged alphabetically according to the chemical formula), see also Table 35 [26]

Table 8: Continued

Concentration, %	*Temperature K (°C)*	*Chemical composition of the steels, %*												
		Cr	*–*	*13*	*17*	*18*	*18*	*17*	*18*	*17*	*20*	*25*	*Ti*	*Ti-Pd*
		Ni	*–*	*–*	*–*	*–*	*9*	*12*	*14*	*15*	*25*	*5*		
		Mo	*–*	*–*	*–*	*2*	*–*	*2.5*	*3.5*	*4.5*	*4.5*	*1.5*		
		Cu	*–*	*–*	*–*	*–*	*–*	*–*	*–*	*–*	*1.5*	*–*		
0.5	373 (100)		2	2	2	2	2	1	1	1	1	1	1	0
1	293 (20)		2	2	2	0	0	0	0	0	0	0	0	0
1	323 (50)		2	2	2	2	1	0	0	0	0	0	0	0
1	343 (70)		2	2	2	2	1	0	0	0	0	0	1	0
1	358 (85)		2	2	2	2	2	1	1	0	0	0	1	0
1	373 (100)		2	2	2	2	2	1	1	1	1	1	1	0
2	293 (20)		2	2	2	0	0	0	0	0	0	0	0	0
2	323 (50)		2	2	2	2	1	0	0	0	0	0	0	0
2	333 (60)		2	2	2	2	1	0	0	0	0	0	1	0
3	293 (20)		2	2	2	0	0	0	0	0	0	0	0	0
3	308 (35)		2	2	2	2	1	0	0	0	0	0	0	0
3	323 (50)		2	2	2	2	1	0	0	0	0	0	1	0
3	358 (85)		2	2	2	2	2	1	1	1	0	0	1	1
3	373 (100)		2	2	2	2	2	2	2	2	1	2	2	1
5	293 (20)		2	2	2	2	1	0	0	0	0	0	0	0
5	308 (35)		2	2	2	2	1	0	0	0	0	0	1	0
5	333 (60)		2	2	2	2	2	1	0	0	0	0	1	0
5	348 (75)		2	2	2	2	2	1	1	1	0	0	2	1
5	358 (85)		2	2	2	2	2	2	2	2	1	0	2	1
5	BP		2	2	2	2	2	2	2	2	2	1	2	1

0: corrosion rate < 0.1 mm/a (< 3.94 mpy)
1: corrosion rate 0.1–1 mm/a (3.94–39.4 mpy)
2: corrosion rate > 1 mm/a (> 39.4 mpy)
P pitting and crevice corrosion possible
S stress corrosion cracking
BP boiling point

Table 8: Compilation of corrosion rates on steels in sulfuric acid and in sulfuric acid solutions (arranged alphabetically according to the chemical formula), see also Table 35 [26]

Table 8: Continued

Concentration, %	*Temperature K (°C)*	*Chemical composition of the steels, %*											
		Cr	–	13	17	18	18	17	18	17	20	25	Ti	Ti-Pd
		Ni	–	–	–	–	9	12	14	15	25	5		
		Mo	–	–	–	2	–	2.5	3.5	4.5	4.5	1.5		
		Cu	–	–	–	–	–	–	–	–	1.5	–		
10	293 (20)		2	2	2	2	2	0	0	0	0	0	1	0
10	323 (50)		2	2	2	2	2	1	1	0	0	0	2	0
10	333 (60)		2	2	2	2	2	1	1	1	0	0	2	1
10	353 (80)		2	2	2	2	2	2	2	2	1	0	2	1
10	BP		2	2	2	2	2	2	2	2	2	2	2	2
20	293 (20)		2	2	2	2	2	0	0	0	0	0	2	0
20	313 (40)		2	2	2	2	2	1	1	1	0	1	2	0
20	323 (50)		2	2	2	2	2	1	1	1	0	1	2	1
20	333 (60)		2	2	2	2	2	2	1	1	1	2	2	1
20	373 (100)		2	2	2	2	2	2	2	2	2	2	2	2
30	293 (20)		2	2	2	2	2	1	1	1	0	1	2	0
30	313 (40)		2	2	2	2	2	2	1	1	0	2	2	1
30	333 (60)		2	2	2	2	2	2	2	2	1	2	2	2
40	293 (20)		2	2	2	2	2	2	2	2	0	2	2	0
40	313 (40)		2	2	2	2	2	2	2	2	0	2	2	1
40	333 (60)		2	2	2	2	2	2	2	2	1	2	2	2
40	363 (90)		2	2	2	2	2	2	2	2	2	2	2	2
50	293 (20)		2	2	2	2	2	2	2	2	0	2	2	1
50	313 (40)		2	2	2	2	2	2	2	2	0	2	2	2
50	343 (70)		2	2	2	2	2	2	2	2	2	2	2	2

0: corrosion rate < 0.1 mm/a (< 3.94 mpy)
1: corrosion rate 0.1–1 mm/a (3.94–39.4 mpy)
2: corrosion rate > 1 mm/a (> 39.4 mpy)
P pitting and crevice corrosion possible
S stress corrosion cracking
BP boiling point

Table 8: Compilation of corrosion rates on steels in sulfuric acid and in sulfuric acid solutions (arranged alphabetically according to the chemical formula), see also Table 35 [26]

Table 8: Continued

Concentration, %	*Temperature K (°C)*		*Chemical composition of the steels, %*											
		Cr	–	13	17	18	18	17	18	17	20	25	Ti	Ti-Pd
		Ni	–	–	–	–	9	12	14	15	25	5		
		Mo	–	–	–	2	–	2.5	3.5	4.5	4.5	1.5		
		Cu	–	–	–	–	–	–	–	–	1.5	–		
60	293 (20)		2	2	2	2	2	2	2	2	0	2	2	2
60	313 (40)		2	2	2	2	2	2	2	2	1	2	2	2
60	343 (70)		2	2	2	2	2	2	2	2	2	2	2	2
70	293 (20)		2	2	2	2	2	2	2	2	0	2	2	2
70	313 (40)		2	2	2	2	2	2	2	2	1	2	2	2
70	343 (70)		2	2	2	2	2	2	2	2	2	2	2	2
80	293 (20)		2	2	2	2	2	1	1	1	0	1	2	2
80	313 (40)		2	2	2	2	2	2	2	2	1	2	2	2
80	333 (60)		2	2	2	2	2	2	2	2	2	2	2	2
85	293 (20)		0	1	1	1	1	1	1	1	0	1	2	2
85	303 (30)		1	1	1	1	1	1	1	1	0	1	2	2
85	313 (40)		2	2	2	1	1	1	1	1	1	2	2	2
85	323 (50)		2	2	2	2	2	2	2	2	1	2	2	2
90	293 (20)		0	0	0	0	0	0	0	0	0	0	2	2
90	303 (30)		1	1	1	1	0	0	1	1	0	1	2	2
90	313 (40)		2	2	2	2	2	1	1	1	1	2	2	2
90	343 (70)		2	2	2	2	2	2	2	2	2	2	2	2
94	293 (20)		0	0	0	0	0	0	0	0	0	0	2	2
94	303 (30)		2	1	1	0	0	0	0	0	0	0	2	2
94	313 (40)		2	2	2	2	1	0	1	1	1	1	2	2

0: corrosion rate < 0.1 mm/a (< 3.94 mpy)
1: corrosion rate 0.1–1 mm/a (3.94–39.4 mpy)
2: corrosion rate > 1 mm/a (> 39.4 mpy)
P pitting and crevice corrosion possible
S stress corrosion cracking
BP boiling point

Table 8: Compilation of corrosion rates on steels in sulfuric acid and in sulfuric acid solutions (arranged alphabetically according to the chemical formula), see also Table 35 [26]

Table 8: Continued

Concentration, %		*Temperature K (°C)*	*Chemical composition of the steels, %*												
			Cr	**–**	**13**	**17**	**18**	**18**	**17**	**18**	**17**	**20**	**25**	***Ti***	***Ti-Pd***
			Ni	**–**	**–**	**–**	**–**	**9**	**12**	**14**	**15**	**25**	**5**		
			Mo	**–**	**–**	**–**	**2**	**–**	**2.5**	**3.5**	**4.5**	**4.5**	**1.5**		
			Cu	**–**	**–**	**–**	**–**	**–**	**–**	**–**	**–**	**1.5**	**–**		
94		323 (50)		2	2	2	2	1	1	1	1	1	1	2	2
96		293 (20)		0	0	0	0	0	0	0	0	0	0	2	2
96		303 (30)		1	1	0	0	0	0	0	0	0	0	2	2
96		313 (40)		2	2	1	1	0	0	1	1	1	1	2	2
96		323 (50)		2	2	2	2	1	1	1	1	1	2	2	2
98		303 (30)		1	1	0	0	0	0	0	0	0	0	2	2
98		313 (40)		1	1	1	1	0	0	0	0	1	1	2	2
98		323 (50)		2	2	2	2	2	0	1	1	1	1	2	2
98		353 (80)		2	2	2	2	2	2	2	2	2	2	2	2
100		343 (70)						0	0				0		
$H_2SO_4 + Al_2(SO_4)_3$ (+ aluminium sulfate)															
42	1.5	318 (45)		2	2	2	2	2	2	2	2	1	2	2	2
$H_2SO_4 + CH_3COOH$ (+ acetic acid)															
1	1	293 (20)				0	0	0	0	0	0	0	0	0	0
1	1	BP				2	1	1	1	1	1	0	1	1	0
1	25	BP							1	1	1	1	1	1	0
2	0.5	BP			2	2	2	2	2	1	1	1	2	1	0
2	25	353 (80)			2	2	2	2	1	1	1	0	1	1	0
2	0.2	393 (120)		2	2	2	2	2	2	1	1	1	2	2	2
5	90	293 (20)						1	0	0	0	0	0	0	0

0: corrosion rate < 0.1 mm/a (< 3.94 mpy)
1: corrosion rate 0.1–1 mm/a (3.94–39.4 mpy)
2: corrosion rate > 1 mm/a (> 39.4 mpy)
P pitting and crevice corrosion possible
S stress corrosion cracking
BP boiling point

Table 8: Compilation of corrosion rates on steels in sulfuric acid and in sulfuric acid solutions (arranged alphabetically according to the chemical formula), see also Table 35 [26]

Table 8: Continued

Concentration, %			*Temperature K (°C)*	*Chemical composition of the steels, %*												
				Cr	–	13	17	18	18	17	18	17	20	25	Ti	Ti-Pd
				Ni	–	–	–	–	9	12	14	15	25	5		
				Mo	–	–	–	2	–	2.5	3.5	4.5	4.5	1.5		
				Cu	–	–	–	–	–	–	–	–	1.5	–		
10	2		BP		2	2	2	2	2	2	2	2	2	2	2	2
10	90		293 (20)						1	0	0	0	0	0	1	0
$H_2SO_4 + CH_3COOH + (CH_3CO)_2O$ (+ acetic acid + acetic anhydride)																
0.4	71.3	28.3	408 (135)		2	2	2	2	2	1				1		
5	47.5	47.5	293–313 (20–40)		2	2	2	0	0	0	0	0		0		
5	47.5	47.5	353 (80)		2	2	2	2	2	1				1		
$H_2SO_4 + C_2H_4(OH)COOH$ (lactic acid)																
10–50	25		BP		2	2	2	2	2	2	2	2	1	2	2	2
$H_2SO_4 + Cl_2$ (+ chlorine) (saturated solution)																
40–50			298 (25)		2	2	2	2	2	2	2	2	2	2	0	0
60			313 (40)		2	2	2	2	2	2	2	2	2	2	2	2
82			323 (50)							1	1	1	1	2	2	2
96			323 (50)							1						
$H_2SO_4 + (COOH)_2 + HNO_3$ (+ oxalic acid + nitric acid)																
5	2	0	333 (60)							2	2	2				
5	2	0.5	333 (60)							0	0	0	0			
5	2	1	333 (60)							0	0	0	0			
$H_2SO_4 + CrO_3$ (+ chromic acid)																
1	3.5		308 (35)		0	0	0	0	0	0	0	0	0	0	0	0
1.5	1.5		BP		2	2	1	0	0	0	0	0	0	0	0	0

0: corrosion rate < 0.1 mm/a (< 3.94 mpy)
1: corrosion rate 0.1–1 mm/a (3.94–39.4 mpy)
2: corrosion rate > 1 mm/a (> 39.4 mpy)
P pitting and crevice corrosion possible
S stress corrosion cracking
BP boiling point

Table 8: Compilation of corrosion rates on steels in sulfuric acid and in sulfuric acid solutions (arranged alphabetically according to the chemical formula), see also Table 35 [26]

Table 8: Continued

Concentration, %		*Temperature K (°C)*	*Chemical composition of the steels, %*												
			Cr	–	13	17	18	18	17	18	17	20	25	Ti	Ti-Pd
			Ni	–	–	–	–	9	12	14	15	25	5		
			Mo	–	–	–	2	–	2.5	3.5	4.5	4.5	1.5		
			Cu	–	–	–	–	–	–	–	–	1.5	–		
5–10	3.5	308 (35)		2	0	0	0	0	0	0	0	0	0	0	0
10	7	323 (50)						0	0	0	0	0	0	0	0
20	2	323 (50)						0	0	0	0	0	0	0	0
20	4	333 (60)						1							
25	24	381 (108)		2	2	2	2	2	2	2	2	2	2	0	0
32	20	363 (90)		2	2	2	2	2	2	2	2	2	2	0	0
46	18	373 (100)		2	2	2	2	2	2	2	2	2	2	0	0
51	4	343 (70)		2	2	2	2	2	2	2	2	2	2	0	0
80	0.5	353 (80)						2							
80	5.5	298 (25)						0	0						
96	0.3	353 (80)						0							
$H_2SO_4 + CuSO_4$ (+ copper sulfate)															
4	1	293 (20)			2	2	0	0	0	0	0	0	0	0	0
4–5	approx. 5														
	sat. sol.	293 (20)		2	0	0	0	0	0	0	0	0	0	0	0
8	0.05	353 (80)			2	2	2	2	0	0	0	0	1	2	1
8	1	293 (20)			2	2	0	0	0	0	0	0	0	0	0
10	10	BP			2	1	0	0	0	0	0	0	0	0	0
13	1.3	313 (40)						0	0	0	0	0	0	0	0
14	1.5	313 (40)							0	0	0	0	0	0	0

0: corrosion rate < 0.1 mm/a (< 3.94 mpy)
1: corrosion rate 0.1–1 mm/a (3.94–39.4 mpy)
2: corrosion rate > 1 mm/a (> 39.4 mpy)
P pitting and crevice corrosion possible
S stress corrosion cracking
BP boiling point

Table 8: Compilation of corrosion rates on steels in sulfuric acid and in sulfuric acid solutions (arranged alphabetically according to the chemical formula), see also Table 35 [26]

Table 8: Continued

Concentration, %			Temperature K (°C)	Chemical composition of the steels, %												
				Cr	–	13	17	18	18	17	18	17	20	25	Ti	Ti-Pd
				Ni	–	–	–	–	9	12	14	15	25	5		
				Mo	–	–	–	2	–	2.5	3.5	4.5	4.5	1.5		
				Cu	–	–	–	–	–	–	–	–	1.5	–		
16	12		393 (120)							1	1	1	1	1	1	1
16	13		363 (90)						0	0	0	0	0	0	0	0
65	0.05		311 (38)							0	0	0	0	0	0	0
65	1		311 (38)							0	0	0	0	0	0	0
$H_2SO_4 + FeSO_4$ (+ ferrous sulfate)																
5	0.05		343 (70)		2	2	2	2	2	0	0	0	0	0	2	0
5	5		313 (40)		2	2	2	0	0	0	0	0	0	0	0	0
8	20		293 (20)		2	2	0	0	0	0	0	0	0	0	0	0
10	0.2		BP		2	2	2	2	2	2	2	2	1	2	0	0
17	7		333 (60)		2	2	2	1	1	0	0	0	0	0	0	0
25	sat. at 298 K (25 °C)		BP		2	2	2	2	2	2	2	2	2	2	2	2
$H_2SO_4 + Fe_2(SO_4)_3$ (+ ferric sulfate)																
2	0.02		BP						2	0	0	0	0	0	0	0
2	10		373 (100)						0	0	0	0	0	0	0	0
7	0.05		353 (80)						0	0	0	0	0	0	1	0
7	10		353 (80)						0	0	0	0	0	0	0	0
8	0.05		353 (80)		2	2	2	0	0	0	0	0	0	0	2	0
10	2		BP			2	2	2	2	2	2	2	0	2	2	0
$H_2SO_4 + HF + KClO_3$ (+ hydrogen fluoride + potassium chlorate)																
9	1	3	333 (60)		2	2	2	1	1^{4S}	1^{4S}	1^{4S}	1^{4S}	0^{4S}	1	2	2

0: corrosion rate < 0.1 mm/a (< 3.94 mpy)
1: corrosion rate 0.1–1 mm/a (3.94–39.4 mpy)
2: corrosion rate > 1 mm/a (> 39.4 mpy)
P pitting and crevice corrosion possible
S stress corrosion cracking
BP boiling point

Table 8: Compilation of corrosion rates on steels in sulfuric acid and in sulfuric acid solutions (arranged alphabetically according to the chemical formula), see also Table 35 [26]

Table 8: Continued

Concentration, %		*Temperature K (°C)*	*Chemical composition of the steels, %*												
			Cr	–	13	17	18	18	17	18	17	20	25	Ti	Ti-Pd
			Ni	–	–	–	–	9	12	14	15	25	5		
			Mo	–	–	–	2	–	2.5	3.5	4.5	4.5	1.5		
			Cu	–	–	–	–	–	–	–	–	1.5	–		
$H_2SO_4 + HNO_3$ (+ nitric acid)															
5	1	298 (25)						0	0	0	0	0	0	0	0
5	1	323 (50)						0	0	0	0	0	0	1	0
10	1	298 (25)						0	0	0	0	0	0	1	0
10	1	353 (80)						1	0	0	0	0	0	2	0
17	1	373 (100)				2		1							
95	1	323 (50)						1	0	0	0	0	0	2	2
99	1	308 (35)						0	0	0	0	0	0	1	1
10	3	298 (25)						0	0	0	0	0	0	0	0
10	3	353 (80)						1	0	0	0	0	0	1	0
50	3	298 (25)						0	0	0	0	0	0	0	0
20	5	298 (25)						0	0	0	0	0	0	0	0
20	5	323 (50)						0	0	0	0	0	0	0	0
60	5	298 (25)						0	0	0	0	0	0	0	0
60	5	323 (50)						0	0	0	0	0	0	0	0
60	5	353 (80)							1	1	1	1	1	2	2
17	7	373 (100)				2		0							
60	10	333 (60)		2	2			0	0	0	0	0	0	1	1
60	10	353 (80)		2	2			1	1	1	1	1	1	2	2
80	10	323 (50)						0	0	0	0	0	0	1	1

0: corrosion rate < 0.1 mm/a (< 3.94 mpy)
1: corrosion rate 0.1–1 mm/a (3.94–39.4 mpy)
2: corrosion rate > 1 mm/a (> 39.4 mpy)
P pitting and crevice corrosion possible
S stress corrosion cracking
BP boiling point

Table 8: Compilation of corrosion rates on steels in sulfuric acid and in sulfuric acid solutions (arranged alphabetically according to the chemical formula), see also Table 35 [26]

Table 8: Continued

Concentration, %		*Temperature K (°C)*	*Chemical composition of the steels, %*												
			Cr	**–**	**13**	**17**	**18**	**18**	**17**	**18**	**17**	**20**	**25**	***Ti***	***Ti-Pd***
			Ni	**–**	**–**	**–**	**–**	**9**	**12**	**14**	**15**	**25**	**5**		
			Mo	**–**	**–**	**–**	**2**	**–**	**2.5**	**3.5**	**4.5**	**4.5**	**1.5**		
			Cu	**–**	**–**	**–**	**–**	**–**	**–**	**–**	**–**	**1.5**	**–**		
90	10	308 (35)						0	0	0	0	0	0	2	2
16	13	373 (100)				1		0							
80	20	293 (20)					0	0	0	0	0	0	0		
80	20	333 (60)			1	1	1	1	0	0	0	0	0	2	2
80	20	373 (100)						1	1	1	1	1	1	2	2
15	25	373 (100)				1		0							
20	30	353 (80)						1	0	0	0	0	0	1	1
40	30	353 (80)						1	1	1	1	1	1	1	1
70	30	308 (35)						0	0	0	0	0	0	1	1
14	47	373 (100)				1		0							
20	50	353 (80)						1	1	1	1	1	1	1	1
50	50	333 (60)		2	1	1	1	1	1	0	0	0	0	0	0
67	54	348 (75)						0							
67	54	BP						2							
95	54	293 (20)				1									
95	54	333 (60)				2		1							
14	56	373 (100)				1		0							
35	65	308 (35)					0	0	0	0	0	0	0	0	0
10	90	308 (35)						0	0	0	0	0	0	0	0

0: corrosion rate < 0.1 mm/a (< 3.94 mpy)
1: corrosion rate 0.1–1 mm/a (3.94–39.4 mpy)
2. corrosion rate > 1 mm/a (> 39.4 mpy)
P pitting and crevice corrosion possible
S stress corrosion cracking
BP boiling point

Table 8: Compilation of corrosion rates on steels in sulfuric acid and in sulfuric acid solutions (arranged alphabetically according to the chemical formula), see also Table 35 [26]

Table 8: Continued

Concentration, %				*Temperature K (°C)*	*Chemical composition of the steels, %*												
					Cr	*–*	*13*	*17*	*18*	*18*	*17*	*18*	*17*	*20*	*25*	*Ti*	*Ti-Pd*
					Ni	*–*	*–*	*–*	*–*	*9*	*12*	*14*	*15*	*25*	*5*		
					Mo	*–*	*–*	*–*	*2*	*–*	*2.5*	*3.5*	*4.5*	*4.5*	*1.5*		
					Cu	*–*	*–*	*–*	*–*	*–*	*–*	*–*	*–*	*1.5*	*–*		
$H_2SO_4 + HNO_3 + C_2H_5OH$ (+ nitric acid + ethyl alcohol)																	
65	5	7		403 (130)		2	2	2	2	2	2	2	2	2	2	2	2
$H_2SO_4 + H_3PO_4$ (+ phosphoric acid)																	
2	40			BP		2	2	2	2	2	2	2	2	2	2	2	2
2	41.4			353 (80)		2	2	2	2	2	0	0	0	0	1	2	2
3.5	41.4			353 (80)		2	2	2	2	2	1	0	0	0	1	0	2
47	43			343 (70)		2	2	2	2	2	1	1	1	1	2	2	2
15	53			333 (60)		2	2	2	2	2	2	2	2	1	2	2	2
3.5	76			353 (80)		2	2	2	2	2	1	0	0	0	1	2	2
$H_2SO_4 + H_3PO_4 + CaSO_4$ (+ phosphoric acid + calcium sulfate)																	
2	4	50		323 (50)		2	2	2		1	0	0	0	0	0	1	0
1	22	traces				2	2	1		1	0	0	0	0	0	2	1
$H_2SO_4 + H_3PO_4 + CrO_3$ (+ phosphoric + chromic acid)																	
14	57	9		353 (80)							2	2	2				
$H_2SO_4 + H_3PO_4 + HF + HNO_3$ (+ phosphoric acid + hydrogen fluoride + nitric acid)																	
25.8	7.9	0.1	12	363 (90)		2	2	2	2	2	1	1	1	1	1	2	2
$H_2SO_4 + H_3PO_4 + HNO_3$ (+ phosphoric + nitric acid)																	
45	43	2		373 (100)							1	1	1				
45	43	2.2		378 (105)							1	1	1	1			
23	64	1.9		363 (90)						0	0	0	0	0			

0: corrosion rate < 0.1 mm/a (< 3.94 mpy)
1: corrosion rate 0.1–1 mm/a (3.94–39.4 mpy)
2: corrosion rate > 1 mm/a (> 39.4 mpy)
P pitting and crevice corrosion possible
S stress corrosion cracking
BP boiling point

Table 8: Compilation of corrosion rates on steels in sulfuric acid and in sulfuric acid solutions (arranged alphabetically according to the chemical formula), see also Table 35 [26]

Table 8: Continued

Concentration, %			*Temperature K (°C)*	*Chemical composition of the steels, %*												
				Cr	–	13	17	18	18	17	18	17	20	25	Ti	Ti-Pd
				Ni	–	–	–	–	9	12	14	15	25	5		
				Mo	–	–	–	2	–	2.5	3.5	4.5	4.5	1.5		
				Cu	–	–	–	–	–	–	–	–	1.5	–		
10.5	66	7.2	373 (100)							1	1	1	1			
18	78	3	368 (95)							1	1	1	1			
$H_2SO_4 + H_3PO_4 + H_2SiF_6$ (+ phosphoric + fluorosilic acid)																
3	30	1	343 (70)		2	2	2	2	2	0	0	0	0	1	2	2
1	55	1	363 (90)		2	2	2	2	2	1	1	1	1	2	2	2
$H_2SO_4 + H_3PO_4 + (NH_4)_2SO_4$ (+ phosphoric acid + ammonium sulfate)																
1.5	10	25	363 (90)							0	0	0	0	0		
1	15	25	373 (100)						2	0	0	0	0			
3	15	25	373 (100)						2	0	0	0	0			
3	15	30	363 (90)						2	0	0	0	0			
20	15	20	293 (20)						2	0	0	0	0			
1	16	9	353 (80)						0	0	0	0	0	0		
$H_2SO_4 + K_2Cr_2O_7$ (+ potassium dichromate)																
1	5		308 (35)		0	0	0	0	0	0	0	0	0	0	0	0
1.5	2.5		BP		2	2	1	0	0	0	0	0	0	0	0	0
5–10	5		308 (35)		2	0	0	0	0	0	0	0	0	0	0	0
51	6		343 (70)		2	2	2	2	2	2	2	2	2	2	0	0
80	0.6		353 (80)						2							
80	8		298 (25)						0	0						
96	0.5		353 (80)						0							

0: corrosion rate < 0.1 mm/a (< 3.94 mpy)
1: corrosion rate 0.1–1 mm/a (3.94–39.4 mpy)
2: corrosion rate > 1 mm/a (> 39.4 mpy)
P pitting and crevice corrosion possible
S stress corrosion cracking
BP boiling point

Table 8: Compilation of corrosion rates on steels in sulfuric acid and in sulfuric acid solutions (arranged alphabetically according to the chemical formula), see also Table 35 [26]

Table 8: Continued

Concentration, %		*Temperature K (°C)*	*Chemical composition of the steels, %*												
			Cr	**–**	**13**	**17**	**18**	**18**	**17**	**18**	**17**	**20**	**25**	***Ti***	***Ti-Pd***
			Ni	**–**	**–**	**–**	**–**	**9**	**12**	**14**	**15**	**25**	**5**		
			Mo	**–**	**–**	**–**	**2**	**–**	**2.5**	**3.5**	**4.5**	**4.5**	**1.5**		
			Cu	**–**	**–**	**–**	**–**	**–**	**–**	**–**	**–**	**1.5**	**–**		
$H_2SO_4 + N_2$ Sat. sol.															
*	0.2	BP						1	1	1	1				
	0.3	BP							1	1	1				
	2	333 (60)							2		1		1		
	20	343 (70)							2		1				
$H_2SO_4 + NaCl$ (+ sodium chloride)															
0.5	1	BP		2	2	2	2	2	2	2	2	2	2	2	0
1	0	323 (50)						0	0			0		0	
1	0.02	323 (50)						0	0			0		0	
1	0.1	323 (50)						0	0			0		0	
1	0.1	333 (60)									0	0		0	
1	0.2	323 (50)						0	0			0		0	
2	0	323 (50)						0	0			0		0	
2	0.02	323 (50)						0	0			0		1	
2	0.1	323 (50)						0	0			0		1	
2	0.1	333 (60)									0	0		1	
2	0.2	323(50)						0	0			0		1	
4	0.1	333 (60)									1			2	0
5	0	323 (50)						0	0			0		2	
5	0.02	323 (50)					0	0			0		2		

0: corrosion rate < 0.1 mm/a (< 3.94 mpy)
1: corrosion rate 0.1–1 mm/a (3.94–39.4 mpy)
2: corrosion rate > 1 mm/a (> 39.4 mpy)
P pitting and crevice corrosion possible
S stress corrosion cracking
BP boiling point
* Sat. sol. saturated solution

Table 8: Compilation of corrosion rates on steels in sulfuric acid and in sulfuric acid solutions (arranged alphabetically according to the chemical formula), see also Table 35 [26]

Table 8: Continued

Concentration, %			*Temperature K (°C)*	*Chemical composition of the steels, %*												
				Cr	**–**	**13**	**17**	**18**	**18**	**17**	**18**	**17**	**20**	**25**	***Ti***	***Ti-Pd***
				Ni	**–**	**–**	**–**	**–**	**9**	**12**	**14**	**15**	**25**	**5**		
				Mo	**–**	**–**	**–**	**2**	**–**	**2.5**	**3.5**	**4.5**	**4.5**	**1.5**		
				Cu	**–**	**–**	**–**	**–**	**–**	**–**	**–**	**–**	**1.5**	**–**		
5	0.1		323 (50)						1	1			0		2	
5	0.2		323 (50)						1	1			0		2	
8	0.1		333 (60)										1			0
10	5		333 (60)						2			2	2	2	2	0
10	5		353 (80)						2			2				1
16	0.1		333 (60)													0
32	0.1		333 (60)										2			
$H_2SO_4 + NaCl + NaF$ (+ sodium chloride + sodium fluoride)																
96	0.1	0.1	316 (43)								0					
$H_2SO_4 + Na_2Cr_2O_7$ (+ sodium dichromate)																
10	9		323 (50)						0	0	0	0	0	0	0	0
20	2.6		323 (50)						0	0	0	0	0	0	0	0
20	5		333 (60)						1							
25	30		381 (108)		2	2	2	2	2	2	2	2	2	2	0	0
32	26		363 (90)		2	2	2	2	2	2	2	2	2	2	0	0
46	23		373 (100)		2	2	2	2	2	2	2	2	2	2	0	0
$H_2SO_4 + Na_2SO_4$ (+ sodium sulfate)																
0.5	1		BP		2	2	2	1	1	1	1	1	0	1	2	0
0.5	4		BP		2	2	2	1	1	1	1	1	0	1	0	0
3	2		BP		2	2	2	2	2	2	2	2	1	2	2	1

0: corrosion rate < 0.1 mm/a (< 3.94 mpy)
1: corrosion rate 0.1–1 mm/a (3.94–39.4 mpy)
2: corrosion rate > 1 mm/a (> 39.4 mpy)
P pitting and crevice corrosion possible
S stress corrosion cracking
BP boiling point

Table 8: Compilation of corrosion rates on steels in sulfuric acid and in sulfuric acid solutions (arranged alphabetically according to the chemical formula), see also Table 35 [26]

Table 8: Continued

Concentration, %			*Temperature K (°C)*	*Chemical composition of the steels, %*												
				Cr	**–**	**13**	**17**	**18**	**18**	**17**	**18**	**17**	**20**	**25**	**Ti**	**Ti-Pd**
				Ni	**–**	**–**	**–**	**–**	**9**	**12**	**14**	**15**	**25**	**5**		
				Mo	**–**	**–**	**–**	**2**	**–**	**2.5**	**3.5**	**4.5**	**4.5**	**1.5**		
				Cu	**–**	**–**	**–**	**–**	**–**	**–**	**–**	**–**	**1.5**	**–**		
5	15		368 (95)		2	2	2	2	2	2	2	2	1	2	1	0
13	20		323 (50)		2	2	2	2	2	1	1	1	0	1	1	0
25	24		303 (30)		2	2	2	2	2	0	0	0	0	0	1	0
$H_2SO_4 + Na_2SO_4 + ZnSO_4$ (spinning bath solution)																
10	20	1	323 (50)							1		0	0		1	0
Sat. with H_2S										1		0	0		2	2
10	20	3	BP		2	2	2	2	2	2	2	2	2	2	2	2
12	10	3	333 (60)							1		0	0		1	0
12	21	0.3	333 (60)													
Sat. with H_2S										2		2	1		2	2
15	11	3	BP		2	2	2	2	2	2	2	2	2	2	2	2
$H_2SO_4 + (NH_4)_2SO_4$ (+ ammonium sulfate)																
0.2	42		373 (100)		2	2	2	1	1	0	0	0	0	0	1	0
1	20		BP		2	2	2	1	1	1	1	0	0	1	2	0
1	40		353		2	2	2	1	1	1	1	0	0	1	1	0
1	40		BP		2	2	2	2	2	2	2	1	0	2	2	0
2	40		353 (80)		2	2	2	1	1	1	1	0	0	1	1	0
2	40		BP		2	2	2	2	2	2	2	1	0	2	2	1
5	10		313 (40)		2	2	1	0	0	0	0	0	0	0	1	0
5	20		313 (40)		2	2	1	0	0	0	0	0	0	0	1	0

0: corrosion rate < 0.1 mm/a (< 3.94 mpy)
1: corrosion rate 0.1–1 mm/a (3.94–39.4 mpy)
2: corrosion rate > 1 mm/a (> 39.4 mpy)
P pitting and crevice corrosion possible
S stress corrosion cracking
BP boiling point

Table 8: Compilation of corrosion rates on steels in sulfuric acid and in sulfuric acid solutions (arranged alphabetically according to the chemical formula), see also Table 35 [26]

Table 8: Continued

Concentration, %		*Temperature K (°C)*	*Chemical composition of the steels, %*												
			Cr	–	13	17	18	18	17	18	17	20	25	Ti	Ti-Pd
			Ni	–	–	–	–	9	12	14	15	25	5		
			Mo	–	–	–	2	–	2.5	3.5	4.5	4.5	1.5		
			Cu	–	–	–	–	–	–	–	–	1.5	–		
5	20	333 (60)		2	2	2	1	1	0	0	0	0	0	1	0
5	20	353 (80)		2	2	2	2	2	2	2	1	0	2	2	1
5	20	BP		2	2	2	2	2	2	2	1	1	2	2	2
5	40	333 (60)		2	2	2	1	1	0	0	0	0	0	1	0
5	40	BP		2	2	2	2	2	2	2	1	1	2	2	2
10	20	313 (40)		2	2	2	1	1	0	0	0	0	0	2	0
10	20	353 (80)		2	2	2	2	2	1	1	1	0	1	2	1
10	20	BP		2	2	2	2	2	2	2	1	1	2	2	2
10	40	313 (40)		2	2	2	1	1	0	0	0	0	0	2	0
10	40	353 (80)		2	2	2	2	2	1	1	1	1	1	2	1
10	51	373 (100)		2	2	2	2	2	2	2	1	1	2	2	2
$H_2SO_4 + O_2$ (+ oxygen) sat. solution															
*	0.2	BP					2								
	0.3	BP						0		0					
		BP						0		1					
	2	343 (70)						1							
	3	343 (70)						2							
	95	323 (50)						0							
	95	333 (60)						1							

0: corrosion rate < 0.1 mm/a (< 3.94 mpy)
1: corrosion rate 0.1–1 mm/a (3.94–39.4 mpy)
2: corrosion rate > 1 mm/a (> 39.4 mpy)
P pitting and crevice corrosion possible
S stress corrosion cracking
BP boiling point
* Sat. sol. saturated solution

Table 8: Compilation of corrosion rates on steels in sulfuric acid and in sulfuric acid solutions (arranged alphabetically according to the chemical formula), see also Table 35 [26]

Table 8: Continued

Concentration, %	*Temperature K (°C)*	*Chemical composition of the steels, %*												
		Cr	*–*	*13*	*17*	*18*	*18*	*17*	*18*	*17*	*20*	*25*	*Ti*	*Ti-Pd*
		Ni	*–*	*–*	*–*	*–*	*9*	*12*	*14*	*15*	*25*	*5*		
		Mo	*–*	*–*	*–*	*2*	*–*	*2.5*	*3.5*	*4.5*	*4.5*	*1.5*		
		Cu	*–*	*–*	*–*	*–*	*–*	*–*	*–*	*–*	*1.5*	*–*		
$H_2SO_4 + SO_2$ (+ sulfur dioxide) sat. solution														
* 0.5	363 (90)		2	2	2	2	2	1	0	0	0	1	0	0
0.5	BP		2	2	2	2	2	2	2	2	1	2	0	0
2	343 (70)									0				
5	343 (70)									1				
10	323 (50)		2	2	2	2	2	1	1	1	0	2	0	0
20	313 (40)		2	2	2	2	2	1	1	1	0	2	0	0
20	333 (60)		2	2	2	2	2	2	2	2	2	2	1	1
30	293 (20)		2	2	2	2	2	1	1	1	0	2	0	0
30	353 (80)		2	2	2	2	2	2	2	2	2	2	2	1
50	293 (20)		2	2	2	2	2	2	1	1	0	2	0	0
60	293 (20)		2	2	2	2	2	2	2	2	0	2	2	2
90	313 (40)										1			
95	323 (50)									1		0		
95	333 (60)									1		1		
96	313 (40)										1			
96	328 (55)						1	1	1	1	1	1	2	2
98	353 (80)		2				1	1	1	1	1	1	2	2
$H_2SO_4 + SO_3$ (+ sulfur trioxide (oleum))														
100 7	333 (60)		0	0	0	0	0	0	0	0	0	0	2	2
100 11	333 (60)		0	0	0	0	0	0	0	0	0	0	2	2

0: corrosion rate < 0.1 mm/a (< 3.94 mpy)
1: corrosion rate 0.1–1 mm/a (3.94–39.4 mpy)
2: corrosion rate > 1 mm/a (> 39.4 mpy)
P pitting and crevice corrosion possible
S stress corrosion cracking
BP boiling point
* Sat. sol. saturated solution

Table 8: Compilation of corrosion rates on steels in sulfuric acid and in sulfuric acid solutions (arranged alphabetically according to the chemical formula), see also Table 35 [26]

Table 8: Continued

Concentration, %		*Temperature K (°C)*	*Chemical composition of the steels, %*												
			Cr	–	13	17	18	18	17	18	17	20	25	*Ti*	*Ti-Pd*
			Ni	–	–	–	–	9	12	14	15	25	5		
			Mo	–	–	–	2	–	2.5	3.5	4.5	4.5	1.5		
			Cu	–	–	–	–	–	–	–	–	1.5	–		
100	11	373 (100)		2	2	2		1	0				0	2	2
100	60	293–343 (20–70)						0	0	0	0	0	0	2	2
100	60	353 (80)			2	2		0	0				0	2	2
$H_2SO_4 + ZnSO_4$ (+ zinc sulfate)															
0.5	30	BP		2	2	2	2	2	1	1	1	0	1	1	0
1	1	338 (65)		2	2	2	1	1	0	0	0	0	0	0	0
2	30–45	353 (80)		2	2	2	2	2	1	1	1	0	1	2	0
10	5	323 (50)		2	2	2	2	2	1	1	1	0	1	2	0

0: corrosion rate < 0.1 mm/a (< 3.94 mpy)
1: corrosion rate 0.1–1 mm/a (3.94–39.4 mpy)
2: corrosion rate > 1 mm/a (> 39.4 mpy)
P pitting and crevice corrosion possible
S stress corrosion cracking
BP boiling point

Table 8: Compilation of corrosion rates on steels in sulfuric acid and in sulfuric acid solutions (arranged alphabetically according to the chemical formula), see also Table 35 [26]

– Stress corrosion cracking in sulfuric acid solutions –

According to [27], stress corrosion cracking (SCC) may occur in sulfuric acid solutions. The consequence of intermittent operation (start-up, shut-down and preheating periods) in power stations, especially those heated with fossil fuels which are operated mainly during medium and peak loads, is condensate formation in the entire turbo-generator with accumulation of the impurities present in the steam, for example sulfates and chlorides. The alternating operating states lead to concentration of the condensates and stress fluctuations in the material. The prestressing applied to screws in particular provides conditions for inducing stress corrosion cracking on screw materials. The MoS_2 (molybdenum disulfide) lubricating pastes applied may lead to decomposition of MoS_2 at operating conditions above 623 K (350 °C), to form SO_2, H_2SO_3 and finally H_2SO_4.

Screw materials shown in Table 9 were tested for susceptibility to stress corrosion cracking in aqueous solutions containing sulfite, sulfate and chloride under conditions similar to in practice [28, 29]. The test temperature was 363 K (90 °C), and in isolated instances 353 K (80 °C). The test stress was between 0.62 and 1.45 $R_{p0.2}$, i.e. below and above the 0.2 % yield strength. Figure 17 shows the test results for the low-alloy screw steel 21 CrMoV 5 7. No SCC was found below a test stress of $R_{p0.2}$. Premature fractures were, without exception, a consequence of general corrosion (reduction in cross-section). The specimens exhibited a severe reduction in area after fracture, but no cracking was found [28, 29].

– Influence of hydrogen uptake on corrosion behavior –

As well as corrosion in sulfuric acid solutions, hydrogen uptake may occur as a secondary effect. This is possible, for example, during electroplating or pickling [30–34]. The hydrogen uptake is associated with a reduction in the deformability. Hydrogen can be taken up by the material in several forms:

- in atomic form, dissolved in the lattice
- in molecular form, accumulated (recombined) in slag lines, pores and microcracks
- in embedded form at dislocations, phase boundaries, lattice defects and
- in chemically bound form as hydride.

Material	*Section*	*Melt*	*Chemical composition, %*											
			C	*Si*	*Mn*	*P*	*S*	*Al*	*Mo*	*Cr*	*V*	*Ti*	*Nb*	*N*
NiCr20TiAl		1	0.07	0.24	0.05	–	–	1.20	–	19.65	–	2.48	–	–
		2	0.07	0.02	0.02	0.003	0.003	1.32	–	19.65	–	20.07	–	–
21 CrMoV 5 7	A	1	0.20	0.21	0.63	0.019	0.018	–	0.69	1.39	0.27	–	–	–
		2	0.20	0.24	0.41	0.008	0.005	–	0.76	1.29	0.32	–	–	–
X 19 CrMoVNbN 11 1	D	1	0.19	0.22	0.57	0.019	0.005	–	0.69	10.41	0.16	–	0.49	0.059

A Unalloyed steels and cast steel
D Structural steels with up to 12 % chromium

Table 9: Chemical composition of heat-resistant screw materials investigated for susceptibility to stress corrosion cracking in aqueous solutions containing sulfite, sulfate and chloride (temperature 363 K (90 °C), in isolated instances 353 K (80 °C), test stress 0.62–1.45 $R_{p0.2}$, see Figures 17 and 43) [28, 29]

The following types of hydrogen attack result:

- Hydrogen absorption: formation of internal cracks, H-induced stress corrosion cracking, blistering on unalloyed steels, reduction in deformability
- internal chemical reaction: decarburization by pressurized hydrogen at temperatures above 473 K (200 °C)
- external hydride formation with general corrosion
- internal hydride formation with embrittlement of the material and the risk of cracking under load [35].

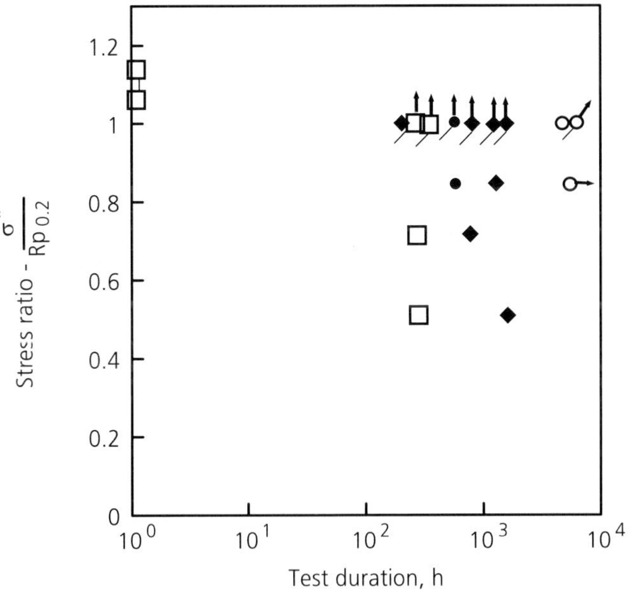

Figure 17: Stress corrosion cracking behavior of the heat-resistant screw steel 21 CrMoV 5 7 (see Table 9) in various solutions (simulated turbine condensates) at 363 K (90 °C) [28, 29]
* test stress
○ 1 % NaCl
● 0.05 mol/l H_2SO_4
◆ 1 % NaCl + 0.05 mol/l H_2SO_4
□ $H_2O + SO_2 \uparrow$
/ = corrected

Figure 18 shows the dependence of the hydrogen uptake of iron and iron alloys on the duration of corrosion in 1 mol/l sulfuric acid. The graph shows the hydrogen uptake as a function of the alloying elements and the heat treatment independent of the use of iron materials in this medium [32]. Figure 19 shows the weight loss under such conditions. Anodic dissolution is greatly inhibited by alloying elements such as copper, tin or nickel, and, as can be seen in the case of copper and nickel, the hydrogen uptake and therefore the risk of possible embrittlement is also reduced. An increase in the weight loss is observed in the copper-rich iron alloy containing 3.4 % Cu as compared with the alloy containing 0.55 % Cu. This is attributed to the formation of microcells by Cu-precipitates in the iron matrix. Sulfur, phosphorus and manganese increase the corrosion rate and hydrogen activity of iron. Similarly, 3.1 % Si or 10.4 % Cr do not provide protection from corrosion and hydrogen uptake in sulfuric acid [33].

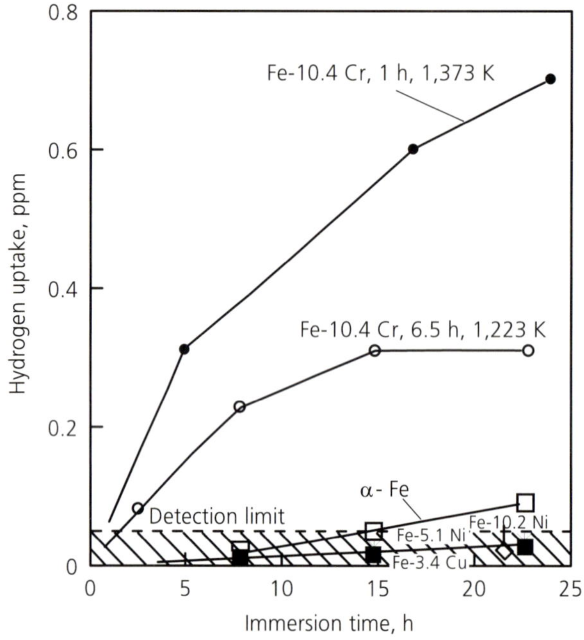

Figure 18: Hydrogen uptake of iron and iron alloys in sulfuric acid (1 mol/l) at 298 K (25 °C) as a function of the immersion time [32]

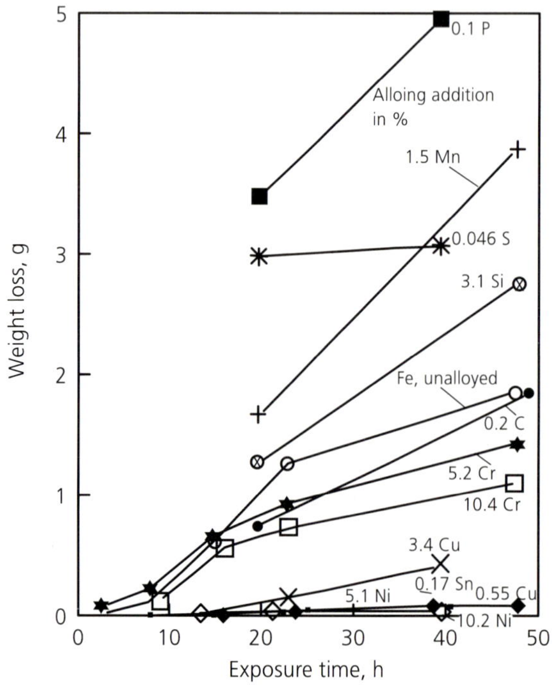

Figure 19: Weight loss of iron and iron alloys during corrosion in sulfuric acid (1 mol/l) at 298 K (25 °C) (see also Figure 18) [33]

– Hydrogen-induced stress corrosion cracking (HSCC) in sulfuric acid solutions –

Corrosion studies on hydrogen-induced stress corrosion cracking (HSCC) in 1 mol/l sulfuric acid have demonstrated that the sensitivity of low-strength steels (120 to 265 MPa yield strength) increases as the phosphorus content increases. In contrast, the sensitivity in solutions containing hydrogen sulfide is independent of the phosphorus content. Whereas the sensitivity of these steels to HSCC does not depend on the intercrystalline segregation of phosphorus at concentrations of up to 20 atomic percent, the sensitivity of high-strength steels (875 to 960 MPa) does. In the first case transcrystalline fracture was found in a tensile test at a low elongation rate (10^{-6} 1/s) at 298 K (25 °C), and in the second case intercrystalline fracture was observed [36].

Corrosion studies on susceptibility to H-induced stress corrosion cracking were performed on high-strength piping steels (Table 10) with $R_{p0.2}$ = 430–700 MPa [37]. H-induced stress corrosion cracking may be facilitated by increasing load. This is particularly the case if the components (pipes) are cathodically protected in the region of hydrogen deposition and if they have sharp notches. Hydrogen deposition alone is not adequate, even under a high static load, if no promoters of hydrogen uptake are present in the medium. Hydrogen sulfide and sulfides in the soil are the most important promoters. The specific action of the cathodic protective current, which reduces the oxygen at the steel surface and provides an anaerobic environment, should be noted here. In such an environment, sulfates can be reduced to sulfide by sulfate-reducing bacteria. The pH values in the electrolyte may be quite low under these conditions.

For this reason, the corrosion studies were performed mainly at pH 5.5 in an electrolyte called a "synthetic soil solution" (see Table 10).

The susceptibility of unalloyed and low-alloy steels to H-induced stress corrosion cracking depends on the temperature. As the temperature increases, the sensitivity to HSCC decreases, and at 333 K (60 °C) is non-existent in the steels shown in Table 10. The studies were performed as a result of the requirement for risk-free cathodic protection of pipelines from high-strength steels, i.e. H-induced stress corrosion cracking had to be avoided.

It was thus found that the essential influencing parameters are not the strength and cathodic protection, but the presence of sulfide, mechanical stresses and the pH value. Under these conditions, cathodic protection can be applied whenever sulfides are absent and the pH value at the pipe wall is sufficiently high, which ought to be the case in neutral media. In acid media, if sulfides are present there is the risk of H-induced stress corrosion cracking even in soft steels. In neutral media, there is no risk of HSCC in the case of free corrosion because the steady state potential is not sufficiently negative. The worst that can happen here is shallow pitting and pitting corrosion.

The potential must be below the protection potential $U_{Cu/CuSO4}$ = –0.85 (U_H = –0.53 V) for the cathodic protection of unalloyed steels. For this reason, the studies were performed at U_H = –0.55 V.

The results of the studies on high-strength piping steels of $R_{p0.2}$ = 430–700 MPa have shown that in weakly acidic, sulfide-containing media (pH 5.5) H-induced stress corrosion cracking occurs above a critical threshold stress of about 60 % $R_{p0.2}$. The critical stress decreases as the pH value decreases (30 % $R_{p0.2}$ at pH 3) and increases as the pH increases. If the pH value is sufficiently high, the critical stress is identical to the yield strength. HSCC cannot be concluded from the appearance of H-induced cracks (internal separation of the material). Similarly, high-strength steels (yield strength up to 700 MPa) are no more susceptible to HSCC than those of lower strength. The corrosion behavior with this type of corrosion in sulfide-containing solutions at pH 5.5 basically corresponds to that in stronger acid solutions of pH 3. The threshold stresses of course also depend on the potential. At pH 5.5, the threshold potential U_{HSCC} for the occurrence of HSCC at 60 % $R_{p0.2}$ which depends on the tensile stress, is between U_H = –0.42 and 0.55 V. Below this, at U_H = –0.8 and –1.0 V, further HSCC is unlikely [37].

Chemical composition, %												
C	*Si*	*Mn*	*P*	*S*	*Al*	*Cu*	*Cr*	*Ni*	*Mo*	*Nb*	*V*	*N*
0.18	0.39	1.39	0.014	0.012	0.035	n.g.	n.g.	n.g.	n.g.	0.01	0.10	0.0165
0.14	0.37	1.49	0.022	0.012	0.023	n.g.	n.g.	n.g.	n.g.	0.031	0.06	0.0140
0.15	0.29	1.39	0.017	0.015	0.032	0.09	0.03	0.06	0.02	0.032	0.00	0.0070
0.16	0.37	1.36	0.015	0.013	0.034	0.06	0.05	0.06	0.01	0.031	0.00	0.0060
0.07	0.32	1.26	0.013	0.008	0.033	0.02	0.02	0.00	0.00	0.039	0.00	0.0070
0.14	0.43	1.36	0.015	0.010	0.044	0.02	0.02	0.00	0.00	0.028	0.07	0.0130
0.16	0.49	1.40	0.014	0.010	0.036	n.g.	n.g.	n.g.	n.g.	0.045	0.08	0.0160
0.16	0.40	1.45	0.015	0.008	0.032	0.05	0.07	0.04	n.g.	0.035	n.g.	n.g.
0.03	0.19	2.19	0.021	0.008	0.027	0.08	0.04	0.01	0.33	0.115	0.00	0.0070
0.07	0.37	1.87	0.017	0.012	0.055	0.05	0.02	0.01	0.01	0.050	n.g.	0.0150
0.18	0.42	1.53	0.014	0.016	0.043	n.g.	n.g.	n.g.	n.g.	n.g.	0.10	0.0190

n.g. = not given

Table 10: Chemical composition of high-strength piping steels in corrosion studies in a "synthetic soil solution" (0.025 mol/l (COONa)-(CH·OH)$_2$-(COONa) + 0.05 mol/l K_2SO_4 + x ml 1 mol/l H_2SO_4 + y ml 0.1 mol/l NaOH containing 0.2 g/l Na_2S) for susceptibility to H-induced stress corrosion cracking (the pH was established with H_2SO_4 at ≤ and with NaOH at ≥ in the pH range 5.0–9.0 at temperatures between 275 and 313 K (2 and 40 °C) [37]

Hydrogen-induced stress corrosion cracking can occur in prestressed steels in non-pressed encasement tubes due to atmospheric corrosion, as well as corrosion in encasement tube water formed in the prestressed channels after concreting. In this case, hydrogen uptake during corrosion is possible both in weakly acidic and in neutral to alkaline solutions. The direct hydrogen uptake in acid solutions by discharge of hydrogen ions loses significance as the pH value of the electrolyte increases. Hydrogen can be deposited and taken up on zinc-coated steels with a damaged zinc layer as a result of cathodic polarization of the exposed steel surface by zinc. The H-induced crack propagation is more transcrystalline in character in soft steels, whereas in high-strength steels it is more intercrystalline in character. Grain and phase boundaries perpendicular to the attacking tensile stress facilitate crack propagation. Cold-drawn steels in which the internal boundaries are aligned predominantly axially and parallel to the attacking tensile stresses are less sensitive since they can bind large quantities of hydrogen at these surfaces and render them practically harmless. If tensile stresses occur perpendicularly to the deformation direction in these steels, cracks can also occur in these steels, but they run axially. In the case of prestressed steels, tensile strength plays a minor role alongside the parameters of H-concentration, tensile stress and steel structure. The difference in structure manifests itself in particular in steels of similar chemical composition and strength but different pretreatment (e.g. heat treatment or patented drawing) [38].

– *Electrochemical corrosion protection processes* –

In the electrochemical corrosion protection processes, a potential at which the corrosion rate is practically negligible and, for example, local corrosion phenomena (pitting and stress corrosion cracking) are avoided is in general imposed on the metallic surface.

The adjustment of the potential is achieved by direct current polarization and is called cathodic or anodic protection, depending on the direction of the change in potential.

Knowing the current density potential curve of the corrosion system is the most important criterion for deciding which protection process can be used or is more suitable.

– *Cathodic protection* –

In this process nowadays, the component to be protected is connected chiefly to the negative pole of a direct current source. A resistant anode serves as the counter-electrode.

The weight losses at certain potentials, taking into account, for example, pitting corrosion or hydrogen uptake (see in this context the influence of hydrogen and [37]) are a precondition for determination of the protection potential. The protection current density must be present over the entire surface.

For iron, a protection potential more negative than $U_H = -0.53$ V is necessary for adequate cathodic protection in neutral solutions. For unalloyed steels, a polarization potential slightly below this can be applied (or is necessary) in weak sulfuric acid media.

If the current density potential curve exhibits a pronounced passive region in the corrosion medium, the use of anodic protection is to be recommended. This is possible at high concentrations of sulfuric acid, since under these conditions sulfate layers, which inhibit further corrosion, are formed.

– Anodic protection –

As already mentioned under cathodic protection, a passive region in the current density potential curve is the fundamental precondition for use of this protection process in sulfuric acid.

In this process, after the relatively high passivation current density has been overcome, the corrosion potential is shifted into the passive range, in which unalloyed steel (as the material in our case) corrodes at a substantially lower current density (see Table 11) [39]. In contrast to the cathodic protection process in which corrosion can be completely suppressed, the material corrodes under the conditions of passivity with very small corrosion losses. The presence of chlorides has a very adverse effect on this process. As in all cases, chlorides cause pitting, and, to avoid this, the material must remain in the passive state below certain potential limits or anodic protection must be dispensed with completely.

For anodic protection, storage tanks for 93 % sulfuric acid of unalloyed steel, for example, are suitable at temperatures of up to 313 K (40 °C) as a guarantee of reliable protection [54].

Table 11 shows the effect of anodic protection in the region of passivatability of unalloyed steel in various concentrations of sulfuric acid at 313 K (40 °C). As a result of the anodic protection potential, corrosion rates are achieved on unalloyed steel in highly concentrated sulfuric acid which render it practically resistant under these conditions.

The decline in the material consumption rate can be seen particularly clearly in 85 % sulfuric acid, where it drops from 3.3 to 0.1 g/m^2 h (0.11 mm/a (4.33 mpy)) at a protection potential of 1.5 V and a very low corrosion current density in the passive state of only 4.5 $\mu A/cm^2$ [39]. Anodic protection is a reliable means of avoiding breakdowns [54].

H_2SO_4 concentration %	Anodic protection		Corrosion current density in the passive state $\mu A/cm^2$
	Without	With (1.5 V)	
	Material consumption rate, g/m^2 h*		
70	0.86	1.34	130.0
75	0.53	0.38	39.0
80	0.73	0.20	20.0
85	3.30	0.10	4.5
90	0.63	0.11	7.9
93	0.84	0.14	15.0

* × 1.1 mm/a

Table 11: Influence of anodic protection on the corrosion rate of unalloyed steel in various concentrations of sulfuric acid at 313 K (40 °C) [39]

– Corrosion inhibitors –

The corrosion rate of the steel Ch 40 (about 1.7035; 41Cr4) is about 0.22 mm/a (8.66 mpy) at room temperature in artificial seawater and in pit water (0.35 mol/l H_2SO_4 + 0.5 mol/l NaCl), regardless of whether the inhibitor used is ChOSp-10 or Katapin BPV. Under dynamic loading (0.83 Hz, 2×10^5 load cycles), the first inhibitor proves to be the more effective, whilst the second has no effect. The protective action of the inhibitor increases as the concentration increases, and to a greater degree in pit water than in artificial seawater. The service life can even reach that in air. Inhibitor additions of more than 1.5 g/l are quoted. The best action is achieved with concentrations of about 6 g/l in pit water [40].

Inhibitors which largely suppress or reduce corrosion are used to protect unalloyed and low-alloy steel from corrosion by media containing sulfuric acid [40–51]. If used in pickling solutions, they should reduce the formation of hydrogen with its associated consequences (see also the influence of hydrogen), as well as the severe material consumption.

In the use of unalloyed steels in sulfuric acid at concentrations above 65 %, inhibitors should drastically reduce material consumption due to corrosion. Sorbic acid occurs in rowan fruits and is used as a natural, physiologically acceptable preservative as a result of its fungistatic and anti-microbial action. Since long-chain, aliphatic carboxylic acids can have an inhibitory effect on metal corrosion (see also [48]) and sorbic acid (CH_3-CH=CH-CH=CH-COOH) is used as a preservative in sour preserves, experiments have been performed on its inhibitory action in acid solutions. In a sulfuric acid concentration of 0.05 mol/l, sorbic acid additions of 0.01 and 0.1 g/l slightly stimulate corrosion both in oxygen-containing and in oxygen-free solutions. When 1.0 g/l is added, a slight inhibiting action is observed at 294 K (21 °C), with a protective action of 19 % in oxygen-containing solution and 24 % in oxygen-free solution [41].

Table 12 shows the protective action of some inhibitors in 10 % sulfuric acid at various temperatures on static and agitated specimens. Table 13 shows that the inhibitors also have a very good action in other acids, with an inhibition efficiency above 90 % [42]. As well as the inhibiting action of the individual inhibitors, the hydrogen uptake in the steel should also be noted, as can be seen from Table 14. This summary shows that a good inhibiting action regarding weight loss does not necessarily mean a good inhibition of hydrogen uptake [31, 42].

In some cases, the hydrogen absorption is even stimulated by inhibitors. For this reason it is also important to know the minimum concentration of hydrogen in the material at which H-induced damage no longer occurs, as well as the reduction in the weight loss [37, 51].

In spite of its adverse effect on the mechanical properties of steel (H-embrittlement), thiourea is still often used as an inhibitor in acid corrosion of steels [44]. Corrosion studies on steels of varying carbon content (0.002 %, 0.17 %, 0.36 %, 0.61 %, 1.05 %) but the same content of the following elements: 0.015 % S, 0.008 % P, 0.22 % Si, 0.24 % Mn, 0.20 % Al, 0.04 % Ni, 0.02 % Cu, 0.04 % Cr, 0.01 % Mo, have demonstrated that the rate of acid corrosion passes through a minimum according to the thiourea concentration. The minima are shifted more in the direction of higher thiourea concentrations the higher the carbon content of the steels. Oxygen-free 2 mol/l sulfuric acid at 298 K (25 °C) was used as the test solution. At concentrations above 0.1 % thiourea, the inhibitor becomes more effective the higher the carbon content of the steel [44].

Inhibitor	*Temperature, K (°C)*		
	328 (55)	*358 (85)*	*328 (55)*
	Static sample		*Rotating disk, 1 m/s*
0.05 %	*Inhibition efficiency, %*		
None	–	–	–
DBSO[1)]	99	99	81
EMBI[2)]	99	99	86
BUD[3)]	96	76	61
DOT[4)]	98	96	93

1) dibenzyl sulfoxide (Preventol® Cl–5)
2) ethoxylated mercaptobenzimidazole (Preventol® C1–6)
3) 1,4-butynediol
4) di-o-tolylthiourea

Table 12: Inhibition efficiency of inhibitors in unalloyed steel RSt 14 in 10 % sulfuric acid [42]

Inhibitor	*Inhibition efficiency, %*			
0.05 %	*Hydrochloric acid*	*Sulfuric acid*	*Phosphoric acid*	*Formic acid*
DBSO[1)]				
328 K (55 °C)	99	99	99	90
358 K (85 °C)	99	99	99	97
EMBI[2)]				
328 K (55 °C)	97	99	99	98
358 K (85 °C)	99	99	99	99

1) dibenzyl sulfoxide (Preventol® C1–5)dibenzyl sulfoxide
2) ethoxylated mercaptobenzimidazole (Preventol® C1–6)

Table 13: Inhibition efficiency of commercially available inhibitors on unalloyed steel RSt 14 in static 10 % sulfuric acid in comparison with other acids [42]

Inhibitor	*Inhibition efficiency*	*Hydrogen uptake*
Mol/l	*%*	*ml per 100 g*
None		8.93
Thiourea		
10^{-1}	53	10.8
10^{-2}	90	9.8
10^{-3}	95	8.0
10^{-4}	92	9.0
10^{-5}	44	9.2
Monomethylthiourea		
10^{-2}	93	8.7
10^{-3}	97	7.4
10^{-4}	93	7.3
10^{-5}	27	8.0
Tetramethylthiourea		
10^{-1}	91	7.4
10^{-3}	97	4.7
10^{-4}	94	7.1
10^{-5}	41	9.0
Diethylthiourea		
10^{-3}	99	3.3
10^{-4}	93	6.8
10^{-5}	9	10.0
Monophenylthiourea		
10^{-3}	98	3.1
10^{-4}	91	7.2
10^{-5}	26	8.7

Table 14: Inhibition efficiency and hydrogen uptake of unalloyed steel during corrosion in sulfuric acid (10 mol/l) at 323 K (50 °C) due to inhibitors [31, 42]

The influence of alkynols on the temperature dependence of the H-permeation current density on unalloyed steel in H_2S-free and H_2S-containing 2.5 mol/l sulfuric acid is investigated in [45]. According to this work, the H-permeation rate increases as the temperature rises. Above 323 K (50 °C), noticeable inhibition is achieved in H_2S-containing and H_2S-free acid only by 1-hexyne-3-ol and 1-octyne-3-ol.

The inhibitory activity of sulfoxides, as well as other substances, in H_2S-containing and H_2S-free sulfuric acid is investigated. Table 15 contains the results. The best action is achieved with dibenzyl sulfoxide at a concentration of 10^{-3} mol/l at 298 K (25 °C) in 2.5 mol/l sulfuric acid. The corrosion rate drops to 0.08 mm/a (3.15 mpy), with a corrosion rate in the inhibitor-free solution of 178 mm/a (7,008 mpy) at H_2S saturation [45].

Inhibitor	*N_2-flushed*		*H_2S-saturated*	
	mol/l		*mol/l*	
	10^{-4}	*10^{-3}*	*10^{-4}*	*10^{-3}*
	Corrosion rate, mm/a (mpy)			
None	22 (866)	22 (866)	178 (7,008)	178 (7,008)
Dimethyl sulfoxide	23 (906)	21 (827)	205 (8,071)	192 (7,559)
Di-n-butyl sulfoxide	17 (669)	1 (39.4)	9 (354)	1.1 (43.3)
Tetramethylene sulfoxide	16 (630)	9 (354)	200 (7,874)	12 (472)
Diphenyl sulfoxide	22 (866)	6 (236)	165 (6,496)	0.55 (21.7)
Di-p-tolyl sulfoxide	23 (905)	13 (512)	8 (315)	0.15 (5.91)
Dibenzyl sulfoxide	1.2 (47.2)	0.03 (1.18)	0.31 (12.2)	0.08 (3.15)

Table 15: Corrosion of unalloyed steel in sulfuric acid (2.5 mol/l) at 298 K (25 °C) with and without inhibitor addition (sulfoxides) [45]

Table 16 shows the inhibiting action of pyrrole with and without halide addition.

Inhibitor	*Temperature, K (°C)*		
	311 (38)	*339 (66)*	*366 (93)*
	Corrosion rate, mm/a (mpy)*		
None	35.8 (1,409)	188.0 (7,402)	753.0 (29,645)
Pyrrole (0.1 %)	8.2 (323)	44.6 (1,756)	155.0 (6,102)
NaCl (0.017 mol/l)	29.2 (1,150)	180.6 (7,110)	586.0 (23,071)
NaI (0.017 mol/l)	1.2 (47.2)	6.4 (252)	23.7 (933)
Pyrrole + NaCl	1.6 (63)	8.7 (343)	22.2 (874)
Pyrrole + NaI	0.1 (3.94)	0.5 (19.7)	1.6 (63.0)

* conversion from mg/cm^2 h

Table 16: Corrosion on low-carbon steel (for analysis, see text) in sulfuric acid (1 mol/l) as a function of the temperature and the addition of pyrrole as an inhibitor [43]

The corrosion rate increases sharply with and without inhibitor. Whereas the unalloyed steel ((%) Fe-0.022 C, 0.14 Mn, 0.007 P, 0.017 S, 0.004 Si, 0.015 Cu, < 0.002 Ni, 0.010 Cr) is already destroyed at 311 K (38 °C) without an inhibitor, it can easily be used at 311 K (38 °C) if 0.1 % pyrrole and 0.017 mol/l NaI are present, and under certain circumstances is worth considering at higher temperatures [43].

Substances which can be used as inhibitors are obtained as by-products in the production of corn starch. They consist of 40–52 % protein, 22–27 % soluble hydrocarbons, 1–3 % fat, up to 0.5 % starch and 0.7–1.1 % lactic acid (EK), as well as ethanol octacetylgentiobiose and sodium chloride (EK-1).

The steel St 3 is dissolved in 20 % H_2SO_4, as already described above. The corrosion rates, which decrease with time as a result of the high initial corrosion, were found to be 45 mm/a (1,772 mpy) at 293 K (20 °C) after a test duration of 6 h (144 mm/a (5,669 mpy) after 1 h). Additions of the inhibitors EK and EK-1 led to the corrosion rates in Table 17. A corrosion rate of 2.2 mm/a (86.6 mpy) is found after an exposure time of 1 h by addition of 6 ml/l [46]. On the basis of the short test durations, these figures can only give an initial overview of the action of the inhibitors and a clear conclusion on the temperature dependence can be made only with reservation.

Benzylhexamethylenetetramine iodide in concentrations of about 0.003 mol/l has proved to be a very good inhibitor of corrosion on the steel St 3 in 20 % sulfuric acid. The inhibition increases as the temperature rises [47].

The acid media used in industry, in this case sulfuric acid media in particular, are rarely pure. They contain substances which may influence the activity of inhibitors in various ways. According to [50], an inhibitor which protects both unalloyed and high-alloy steel from corrosion in acid media has been developed. The starting substance for its preparation is a higher aldehyde. This is reacted with liquid ammonia at 353 K (80 °C) for 30 min. Two phases then form, of which only the upper phase is used. Ammonium carbonate is added to this in a CO_2 atmosphere at 383 K (110 °C). The reaction takes place in a reflux condenser over a period of between 1 and 4 h. After this treatment, the solution is deep red and viscous. The solution obtained after a reaction of 4 h has been called DX-A, and exhibited the best inhibitor action at a concentration of 1.0 % in 2.5 mol/l H_2SO_4 at room temperature, with an inhibition efficiency of 99.23 %. The corrosion rate (no test duration stated) of 46 mm/a (1,811 mpy) without DX-A drops to 0.4 mm/a (15.7 mpy) when 0.1 % is added (Table 55). Figure 20 shows the influence of DX-A on the corrosion behavior of the unalloyed steel PCRCA in sulfuric acid as a function of the temperature at room temperature. A marked decrease in the corrosion rates of unalloyed steel in sulfuric acid is detectable here, which makes its use in dilute acid and at high concentrations worth considering.

Figure 21 shows the influence of inhibitor DX-A activity on unalloyed steel in 2.5 mol/l H_2SO_4 at room temperature as a function of the test duration.

Whereas without the inhibitor the corrosion rate in 2.5 mol/l sulfuric acid rises virtually continuously as the immersion time increases, it remains virtually constant or rises only slightly in the given test duration of 24 h. Only a slight increase is observed as the temperature rises to 333 K (60 °C); this intensifies up to 353 K (80 °C) and above this temperature increases rapidly (Figure 22) [50]. For corrosion protection by inhibitors see also [3].

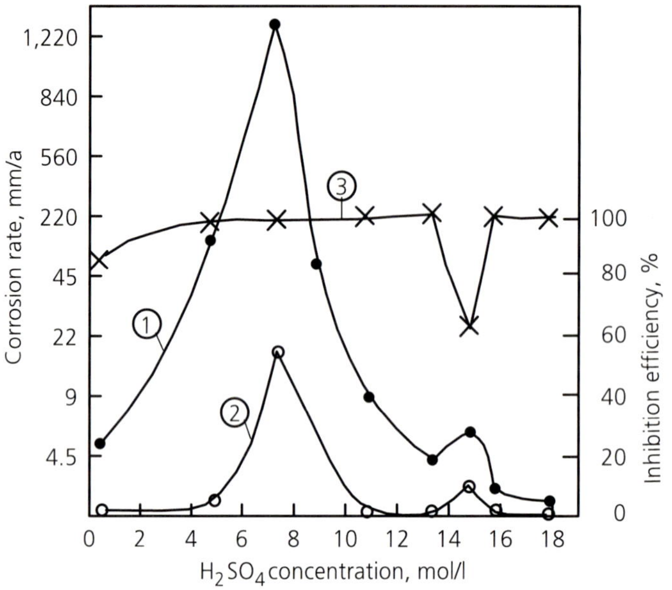

Figure 20: Influence of the inhibitor DX-A on the corrosion behavior of unalloyed steel in sulfuric acid at room temperature [50]
① H_2SO_4 without DX-A
② H_2SO_4 + 1.0 % DX-A
③ inhibition efficiency

Inhibitor addition ml/l	*Temperature K (°C)*	*Test duration h*	*Material consumption rate g/m² h*	*Corrosion rate mm/a (mpy)*
	293 (20)	1	130	144 (5,669)
	293 (20)	2	80	87 (3,425)
	293 (20)	4	56	62 (2,441)
	293 (20)	6	41	45 (1,772)
6 EK[1)]	293 (20)	3	2	2.2 (86.6)
2 EK	353 (80)	1	52	58 (2,283)
4 EK-1[2)]	363 (90)	1	78	87 (3,425)

1) 40–52 % protein, 22–27 % soluble hydrocarbons, 1–3 % fat, up to 0.5 % starch, 0.7–1.1 % lactic acid
2) ethanol, octacetylgentiobiose, sodium chloride

Table 17: Inhibitors obtained as by-products in the production of corn starch and their effect on the corrosion behavior of unalloyed steel St 3 in 20 % sulfuric acid at various temperatures [46]

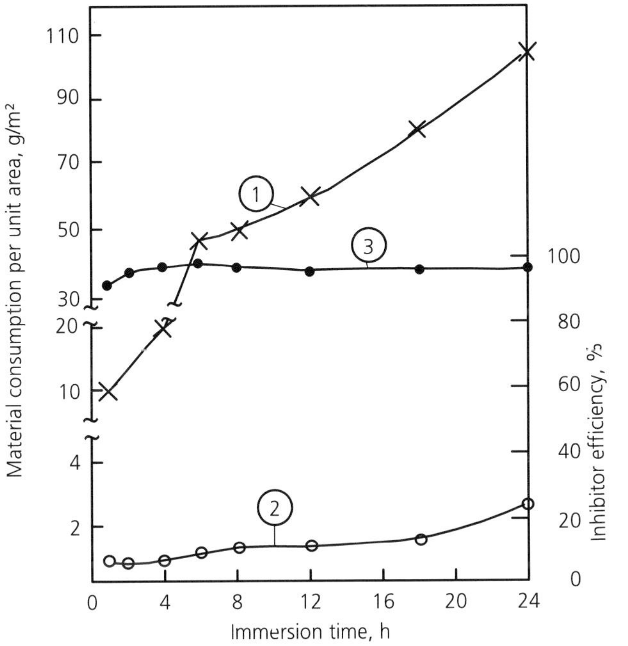

Figure 21: Influence of the immersion time on the inhibition efficiency of the inhibitor DX-A during corrosion of unalloyed steel in 2.5 mol/l H_2SO_4 at room temperature [50]
① H_2SO_4 without DX-A; ② H_2SO_4 + 1.0 % DX-A; ③ inhibition efficiency

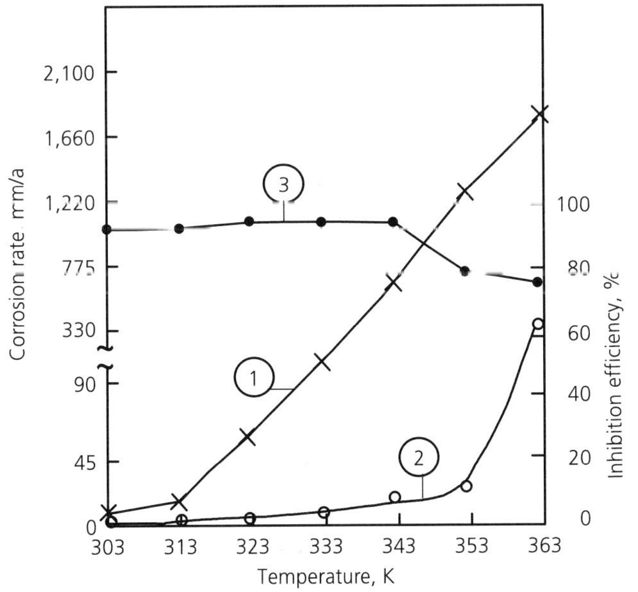

Figure 22: Influence of temperature on the inhibition efficiency of the inhibitor DX-A during corrosion of unalloyed steel in 2.5 mol/l H_2SO_4 at room temperature [50]
① H_2SO_4 without DX-A; ② H_2SO_4 + 1.0 % DX-A; ③ inhibition efficiency

– Areas of application –

According to [52, 53], unalloyed steel is suitable for the production of storage and transportation tanks for sulfuric acid above a concentration of 85 % at room temperature (see Figures 1–5). The reason for damage due to cracking of steel tanks (in the region between the acid intake and the bottom during storage of highly concentrated sulfuric acid in the concentration range between 93 and 98 %) is the change in the corrosive properties of the acid resulting from the change in concentration or temperature. The best protection against such influences is application of a corrosion protection system, for example a paint film, the use of anodic protection [54] or the use of corrosion-resistant linings. Unalloyed or low-alloy steel is proposed as the basic material in this context, with high-alloy steel as the lining material (see Sections Austenitic chromium-nickel steels, Austenitic chromium-nickel-molybdenum steels, Austenitic chromium-nickel steels with special alloying additions and Special iron-based alloys) [55].

According to this work, corrosive attack on unalloyed steel in 93 % sulfuric acid is too high at room temperature, especially in the region of the phase boundary, compared with the higher-alloyed material. The austenitic CrNiMo-steels 10Ch17N13M2T (cf. 1.4571, SAE 316 Ti) or 06ChN28MDT are suggested as lining materials [55].

Further possibilities consist of lining with glass fiber-reinforced polyester resins [53].

In 20 % sulfuric acid at boiling point, the alloy S-CuA18 is particularly suitable for fusion welding [56].

Through boriding it is possible to improve both the acid resistance and the wear resistance of unalloyed and low-alloy steels. Table 18 contains steels of this group suitable for boriding. Figure 23 shows the improvement in the acid resistance of Ck 45 in various acids and 10 % H_2SO_4. This graph shows a very sharp drop in the material consumption of the borided specimens in acid. The material consumption in 10 % H_2SO_4 at 329 K (56 °C) falls from 156 to 4.9 mg/cm^2 after a test duration of 8 h due to boriding. This corresponds to corrosion rates of 216 and 6.8 mm/a (8,504 and 268 mpy) respectively [57].

Steel	*DIN-Mat. No.*	*cf. SAE / UNS*
St 00 †	1.0030 †	–
S235JRC (ST 37 †)	1.0120	–
C15	1.0401	1015, 1017
C45	1.0503	1045
C50	1.0540	1049, 1050

†: old designation

Table 18: Selection of steels suitable for boriding with improved corrosion properties in acids (see Figures 23 and 78) [57]

Table 18: Continued

Steel	*DIN-Mat. No.*	*cf. SAE / UNS*
C60	1.0601	1060
(or also the corresponding Ck grades)		
16MnCr5	1.7131	G51150
40CrMnMo7	1.2311	–
42CrMo4	1.7225	K14248
60WCrV8	1.2550	–
90MnCrV8	1.2842	O 2
100Cr6	1.3505	G52986
115CrV3	1.2210	L 2
X6CrNiTi18-10	1.4541	321
X 12 CrNi 18 8 †	1.4300 †	–
30WCrV17-2	1.2567	–
X37CrMoV5-1	1.2343	H 11
X40CrMoV5-1	1.2344	H 13
X90CrMoV18	1.4112	440 B

†: old designation

Table 18: Selection of steels suitable for boriding with improved corrosion properties in acids (see Figures 23 and 78) [57]

According to [58], unalloyed steel is used for storage and transportation tanks for sulfuric acid of concentrations from 68 to 100 % at low temperatures. It can also be used in plants for the contact process in the acid range between 93 and 98.5 %. Its resistance drops sharply in reaction equipment where the temperature increases, for example in the range 323 to 353 K (50 to 80 °C), so that the steel surface is preferably protected by an acid-resistant brick lining, especially where a turbulent flow exists. An additional film of plastic between the steel jacket and brick lining increases the reliability. Unalloyed steel cannot be used above 353 K (80 °C) [58] and is no longer used as the construction material in the nitric oxide process for the production of sulfuric acid developed by Ciba-Geigy. Since this process operates at temperatures below 353 K (80 °C), plastics are predominantly used [5].

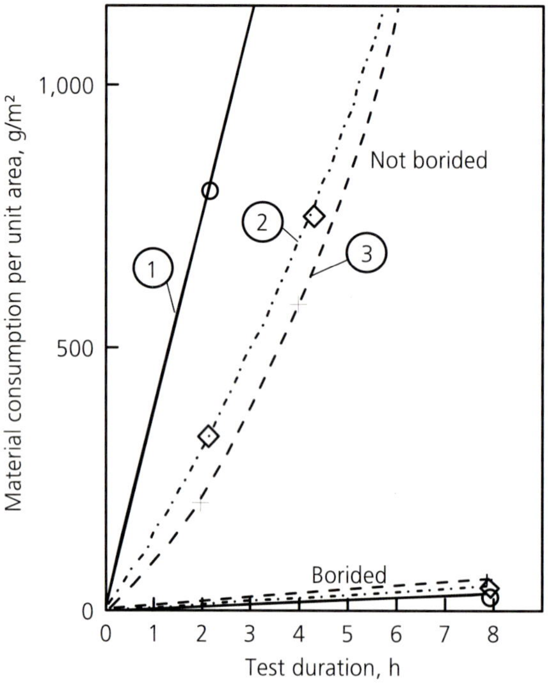

Figure 23: Material consumption of the steel C45E (DIN-Mat. No. 1.1191, cf. SAE 1042) in acid solutions (329 K (56 °C)) as a function of the surface condition [57]
① 20 % HCl
② 30 % H_3PO_4
③ 10 % H_2SO_4

Unalloyed cast iron

The corrosion behavior of unalloyed cast iron in sulfuric acid is similar to that of the unalloyed steels (see also Section Unalloyed steels and cast steel) [2, 4, 7, 9, 22, 59–67]. Table 19 contains the corrosion rates in mixed acids (H_2SO_4 containing HNO_3 and H_2O) [22]. Figure 24 shows the corrosion rate of unalloyed cast iron in sulfuric acid as a function of the temperature [61]. This graph shows that in the concentration range above 62 % sulfuric acid, corrosion rates < 0.1 mm/a (< 3.94 mpy) are to be expected only below about 293 K (20 °C). Above this temperature, the corrosion increases rapidly [61]. Figure 9 gives a comparison of the corrosion rates of unalloyed steel in sulfuric acid solutions > 60 % (possible area of use) with those of unalloyed cast iron. Figure 16 shows the corrosion resistance of unalloyed iron materials in sulfuric acid in comparison with higher-alloy steels.

Concentration, %			***Cast iron***	***Steel***
H_2SO_4	HNO_3	H_2O	***Corrosion rate, mm/a (mpy)***	
40	10		0.25 (9.84)	0.13 (5.12)
85	15	–	0.16 (6.30)	0.10 (3.94)
75	25	–	0.14 (5.51)	0.10 (3.94)
72.5	27	0.5		0.42 (16.5)
65	23.5	11.5		0.30 (11.8)
55	20	25	0.31 (12.2)	0.43 (16.9)
53	46	1.0		0.08 (3.15)
50	25	25	0.30 (11.8)	0.36 (14.2)
45	30	25	0.29 (11.4)	0.35 (13.8)
70	5	25	1.31 (51.6)	unsuitable
10	40	50	unsuitable	unsuitable
10	70	20		0.04 (1.57)

Table 19: Corrosion on unalloyed steel and cast iron in acid mixtures at room temperature [22]

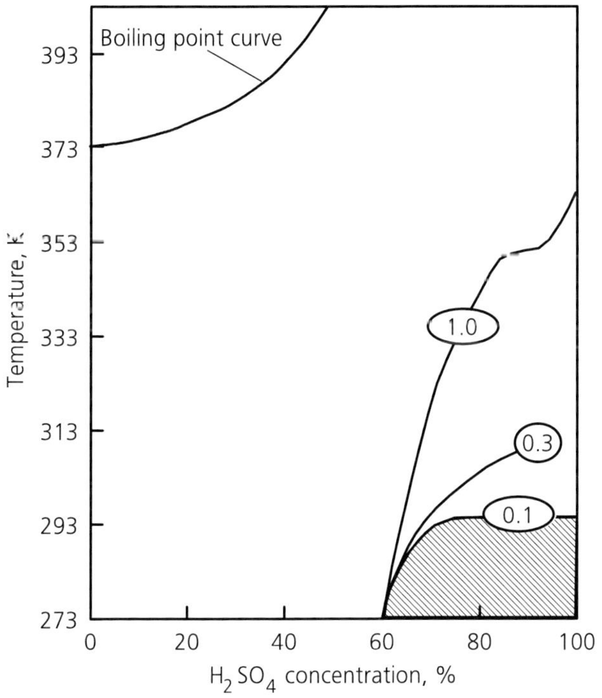

Figure 24: Isocorrosion curves (mm/a) for unalloyed cast iron in sulfuric acid [61]

Gray cast iron is severely attacked by dilute sulfuric acid at room temperature and not resistant. Above a concentration of 65 % sulfuric acid, a firmly adhering layer of corrosion products and graphite residues forms and this greatly inhibits further attack. From concentrations of about 80 %, cast iron is resistant even at temperatures up to the boiling point. Since the flow rate has less influence on the corrosion of cast iron than on that of steel, the former is used for components such as pumps, impellers, valves and fittings [60, 66].

Figure 25 shows the corrosion rates of unalloyed cast iron at low acid concentrations of up to about 40 % sulfuric acid. The decrease in the corrosion rate after a test duration of 14 days is due to the accumulation of graphite corrosion products on the surface.

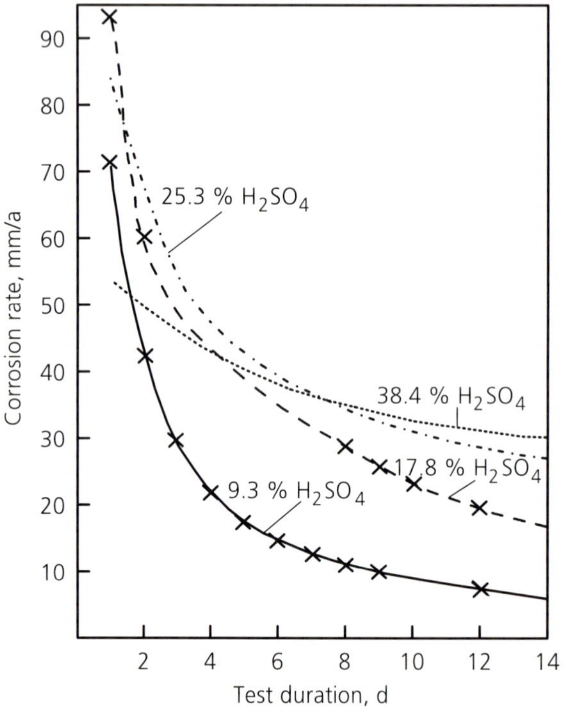

Figure 25: Corrosion behavior of unalloyed cast iron in dilute sulfuric acid at 297 to 299 K (24 to 26 °C) [60, 66]

Table 20 gives a summary of corrosion rates at 293 K (20 °C) in sulfuric acid of low concentration [60] and illustrates the high corrosive attack, as expected, at lower sulfuric acid concentrations. If these materials are used in highly concentrated sulfuric acid, it must be ensured that dilution of the acid, for example by water, is excluded.

Material	H_2SO_4 concentration, %			
	0.5	5.0	10	33
	Corrosion rate, mm/a (mpy)			
Nodular cast iron	32 (1,260)	76 (2,992)	143 (5,630)	113 (4,449)
Nodular cast iron with 1.5 % Ni	25 (984)	58 (2,283)	130 (5,118)	114 (4,488)
Gray cast iron	41 (1,614)	79 (3,110)	175 (6,890)	162 (6,378)
Steel, unalloyed	47 (1,850)	115 (4,528)	192 (7,559)	192 (7,559)

Table 20: Corrosion resistance of cast iron and unalloyed steel in dilute sulfuric acid at 293 K (20 °C) [60]

As can also be seen from Table 20, the structure of the graphite in cast iron has relatively little influence on its corrosion resistance. Cast iron generally proves to be somewhat more resistant than unalloyed steel at low acid concentrations, nodular cast iron being slightly more resistant than gray cast iron [60, 66].

The influence of the basic structure (perlite and ferrite) in 5 % sulfuric acid at 303 K (30 °C) manifests itself in the following corrosion rates [60]:

- Nodular cast iron
  annealed, ferritic structure 81 mm/a (3,189 mpy)
  cast condition, perlitic structure 134 mm/a (5,276 mpy)
- Gray cast iron
  cast condition, perlitic structure 167 mm/a (6,575 mpy)
  annealed, ferritic structure 202 mm/a (7,953 mpy)

Table 21 shows the corrosion behavior, with relatively low corrosion rates in sulfuric acid (83 %) at higher temperatures [60, 66]. With a low corrosion rate of 0.13 mm/a (5.12 mpy) at 394 K (121 °C), nodular cast iron is particularly suitable for use in 99.6 % H_2SO_4. Cast iron is more resistant than steel when used in oleum (Figure 9).

H_2SO_4 concentration	Temperature	Corrosion rate
%	K (°C)	mm/a (mpy)
83.4	305 (32)	0.77 (30.3)
83.4	323 (50)	2.15 (84.6)
93.4	305 (32)	0.30 (11.8)
93.4	323 (50)	0.45 (17.7)
99.6	377 (104)	0.28 (11.0)
99.6	394 (121)	0.13 (5.12)

Table 21: Corrosion on nodular cast iron in concentrated sulfuric acid at high temperatures [60, 66]

Figure 26 shows the marked increase in the corrosion rate of unalloyed gray cast iron in 96 % sulfuric acid at high temperatures [60, 66, 68]. After a sharp rise in the temperature range between 273 and 473 K (0 and 200 °C) to more than 50 mm/a (1,969 mpy), the corrosion rate falls to less than 2.0 mm/a (78.7 mpy) at 573 K (300 °C). These corrosion rates are acceptable for short operating times or in pumps. By comparison, the corrosion rate for gray cast iron in 96 % sulfuric acid at 292 to 293 K (19 to 20 °C) is quoted as 0.1 mm/a (3.94 mpy) [68].

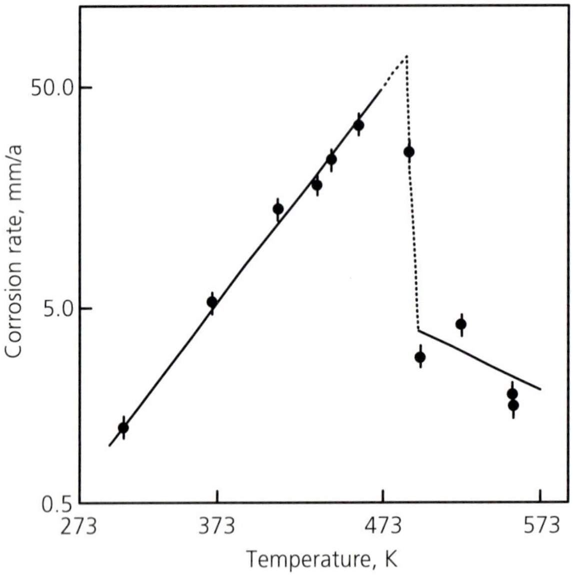

Figure 26: Corrosion rate of unalloyed gray cast iron (Fe-3.2 % C, 2.0 % Si, 0.6 % Mn, 0.2 % P) in 96 % sulfuric acid in the temperature range 273 to 573 K (0 to 300 °C) [60, 66, 68]

At this point, it should be pointed out that downtime corrosion and increasing corrosiveness of sulfuric acid due to dilution with moisture are possible. According to [63], corrosion products cause "explosions" of the structure of gray cast iron under the action of oleum, but these were not found with nodular cast iron [69].

Because of its good resistance in hot concentrated sulfuric acid, gray cast iron is often used under these conditions. Its corrosion behavior in boiling concentrated sulfuric acid is particularly important in the Pauling process [5] for the regeneration of waste sulfuric acid. "Gray cast iron containing finely divided graphite and having a perlitic matrix has shown a more favorable corrosion behavior than gray cast iron with a different structure in many instances" [70].

Gray cast iron can be in the active or passive state in boiling concentrated sulfuric acid. If pure concentrated sulfuric acid is heated to boiling point, cleaned gray cast iron specimens then remain active within a period of 24 h and the corrosion mixture rapidly becomes cloudy as a result of precipitated ferric sulfate. If passivators such as nitric acid, vanadium acids, chromic acid or noble metals are added to this acid before the start of the experiment, the corrosion rate is reduced to less than 5 % of that determined on gray cast iron in the active state. Boiling 98 % sulfuric acid remains clear under such conditions. Gray cast iron can also be passivated electrochemically (see Section Unalloyed steels and cast steel – Anodic protection – and also in this section – Areas of application –). In [70], gray cast iron specimens of varying composition in the active or passive state were exposed to corrosion in boiling 90 to 98 % H_2SO_4. Platinum oxide was used as the passivator. Specimens about 30 mm in diameter and 5 mm in thickness were used to determine the weight losses. They were boiled under reflux in 1 l round-bottomed flasks in 1 kg analytically pure sulfuric acid of various concentrations for 24 h. The cast irons used and their structures are summarized in Table 22. The corrosion rates of the individual alloys show substantial differences in the active and passive state. Alloy No. 4 is a good example of this. The corrosion rates contained in Table 23 illustrate the difference. Corrosion rates of 79.8 mm/a (3,142 mpy) are measured in the active state in 97 % H_2SO_4 at boiling point, and rates of about 2 mm/a (78.7 mpy) in the passive state. No clear dependence of the corrosion rate on the amount of passivator was found in these studies. According to the results in Table 23, 5 mg per kg H_2SO_4 are already sufficient to force passivity in concentrated boiling H_2SO_4. Table 24 contains the corrosion rates in the passive state for the alloys in Table 22.

The results of the studies are summarized according to [70] as follows: Unalloyed gray cast iron in boiling concentrated sulfuric acid exhibits a corrosion minimum in the active state at 96 % H_2SO_4 and a corrosion maximum at 97 % H_2SO_4. The latter is associated with a Si-content of more than 1.4 to 1.5 % and occurs in the case of both unalloyed gray cast iron and low-alloy gray cast iron containing Cu and/or Sn.

Cu/Sn-alloyed gray cast iron can be used in a wider H_2SO_4 concentration range, whereas the corrosion minimum is limited to a concentration of 96 % in the case of unalloyed gray cast iron.

High Si-contents in gray cast iron reduce the corrosion resistance in the active and passive regions.

The corrosion resistance of Cu/Sn-alloyed gray cast iron in the passive state increases uniformly as the Si-content decreases and the Sn-content increases, and as the H_2SO_4 concentration increases (see Tables 22 and 24) [70].

The best corrosion bahavior under the test conditions mentioned in boiling con centrated sulfuric acid is shown by the Ni-containing Alloy No. 10, which exhibits corrosion rates of less than 2 mm/a (78.7 mpy) (mean value of 2 specimens) in the passive state in the entire concentration range studied (93 to 98 % H_2SO_4).

In the Pauling process for the reconcentration of waste sulfuric acid (not less than 70 %), a gray cast iron tank heated with flue gases can be used in the simplest case. This is filled with hot, boiling 96 % H_2SO_4 at normal pressure.

The tank contains a slow-running anchor stirrer. The waste acid to be concentrated is introduced via a distillation head. For corrosion reasons, the sulfuric acid concentration in the tank should not be below 95 %. The 96 to 97 % acid in the tank has a temperature of 593 K (320 °C) [5].

Alloy No.	Chemical composition, %										Graphit structure according ASTM	Matrix
	C	Si	Mn	P	S	Cu	Ni	Cr	Sn	V		
1	3.47	2.31	0.47	0.61	0.14	0.056	0.032	0.075	0.041	0.060	I D7	mainly perlitic (about 15 % by volume ferrite in the structure), phosphide eutectic
2	3.25	1.87	0.52	< 0.03	< 0.03	< 0.03	< 0.03	0.02–0.03	0.011	0.21	I A3-4, D8	perlitic, traces of ferrite
3	3.53	1.65	0.28	0.15	0.080	0.12	0.080	0.055	0.013		I A-B 4-6	mainly perlitic, ferritic regions (about 25 % by volume in total in the structure)
4	3.17	1.79	0.19	0.05	0.02	0.94	0.03–0.05	0.02–0.06	0.01		I A4-6	perlitic, dendritic solidification
5	3.15	1.67	0.3	0.04	0.02	0.04	0.04	0.03	0.09		I A5-6	perlitic, dendritic solidification
6	3.24	1.82	0.2	0.04	0.02	0.96	0.05	0.04	0.95		I A4-5	perlitic, slightly dendritic solidification
7	3.28	0.76	0.28	0.05	0.02	0.98	0.03–0.05	0.02–0.06	0.086		I A4-5	perlitic, small amounts of ferrite
8	3.02	1.32	0.41	0.04	0.03	0.97	0.03–0.05	0.02–0.06	0.064		I D7, E5	perlitic
9	3.84	1.48	0.3	0.06	0.02	0.98	0.03	0.04	0.95		I B5-6, A3-4	perlitic
10	3.55	0.38	0.21	0.010	0.009	1.05	2.13	0.01	< 0.001		I A3-4	perlitic, dendritic solidification

Table 22: Chemical analysis and structure of gray cast iron of different alloying contents for corrosion studies in boiling concentrated sulfuric acid (see Tables 23 and 24) [70]

Passivator PtO_2 (82 % Pt) mg/kg H_2SO_4	*Corrosion rate mm/a* (mpy)*
0	79.8 (3,142)
4.63	2.2 (86.6)
10.21	1.6 (63.0)
14.93	1.4 (55.1)
21.38	1.6 (63.0)

* conversion from mg/cm^2 d (test duration 24 h)

Table 23: Dependence of the corrosion of gray cast iron alloy No. 4 (see Table 22) in boiling 97 % H_2SO_4 (initial concentration) under reflux on the amount of passivator [70]

The corrosion rates of 8 to 10 mm/a (315 to 394 mpy) determined on such a tank are in the borderline region between the active and passive state of the cast iron [71].

A mixture of sodium nitrite and sodium nitrate was used in corrosion studies to determine the influence of the stirrer speed on the corrosion loss of cast iron [71].

In a model experiment, the dependence of the corrosion rate on the stirrer speed was determined in an electrically heated gray cast iron tank of about 1 l capacity. The tank had a diameter of 120 mm, a height of 125 mm and a wall thickness of 10 mm. The following gray cast iron was used:

- Analysis: 3.56 % C, 1.91 % Si, 0.48 % Mn, 0.064 % P, 0.041 % S,
- Structure: Lamellar graphite A4–5 according to ASTM, perlitic matrix, few ferrite halos.

A scale model of the anchor stirrer in Pauling tanks used in industry was used as the stirrer. In such a tank with a capacity of 22 t SO_3/d and internal diameter of 2,500 mm, a path velocity of 3.8 m/s would correspond to a stirrer speed of 29 rpm, which is close to the operating speed of 32 rpm. 1.5 g concentrated sulfuric acid with additions of 7 g $NaNO_2$ and 49 g $NaNO_3$ was used as the corrosion medium. The experiments were carried out at boiling point, water present in the acid or formed by the corrosion reaction being distilled off continuously as dilute sulfuric acid. The corrosion rates under flow conditions for the container wall are summarized in Table 25 and Figure 27. The increase in the corrosion rate as the flow rate at the container wall increases can be attributed to the accelerated material consumption of the passive layer. Under heat transfer conditions, the corrosion rate increases as the temperature difference between the tank wall and the liquid (corrosion medium) increases. However, since faster stirring (increasing flow rate) intensifies the transfer of matter and reduces the difference in temperature, the corrosion decreases. The overlapping of the two effects results in the minimum in the curve in Figure 27 [71].

Alloy No.	*H_2SO_4 concentration, %*						
	93	*94*	*95*	*96*	*97*	*98*	*Others*
	Corrosion rate, mm/a (mpy)						
1	16.5 (650)	12.8 (504)	7.8 (307)	5.7 (224)	1.9 (74.8)	1.7 (66.9)	a)
2	10.7 (421)	9.5 (374)	7.0 (276)	2.8 (110)	2.1 (82.7)	1.6 (63.0)	
3	6.7 (264)	5.1 (201)	3.7 (146)	1.9 (74.8)	1.3 (51.2)	0.9 (35.4)	
4	8.7 (343)	7.5 (295)	4.7 (185)	3.4 (134)	1.6 (63.0)	1.3 (51.2)	
5	5.7 (224)	5.7 (224)	2.7 (106)	2.3 (90.6)	1.8 (70.9)	1.2 (47.2)	b)
6	7.9 (311)	5.1 (201)	4.4 (173)	2.5 (98.4)	1.8 (70.9)	1.1 (43.3)	
7	2.4 (94.5)	2.2 (86.6)	2.5 (98.4)	2.0 (78.7)	1.4 (55.1)	1.1 (43.3)	
8	8.5 (335)	6.6 (260)	4.2 (165)	2.8 (110)	2.1 (82.7)	1.6 (63.0)	
9	5.7 (224)	5.7 (224)	3.8 (150)	2.6 (102)	1.8 (70.9)	1.7 (66.9)	
10	1.9 (74.8)	1.9 (74.8)	1.4 (55.1)	1.3 (51.2)	0.7 (27.6)	0.7 (27.6)	c)

a) 94.5 % H_2SO_4 7.6 mm/a (299 mpy), 96.7 % H_2SO_4 3.9 mm/a (154 mpy)
b) 93 % H_2SO_4 6.7 mm/a (263 mpy)
c) 92 % H_2SO_4 2.9 mm/a (114 mpy), 93.5 % H_2SO_4 1.8 mm/a (70.9 mpy), 96.6 % H_2SO_4 1.2 mm/a (47.2 mpy)

Table 24: Corrosion rates on gray cast iron in the passive state in boiling, concentrated sulfuric acid (test duration 24 h, conversion from mg/cm^2 d) [70]

Stirrer speed	*Path velocity of the liquid close to the tank wall of the test tank*	*Material consumption rate*	*Corrosion rate*
rpm	*m/s*	*mg/cm^2 d*	*mm/a (mpy)*
0	0	27.5	13.8 (543)
200	1.26	26.3	13.2 (520)
300	1.88	25.7	12.9 (508)
400	2.51	26.3	13.2 (520)
500	3.14	17.3	8.7 (343)
600	3.77	16.1	8.1 (319)
700	4.40	17.9	9.0 (354)
800	5.02	22.3	11.2 (441)
950	5.97	21.7	10.9 (429)

Table 25: Dependence of the corrosion rate of a gray cast iron tank in boiling, concentrated H_2SO_4 with additions of 7 g $NaNO_2$ and 49 g $NaNO_3$ per 1.5 kg conc. H_2SO_4 on the flow rate at the tank wall (test duration 24 h), see also Figure 27 [71]

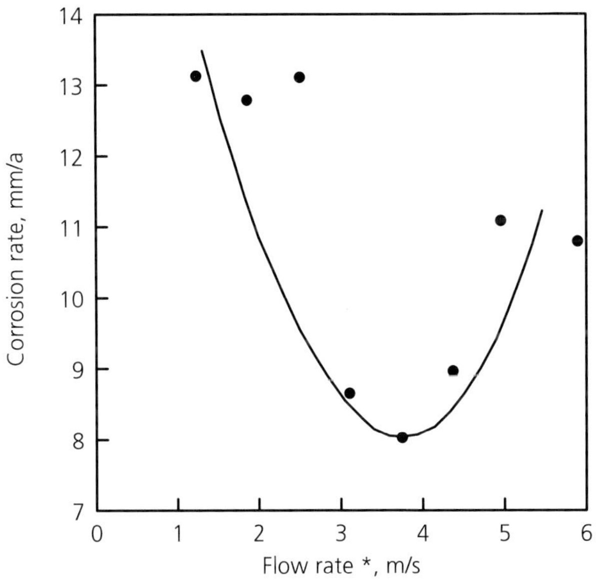

Figure 27: Corrosion rate as a function of the flow rate in a gray cast iron tank for reconcentrating waste sulfuric acid. Model experiment in boiling concentrated H_2SO_4 in accordance with the Pauling process (see also Table 25) [71]

The material consumption rate in the experiments with heat transfer and stirring were higher than those in the experiments under isothermal conditions without stirring. The difference was at least 12 mg/cm^2 d (6 mm/a (236 mpy)) [50] (see – Areas of application –).

According to [72], cast iron is more corrosion-resistant than steel, and the perlitic Meehanite® cast irons have better metallurgical properties than conventional cast iron.

Meehanite® CB3 was developed for use in sulfuric acid above a concentration of 77 % and at temperatures up to 368 K (95 °C).

As has already been seen from Figure 9, increased corrosion of cast iron must be expected in fuming sulfuric acid (oleum) in the range of about 15 to 20 % free SO_3. Investigations into the damage caused to a stirred tank by oleum (65 % oleum at 423 K (150 °C)) showed that free SO_3 diffuses into the material particularly at regions of coarse graphite lamellae in the cast iron. As a result stresses build up due to chemical reactions and the associated increase in volume. Cracks then form which are further exaggerated by operating stresses [69].

Pilot experiments using saturated ammonium sulfate solution at 311 to 320 K (38 to 47 °C) over a period of 33 days caused a corrosion rate of 1.13 mm/a (44.5 mpy) on cast iron in the presence of 5 % free sulfuric acid (Table 26) [73]. Cast iron is destroyed in an ammonium sulfate crystal slurry containing sulfuric acid at 373 K (100 °C).

Material	*Conditions*	
	Pilot experiment (33 days) at 311 to 320 K (38 to 47 °C) to using saturated ammonium sulfate solution which contained 5 % free sulfuric acid	*Pilot experiment (36 days) at 373 K (100 °C), aerated, using ammonium sulfate crystal slurry containing sulfuric acid in the centrifuge during centrifugation and washing*
	Corrosion rate, mm/a (mpy)	
Monel® 1)	0.14 (5.51)	0.175 (6.89)
Nickel	0.13 (5.12)	0.200 (7.87)
Inconel® 2)	0.09 (pitting depth 0.3 mm) (3.54)	0.005 (0.20)
Ni-Resist® 3)	0.14 (5.51)	0.475 (18.7)
Cast iron	1.13 (44.5)	

1) 67 % Ni, 30 % Cu
2) 79.5 % Ni, 13 % Cr, 6.5 % Fe
3) No precise data, probably Ni-Resist®1 (16 % Ni, 6 % Cu, 2 % Cr)

Table 26: Corrosion of metallic materials in ammonium sulfate solutions ($(NH_4)_2SO_4$) containing sulfuric acid [73]

– *Electrochemical corrosion protection processes* –

Since anodic protection of unalloyed steel in concentrated sulfuric acid at low temperatures is an established process and steels can also be passivated in hot concentrated H_2SO_4, corrosion studies were also performed on gray cast iron, which exhibits a behavior similar to that of steel under such conditions, in boiling 95 % H_2SO_4. The curves plotted in Figure 28 were obtained as a result of the electrochemical studies.

A line of the optimum values for the protection potential was determined from the position of the individual potentials and the course of the corrosion potential. According to this graph, the resulting protection potential in boiling 95 % H_2SO_4 is 1.3 V (SHE). A cast iron of composition (%) Fe-3.21C-1.41Si-0.6Mn-0.09P-0.4Ni was used as the test material. Figure 29 shows the drop in the current density for boiling 96 % H_2SO_4 over a period of 3 days as the test duration increases. Under these conditions, an average corrosion rate of 2.5 mm/a (98.4 mpy) was measured in the anodically protected state and a rate of 62 mm/a (2,441 mpy) was measured in the unprotected state. Practical experiments on operating plants have shown encouraging results after one year [74].

According to detailed studies, it may be possible to use this process in individual cases. The use of other materials is more reliable however (see Figures 47, 110, 132, 141 and Table 54). Given the high protection potentials required, limited local pitting may also occur in the absence of chlorides. High-silicon cast iron may serve as a cathode in the plant during use [74]. Figure 30 shows the dependence of the corrosion potential of

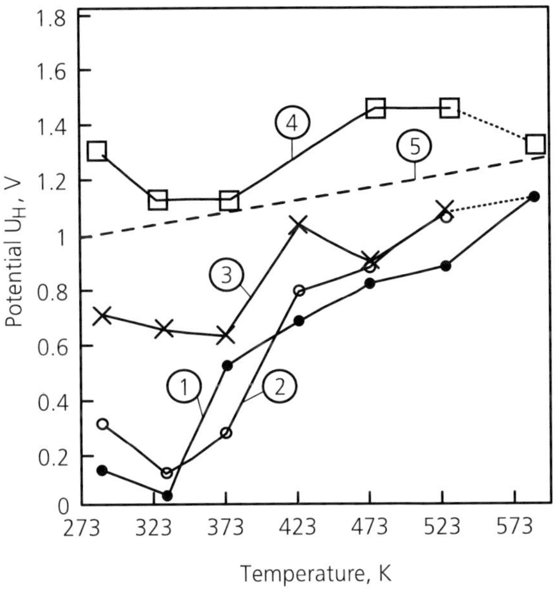

Figure 28: Changes in the potential of unalloyed cast iron in 95 % sulfuric acid as a function of the temperature [74]
① steady state potential
② corrosion potential
③ passivation potential
④ breakdown potential
⑤ protection potential (optimum values)

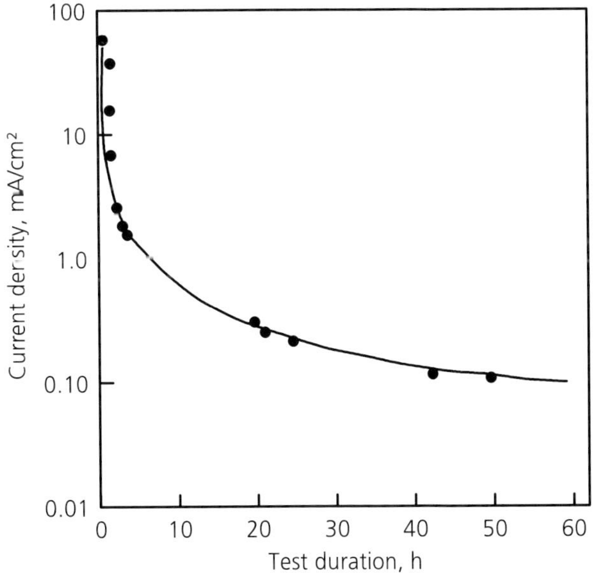

Figure 29: Change in the current density in unalloyed cast iron in boiling 96 % sulfuric acid at +1.3 V (see Figure 28) [74]

gray cast iron with lamellar graphite (for corrosion rates see Figure 26) in various concentrations of sulfuric acid at various temperatures (see also Figure 28). In this graph, the passive state for 98 % sulfuric acid at about 393 K (120 °C) is reached at potentials above +700 mV. Passivation is not yet present in 96 % acid at 503 K (230 °C) and in 93 % sulfuric acid at the boiling point. In 97 % H_2SO_4, passivation is prevented by liquid paraffin.

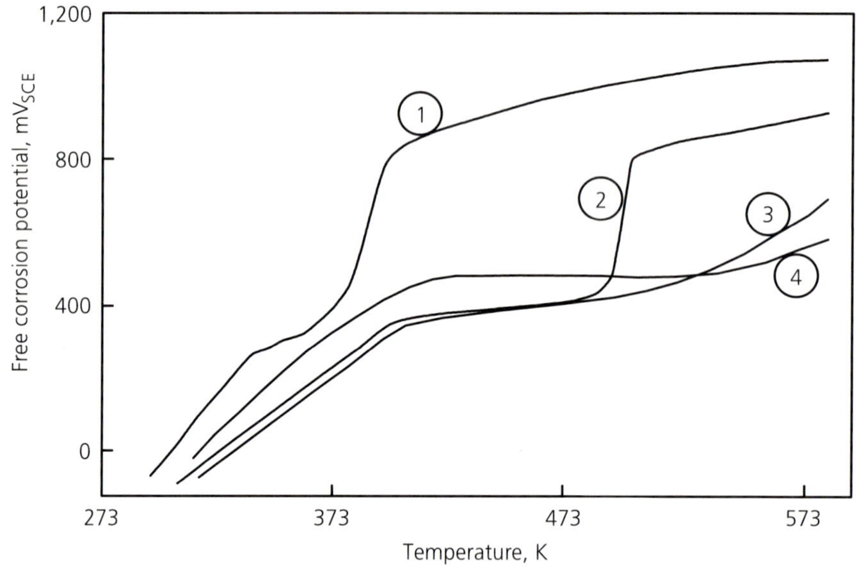

Figure 30: Change in the corrosion potential of cast iron (Fe-3.2 % C, 2.0 % Si, 0.6 % Mn, 0.2 % P) in various concentrations of sulfuric acid as a function of the temperature [68]
Sulfuric acid concentration:
① 98 %
② 96 %
③ 93 %
④ 97 % with 2 % by volume liquid paraffin

It follows that an acid concentration above 93 % and the absence of reducing agents are necessary to achieve passivity at the boiling point. Figure 31 shows the dependence of the corrosion potential in boiling sulfuric acid containing no impurities on the H_2SO_4 concentration.

Figure 32 shows the material consumption rate (or corrosion rate) to be expected of cast iron in boiling concentrated sulfuric acid as a function of some essential factors. Table 27 contains a summary of corrosion rates in the active and passive state [68]. The latter shows how difficult it can be in practice to obtain the passive state.

According to [5], tanks of perlitic cast iron show a good corrosion resistance in sulfuric acid above a concentration of 95 % in the temperature range from 593 to 603 K (320 to 330 °C) due to formation of passive layers, and this resistance can be monitored by measurement of the potential. It can also be established by addition of oxidizing agents.

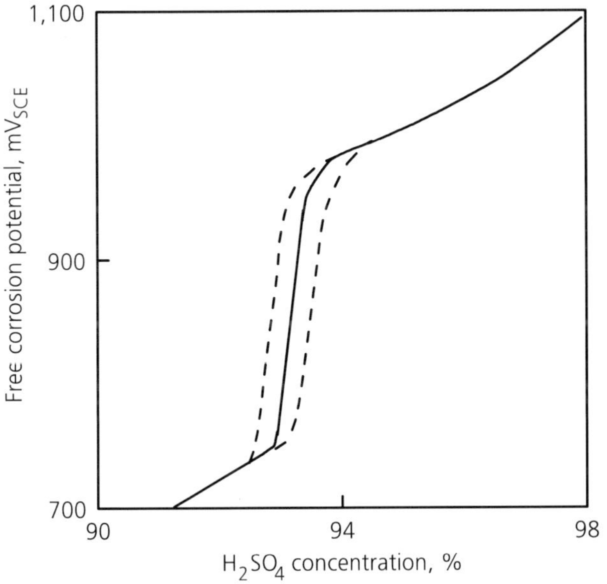

Figure 31: Dependence of the corrosion potential of cast iron (Table 27) on the concentration of boiling sulfuric acid [68]

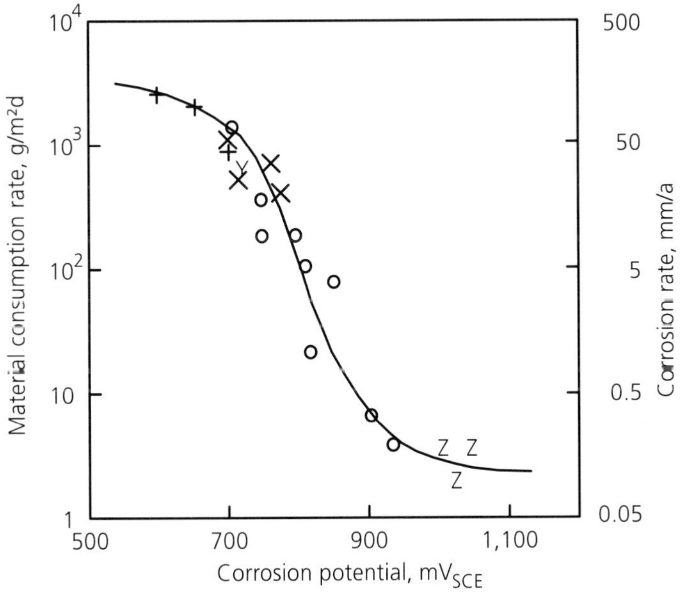

Figure 32: Corrosion of cast iron (compare Table 27) in boiling concentrated sulfuric acid as a function of the potential [68]

+ addition of liquid paraffin

○ dilution with water

x cathodic polarization

y H_2S passed through

z anodic polarization

The corrosion resistance described as "good" here is based on corrosion rates of 8 to 10 mm/a (315 to 394 mpy) [5]. A decision on the use of cast iron must be made in the individual case (see Figure 33).

Table 28 shows another possible use for anodic protection – for cast iron in 96 % sulfuric acid below 473 K (200 °C) [68].

– *Areas of application* –

Tanks of unalloyed gray cast iron with lamellar graphite are used in contact with boiling sulfuric acid at 583 K (310 °C) during recovery of sulfuric acid. The tanks have a diameter of 2.5 m and are 2 m high. The wall thickness is 90 mm and they contain 10 t acid.

The tank lid is also made of unalloyed cast iron and the feed and discharge lines have high-silicon cast iron inserts. The stirrer is made of high-silicon cast iron and its speed is 40 rpm [50].

The service life of these tanks varies and ranges from 6 months up to 3 years. Failure is usually due to the corrosion rate on the tank bottom which is 35 mm/a (1,378 mpy). Figure 33 shows the influence of the concentration of boiling sulfuric acid on the corrosion rate under these conditions.

Free corrosion (active state)			*Corrosion with anodic polarization (passive state)*			*Inhibition efficiency %*
Activator	*Corrosion potential mV_{SCE}*	*Corrosion rate mm/a (mpy)*	*Potential mV_{SCE}*	*Current density mA/cm^2*	*Corrosion rate mm/a (mpy)*	
H_2O	750	9.75 (384)	900	0.5	3.9 (154)	60
H_2O	750	18.5 (728)	900	0.8	6.5 (256)	65
H_2O	735	31.5 (1,240)	900	0.7	5.4 (213)	83
Paraffin oil	700	47.5 (1,870)	900	0.3	3.1 (122)	93
H_2O	820	1.15 (45.3)	1,000	0.1	0.6 (23.6)	48

Table 27: Corrosion of gray cast iron (Fe-3.2 % C, 2.0 % Si, 0.6 % Mn, 0.2 % P) in boiling concentrated H_2SO_4 in the active and passive state (see Figure 32) [68]

Heating is by oil burners. The process runs continuously by addition of 70 % sulfuric acid and discharge of boiling 95 % acid. The sulfuric acid concentrated in this plant contains tar formed by decomposition of organic substances. These and other circumstances give rise to the aggressiveness of the acid and variations in its concentration around 95 %, with the associated corrosion rates. There are also differences in the structural formation of the cast iron.

Laboratory corrosion studies on cast iron filings from the tanks showed corrosion rates of between 3 and 30 mm/a (118 and 1,181 mpy) under the conditions described. The microstructure of the specimens varied from perlitic to ferritic. No correlation was found between the microstructure and the chemical composition of the cast iron.

Little advantage was gained by changing to a more expensive material, since even tantalum exhibited a material loss of 1.5 mm/a (59.1 mpy) under these conditions [74].

Temperature K (°C)	*Corrosion in the active state*		*Anodic polarization (anodic protection)*		*Inhibition efficiency %*
	Corrosion potential mV_{SCE}	*Corrosion rate mm/a (mpy)*	*Potential mV_{SCE}*	*Corrosion rate mm/a (mpy)*	
303 (30)	−100	1.2 (50.0)	+800	0.26 (10.2)	78
363 (90)	+50	5.2 (205)	+850	0.22 (8.66)	96
403 (130)	+400	17.5 (685)	+850	0.16 (6.30)	99
433 (160)	+480	24.5 (965)	+850	0.43 (16.9)	98
443 (170)	+425	30.7 (1,209)	+800	0.07 (2.76)	99
460 (187)	+425	48.5 (1,909)	+900	0.31 (12.2)	99

Table 28: Corrosion of gray cast iron (Table 27) in 96% H_2SO_4 at temperatures below 473 K (200 °C) [68]

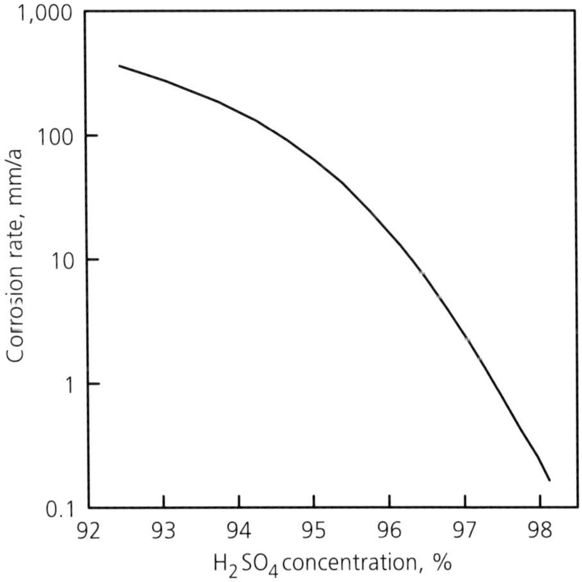

Figure 33: Corrosion rate of gray cast iron in various concentrations of boiling sulfuric acid during concentration of waste sulfuric acid [74]

See Section Unalloyed steels and cast steel for the corrosion behavior of cast iron on exposure to flue gas.

High-alloy cast iron, high-silicon cast iron

Cast iron with a high silicon content has a very good resistance in all concentrations of sulfuric acid up to the boiling point [5, 58, 74–80]. The good resistance of this material in sulfuric acid is already well-known. Figure 34 shows the increase in the corrosion resistance of iron with increasing Si-content in sulfuric acid at room temperature (no further information available). This graph shows that the good corrosion resistance of high-silicon cast iron in sulfuric acid is present only above about 14.5 % Si [81].

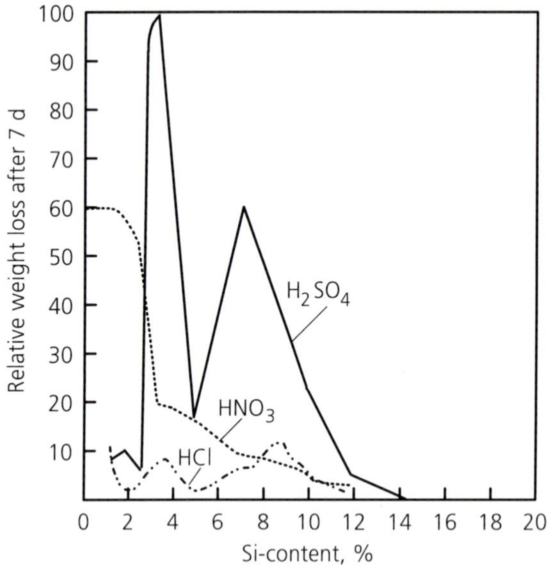

Figure 34: Influence of the Si-content of FeSi-alloys on the corrosion resistance in sulfuric acid at room temperature [81]

Corrosion studies on iron-silicon alloys of varying Si-content (Table 29) confirm this trend, as can be seen from Figure 35 [82–84]. Hardly any weight loss is detectable in the alloy containing 14.5 % Si. The specimen size was 25 × 25 × 3–8 mm. Table 30 [82, 83] summarizes the corrosion rates for sulfuric acid at 323 K (50 °C). According to this summary, the corrosion rate of 14.5 % high-silicon cast iron is only 0.088 mm/a (3.46 mpy) in 5 % H_2SO_4 and is virtually zero in 70 % acid at 323 K (50 °C) [82].

Figure 36 shows the sudden drop in the corrosion rate in 10 % sulfuric acid at 353 K (80 °C) as the Si-content increases [85]. Table 31 shows the influence of the temperature, concentration and Si-content. Here also, the threshold value of 14.5 % Si can be clearly recognized [82, 83]. Figure 37 shows the corrosion rate of Duriron® (14.5 % Si) in various concentrations of sulfuric acid at the boiling point. From this graph, the maximum corrosion rates are in the concentration range between about 5 and 50 % boiling H_2SO_4 [86].

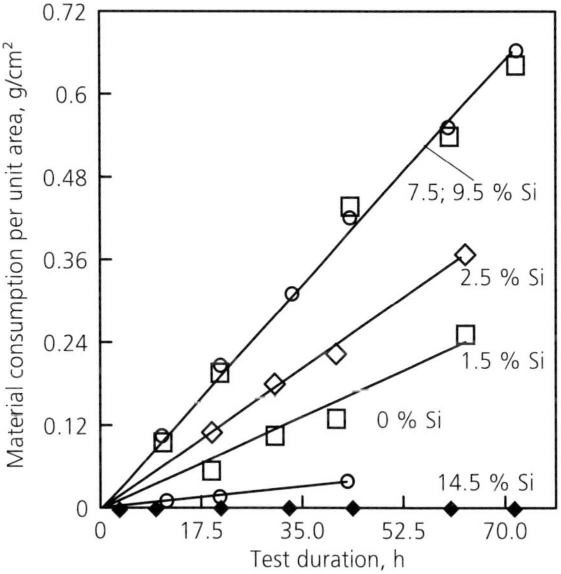

Figure 35: Material consumption of FeSi-alloys (Table 29) in 30 % sulfuric acid at 323 K (50 °C) as a function of the Si-content and exposure time [82, 83, 84]

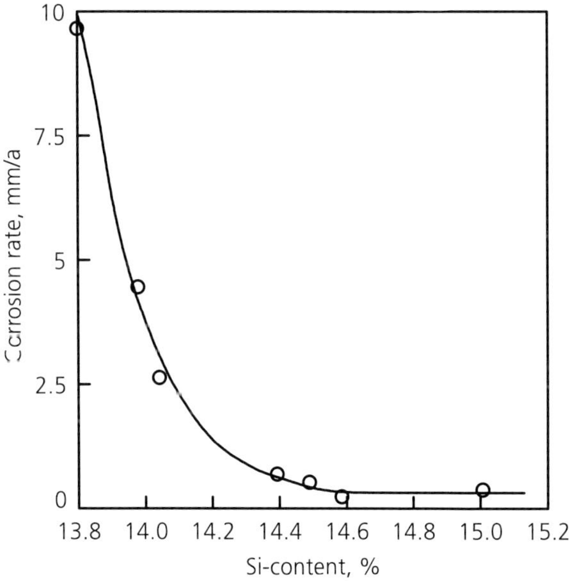

Figure 36: Influence of the Si-content in iron on its corrosion behavior in 10 % sulfuric acid at 353 K (80 °C) [85]

Alloy	*Chemical analysis, %*						
	Si	*Cr*	*C*	*Mn*	*S*	*Al*	*Fe*
0.0 % Si	0.01	0.01	0.02	0.08	0.005	–	balance
1.5 % Si	1.48	–	0.02	0.23	0.001	0.33	balance
2.5 % Si	2.50	–	0.02	0.23	0.001	0.40	balance
7.6 % Si	7.61	–	0.05	–	–	–	balance
9.5 % Si	9.51	–	0.02	–	–	–	balance
11.5 % Si	11.56	–	–	–	–	–	balance
14.5 % Si	14.79	0.05	0.02	0.07	0.004	0.03	balance

Table 29: Alloy composition of iron-silicon alloys (see Figures 35, 38 Tables 30, 31) [82–84]

Alloy[1)]	*H_2SO_4 concentration, %*					
	5	*10*	*20*	*35*	*50*	*70*
	Corrosion rate, mm/a (mpy)					
0.0 % Si	19 (748)	17 (669)	54 (2,126)	144 (5,669)	126 (4,960)	0.95 (37.4)
1.5 % Si	124 (4,881)	116 (4,567)	117 (4,606)	145 (5,709)	200 (7,874)	0.99 (39.0)
2.5 % Si	224 (8,819)	197 (7,756)	200 (7,874)	250 (9,843)	278 (10,945)	1.00 (39.4)
7.5 % Si	249 (9,803)	353 (13,898)	506 (19,921)	545 (21,457)	182 (7,165)	2.17 (85.4)
9.5 % Si	274 (10,787)	365 (14,370)	711 (27,992)	374 (14,724)	133 (5,236)	0.87 (34.3)
11.5 % Si	–	767 (30,197)	–	–	–	0.50 (19.7)
14.5 % Si	0.088 (3.46)	0.075 (2.95)	–	–	–	0.00 (0.0)

1) see Table 29

Table 30: Corrosion on iron-silicon alloys in various concentrations of sulfuric acid at 323 K (50 °C) [82–84]

Alloy [1]	***H_2SO_4 concentration, %***				
	5	***10***	***20***	***35***	***50***
	373 K [2] ***(100 °C)***	***375 K (102 °C)***	***377 K (104 °C)***	***383 K (110 °C)***	***396 K (123 °C)***
	Corrosion rate, mm/a (mpy)				
0.0 % Si	422 (16,614)	551 (21,693)	837 (32,952)	2,000 (78,740)	6,836 (269,134)
1.5 % Si	4,753 (187,126)	4,372 (172,126)	5,024 (197,795)	7,026 (276,614)	5,258 (207,007)
2.5 % Si	6,620 (260,630)	5,831 (229,567)	6,487 (255,394)	10,440 (411,024)	2,746 (108,110)
7.5 % Si	6,684 (263,150)	7,372 (290,236)	9,153 (360,354)	11,746 (462,441)	1,100 (43,307)
9.5 % Si	6,570 (258,661)	9,377 (369,173)	10,786 (424,646)	6,991 (275,236)	869 (34,213)
11.5 % Si	–	–	7,290 (287,007)	5,543 (218,228)	24,879 (979,488)
14.5 % Si	0.016 (0.63)	0.019 (0.75)	0.069 (2.72)	0.048 (1.89)	0.09 (3.54)

1) see Table 29
2) boiling point

Table 31: Corrosion on iron-silicon alloys in various concentrations of sulfuric acid at the boiling point [82–84]

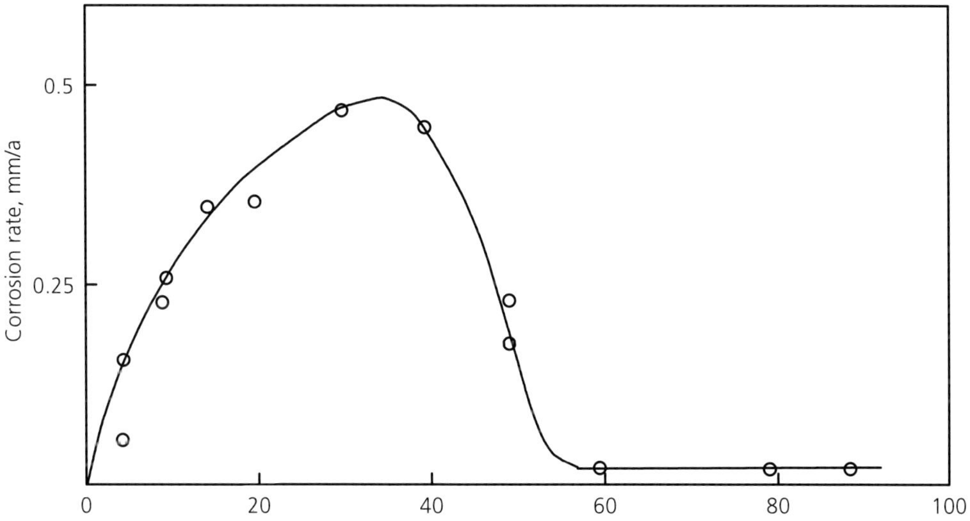

Figure 37: Corrosion rate of Duriron® (14.5 % Si) in various concentrations of boiling sulfuric acid [86]

Accordingly a value of 0.5 mm/a (19.7 mpy) is given for approximately 33% H_2SO_4.

High-silicon cast iron achieves good corrosion resistance in sulfuric acid by the formation of a protective SiO_2 layer on the surface. Associated with this is a drop in the corrosion rate as the acid exposure time increases. Figure 38 shows this reduction in the corrosion rate [82–84]. The final values for the corrosion rates are not reached until an exposure time of hours or weeks [87].

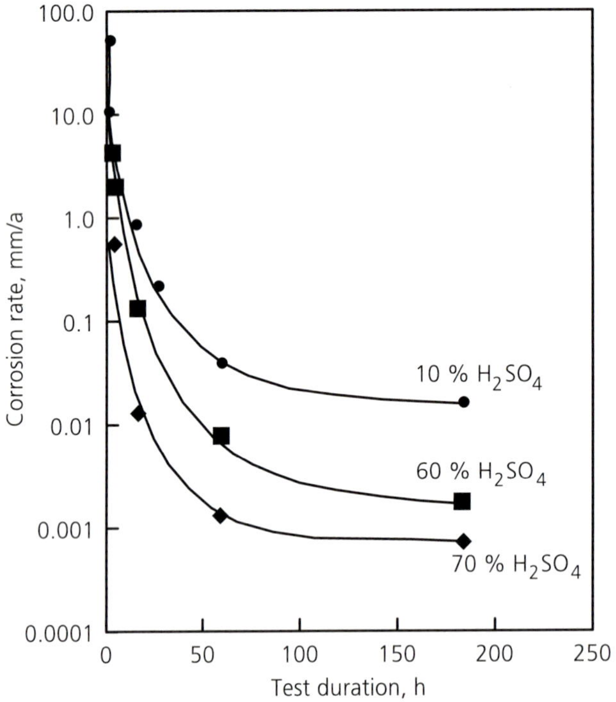

Figure 38: Influence of the test duration on the corrosion rate of the FeSi-alloy containing 14.5% Si (Table 29) in various concentrations of boiling sulfuric acid [82–84]

Graphs of the areas of use showing the corresponding corrosion rates are contained in [1, 2, 4, 6, 7, 60, 61, 63, 66, 86, 88, 89].

Figure 39 contains a detailed reproduction of isocorrosion curves for high-Si cast iron [61].

This figure shows the excellent corrosion resistance of high-Si cast iron in all concentrations of sulfuric acid up to the boiling point. Corrosion rates of about 2 mm/a (78.7 mpy) are shown in the "critical" concentration range between about 5 and 50% H_2SO_4 [61]. According to [82], the maximum corrosion rates are 0.1 mm/a (3.94 mpy). In this context, see Table 31 [82] and Figure 40 [82].

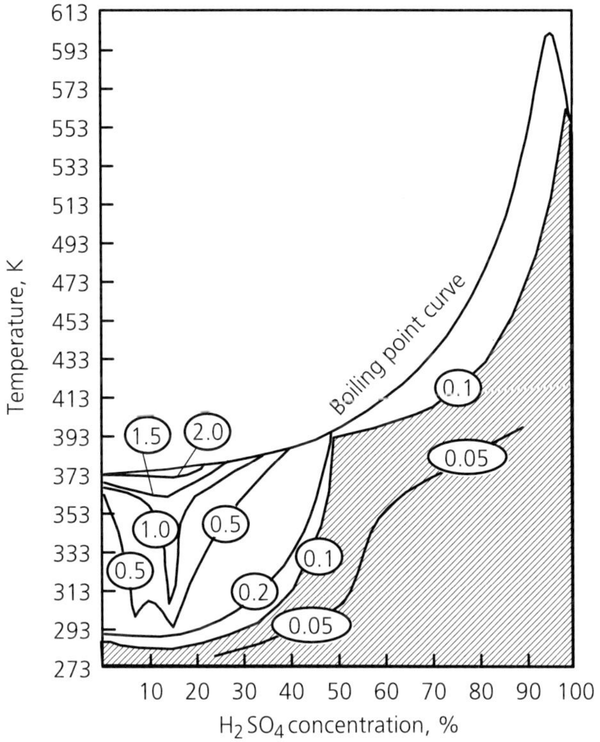

Figure 39: Isocorrosion curves for high-silicon cast iron (14.5 % Si) in sulfuric acid [61]

Cast irons containing 14 to 18 % Si are used in practice. The corrosion resistance of this material in sulfuric acid increases as the Si-content increases as a consequence of the formation of a protective SiO_2 layer; however, it is more susceptible to hydrogen fluoride as a result.

Since pure acid is hardly ever encountered in chemical practice, attention should be paid to the side effects of impurities. The deterioration in the corrosion behavior of high-silicon cast iron due to hydrogen fluoride has already been mentioned. The presence of oxidizing salts (copper sulfate, sodium bichromate) and sodium chloride, ammonium sulfate and sodium sulfate in aqueous solutions of sulfuric acid does not increase the attack of high-silicon cast iron containing 14.5 % Si [60, 66].

Very high-silicon cast iron having Si-contents of about 65 % showed no weight loss in sulfuric acid at the boiling point. According to this work, the corrosion rate in boiling 75 % sulfuric acid dropped to 0.005 mm/a (0.20 mpy) after a test duration of 5 days. The initial corrosion rate was 1.57 mm/a (61.8 mpy) after a test duration of one day [90].

The resistance of high-Si cast iron containing less than 14.5 % Si in sulfuric acid can be increased by modification. Good resistance of an alloy containing 8 to 9 % Si in sulfuric acid in the concentration range between 25 and 100 % can be achieved by addition of Mo [80].

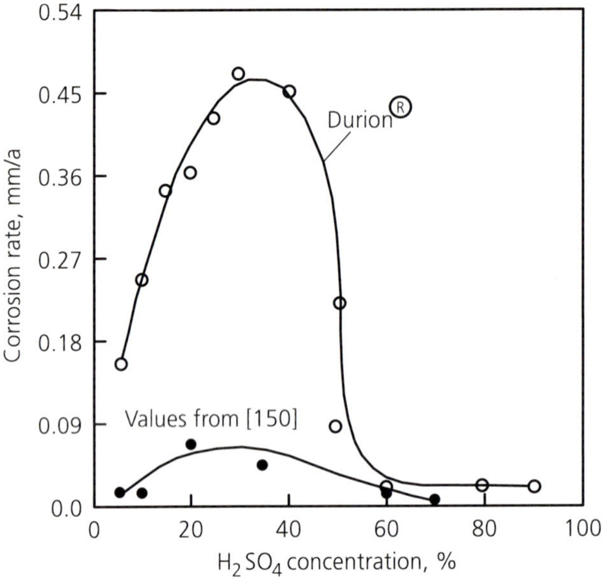

Figure 40: Influence of various concentrations of boiling sulfuric acid on the corrosion of high-Si cast iron (14.5 % Si) (see also Tables 29 and 31), test duration 175 h [82–84]

The mechanical and technological properties of high-Si cast iron can be improved by modifying the alloy composition. Table 32 contains a summary of such alloys. A protective surface layer is formed on the surface of these alloys after an exposure time of only 20 h in 40 % H_2SO_4 at 333 K (60 °C), and in association with this a very marked decrease in the corrosion rate occurs. Figure 41 shows the course of the corrosion and the Si-content in the surface layer [91].

Table 33 contains a summary of corrosion rates for high-silicon cast iron with molybdenum [22]. Molybdenum contents reduce the corrosion rate of high-silicon cast iron in warm and hot sulfuric acid even at lower concentrations.

Alloy designation	*Chemical composition, %*[1)]							
	C	*Si*	*Mn*	*P*	*S*	*Cu*	*Cr*	*Rare metals*[2)]
ChS15	0.44	15.2	0.44	0.08	0.23	–	–	–
ChS15Ch	0.42	15.1	0.40	0.07	0.021	–	–	0.06
ChS17	0.30	17.0	0.50	0.08	0.04	–	–	–
ChS17Ch	0.43	16.9	0.46	0.07	0.033	–	–	0.08
ChS17D2KhCh	0.42	16.8	0.40	0.07	0.032	1.95	1.05	0.05

1) balance Fe
2) rare metals (Ce, Y, Pr, Nd, Tb, Gd)

Table 32: Chemical composition of modified FeSi-alloys (see Figure 41) [91]

High-Si cast iron pumps are used up to a maximum of 413 K (140 °C) with sulfuric acid concentrations up to 75 %. Special constructions are used for more severe stresses [58].

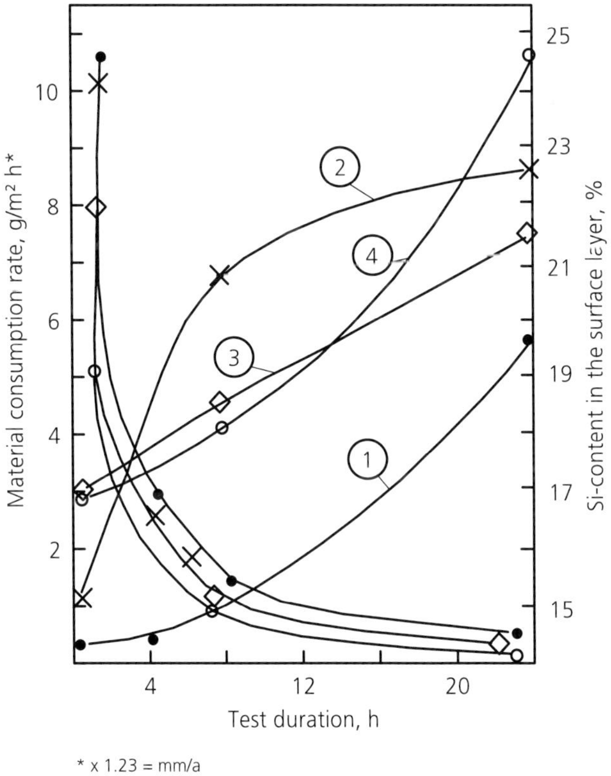

Figure 41: Material consumption rate of FeSi-alloys in 40 % sulfuric acid at room temperature as a function of the test duration and Si-content in the surface layer (see Table 32) [91]
① ChS15
② ChS15Ch
③ ChS17Ch
④ ChS17D2KhCh

Distillation columns on which sulfuric acid acts in concentrations of 65 to 95 % at temperatures between 413 and 573 K (140 and 300 °C) are made of high-silicon cast iron containing 15 % Si [5]. High-silicon cast iron is used also for stirrers in sulfuric acid.

High-Si cast iron is not used in oleum, since cracking and bursting may occur along the grain boundaries of the material due to oxidation of the silicides with subsequent salt formation [5, 58].

If anodic protection of unalloyed cast iron is used in hot concentrated sulfuric acid, high-silicon cast iron can be used as the cathode [74]. See also Section Unalloyed steels and cast steel.

H_2SO_4 concentration, %	Temperature K (°C)	without Mo	with 3 % Mo
		Corrosion rate, mm/a (mpy)	
40	343 (70)	0.275 (10.8)	0.075 (2.95)
40	boiling temp.	0.350 (13.8)	0.170 (6.69)
50	343 (70)	0.085 (3.35)	0.008 (0.31)
50	boiling temp.	0.150 (5.91)	0.040 (1.57)
75	boiling temp.	0.007 (0.28)	0.017 (0.67)
95	343 (70)	0.001 (0.04)	0.007 (0.28)
95	343 (70)	0.009[1)] (0.35)	
88	553 (280)	< 0.1[2)] (< 3.94)	< 0.1[3)] (< 3.94)
90	573 (300)	< 0.1[2)] (< 3.94)	< 0.1[3)] (< 3.94)
95	593 (320)	< 0.1[2)] (< 3.94)	< 0.1[3)] (< 3.94)

1) aerated
2) Duracid® containing 15 % Si
3) Duracid® containing 15 % Si + 3 % Mo

Table 33: Corrosion rates of high-silicon cast iron in various concentrations of sulfuric acid at various temperatures, with and without addition of molybdenum [22]

Structural steels with up to 12 % chromium

These steels hardly differ from unalloyed and low-alloy steels in their corrosion resistance in sulfuric acid and sulfuric acid solutions (see also Section Unalloyed steels and cast steel).

For steels containing 12 and 17 % chromium corrosion rates in excess of 1.27 mm/a (50 mpy) at room temperature are given over the entire concentration range of sulfuric acid [59]. In comparison with higher-alloyed materials or, for example, with high-silicon cast iron, the corrosion rates are in fact far in excess of the given value of 1.27 mm/a (50 mpy).

According to [76], these steels are only slightly more corrosion-resistant than unalloyed steels under weakly oxidizing conditions. Under atmospheric conditions (SO_2 atmosphere), a better resistance than unalloyed steels can also be expected.

If the material does not have a natural resistance in certain media (cf. the noble metals), as in sulfuric acid here, this resistance can be achieved only by passivation – if the material is intended for use. Low-alloy iron materials do not have these properties initially. As has already been described for unalloyed steels (Section Unalloyed steels and cast steel), this resistance can be achieved only under certain conditions by anodic polarization (anodic protection), or by alloying.

If iron is alloyed with chromium, its passive range is extended to negative potentials, and the initiating current density is lowered. Figure 42 shows the reduction in the current density and the associated reduced weight loss in 10 % sulfuric acid at room temperature. See also Figures 18 and 19 and Section Unalloyed steels and cast steel for the influence of alloying elements on iron. As Figure 42 shows, the passive range of iron is shifted to highly negative potentials only at above 10 % chromium. This is also illustrated by Figures 18 and 19. Iron-chromium alloys containing more than 13 % chromium behave passively and are resistant under mild corrosive conditions [92].

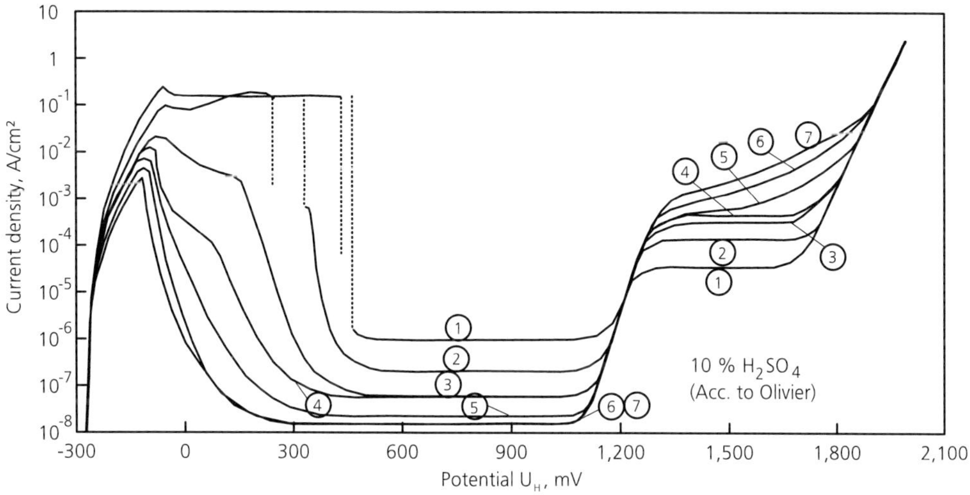

Figure 42: Anodic current density/potential curves for FeCr-alloys of varying Cr-content in 10 % sulfuric acid at 298 K (25 °C) [92]
① 2.8 %
② 6.7 %
③ 9.5 %
④ 12 %
⑤ 14 %
⑥ 16 %
⑦ 18 %

Table 8 shows the low and approximately identical corrosion resistance of steels containing about 13 % Cr with unalloyed steels in sulfuric acid and sulfuric acid solutions. Tables 34 [93] and 50 [94] give a rough summary of corrosion rates on chromium steels and chromium-nickel steels in sulfuric acid and sulfuric acid solutions for certain steels.

Table 35 contains the corresponding material designations [93]. The steel 1.4512 (X2CrTi12, cf. SAE 409) from this table belongs to the materials dealt with in this section. In summary, the area of use of steels with less than, or a maximum of, not more than 12 % chromium can be located in the area for unalloyed steels, as Figure 16 show. According to Figure 16, these steels are less resistant than those containing 17 % Cr. From Figures 18 and 19, at chromium contents of up to 10.4 % an even poorer corrosion resistance than that of low-alloy steels without added chromium can be expected.

Table 36 shows the corrosion behavior of chromium steels with various Cr-contents at 293 K (20 °C) in 6 % sulfuric acid flushed with O_2 or H_2. Hardly any difference is detectable under these conditions. The corrosion resistance of these steels decreases as the Cr-content increases. Whereas the 8.3 % Cr-steel corrodes at 57 mm/a (2,244 mpy) in 6 % sulfuric acid at 293 K (20 °C), the corrosion rate on 16 % steel rises to 135 mm/a (5,315 mpy) with O_2 flushing. The same values were determined with H_2 flushing. By comparison, the corrosion rates on unalloyed steel under the same conditions are 9.1 mm/a (358 mpy) with O_2 flushing, and 0.79 mm/a (31.1 mpy) with H_2 flushing [12]. See also Tables 1 and 2.

Medium	***Concentration %***	***Temperature K (°C)***	***Material (DIN-Material No.)***						
			1.4000 1.4002 1.4005 1.4006 1.4021 1.4024 1.4034 1.4104 1.4116 1.4117 1.4125 1.4512	***1.4016 1.4057 1.4112 1.4120 1.4122 1.4305 1.4510 1.4511***	***1.4113 1.4542 1.4568***	***1.4301 1.4303 1.4306 1.4310 1.4311 1.4541 1.4550***	***1.4521 1.4522***	***1.4401 1.4404 1.4406 1.4429 1.4435 1.4436 1.4460 1.4571 1.4580***	***1.4438 1.4439 1.4465 1.4505***
Sulfuric acid[1)] H_2SO_4	1	293 (20)	3	3	2	1	0	0	
		343 (70)	3	3	2	1	2	0	
		boiling	3	3	3	1	–	1	0
	2.5	293 (20)	3	3	3	1	–	0	
		343 (70)	3	3	3	1	–	0	
		boiling	3	3	3	2	–	2	0
	5	293 (20)	3	3	3	1	–	0	
		343 (70)	3	3	3	1	–	1	0
		boiling	3	3	3	3	–	2	1
	7.5	293 (20)	3	3	3	1	–	0	

1) Under oxidizing conditions, it is possible to use the steels 1.4438 and 1.4439 at low H_2SO_4 concentrations (up to about 10 %) – consult the producer –

Corrosion rates:

0: < 0.11 mm/a (< 4.33 mpy)

1: 0.11–1.1 mm/a (4.33–43.3 mpy)

2: 1.1–11.0 mm/a (43.3–433 mpy)

3: > 11.0 mm/a (> 433 mpy)

Table 34: Possible uses for chromium steels and chromium-nickel steels (see Table 35) in sulfuric acid and solutions containing sulfuric acid (see also Table 8) [93]

Table 34: Continued

Medium	*Concentration %*	*Temperature K (°C)*	*Material (DIN-Material No.)*						
			1.4000 1.4002 1.4005 1.4006 1.4021 1.4024 1.4034 1.4104 1.4116 1.4117 1.4125 1.4512	*1.4016 1.4057 1.4112 1.4120 1.4122 1.4305 1.4510 1.4511*	*1.4113 1.4542 1.4568*	*1.4301 1.4303 1.4306 1.4310 1.4311 1.4541 1.4550*	*1.4521 1.4522*	*1.4401 1.4404 1.4406 1.4429 1.4435 1.4436 1.4460 1.4571 1.4580*	*1.4438 1.4439 1.4465 1.4505*
	7.5	343 (70)	3	3	3	1	–	1	0
		boiling	3	3	3	2	–	2	1
	10	293 (20)	3	3	3	2	–	1	0
		343 (70)	3	3	3	2	–	2	1
		boiling	3	3	3	3	–	2	0
	20	293 (20)	3	3	3	1	–	1	0
		343 (70)	3	3	3	2	–	2	1
		boiling	3	3	3	3	–	3	1
	40	293 (20)	3	3	3	1	–	1	0
		343 (70)	3	3	3	2	–	2	1
		boiling	3	3	3	3	–	3	2
	60	293 (20)	3	3	3	3	–	2	0
		343 (70)	3	3	3	3	–	3	1
		boiling	3	3	3	3	–	3	–
	80	293 (20)	3	3	3	1	–	1	0
		343 (70)	3	3	3	3	–	2	1
		boiling	3	3	3	3	–	3	–

1) Under oxidizing conditions, it is possible to use the steels 1.4438 and 1.4439 at low H_2SO_4 concentrations (up to about 10%) – consult the producer –
Corrosion rates:
0: < 0.11 mm/a (< 4.33 mpy)
1: 0.11–1.1 mm/a (4.33–43.3 mpy)
2: 1.1–11.0 mm/a (43.3–433 mpy)
3: > 11.0 mm/a (> 433 mpy)

Table 34: Possible uses for chromium steels and chromium-nickel steels (see Table 35) in sulfuric acid and solutions containing sulfuric acid (see also Table 8) [93]

Table 34: Continued

Medium	*Concentration %*	*Temperature K (°C)*	*Material (DIN-Material No.)*						
			1.4000 1.4002 1.4005 1.4006 1.4021 1.4024 1.4034 1.4104 1.4116 1.4117 1.4125 1.4512	*1.4016 1.4057 1.4112 1.4120 1.4122 1.4305 1.4510 1.4511*	*1.4113 1.4542 1.4568*	*1.4301 1.4303 1.4306 1.4310 1.4311 1.4541 1.4550*	*1.4521 1.4522*	*1.4401 1.4404 1.4406 1.4429 1.4435 1.4436 1.4460 1.4571 1.4580*	*1.4438 1.4439 1.4465 1.4505*
	98 (concentrated)	293 (20)	–	0	0	0	–	0	0
		343 (70)	2	2	2	2	–	2	1
		423 (150)	3	3	3	2	–	2	–
		boiling	3	3	3	3	–	3	–
Fuming (11 % free SO_3)		293 (20)	–	0	0	0	0	0	
		373 (100)	3	3	3	1	–	0	
Fuming (60 % free SO_3)		293 (20)	–	0	0	0	0	0	
		353 (80)	3	3	3	0	–	0	
Mixed acids (nitrating acids)	50 % H_2SO_4 + 50 % HNO_3	323 (50)	3	2	1	0	1	0	
		363 (90)	3	3	2	1	2	1	
		393 (120)	3	3	3	2	2	2	

1) Under oxidizing conditions, it is possible to use the steels 1.4438 and 1.4439 at low H_2SO_4 concentrations (up to about 10 %) – consult the producer –

Corrosion rates:

0: < 0.11 mm/a (< 4.33 mpy)

1: 0.11–1.1 mm/a (4.33–43.3 mpy)

2: 1.1–11.0 mm/a (43.3–433 mpy)

3: > 11.0 mm/a (> 433 mpy)

Table 34: Possible uses for chromium steels and chromium-nickel steels (see Table 35) in sulfuric acid and solutions containing sulfuric acid (see also Table 8) [93]

Table 34: Continued

Medium	***Concentration %***	***Temperature K (°C)***	***Material (DIN-Material No.)***						
			1.4000 1.4002 1.4005 1.4006 1.4021 1.4024 1.4034 1.4104 1.4116 1.4117 1.4125 1.4512	***1.4016 1.4057 1.4112 1.4120 1.4122 1.4305 1.4510 1.4511***	***1.4113 1.4542 1.4568***	***1.4301 1.4303 1.4306 1.4310 1.4311 1.4541 1.4550***	***1.4521 1.4522***	***1.4401 1.4404 1.4406 1.4429 1.4435 1.4436 1.4460 1.4571 1.4580***	***1.4438 1.4439 1.4465 1.4505***
	75% H_2SO_4 + 25% HNO_3	323 (50)	3	2	1	1	1	0	
		363 (90)	3	3	1	1	1	1	
		430 (157)	3	3	3	3	3	3	
	20% H_2SO_4 + 15% HNO_3	323 (50)	3	3	1	0	1	0	
		353 (80)	3	3	2	1	2	0	
	70% H_2SO_4 + 10% HNO_3	323 (50)	3	3	1	0	1	0	
		363 (90)	3	3	3	1	3	0	
		441 (168)	3	3	3	3	3	3	
	30% H_2SO_4 + 5% HNO_3	363 (90)	3	3	1	0	1	0	
		383 (110)	3	3	2	1	2	0	
	15% H_2SO_4 + 5% HNO_3	407 (134)	3	3	2	1	2	1	1

1) Under oxidizing conditions, it is possible to use the steels 1.4438 and 1.4439 at low H_2SO_4 concentrations (up to about 10%) – consult the producer –

Corrosion rates:

0: < 0.11 mm/a (< 4.33 mpy)

1: 0.11–1.1 mm/a (4.33–43.3 mpy)

2: 1.1–11.0 mm/a (43.3–433 mpy)

3: > 11.0 mm/a (> 433 mpy)

Table 34: Possible uses for chromium steels and chromium-nickel steels (see Table 35) in sulfuric acid and solutions containing sulfuric acid (see also Table 8) [93]

Table 34: Continued

Medium	*Concentration %*	*Temperature K (°C)*	*Material (DIN-Material No.)*						
			1.4000 ***1.4002*** ***1.4005*** ***1.4006*** ***1.4021*** ***1.4024*** ***1.4034*** ***1.4104*** ***1.4116*** ***1.4117*** ***1.4125*** ***1.4512***	***1.4016*** ***1.4057*** ***1.4112*** ***1.4120*** ***1.4122*** ***1.4305*** ***1.4510*** ***1.4511***	***1.4113*** ***1.4542*** ***1.4568***	***1.4301*** ***1.4303*** ***1.4306*** ***1.4310*** ***1.4311*** ***1.4541*** ***1.4550***	***1.4521*** ***1.4522***	***1.4401*** ***1.4404*** ***1.4406*** ***1.4429*** ***1.4435*** ***1.4436*** ***1.4460*** ***1.4571*** ***1.4580***	***1.4438*** ***1.4439*** ***1.4465*** ***1.4505***
	2% H_2SO_4 + 1% HNO_3	boiling	3	3	2	2	2	0	
Fatty acid + traces of H_2SO_4		hot	–	–	3	2	–	1	0
Linseed oil + 3% H_2SO_4		293 (20)	0	0	0	0	0	0	
		473 (200)	–	–	1	0	0	0	
Super phosphate $Ca(H_2PO_4)_2$ + $CaSO_4$ + 3% H_2SO_4		293 (20)	–	–	–	0	–	0	

1) Under oxidizing conditions, it is possible to use the steels 1.4438 and 1.4439 at low H_2SO_4 concentrations (up to about 10%) – consult the producer –

Corrosion rates:

0: < 0.11 mm/a (< 4.33 mpy)

1: 0.11–1.1 mm/a (4.33–43.3 mpy)

2: 1.1–11.0 mm/a (43.3–433 mpy)

3: > 11.0 mm/a (> 433 mpy)

Table 34: Possible uses for chromium steels and chromium-nickel steels (see Table 35) in sulfuric acid and solutions containing sulfuric acid (see also Table 8) [93]

Material No.	*Material DIN-designation*	*Comparable SAE type (approximate)*	*Material No.*	*Material DIN-designation*	*Comparable SAE type (approximate)*
1.4000	X6Cr13	403, 410 S[1)]	1.4401	X5CrNiMo17-12-2	316
1.4002	X6CrAl13	405	1.4404	X2CrNiMo17-12-2	316 L
1.4005	X12CrS13	416	1.4406	X2CrNiMoN17-11-2	316 LN
1.4006	X12Cr13	410[1)]	1.4429	X2CrNiMoN17-13-3	316 LN
1.4016	X6Cr17	430	1.4435	X2CrNiMo18-14-3	316 L
1.4021	X20Cr13	420[1)]	1.4436	X3CrNiMo17-13-3	316
1.4024	X15Cr13	–	1.4438	X2CrNiMo18-15-4	317 L
1.4034	X46Cr13	–	1.4439	X2CrNiMoN17-13-5	317 LMN
1.4057	X17CrNi16-2	431	1.4460	X3CrNiMoN27-5-2	329
1.4104	X14CrMoS17	430 F	1.4465	X1CrNiMoN25-25-2	310 MoLN
1.4112	X90CrMoV18	440 B	1.4505	X4NiCrMoCuNb20-18-2	–
1.4113	X6CrMo17-1	434	1.4510	X3CrTi17	430 Ti
1.4116	X50CrMoV15	–	1.4511	X3CrNb17	–
1.4117	X38CrMoV15	–	1.4512	X2CrTi12	409
1.4120	GX20CrMo13	–	1.4521	X2CrMoTi18-2	443, 444
1.4122	X39CrMo17-1	–	1.4522	X2CrMoNb18-2	443
1.4125	X105CrMo17	440 C	1.4541	X6CrNiTi18-10	321
1.4301	X5CrNi18-10	304	1.4542	X5CrNiCuNb16-4	630
1.4303	X4CrNi18-12	305	1.4550	X6CrNiNb18-10	347, 348
1.4305	X8CrNiS18-9	303	1.4568	X7CrNiAl17-7	631
1.4306	X2CrNi19-11	304 L	1.4571	X6CrNiMoTi17-12-2	316 Ti
1.4310	X10CrNi18-8	301, 302	1.4580	X6CrNiMoNb17-12-2	316 Cb
1.4311	X2CrNiN18-10	304 LN			

1) If SAE materials are required, consult the producer since several DIN materials correspond to these

Table 35: Material designation of the steels listed in Table 34 [93]

Cr-content %	Flushing with	
	O_2	H_2
	Corrosion rate, mm/a (mpy)	
8.3	57 (2,244)	47 (1,850)
9.2	69 (2,717)	64 (2,520)
13.0	109 (4,291)	112 (4,409)
16.0	135 (5,315)	137 (5,394)

Table 36: Comparison of the corrosion rates of Cr-steels in 6 % H_2SO_4 at 293 K (20 °C) on flushing with O_2 or H_2 [12]

Hence chromium steels are the least resistant of the materials studied (unalloyed steel, aluminium, copper, lead, nickel, tin and Monel®) [12]. As Figure 42 shows, these steels are suitable for anodic protection. The passive current density required for the particular use is important here. Steels with a higher Cr-content are more suitable [95].

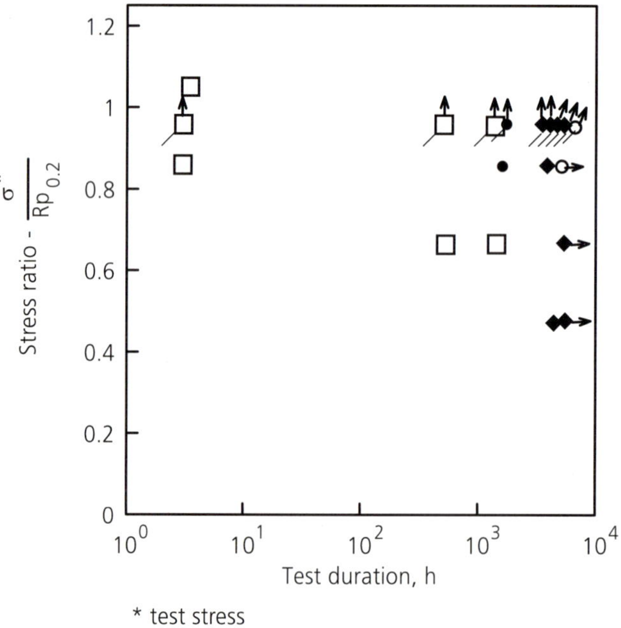

Figure 43: Stress corrosion cracking behavior of the heat-resistant screw steel X19CrMoNbVN11-1 (Table 9) in aqueous solutions (simulated turbine condensate) at 363 K (90 °C) [28, 29]
○ 1 % NaCl
● 0.05 mol/l H_2SO_4
◆ 1 % NaCl + 0.05 mol/l H_2SO_4
□ $H_2O + SO_2$ ↑
/ = corrected

In the case of chromium steels and chromium-nickel steels, an anodic current density of 1 mA/cm^2 corresponds to a corrosion rate of 0.01 mm/a (0.39 mpy) [92].

Ion implantations on the steel 3CR12 ((%) Fe-0.015C-0.018S-0.015P-0.847Mn-0.50Si-0.11V-0.27Ti-0.23Mo-11.55Cr-1.50Ni) exhibited a favorable influence on the corrosion properties of this material in 0.5 mol/l H_2SO_4 at room temperature. However, this measure has no long-term effect [96].

Electrochemical studies have demonstrated that the steel SAE M 50 ((%) Fe-0.85C-4.0Cr-0.27Mn-0.17Si-4.25Mo-1.0V, cf: 1.3551) exhibits a considerably improved corrosion behavior in 1 mol/l H_2SO_4 with Ta implantation [97].

10.4 % chromium in iron does not protect against corrosion and hydrogen uptake in sulfuric acid. On the contrary, chromium forms so-called hydrogen traps which increase the uptake of hydrogen [33]. See Section Unalloyed steels and cast steel for the influence of hydrogen in steel.

The screw materials in Table 9 were investigated for susceptibility to stress corrosion cracking in aqueous solutions containing sulfite, sulfate and chloride under conditions similar to in practice. For details, see Section Unalloyed steels and cast steel. Figure 43 shows the test results for the high-alloy, heat-resistant screw material X19CrMoNbVN11-1 (1.4913). In the certified delivery condition, no stress corrosion cracking was detected in the test media at stresses below the $R_{p0.2}$ limit. Above this limit, phenomena similar to stress corrosion cracking (time of failure 3 h) were found on this steel in solutions saturated with SO_2 at 363 K (90 °C).

Longer-running specimens fractured as a result of a reduction in cross-section (see corrected value), without displaying features of stress corrosion cracking [28, 29].

A steel containing 2.25 % Cr and 1 % Mo in the tempered state exhibited a better corrosion behavior in 0.025–2.5 mol/l H_2SO_4 than the cold worked material in corrosion studies. No corrosion rates are given in this work. The conclusions are based on electrochemical studies [98].

Ferritic chromium steels with more than 12 % chromium

As with steels alloyed with less than 12 % chromium (see Section Structural steels with up to 12 % chromium), those containing more than 12 % chromium can be used under corrosion conditions only if they are in the passive state, or can be passivated in the corrosion medium. Since sulfuric acid has a reducing action over wide ranges, adequate corrosion resistance of these materials in sulfuric acid can be expected only under quite specific conditions.

From [59], corrosion rates in excess of 1.27 mm/a (50 mpy) can be expected already at room temperature in all concentrations of aerated sulfuric acid. The corrosion rate can, in fact, be a multiple of this value.

As can be seen from Figure 19, the passivatability of steels increases as the chromium content increases. Above 10 % Cr, the corrosion currents in the passive range in 10 % sulfuric acid show a marked decrease. According to [92], iron-chromium alloys containing more than 13 % Cr can be passive and stable. This means that oxidizing agents present in the corrosion medium, in this case in sulfuric acid, may reduce the aggressivity towards chromium steels. Thus oxidizing acids attack these materials only slightly or hardly at all.

Figure 16 confirms this behavior. 17 % chromium steels have an extended range of use in oxidizing, concentrated sulfuric acid in comparison with unalloyed steels.

See Table 49 for the possible uses of 1.4015 (X8Cr18, cf. UNS S43080) in the production of caprolactam.

As a result of their passivatability, the use of anodic protection is suitable for these steels [95].

Corrosion studies on stainless steels over a period of up to 10 years in an industrial and in a sea atmosphere have shown that the chromium steel X8Cr18 is not suitable in this context. When exposed to the atmosphere, sulfurous acid (H_2SO_3) and then sulfuric acid (H_2SO_4) form as a result of the exposure to SO_2, and these products, together with other pollutants, initiate severe corrosion on this steel [99].

Vigorous corrosive attack occurs in aqueous solutions of sulfuric acid, as summarized in Table 36 for 6 % acid. As the chromium content increases, the corrosion rates rise regardless of the gas used for flushing (O_2 or H_2). At a corrosion rate of 135 mm/a (5,315 mpy) at 293 K (20 °C), the steel containing 16 % Cr is least resistant. Under these conditions, the steels listed in this table are less corrosion-resistant than unalloyed steels [12].

Table 8 gives a summary of corrosion rates in sulfuric acid for various groups of steels. Columns 2 to 4 show values for chromium steels. Table 34 contains corrosion rates for commercially available steels (see Table 35). These two summaries clearly show again the narrow area of use of these steels in H_2SO_4. Table 37 contains a further classification of the resistance of chromium-containing materials (cast materials) [73]. Although the Cr-content is high, the iron alloy containing 30 % Cr is of little use in sulfuric acid, that is to say at low temperatures and high concentrations.

Figure 44 shows isocorrosion curves for chromium cast steel containing 0.6 % C and 30 % Cr [61]. This graph illustrates that the material is practically resistant in up to 10 % sulfuric acid at 333 K (60 °C) and in 75 to 100 % acid from 293 to 313 K (20 to 40 °C). Alloying with molybdenum improves the corrosion resistance in sulfuric acid.

The current development of chromium steels is being concentrated on specific, ferritic chromium steels, and is aimed at improving the general corrosion resistance, providing less sensitivity towards stress corrosion cracking and the action of chlorides, and thus increasing the resistance to pitting corrosion, crevice corrosion and stress corrosion cracking.

These conditions are met largely by the ferritic chromium steels. Table 38 gives a summary of those steels with Cr-contents above 18 %. From this compilation, a distinction may be made between three groups of chromium steels which are used chiefly today:

- 18 % Cr + Mo of low C- and N-content, stabilized with Ti or Nb.
- 25–28 % Cr, nickel-free, partly alloyed with Mo and of low C- and N-content, stabilized and
- stabilized Mo- and Ni-containing ferritic steels of high Cr-content [100]

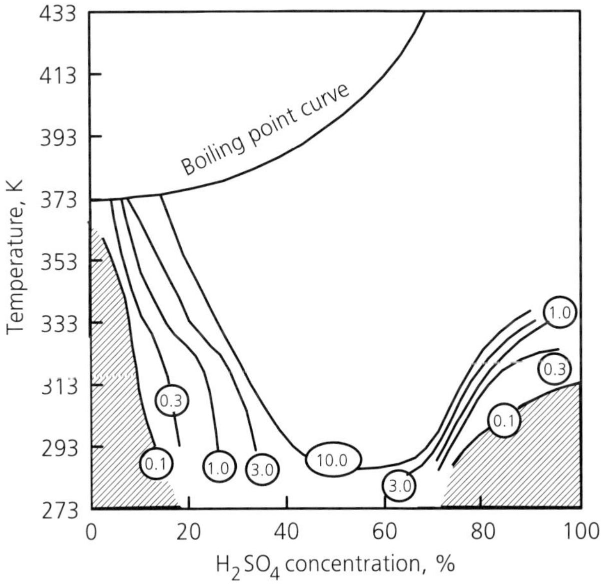

Figure 44: Isocorrosion curves for chromium cast steel containing 0.6% C and 30% Cr [61]

Material	*Temperature K (°C)*	*Sulfuric acid concentration, %*					
		5	*10*	*30*	*50*	*75*	*98*
Fe-30Cr	293 (20)	⊗	–	–	–	+	+
	323 (50)	–	–	–	–	–	–
	373 (100)	–	–	–	–	–	–
Fe-30Cr + Mo	293 (20)	⊗	⊕	–	–		+
	323 (50)		–	–	–		+
	373 (100)	–	–	–	–	–	⊕
Fe-18Cr 3Ni	293 (20)	+	⊕	–	–		+
	323 (50)	⊕	⊗	–	–	–	⊕
	373 (100)	–	–	–	–	–	⊗

Material consumption rate per unit area
+ 0 to 0.1 g/m^2 h
⊕ 0.1 to 1 g/m^2 h
⊗ 1.0 to 10.0 g/m^2 h
– > 10.0 g/m^2 h

Table 37: Corrosion behavior of high-alloy cast steel and nickel cast alloys in sulfuric acid at various temperatures [73]

Table 37: Continued

Material	*Temperature K (°C)*	*Sulfuric acid concentration, %*					
		5	*10*	*30*	*50*	*75*	*98*
Fe-18Cr-8Ni + Mo	293 (20)	+	+	⊕	⊗	⊕	+
	323 (50)	⊕	⊕	⊗	–	⊗	⊗
	373 (100)	+	⊗	⊗	–	–	⊗
Fe-20Cr-25Ni + Mo + Cu	293 (20)	+	+	+	+	+	+
	323 (50)	⊕	⊕	⊕	⊕	⊕	+
	373 (100)	⊗	⊕	⊕	⊕	⊗	⊕
Fe-56Ni-22Mo	293 (20)	+	+	+	+	+	+
	323 (50)	+	+	+	+	+	+
	373 (100)	⊕	⊕	⊕	⊕	⊗	⊕
Fe-62Ni-32Mo	293 (20)	+	+	+	+	+	+
	323 (50)	+	⊕	⊕	+	+	+
	373 (100)	⊕	⊕	⊕	⊕	⊗	⊕
Fe-55Ni-19Mo-17Cr	293 (20)	+	⊕	+	+	+	+
	323 (50)	⊕	⊕	⊕	⊕	+	+
	373 (100)	⊕	⊕	⊕	⊕	–	⊕
Fe-14.5Si	293 (20)	⊕	⊕	+	+	+	+
	323 (50)	⊗	⊕	+	+	+	+
	373 (100)	⊗	⊗	⊕	⊕	⊕	+

Material consumption rate per unit area
+ 0 to 0.1 g/m^2 h
⊕ 0.1 to 1 g/m^2 h
⊗ 1.0 to 10.0 g/m^2 h
– > 10.0 g/m^2 h

Table 37: Corrosion behavior of high-alloy cast steel and nickel cast alloys in sulfuric acid at various temperatures [73]

Trade name/ Mat. No.	Alloy type	Chemical composition, %							
		C	Si	Mn	Cr	Mo	Ni	N	Others
1.4521	X2CrMoTi18-2	0.01	+	+	18	2	–	0.02	Ti
18-3	CrMo 18 3	0.01	+	+	18	3	–	0.02	Ti
E-Brite® 16–1	CrMo 26 1	0.002	0.25	0.10	26	1	–	0.01	–
26–1 S	CrMoTi 26 1	0.04	0.20	0.30	26	1	–	0.02	Ti
26–1 Cb	CrMoNb 26 1	0.04	0.20	0.30	26	1	–	0.02	Nb
Remanit® 4133	X 1 CrMo 28 2	0.005	0.10	0.10	28	2	–	0.005	–
Remanit® 4575	X1CrNiMoNb28-4-2	0.01	0.20	0.4	28	2.5	4	0.02	Nb
Monit®	CrNiMoTi 25 4 4	0.015	+	+	25	4	4	0.02	Ti
Sea-cure® SC-1	CrNiMoTi 26 3 3	0.02	0.30	0.5	26	3	3	0.03	Ti
AL 29-4	CrMo 29 4	0.01	0.02	0.1	29	4	–	0.01	–
AL 29-4-2	CrMoNi 29 4 2	0.01	0.02	0.1	29	4	2	0.01	–
AL 29-4C	CrMoNb 29 4	0.02	+	+	29	4	–	0.02	Ti
S-Mac® 302	CrMo 30 2	0.005	+	+	30	2	–	0.01	–

Table 38: Chemical composition of ferritic chromium steels containing at least 18 % Cr [100]

These materials are of particular interest due to their good mechanical and technological properties and because they are economical.

Figure 45 shows isocorrosion curves of the steel X 2 CrNiMoTi 25 4 4 (Monit®) in comparison with those of a CrNi-cast steel. This graph illustrates that the ferritic steel X 2 CrNiMoTi 25 4 4 is better for use at concentrations of up to 40 % sulfuric acid at higher temperatures than the austenitic steel GX7NiCrMoCuNb25-20 (1.4500, cf. CN-7M). Whereas the high-alloy steel X6CrNiMoTi17-12-2 failed after use as a heating coil material (3 months in sodium bisulfate solution at 368 K (95 °C) and pH 1 (H_2SO_4)) as a result of pitting and stress corrosion cracking, the ferritic steel X 2 CrNiMoTi 25 4 4 mentioned exhibited no corrosive attack after 20 months.

For alloying and processing reasons, the maximum wall thickness for this material is limited to 3 mm. On the other hand, the ferritic steel Remanit® 4575 (X1CrNiMoNb28-4-2 or 1.4575) is obtainable in any semi-finished form and size, and for this reason need not be combined with other materials. Nevertheless, processing of this material is exacting (welding guidelines as for the processing of titanium) and makes production of equipment expensive [100].

The corrosion resistance of X1CrNiMoNb28-4-2 in sulfuric acid is comparable to that of the austenitic steel X1CrNiMoN25-25-2 (1.4465, cf. SAE 310 MoLN). Figure 46 contains the weight losses of the ferritic steel in sulfuric acid as a function of the temperature and concentration. It should be noted that the resistance of this steel to sulfuric acid can be substantially improved by addition of copper salts [101].

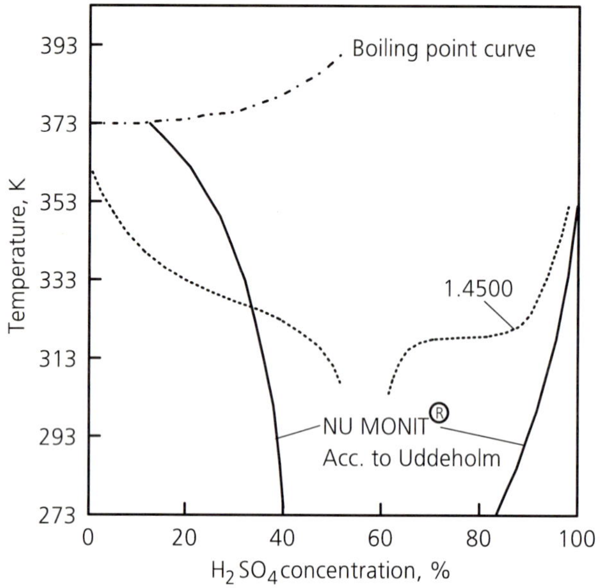

Figure 45: Isocorrosion curves (0.1 mm/a (3.94 mpy)) of the ferritic steel Monit® (X 2 CrNiMoTi 25 4 4) in comparison with the austenitic cast steel GX7NiCrMoCuNb25-20 (DIN-Mat. No. 1.4500) in sulfuric acid [100]

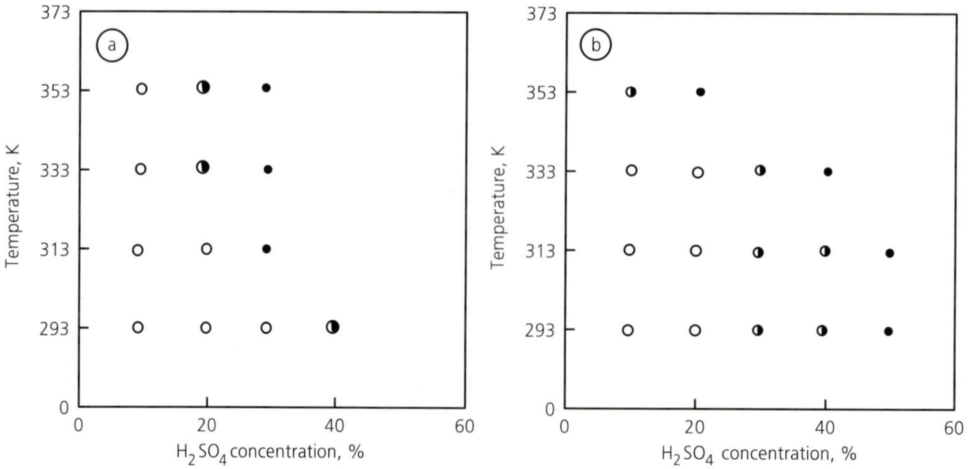

Figure 46: Corrosion resistance of high-alloy steels in sulfuric acid [101]
a) X1CrNiMoNb28-4-2 (Remanit® 4575)
b) X1CrNiMoN25-25-2 (Remanit® 4465)
○ < 0.1 g/m^2 h
◑ 0.1–1 g/m^2 h
● > 1 g/m^2 h

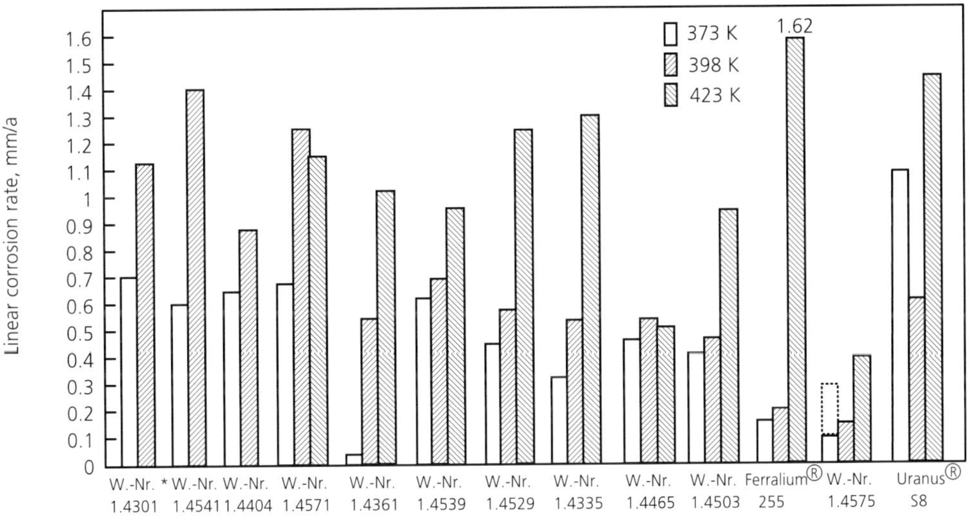

* W.-Nr. = DIN-Mat. No.

Figure 47: Corrosion behavior of high-alloy steels in static, non-aerated, hot 95% sulfuric acid at various temperatures (compare also Figures 101, 135) [102, 103, 109] test duration 21 d

This steel shows a good resistance, in comparison with other steels, in hot 95% sulfuric acid, particularly at 423 K (150 °C), with a corrosion rate of not more than 0.35 mm/a (13.8 mpy) (Figure 47) [102, 103]. This corrosion rate is of particular interest for thick-walled components.

Figure 48 shows a summary of the action of alloying elements in high-alloy chromium steels containing 20 and 26% Cr in various concentrations of sulfuric acid at various temperatures [104, 105]. The specimens used for this graph were activated with a zinc rod before being introduced into the acid. In these graphs, the shaded area characterizes the passivation established under these conditions. The corrosion rates of the specimens in the passive state are so low that no discoloration of the solution occurs [104]. Chromium steels are given a considerable increase in corrosion resistance in sulfuric acid only by Ni-additions. The beneficial influence of chromium and molybdenum on the corrosion resistance can also be seen. An addition of 1% Cu extends the resistance range to higher temperatures. The steel investigated in this study contains (%) Fe-28Cr-2Mo-4Ni-1Cu and is superior to the austenitic CrNiMo-steels in acid concentrations of up to 30% H_2SO_4 at 333 K (60 °C). It can be used in boiling sulfuric acid in concentrations below 30% and is superior to the copper-containing CrNiMo-steels under these conditions [105]. Similar studies were performed in [106].

These graphs explain the good corrosion resistance of the ferritic steels containing certain alloying additions in comparison with other grades. Table 39 contains the weight losses of the cast steel 90Ch28MFTAL ((%) Fe-28Cr-max.1C-3.5Mo-1Si-1V-1Mn-0.5Ti-0.4N) in various concentrations of sulfuric acid at various tempera-

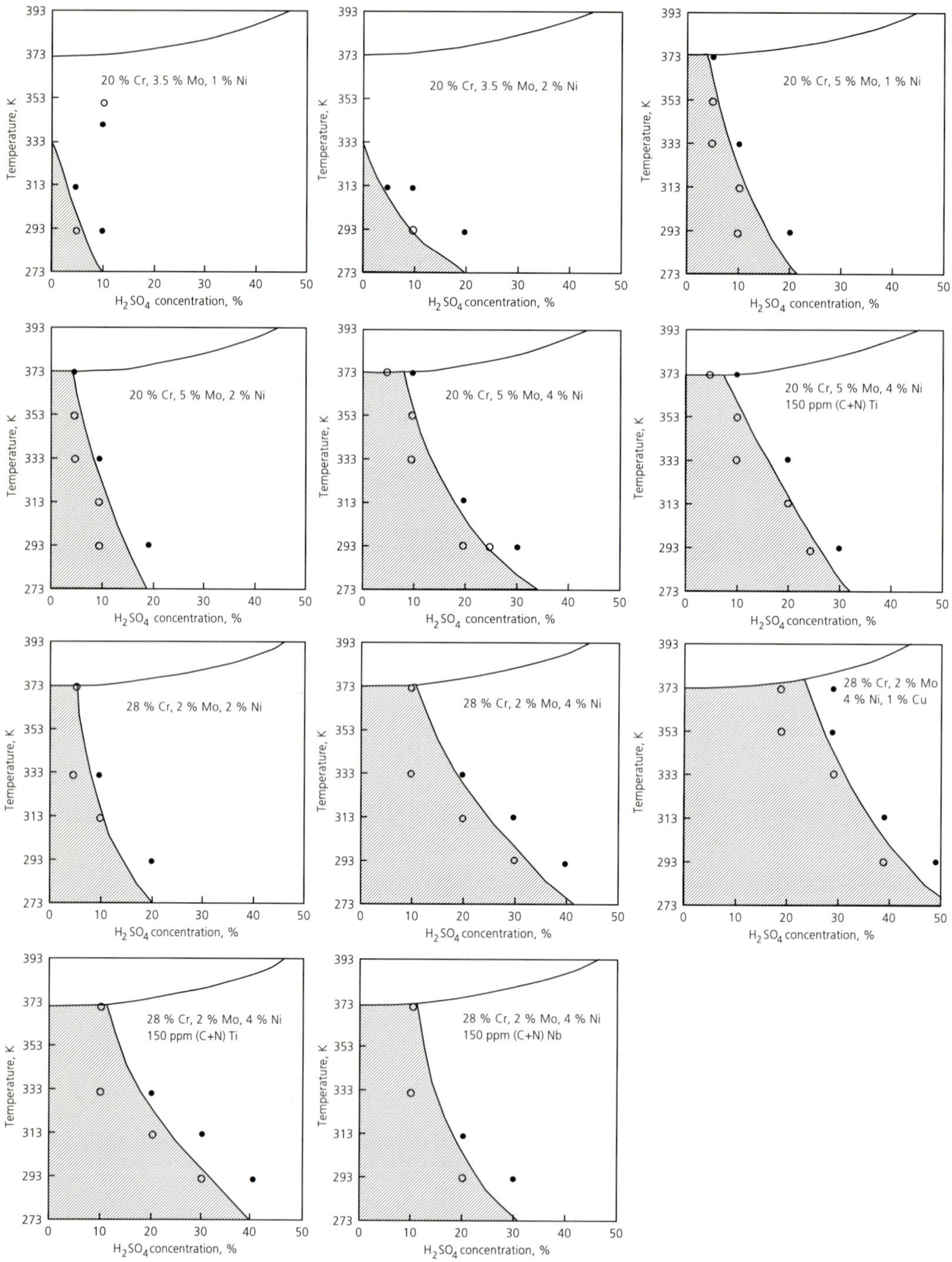

Figure 48: Influence of alloying elements in high-Cr steels on the corrosion behavior in various concentrations of sulfuric acid (compare also Figure 16) [104, 105]

○ resistant

● not resistant

tures. As was to be expected from Figures 45, 46 and 47, these materials can be used only up to concentrations of 15 % H_2SO_4, as they are not resistant at higher concentrations. As expected, the corrosion rate drops in concentrated sulfuric acid and is 0.4 mm/a (15.7 mpy) in 93 % acid at 353 K (80 °C) [107].

According to [108], copper additions in the cast steel 90Ch28MFTAL ((%) Fe-0.7C-31.1Cr-2.8Mo-0.8Si-0.8Mn-0.3Ti-0.6V) improved the corrosion resistance in sulfuric acid. Table 40 shows this influence of copper [73]. Although the corrosion rates in 35 % H_2SO_4 are very high in this summary, a reduction in the material consumption is also clearly detectable under these conditions. With a Cu-addition of 3.25 %, this cast steel can be regarded as resistant in 15 and 93 % sulfuric acid at 353 K (80 °C). The action of copper here is based on its accumulation on the surface of the material, which makes passivation easier [108].

A comparison of the results in Tables 39 and 40 shows the influence of copper, as can be seen also from Figure 48. According to Table 39, the corrosion behavior of the steel with the same designation but without copper is no worse. The corrosion resistance deviates, above all, in 15 and 35 % sulfuric acid. Needless to say, the steels also differ in their chemical composition.

Material	Temperature, K (°C)	H_2SO_4 concentration, %				
		15	35	55	70	93
		Material consumption rate, g/m^2 h				
90Ch28MFTAL[1)]	333 (60)	0.005	0.01	10.0	4.2	0.23
90Ch28MFTAL	353 (80)	0.010	5.9	13.7	7.7	0.38
06ChN28MDTL[2)]	333 (60)	0.03	0.052	1.78	0.627	0.01
06ChN28MDTL	353 (80)	0.126	0.271	3.4	0.950	0.15

1) 28 Cr, max. 1 C, 3.5 Mo, 1 Si, 1 V, 1 Mn, 0.5 Ti, 0.4 N
2) ≤ 0.06 C, ≤ 0.8 Si, ≤ 0.8 Mn, 22–25 Cr, 2.5–3.0 Mo, 26–29 Ni, 0.5–0.9 Ti, 2.5–3.5 Cu

Table 39: Material consumption rates of high-alloy steels in various concentrations of sulfuric acid at various temperatures, test duration 100 h [107]

Cu-content	H_2SO_4 concentration, %			
	15	35	55	93
	Corrosion rate, mm/a (mpy)			
0	1.44 (56.7)	20.19 (795)	> 100 (> 3,94)	0.41 (16.1)
1.5	1.21 (47.6)	20.24 (797)	> 100 (> 3,94)	0.26 (10.2)
2.0	0.97 (38.2)	10.33 (407)	> 100 (> 3,94)	0.20 (7.87)
3.25	0.15 (5.91)	9.35 (368)	> 100 (> 3,94)	0.11 (4.33)

Table 40: Corrosion rates of 90Ch28MFTAL cast steel ((%) Fe-0.7-31.1Cr-2.8Mo-0.8Si-0.8Mn-0.3Ti-0.6V) of various Cu-contents in various concentrations of sulfuric acid at 353 K (80 °C), test duration 100 h [108]

A poorer corrosion behavior can be expected on principle from lower-alloy steels on exposure to sulfuric acid. For a specific use, the manufacturer's instructions on possible suitability should be followed.

As already described above, the corrosion resistance of Cr-steels is based on their passivity, and they must be self-passivatable under the appropriate corrosion conditions. This self-passivation can be intensified in two possible ways:

1. Alloying of the material with a relatively large amount of a readily passivatable metal (see also Figure 42) and hence inhibition of the anodic process.
2. Alloying of a material which tends towards passivation with small amounts of those metals which display a low overpotential regarding cathodic processes (usually evolution of hydrogen) (cathodically active alloying) [110–113].

A cathodically active alloying component (which corresponds to anodic protection) must meet the following conditions:

The equilibrium potential must be more positive than the corrosion potential of the basic material, and the overpotential of the cathodic process should have the lowest possible value (In acid media, discharge of hydrogen usually takes place. The overpotential is a measure of the inhibition of the electrode reaction).

These conditions are met, for example, by the metals Pd, Ru, Pt, Rh, Re etc. Nickel, tungsten and molybdenum may appear as cathodic additions if the medium has a low aggressivity, due to their low hydrogen overpotential and adequate resistance in many media. Intermetallic compounds, such as carbides, borides, nitrides and silicides, are also of interest for this purpose.

A protective action can be achieved with much smaller amounts in the case of cathodic alloying in contrast to alloying with readily passivatable components.

After titanium and its alloys, FeCr-alloys show the greatest tendency towards self-passivation. They are particularly suitable for cathodic alloying. However, the process encounters difficulties because of the low resistance of the passive state to the influence of chloride. Figure 49 shows the influence of small amounts of palladium on the self-passivation of FeCr-alloys and the associated drop in the corrosion rate at 373 K (100 °C) in 20 % sulfuric acid [110, 111, 114]. On severe exposure to corrosion, the corrosion rate of the steel 25Cr8MnN drops to about 10 mm/a (394 mpy), whereas without palladium it is destroyed immediately. Ni and Mn prove to be very effective alloying additions to these steels, in addition to the cathodic components. Figure 50 shows the influence of an alloying addition of 0.5 Pd on the corrosion behavior of a Cr-steel containing 25 % Cr and a Cr-cast steel containing 28 % Cr in various concentrations of sulfuric acid at 373 K (100 °C) [111, 112]. This graph shows that the material consumption rate also drops considerably as the concentration increases. In the passive state in 20 % sulfuric acid, the material consumption rate was less than 0.1 g/m^2 h (below 0.1 mm/a (3.94 mpy)) [111].

Table 41 contains details of the chromium steels used for "cathodic alloying". It shows that the majority of steels had a mixed structure of austenite and ferrite and only the pure Cr-steels had a purely ferritic structure. The steel 25Cr8MnN showed traces of ferrite [110]. See also Section Ferritic-austenitic steels with more than 12 % chromium.

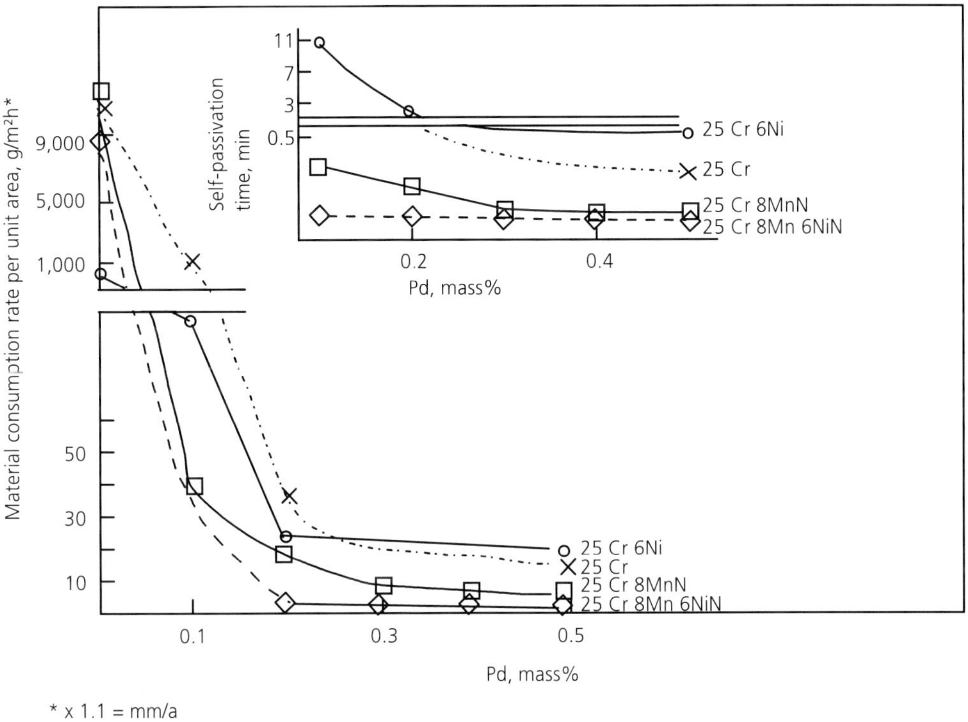

Figure 49: Influence of palladium as a cathodically acting alloying component on the self-passivation and corrosion rate of chromium steels (see Table 41) in 20% sulfuric acid at 373 K (100 °C), test duration 5 h [110, 111, 114]

Table 42 shows the material consumption rate of an iron-chromium alloy containing 40% Cr + 0.2% Pd in various concentrations of sulfuric acid at 373 K (100 °C) [110, 111]. At a corrosion rate of only 0.07 mm/a (2.76 mpy) in 50% H_2SO_4, this alloy is resistant. However, the corrosion rate rises to 22 mm/a (866 mpy) at the boiling point. Under these conditions, the material can be used only in up to 20% sulfuric acid [110].

In comparison with this corrosion behavior in hot sulfuric acid (10%), Table 43 contains the corrosion rates of the Pd-free alloys and those of a few other alloys which are often used [111, 113].

The corrosion resistance of the free-cutting steel Sandvik® 1802 ((%) Fe-18Cr-2.3Mo-0.5Si-0.1Mn-0.6Ti-0.03C + N-0.3S, X2CrMoTiS18-2, 1.4523) in dilute sulfuric acid is approximately comparable to that of the steel X5CrNiMo17-12-2 (1.4401) [115].

The machinability of stainless steels can be improved noticeably also by Ti- and S-additions, as well as by Mn-sulfides. In addition, the Ti-sulfides present in the material lead to lower cutting forces and short chips. In contrast to manganese sulfide, titanium sulfide does not impair the corrosion resistance of materials. The latter is still resistant in strongly acidic electrolytes, whereas manganese sulfide is attacked already in neutral salt solutions [116].

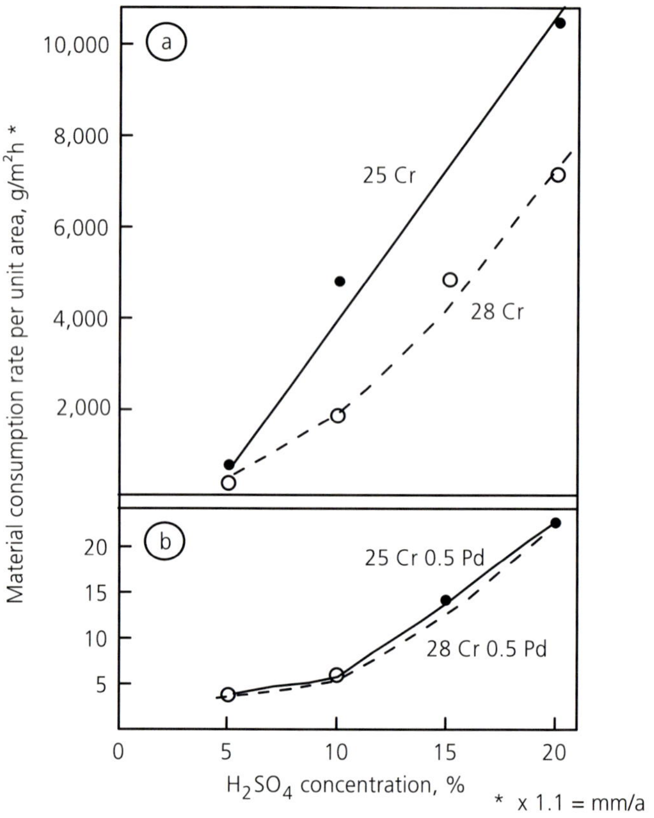

Figure 50: Influencing the average corrosion rate of ductile 25 % Cr-steel and 28 % Cr-cast steel in sulfuric acid at 373 K (100 °C) [111, 112]
a) without cathodically acting Pd-addition
b) with Pd-addition

As expected, the free-cutting steel X 3 CrMoTiS 17 ((%) Fe-0.015C-1.35Si-0.4Mn-0.011P-0.23S-17.11Cr-1.10Mo-0.25Ni-1.20Ti) can hardly be used at all in sulfuric acid. The material consumption rates in 5 % H_2SO_4 are 0 g/m^2 h at room temperature, 51.9 g/m^2 h (57 mm/a (2,244 mpy)) at 313 K (40 °C) and 141.4 g/m^{2}h (156 mm/a (6,142 mpy)) at 333 K (60 °C). In 10 % acid, the material consumption rates are 16.74 g/m^2 h (19 mm/a (748 mpy)) at room temperature, 61.26 g/m^{2}h (67 mm/a (2,638 mpy)) at 313 K (40 °C) and 174.87 g/m^{2}h at 333 K (60 °C) (192 mm/a (7,559 mpy)). This steel is thus considerably less resistant than CrNi-steel under these conditions (see also Section Austenitic chromium-nickel steels and Austenitic chromium-nickel-molybdenum steels). In contrast to the austenites, it is not attacked in a sulfuric acid solution of sodium chloride at pH 0 [116, 117].

The corrosion rate of 18Cr2Mo steels is quoted as 10.2 mm/a (402 mpy) in 2 % sulfuric acid at 303 K (30 °C) [118, 119] and 823 mm/a (32,402 mpy) in boiling 19 % acid [119]. Both values show the low corrosion resistance of these materials in sulfuric acid solution, the latter value being more of theoretical interest since these steels are destroyed under these conditions.

Steel	Melting	Working	Heat treatment	Alloying content, %*						Structure**
				Cr	Ni	Mo	Mn	Pd	N	
25Cr	1	2	3	25.2	–	–	0.6	–	0.03–0.04	F
25Cr0.2Pd				25.6	–	–	0.6	0.2	–	F
25Cr0.5Pd				25.6	–	–	0.6	0.5	–	F
25Cr6Ni				24.7	6.1	–	0.6	–	–	A + 40 % F
25Cr6Ni0.2Pd				24.4	6.2	–	0.6	0.2	–	A + 40 % F
25Cr6Ni0.5 Pd				24.4	6.2	–	0.6	0.5	–	A + 40 % F
25Cr8MnN	1a	2a	3a	23.0	–	–	8.4	–	0.94	A + F (traces)
25Cr8MnN0.2Pd				22.2	–	–	7.7	0.2	0.69	A + 8 % F
25Cr8MnN0.5Pd				22.0	–	–	8.3	0.5	0.91	A + 50 % F
25Cr8Mn6NiN				25.0	5.4	–	7.0	–	0.77	A + 9 % F
25Cr8Mn6NiN0.2Pd				26.5	5.6	–	7.1	0.2	0.97	A
25Cr8Mn6NiN0.5Pd				26.5	5.1	–	9.0	0.5	0.83	A + 10 % F
25Cr2Mn6Ni2MoN				24.3	6.8	2.1	2.2	–	0.53	A + 11 % F
25Cr2Mn6Ni2-MoN0.2Pd				27.0	8.1	2.0	0.8	0.2	0.92	A + 13 % F
25Cr8Mn2MoN				18.0	–	1.9	10.0	–	0.77	A + 40 % F
25Cr8Mn2MoN0.2Pd				23.5	–	1.7	9.5	0.2	0.77	A + 50 % F
25Cr8Mn2MoN0.5Pd				24.5	–	2.0	9.3	0.5	0.97	A + 36 % F

* 0.06 % C
** A = austenite, F = ferrite
1. Under-slag smelting in a high frequency furnace
1a: Raw material smelting in an argon atmosphere, ingot smelting in a nitrogen atmosphere under a partial pressure of 0.4 MPa
2: Steel ingot forging in air and sheet milling after heating to 1,373 K (1,100 °C)
2a: Steel ingot forging in air after heating to 1,473 K (1,200 °C)
3: Annealing at 1,043 K (770 °C) (30 min) with water cooling, hardening from 1,273 K (1,000 °C) (10 min) with water cooling
3a: Hardening from 1,423 K (1,150 °C) (30 min) in water

Table 41: Summary of steels containing about 25 % chromium [110]

Temperature, K (°C)	H_2SO_4 concentration, %				Literature
	20	30	40	50	
	Material consumption rate, $g/m^2\,h$				
373 (100)	0.05	0.05	0.05	0.06	[110, 111]
boiling	0.08	20.00	20.00	20.00	[110]

Table 42: Corrosion of the alloy Fe-40Cr-0.2Pd in the passive state in various concentrations of sulfuric acid at various temperatures after a test duration of 10 h [110, 111]

Alloy	Corrosion rate, mm/a (mpy)
Fe-25Cr	10,000 (393,700)
Fe-40Cr	
Fe-60Cr	
0Ch23N28M3D3T[1)]	3.0 (118)
Hastelloy® A[2)] and B[3)]	0.2 (7.87)
Hastelloy® G[4)]	3.0 (118)
Ti-30Mo	0.18 (7.08)
Fe-40Cr0.2Pd	0.05 (1.97)

1) 0.06 C, 26–29 Ni, 22–25 Cr, 2.4–3.0 Mo, 2.5–3.5 Cu (see Section Austenitic chromium-nickel steels with special alloying additions)
2) 0.15 C, 23 Fe, max. 23 Mo, 2 Mn, 1 Si, balance Ni
3) NiMo-alloy
4) NiMoCr-alloy

Table 43: Linear corrosion rates of high-alloy materials in 40% sulfuric acid at 373 K (100 °C) in comparison with Pd-alloyed FeCr-alloy [111, 113]

Severe intercrystalline corrosion occurs in the steel 26Cr1Mo in an aqueous solution containing 49% H_2O, 49% H_2SO_4 and 2.2% $Fe_2(SO_4)_2$ [120].

Extensive corrosion studies have been performed in Cr-steels with Co-additions [121, 122]. The results of this work have confirmed that Co-additions to rust- and acid-resistant steels increase their resistance in sulfuric acid. As electrochemical studies have shown, the Co-addition reduces the corrosion current in the active range (see also Figure 50). In chromium steels containing 18 and 27% Cr, a passive region develops in the current density potential curve in 10% sulfuric acid at 298 K (25 °C). Figure 51 shows the influence of the Co-content on the material consumption per unit area in 10% aerated sulfuric acid at room temperature as a function of the heat treatment over various test durations. The graph shows that the material consumption decreases as the Co-content increases. A material consumption of 250 mg/cm^2 over 90 h corresponds to a corrosion rate of 31 mm/a (1,220 mpy) and is therefore of no interest in industry [121].

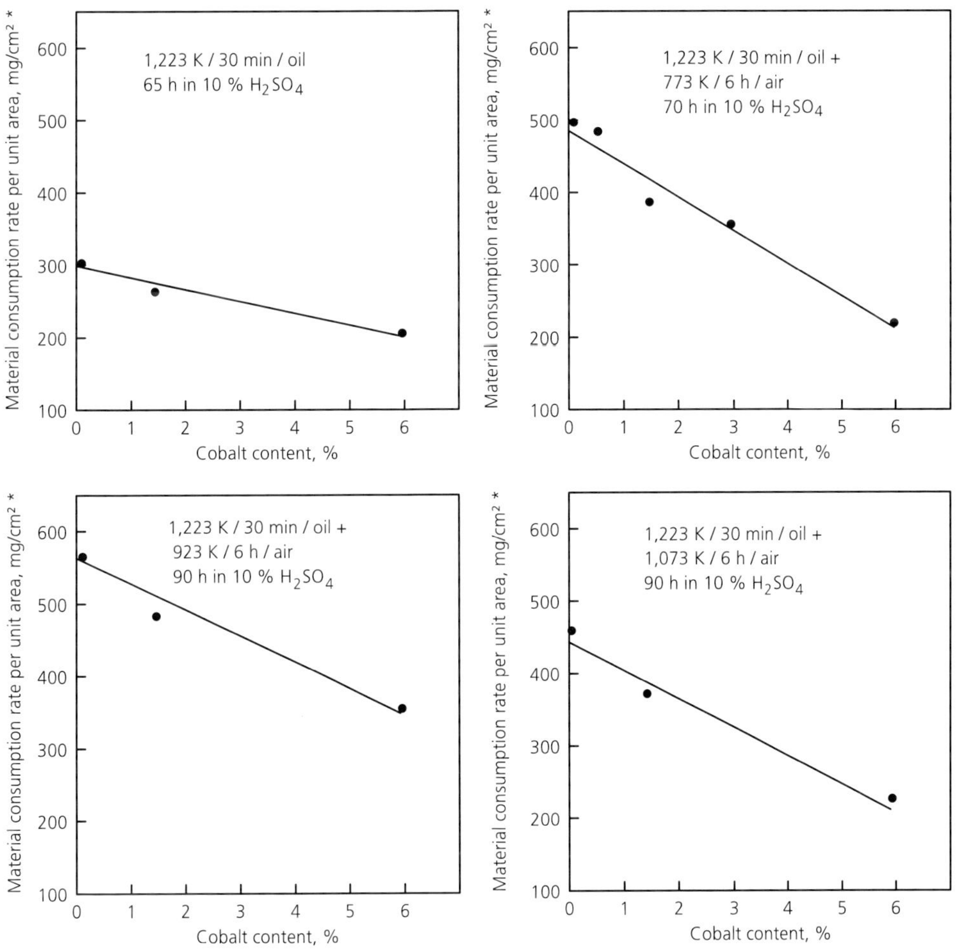

* 250 mg/cm² over 90 h corresponds to about 31 mm/a

Figure 51: Influence of cobalt on the material consumption rate per unit area of an 18 % Cr-steel in aerated 10 % sulfuric acid at 298 K (25 °C) [121]

The following corrosion rates occur on chromium steels of the type 26Cr1Mo and other Cr-steels in boiling butyl-acetate containing 0.3 % H_2SO_4 (test duration 48 h): EB-26-1 4.3 mm/a (169 mpy), EB-26-1 W 24.0 mm/a (945 mpy), AOD 26-1 13.0 mm/a (512 mpy), 20Cr-3Mo 1.6 mm/a (63.0 mpy) and 16Cr-1Ni 11.0 mm/a (433 mpy). The better resistance of the steel 20Cr3Mo is due to the higher Mo-content. Under the same conditions, the steel SAE 316 (X5CrNiMo17-12-2 or 1.4401) has a corrosion rate of less than 0.025 mm/a (0.98 mpy) [123].

– Stress corrosion cracking and pitting in sulfuric acid solutions –

As already mentioned, chromium steels have an adequate corrosion resistance only in the passive range. Whereas the influence of chlorides on the higher-alloy steels is relatively low, it is significant in the lower-alloy steels. Figure 52 shows this influence.

In this graph, the transition from the active to the passive state is characterized in the current density potential curve by a sharp drop in the current density after a critical value, the passivation current density, is exceeded. The higher this value, the more difficult spontaneous passivation is (see also Figure 50). As can be seen from Figure 52, the passivation current density increases as the content of alloying elements decreases and the chloride content increases, as is shown for the steel containing 13 % Cr and 1 % Ni [124]. For the test methods, see [125].

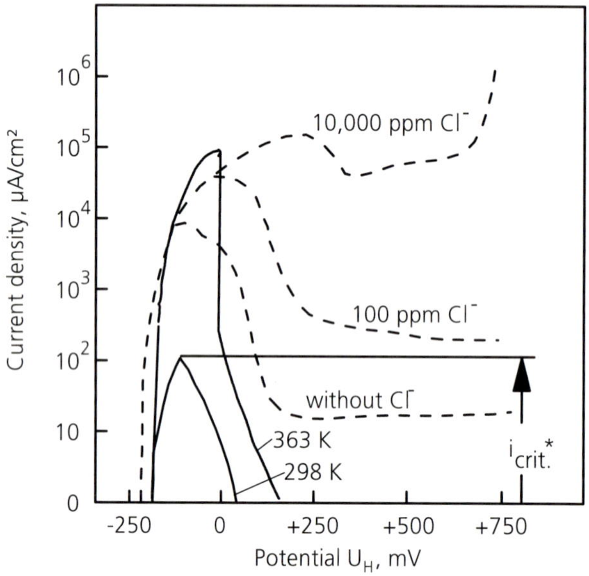

* passivation current density

Figure 52: Influence of the temperature and the chloride content on the passivation current density of Cr-steel and CrNi-steel in 1 mol/l sulfuric acid [124]
——— steel containing 19 % Cr, 8 % Ni
- - - - - steel containing 13 % Cr, 1 % Ni

Chlorides can also impede the passivity. In austenitic steels, a local disturbance in passivity leads to pitting.

The pitting and stress corrosion cracking resistance of Cr-steels and CrNi-steels are determined chiefly by the Cr- and Mo-content. The improving effect of these alloying elements on the pitting corrosion resistance is characterized by the "pitting index" = 1 ×% Cr + 3 ×% Mo. This allows relevant weighting of the improving action. The increase in resistance is characterized also by a rise in the threshold potentials for initiating pitting. Table 44 shows the increase in the pitting resistance equivalent for, inter alia, ferritic steels. According to this table, the pitting corrosion resistance of the steel Monit® is very good in comparison with the usual CrNi-MoCu-steels [100].

Nickel-free chromium steels with Cr-contents of between 18 and 25 % exhibit no susceptibility to stress corrosion cracking, whereas the sensitivity to this corrosion increases greatly with only low Ni-contents. Whereas pitting resistance is not influenced by the Ni-content, Ni is a deciding factor for resistance to chloride-induced stress corrosion cracking [126].

Corrosion studies on the susceptibility to intercrystalline stress corrosion cracking in the Strauss test (10 % $CuSO_4$ + 10 % H_2SO_4, pH 2.5 at 562 K (289 °C)) showed no susceptibility of CrMo-steels 26 1, unless they were sensitized [127].

DIN-Mat. No.	***Trade names***	***Average value of alloying components, %***					***Pitting resistance equivalent***	***U_{1p} 1)***
		Cr	***Ni***	***Mo***	***Cu***	***N***	***% Cr + 3 × % Mo***	***mV***
	Sea-Cure®	26	2	3	–	–	35	550
	Superferrit	28	3.5	2.5	–	–	35.5	575
	Monit®	25	4	4	–	–	37	530
1.4592	AL 29-4C	29	–	4	–	–	41	n.g.
	NSCD	17	16	5.5	3	–	33.5	n.g.
	VEW A 963	17	16	6.3	1.6	+	35.9	610
	AL-6 X	20	24	6	–	–	38	n.g.
1.4547	254 SMO	20	18	6.1	0.7	+	38.3	590
1.4529	Cronifer® 1925 h Mo	20	25	6.25	1.15	+	38.75	400
1.4563	Nicrofer® 3127 and Sanicro® 28	27	31	3.5	1.15	–	37.5	430
2.4614	Nicrofer® 4221 h Mo	21	42.5	6.25	2.25	–	39.7	420

1) U_{1p} = pitting passivation potential in 0.1 mol/l NaCl at 353 K (80 °C), pH 7.5
n.g. not given

Table 44: Material parameters of selected chemical apparatus materials [100]

– Areas of application –

Low-coercivity materials with a high initial and maximum permeability and good magnetizability are used for the construction of magnetic systems in the electrical industry. The ferritic steel X1CrNiMoNb28-4-2 (Remanit® 4575) exhibits good corrosion resistance and good magnetic properties. Its magnetic properties are comparable to those of 17 % Cr-steels (X8Cr18; 1.4015 and X6CrMo17-1; 1.4113) and significantly better than those of the ferritic-austenitic steel X3CrNiMoN27-5-2 (1.4460). In 20 % H_2SO_4 at the boiling point, corrosion rates in excess of 10 mm/a (394 mpy) are given for the steels X8Cr18, X6CrMo17-1 and X3CrNiMoN27-5-2, and rates of less than 0.1 mm/a (3.94 mpy) for the steel X1CrNiMoNb28-4-2 (1.4575) [122].

Ferritic-austenitic steels with more than 12 % chromium

These materials, called duplex steels, have a relatively high resistance to stress corrosion cracking in addition to a good corrosion resistance. Because of their high yield strength of at least 450 N/mm^2, they are used for components in the chemical industry which, in addition to being exposed to corrosion by the medium, are also subjected to wear and vibration [100, 126, 127].

Whereas earlier steels of this group still have the disadvantages of lack of strength and the tendency towards intercrystalline corrosion, especially after welding after rapid cooling, these problems have been virtually eliminated in more recent grades.

During testing of the two-phase alloy CrNi 24 13 (used for weld-claddings) for resistance to intercrystalline corrosion in copper sulfate-sulfuric acid solution in accordance with DIN EN ISO 3651-2, corrosion at the austenite-ferrite grain boundaries is detected only after subsequent bending. This corrosion can be found on specimens which have not been bent after considerably longer testing (in some cases several hundred hours) [128].

Under the same test conditions (Strauss test), the steel AF 22 (1.4462) containing (%) Fe-≤ 0.03C-≤ 1Si≤ 2Mn-21–23Cr-4.5–6.5Ni-2.5–3.5Mo-0.08–0.20N exhibited no corrosion effect [129].

Table 45 contains a summary of the steels with a ferritic-austenitic structure used today [100, 130–137]. In accordance with their Cr-, Ni- and Mo-contents, these steels can be expected to have a better corrosion resistance in sulfuric acid (compare Figure 48) than is the case with the ferritic steels of high Cr-content (see Section Ferritic chromium steels with more than 12 % chromium).

Table 8 gives a comparison of the corrosion resistance of these steels with that of other high-alloy steels. The ferritic-austenitic steels in column 10 have a wider area of use under exposure to sulfuric acid in comparison with lower-alloy chromium steels, and under certain conditions are equivalent in their resistance to high-alloy chromium-nickel steels. Table 34 shows the corrosion resistance of the steel X3CrNiMoN27-5-2 (1.4460) in sulfuric acid solutions.

These steels can be used chiefly in sulfuric acid at low or high concentrations and predominantly at low temperatures. Because of the relatively wide range of these materials, a noticeable difference in their corrosion resistance is to be expected also. Figure 53 shows the corrosion behavior of the steel 3RE60 (1.4424, X2CrNiMoSi18-5-3, Table 45) in sulfuric acid at various temperatures and concentrations. With a corrosion rate of only 0.1 mm/a (3.94 mpy) in the concentration range up to 25 % H_2SO_4 at 303 K (30 °C) and at a decreasing concentration up to almost 363 K (90 °C), this steel is quite resistant in sulfuric acid and is even superior to the chromium-nickel steel SAE 316 L (1.4404). On the other hand, it does not have the resistance of

Trade name/ DIN-Mat. No.	*Alloy type*	*Chemical composition, %*								*Literature*
		C	*Si*	*Mn*	*Cr*	*Mo*	*Ni*	*Cu*	*N*	
3RE60 (1.4424)	X2CrNiMoSi18-5-3	0.03	2.0	2.0	18.5	2.7	4.9	–	–	[130]
SAF® 2205 (1.4462)	X2CrNiMoN22-5-3	0.03	1.0	2.0	22.0	3.0	5.5	–	0.74	[130]
Ferralium® 255 (cf. 1.4507)	X2CrNiMoCuN25-6-3	0.02	1.4	1.4	26.0	3.3	5.5	3.0	0.20	[100, 137]
Ferralium® 255-3SC		0.05			25.0	3.0	6.0	2.5	0.18	[131]
Fermanel®		0.06			27.0	3.1	8.5	1.0	0.23	[132]
1.4530	X1CrNiMoAlTi12-9-2	0.03			25	3.5	5	1.5		[134]
A 905 (1.4467)	X2CrMnNiMoN26-5-4	0.04	0.35	5.8	25.5	2.3	3.7	–	0.37	[100, 135]
Noridur® 9.4460	G-X 3 CrNiMoCuN 24 6	0.03	1.5	1.5	24	2.5	6	3	0.15	[100, 133]
1.4515	GX3CrNiMoCuN26-6-3	0.04	0.7	1.3	27.5	3.6	7.5	1.25	0.2	[136, 138]

Table 45: Chemical composition of selected ferritic-austenitic CrNi-steels (see also Table 41)

the ferritic steel X 2 CrNiMoTi 25 4 4, which can be used up to a concentration of 40% H_2SO_4 (see Figure 45). The high resistance of the steel 3RE60 towards stress corrosion cracking is emphasized [118, 130].

The steel SAF® 2205 (1.4462, X2CrNiMoN22-5-3) is particularly suitable for use in solutions containing chlorides and hydrogen sulfide, and in sulfuric acid solutions of less than 25%. Figure 54 shows its corrosion behavior in pure sulfuric acid. It is more suitable for use in this medium at a concentration of 20% at 333 K (60 °C) than the steel 3RE60 [130].

An increase in the corrosion resistance can be achieved by increasing the alloying content or by improving the alloys, especially in the middle concentration range of sulfuric acid, as can be seen from Figure 55 for the two duplex steel grades of Ferralium® (see composition in Table 45) [131].

Of these two steels, the cast steel Ferralium® 255 should be used in preference in sulfuric acid. Whilst it exhibits corrosion rates of less than 0.003 mm/a (0.12 mpy) in 20% H_2SO_4 at 336 K (63 °C), these rise to 3.05 mm/a (120 mpy) for 3RE60 under the same conditions [139]; this can be concluded from Figures 53–55 and 56. The corrosion rate in boiling 0.05 mol/l H_2SO_4 and 5% NaCl solution is 0.025 mm/a (0.98 mpy) for Ferralium® 255, and 1.01 mm/a (39.8 mpy) for the steel 3RE60. Under these conditions, the duplex steels are superior to the austenites SAE 317 L (X2CrNiMo18-15-4) with a corrosion rate of 3.8 mm/a (150 mpy) and Carpenter® 20Cb-3 (NiCr20CuMo) with a rate of 3.3 mm/a (130 mpy). The pitting and crevice corrosion resistance of these materials in a solution of 0.01 mol/l H_2SO_4 + 0.1% $Fe_2(SO_4)_3$ + 1,000 ppm Cl^- was investigated in the same study. According to these studies, the pitting and crevice corrosion temperatures given are both 328 K (55 °C) for Ferralium®, 313 and 303 K (40 and 30 °C) for 3RE60, 308 and 303 K (35 and 30 °C) for SAE 317 L, and 303 and 293 K (30 and 20 °C) for Carpenter® 20Cb-3

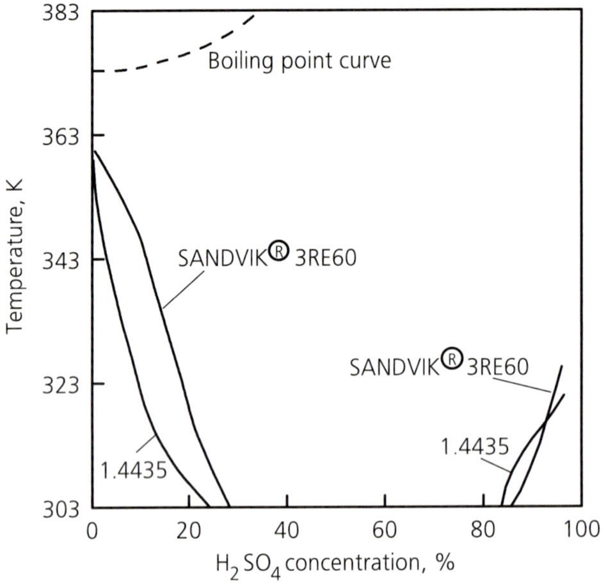

Figure 53: Isocorrosion curves (0.1 mm/a (3.94 mpy)) of the ferritic-austenitic steel 3RE60 (X2CrNiMoSi18-5-3) in comparison with the austenitic steel 1.4435 (X2CrNiMo18-14-3) [118, 130]

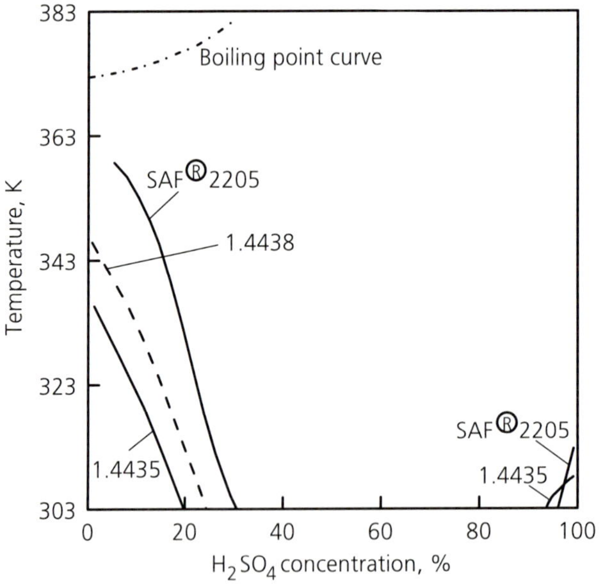

Figure 54: Isocorrosion curves (0.1 mm/a (3.94 mpy)) of the ferritic-austenitic steel SAF® 2205 (X2CrNiMoN22-5-3) in comparison with the austenitic steels 1.4435 (X2CrNiMo18-14-3) and 1.4438 (X2CrNiMo18-15-4) [130]

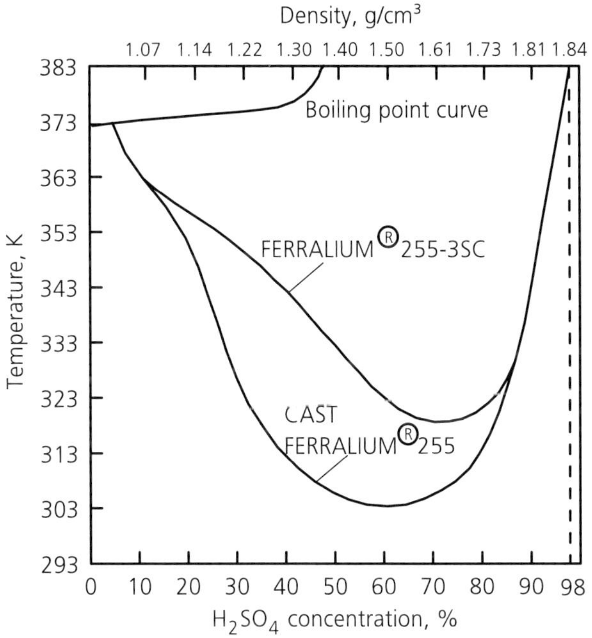

Figure 55: Isocorrosion curves (0.1 mm/a (3.94 mpy)) for the duplex steel Ferralium® (see Table 45) in sulfuric acid [131]

(Table 53). The steel Ferralium® Alloy 255-3SC has the best corrosion resistance in sulfuric acid of the steels with the same trade name, such as Ferralium® 255 and 288 ((%) Fe-28Cr-8Ni-2.5Mo-1.25Cu-max.0.08C) [131, 140, 141].

Its resistance passes through a maximum at 320 K (47 °C) and a sulfuric acid concentration of 70 % with a corrosion rate of 0.1 mm/a (3.94 mpy). The maximum corrosion rates for Ferralium® 255 of 0.5 mm/a (19.7 mpy) are given in the same study [131] and in [140, 141], as can be seen from Figure 56 [131], although the rates relate to the cast alloy in one case and the forged alloy in the other. Ferralium® 255-3SF shows the same curve shape as Ferralium® 255-3SC in Figure 56, but at 0.5 mm/a (19.7 mpy) (see also Figure 55 and Figures 58 and 59, regions "B").

The graph in Figure 56 also shows the superiority of duplex steels of this composition in comparison with the steel SAE 316 (cf. X5CrNiMo17-12-2), which cannot be used in sulfuric acid in the concentration range between 30 and 90 % (see Section Austenitic chromium-nickel-molybdenum steels).

According to [137], corrosion rates of not more than 0.15 mm/a (5.91 mpy) are given for Ferralium® in sulfuric acid in the concentration range between 0 and 20 % at 353 K (80 °C) and in 98 % H_2SO_4 up to 353 K (80 °C).

A further development of the steel Ferralium® 255 is the steel Fermanel®, the isocorrosion curve of which (0.1 mm/a (3.94 mpy)) is plotted in Figure 57 in comparison with other high-alloy steels. See Table 45 for the alloy composition. This steel is still resistant (for metallic materials) even in the critical concentration range

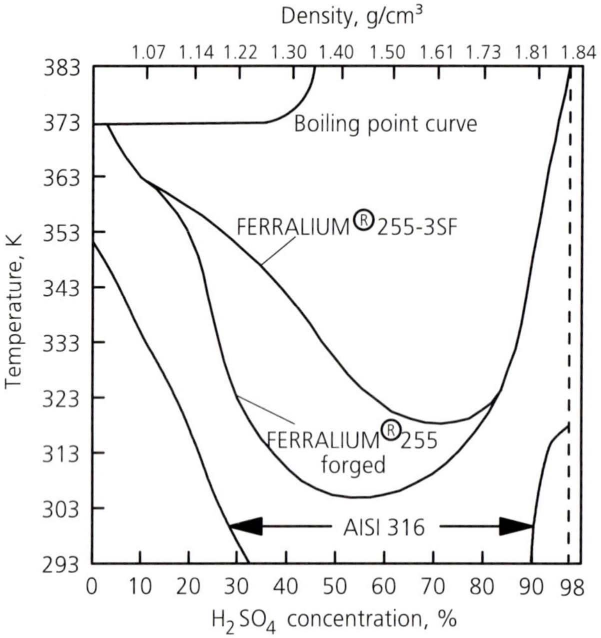

Figure 56: Isocorrosion curves (max. 0.5 mm/a (19.7 mpy) for the duplex steel Ferralium® (see Table 45) in comparison with the steel SAE 316 (cf. X5CrNiMo17-12-2) [131]

in sulfuric acid between 50 and 70 % at 333 K (60 °C). If the sulfuric acid contains traces of highly oxidizing substances, such as, for example, nitric acid, the application range of Fermanel® is extended [132].

The ferritic-austenitic cast steel G-X 3 CrNiMoCuN 24 6 (see Table 45) used as a pump material exhibits a relatively good corrosion resistance in sulfuric acid solutions [133, 142–144].

Specimens were subjected to a long-term immersion test in accordance with DIN 50905 to determine the corrosion resistance of this material in sulfuric acid. The specimens were first activated in the corrosion medium with a Zn-rod until clear evolution of gas occurred. To obtain results which can be used in practice, test durations of 21 days were chosen. Some of the corrosion rates showed very marked variations in the corrosion studies.

Corrosion studies in 10 % H_2SO_4 at 353 K (80 °C) have shown that steady-state corrosion is established only after a test duration of 24 h. After this time, the corrosion rate dropped from more than 1 mm/a (39.4 mpy) to less than 0.01 mm/a (0.39 mpy). Figure 58 shows the corrosion behavior of G-X 3 CrNiMoCuN 24 6 (Noridur® 9.4460) in aerated sulfuric acid with a stationary specimen. In this graph, corrosion rates of less than 0.1 mm/a (3.94 mpy) can be expected in the region "A". The region "B" is less certain and very high corrosion rates were determined occasionally on ferritic-austenitic steels. According to these studies, it is significant that this cast steel can be severely corroded in the concentration range between 50 and

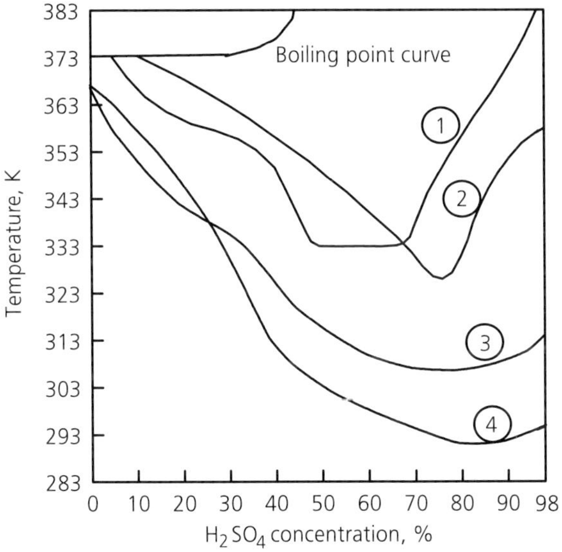

Figure 57: Comparison of the isocorrosion curves (0.1 mm/a (3.94 mpy)) of the duplex steel Fermanel® with austenites in various concentrations of sulfuric acid [132]
① Fermanel®
② CN-7M (Fe-20Cr-29Ni-3.5Mo-2.5Cu, cf. 1.4500, GX7NiCrMoCuNb25-20)
③ N08904 – 1.4539 (X1NiCrMoCu25-20-5)
④ S31254 (Fe-20Cr-18Ni-6Mo, cf. 1.4547, X1CrNiMoCuN20-18-7)

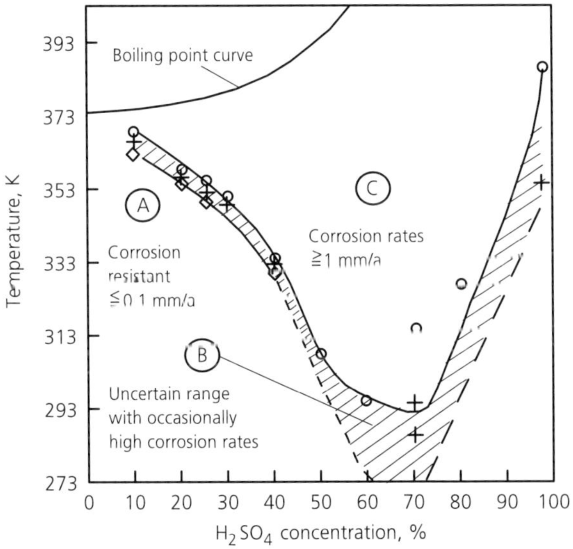

Figure 58: Isocorrosion curves of the duplex steel G-X 3 CrNiMoCuN 24 6 in static, aerated sulfuric acid; specimens activated [133, 143]
◇ 0.1 mm/a (3.94 mpy)
+ 0.4 mm/a (15.7 mpy)
○ < 1.0 mm/a (< 39.4 mpy)

80% H_2SO_4 even at room temperature [133, 143]. This steel shows a somewhat more favorable corrosion behavior in flowing sulfuric acid (10 m/s), as is shown in Figure 59. Under these conditions, the resistance region is extended in comparison with static sulfuric acid solutions. Under flow conditions, in contrast to static acid, this material can be used in the concentration range between 50 and 80% up to 313 K (40 °C). However, the problems of steady-state corrosion should be noted here [143, 144].

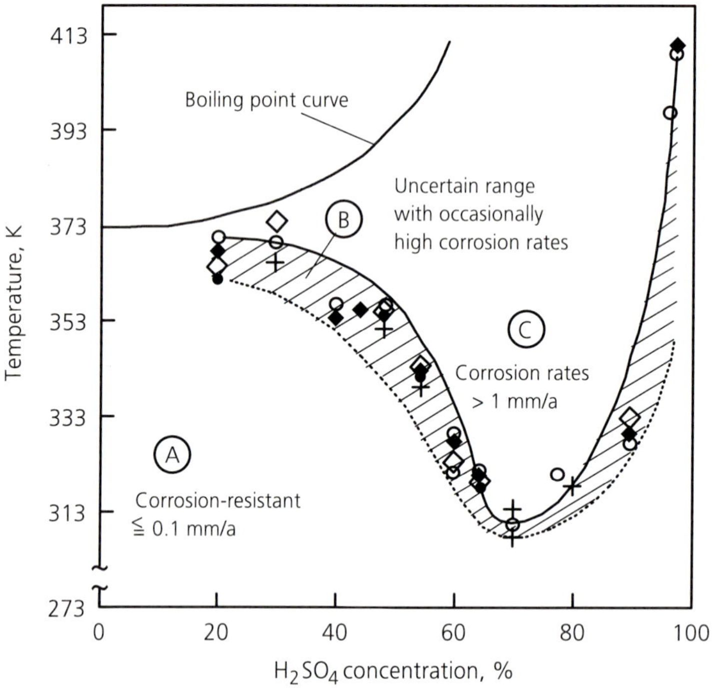

Figure 59: Isocorrosion curves of the duplex steel G-X 3 CrNiMoCuN 24 6 in flowing (10 m/s) sulfuric acid [143, 144]

- \+ 0.01 mm/a (0.39 mpy)
- ● 0.005 mm/a (0.20 mpy)
- ◇ 0.1 mm/a (3.94 mpy)
- ◆ 0.5 mm/a (19.7 mpy)
- ○ < 1.0 mm/a (< 39.4 mpy)

40 mm long specimens with an internal diameter of 8 mm and an effective area of 10 cm^2 were used for corrosion studies on pipes containing a flowing medium. Studies in 70% H_2SO_4 at various temperatures gave the results summarized in Figure 131 in comparison with an austenitic steel. The flow rate could be varied in the range up to 21 m/s and the test duration was 42 days.

In the groups of 5 specimens from the small melt, some were found to be passive whereas others were immediately active. All specimens were subjected to the same pretreatment, exposed to stress simultaneously and passivated in air before being introduced into the medium. For this reason, all the specimens were activated with a zinc strip. Nevertheless, the active-passive phenomenon just described reoccurred. The resulting corrosion rates can be explained with the aid of Figure 134. According to this graph, in contrast to the austenites, the ferritic-austenitic steel G-X 3 CrNiMo-CuNb 24 6 is no longer completely passive at 323 K (50 °C). As the temperature increases, attack by the acid increases as a consequence of decreasing passivity. In the case of active corrosion, selective corrosion is found almost exclusively on this steel. Only on exposure to concentrated sulfuric acid is there uniform attack on both constituents of the structure. The austenitic structural constituent is preferentially dissolved in about 10 % H_2SO_4, whereas the ferritic structural constituent is preferentially dissolved at higher concentrations. This also applies to flowing sulfuric acid of the same concentration [144].

The corrosion resistance of the stainless steels in 20 % sulfuric acid can be improved by stirring. The passive state is retained under these conditions. Another possibility is passivation prior to use, for example as pipes in heat exchangers. Treatment with a polarization voltage of 0.15 V at 353 K (80 °C) for 10 min in 20 % H_2SO_4 is sufficient here to retain the passivity even in static acid. The corrosion rates can reach 0.5 mm/a (19.7 mpy) [145].

A pump impeller of G-X 3 CrNiMoCuN 24 6 exhibited no corrosion after an operating time of 1,000 h in 25 % sulfuric acid at 318 K (45 °C). The speed of rotation was 1,450 rpm. An impeller of the austenitic material GX7NiCrMoCuNb25-20 (CN-7M) was completely destroyed under the same conditions.

An impeller made of the duplex steel showed an equally good corrosion resistance in 25 % sulfuric acid at 353 K (80 °C) and a rotation speed of 2,900 rpm (operating time 1,500 h). Whereas the steel G-X 3 CrNiMoCuN 24 6 was attacked hardly at all under these conditions, the ferritic-austenitic steel 1.4464 (GX40CrNiMo27-5) exhibited noticeable corrosion attack in the areas of high flow. The cobalt alloy G-X 200 CoCrNiMo 47 29 lay between the two duplex steels in corrosion resistance. The corrosion medium contained 200 g/l titanium dioxide [133].

The steel GX3CrNiMoCuN26-6-3 (1.4515) shows a different corrosion behavior in flowing sulfuric acid in comparison with the steel G-X 3 CrNiMoCuN 24 6 (9.4460), as can be seen from Figure 60. The corrosion maximum of this steel occurs in 40 % sulfuric acid. Under these conditions, the corrosion rate is less than 0.1 mm/a (3.94 mpy) only at room temperature. To determine the corrosion rates, rotating disk specimens 40 mm in diameter and 8 mm thick were exposed isothermally to various concentrations of sulfuric acid so that a flow rate of 0.6 m/s was established on their edge. The test duration was 48 h. This was increased to 5 days in the temperature or concentration ranges in which weight losses were below 0.5 mg/d. Under these conditions, the corrosion diagram in Figure 60 is more comparable with Figure 59.

The steel X 3 CrMnNiMoN 25 6 4 is used to advantage in the chemical industry as a weldable structural steel for plants in instances where ferritic 17 % chromium steels can no longer be used due to low corrosion resistance and CrNiMo-steels of the 18 10 type are unsuitable because of the possible risk of stress corrosion cracking. This applies mainly to heat exchangers in which service water containing

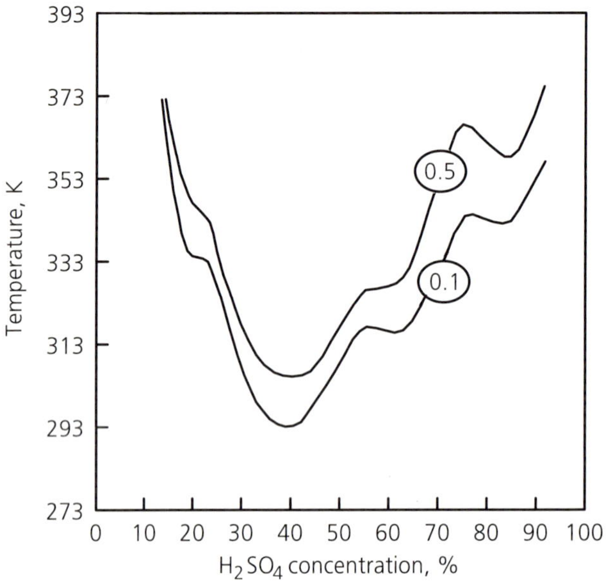

Figure 60: Isocorrosion curves (mm/a) of the steel GX3CrNiMoCuN26-6-3 (1.4515) in sulfuric acid [136]

chlorides is used for cooling. This steel has a better corrosion resistance in sulfuric acid than the steel X2CrNiMoN17-13-5 (Figure 61) [135].

Whereas there is no information on the type of corrosion test for Figure 61, Figure 62 shows isocorrosion curves for the steel X 3 CrMnNiMoN 25 6 4 in sulfuric acid flushed with air. This graph does not contain the pronounced corrosion maximum of Figure 61. The corrosion rates which lie very close together at between 30 and 70 % sulfuric acid are, without doubt, due to the uncertain resistance region (see region "B" in Figure 58) [133].

Figure 47 shows the corrosion behavior of Ferralium® 255 (see Table 45) in hot sulfuric acid. In comparison with other high-alloy materials, this steel exhibits a relatively low corrosion rate in 95 % H_2SO_4 – about 0.1 mm/a (3.94 mpy) at 373 K (100 °C) and 1.5 mm/a (59.1 mpy) at 398 K (125 °C). It is attacked the most severely of the steels listed at 423 K (150 °C) (1.62 mm/a (63.8 mpy)). Under these conditions, it is better to use the ferritic steel 1.4575 (X1CrNiMoNb28-4-2) (see Section Ferritic chromium steels with more than 12 % chromium) [102, 103]. In this context, see also "Hot concentrated sulfuric acid" in Section Austenitic chromium-nickel steels with special alloying additions and Figure 135 and Table 61.

Figure 63 shows the course of corrosion on the cast steel 9.4460 (Table 45) in 98.5 % aerated H_2SO_4 at 423 K (150 °C). As this graph shows, a steady corrosion rate of 1.7 mm/a (66.9 mpy) is established only after a test duration of 4 days [133].

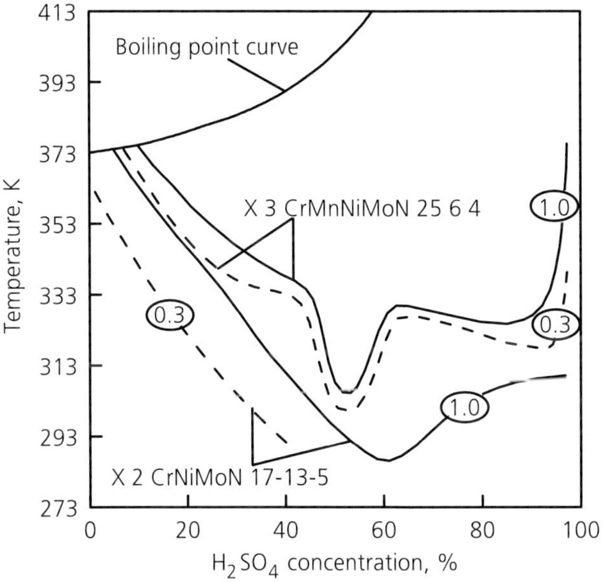

Figure 61: Isocorrosion curves (g/m^2 h) of the steel X 3 CrMnNiMoN 25 6 4 in comparison with the steel X2CrNiMoN17-13-5 in sulfuric acid [135]

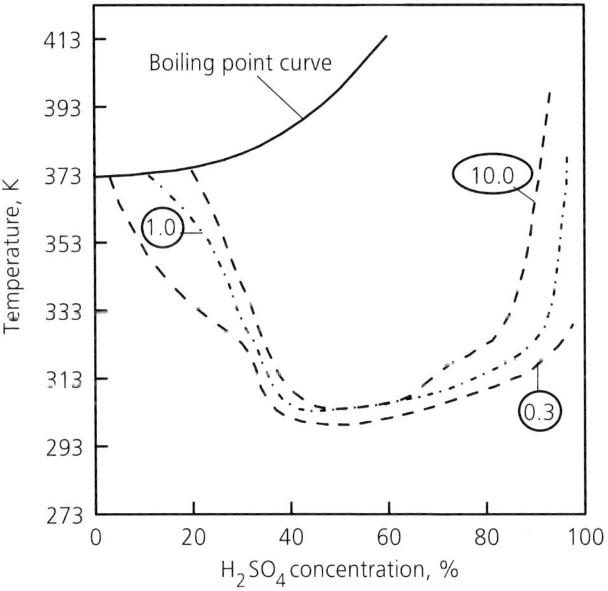

Figure 62: Isocorrosion curves (g/m^2 h) of the steel X 3 CrMnNiMoN 25 6 4 in sulfuric acid flushed with air (static specimens) [133]

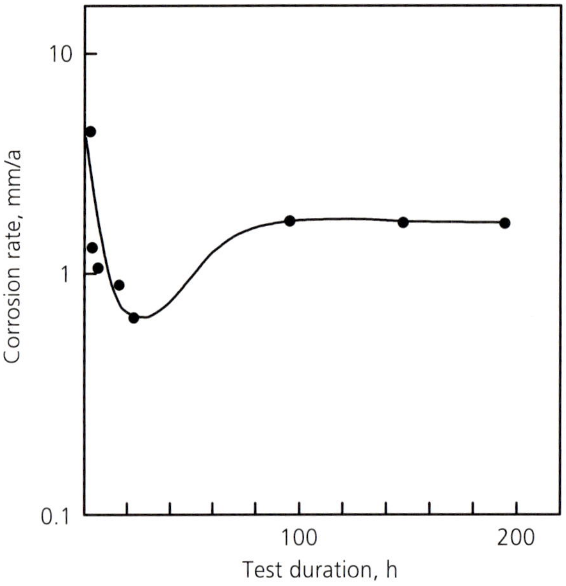

Figure 63: Influence of the test duration on the corrosion rate of the cast steel G-X 3 CrNiMoCuN 24 6 (9.4460) in aerated 98.5 % sulfuric acid at 423 K (150 °C) [133]

In corrosion studies in chamber acid (73 to 78 % sulfuric acid containing 0.02 to 0.05 % HNO_3), the steel 08Ch22N6T (Table 6, Figure 65) showed a different corrosion behavior. It can be used in corrosion solution B upstream of the cooler (344 to 387 K (71 to 114 °C)), but suffers very severe corrosion under the conditions A also upstream of the cooler [23].

For self-passivation by cathodically active alloying elements, see Section Ferritic chromium steels with more than 12 % chromium. A number of 25 % Cr steels of different alloying contents and the resulting different structural compositions are summarized in Table 41. As Figure 49 shows, an improvement in the corrosion behavior of CrMn-steels of mixed structure is found in 20 % sulfuric acid at boiling point, as a result of alloying with small amounts of palladium. This addition has a greater effect in the higher-alloy steels 25Cr8MnN, and even more so in the steel 25Cr8Mn6NiN containing 9 % ferrite. Figure 64 summarizes the material consumption rates of these steels in various concentrations of sulfuric acid at 293 K (20 °C) [110, 114]. From this graph, corrosion rates acceptable in industry are achieved with a Pd-content of 0.3 %. The corrosion resistance of CrMn-steels in sulfuric acid increases considerably on alloying with Ni or Mo. Steels which contain Pd as well as Mn and Ni exhibit the best corrosion resistance [110].

Instead of the ferritic steel X6CrMo17-1 (1.4113), the steel X3CrNiMoN27-5-2 (1.4460) with a ferritic-austenitic structure, for example, is used for cores in solenoid valves, especially if exposure to corrosion is relatively high. Because of its mixed structure, its magnetic properties are often inadequate, since they vary according to the processing and composition of the steel. The two steels mentioned and X8Cr18

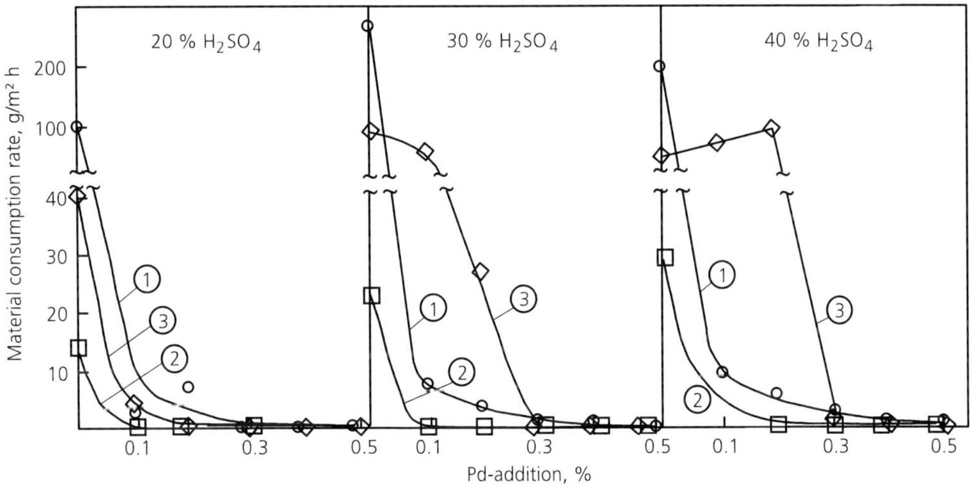

Figure 64: Influence of Pd in CrMn-steels on the corrosion behavior in sulfuric acid solutions (293 K (20 °C)), test duration 24 h [110, 114]
① 25Cr8MnN
② 25Cr8Mn6NiN (Table 41)
③ 25Cr8Mn2MoN

(1.4015) are corroded to an extent of more than 10 mm/a (394 mpy) in 20 % boiling sulfuric acid. The steel X1CrNiMoNb28-4-2 (1.4575) is technically resistant under these conditions (see Section Ferritic chromium steels with more than 12 % chromium) [122].

– Corrosion behavior in sulfuric acid-acid mixtures –

On evaporation of phosphoric acid, the steel 08Ch22N6T (Figure 65) showed the following corrosion rates in a mixture of 52.35 % P_2O_5, 0.45 % F and H_2SO_4 between 353 and 363 K (80 and 90 °C), 0.098 mm/a (3.86 mpy) at 3.1 % SO_4^{2-} and 0.085 mm/a (3.35 mpy) at 8.2 % SO_4^{2-}. It is superior to the austenitic steels, such as 03Ch21N21M4GB under these conditions (see Section Austenitic chromium-nickel-molybdenum steels). The steel 08Ch21N6M2T is even more resistant under these conditions, with corrosion rates in the sequence mentioned of 0.06 and 0.07 mm/a (2.36 and 2.76 mpy) [146]. In the production of sodium hexafluorosilicate, the steel 08Ch22N6T exhibits a corrosion rate of 0.73 mm/a (28.7 mpy) in the sulfate process in a filtrate of 1.5 % Na_2SiF_6 and 6–8 % H_2SO_4 at 303 to 313 K (30 to 40 °C), and a corrosion rate of 6.5 mm/a (256 mpy) in the collecting tank for an industrial effluent consisting of 1 % Na_2SiF_6 and 1 to 2 % H_2SO_4 (Table 7) [24]. Under these conditions, it can be used only with reservations, or not at all.

The corrosion of CrNi-steels in sulfuric acid is reduced by additions of nitric acid [147, 148]. The acid acts as an oxidizing agent and promotes the passivatability of these materials. Figure 65 shows the influence of HNO_3 additions to 10 % sulfuric acid on these steels. According to this graph, no corrosive attack is found on these

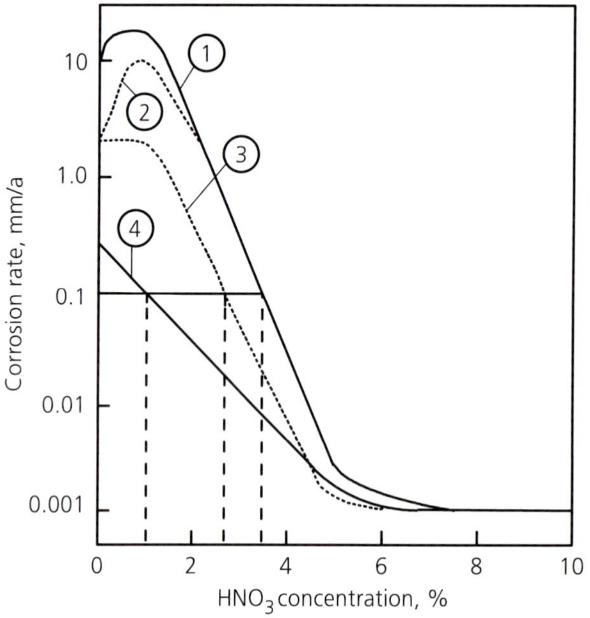

Figure 65: Influence of nitric acid in 10 % sulfuric acid at room temperature on the corrosion behavior of various CrNi-steel [147]
① 08Ch22N6T (steel containing 21 – 23 % Cr, 5.3–6.3 % Ni)
② 12Ch18N10T (cf. 1.4541, X6CrNiTi18-10)
③ 08Ch21N6M2T (containing 20–22 % Cr, 5.5–6.5 % Ni, 1.8–2.5 % Mo)
④ 06ChN28MDT (containing 22–25 % Cr, 26–29 % Ni, 2.5–3.5 % Mo, 2.5–3.5 % Cu)

steels with an 8 % HNO_3 addition to 10 % sulfuric acid. A sharp drop in the corrosion rate is found in particular with the Mo-free duplex steel 08Ch22N6T, the rate falling from more than 10 mm/a (394 mpy) in pure 10 % H_2SO_4 to 0.1 mm/a (3.94 mpy) if about 3.5 % HNO_3 is added. It is more advantageous for the materials used in the production of sulfuric acid if nitrous gases are washed out with an acid mixture containing up to 5 % HNO_3 instead of with 10 to 45 % H_2SO_4 [147]. Table 46 contains a summary of corrosion rates in such an acid mixture.

The effect of nitric acid is particularly striking at medium H_2SO_4 concentrations, in which these steels cannot be used without the addition of HNO_3. A mere 0.1 % HNO_3 is sufficient in 50 % H_2SO_4 at 333 K (60 °C) for these steels (including the ferritic-austenitic steels) to become practically resistant, as Table 47 shows [148].

CrNi-steels are used for the extraction of phosphoric acid. As well as depending on the corrosion rate in phosphoric acid, their corrosion rate depends chiefly on the concomitant substances present in the acid. Table 48 contains a summary of corrosion rates in phosphoric acid with various additions. As this summary shows, the corrosion resistance of the ferritic-austenitic steels 08Ch22N6T and 08Ch21N6MT is impaired only insignificantly or not at all by sulfuric acid additions. Additional amounts of hydrofluoric acid greatly impair the corrosion resistance, more so than in a mixture of only phosphoric acid and hydrofluoric acid [149]. The ferritic-austenitic steels can no longer be used if sulfuric and hydrofluoric acid are present.

Material	Temperature K (°C)	10 % H_2SO_4 + 1 % HNO_3	10 % H_2SO_4 + 5 % HNO_3	30 % H_2SO_4 + 1 % HNO_3	30 % H_2SO_4 + 5 % HNO_3	50 % H_2SO_4 + 1 % HNO_3	50 % H_2SO_4 + 5 % HNO_3
		Corrosion rate, mm/a (mpy)					
08Ch22N6T	295 (22)	32 (1,260)	0.003 (0.12)	0.004 (0.16)	0.003 (0.12)	0.004 (0.16)	0.001 (0.04)
	328 (55)	0.3 (11.8)	0.009 (0.35)	0.01 (0.39)	0.02 (0.79)	0.009 (0.35)	0.02 (0.79)
08Ch21N6M2T	295 (22)	3.0 (118)	0.002 (0.08)	0.002 (0.08)	0.001 (0.04)	0.004 (0.16)	0.001 (0.04)
	328 (55)	0.03 (1.18)	0.009 (0.35)	0.01 (0.39)	0.008 (0.31)	0.002 (0.08)	0.008 (0.31)
12Ch18N10T (cf. 1.4541)	295 (22)	2.1 (82.7)	0.004 (0.16)	0.002 (0.08)	0.001 (0.04)	0.002 (0.08)	0.001 (0.04)
	328 (55)	0.3 (11.8)	0.009 (0.35)	0.01 (0.39)	0.01 (0.39)	0.008 (0.31)	0.02 (0.79)
06ChN28MDT	295 (22)	0.2 (7.87)	0.003 (0.12)	0.001 (0.04)	0.001 (0.04)	0.004 (0.16)	0.001 (0.04)
	328 (55)	0.01 (0.39)	0.005 (0.20)	0.006 (0.24)	0.006 (0.24)	0.003 (0.12)	0.002 (0.08)

Table 46: Influence of nitric acid in sulfuric acid on the corrosion behavior of CrNi-steels (see Figure 65) [147]

Table 49 shows a possible use of ferritic-austenitic steels under the conditions of caprolactam production [150].

Instead of the austenite 1Ch18N10T, the ferritic-austenitic steel 0Ch22N5T (%, 0.05C, 0.59Mn, 0.78Si, 0.02P, 20.96Cr, 5.70Ni, 0.019S, 0.48Ti analytical values) can be used in oleum of a concentration of about 20 %. Even at 368 K (95 °C) the corrosion rate does not exceed 0.09 mm/a (3.54 mpy). It is considerably below this value at lower temperatures. The same also applies to the phase boundary [151].

– Intercrystalline corrosion and corrosion fatigue in sulfuric acid solutions –

Corrosion studies on susceptibility towards intercrystalline corrosion have shown that no intercrystalline corrosion occurs in the steel 1.4527 (GX4NiCrCuMo30-20-4) below 0.04 % C in the more stringent Strauss test (30 % H_2SO_4). At higher C-contents (between 0.04 and 0.06 %), corrosion depths of more than 0.05 mm were found after a test duration of 48 h [134].

The corrosion resistance of the steels is impaired by corrosion fatigue in sulfuric acid [152–154].

HNO_3, %	08Ch22N6T[1)]	12Ch18N10T (cf. 1.4541)[1)]	08Ch21N6M2T[1)]	10Ch17N13M2T (cf. 1.4571)
	Corrosion rate, mm/a (mpy)			
–	284 (11,181)	195 (7,677)	361 (14,213)	171 (6,732)
0.025	320 (12,598)	219 (8,622)	0.01 (0.39)	83 (3,268)
0.050	0.02 (0.79)	0.01 (0.39)	0.02 (0.79)	0.02 (0.79)
0.075	323 (12,717)	0.01 (0.39)	0.02 (0.79)	0.01 (0.39)
0.100	0.03 (1.18)	0.01–317 (0.39 –12,480)	0.01 (0.39)	0.03 (1.18)
0.500	0.03 (1.18)	0.02 (0.79)	0.02 (0.79)	0.02 (0.79)
1.000	0.02 (0.79)	0.02 (0.79)	0.01 (0.39)	0.02 (0.79)

1) see Figure 65

Table 47: Influence of nitric acid in 50% sulfuric acid at 333 K (60 °C) on the corrosion behavior of CrNi-steels [148]

Addition	06ChN28MDT[2)]	03Ch21N21M4GB[2)]	08Ch21N6M2T[1)]	08Ch22N6T[1)]
	Corrosion rate, mm/a (mpy)			
None	0.004 (0.16)	0.006 (0.24)	0.004 (0.16)	0.004 (0.16)
1% H_2SO_4	0.003 (0.12)	0.005 (0.20)	0.004 (0.16)	0.008 (0.31)
5% H_2SO_4	0.003 (0.12)	0.001 (0.04)	0.003 (0.12)	0.003 (0.12)
0.5% HF	0.077 (3.03)	7.5–11.4 (295–449)	0.96 (37.8)	3.34 (131)
1.5% HF	0.89 (35.0)	0.19–0.75 (7.48–29.5)	3.05 (120)	7.41 (292)
1% H_2SO_4 + 0.5% HF	0.14 (5.51)	1.61 (63.4)	2.54 (100)	6.41 (252)
1% H_2SO_4 + 1.5% HF	0.87 (34.3)	0.90 (35.4)	5.07 (200)	12.6 (496)
5% H_2SO_4 + 0.5% HF	0.20 (7.87)	0.36 (14.2)	7.81 (307)	21.2 (835)
5% H_2SO_4 + 1.5% HF	1.57 (61.8)	1.77 (69.7)	21.5 (846)	48.6 (1,913)
Production acid (28.6% P_2O_5, 2.26% SO_3, 1.82% F)	0.04 (1.57)	0.02 (0.79)	0.05 (1.97)	0.01 (0.39)

1) see Figure 65
2) 20–22 Cr, 20–22 Ni, 3.4–7 Mo, 1.8–2.5 Mn

Table 48: Corrosion rates (converted from g/m^2 h) of CrNi-steels in a 41.4% H_3PO_4 solution with various additions and in phosphoric acid (production acid) (test duration 100 h, 343 K (70 °C), specimen size 40 × 20 × 2 mm) [149]

Material	See section	Collector, 63 % H_2SO_4 318–323 K (45–50 °C)	Collector, 820 g/l NH_4HSO_3, pH 4.0–4.5 up to 333 K (60 °C)	Gas flow, 12 % SO_2, 0.1 g/m^3 H_2SO_4, 3 g/l H_2O, up to 323 K (50 °C)	Column, 120–150 g/l $HON(SO_3NH_4)_2$ up to 2 g/l H_2SO_4, 268–273 K (–5–0 °C)	Hydrolysis tower, 120–150 g/l $(NH_2OH)_2$ x H_2SO_4, 120 g/l H_2SO_4, up to 383 K (110 °C)	Collector, $(NH_2OH)_2$ x H_2SO_4, up to 353 K (80 °C)	Preoxidation, 8 % $(NH_2OH)_2$ x H_2SO_4, 13 % $C_5H_{10}CO$, 5 % $C_5H_{10}CNOH$, 15 % H_2SO_4, up to 318 K (45 °C)	Oxidation, 8 g/l $(NH_2OH)_2$ x H_2SO_4, 5 % $C_5H_{10}CO$, 12 % $C_5H_{10}CNOH$, 2 g/l H_2SO_4, up to 353 K (80 °C)	Rearrangement reactor, Lactam up to 40 %, H_2SO_4 up to 58 %, 368–378 K (95–105 °C)	Collector, 45 % $(NH_4)_2SO_4$, 15 g/l H_2SO_4, up to 343 K (70 °C)
		Corrosion rate, mm/a (mpy)									
Ch18N10T	G	1.150 (45.3)	0.010 (0.39)	5.700 (224)	1.100 (43.3)	3.500 (138)	0.570 (22.4)	0.678 (26.7)	0.050 (1.97)	–	0.005 (0.20)
Ch17N13M2T	H	–	0.008 (0.31)	1.080 (42.5)	0.022 (0.87)	1.800 (70.9)	0.050 (1.97)	0.507 (20.0)	0.010 (0.39)	0.023 (0.91)	0.002 (0.08)
08Ch22N6T	F	1.090 (42.9)	0.015 (0.59)	–	1.310 (51.6)	4.310 (170)	0.420 (16.5)	0.725 (28.5)	0.086 (3.39)	–	0.003 (0.12)
08Ch21N6M2T	F	0.019 (0.75)	0.002 (0.08)	0.890 (35.0)	0.009 (0.35)	1.190 (46.9)	0.030 (1.18)	0.482 (19.0)	0.015 (0.59)	0.018 (0.71)	0.001 (0.04)
X8Cr18	E	–	0.005 (0.20)	–	–	–	–	–	–	–	0.003 (0.12)
10Ch14AG15	K	1.225 (48.2)	0.015 (0.59)	12.800 (504)	–	–	–	–	–	5.863 (231)	0.003 (0.12)
00Ch23N28M3D3T	I	0.005 (0.20)	0.001 (0.04)	0.004 (0.16)	0.001 (0.04)	0.004 (0.16)	0.001 (0.04)	0.013 (0.51)	0.006 (0.24)	0.009 (0.35)	–
08Ch17N16M3T	H	–	0.005 (0.20)	1.150 (45.3)	–	0.014 (0.55)	0.025 (0.98)	0.186 (7.32)	–	–	–
0Ch18N20S3M3D3B	K	0.007 (0.28)	0.001 (0.04)	0.008 (0.31)	0.001 (0.04)	0.007 (0.28)	0.001 (0.04)	0.062 (2.44)	0.007 (0.28)	0.012 (0.47)	–
VT1-1		0.530 (20.9)	0.002 (0.08)	0.800 (31.5)	0.002 (0.08)	0.019 (0.75)	0.007 (0.28)	0.013 (0.51)	0.002 (0.08)	–	0.008 (0.31)
VT5-1		0.720 (28.3)	0.002 (0.08)	0.980 (38.6)	0.002 (0.08)	0.011 (0.43)	0.012 (0.47)	0.017 (0.67)	0.002 (0.08)	0.820 (32.3)	0.010 (0.39)
AT 3		0.250 (9.84)	0.005 (0.20)	0.500 (19.7)	0.005 (0.20)	0.10 (3.94)	0.017 (0.67)	0.010 (0.39)	0.008 (0.31)	0.120 (4.72)	0.001 (0.04)
AT 4		0.890 (35.0)	0.002 (0.08)	0.654 (25.7)	0.008 (0.31)	0.020 (0.79)	0.020 (0.79)	0.015 (0.59)	0.010 (0.39)	0.123 (4.84)	0.002 (0.08)
Aluminium		–	0.350 (13.8)	0.856 (33.7)	–	2.360 (92.9)	1.060 (41.7)	–	0.070 (2.76)	–	0.016 (0.63)
Lead		0.182 (7.17)	–	0.045 (1.77)	–	1.210 (47.6)	0.518 (20.4)	0.76 (29.9)	0.734 (28.9)	–	0.518 (20.4)
Copper		–	–	–	–	1.365 (53.7)	0.286 (11.3)	–	–	–	0.156 (6.14)

E: see section Ferritic chromium steels with more than 12 % chromium
F: see section Ferritic-austenitic steels with more than 12 % chromium
G: see section Austenitic chromium-nickel steels
H: see section Austenitic chromium-nickel-molybdenum steels
I: see section Austenitic chromium-nickel steels with special alloying additions
K: see section Special iron-based alloys

Table 49: Corrosion rates of metallic materials in the operating solutions of some apparatuses in caprolactam production [150]

Corrosion studies on the ferritic-austenitic steel 1.4462 (X2CrNiMoN22-5-3, cf. 2205) in 90 and 96 % H_2SO_4 at 353 K (80 °C) have shown that under these conditions the steel has no endurance strength but only a corrosion fatigue strength (see Figures 117 and 118). In 96 % H_2SO_4 at 353 K (80 °C) this amounts to only 49 % of the fatigue strength in air under a stress cycle number of 10^6. For details in this context, see – Corrosion fatigue – in Section Austenitic chromium-nickel-molybdenum steels.

– Electrochemical protection –

As already described under corrosion fatigue, upon exposure to the latter in hot sulfuric acid, the steel 1.4462 (cf. 2205) has a corrosion resistance dependent on the stress cycle number. Anodic protection (see Figures 116–118) at a potential of +600 mV (U_H) improves the corrosion behavior, and indeed to a greater extent than in the case of austenitic steels. However, anodic protection is not sufficient for the fatigue strength of the material in air to be achieved; although it does enable the corrosion fatigue of the steel to be calculated. In this context, see also – Corrosion fatigue – in Section Austenitic chromium-nickel-molybdenum steels.

Austenitic chromium-nickel steels

These steels achieve their corrosion resistance by the formation of a passive layer on their surface. This layer can also develop under oxidizing conditions in sulfuric acid, and consists of iron oxide and chromium oxide with incorporated sulfates, which improve its stability. At higher acid flow rates and under reducing conditions, the protective layer is destroyed or its formation inhibited. A sometimes quite considerable increase in corrosion is associated with this. In addition to the largely reducing character of sulfuric acid, such conditions also arise, for example, in non-agitated and non-aerated sulfuric acid during relatively long downtimes without emptying (downtime corrosion). As well as these adverse conditions, these steels are also at risk of depassivation by chlorides.

An unsatisfactory corrosion resistance of the austenitic CrNi-steels can be concluded from these comments. However, it should be remembered that different corrosion behaviors are to be expected from the large number of steels available on the market. According to [60, 155], cast and wrought materials differ only insignificantly in their corrosion behavior. Figures 66, 67 and 68 [9, 63, 156] give an overview of the corrosion behavior of the austenitic steels as a function of their alloy composition. These graphs show that the CrNi-steel 18 8 has the lowest resistance in sulfuric acid, and only a limited increase in the corrosion resistance is possible by alloying. A noticeable improvement can be achieved by higher Ni-contents (Figure 66).

Figure 69 shows isocorrosion curves for a steel containing 0.1 % C, 18 % Cr, 8 % Ni [61]. The graph shows that a corrosion rate of more than 1 mm/a (39.4 mpy) is to be expected even at room temperature in the concentration range between 20 % and 70 % sulfuric acid, which excludes use of these materials. They can be used only at very low temperatures, and at high concentrations and low temperatures. In boiling

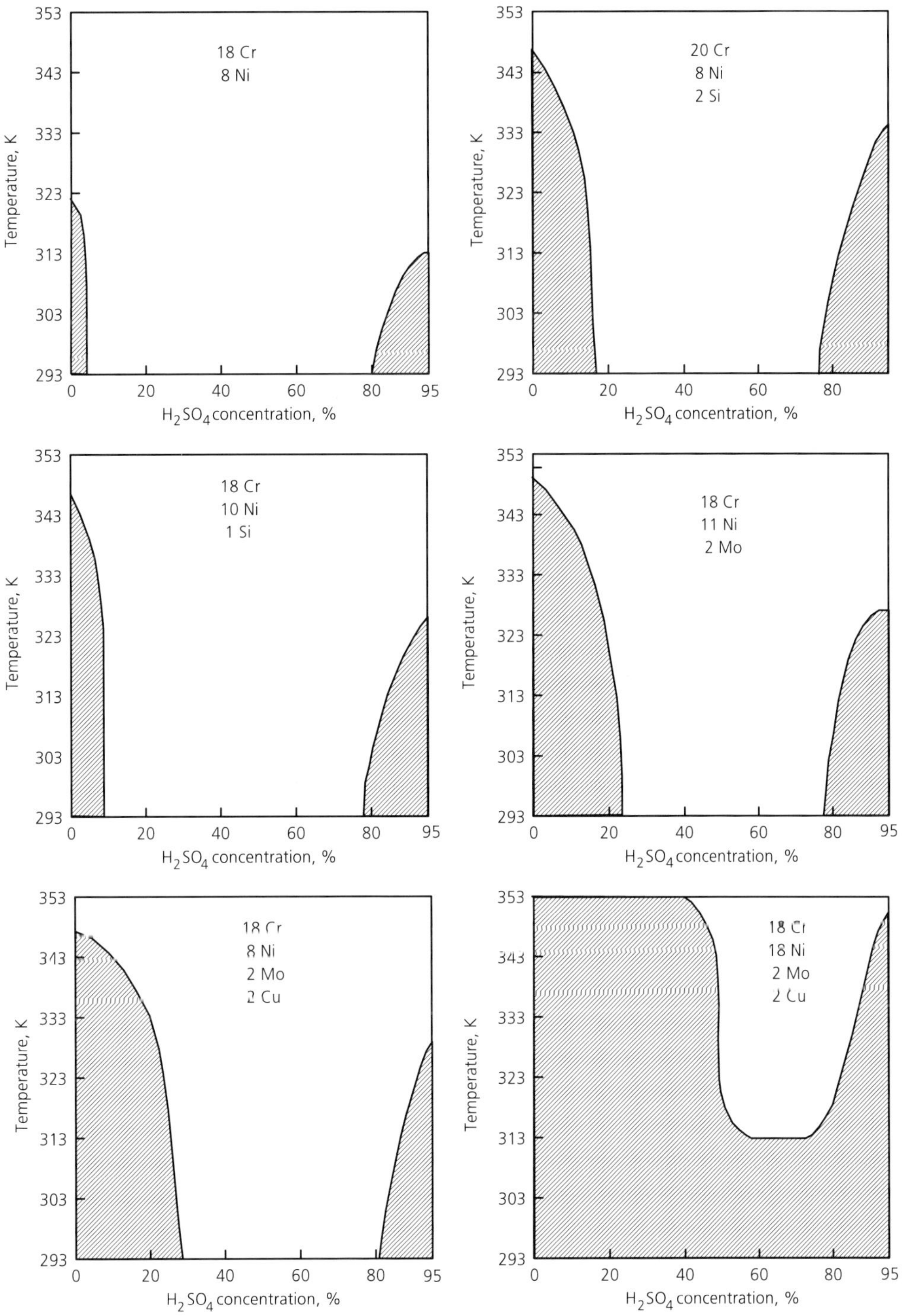

Figure 66: Influence of various alloying elements on the corrosion behavior of some CrNi-steels in sulfuric acid [9, 63, 156]
/// corrosion rate < 0.1 mm/a (< 3.94 mpy)

sulfuric acid the corrosion resistance can be improved by increased Ni-content and addition of Cu. Figure 70 shows this dependence [3, 60]. It can be seen from this figure that Mo-contents alone do not help to increase the corrosion resistance of these steels in sulfuric acid.

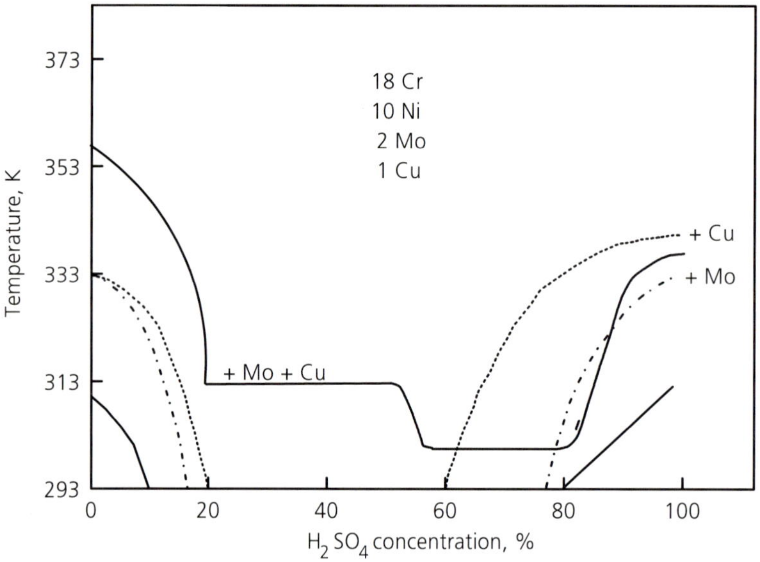

Figure 67: Influence of various alloying elements on the corrosion behavior of the CrNi-steel 18 10 in sulfuric acid [9, 63, 92, 156]

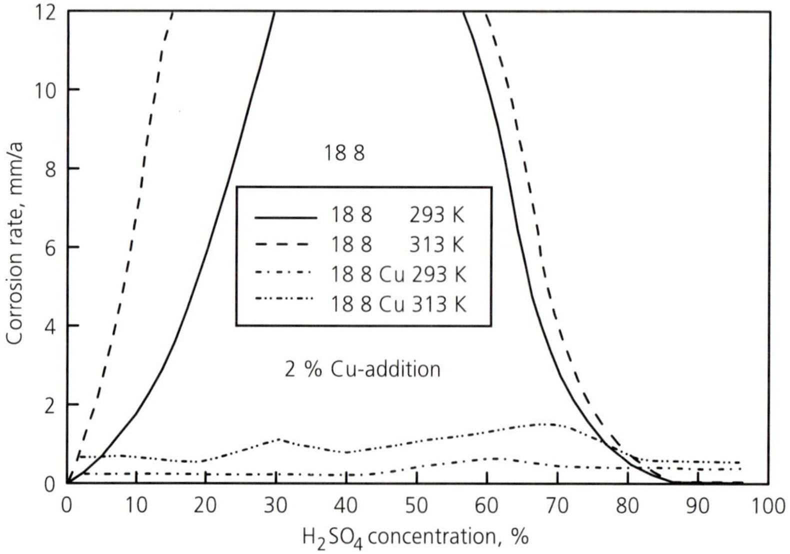

Figure 68: Corrosion rates of a CrNi-steel 18 8 with 2 % Cu as a function of the temperature and concentration of the sulfuric acid [63]

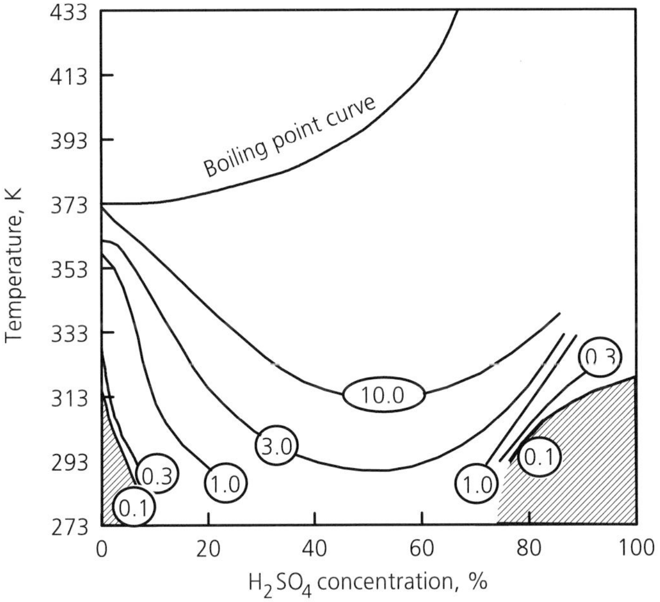

Figure 69: Isocorrosion curves of a CrNi-steel containing 0.1 % C, 18 % Cr, 8 % Ni in various concentrations of sulfuric acid [61]

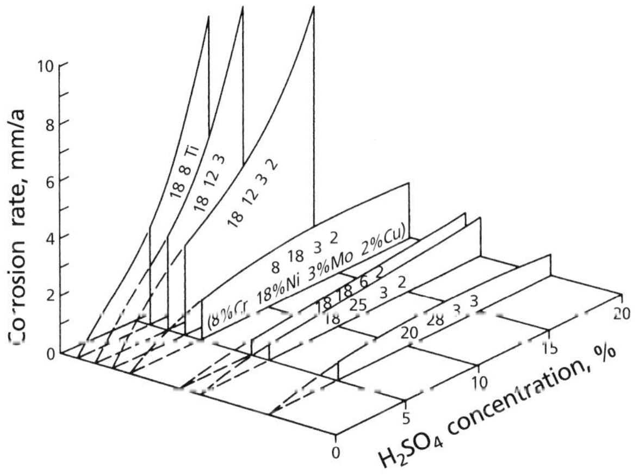

Figure 70: Influence of alloying elements in austenitic CrNi-steels on their corrosion resistance in up to 20 % boiling sulfuric acid [3, 60]

The chromium-nickel steels of the type 18 10 can be used with satisfactory corrosion rates only in cold sulfuric acids up to 5 % and above 80 %. As the temperatures rise, the corrosion resistance decreases rapidly. For more about the low resistance of these steels in sulfuric acid, see also [2, 4, 11, 22, 26, 59, 60, 87, 92, 93, 119, 137, 155, 157–164], Figures 15, 16, 47, 52, and 65 and Tables 6–8 (column 9), 34 (1.4301–1.4311, 1.4541, 1.4550), 37, 46, 47, 56 and 66.

In very dilute sulfuric acid in the concentration range from 0.005 to 0.4 mol/l there are differences in the corrosion behavior of the Mo-free CrNi-steels. The Ti-stabilized steel Ch18N10T ((%) Fe-0.11C-18.4Cr-10.5Ni-0.70Ti) is thus more resistant at room temperature in 0.4 mol/l H_2SO_4 with a corrosion rate of 0.01 mm/a (0.39 mpy) than the non-stabilized steel 000Ch18N11 ((%) Fe-0.03C-19.2Cr-10.9Ni; cf. 1.4306, X2CrNi19-11), which has a corrosion rate of 0.19 mm/a (7.48 mpy). Ch18N9 ((%) Fe-0.15C-17.65Cr-9.3Ni) corrodes at a rate of 0.82 mm/a (32.3 mpy) and is practically unusable. The test duration was 24 hours [165].

The influence of manganese on the corrosion behavior of CrNi-steels 18 10 was investigated in [166]. From this study, a tendency towards improved passivatability and increased resistance to pitting corrosion is observed for Ti-stabilized steels as the Mn-content increases (content of these steels: 0.2 % Mn). A significant deterioration in passivation and pitting corrosion behavior is found with non-stabilized steels, in particular when the Mn-content rises from less than 0.7 % to more than 1.0 %. 0.05 mol/l H_2SO_4 with additives was used as the test solution.

The steel X5CrNiN19-7 was developed to save nickel. Analysis of the test material showed the following composition: (%) Fe-0.03C-0.36Si-19.5Cr-6.72Ni-0.18N [94].

After test durations of 2,000 h in 5 % H_2SO_4 at 353 K (80 °C), the corrosion rates were somewhat more than 2 mm/a (78.7 mpy) and were therefore twice as high as those of the X8CrNiTi18-10 steels and still somewhat higher than in the case of X8CrNiTi18-11. In 10 % H_2SO_4, also at 353 K (80 °C), the corrosion rates were, in the above sequence, about 3.5, 2.5 and 1.5 mm/a (138, 98.4 and 59.1 mpy). Figure 71

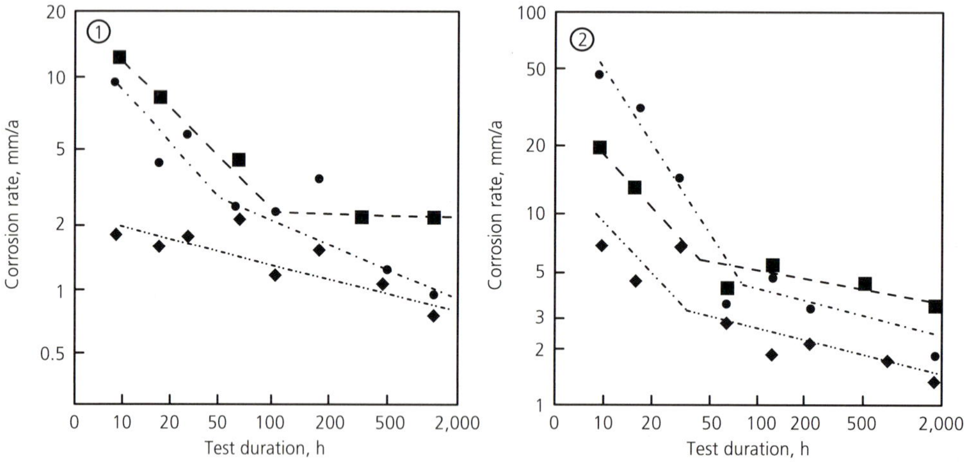

Figure 71: Influence of the test duration on the corrosion rates of CrNi-steels, especially on the Ni-saving steel X 5 CrNiN 19.7 [94]
① in 5 % H_2SO_4, 357 K (84 °C)
② in 10 % H_2SO_4, 357 K (84 °C)
■ X 5 CrNiN 19.7 (%, 0.03C, 0.36Si, 19.5Cr, 7.72Ni, 0.18N)
● X 8 CrNiTi 18.10 (%, 0.06C, 18.36Cr, 10.35Ni, 0.43Mo, 0.61Ti)
◆ X 8 CrNiMoTi 18.11 (%, 0.06C, 16.89Cr, 11.93Ni, 2.14Mo, 0.51Ti)

illustrates the corrosion rates as a function of the test duration [94]. This nickel-saving steel shows hardly any difference in corrosion behavior to the steel X8CrNiMoTi18-11 in the concentration range up to 5 % H_2SO_4 at room temperature. Under these conditions, the corrosion rates are less than 0.1 mm/a (3.94 mpy), but in 5 % acid at room temperature they reach 0.47 and 0.37 mm/a (18.5 and 14.6 mpy) respectively in the sequence X5CrNiN19-7, X8CrNiMoTi18-11. Table 50 shows a comparison of corrosion rates. X5CrNiN19-7 is said to be superior to the steel X8CrNiMoTi18-11 in its pitting corrosion resistance and only less resistant than the steel X 5 CrNiMo 17.13. It is not resistant to stress corrosion and the resistance to intercrystalline corrosion is described as adequate [94].

The chromium-nickel steels of the type 18 10 are no longer usable in 4 % sulfuric acid containing up to 6 % SO_2. At 308 K (35 °C), the corrosion rates are about 6.5 mm/a (256 mpy). Higher-alloy steels are more resistant, with corrosion rates of about 0.03 mm/a (1.18 mpy) (see Sections Austenitic chromium-nickel-molybdenum steels, Austenitic chromium-nickel steels with special alloying additions) [167]. The resistance of the CrNi-steels can be increased by agitating the corrosion solution, and the passive state, if present, can be stabilized or its range extended (see also – Electrochemical corrosion protection processes –) [145].

Figure 47 shows the corrosion rates of chromium-nickel steels (1.4301, 1.4541, 1.4335) in hot sulfuric acid at temperatures of 373, 398 and 423 K (100, 125 and 150 °C). Up to 398 K (125 °C) the corrosion rates lie within a range which is usable under certain circumstances. A ferritic steel is more suitable here (see Section Ferritic chromium steels with more than 12 % chromium and Section Austenitic chromium-nickel steels with special alloying additions and also Figure 135 and Table 61).

Figure 72 shows the corrosion rates of 1.4541 (X6CrNiTi18-10, cf. SAE 321) in sulfuric acid in the concentration range 67 to 98 % at temperatures above 353 K

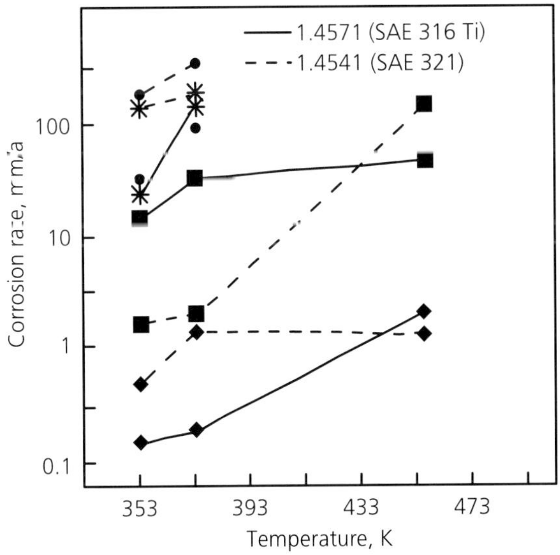

Figure 72: Influence of the temperature and concentration of sulfuric acid on the corrosion rate of the steels 1.4541 (X6CrNiTi18-10) and 1.4571 (X6CrNiMoTi17-12-2) [168]
Sulfuric acid concentration:
✳ 67 %; ■ 90 %; ● 78 %; ◆ 97–98 %

(80 °C) [168]. The corrosion rates decrease as the concentration increases and increase as the temperature rises. In practice, this material cannot be used in hot sulfuric acid under long-term exposure (see – Electrochemical corrosion protection processes –). Figure 73 shows the dependence of the corrosion rates on the acid concentration at 373 K (100 °C). At an acid concentration of 100 % and 373 K (100 °C), the material is practically resistant, with a corrosion rate of 0.03 mm/a (1.18 mpy) [102, 103, 168].

Temperature K (°C)	*Test duration d*	*H_2SO_4 concentration %*	*X8CrTi17*	*X5CrNiN19-7*	*X8CrNiTi18-10*	*X8CrNiMoTi18-11*
			Corrosion rate, mm/a (mpy)			
RT	n.d.[1)]	0.2	–	0.016 (0.63)	–	0.013 (0.51)
RT	n.d.	0.5	–	0.052 (2.05)	–	0.019 (0.75)
RT	n.d.	1.0	–	0.080 (3.15)	–	0.060 (2.36)
RT	n.d.	5.0	–	0.47 (18.5)	–	0.37 (14.6)
RT	n.d.	10.0	–	1.1 (43.3)	–	1.0 (39.4)
RT	n.d.	30	–	3.5 (138)	–	2.4 (94.5)
RT	n.d.	90	–	7.1 (280)	–	4.5 (177)
RT	n.d.	98	–	0.1 (3.94)	–	0.12 (4.72)
313 (40)	64	2	0.0003 (0.01)	0.0002 (0.008)	0.0002 (0.008)	0.0001 (0.004)
313 (40)	64	5	0.76 (29.9)	0.0008 (0.03)	0.0022 (0.09)	0.0015 (0.06)
313 (40)	64	10	2.85 (112)	0.41 (16.1)	0.022 (0.87)	0.0003 (0.01)
313 (40)	64	20	5.17 (204)	3.19 (126)	0.49 (19.3)	0.19 (7.48)
353 (80)	64	2	0.41 (16.1)	0.56 (22.0)	0.17 (6.69)	0.31 (12.2)
353 (80)	64	5	1.08 (42.5)	2.22 (87.4)	0.93 (36.6)	0.77 (30.3)
353 (80)	64	10	3.34 (131)	3.22 (127)	1.74 (68.5)	1.86 (73.2)
353 (80)	64	20	> 50 (> 1,969)	5.6 (220)	8.4 (331)	3.8 (150)

1) no data available

Table 50: Corrosion rates of Cr-steels and CrNi-steels in sulfuric acid, in particular in relation to the corrosion behavior of the Ni-saving steel X5CrNiN19-7 [94]

In highly concentrated hot sulfuric acid, the corrosion rates fall as the metal sulfate content of the acid increases. Figure 74 shows the influence of the metal sulfate content in 90 % H_2SO_4 on the corrosion rate of 1.4541 at 373 K (100 °C) [102, 103].

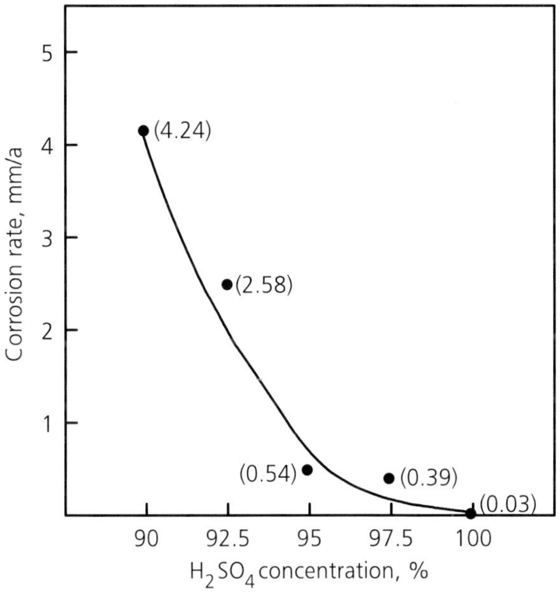

Figure 73: Influence of the H_2SO_4 concentration on the corrosion rate of the steel 1.4541 (X6CrNiTi18-10, cf. SAE 321) at 373 K (100 °C) over a test duration of 21 d [102, 103]

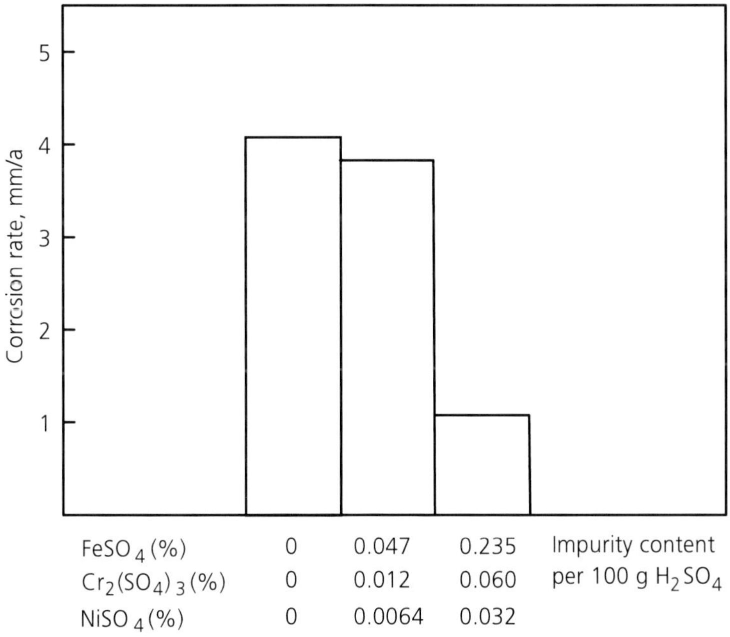

Figure 74: Influence of the metal sulfate content in 90% sulfuric acid at 373 K (100 °C) on the corrosion rate of the steel 1.4541 (X6CrNiTi18-10, cf. SAE 321) [102, 103]

This fact is of interest only in a static medium, because in agitated or flowing acid, fresh acid is constantly supplied.

The decreasing corrosion rate of chromium-nickel steels with increasing concentration of metal sulfates and impurities resultant thereof, is due to the diffusion polarization of both the cathodic and anodic reactions [102, 103, 168].

For the influence of nitric acid in concentrated sulfuric acid on CrNi-steels at 393 K (120 °C), see Figure 102 and Section Austenitic chromium-nickel-molybdenum steels.

In the atmosphere (Munich, Frankfurt, Düsseldorf), samples of the steel 1.4301 (X5CrNi18-10, SAE 304) showed no corrosion attack on their surfaces after a test duration of 4 years. This was due to the constant cleaning of the surface by rain. The underside of the samples lacked this cleaning and were attacked by corrosion. The corrosive attack is caused by the SO_2 content of the atmosphere and the associated formation of H_2SO_4, and manifests itself in the form of pitting corrosion [99].

The corrosion resistance of the steel 1.4301 in H_2SO_4 deteriorates on addition of sulfur. Additions of sulfur and Ti give the steel a good resistance in H_2SO_4 up to a concentration of 2.5 %, with corrosion rates which are substantially below 0.1 mm/a (3.94 mpy). The corrosion rates rise to 0.5 mm/a (19.7 mpy) in 5 % H_2SO_4 and to 0.9 mm/a (35.4 mpy) in 10 % acid at room temperature. At higher temperatures, the corrosion rates sometimes rise to considerably more than 1 mm/a (39.4 mpy) [117]. Corrosion tests on the steel X 2 CrNiTiS 18 9 from a 20 t melt gave similar results (see also Tables 56–59) [116, 117].

– Corrosion behavior in sulfuric acid mixtures –

During the production of sulfuric acid, corrosion studies were performed in chamber acid at a concentration of 75.3 to 75.6 % H_2SO_4 and a flow rate of 0.7 and 1.4 m/s at various temperatures. The chamber acid contained small amounts of HNO_3. Figure 14 shows the experimental results. The test durations were between 44 and 160 h at 383 K (110 °C) and 260 h in the temperature range between 393 and 403 K (120 and 130 °C). Under the experimental conditions described in Figure 15, the steel 12Ch18N10T ((%)Fe-≤ 0.12C-≤ 0.80Si-≤ 2.00Mn-≤ 0.035P-≤ 0.020S-17–19Cr-≤ 0.30Mo-9–11Ni-≤ 0.20W-≤ 0.30Cu-0.80Ti; cf. 1.4541) shows corrosion rates of 0.2 to 0.3 mm/a (7.87 to 11.8 mpy) in the temperature range between 365 and 373 K (92 and 100 °C), and under these conditions shows about the same corrosion resistance as the molybdenum-containing steel 10Ch17NMo2T (see Section Austenitic chromium-nickel-molybdenum steels). Plant studies (Table 6) show that 12Ch18N10T cannot be used under the conditions prevailing upstream of the cooler (as shown in Table 6). The samples were in flowing acid. After a test duration of 92 h the material samples of the material mentioned had dissolved [23].

See also Figure 102 in Section Austenitic chromium-nickel-molybdenum steels for the influence of small amounts of HNO_3 in 96 % sulfuric acid on CrNi-steels at 393 K (120 °C).

In the hydrolysis of sulfuric acid from titanium dioxide production (10 to 15 % H_2SO_4, 0.8 to 1.4 % $TiOSO_4$, 11 to 17 % $FeSO_4$) at 313 K (40 °C), specimens of

Ch18N10T showed material consumption rates of 0.05 $g/m^2 h$ (0.06 mm/a (2.36 mpy)) after test durations of 1,440 h in a laboratory experiment, and material consumption rates of 0.32 g/m^2 h (0.35 mm/a (13.8 mpy)) under operating conditions. Ch18N10T cannot be used under these conditions [169].

During production of tartaric acid, in the cracking apparatus at 338 K (65 °C) under a pressure of 0.1 MPa, the steel Ch18N10T corrodes in a solution of 250 g/l $H_2C_4H_4O_6$ + 300 g/l H_2SO_4 at a material consumption rate of 3.405 g/m^2 h (3.7 mm/a (146 mpy)). In the vacuum evaporator in a solution of 900 g/l $H_2C_4H_4O_6$ + 100 g/l H_2SO_4 at 353 K (80 °C), the material consumption rate under a pressure of 0.01 MPa is 2.18 $g/m^2 h$ (2.4 mm/a (94.5 mpy)). The Mo-free steel is also unusable under such conditions [170]. Titanium alloys are more suitable. The material consumptions originate from test durations of 1,160 h.

In the production of sodium hexafluorosilicate, the steel 12Ch18N10T exhibits a corrosion rate of 0.41 mm/a (16.1 mpy) in a filtrate comprising 1.5 % Na_2SiF_6 and 6 to 8 % H_2SO_4 at 303 to 313 K (30 to 40 °C) in the sulfate process, and a rate of 0.57 mm/a (22.4 mpy) in a collecting tank containing industrial effluent (consisting of 1 % Na_2SiF_6 and 1 to 2 % H_2SO_4) (see also Table 7 and Figure 65). Crevice and pitting corrosion have been found and hence this steel cannot be used [24].

In the production of caprolactam, the resistance of the steel Ch18N10T varies according to the composition of the media. Table 49 contains the corrosion rates determined after a test duration of 3,600 h under operating conditions. At the same time, it gives a comparison with other materials. This table shows that the simple chromium-nickel steel can be used only to a limited degree for caprolactam production [171].

Studies on the choice of materials for use in centrifuges in contact with thiourea solutions containing sulfuric acid, have shown that simple chromium-nickel steel Ch18N10T is not suitable. In production solutions, which apart from solids also contain 20 g/l H_2SO_4 and 100 g/l urea, material consumption rates of 4 g/m^2 h (4.4 mm/a (173 mpy)) were measured (test duration 100 h) [172]. The corrosion under these conditions was uniform, in contrast to higher-alloy steels, on which crevice corrosion was found (see also Sections Austenitic chromium-nickel-molybdenum steels and Austenitic chromium-nickel steels with special alloying additions).

For the production of iodine, the steel 12Ch18N9T can be used only under oxidizing conditions in the desorption stage. The corrosion rates for this exposure are quoted as 0.09 mm/a (3.54 mpy). Indeed only titanium is adequately resistant [173] under all operating conditions.

In the production of alkylsulfonates (in which solutions containing 13.3 % alkylsulfonic acid, 29.6 % hydrocarbons, 3.7 % sulfuric acid, 2.1 % sulfur trioxide and 51.3 % water are obtained), it is not possible to use CrNi-steels without further alloying additions. Temperatures can rise to 393 K (120 °C). In an argon atmosphere, a CrNi-steel 18 10 corrodes at a rate of 0.83 mm/a (32.7 mpy). It is almost destroyed, at 7.9 mm/a (311 mpy), if oxygen is present [174]. A high-alloy steel containing Mo and Cu is also suitable here (see Section Austenitic chromium-nickel steels with special alloying additions).

In the production of anhydrous hydrogen fluoride, corrosion rates of up to or more than 5 mm/a (197 mpy) occur on the steel Ch18N10T in the reaction apparatuses.

Gases containing up to 80 % HF, 5 to 10 % H_2SO_4 and 2 to 3 % H_2O are formed in the decomposition of fluorspar. The condensates contain up to 40 % water, 18 to 40 % hydrogen fluoride and 35 to 70 % sulfuric acid. Under the conditions of the gas intake at 393 to 423 K (120 to 150 °C), the steel Ch18N10T corrodes at a rate of between 3.7 and 5 mm/a (146 and 197 mpy), and in the condensates the rate is more than 5 mm/a (197 mpy) [25]. Corrosion rates which may still be acceptable are to be expected with steels having a high alloy content and Mo- and Cu-additions (see Section Austenitic chromium-nickel steels with special alloying additions), although the corrosion rates are also about 1 mm/a (39.4 mpy) or more.

As already mentioned, the corrosion behavior of chromium-nickel steels is determined by their passivity in the corrosion medium. This is of particular importance in acid mixtures of H_2SO_4, HNO_3 and HCl [147, 148, 175, 176]. As the HNO_3 concentration increases the resistance of the steel 12Ch18N10T (cf. 1.4541) in sulfuric acid increases, as can be seen from Figure 65. In 10 % H_2SO_4 an addition of 5 % HNO_3 is sufficient to almost prevent corrosion. The influence of nitric acid in various H_2SO_4-HNO_3 mixtures is shown in Table 46, and in Table 47 for 50 % sulfuric acid at 333 K (60 °C). These data show that nitric acid inhibits sulfuric acid corrosion, and the otherwise very high corrosion rates of more than 1 mm/a (39.4 mpy) are shifted to the region of 0.03 mm/a (1.18 mpy), making these steels still usable in practice. The higher corrosion rates in Table 46 at 295 K (22 °C) in an acid mixture of 10 % H_2SO_4 with 1 % HNO_3 are to be attributed to the less potent oxidizing action of nitric acid at lower temperatures. Additions of fluorides have a similar effect. Additions of 0.1 to 0.5 % to 85 to 93 % sulfuric acid at 343 to 393 K (70 to 120 °C) reduce corrosion considerably. The reason for this is the increasing self-passivation of the steel 08Ch18N10T (cf. 1.4541) [177].

While corrosion is reduced by nitric acid, the presence of HCl promotes corrosion in all cases, in particular by destroying any passive layer present, resulting in pitting corrosion under adverse conditions [175, 176]. Corrosion in sulfuric acid solutions containing hydrochloric acid is also reduced by additions of nitric acid. Figures 75 and 76 show the influence of the total concentration, the composition of the mixture and the temperature on the corrosion behavior [175]. A corrosion minimum is found at a ratio of the quantities of H_2SO_4:HNO_3:HCl of 90:6:4. If the HNO_3 concentration is increased further to an H_2SO_4:HNO_3:HCl ratio of 90:8:2, the corrosion is increased again, as it is if the total acid concentration and the temperature are increased. The experimental results are obtained from short-term experiments over a maximum of 24 h. The specimens were activated with zinc before the start of the experiment [175, 176].

The steel 1Ch18N10T ((%) Fe-0.06C-0.31Mn-0.61Si-0.027P-18.04Cr-10.22Ni-0.53Ti) has a very good resistance in oleum in the temperature range up to 368 K (95 °C). A summary of corrosion rates in 20 % oleum is shown in Table 51, and as a function of the flow rate in Table 52 [151, 178].

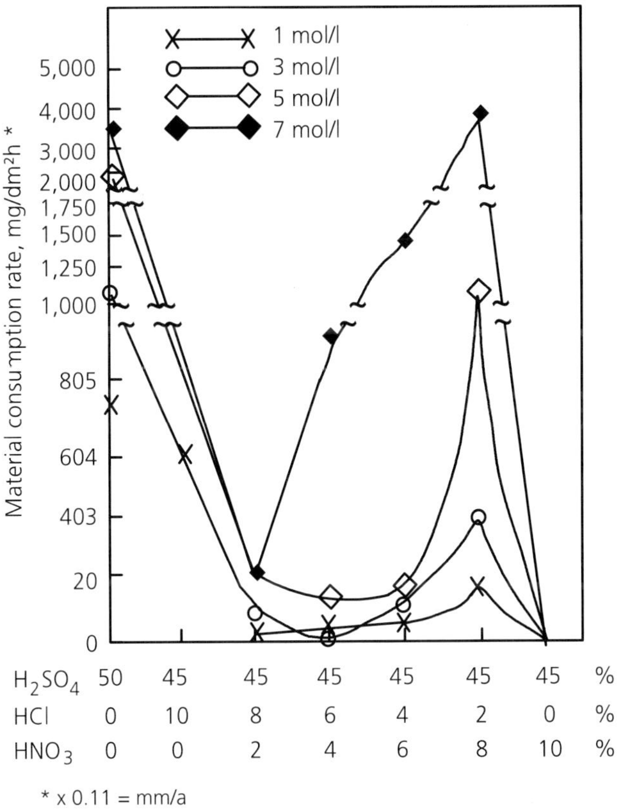

Figure 75: Influence of the concentration of H_2SO_4, HCl and HNO_3 acid mixtures on the material consumption rate of the steel SAE 304 (X5CrNi18-10) at room temperature [175]

Laboratory studies in a pickling solution for cast iron containing 5 % H_2SO_4 + 5 % HF + 10 % $FeSO_4$ at 322 K (49 °C) showed a corrosion rate of 10 mm/a (394 mpy) on the steel X5CrNi18-10 (Table 62). The total test duration of the alternate immersion tests was 40 h [16].

The corrosion resistance of CrNi-steel can be improved considerably by additions with a cathodic action. Large-scale industrial use is a current problem [110, 111, 179–181]. Corrosion studies in the laboratory have shown that the corrosion behavior of the austenitic steels varies. The steel Fe-18Cr8Mn2Ni is significantly inferior to the steel Fe-18Cr-10Ni in its corrosion behavior in sulfuric acid, as can be seen from Figure 77. An alloying addition of 0.2 % Pd in the material surface reduces the corrosion rates in 20 % H_2SO_4 at 293 K (20 °C) to values which render both materials usable in this medium. According to Figure 77, the corrosion rates for both steels are below 1 mm/a (39.4 mpy).

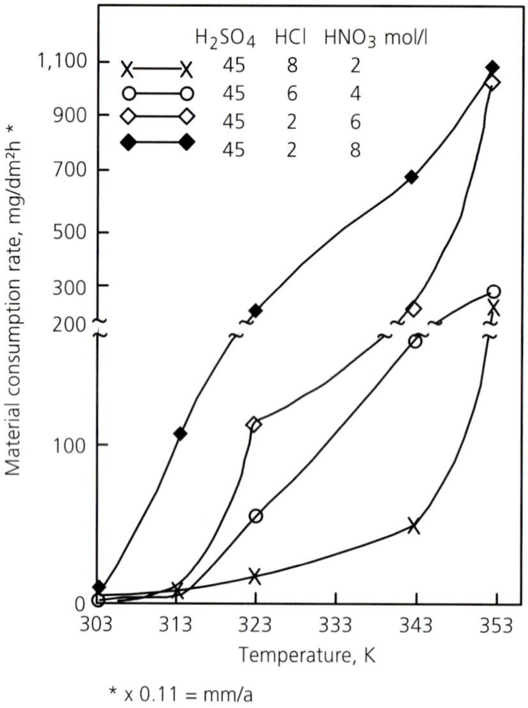

Figure 76: Influence of the temperature on the material consumption rate of the steel SAE 304 (X5CrNi18-10) in acid mixtures [175]

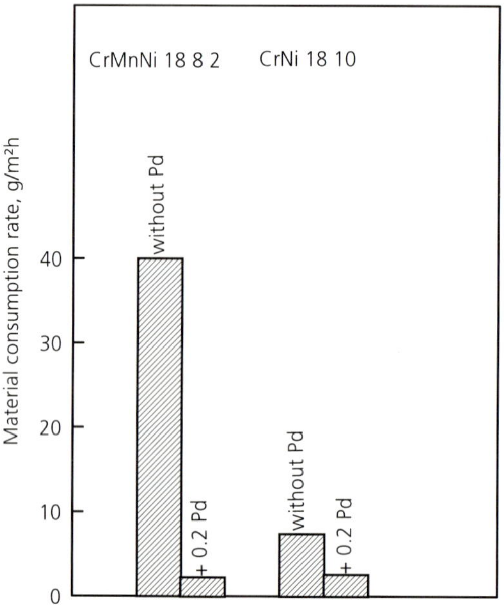

Figure 77a: Corrosion behavior of CrNi-steel 18 10 and CrMnNi-steel 18 8 2 with and without a Pd-content in 20 % sulfuric acid at 293 K (20 °C) [111]

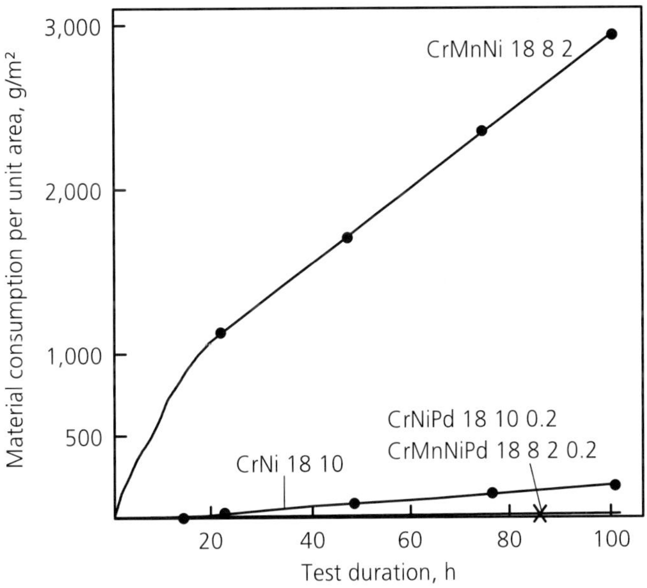

Figure 77b: Corrosion behavior of CrNi-steel 18 10 and CrMnNi-steel 18 8 2 with and without Pd in 20 % sulfuric acid at 293 K (20 °C) [111]

Figure 49 shows the corrosion behavior of Mn-containing, predominantly ferritic-austenitic CrNi-steels (for their composition, see Table 41) in 20 % sulfuric acid as a function of the Pd-content. The austenitic steel 25Cr8Mn6NiN is to be regarded as being largely resistant in 20 % sulfuric acid at 373 K (100 °C). No exact corrosion rates are quoted [111].

Ion implantation (introduction of an alloy partner into the layer close to the surface of a base metal) results in the formation of very thin "heat treatment layers" (10 to 100 nm) in some cases. The low heat treatment temperature compared with other surface heat treatment processes, such as the CVD and most PVD processes, is an advantage. Although the very thin layer thicknesses have the disadvantage of corroding through more quickly, on the other hand they have the advantage that the dimensions of work-pieces and therefore the production tolerances are not changed. In addition to this process, ion beam mixing, amongst others, has also been investigated [180].

The CrNi-steel 18 8 was bombarded with 5 × 10^{16} Pt-ions/cm^2 with an energy of 80 keV. A 50 nm thick surface alloy layer containing about 5 % Pt was formed, or rather developed on this steel.

The specimens treated in this way were investigated for their corrosion behavior at room temperature in 20 % H_2SO_4 (O_2-saturated), together with untreated specimens. The untreated specimens started to corrode actively after a few hours. After a longer period of time (not specified), the corrosion rate decreased as a consequence of the formation of a surface layer, although this had little adhesion and could easily be wiped off. The specimen irradiated with Pt-ions remained passive over the entire

Temperature K (°C)	*Test duration h*	*In the gas phase (over the medium)*	*On the phase interface*	*In the liquid*
		Corrosion rate, mm/a (mpy)		
283–286 (10–13)	93.0	0.001 (0.04)	0.001 (0.04)	0.001 (0.04)
		0.002 (0.08)	0.003 (0.12)	0.005 (0.20)
283–289 (10–16)	98.0	0.009 (0.35)	0.006 (0.24)	0.004 (0.16)
		0.007 (0.28)	0.015 (0.59)	0.010 (0.39)
293 (20)	98.5	0.000 (0)	0.000 (0)	0.000 (0)
		0.000 (0)	0.000 (0)	0.000 (0)
303 (30)	95.0	0.002 (0.08)	0.001 (0.04)	0.001 (0.04)
		0.001 (0.04)	0.007 (0.28)	0.001 (0.04)
318 (45)	116.5	0.000 (0)	0.000 (0)	0.000 (0)
		0.000 (0)	0.000 (0)	0.000 (0)
348 (75)	240.0	0.000 (0)	0.008 (0.31)	0.002 (0.08)
		0.000 (0)	0.010 (0.39)	0.002 (0.08)
348 (75)	523.0	0.0007 (0.03)	0.003 (0.12)	0.010 (0.39)
		0.0002 (0.008)	0.006 (0.24)	0.011 (0.43)
368 (95)	95.0	0.006 (0.24)	0.016 (0.63)	0.065 (2.56)
		0.007 (0.28)	0.018 (0.71)	0.064 (2.52)

Table 51: Corrosion behavior of 1Ch18N10T ((%) Fe-0.06C-1.31Mn-0.61Si-0.027P-18.04Cr-10.22Ni-0.53Ti) in oleum (19.1 %) at various temperatures [178]

Temperature K (°C)	*Specimen speed (rotating disk) m/s*	*Test duration h*	*Corrosion rate mm/a (mpy)*
293 (20)	1.0	69.0	no measurable corrosion
293 (20)	5.0	49.0	no measurable corrosion
293 (20)	11.5	47.5	no measurable corrosion
343 (70)	1.0	37.0	0.012 (0.47)
343 (70)	5.0	24.5	0.009 (0.35)
343 (70)	11.5	21.5	0.015 (0.59)

Table 52: Corrosion of the steel 1Ch18N10T in oleum (19.1 %) as a function of the flow rate and temperature [178]

test duration of 80 days. The surface layer which forms proved to have a very high mechanical stability. The corrosion rates, determined by neutron activation analysis of the Cr-dissolving, were lower than those of the non-irradiated specimens by 10^3 [179–181].

This type of corrosion protection can be used in practice on components where only low material losses are expected during their service life, or under conditions under which local corrosion or cracking should be prevented [181].

The relatively soft austenitic steels can be protected from wear by boriding. An additional improvement in corrosion resistance is associated with the boriding. This is particularly effective against hydrochloric acid, as can be seen from Figure 78. The material consumption is considerably lower in 10 % sulfuric acid than in 20 % HCl, and the resistance in sulfuric acid is not as radically improved by boriding as in hydrochloric acid. After a test duration of 8 h, a drop in the corrosion rates in 10 % H_2SO_4 at 329 K (56 °C) from 7.6 mm/a (299 mpy) in the untreated state to 5.3 mm/a (209 mpy) in the borided state is quoted for the steel 1.4541 (X6CrNiTi18-10). In 20 % HCl the corrosion rate falls drastically from 198 mm/a (7,795 mpy) to 1.2 mm/a (47.2 mpy) [57].

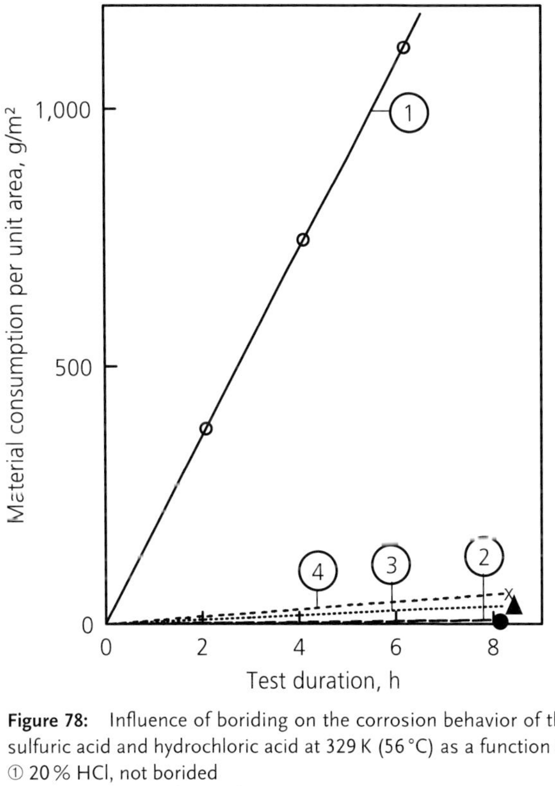

Figure 78: Influence of boriding on the corrosion behavior of the steel 1.4541 (X6CrNiTi18-10) in sulfuric acid and hydrochloric acid at 329 K (56 °C) as a function of the test duration [57]
① 20 % HCl, not borided
② 10 % H_2SO_4, not borided
③ 10 % H_2SO_4, borided
④ 20 % HCl, borided

– *Stress corrosion cracking in sulfuric acid solutions* –

[182] contains a summary and overview of the occurrence of stress corrosion cracking in sulfuric acid amongst other media. According to this reference, stress corrosion cracking is found only in relatively highly concentrated H_2SO_4 solutions in the presence of chlorides. These include, for example, solutions of 2.5 to 5 mol/l H_2SO_4 + 0.1 mol/l NaCl [182].

The relationship between stress corrosion cracking and hydrogen embrittlement was investigated in [183, 184]. Specimens were charged with hydrogen by electrolytic charging experiments on flat specimens for the tensile test in 0.5 mol/l H_2SO_4 with an addition of 40 mg/l arsenous acid as a promoter. The charging current density was 100 mA/cm^2. After the charging, the samples were subjected to tensile testing at a strain rate of 5×10^{-4} 1/s. Hydrogen damage occurs after a charging time of about 25 h on the steels X5CrNi18-10 (1.4301) and X6CrNiTi18-10 (1.4541). These steels, however, are among those less at risk from hydrogen embrittlement. To determine whether the influence of hydrogen can be reversed, specimens were then heat-treated at 293 to 873 K (20 to 600 °C) for 5 h after the hydrogen charging. These experiments showed that CrNi-steels were not irreversibly damaged by hydrogen charging, which is the case with corrosion in sulfuric acid. After heat treatment at 423 K (150 °C), the steel X 10 CrNi 18 9 (cf. 1.4319) showed no effects of the hydrogen charging during subsequent deformation. This cannot be attributed to a change in the structure of the material, since martensite can only be changed reversibly at about 673 K (400 °C). In contrast, if the material is worked immediately after the hydrogen charging, irreversible damage occurs to X 10 CrNi 18 9. The ductility values of the starting state were reached after heat treatment at 523 K (250 °C) after 5 h. This leads one to the conclusion that hydrogen charging of the material without simultaneous stress action does not cause damage to the material [184]. Hydrogen charging has less influence on the deformation capacity of the material in the case of the Mo containing steels (see Section Austenitic chromium-nickel-molybdenum steels).

– *Intercrystalline corrosion in sulfuric acid solutions* –

The most frequent cause of intercrystalline corrosion of high-alloy chromium-containing steels is the precipitation of chromium-rich carbides, predominantly in the form of $M_{23}C_6$, and the associated decrease in the Cr-content in the regions adjacent to the precipitates. Of the known chromium nitrides CrN and Cr_2N, the latter is the only precipitation form present in the chemically resistant steels. Its precipitation also leads to a reduction in the Cr-content in its surroundings. The presence of linked chromium-depleted regions with a defined minimum level of chromium is a prerequisite for susceptibility to intercrystalline corrosion due to chromium-rich precipitates. This minimum content of chromium, called the resistance limit, depends on the corrosive conditions. In trials using the Strauss test (DIN EN ISO 3651-2) it approximates to 14 % Cr. However, a deterioration in the intercrystalline corrosion behavior of the steels is to be expected only at about 0.2 % N. The influence of nitrogen due to the precipitation of Cr_2N is not recorded in the Strauss test (copper sulfate-sulfuric acid solution) [185].

In trials with the more stringent Strauss test under Euronorm conditions 121–72, resistance spot-welded sheets of non-stabilized X 12 CrNi 18 8 (1.4300) steel showed no susceptibility to intercrystalline corrosion even after extremely long welding times [186].

On testing in the Strauss test, for the susceptibility of the material to intercrystalline corrosion in boiling copper sulfate-sulfuric acid solution with a copper addition (in the form of filings), the potential of these steels at +0.34 V is somewhat higher than the activation potential of the grain boundaries. Quenched CrNi-steel is passivated under these conditions; the corrosion rate being low. As a result of the passivating properties of the Strauss solution, the chromium-depleted zones at the resistance limit are passivated in conductive contact with the more noble grain. A steel which is susceptible to intercrystalline corrosion, as a result of chromium precipitates, is attacked mainly at the grain boundaries or grain edges, without substantial general corrosion or pitting corrosion. Stresses have no noticeable influence on the results of this test [186].

Figures 79–82 show the influence of the state of treatment on the corrosion behavior of various stainless steels in the Strauss test [187]. With very low corrosion

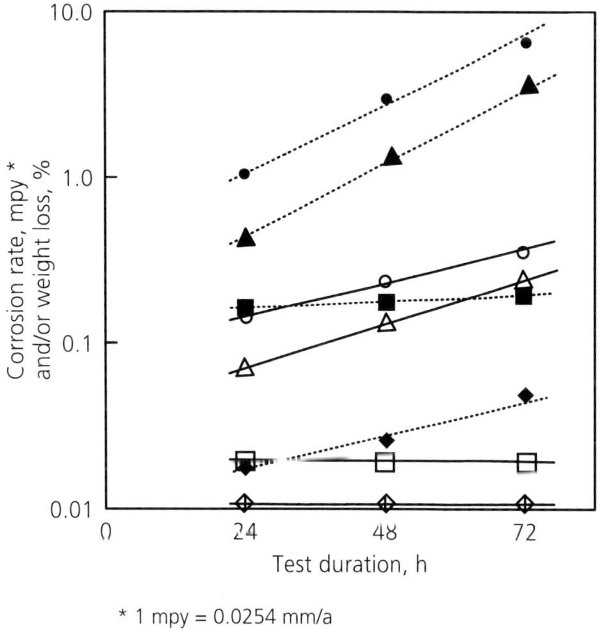

Figure 79: Comparison of the corrosion behavior of the steel SAE 304 (X5CrNi18-10) as a function of the treatment state in the Strauss test [187]
Values in:

mpy	%	
□	■	1,373 K (1,100 °C)/2 h/water
◇	◆	delivery state
△	▲	923 K (650 °C)/3 h/air
○	●	after LCF testing at 923 K (650 °C)

rates, as are to be expected in the Strauss test, the steel SAE 304 shows the poorest corrosion behavior, and not only after low cycle fatigue testing (LCF). The lowest corrosion rate is achieved in the solution-annealed state. Sensitization at 923 K (650 °C)/3 h with subsequent air cooling increases the corrosive attack, mainly in the grain boundary region, as a consequence of chromium depletion due to carbide formation, as has already been explained above. Additional stress (LCF) increases this tendency. Testing under LCF conditions was carried out at 923 K (650 °C) in continuous cycles at a strain rate of 2×10^{-3} 1/s up to a total elongation of the specimens of about 1 %. The increased susceptibility of the materials to corrosion under these conditions can be counteracted by low C-content, high Cr-content and addition of elements which firmly bind to carbide, as can be seen from Figures 80–82. The high-alloy NiCr-steel Incoloy® 800 shows no rise in the corrosion rate, even with increased corrosion testing (Figure 81). The steel SAE 316 L gives an even better corrosion behavior (Figure 82 – see also Section Austenitic chromium-nickel-molybdenum steels) [187].

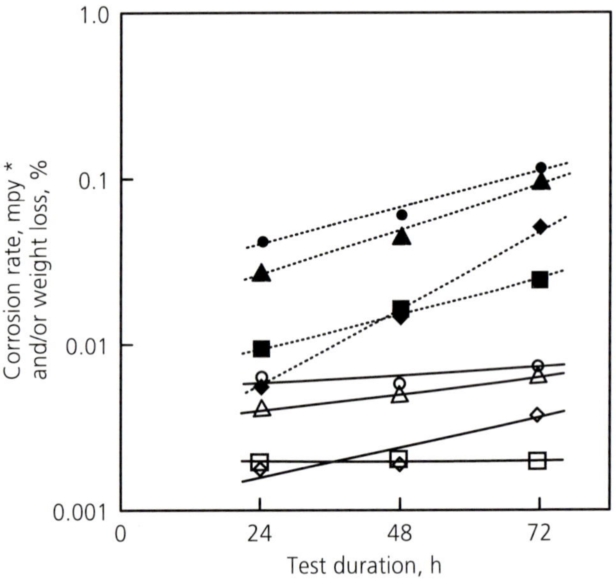

* 1 mpy = 0.0254 mm/a

Figure 80: Comparison of the corrosion behavior of the steel SAE 321 (X6CrNiTi18-10) as a function of the treatment state in the Strauss test [187]

Values in:

mpy	%	
□	■	1,373 K (1,100 °C)/2 h/water
◇	◆	delivery state
△	▲	923 K (650 °C)/3 h/air
○	●	after LCF testing at 923 K (650 °C)

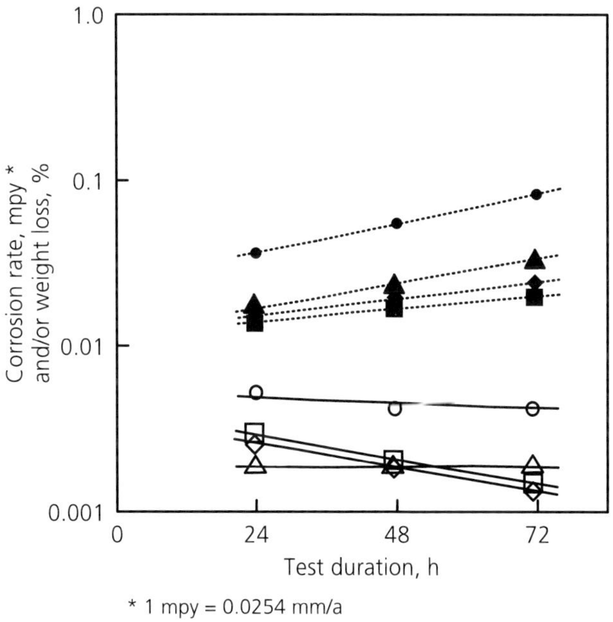

Figure 81: Comparison of the corrosion behavior of Incoloy® 800 (X10NiCrAlTi32-21) as a function of the treatment state in the Strauss test [187]
Values in:

mpy	%	
□	■	1,373 K (1,100 °C)/2 h/water
◇	◆	delivery state
△	▲	923 K (650 °C)/3 h/air
○	●	after LCF testing at 923 K (650 °C)

– Pitting corrosion in sulfuric acid solutions –

The susceptibility of chromium-nickel steels, without molybdenum additions, to pitting corrosion is considerably higher than of those which contain Mo. Corrosion studies on the susceptibility to crevice and pitting corrosion of these steels showed very low critical temperatures. For the steel 304 L, the pitting corrosion temperature is quoted as 288 K (15 °C) and the crevice corrosion temperature as 283 K (10 °C) after testing in 0.01 mol/l H_2SO_4 + 0.1 % $Fe_2(SO_4)_3$ + 1,000 ppm Cl^-. This means that the corrosion phenomena mentioned can be expected under test conditions above this temperature, if chlorides are present [139]. Table 53 shows a comparison with other steels. See also Section Austenitic chromium-nickel-molybdenum steels. The pitting corrosion behavior of the steel 1.4301 (SAE 304) was investigated in 0.5 mol/l sulfuric acid + 1 mol/l potassium bromide solution with and without the addition of 0.5 mol/l sodium nitrate in comparison with sodium chloride solutions containing sulfuric acid in potentiostatic holding tests. In nitrate-free solutions, the steel tends to suffer from pitting corrosion after a critical potential is exceeded. This potential is higher than that of pitting corrosion in chloride-containing solution.

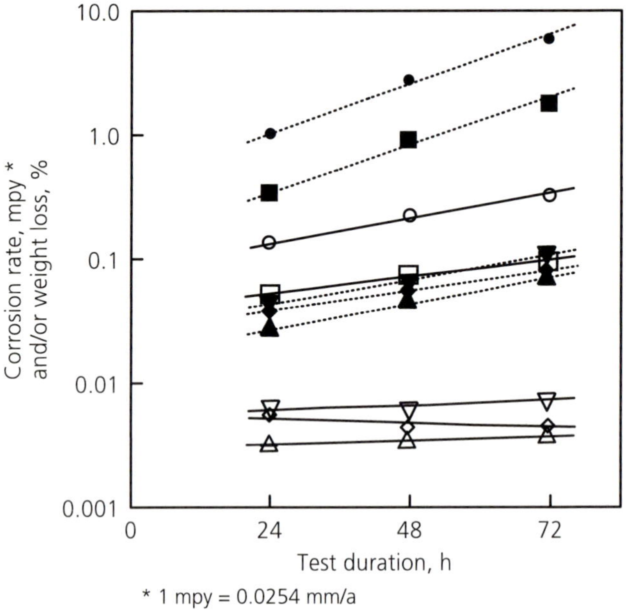

Figure 82: Corrosion behavior of austenitic steels after LCF (low cycle fatigue) testing at 923 K (650 °C) as a function of time in the Strauss test [187]
Values in:

mpy	%	
○	●	Steel SAE 304
△	▲	Steel SAE 316 L
▽	▼	Steel SAE 321
◇	◆	Incoloy® 800
□	■	Incoloy® 800 H

The bromides, in contrast to the chlorides, represent less of a threat, in terms of pitting corrosion. As the pitting corrosion proceeds, the pitting corrosion density is considerably greater under the influence of bromide than has been found with chlorides. Pitting corrosion in solutions containing bromide is inhibited by nitrate additions, as is the case in chloride-containing solutions. An upper threshold potential develops, this likewise being considerably more positive than that in chloride-containing solutions. The pitting corrosion density on the cut and rolled surfaces of the specimens is reduced by subsequent passivation of this steel in nitric acid [188].

– Electrochemical corrosion protection processes –

– Cathodic protection

Figure 83 shows the influence of additions to the corrosion medium. These extend or increase the potential range of active corrosion and the corrosion rate on the steel X5CrNi18-10 in sulfuric acid, in particular by hydrogen sulfide and sulfur dioxide (flush gases).

The marked increase in the rate of corrosion is equivalent here to an increase in the passivation current density and thus to a more difficult transition from the active to the passive state.

Figure 83, furthermore, shows that the material consumption rates of chromium-nickel steels in acid media almost always increase at potentials which are more negative than the free corrosion potential. For this reason, these steels cannot be given cathodic protection in acids [92].

Material	*Pitting Corrosion Temperature, K (°C)*	*Crevice Corrosion Temperature, K (°C)*
SAE 304 L (X2CrNi19-11)	288 (15)	283 (10)
SAE 316 L (X2CrNiMo17-12-2)	303 (30)	293 (20)
SAE 317 L (X2CrNiMo18-15-4)	298 (25)	298 (25)
Alloy 825 (NiCr21Mo)	298 (25)	293 (20)
20Cb-3 Alloy (NiCr20CuMo, cf. 2.4660)	303 (30)	293 (20)
SAE 317 LM (X 2 CrNiMoN 17 13 5)	308 (35)	303 (30)
3RE60 (X2CrNiMoSi18-5-3)	313 (40)	303 (30)
SAE 904 L (X1NiCrMoCu25-20-5)	328 (55)	318 (45)
Ferralium® 255 (X2CrNiMoCuN25-6-3)	328 (55)	328 (55)

Table 53: Pitting and crevice corrosion temperatures, determined in 0.01 mol/l H_2SO_4 + 0.1 % $Fe_2(SO_4)_3$ + 1,000 ppm Cl^- (see also Figure 109) [139]

– *Anodic protection*

The formation of a pronounced passive region in the corrosion medium is a prerequisite for the use of this corrosion protection process. Although complete protection from corrosion, as with cathodic protection, is not possible under these conditions, a drastic reduction in the corrosion current can be achieved, with corrosion rates which can be very much lower than 0.1 mm/a (3.94 mpy). Sulfuric acid is a reducing medium at low concentrations and at all temperatures up to 65 %, has an oxidizing action only at higher temperatures and above 65 % H_2SO_4, and under these conditions can bring about passivity in the materials.

Electrochemical corrosion studies on chromium-nickel steels in 20 % sulfuric acid at 353 K (80 °C) have confirmed the corrosion expected under these conditions (compare also Figure 69). Stirring and the resulting flow rates of little more than 1 m/s lead to the development of a passive state. Passivity can also be maintained, even in non-agitated sulfuric acid, by brief (10 min) polarization with +0.15 V at 353 K (80 °C). Such a pretreatment is recommended for the use of 12Ch18N10T (cf. 1.4541) as a material for tubular heat exchangers [145].

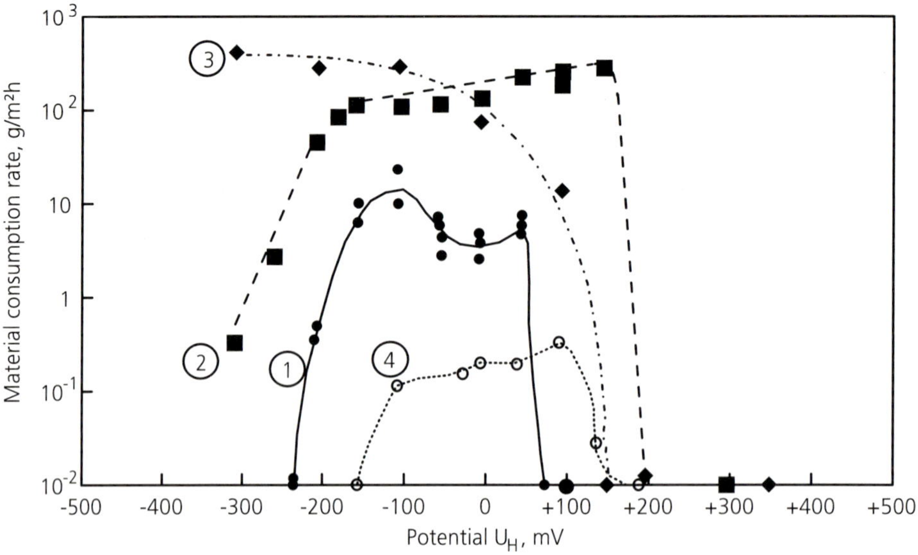

Figure 83: Influence of flush gases on the corrosion behavior of the steel X5CrNi18-10 (1.4301) in sulfuric acid [92]
① H_2
② H_2S
③ SO_2
④ CO

Preliminary studies on the particular application are advisable, if only because these steels have corrosion rates of greater than zero in the passive state. More than 0.2 mm/a (7.87 mpy) is to be expected in this case. The steel 10Ch17N13M2T (cf. 1.4571) proves to be better under such conditions (see Section Austenitic chromium-nickel-molybdenum steels).

Better conditions for anodic corrosion protection on CrNi-steels in sulfuric acid exist at high concentrations, at which the acid has an oxidizing action. Pronounced corrosion phenomena arise on stainless austenitic steels, as a function of their composition, at sulfuric acid concentrations above 90 % and temperatures of about 343 K (70 °C) or more. Under these conditions, these steels are alternately in the passive or active state [168, 189]. If the corrosion chemistry of such specimens is investigated, a constant alternation is found at periodic intervals, the so-called potential oscillations. This can be prevented by anodic polarization.

Figure 84 shows the corrosion behavior of Ch18N10T (corresponding to 1.4541 or X6CrNiTi18-10) in 93.6 % sulfuric acid as a function of temperature and gas content. Up to a temperature of 313 K (40 °C) the corrosion rates are low, but thereafter they increase, especially in aerated sulfuric acid (increasing scatter of the corrosion rates): No scatter of the corrosion rates is observed in SO_2-containing acid. Figure 85 shows the higher variations in potential from 343 K (70 °C), as described above. Up to 373 K (100 °C), the material can be used under these conditions at acid flow rates of up to 2 m/s with constant monitoring of the anodic potential, as, for example, in heat exchangers. Potentials of between +600 and +700 mV have been found to be the optimum protection under these conditions [189].

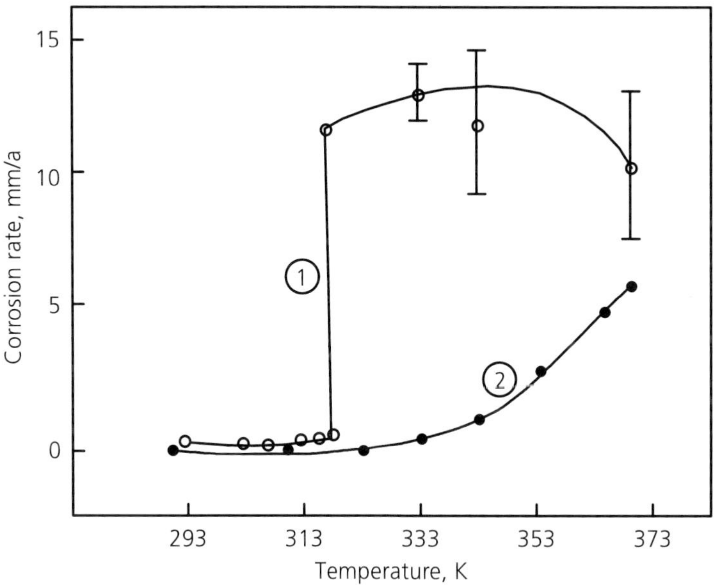

Figure 84: Corrosion of the steel Ch18N10T (X6CrNiTi18-10) in 93.6 % sulfuric acid at a flow rate of 0.85 m/s as a function of the temperature (test duration 5 to 70 h) [189]
① aerated with dry air
② 7 to 10 % SO_2

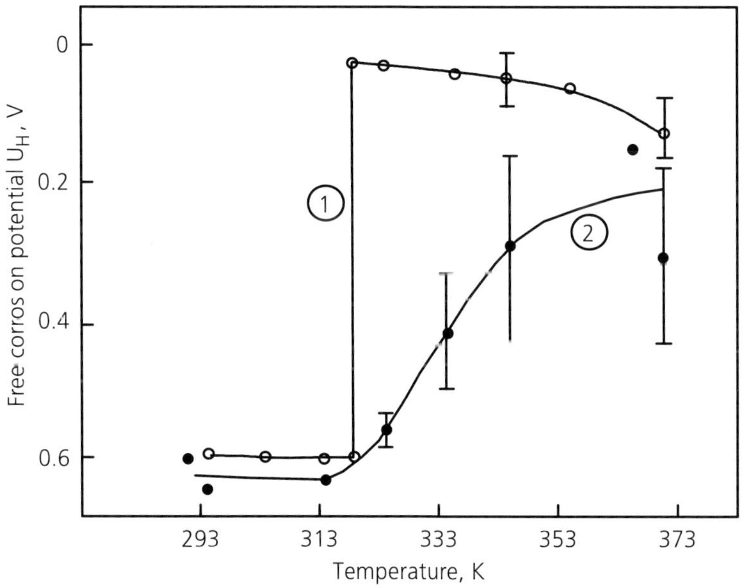

Figure 85: Dependence of the corrosion potential of the steel Ch18N10T (X6CrNiTi18-10) in 93.6 % sulfuric acid on the temperature [189]
① aerated with dry air
② 7 to 10 % SO_2

Figure 72 shows the corrosion rate as a function of the temperature above 373 K (100 °C). The corrosion rates of X6CrNiTi18-10 as a function of the potential are shown in Figure 86 for 90 % H_2SO_4, and in Figure 87 for 97 to 98 % H_2SO_4 [168]. They decrease as the polarization potential increases. As can be seen from these graphs, from a potential of more than 1,000 mV, only very low corrosion rates and no further corrosion damage are to be expected in 90 % and 98 % H_2SO_4 in the temperature range from 373 to 413 K (100 to 140 °C).

The corrosion rates also drop as the concentration rises to about 0.1 mm/a (3.94 mpy) in 90 % acid and 0.01 mm/a (0.39 mpy) in 98 % acid. The material 1.4571 (X6CrNiMoTi17-12-2) shows a better corrosion behavior (see also Figures 88, 110–115). These circumstances were observed in aerated acid. The corrosion rates seen in Figure 88 were found in flowing 98 % H_2SO_4. Under these conditions, the Mo-free steel X6CrNiTi18-10 or X6CrNiTi18-10 proved to be better [168].

This graph shows that the corrosion rate of the two steels in the passive range in flowing acid is lower than that in static acid, and that of the molybdenum-free steel is lower than that of the molybdenum-containing steel by about one power of ten. The lowest corrosion rate of about 0.002 mm/a (0.08 mpy) is achieved by X6CrNiTi18-10 at a U_H of about 1,200 mV [168].

According to [102, 103], Mo-containing and Mo-free stainless austenitic steels are protected equally well in 90 % sulfuric acid at 373 K (100 °C). Table 54 contains a summary of the corrosion rates of anodically protected materials in 98.7 % H_2SO_4 at 373 K (100 °C) and a flow rate of about 1 m/s in a pilot plant under semi-industrial conditions. For details of the experimental conditions see [102].

According to this summary, the corrosion rate on the steel 1.4541 (after corrosion testing under the conditions in Table 54) is less than 0.01 mm/a (0.39 mpy) as a result of anodic protection. Without this protection, the corrosion rate could be as high as 0.5 mm/a (19.7 mpy) at a flow rate of 1.3 m/s.

By using anodic corrosion protection in sulfuric acid plants with a concentration of more than 90 % and at temperatures of up to or about 413 K (140 °C), inexpensive CrNi-steels such as 1.4301 (X5CrNi18-10) and 1.4541 (X6CrNiTi18-10) can also be used [190].

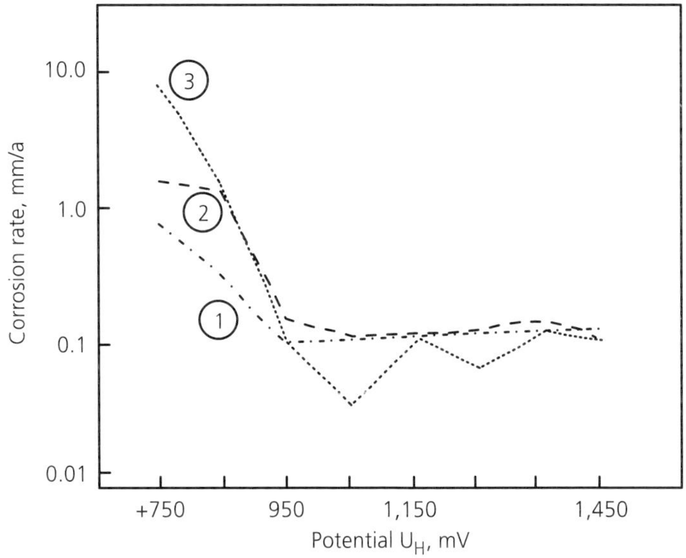

Figure 86: Corrosion rate of the steel X6CrNiTi18-10 (1.4541, SAE 321) in aerated sulfuric acid (90 %) as a function of the temperature and potential (test duration 5.3 d) [168]
① 373 K (100 °C)
② 393 K (120 °C)
③ 413 K (140 °C)

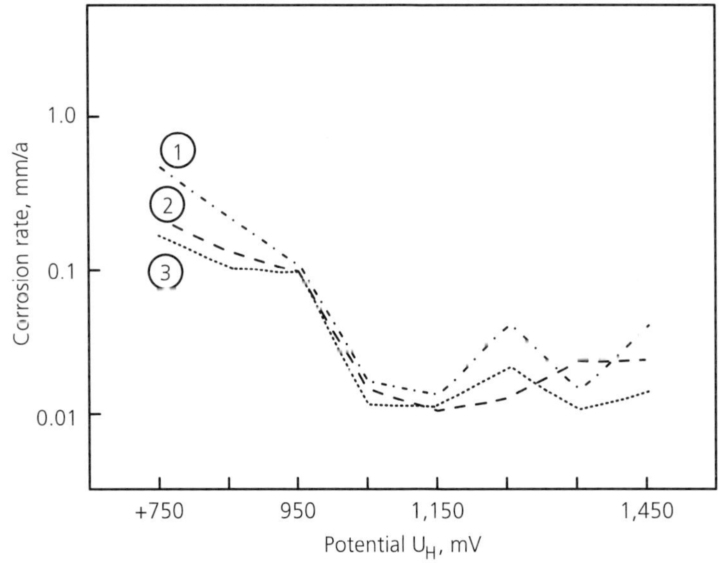

Figure 87: Corrosion rate of the steel X6CrNiTi18-10 (1.4541, SAE 321) in aerated sulfuric acid (97 to 98 %) as a function of the temperature and potential (test duration 5.3 d) [168]
① 373 K (100 °C)
② 393 K (120 °C)
③ 413 K (140 °C)

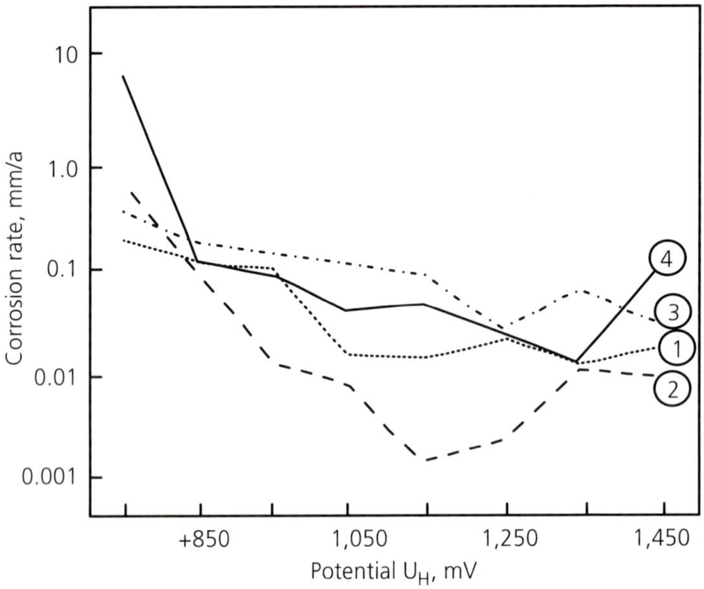

Figure 88: Comparison of the corrosion behaviour of the steels X6CrNiTi18-10 (1.4541) and X6CrNiMoTi17-12-2 (1.4571) in aerated 98 % sulfuric acid at 413 K (140 °C) as a function of the potential and flow rate (test duration 10 d) [168]
① X6CrNiTi18-10 (SAE 321), static
② X6CrNiTi18-10 (SAE 321), 0.8 m/s
③ X6CrNiMoTi17-12-2 (SAE 316 Ti), static
④ X6CrNiMoTi17-12-2 (SAE 316 Ti), 0.8 m/s

The corrosion of austenitic steels in sulfuric acid is reduced by additions of oxidizing substances (e.g. HNO_3), or the unwanted contamination resulting from the latter.

Table 55 [50] shows the protective action of an inhibitor called DX-A (for its composition, see Section Unalloyed steels and cast steel) on the steel SAE 304 (X5CrNi18-10) in 2.5 mol/l H_2SO_4 at room temperature. The corrosion rate of this steel falls from 45 mm/a (1,772 mpy) in pure acid to 0.4 mm/a (15.7 mpy) on addition of inhibitor (1.0 %). As the table shows, there are hardly any differences in the corrosion behaviors of unalloyed steel and the relatively high-alloy steel in sulfuric acid under these conditions. According to [50], the optimum inhibitor concentration was determined as 1 %. Figure 89 [50] shows the influence of the sulfuric acid concentration on the corrosion rate of SAE 304. At room temperature, the corrosion maximum without an inhibitor (at a corrosion rate of 66 mm/a (2,598 mpy) in 8 mol/l H_2SO_4) is shifted to a value of approximately 2 mm/a (78.7 mpy) in 4 mol/l H_2SO_4 by the inhibitor DX-A. In this range, a considerably lower corrosion rate is to be expected compared with the unalloyed steel (see Figure 20). Overall, the inhibitor action can be considered as being very good. Figure 90 shows the influence of the inhibitor on the weight loss as a function of the test duration. According to this graph, after initially little corrosion, the material consumption rises after a test duration of 12 h, reaching a maximum value of about 10 g/m after about 16 h.

Material DIN-designation	*DIN-Mat. No.*	*Flow rate m/s*	*Corrosion rate mm/a (mpy)*		*Protection potential U_H mV*	*Protective current density mA/cm²*	
			Unprotected	*Protected*		*Start*	*End*
X6CrNiTi18-10	1.4541	1.3	0.50 (19.7)	0.006 (0.24)	1,200	0.18	0.01
		1.0	0.16 (6.30)	0.003 (0.12)			
X6CrNiMoTi17-12-2	1.4571	1.3	3.44 (135)	0.002 (0.08)	1,200	0.16	0.01
		0.9	1.42 (55.9)	0.002 (0.08)			
X1NiCrMoCu25-20-5	1.4539 1)	1.2	3.30 (130)	0.003 (0.12)	950	0.19	0.006
X1NiCrMoCuN25-20-7	1.4529	1.2	0.31 (12.2)	0.001 (0.04)	1,200	0.41	0.005
X3NiCrCuMoTi27-23	1.4503	0.9	2.50 (98.4)	0.001 (0.04)	1,200	0.58	0.002
X1NiCrMoCu31-27-4	1.4563	1.0	0.73 (28.7)	0.002 (0.08)	1,150	0.59	0.004
NiCr21Mo	2.4858	0.9	4.94 (194)	0.004 (0.16)	1,200	0.11	0.003
X1NiCrMoCuN25-20-7 2)	1.4529	0.9	>10 (> 394)	0.004 (0.16)	1,200		

1) see also Figures 139 and 140
2) test temperature 393 K (120 °C)

Table 54: Results of corrosion studies in a semi-industrial plant using 98.7% H_2SO_4 at 373 K (100 °C), test duration 20 days [102, 103]

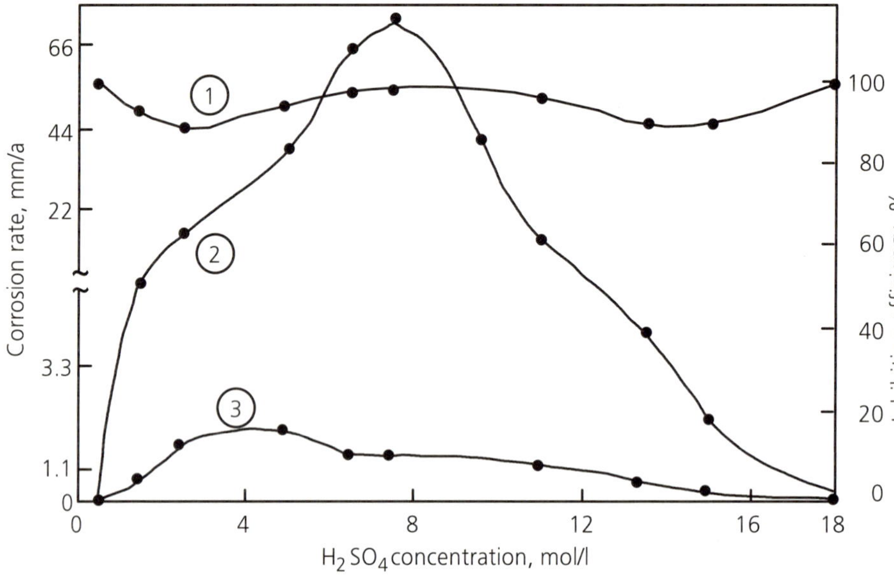

Figure 89: Corrosion of the CrNi-steel SAE 304 (X5CrNi18-10) as a function of the sulfuric acid concentration with and without addition of inhibitor DX-A at room temperature [50]
① inhibition efficiency
② H_2SO_4 without DX-A
③ H_2SO_4 + 10 % DX-A

Inhibitor concentration %	*Corrosion rate[1), mm/a (mpy)*		*Inhibition efficiency, %*	
	Unalloyed steel	*SAE 304*	*Unalloyed steel*	*SAE 304*
0	46.0 (1,811)	45.0 (1,772)	–	–
0.1	28.7 (1,130)	34.2 (1,346)	37.6	24.0
0.2	1.4 (55.1)	16.5 (650)	96.9	63.3
0.3	1.2 (47.2)	1.7 (66.9)	97.4	96.2
0.5	0.8 (31.5)	0.9 (35.4)	98.3	98.0
1.0	0.4 (15.7)	0.4 (15.7)	99.1	99.1

1) converted from mg/dm^2 h

Table 55: Influence of the concentration of the inhibitor DX-A (see composition in Section Unalloyed steels and cast steel) on the corrosion of the unalloyed steel PCRCA and the CrNi-steel SAE 304 (X5CrNi18-10) in 2.5 mol/l H_2SO_4 at room temperature [50]

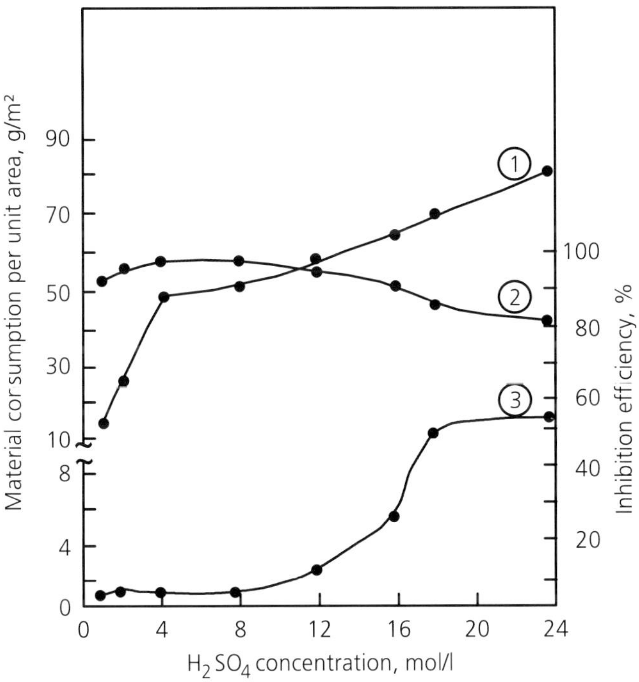

Figure 90: Influence of the immersion time on the corrosion behavior of the CrNi-steel SAE 304 (X5CrNi18-10) in 2.5 mol/l sulfuric acid with and without addition of inhibitor DX-A at room temperature [50]
① inhibition efficiency
② H_2SO_4 without DX-A
③ H_2SO_4 + 10 % DX-A

After a test duration of 24 h, this value is only slightly exceeded. A material consumption per unit area of 10 g/m^2 after 24 h corresponds to a corrosion rate of 0.5 mm/a (19.7 mpy). The graph shows a trend towards lower values as the test duration increases (no data on longer test durations are available).

Figure 91 shows the influence of temperature on the corrosion behavior of SAE 304 in 2.5 mol/l H_2SO_4 in the range 303 to 363 K (30 to 90 °C); good inhibition efficiency is also evident [50].

Corrosion studies on the steel SAE 321 (X6CrNiTi18-10) in various concentrations of sulfuric acid showed the good inhibiting action of a low concentration of potassium iodide, as can be seen from Figure 92 (for specimen dimensions see Figure 84). According to this graph, almost 100 % protection of the steel is achieved with a KI concentration of 10^{-4} mol/l up to sulfuric acid concentrations of 5 mol/l H_2SO_4 (about 38.3 %). Higher KI concentrations also protect the steel in more highly concentrated acid. The test duration was only 1 h.

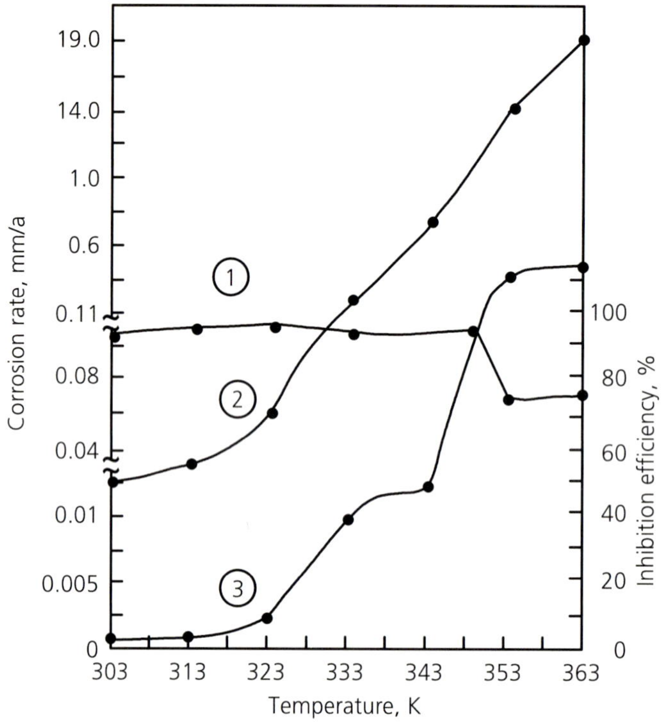

Figure 91: Influence of the inhibitor DX-A in 2.5 mol/l sulfuric acid on the CrNi-steel SAE 304 (X5CrNi18-10) as a function of temperature [50]
① inhibition efficiency
② H_2SO_4 without DX-A
③ H_2SO_4 + 10 % DX-A

Figure 93 shows the corrosion behavior of this steel in 0.5 mol/l H_2SO_4 at 305 K (32 °C) as a function of the test duration.

According to this graph, complete inhibition is to be expected at a KI concentration of between 10^{-4} and 10^{-1} mol/l in 0.5 mol/l H_2SO_4 at 305 K (32 °C). The inhibiting effect is cancelled out at a concentration of 1 mol/l KI. After a test duration of 14 days, pitting corrosion was found on the specimen surface, leading to perforation of the 1.2 mm thick specimens after a test duration of 28 days. Similar studies with NaBr and other halides with similar results are reported in this reference [191].

Figure 94 shows the influence of the inhibitor concentration and temperature in 0.5 mol/l H_2SO_4. Whilst, as expected, the material consumption rate increases as the temperature of the sulfuric acid rises, inhibition of corrosion occurs as the inhibitor addition increases. Complete corrosion protection can be expected only above 10^{-3} mol/l KI (in the range between 10^{-3} and 10^{-1} mol/l KI) in the temperature range between 303 and 353 K (30 and 80 °C) [191] (See also Figure 107).

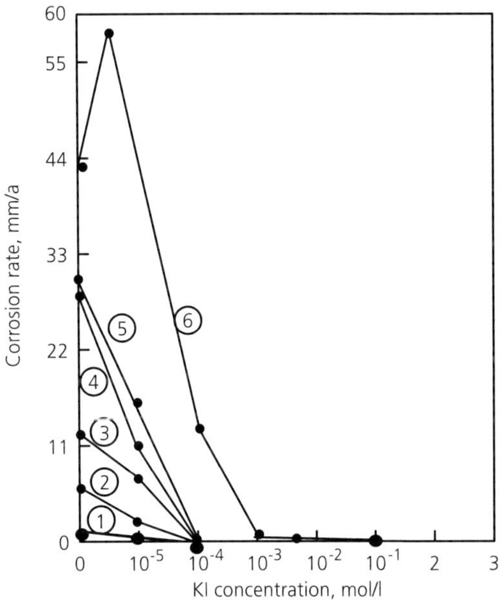

Figure 92: Influence of the potassium iodide concentration on the corrosion rate of the steel SAE 321 (X6CrNiTi18-10) in various concentrations of sulfuric acid at 304 K (31 °C) (test duration 1 h) [191]
Sulfuric acid concentration:
① 0.5 mol/l; ② 1.5 mol/l; ③ 2.5 mol/l; ④ 3.5 mol/l; ⑤ 5.0 mol/l; ⑥ 7.5 mol/l

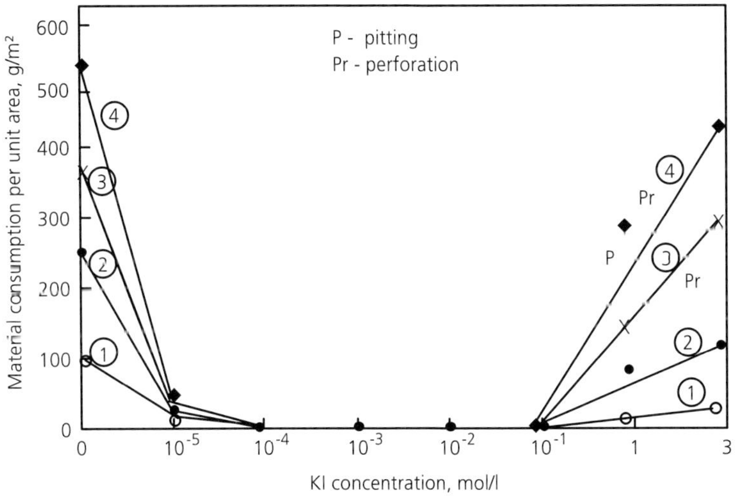

Figure 93: Influence of the potassium iodide concentration and the test duration in 0.5 mol/l H_2SO_4 at 305 K (32 °C) on the corrosion behavior of SAE 321 (X6CrNiTi18-10) (specimen dimensions 50 × 25 × 1.2 mm, 100 ml solution in open glass beakers) [191]
Test duration:
① 24 h, ② 7 d, ③ 14 d, ④ 28 d

Austenitic chromium-nickel-molybdenum steels

The corrosion resistance of Mo-free steels in sulfuric acid solutions is relatively low, as can be seen in Section Austenitic chromium-nickel steels. From Figure 66 it is evident that an increase in the corrosion resistance is achieved by increasing the alloying content. Figure 70 shows the classification of the resistance to be expected from chromium-nickel steels in sulfuric acid as a function of the alloying content. For the increased corrosion resistance of Mo-containing steels, see also Figures 15, 16, 46, 47, 53–57, 61, 65, 71, 72, 82, 88 and Tables 6–8, 34, 37, 48–50, 53, 54 and 72. It should be pointed out once more here that the resistance data are to be interpreted only under these conditions. Very small amounts of other components or oxidants in the medium modify the corrosion resistance of the materials to a considerable degree. In the case of these steels, oxidizing additions have a positive effect on their corrosion resistance. Gassing of the acid and agitation thereof have a further influence. CrNiMo-steels are used in contact with sulfuric acid almost exclusively in the passive state.

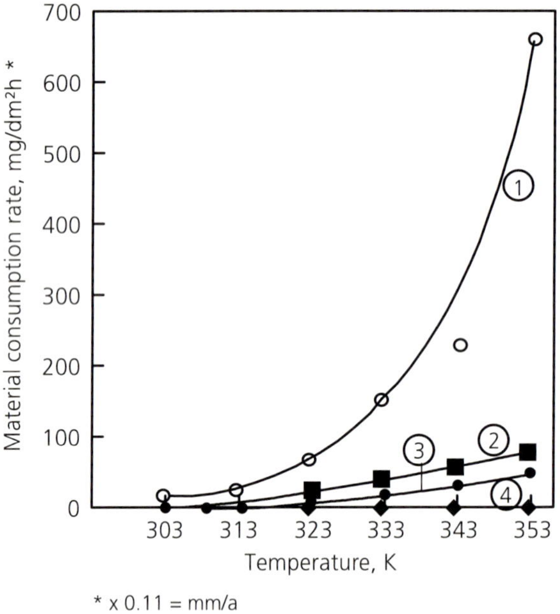

Figure 94: Influence of temperature and potassium iodide concentration on the corrosion behavior of the steel SAE 321 (X6CrNiTi18-10) in 0.5 mol/l sulfuric acid [191]
① 0.5 mol/l H_2SO_4
② 0.5 mol/l $H_2SO_4 + 10^{-5}$ mol/l KI
③ 0.5 mol/l $H_2SO_4 + 10^{-4}$ mol/l KI
④ 0.5 mol/l $H_2SO_4 + 10^{-3}$ to 10^{-1} mol/l KI

Figure 95 is a diagram of the influence of various alloying elements on the passive nature of stainless steels. The dissolution current density in the passive state is reduced chiefly by Cr and the passivation current density is reduced by Mo, the transition from the active to the passive state thus being facilitated. Copper also reduces the passivation current density.

Figure 96 shows the extended range of use of CrNi-steels with additional alloying elements, such as, for example, Mo or Cu.

The steel 1.4401, which contains approximately 2 % Mo and gives a maximum corrosion rate of 0.1 mm/a (3.94 mpy) in pure sulfuric acid, has a greater application range than 1.4301 at room temperature and acid concentrations of slightly above 10 %. At 313 K (40 °C), however, its range of application is smaller.

The maximum permitted temperature in concentrated sulfuric acid is extended to about 328 K (55 °C). For more about the better corrosion resistance of Mo-containing CrNi-steels, see also [119, 137, 157, 161, 167]. In 4 % H_2SO_4 containing up to 6 % gaseous SO_2, the corrosion resistance of the CrNi-steels increases as the alloying content of Mo and Cu increases [167].

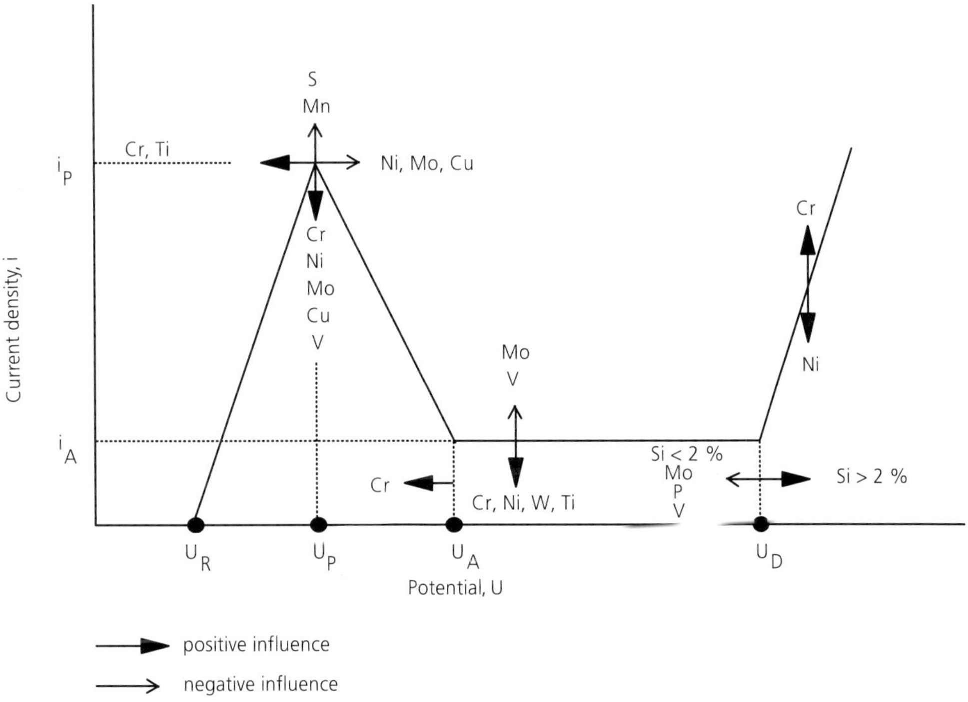

Figure 95: Influence of alloying elements on the passivity of stainless steels [92]
i_P = passivation current density
i_A = passive dissolution current density
U_R = free corrosion potential
U_P = passivation potential
U_A = activation potential
U_D = transpassive breakdown potential

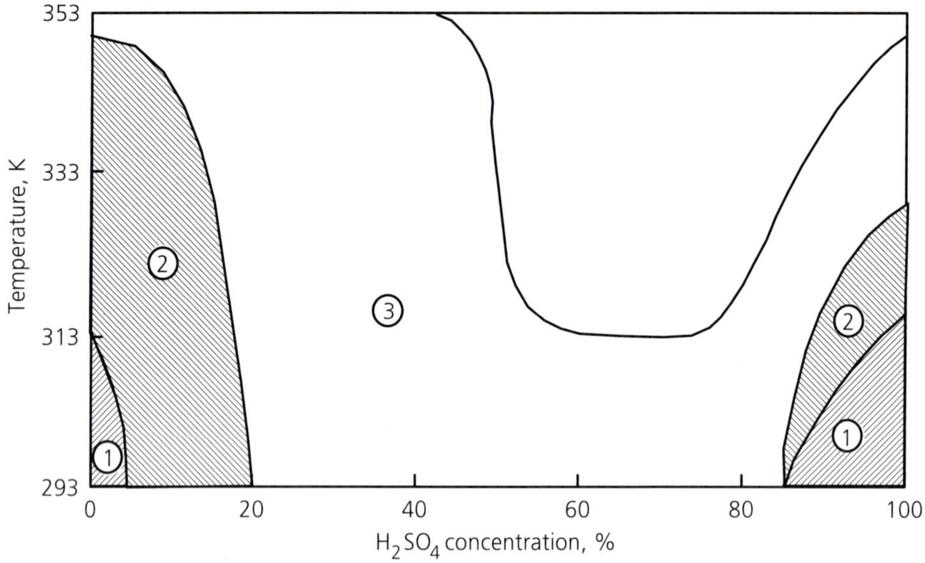

Figure 96: Temperature/concentration diagram of the resistance (corrosion rate 0.1 mm/a (3.94 mpy)) of CrNi-steels in sulfuric acid [92, 160, 163]
1) 1.4301 (X5CrNi18-10)
2) 1.4401 (X5CrNiMo17-12-2)
3) 1.4505 (X4NiCrMoCuNb20-18-2)

Sulfur additions to stainless steels considerably improve their machinability, as with low-alloy steels. On the other hand, the corrosion resistance is reduced by addition of sulfur. Whilst Mn-sulfides are attacked even in weakly acid solutions, titanium sulfides are resistant in strongly acid media. The corrosion resistance of such steels should be improved again by addition of titanium. Table 56 contains the chemical composition of the steels investigated. Table 57 shows the corrosion behavior of these steels in various concentrations of sulfuric acid at various temperatures as a function of the S- and Ti-content. This table shows that CrNiMo-free-cutting steels with a sulfur and titanium addition are particularly suitable for use in 10% H_2SO_4 up to 313 K (40 °C). The corrosion rates are below 0.1 mm/a (3.94 mpy). The corrosion-improving influence of molybdenum can also be seen [117].

Studies in this connection were performed on free-cutting steels having the composition shown in Table 58. Table 59 summarizes the corrosion resistance of experimental melts in sulfuric acid. The best corrosion resistance in sulfuric acid at concentrations up to 10% and temperatures up to 313 K (40 °C) is shown by CrNiMo-steels, for which short-term use at 333 K (60 °C) can also be considered under certain conditions [116].

The standard materials summarized in Table 58 were alloyed with sulfur and titanium, the titanium content being coordinated with the sulfur content. Materials to which only sulfur was alloyed show better machinability than those containing titanium and sulfur.

Steel type	Chemical composition, %								
	C	Si	Mn	P	S	Cr	Mo	Ni	Ti
CrMo-steel 17 0.8	0.070	0.42	1.26	0.026	0.008	16.98	0.83	0.10	< 0.01
+ S	0.066	0.40	1.32	0.010	0.240	17.24	0.84	0.01	< 0.01
+ S + Ti	0.063	0.42	1.40	0.007	0.260	17.23	0.84	0.06	1.31
CrNi-steel 18 10	0.029	0.31	1.97	0.007	0.014	18.39	0.02	9.56	< 0.01
+ S	0.030	0.40	1.82	0.009	0.264	18.88	< 0.01	10.33	< 0.01
+ S + Ti	0.026	0.50	1.82	0.005	0.226	17.67	< 0.01	10.40	1.30
+ S + Zr	0.029	0.45	1.68	0.006	0.250	17.53	0.05	10.90	< 0.01
CrNiMo-steel 18 14 2.8	0.016	0.43	1.68	0.009	0.017	17.83	2.79	14.16	< 0.01
+ S	0.038	0.43	1.74	0.007	0.230	17.84	2.78	14.53	< 0.01
+ S + Ti	0.018	0.44	1.88	0.010	0.235	17.65	2.81	14.02	1.37

Table 56: Chemical composition of high-alloy free-cutting steels (experimental batches) for corrosion studies in H_2SO_4 (see Tables 57, 59) [117]

Temperature K (°C)	H_2SO_4 concentration %	Material consumption rate, g/m² h*					
		CrNi-steel 18 10			CrNiMo-steel 18 10 2.8		
		–	+ S	+S + Ti	–	+ S	+ S + Ti
Room temperature	1.25	0.6	3.7	< 0.1	< 0.1	< 0.1	< 0.1
	2.50	0.7	4.4	< 0.1	< 0.1	< 0.1	< 0.1
	5.00	1.1	5.1	0.5	0.1	0.1	< 0.1
	10.00	1.5	6.9	0.9	0.2	0.5	< 0.1
313 (40)	1.25	1.3	–	1.4	0.1	0.8	< 0.1
	2.50	1.8	–	1.6	0.1	1.3	< 0.1
	5.00	–	–	1.7	0.4	1.8	< 0.1
	10.00	–	–	3.7	0.4	2.2	< 0.1
333 (60)	1.25	–	–	–	0.5	–	0.5
	2.50	–	–	–	0.5	–	0.5
	5.00	–	–	–	–	–	1.1
	10.00	–	–	–	–	–	1.4

* value × 1.1 mm/a

Table 57: Corrosion behavior of CrNi-free-cutting steels (experimental batches, see Table 56) in sulfuric acid as a function of their alloying additions [117]

Steel type	Chemical composition, %								
	C	Si	Mn	P	S	Cr	Mo	Ni	Ti
X 3 CrMoTiS 17 (Remanit® 4193 X)	0.015	1.35	0.41	0.011	0.23	17.11	1.10	0.25	1.20
X 2 CrNiTiS 18 9 (Remanit® 4306 X)	0.023	0.45	1.20	0.017	0.22	18.32	0.12	11.15	1.33
X 2 CrNiMoTiS 18 12 (Remanit® 4435 X)	0.012	0.40	1.28	0.018	0.23	18.07	2.62	14.05	1.03

Table 58: Chemical composition of high-alloy free-cutting steels (experimental batches) for corrosion studies in acids (see Table 59) [116]

Medium	Temperature K (°C)	X 3 CrMoTiS 17 (Remanit® 4193X)	X 2 CrNiTiS 18 9 (Remanit® 4306 X)		X 2 CrNiMoTiS 18 12 (Remanit® 4435 X)	
		164 × 164 mm	164 □	212 mm ∅	164 □	212 mm ∅
		Material consumption rate, g/m^2 h*				
H_2SO_4						
5 %	room temperature	0	0.92	0.69	0.02	0.01
5 %	313 (40)	51.93	0.97	1.71	0	0.06
5 %	333 (60)	141.40	3.96	4.65	0	0
10 %	room temperature	16.74	1.12	0.87	0.01	0.01
10 %	313 (40)	61.26	1.65	2.58	0.04	0.14
10 %	333 (60)	174.87	8.24	7.02	0.61	0.62
HNO_3						
2.5 %	room temperature	0	0	–	0	–
2.5 %	boiling	0	0.01	0.01	0.01	0.01
65 %	room temperature	0	0	–	0	–
65 %	boiling	1.96	0.22	0.26	0.41	0.63
5 % Lactic acid	boiling	0	10.66	13.84	0.01	0.01
20 % Acetic acid	boiling	0	4.48	6.04	0.01	0.01
5 % Formic acid	333 (60)	0.02	0	0	0	0
5 % Formic acid	boiling	0	10.92	13.79	0.70	1.03

* value × 1.1 mm/a, □ square bar steel

Table 59: Corrosion behavior of square and pipe material from free-cutting steels in acids [116]

Only sulfur-containing free-cutting steels show a significantly lower corrosion resistance than low-sulfur standard types. On the other hand, the sulfur- and titanium-containing free-cutting grades exhibit a distinctly better corrosion resistance than the standard materials in this study. On the basis of these good results, smelting of free-cutting steels has begun [117].

The corrosion-resistant free-cutting steel Sandvik® 1802 ((%) Fe-≤ 0.030C + N-18Cr-2.3Mo-0.6Ti-0.3S-0.5Si-0.1Mn; cf. 1.4523) is comparable in its corrosion resistance in dilute sulfuric acid to the CrNiMo-steel X5CrNiMo17-12-2 [115].

Specimens produced from the steel SAE 316 by powder metallurgy showed a poorer corrosion behavior than the compact material in 0.5 mol/l H_2SO_4. The causes of this were intermetallic phases formed as a result of grain size and density differences [192].

Under conditions as in [192], specimens of the same steel with additions of 0.5 to 3.0 % Sn showed a corrosion behavior in 0.5 mol/l H_2SO_4 similar to that of rolled steel. The effect of the Sn-addition in improving the corrosion resistance is due in particular to the reduction in the amount of pores joined to one another [193].

The CrNiMo-steels can be classified between the molybdenum-free and the molybdenum- and copper-containing steels in terms of corrosion resistance in sulfuric acid, as can be seen from Figures 16, 66, 70, 96. They are less resistant than ferritic steels or duplex steels, as a result of their alloying content (compare also Figures 46, 53, 54, 56, 57 and 61).

Steels containing about 20 % Cr, 9 % Ni and 2 to 4 % Mo can be regarded as being resistant in sulfuric acid at room temperature up to a concentration of 10 % and from 85 to 90 %; this is also the case in the concentration range 90 to 100 % H_2SO_4, where they are usable above 323 K (50 °C). Figures 97 and 98 (compare also Figure 69) show the increase in corrosion resistance with increasing alloying content with the aid of isocorrosion curves. Although a visible improvement in corrosion resistance in sulfuric acid is observed, in many cases it is not sufficient for use of these materials. This particularly applies to the middle concentration range between about 20 and 80 % H_2SO_4. For the steel X 8 CrNiMoTi 18 10, this range is quoted as being between 25 and 78 % H_2SO_4 at room temperature. An increase in the Ni-content and Cu-addition improve the corrosion resistance according to expectations. The corrosion rates shown in Figure 99 were determined on specimens (45 × 20 × 3 mm, non-pickled and non-ground, merely degreased with acetone) in 66 % H_2SO_4 after a test duration of 60 days at temperatures of 313 and 323 K (40 and 50 °C).

Whereas the steel X 5 CrNiMoTi 18 10 corroded at a rate of 1.32 mm/a (52.0 mpy) at 313 K (40 °C) and 3.58 mm/a (141 mpy) at 323 K (50 °C), the corrosion rates of the higher-alloy steels reach a maximum of 0.03 mm/a (1.18 mpy) [194].

Figure 100 shows another possible choice of material where it is essential to use metallic materials in sulfuric acid, but the molybdenum-containing grades are no longer usable. See also [195–200] on the development and use of corrosion-resistant materials.

In the production of sulfuric acid (chamber acid), CrNiMo-steels give a corrosion rate of approximately 0.01 mm/a (0.39 mpy) when used as pipelines downstream of

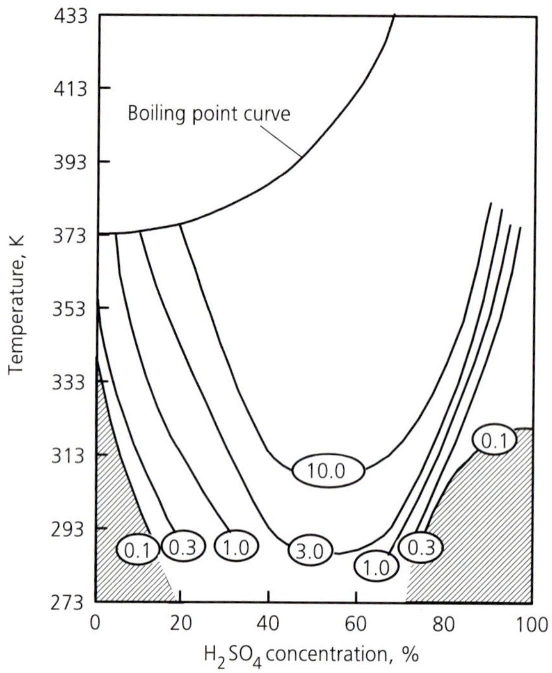

Figure 97: Isocorrosion curves (mm/a) of CrNiMo-steel ((%) Fe-0.05C-18Cr-10Ni-2Mo) in sulfuric acid [61]

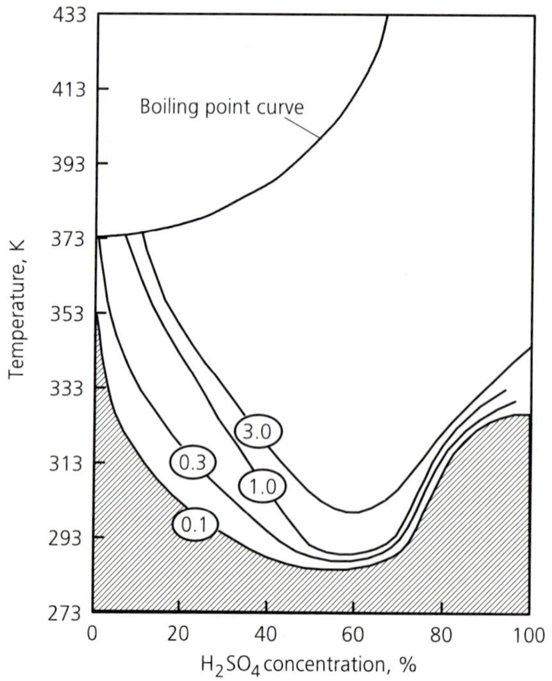

Figure 98: Isocorrosion curves (mm/a) of CrNiMo-steel ((%) Fe-0.05C-17Cr-13Ni-5Mo) in sulfuric acid [61, 84]

the cooler. The conditions under which this figure was recorded were: 72.6 to 77.6 % H_2SO_4 containing 0.02 to 0.05 % HNO_3 in the temperature range 310 to 356 K (37 to 83 °C), and the steel 10Ch17N13M2T, which is similar to 1.4571 (X6CrNiMoTi17-12-2), was used as a typical example of a CrNiMo-steel (see Table 6). Figure 15 shows the corresponding graphs of the corrosion rates of various steels as a function of the temperature [23].

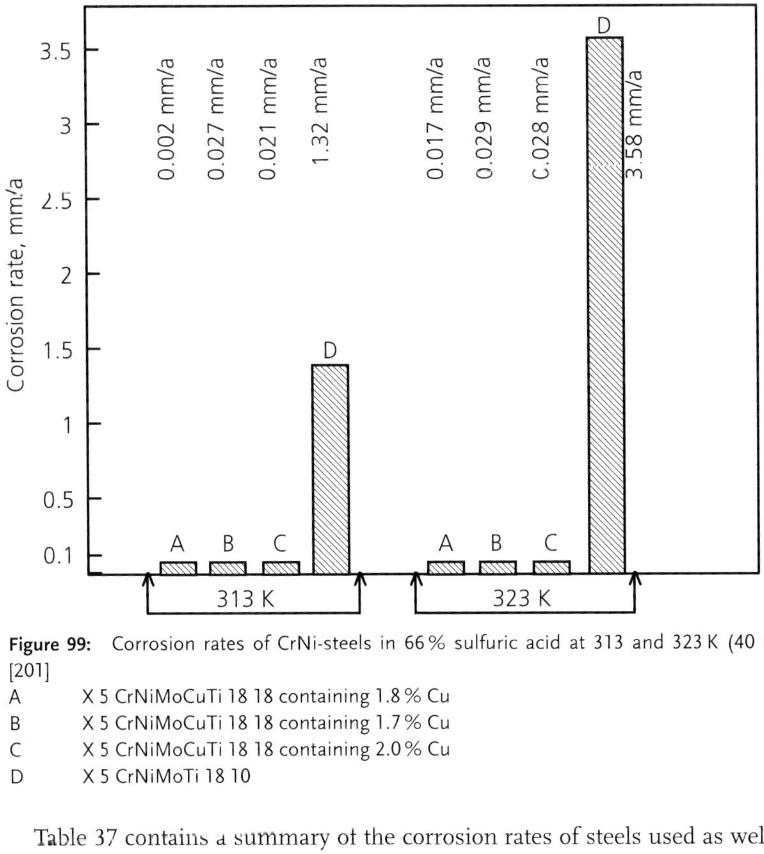

Figure 99: Corrosion rates of CrNi-steels in 66 % sulfuric acid at 313 and 323 K (40 and 50 °C) [201]

A X 5 CrNiMoCuTi 18 18 containing 1.8 % Cu
B X 5 CrNiMoCuTi 18 18 containing 1.7 % Cu
C X 5 CrNiMoCuTi 18 18 containing 2.0 % Cu
D X 5 CrNiMoTi 18 10

Table 37 contains a summary of the corrosion rates of steels used as weld-plating material in sulfuric acid. Further corrosion rates of CrNiMo-steels, in comparison with other steels, are contained in Table 8 for alloying groups, and Table 34 for commercially available materials. For further corrosion rates of the steels SAE 304, 310, 316 and 317 see [4].

Table 60 shows a comparison of the corrosion resistance of the steel SAE 316 (X5CrNiMo17-12-2) in sulfuric acid (10, 19, 50, 75 and 96 %) at temperature 373 K (100 °C). This steel can be used only for short-term operation under these conditions.

Figure 47 shows a comparison of the corrosion resistances of high-alloy materials in 95 % H_2SO_4 at 373, 398 and 423 K (100, 125 and 150 °C). The steels 1.4404 (X2CrNiMo17-12-2), 1.4465 (X1CrNiMoN25-25-2) and 1.4571 (X6CrNiMoTi17-12-2)

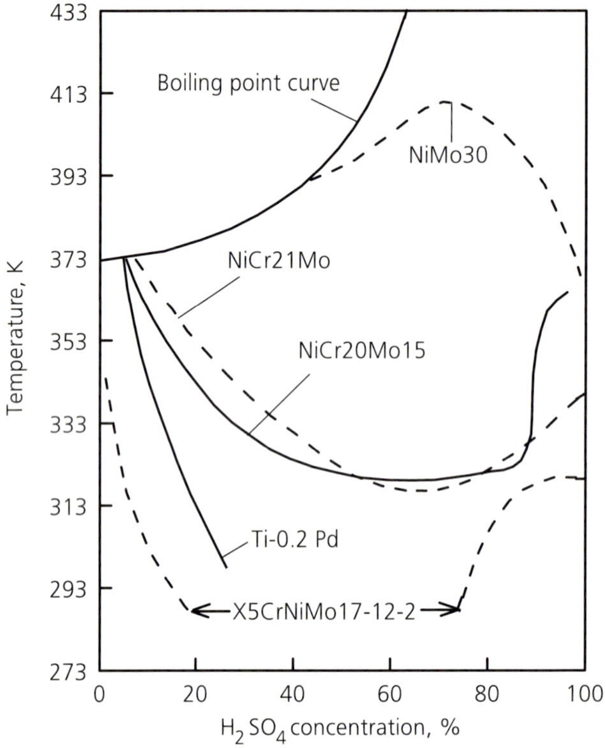

Figure 100: Isocorrosion curves (0.13 mm/a (5.12 mpy)) of high-alloy materials in sulfuric acid [164]

come under this section. Of these materials, the steel 1.4465 shows the best resistance. Its corrosion rate is about 0.5 mm/a (19.7 mpy) at the test temperatures. The maximum corrosion rate is 0.4 mm/a (15.7 mpy) [102, 103]. The corrosion studies were performed under moisture-free conditions. An improvement in the corrosion resistance by anodic protection is possible (see these studies and Table 54).

In this context, see also Section Austenitic chromium-nickel steels with special alloying additions as well as Figure 135 and Table 61.

Figure 101 shows an overview of the corrosion rates of various metallic materials in hot 98 % sulfuric acid. According to this graph, the corrosion resistance of CrNiMo-steels containing 18 % Cr, 10 % Ni and 2.5 % Mo is the lowest [202].

Alloys	Section	Corrosion rate, mm/a (mpy)						
		5 %	10 %	19 %	40 %	50 %	75 %	96 %
		373 K (100 °C)	373 K (100 °C)	377 K (104 °C)	333 K (60 °C)	395 K (122 °C)	455 K (182 °C)	566 K (293 °C)
		10 Days	23 h	23 h	10 Days	20 h	20 h	3 h
Nickel 200 (Ni 99.6)		3.5 (138)	3.0 (118)	2.25 (88.6)	–	25 (984)	23 (906)	25 (984)
Monel® 400 (NiCu 30 Fe)		0.6 (23.6)	0.06 (2.36)	0.19 (7.48)	–	16 (630)	25 (984)	25 (984)
Inconel® 600 (NiCr15Fe)		3.4 (134)	9.8 (386)	16.0 (630)	0.92 (36.2)	25 (984)	25 (984)	21.5 (846)
Inconel® 625 (NiCr22Mo9Nb)		0.15 (5.91)	–	–	< 0.01 (< 0.39)	–	–	–
Incoloy® 800 (X10NiCrAlTi32-21)	K	4.6 (181)	–	–	7.5 (295)	–	–	–
Incoloy® 825 (NiCr21Mo)		0.17 (6.70)	0.5 (19.7)	0.45 (17.7)	< 0.01 (< 0.39)	1.3 (51.2)	25 (984)	1.6 (63.0)
SAE 304 (X5CrNi18-10)	G	8.5 (335)	–	–	3.8 (150)		–	–
SAE 316 (X5CrNiMo17-12-2)	H	2.5 (98.4)	9.2 (362)	25.0 (984)	6.6 (260)		25 (984)	14 (551)

G: see section Austenitic chromium-nickel steels
H: see section Austenitic chromium-nickel-molybdenum steels
K: see sction Special iron based alloys

Table 60: Corrosion behavior of CrNi-steels and nickel alloys according to laboratory studies in various concentrations of sulfuric acid at various temperatures [203]

Material designation	DIN-Material No.	Oscillation temperature* K (°C)	Oscillation range E_H, mV
X5CrNi18-10	1.4301	–	–
X6CrNiMoTi17-12-2	1.4571	353 (80)	370–900
X1CrNiSi18-15-4	1.4361	–	–
X1NiCrMoCuN25-20-7	1.4529	392 (119)	275–780
X3NiCrCuMoTi27-23	1.4503	381 (108)	255–830
X2CrNiMoCuN25-6-3	1.4507	400.5 (127.5)	370–635
X1NiCrMoCu31-27-4	1.4563	383 (110)	260–770
NiCr21Mo	2.4858	377 (104)	270–820
NiMo16Cr15W	2.4819	395.5 (122.5)	450–650

* start of variation in potential in the plot of the current density/potential curves

Table 61: Comparison of the potential oscillation temperatures and oscillation ranges of CrNi-steels and Ni-alloys in 95 % H_2SO_4 [102]

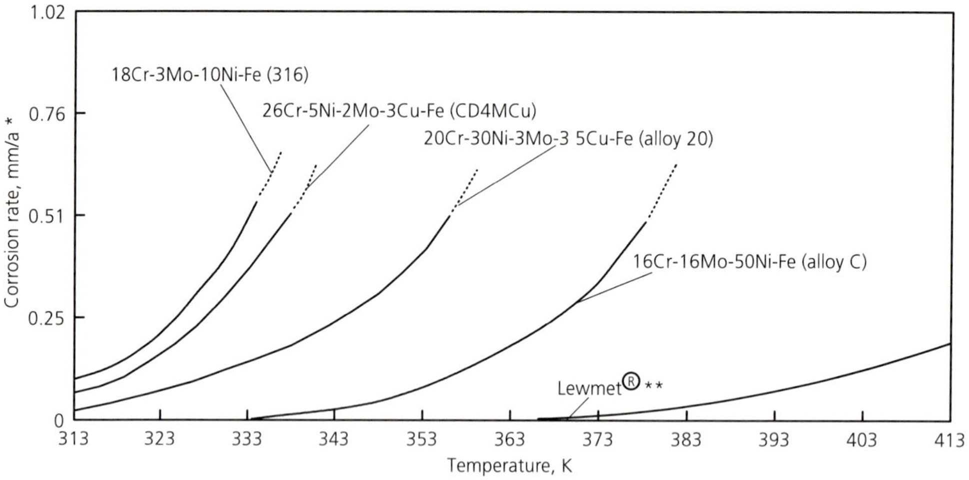

* conversion from mils/year (mpy)

** NiCr-alloys containing 33 % Cr

Figure 101: Corrosion of metallic materials in 98 % sulfuric acid under static conditions (see also Figures 47 and Figures 135) [202, 204]

– *Corrosion behavior in sulfuric acid-acid mixtures* –

A reduction in the corrosion of CrNiMo-steels occurs when nitric acid is added to sulfuric acid due to the oxidizing action of nitric acid [176]. See also Section Austenitic chromium-nickel steels.

Table 47 contains a summary of the corrosion rates of the steel 10Ch17N13M2T (X6CrNiMoTi17-12-2) in 50 % sulfuric acid at 333 K (60 °C). According to this table, corrosion rates of 0.02 mm/a (0.79 mpy) are attained by Cr-steel, CrNi-steel and this CrNiMo-steel when 0.5 % HNO_3 is added. With higher-alloy steels an addition of 0.05 % HNO_3 is already sufficient to reduce the corrosion rate to 0.02 mm/a (0.79 mpy). The corrosion rate without HNO_3 under these conditions is quoted as 171 mm/a (6,732 mpy) [148].

Figure 102 shows the influence of the nitric acid content in hot 96 % sulfuric acid. Sulfuric acid produced by the Müller process has low nitrous content due to nitrogen oxidation in the furnace process. The concentration in the H_2SO_4 can vary in the individual units of the acid part of the plant.

0.008 % N_2O_3 (nitric anhydride) was found in 96 to 98 % absorber sulfuric acid and 0.04 % N_2O_3 was found in the approximately 99 % H_2SO_4 in the waste gas pipelines. Corrosion studies in the two acids have shown that the 96 to 98 % absorber acid containing 0.008 % N_2O_3 was more corrosive than the 99 % H_2SO_4 containing 0.04 % N_2O_3.

The following corrosion rates have been found on the following CrNi-steels in 96 to 98 % acid containing 0.008 % N_2O_3:

- X 8 CrNiTi 18 10 — 0.17 mm/a (6.69 mpy),
- X 8 CrNiMoTi 18 11 — 0.40 mm/a (15.7 mpy) and
- X 5 CrNiMoCuTi 18 18 — 0.21 mm/a (8.27 mpy).

In acid containing 0.04 % N_2O_3, the corrosion rates of all three steels were 0.01 mm/a (0.39 mpy). Figure 102 shows that an N_2O_3 content of 0.03 % or more in 96 % sulfuric acid at 393 K (120 °C) has the effect of a corrosion rate of about 0.1 mm/a (3.94 mpy) on all three steels. Larger additions have no further improving influence. Increased corrosive attack even occurs in the range from 0 to 0.03 % N_2O_3. Under these conditions, the Mo-containing, copper-free steel is most at risk, even more so than in sulfuric acid containing no nitric acid [205]. The corrosive attack is almost exclusively local in character. It may lead to pitting corrosion and is a consequence of the equally hazardous underdosing with passivating agent [205].

Without stating the corrosion rates, the materials X5CrNiMo17-12-2 and GX5CrNiMo19-11-2 are recommended in (%) 50:50, 75:25 and 20:15 sulfuric acid-nitric acid mixtures at 323 K (50 °C); both are also recommended in boiling (%) 2:1 % acid mixtures [206].

Table 62 summarizes the corrosion behavior of iron and Ni-materials in a sulfuric acid pickling solution for cast iron which contains hydrofluoric acid and consists of 5 % H_2SO_4 + 5 % HF + 10 % $FeSO_4$. Although the Mo-containing steel X5CrNiMo17-12-2 is more resistant than unalloyed or CrNi-steel in this solution, with a corrosion rate of 4.85 mm/a (191 mpy), it cannot be considered as being resistant [16].

Material	*Corrosion rate, mm/a (mpy)*
NiCu 30 Fe	4.95 (195)
Ni 99.6	2.00 (78.7)
NiCr15Fe	2.75 (108)
X5CrNi18-10	10.00 (394)
X5CrNiMo17-12-2	4.85 (191)
Unalloyed steel	21.3 (839)

Experimental conditions:
Medium: 5 % HF, 5 % H_2SO_4, 10 % $FeSO_4$
Duration: 40 h
Temperature: 322 K (49 °C)
Aeration: The specimens were alternately immersed in the bath and removed again.

Table 62: Corrosion behavior of iron materials and nickel materials in a pickling solution for cast iron [16]

Hydrofluoric acid in pickling baths containing sulfuric acid can increase the pickling action on steels containing silicon. Corrosion studies in a pilot plant, a pickling tank, gave the corrosion rates shown in Table 63 for metallic materials. The steel X5CrNiMo17-12-2 was completely dissolved in this pickling solution. The test duration was 31 days at an average temperature of 330 K (57 °C) and a maximum temperature of 353 K (80 °C) without aeration in an agitated pickling solution consisting of 25 to 35 % H_2SO_4 containing 4 to 8 % HF [16].

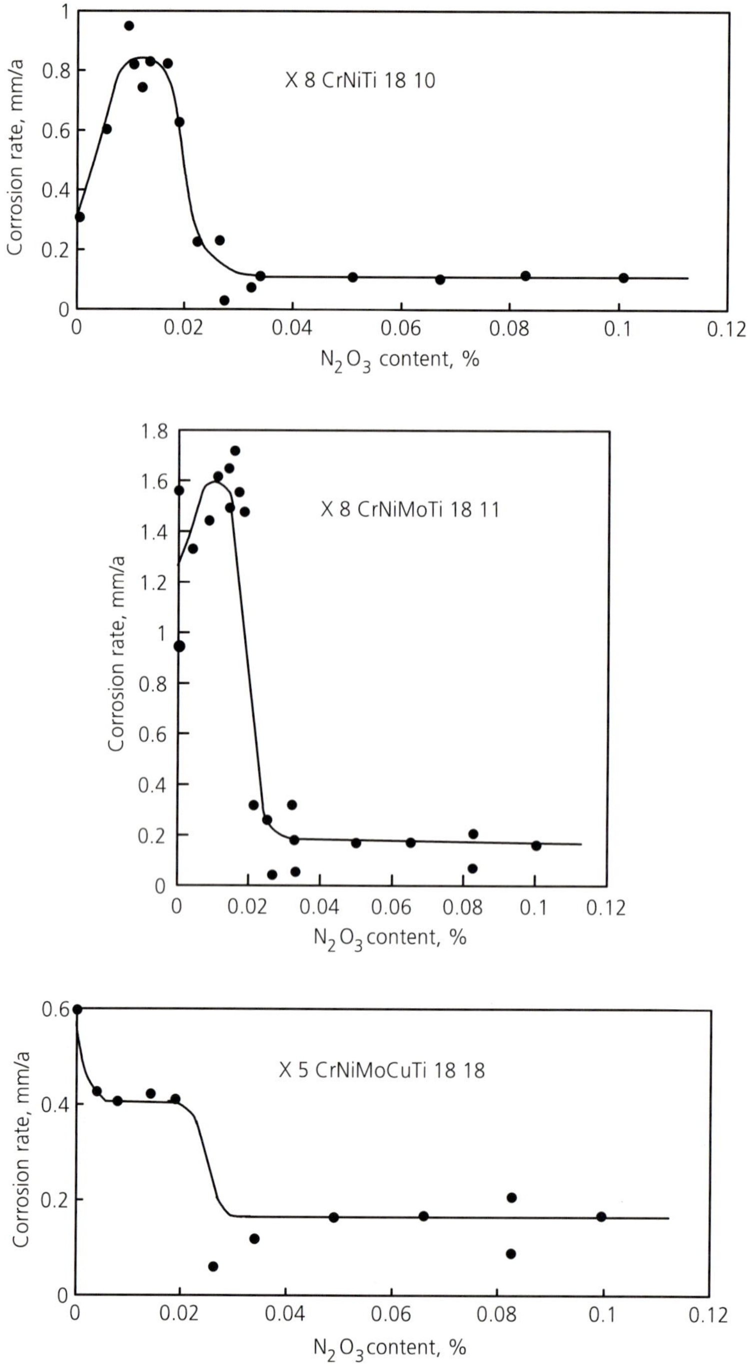

Figure 102: Corrosion rate of CrNi-steels in 96% sulfuric acid at 393 K (120 °C) as a function of the nitric acid content (test duration 32 d) [205]

Material	Corrosion rate mm/a (mpy)	Max. crevice corrosion depth mm
NiCrMoCu 56 23 6 6 (Illium® G)	0.13 (5.12)	0
NiCrMoCu 68 21 5 3 (Illium® R)	0.15 (5.91)	0
G-NiSi 10 Cu (Hastelloy® D)	0.15 (5.91)	0
NiCu30Al	0.18 (7.09)	0
NiCu 30 Fe	0.18 (7.09)	0
High-purity lead	0.18 (7.09)	0
CuNi 30 Fe	0.20 (7.87)	0
NiMo16Cr15W	0.10 (3.94)	0.05
NiMo 30	0.10 (3.94)	0.25
NiCr22Mo9Nb	0.13 (5.12)	0.05
NiCr22Mo6Cu	0.15 (5.91)	0.20
NiCr21Mo	0.15* (5.91)	0.33
X 5 NiCrMoCuNb 29 20 3 3	0.18* (7.09)	0.28
NiCr20CuMo (cf. N08020)	0.18* (7.09)	0.56
X 5 NiCrMoCu 24 20 3 2	0.28 (11.0)	0.13
NiCrMoCu 52 28 8 6	0.64 (25.2)	–
NiCrFe 55 36	2.82 (111)	–
Tantalum	> 90 (> 3,543)	completely destroyed
X5CrNiMo17-12-2	> 180 (> 7,087)	completely destroyed

* Accelerated attack in the region of markings embossed by stresses; no stress corrosion cracking.
Experimental conditions:
Medium: 25 to 35 % H_2SO_4, 4 to 8 % HF
Test location: Pickling tank of a pilot plant
Duration: 31 days
Temperature: 322–353 K (49–80 °C) (average temperature 330 K (57 °C))
Agitation: With the aid of a steel strip
Aeration: none

Table 63: Corrosion behavior of metallic materials in a pickling solution of sulfuric and hydrofluoric acid for steels of relatively high Si-content according to plant studies [16]

Table 48 summarizes the corrosion behavior of CrNi-steels in phosphoric acid with and without H_2SO_4 and HF. Low contents of sulfuric acid in 41 % H_3PO_4 of 1 and 5 % give a corrosion behavior, with corrosion rates of about 0.003 mm/a (0.12 mpy), which is slightly better than, or the same as, that in pure H_3PO_4. Together with hydrofluoric acid, the corrosion behavior deteriorates considerably [149].

Sulfuric acid is usually used for the production of phosphoric acid by the wet breakdown process. The process is based on the reaction of calcium phosphates with sulfuric acid and the separation of the reaction products from one another. Before the ground crude phosphates are broken down with acid, concentrated sulfuric acid is mixed with medium-concentrated phosphoric acid, which is fed to the process, in the acid diluter. Table 64 summarizes the corrosion rates measured on metallic materials in this preliminary diluter. The CrNiMo-steels mentioned last in this table, show the lowest corrosion resistance under these conditions (28 % H_3PO_4 from the wet breakdown from + 20 to 22 % H_2SO_4 + 1 to 1.5 % fluorides) with a corrosion rate of more than 3.8 mm/a (150 mpy). Specimens showed stress corrosion cracking in the region of the markings [207]. According to this summary, higher-alloy Cu-containing steels are more resistant (see Section Austenitic chromium-nickel steels with special alloying additions). Other studies in plant experiments have shown that Mo-containing CrNi-steels are also unusable in phosphoric acids with a low content of sulfuric acid and/or hydrochloric acid. In almost all cases, Mo- and Cu-containing steels are more resistant here [207].

Material	*Corrosion rate mm/a* (mpy)*	*Pitting depth mm*
X 5 NiCrMoCuNb 29 20 3 3	0.028 (1.10)	
NiCr21Mo	0.069 (2.72)	0.13
NiMo 16 Cr	0.071 (2.80)	
NiCrMoCu 56 23 6 6 (Illium® G)	0.076 (3.00)	
X 5 NiCrMoCu 24 20 3 2	0.109 (4.29)	0.08
NiCrMoCu 68 21 5 3 (Illium® R)	0.175; 0.360 (6.89; 14.2)	
NiCu30Al	0.790 (31.1)	≤ 0.08
NiCu 30 Fe	0.970 (38.2)	
NiMo 30	1.98 (78.0)	
X3CrNiMo18-12-3	< 3.80** (< 150)	
X5CrNiMo17-12-2	< 4.50** (< 177)	

* formation of thick layers on all specimens
** stress corrosion cracking in the region of the markings
Experimental conditions:
Medium: 28 % H_3PO_4 (20 % P_2O_5) from the wet breakdown, 20 to 22 % H_2SO_4: approx. 1 to 1.5 % fluorides, probably in the form of H_2SiF_6 (continuous dilution of the conc. sulfuric acid with phosphoric acid fed in from the process)
Test site: at the bottom of the preliminary diluter
Duration: 42 days
Temperature: 355 to 383 K (82 to 110 °C) (average temperature 366 K (93 °C))
Agitation: only by convection
Aeration: moderate

Table 64: Corrosion rates of Ni-containing materials in a sulfuric acid-phosphoric acid mixture in the preliminary diluter of a wet breakdown plant (plant experiments) [207]

In the production of anhydrous hydrogen fluoride, the condensable constituents of the gases formed by decomposition of fluorspar, which consist of 80 % HF, 5 to 10 % H_2SO_4 and 2 to 3 % water, are separated off. The condensates contain 18 to 40 % HF, 35 to 70 % H_2SO_4 and up to 40 % water. Under the gas inlet conditions (393 to 423 K (120 to 150 °C)) the steel Ch17N13M3T (17 % Cr, 13 % Ni, 3 % Mo) shows a corrosion rate of 3.6 to 5 mm/a (142 to 197 mpy). The test duration under operating conditions was 2,200 h. The corrosion rate in the liquid phase at 353 K (80 °C) was 2.2 to 3.7 mm/a (86.6 to 146 mpy). The corrosion rates in the gas vent during prepurification of the gases in the preliminary condenser are quoted as 2.6 mm/a (102 mpy) at temperatures of between 393 and 423 K (120 and 150 °C) (test duration of the plant experiments 1,680 h). The high-alloy Cu-containing steel 0Ch23N28M3D3T had a better resistance under these conditions, but also with a corrosion rate of 0.84 mm/a (33.1 mpy) in the liquid phase [25].

In mixed acid (62 or 75 % H_3PO_4 with a maximum of 4 % H_2SO_4) or contaminated phosphoric acid, the steel 1.4435 (X2CrNiMo18-15-4) is still found to have a relatively good resistance at 353 K (80 °C) with corrosion rates below 0.3 mm/a (11.8 mpy). At 373 K (100 °C), it is virtually destroyed with a rate of about 10 mm/a (394 mpy). It can also no longer be used, even at lower temperatures, in 75 % phosphoric acid contaminated with H_2SO_4, as can be seen from Figure 103. The Cu-containing higher-alloy steels can be used with a relatively good corrosion resistance, in comparison with the Mo-containing steels, under the conditions shown in Figure 103. The steel SAE 904 L (X1NiCrMoCu25-20-5) still has a corrosion rate of less than 0.3 mm/a (11.8 mpy) even at 373 K (100 °C) [208].

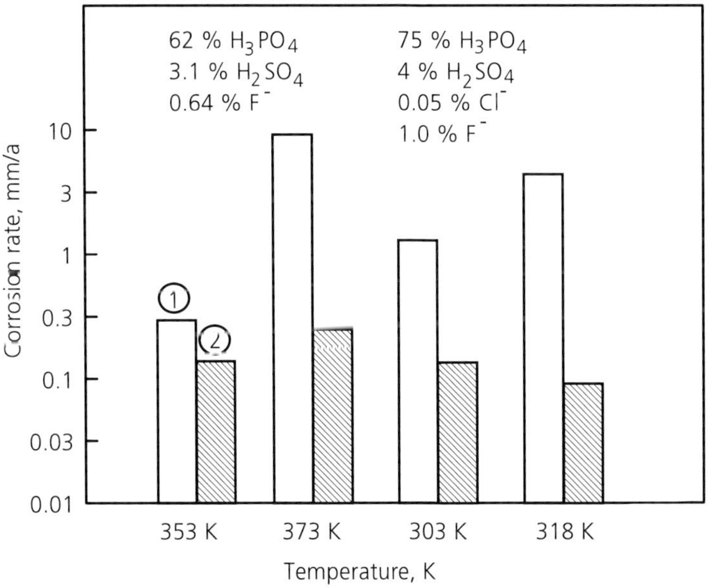

Figure 103: Corrosion rates of CrNiMo-steels in phosphoric acid contaminated by sulfuric acid according to laboratory studies [208]
① X2CrNiMo18-15-4 (1.4435)
② SAE 904 L (X1NiCrMoCu25-20-5)

The corrosion rate of the steel Ch17N13M3T in sulfuric acid solutions of thiourea containing 20 g/l solids, 100 g/l urea and sulfuric acid (pH 0.7, 323 K (50 °C)) is 0.57 mm/a (22.4 mpy) due to crevice corrosion. For this reason, it cannot be used as a centrifuge material. Titanium VT-1 is the preferred material here, with a maximum corrosion rate of only 0.009 mm/a (0.35 mpy) [172].

In the manufacture of sodium hexafluorosilicate in sulfuric acid solutions by the sulfate process, the steel 10Ch17N13M2T (cf. 1.4571) can be used as the collecting tank for the mother solution of 1.5 % Na_2SiF_6 and 6 to 8 % H_2SO_4 at 303 to 313 K (30 to 40 °C). The corrosion rates reach 0.045 mm/a (1.77 mpy). It cannot be used as the collecting tank for the production water containing up to 1 % Na_2SiF_6 and 1 to 2 % H_2SO_4, because of the risk of crevice corrosion and associated pitting corrosion at corrosion rates of 0.383 mm/a (15.1 mpy). The temperature is likewise 303 to 313 K (30 to 40 °C) (Table 7) [24].

The pitting and crevice corrosion temperatures, above which these types of corrosion occur, were determined for the steels SAE 317 L (X2CrNiMo17-12-2) and SAE 317 LM in a solution of 0.01 mol/l H_2SO_4, 0.1 % $Fe_2(SO_4)_3$ and 1,000 ppm Cl^-. For the steel SAE 317 L, they were determined as 298 K (25 °C) in both cases, and for the steel SAE 317 LM as 308 K (35 °C) and 303 K (30 °C) respectively. A comparison with other materials is possible from Table 53.

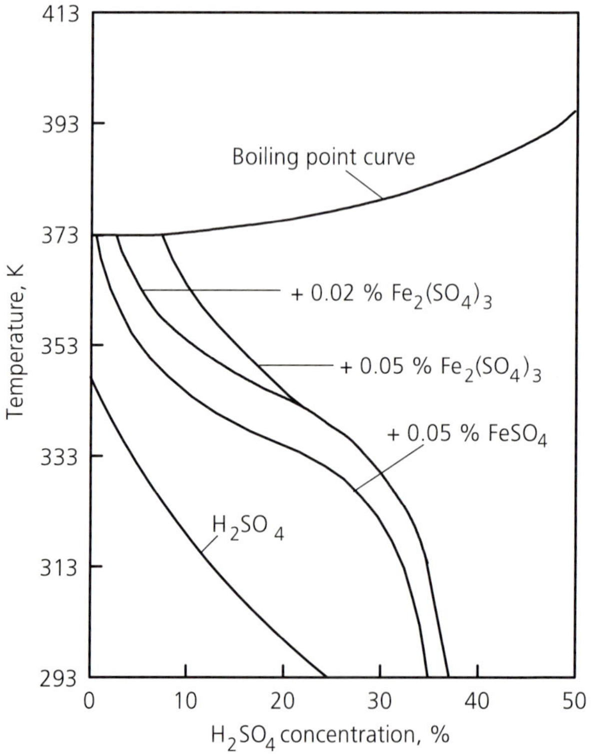

Figure 104: Isocorrosion curves (0.1 mm/a (3.94 mpy)) for the CrNiMo-steel 17 12 2.5 in sulfuric acid with and without sulfate additions [26]

– Corrosion behavior in sulfuric acid with additions –

The corrosion behavior of the CrNi-steels and also the CrNiMo-steels is improved by oxidizing constituents in sulfuric acid. Figures 104 and 105 show the increase in the corrosion resistance of the steel containing 17 % Cr, 12 % Ni and 2.5 % Mo in sulfuric acid containing various sulfates. Figure 106 shows the influence of CrO_3 on this steel. In comparison to this, Figure 107 shows the influence of CrO_3 in H_2SO_4 on the CrNi-steel 18 9, which shows almost the same corrosion resistance as the Mo-containing steel with additions of similar concentration [26].

The examples mentioned indicate once again that additions or impurities in sulfuric acid can considerably influence the corrosion resistance of CrNi-steels, and indeed, as in these examples, improve it. Only a minimal amount is needed to effect a reduction in corrosion by the acid, and if this is exceeded an increase in corrosion can be expected.

– Chlorides

Under the conditions of flue gas purification, increased corrosive attack occurs on metallic materials in particular due to chlorides in aqueous solutions.

Chlorides in most cases cancel out the passivity of CrNi-steels.

If chlorides are present in sulfuric acid, the resistance of the relatively low Mo-alloyed steel is no longer sufficient, as can be seen from Figure 108. Higher-alloy steels or Ni-alloys are the better materials in such cases. As well as a significant improvement in the corrosion resistance of steels with a low copper content, a visible increase in the corrosion resistance of the material in sulfuric acid containing Cl^- is also detectable by increasing the Mo-content to 6.1 % [209, 210].

As the alloying elements increase, a rise in the critical pitting and crevice corrosion temperatures in chloride-containing media is to be expected, as can be seen from Table 53 and Figure 109. Above these temperatures, the particular types of corrosion mentioned can be expected in this medium.

These temperatures rise as the pitting resistance equivalent increases, according to the equation $PRE = 1 \times \%Cr + 3.3 \times \%Mo$ or, with nitrogen, according to $PRE_1 = 1 \times \%Cr + 3.3 \times \%Mo + 30 \times \%N$ [211, 212].

Figure 109 also shows the critical pitting corrosion temperature, as a function of the pitting resistance equivalent for various high-alloy steels. At the same time, the graph contains a comparison of this temperature in various test media, including a simulated flue gas condensate containing H_2SO_4 [211]. For the material SAE 316, see also Table 68.

The critical pitting corrosion temperature (CPT) and crevice corrosion temperature (CCT) can be calculated as follows for 10 % $FeCl_3 \times 6H_2O$ solution:

$$CPT \text{ in } K = 232 + 2.5 \times \%Cr + 7.6 \times \%Mo + 31.9 \times \%N \text{ and}$$
$$CCT \text{ in } K = 192 + 3.2 \times \%Cr + 7.6 \times \%Mo + 10.5 \times \%N.$$

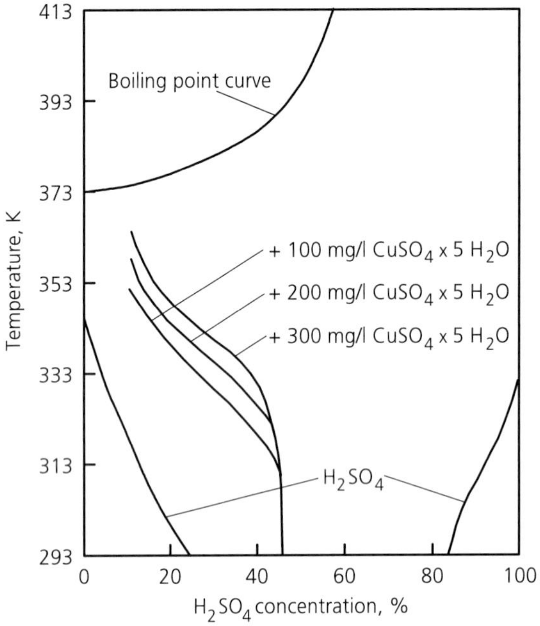

Figure 105: Isocorrosion curves (0.1 mm/a (3.94 mpy)) for the CrNiMo-steel 17 12 2.5 in sulfuric acid with and without addition of $CuSO_4 \times 5\ H_2O$ [26]

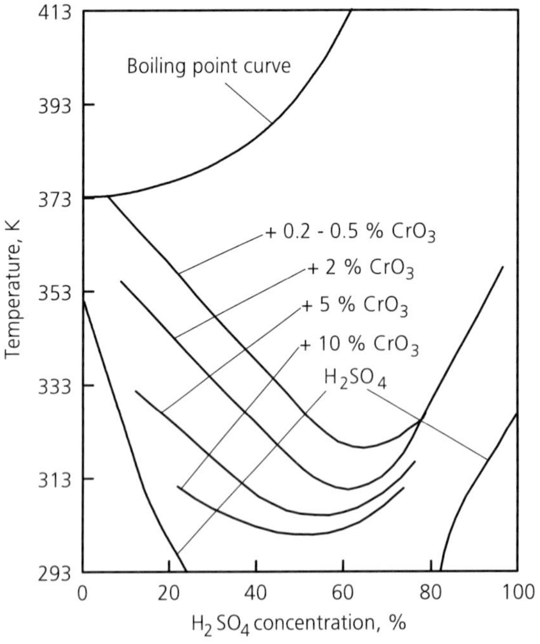

Figure 106: Isocorrosion curves (0.1 mm/a (3.94 mpy)) for the CrNiMo-steel 17 12 2.5 in sulfuric acid with and without addition of CrO_3 [26]

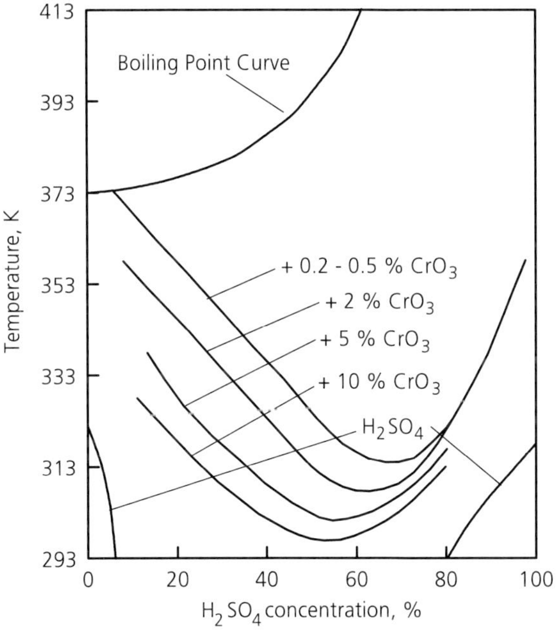

Figure 107: Isocorrosion curves (0.1 mm/a (3.94 mpy)) for the CrNi-steel 18 9 in sulfuric acid with and without addition of CrO_3 [26]

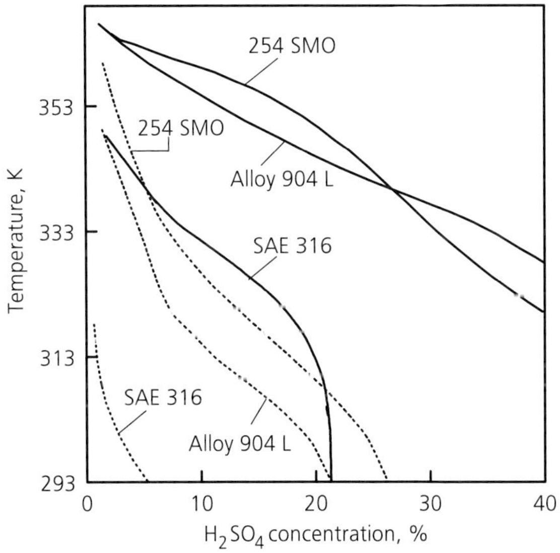

Figure 108: Comparison of the corrosion resistance of CrNiMo-steels (0.1 mm/a (3.94 mpy)) in pure and Cl^--containing sulfuric acid [209, 210]
——— pure sulfuric acid
– – – – sulfuric acid + 200 ppm Cl^-

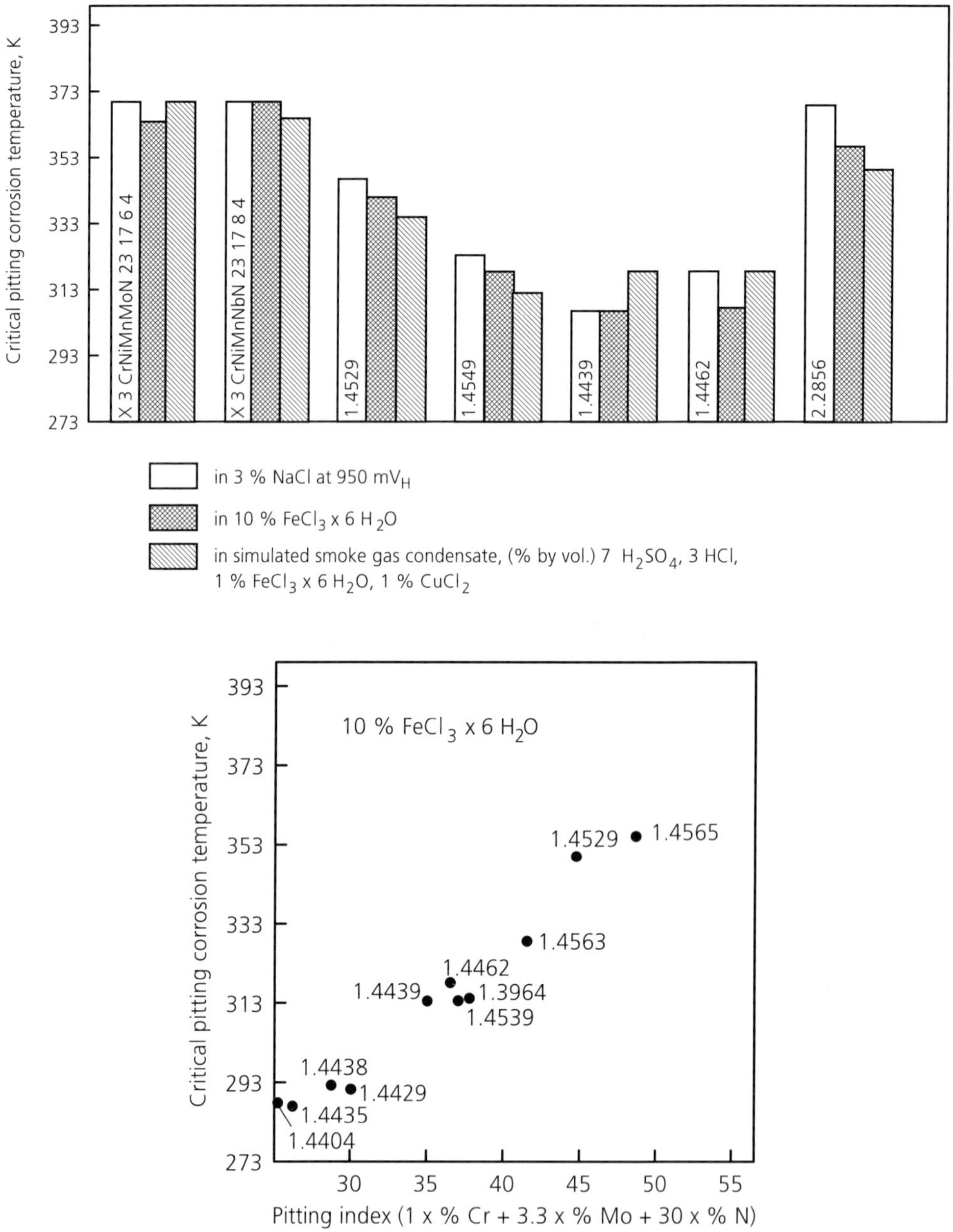

Figure 109: Comparison of the critical pitting corrosion temperatures (CPT) in chloride-containing test solutions with those in solutions containing H_2SO_4 as a function of the pitting resistance equivalent of the high-alloy steels, with nitrogen (test duration 24 h) [211]

If these temperatures are plotted in a coordinate system against the weighted pitting resistance equivalent of the alloying elements, materials alloyed in this way can be joined by a straight line. For the use of CrNiMo-steels in flue gas desulfurization solutions containing H_2SO_4, a ranking of the materials which can be used can be drawn up from the fact that there are hardly any differences between the test solutions [211, 212].

The corrosion behavior of CrNiMo-steels in sulfuric acid solutions (5, 10, 20 and 50 %) without and with chloride ion and copper ion impurities was investigated at 323, 373 and 423 K (50, 100 and 150 °C) [213, 214]. The influence of the structural conditions was also investigated, in particular the delivery condition, the effect of annealing in the laboratory and that of cold working. Table 65 summarizes the results of the immersion experiments after 21 days in a static medium in air in the delivery state [214]. The steels tested include the grades 1.4439 (X2CrNiMoN17-13-5) and 1.4571 (X6CrNiMoTi17-12-2).

A clear decrease in the corrosion resistance as the acid concentration increases at 323 K (50 °C) is to be found only with the steel 1.4571. With the exception of this steel, all the other steels are still readily usable in up to 20 % H_2SO_4. The addition of chloride ions to 20 % acid at 323 K (50 °C) led to poorly reproducible and, in some cases, high corrosion rates on the material 1.4439. The steel 1.4571 showed the same behavior in the 5 % solution of Table 65. For this reason, this steel is not used in the following corrosive media since corrosion rates of 1 mm/a (39.4 mpy) are to be expected. At 373 K (100 °C) the corrosion rates in 5 % H_2SO_4 + 500 ppm Cl^- were 3.81 (150 mpy) for 1.4439 and 4.97 mm/a (196 mpy) for 1.4571. If Cu-ions are present in the acid, even the steel 1.4571 shows a very good resistance in 20 % H_2SO_4 at 373 K (100 °C) with corrosion rates of about 0.02 mm/a (0.79 mpy) (compare also Figures 104–107). Cold working (20 %) leads to a clear deterioration in the corrosion resistance of 1.4571. Electroslag refining of the steel 1.4439 did not improve its corrosion resistance in test solutions 1 to 7 at 323 K (50 °C) and 1 to 3 at 373 K (100 °C). As in the delivery state (fully recrystallized), marked variations in the material consumption rates arise, especially if chlorides are present. The variations are not measurement errors, they are due to unstable behavior of the material, which is in passive-active transition.

A qualitative conclusion on the corrosion behavior under the conditions in Table 65, is possible from the free corrosion potential-time curves. It can be established whether a material becomes spontaneously passive, remains passive, passivates after a relatively long time, activates again after a relatively long time or does not passivate at all. A quantitative conclusion is possible only from knowledge of the anodic partial current density-potential curve of the dissolution of the metal.

Figure 110 shows the course of the free corrosion potential-time curves in 5 % H_2SO_4 (solution 1 in Table 65) at 323 K (50 °C). With a chloride content of 500 ppm, all the steels, including 1.4571, become passive after a test duration of 140 h. This is indicated by the jumps in potential in the range of U_H = about 0 mV to U_H = +500 to 650 mV. By increasing the chloride content to 3,000 ppm, the steel 1.4571, like the Ni-alloy, remains active, and the higher-alloy steels and steel 1.4439 passivate after about 200 h. The corresponding corrosion rates can be seen in Table 65.

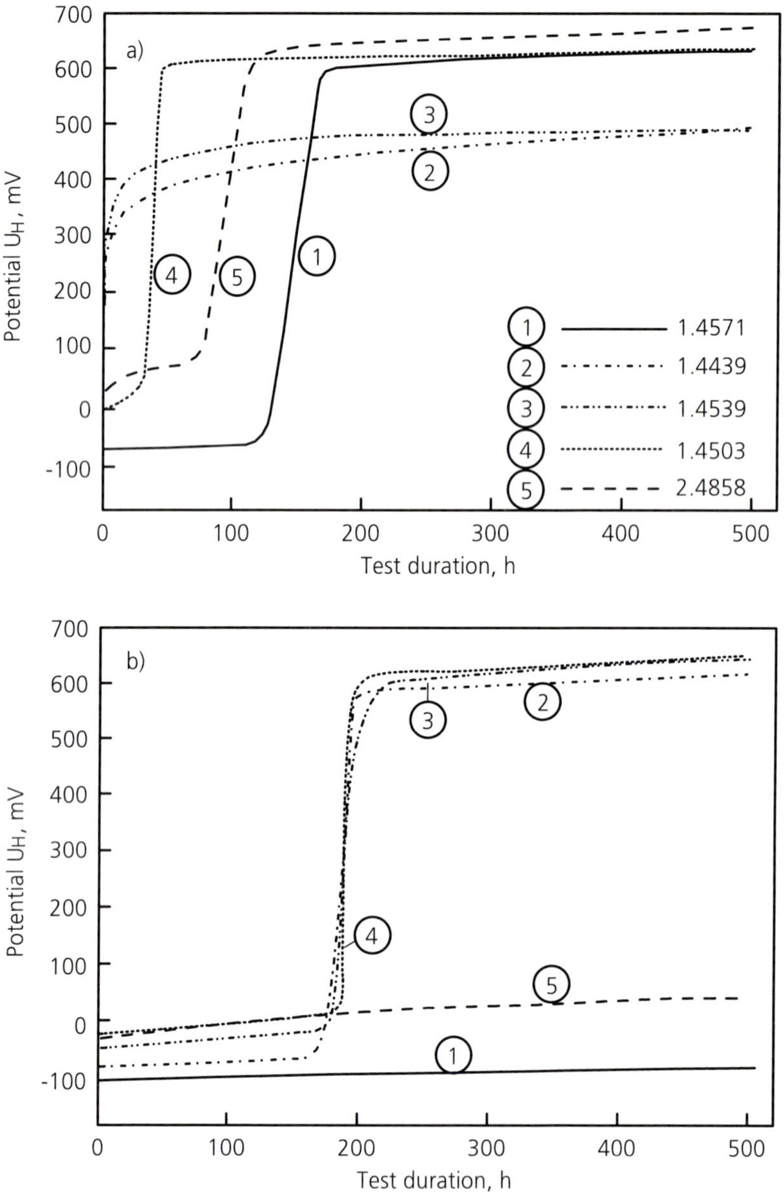

Figure 110: Free corrosion potential/time curves of CrNiMo-steels in 5 % sulfuric acid at 323 K (50 °C), static in air [214]
a) + 500 ppm Cl^-
b) + 3,000 ppm Cl^-

Solution No.	Composition	DIN-Mat. No.*	Temperature, K (°C)							
			323 (50)		373 (100)		423 (150)			
			Material consumption rate, g/m^2 d (Corrosion rate, mm/a)							
1	5 % H_2SO_4 + 500 ppm Cl^-	1.4571	0.03/12.0	(< 0.01/0.55)	109.0	(4.97)	Cl^-	Cl^-		
		1.4439	0.03	(< 0.01)	83.5	(3.81)	without	with	without	with
		1.4539	0.07	(< 0.01)	6.7	(0.30)	207.3	80.8	(9.5)	(3.7)
		1.4503	0.6/1.6	(0.03/0.07)	4.2	(0.19)	167.4	80.7	(7.6)	(3.7)
		2.4858	2.6	(0.12)	3.6	(0.15)				
2	5 % H_2SO_4 + 3,000 ppm Cl^-	1.4571	16.3	(0.74)	–	–				
		1.4439	0.03	(< 0.01)	88.5	(4.04)				
		1.4539	3.2	(0.15)	6.8	(0.31)				
		1.4503	3.4	(0.15)	5.8	(0.26)				
		2.4858	3.5	(0.16)	6.6	(0.27)				
3	10 % H_2SO_4	1.4571	6.8	(0.30)	154.	(7.02)				
		1.4439	0.03	(< 0.01)	105.3	(4.80)				
		1.4539	0.06	(< 0.01)	11.5	(0.52)				
		1.4503	0.3	(0.01)	9.5	(0.43)				
		2.4858	2.2	(0.10)	3.4	(0.14)				
4	10 % H_2SO_4 + 500 ppm Cl^-	1.4571	20.0	(0.91)	–	–				
		1.4439	0.01/5.1	(< 0.01/0.23)	–	–				
		1.4539	0.84	(0.04)	7.4	(0.34)				
		1.4503	1.4	(0.06)	2.9	(0.13)				
		2.4858	3,0	(0.14)	2.9	(0.12)				

* 1.4571 (X6CrNiMoTi17-12-2)
1.4439 (X2CrNiMoN17-13-5)
1.4539 (X1NiCrMoCu25-20-5)
1.4503 (X3NiCrCuMoTi27-23)
2.4858 (NiCr21Mo)

Table 65: Comparison of the corrosion rates in static dilute H_2SO_4 in air with and without impurities (test duration 21 days) [214]

Table 65: Continued

Solution No.	Composition	DIN-Mat. No.*	Temperature, K (°C)					
			323 (50)		373 (100)		423 (150)	
			Material consumption rate, g/m² d (Corrosion rate, mm/a)					
5	20 % H_2SO_4	1.4571	27.9	(1.27)	–	–		
		1.4439	4.4	(0.2)	–	(>> 10)		
		1.4539	0.23	(0.01)	52.7	(2.40)	763.5	(34.8)
		1.4503	0.8	(0.04)	27.5	(1.25)	871.9	(39.8)
		2.4858	2.6	(0.12)	6.5	(0.27)	1951.9	(89)
6	20 % H_2SO_4 + 500 ppm Cl^-	1.4571	90.2	(4.12)	–	–		
		1.4439	0.03/10.5	(< 0.01/0.47)	–	(>> 10)		
		1.4539	2.0	(0.09)	4.5	(0.21)		
		1.4503	1.6/2.4	(0.07/0.11)	4.3	(0.20)		
		2.4858	2.6	(0.12)	3.5	(0.14)		
7	20 % H_2SO_4 + 3,000 ppm Cl^-	1.4571	19.8	(0.9)	–	–		
		1.4439	0.03/15.0	(< 0.01/0.68)	–	–		
		1.4539	3.8	(0.17)	15.7	(0.72)		
		1.4503	3.5	(0.16)	10.2	(0.47)		
		2.4858	2.8	(0.13)	10.4	(0.43)		
8	20 % H_2SO_4 + 2,000 ppm Cl^-	1.4571	0.04	(< 0.01)	0.4	(0.018)		
		1.4439	0.02	(< 0.01)	0.5	(0.023)		
		1.4539	0.03	(< 0.01)	0.4	(0.018)		
		1.4503	0.03	(< 0.01)	0.3	(0.012)		
		2.4858	0.4	(0.02)	0.8	(0.033)		

* 1.4571 (X6CrNiMoTi17-12-2)
1.4439 (X2CrNiMoN17-13-5)
1.4539 (X1NiCrMoCu25-20-5)
1.4503 (X3NiCrCuMoTi27-23)
2.4858 (NiCr21Mo)

Table 65: Comparison of the corrosion rates in static dilute H_2SO_4 in air with and without impurities (test duration 21 days) [214]

Table 65: Continued

Solution No.	***Composition***	***DIN-Mat. No.****	***Temperature, K (°C)***					
			323 (50)		***373 (100)***		***423 (150)***	
			Material consumption rate, g/m^2 d (Corrosion rate, mm/a)					
9	50 % H_2SO_4	1.4571		(>> 10)	–	–		
		1.4439	175	(8.0)	–	–		
		1.4539	0.99	(0.04)	38.5	(1.72)		
		1.4503	1.0	(0.05)	10.8	(0.49)		
		2.4858	1.0	(0.04)	5.3	(0.26)		

* 1.4571 (X6CrNiMoTi17-12-2)
1.4439 (X2CrNiMoN17-13-5)
1.4539 (X1NiCrMoCu25-20-5)
1.4503 (X3NiCrCuMoTi27-23)
2.4858 (NiCr21Mo)

Table 65: Comparison of the corrosion rates in static dilute H_2SO_4 in air with and without impurities (test duration 21 days) [214]

Figures 111 and 112 show the corrosion behavior at 323 K (50 °C) in solutions 5 to 8 in Table 65. In the chloride-free 20 % H_2SO_4 at 323 K (50 °C), all the steels, including 1.4571 and 1.4439, become passive after a relatively short time (Figure 111a). They remain active in the chloride-containing solution (Figure 111b). Although according to these graphs in Figure 111a passivity ought to prevail in the case of the steels 1.4571 and 1.4439, the corrosion rates in Table 65 show that they remain in the active range (corrosion rates of 4.12 and 0.47 mm/a (162 and 18.5 mpy)). The relatively low corrosion rate of 1.4439 is a consequence of the low passivation current density, which allows a transition into the passive range during immersion experiments. Three out of four specimens had assumed a passive free corrosion potential (corrosion rate 0.01 mm/a (0.39 mpy)), and the free corrosion potential of a fourth specimen was in the active range (corrosion rate 0.47 mm/a (18.5 mpy)).

At higher chloride concentrations, all the materials investigated remain active (Figure 112b). The curves in Figure 112a show the corrosion-inhibiting influence of Cu-ions in 20% sulfuric acid solution. Under these conditions, all the steels remain passive. The corrosion rates are below 0.01 mm/a (0.39 mpy) at 323 K (50 °C) and do not exceed 0.03 mm/a (1.18 mpy) at 373 K (100 °C) on the steels 1.4571 and 1.4439.

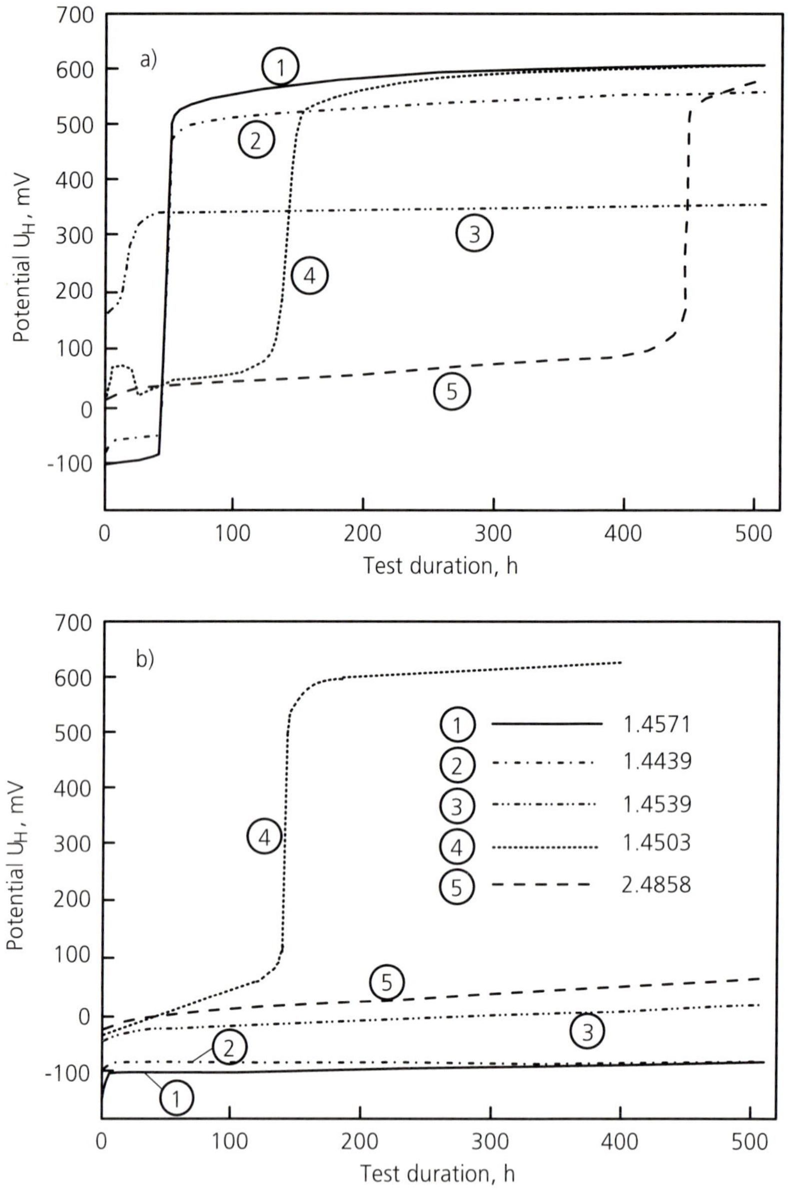

Figure 111: Free corrosion potential/time curves of CrNiMo-steels in 20% sulfuric acid at 323 K (50 °C), static in air [214]
a) without Cl^-
b) + 500 ppm Cl^-

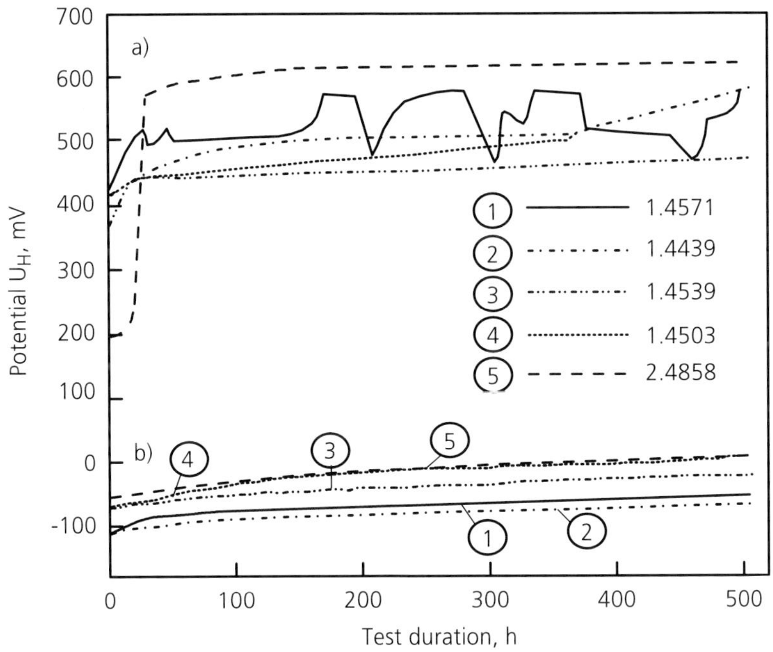

Figure 112: Free corrosion potential/time curves of CrNiMo-steels in 20% sulfuric acid at 323 K (50 °C), static in air [214]
a) + 2,000 ppm Cu^{2+}
b) + 3,000 ppm Cl^{-}

The cyclic-dynamic current density-potential curves ($\Delta U/\Delta t = 20$ mV/min) in 5% H_2SO_4 show curve shapes characteristic of passivatable metallic materials in the case of the steels 1.4571 and 1.4439. The electrochemical parameters, such as passivation current density and dissolution current density in the passive range, are the most unfavorable for the steel 1.4571 (Figure 113), especially at the high chloride level shown in Figure 113b). Figure 114a shows the relatively high passivation current density of this steel in 20% H_2SO_4 at 323 K (50 °C) in the absence of chlorides. This increases very considerably if chlorides are present, the dissolution current density being increased simultaneously (Figure 114b)). The corrosion rate in the last case is 4.12 mm/a (162 mpy) (Table 65 as a comparison). Figure 115b shows the influence of Cu^{2+} with spontaneous passivation of the steels 1.4571 and 1.4439. The cathodic partial current density is so high as a result of the reduction of Cu^{2+}, that all the steels immediately become passive, as is shown by the low corrosion rates of less than 0.01 mm/a (0.39 mpy) at 323 K (50 °C) in 20% H_2SO_4 + 2,000 ppm Cu^{2+} (Table 65).

According to these graphs, anodic protection of these steels would be possible in principle. However, this applies only to chloride-free and chloride-containing dilute sulfuric acid solutions at low temperatures.

Although the steel 1.4571 showed the least favorable corrosion behavior in these studies, the higher-alloy steels are also insufficiently corrosion-resistant in the temperature range of about 423 K (150 °C). No pitting corrosion was found on any of the steels in the studies [214].

Corrosion-inhibiting additions to sulfuric acid can improve the corrosion behavior of even the relatively low-alloy steels 1.4571 and 1.4439 to the extent that they can be used.

– Pitting corrosion in sulfuric acid and sulfuric acid solutions –

Pitting corrosion in mixed acids has already been reported above, see also Table 53. According to this table, Mo-containing CrNi-steels show a better pitting corrosion resistance than Mo-free steels. On the other hand, they are less resistant than Cu-containing steels [139]. The tendency to undergo pitting corrosion is further reduced by N-contents. An alloy (Alloy 30-C) containing (%) Fe-24Cr-20Ni-6Mo-0.44N exhibited hardly any corrosive attack on the surface after a test duration of 24 h in a solution of 11 % H_2SO_4 + 3 % HCl + 1 % $FeCl_3$ + 1 % $CuCl_2$ at boiling point. These specimens were heat-treated at 1,423 K (1,150 °C) for 1 h and then quenched in water. A specimen heat-treated at 1,473 K (1,200 °C) showed a more severe attack in this solution. Heat treatment temperatures below 1,323 K (1,050 °C) also led to an increased attack by pitting corrosion. It follows that 1,423 K (1,150 °C) is the optimum solution annealing temperature for this alloy [215].

The pitting corrosion behavior of the Mo-containing steels 1.4404 (X2CrNiMo17-12-2), 1.4571 (X6CrNiMoTi17-12-2) and 1.4449 (X3CrNiMo18-12-3) was investigated in a solution of 0.5 mol/l H_2SO_4 + 1 mol/l potassium bromide, without and with addition of 0.5 mol/l sodium nitrate, by potentiostatic holding tests. The results are compared with those from the same materials in sulfuric acid solutions of sodium chloride without and with sodium nitrate. In comparison with the Mo-free steels, the pitting corrosion potential in chloride-containing solutions is shifted considerably in the direction of more positive values by addition of molybdenum. This increase in resistance to pitting corrosion by addition of molybdenum is only slight in bromide-containing solutions. The pitting corrosion potential became more positive by only about 0.2 V by alloying 4 % Mo. Mo-contents of about 2 % proved to have an unfavorable effect on pitting corrosion behavior in bromide-containing solutions, since they considerably increased the pit density. As in chloride-containing solutions, pitting corrosion in bromide solutions can be inhibited by nitrates. The threshold potential of pitting corrosion in solutions containing bromide is considerably more positive here than in solutions containing chloride. The inhibition by nitrates in bromide-containing solutions is not influenced by Mo-contents [188].

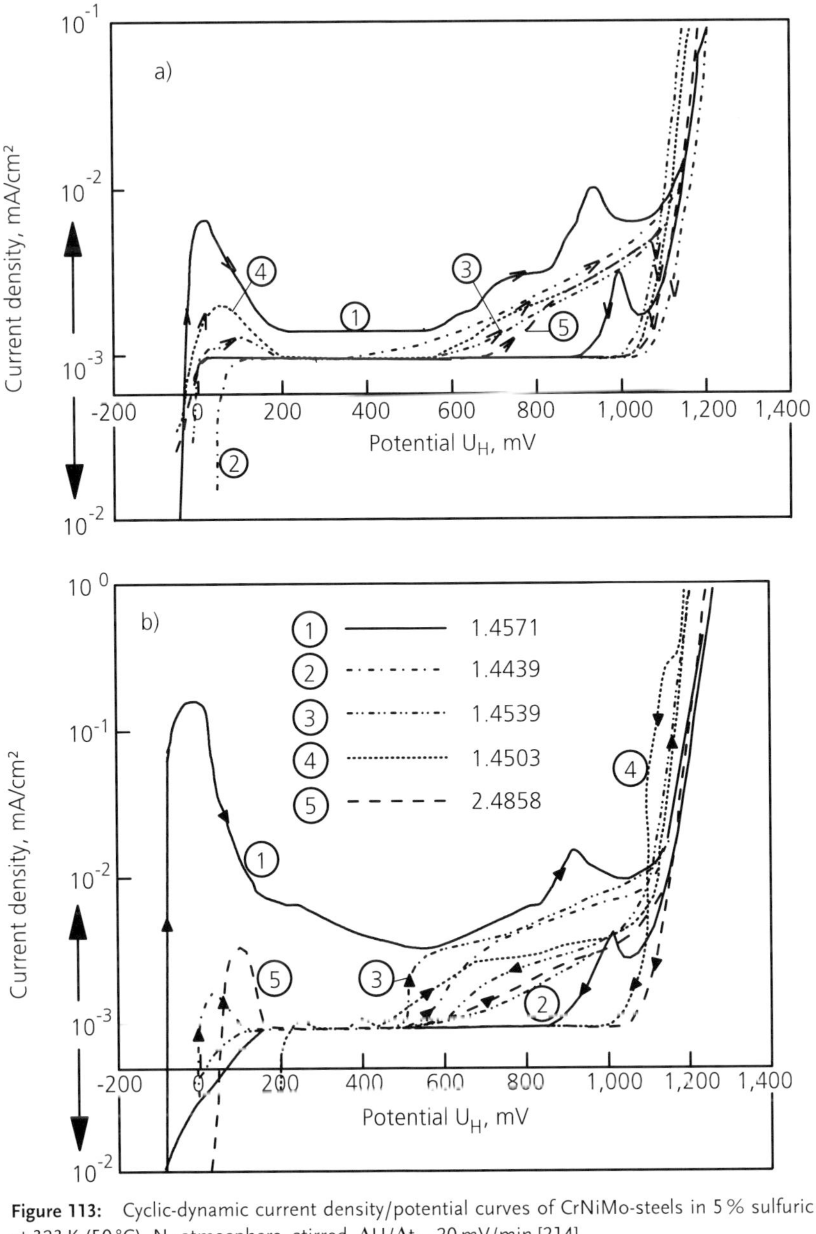

Figure 113: Cyclic-dynamic current density/potential curves of CrNiMo-steels in 5 % sulfuric acid at 323 K (50 °C), N_2 atmosphere, stirred, $\Delta U/\Delta t = 20$ mV/min [214]
a) + 500 ppm Cl^-
b) + 3,000 ppm Cl^-

– Stress corrosion cracking in sulfuric acid and sulfuric acid solutions –

According to [27], media containing sulfuric acid and sulfuric acid with chloride can cause stress corrosion cracking (SCC) on CrNi-steels without and with Mo. No SCC was found in chloride-free solutions. The same also applies to Cu-containing steels [27].

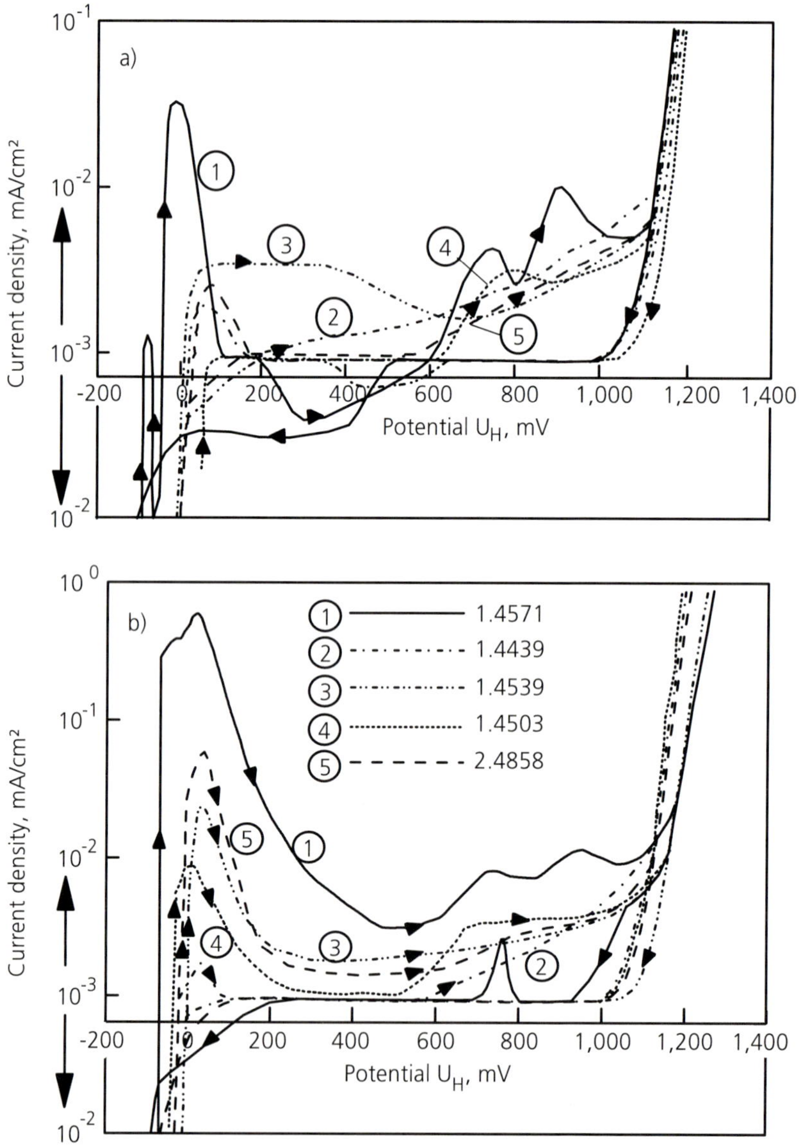

Figure 114: Cyclic-dynamic current density/potential curves of CrNiMo-steels in 20 % sulfuric acid at 323 K (50 °C), N_2-atmosphere, stirred, $\Delta U/\Delta t = 20$ mV/min [214]
a) without Cl^-
b) + 500 ppm Cl^-

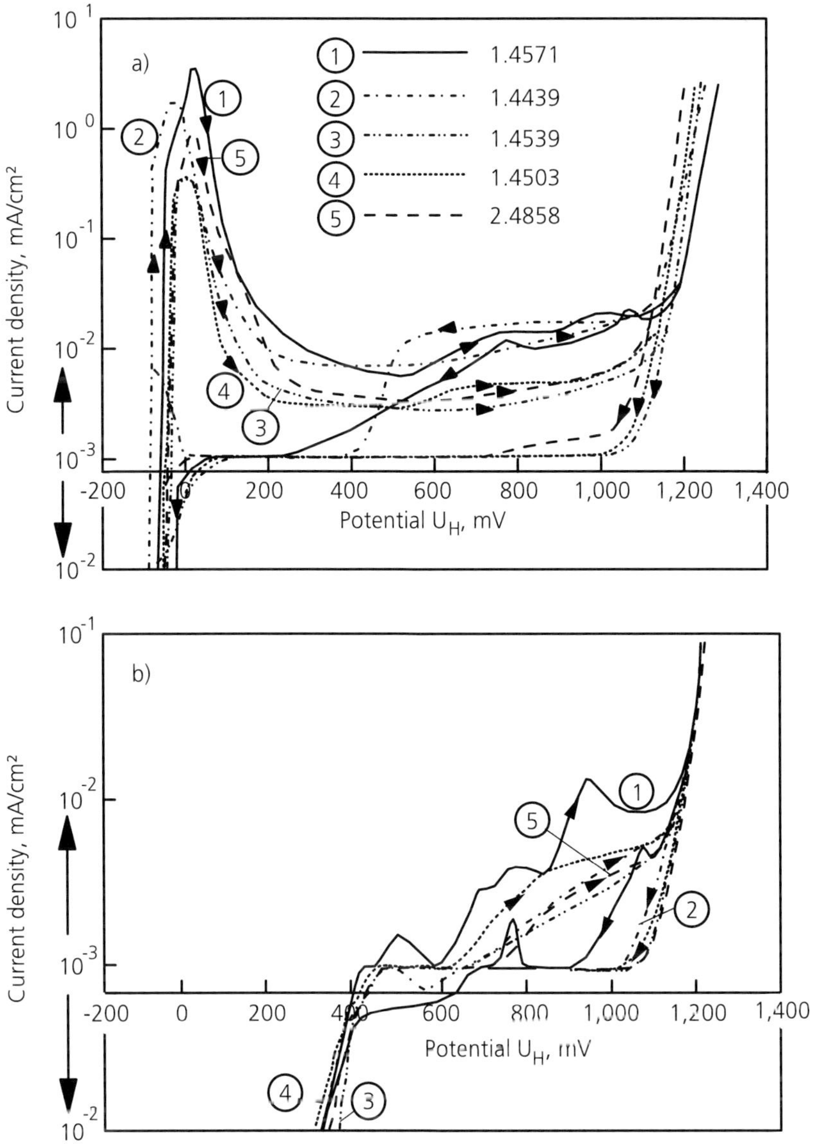

Figure 115: Cyclic-dynamic current density/potential curves of CrNiMo-steels in 20 % sulfuric acid at 323 K (50 °C), N_2-atmosphere, stirred, $\Delta U/\Delta t = 20$ mV/min [214]
a) + 3,000 ppm Cl^-
b) + 2,000 ppm Cu^{2+}

Diverse and sometimes quite corrosive conditions occur in coking plants, necessitating the use of high-alloy steels, especially for the production of ammonium sulfate from ammonia vapors, which are passed into a sulfuric acid solution. These solutions have, for example, an excess of free sulfuric acid of 3 %, a temperature of

367 to 371 K (94 to 98 °C), a pH of 1.5 and a chloride content of 1.2 g/l. The diameters of the ammonia saturation vessels vary between 4.2 and 7.6 m and the capacity between 30 and 100 m^3. One of these saturation vessels, made with the material 1.4577 (X3CrNiMoTi25-25), was found to be damaged by transcrystalline cracks in the region of the weld seam (heat-affected zone) and stress corrosion cracks on a reinforcing strut within the ammonium sulfate saturation vessel, probably caused by parting stresses during cutting on guillotine shears.

The cyclic current density/potential curves of the steels to be used (i.e. 1.4577, 1.4539 (X1NiCrMoCu25-20-5) and 1.4465 (X1CrNiMoN25-25-2)) showed in a similar medium with an increased Cl^--content of 20 g/l, with a change in potential of less than 180 mV/h, that the steel 1.4465 exhibits the better repassivation behavior under these conditions (Figure 138). The repassivation potentials are equally positive for all the steels, and in fact are more positive than the free corrosion potentials in the range between –460 and –515 mV (U_H). Tensile tests with a constant strain rate of 10^{-7} 1/s in the same medium showed a similar corrosion behavior for the steel 1.4465, which is why it is proposed for production of ammonium sulfate saturation vessels. According to the analytical results, it was more suitable than the steel 1.4539 which contains Cu (in this context, see also – Stress corrosion cracking – in Section Austenitic chromium-nickel steels with special alloying additions) [216].

The references [183 and 184] deal with the influence of hydrogen on the mechanical properties.

The steel X 2 CrNiMoN 18 14 was pretreated by solution annealing and then charged electrolytically in an electrolyte containing 5 % H_2SO_4 and 40 mg/l H_2SeO_3 using a current density of 100 mA/cm^2. Directly after charging with hydrogen, the specimens were tensile-tested at a strain rate of 5×10^{-4} 1/s. Supplementary charging experiments were performed on this material before the constant strain rate test. Notched round tensile specimens of 2.5 mm ∅ were used. At the notch, with an angle of 60°, the ∅ was 2.5 mm. The strain rate was 8×10^{-6} 1/s. Constant strain rate tests were further carried out on this material at a hydrogen pressure of 1.0 and 10.0 MPa, and at 1.0 MPa under argon. The specimens were in various states of treatment, that is to say solution-annealed and worked at 1,023 and 1,173 K (750 and 900 °C).

The embrittlement of the Mo-containing steel is not as high as that of the Mo-free steels. However, whereas the process has virtually ended in the last two steels mentioned after about 25 h, it is still increasing in the Mo-containing steel, but is significantly below that of the Mo-free steels.

The specimens tested at a strain rate of 8×10^{-6} 1/s showed a significant loss of ductility in the charged state before working, especially in the case of the specimens worked at 1,023 K (750 °C). No influence of the state of the materials (worked at a high temperature and solution-annealed) on their strength was established. No noticeable embrittlement was detectable on the specimens tensile-tested in hydrogen at a constant strain rate. The mechanical parameters of the material X 2 CrNiMoN 18 14 under argon (1.0 MPa) and hydrogen (1.0 MPa and 10.0 MPa) were about the same. The materials may be better matched to the stress in practice if the influence of hydrogen on the stress corrosion cracking properties is clarified [183].

– Corrosion fatigue in sulfuric acid –

The trials [217–219] contain studies on corrosion fatigue (CF).

The experiments were performed on notched test bars, since this shape of specimen represents more stringent testing of the effect of anodic protection. The notch shape chosen had the notch factor $\alpha_K = 2.3$. In prior studies with cathodic protection, it has been found that, in contrast to smooth round bars, the protective action decreased noticeably on notched specimens [218].

Commercially available concentrated and analytically pure sulfuric acid, with a concentration which varied between 95 and 97 %, was used as the corrosive medium. 90 % dilute acid was also used. The CF studies were carried out with dripping sulfuric acid at 353 K (80 °C). In the electrochemically controlled experiments the specimens were immersed completely in sulfuric acid. The oxygen content established corresponded to the partial pressure at 353 K (80 °C). The investigations were carried out exclusively in rotating-bend fatigue tests, at only one test frequency of 16.7 Hz 1,000 rpm, for the purpose of reducing the scope of the investigation. This stress cycle frequency provides sufficient time for attack by the corrosive medium without the cross-section being noticeably weakened by the corrosive consumption. The test duration for 10^6 stress cycles is about 17 h. During the investigations, the acid dripped directly into the notch. The tensile stress to be applied to the test bar could be determined accurately using strain measurement strips and converted to the nominal stress of the notched specimen. The cyclic-dynamic current density potential curves plotted were used to determine the anodic protection potential within the passive range of these steels.

The anodic protection potential is the same for the steels investigated and is +600 mV (U_H), as can be seen from Figure 116.

Figures 117 and 118 show the results of the investigations. The fatigue curves in Figure 117 show that the bending fatigue strength of the steels (steel 1.4571 in particular) is reduced significantly by dripping sulfuric acid at 353 K (80 °C) in relation to that in air. In this corrosion medium, they no longer have a fatigue strength, but only a corrosion resistance of limited duration. This resistance is somewhat lower for the austenitic steels in 90 % H_2SO_4 than for the semi-austenites. The corrosion fatigue limits after a stress cycle number of 10^6 are compared with the corresponding fatigue limit in air in Figure 118. The numbers on the bar charts indicate the percentage proportion reached under corroding conditions. Under these conditions, the steel 1.4571 reaches only 67 % of its corrosion fatigue limit in air in 96 % sulfuric acid at 353 K (80 °C), and thus reaches the same level as the Cu-containing steel 1.4539 and is better than the semi-austenite at only 49 %. The specimens immersed in 96 % sulfuric acid at 353 K (80 °C) with an anodic protection potential of +600 mV (U_H) show on average a lower corrosion fatigue limit than the "dripped-on" specimens. This is due to a lower oxygen supply in the solution and hence the worsened passivation conditions for the steels.

In Figure 117, the specimens which had become passivated after a relatively short time are identified by an additional circle. In the case of all the steels in this study, the freely passivated specimens have a higher corrosion fatigue limit than those

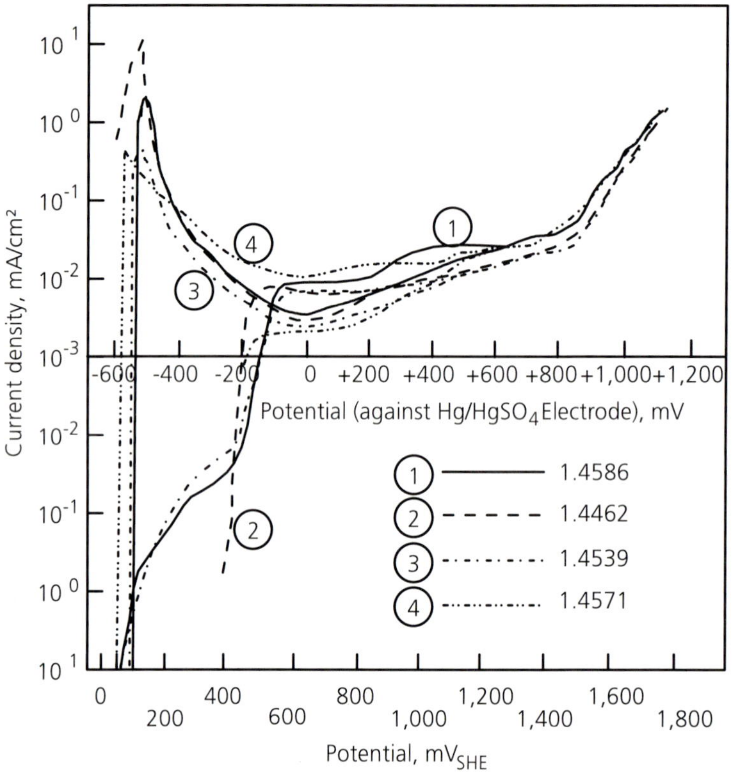

Figure 116: Cyclic-dynamic current density/potential curves of CrNiMo-steels in 90 % sulfuric acid at 353 K (80 °C), in an N_2-atmosphere, stirred, dU/dt = 1,200 mV/h [218]

specimens which remained active or alternated in cycles between the active and passive state.

The specimens tested with anodic protection – especially those under low stresses with correspondingly long test durations – show that the corrosion fatigue limit in 96 % sulfuric acid at 353 K (80 °C) is improved in as much as hardly any corroding takes place in the passive state. This significant improvement can be seen most clearly on the ferritic-austenitic steel, which is attacked the most severely under free corrosion (Figures 117, 118).

The anodic protection of high-alloy steels is not sufficient to obtain the fatigue limit which can be achieved in air. However, it helps to make the CF of these steels calculable [217, 218].

– Corrosion tests with sulfuric acid solutions –

Figure 119 shows the influence of the treatment state of the material on the corrosion behavior in the Strauss test (boiling copper sulfate-sulfuric acid solution) of the steel SAE 316 L (X2CrNiMo17-12-2). As expected, the corrosion rates are low under these corrosion conditions. This test is intended for investigations into the susceptibility of materials to intercrystalline corrosion. As can be seen from Figure 119, the corrosion resistance of the steel shows little dependence on the test duration. The difference between the corrosion rates of the heat-treatment states is only slight and is also more favorable than in the case of the other materials investigated (compare also Figures 79–82). The Mo-containing steel also exhibits a better behavior after LCF testing (see Figure 125 Figure 82). Testing under LCF conditions was carried out at 923 K (650 °C) in continuous cycles with a strain rate of 2×10^{-3} 1/s up to an overall extension of the specimens of about 1 % [187].

– *Electrochemical corrosion protection processes* –

In static 20 % sulfuric acid, the steel 10Ch17N13M2T (X6CrNiMoTi17-12-2) is attacked at a rate of only 0.005 mm/a (0.20 mpy) at 298 K (25 °C). At 323 K (50 °C) the corrosion rates rises to 0.42 (16.5 mpy), and at 353 K (80 °C) to 6.21 mm/a (244 mpy). Under these conditions, this steel is attacked considerably less than 12Ch18N10T (X6CrNiTi18-10) (see Section Austenitic chromium-nickel steels). On anodic polarization of +0.45 V, the corrosion rates fall to 0.0003 mm/a (0.012 mpy) at 298 K (25 °C), to 0.006 mm/a (0.24 mpy) at 323 K (50 °C) and to 0.11 mm/a (4.33 mpy) at 353 K (80 °C), and are therefore below the corrosion rates of the Mo-free steel. CrNiMo-steels are no longer sufficiently resistant in 20 % sulfuric acid at 353 K (80 °C). Under these conditions in aerated acid, corrosion rates of about 0.5 mm/a (19.7 mpy) are expected according to electrochemical studies.

Passivity is established on the steel 10Ch17N13M2T by stirring and the flow rate of 1 m/s generated thereby. This passivity is also subsequently retained in non-agitated acid. Brief polarization (10 min) with 0.15 V at 353 K (80 °C) generates this passivity artificially. This pretreatment is recommended for pipes from this material in tubular heat exchangers [145]. In-house preliminary studies are advisable before use, since the corrosion rate is not zero in the passive range. A corrosion rate of 0.1 mm/a (3.94 mpy) is quoted in this instance.

Anodically protected CrNi-steels can be used in 67 to 90 % H_2SO_4 up to 413 K (140 °C). In 67 % acid at 368 K (95 °C), a current density slightly below 10^{-6} A/cm is established in the passive range between +100 and +500 mV_{cal} after a passivation current density of 10^{-2} A/cm^2 [92]. This corresponds to a corrosion rate of about 0.01 mm/a (0.39 mpy). The steel 1.4571 (X6CrNiMoTi17-12-2) is particularly suitable for anodic corrosion protection in concentrated sulfuric acid at high temperatures, and is employed under these conditions with considerably reduced corrosion rates [95, 102, 168, 220–222]. For anodic corrosion protection, see also Section Austenitic chromium-nickel steels and Figure 96.

It may be anticipated at this point that from studies [168], CrNiMo-steels can be better protected in hot concentrated H_2SO_4 than the Mo-free grades. Mo-containing CrNi-steels could also be given outstanding protection according to [102].

As already described in detail in Section Austenitic chromium-nickel steels, a change occurs in the corrosion behavior of CrNi-steels in hot H_2SO_4 (from about 343 K (70 °C)) at concentrations above 90 %. They can be in either the active or the passive state.

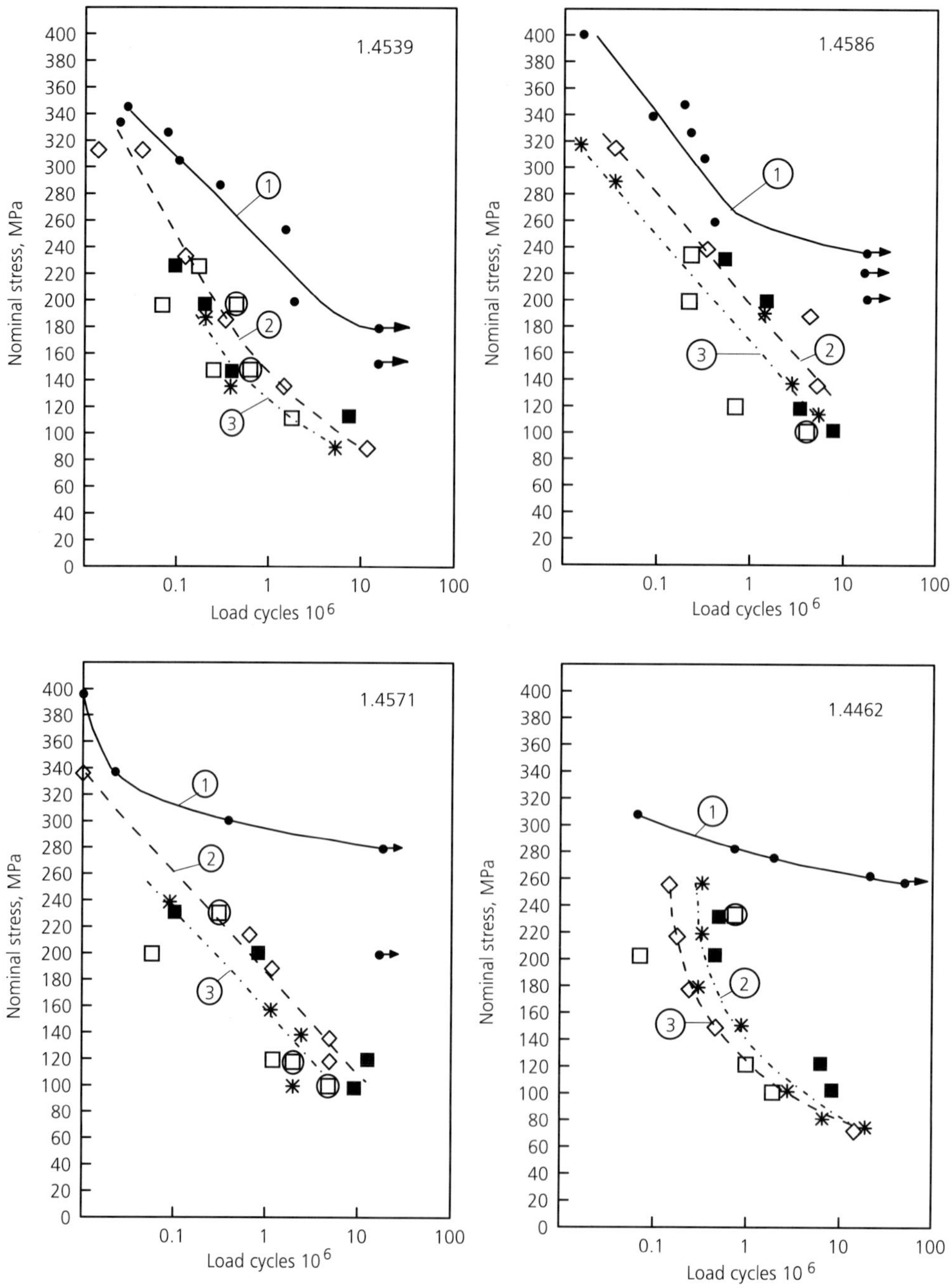

Figure 117: Fatigue curves for the corrosion fatigue cracking of steels in 96 and 90% sulfuric acid at 353 K (80 °C) using notched specimens $\alpha_K = 2.3$ (see also Figure 116) [217, 218]
Test media:
① air
② H_2SO_4, 96%, 353 K (80 °C), dripped
③ H_2SO_4, 90%, dripped
□ H_2SO_4, 96%, immersed
■ H_2SO_4, 96%, anodically protected
○ test bar, passive corrosion potential

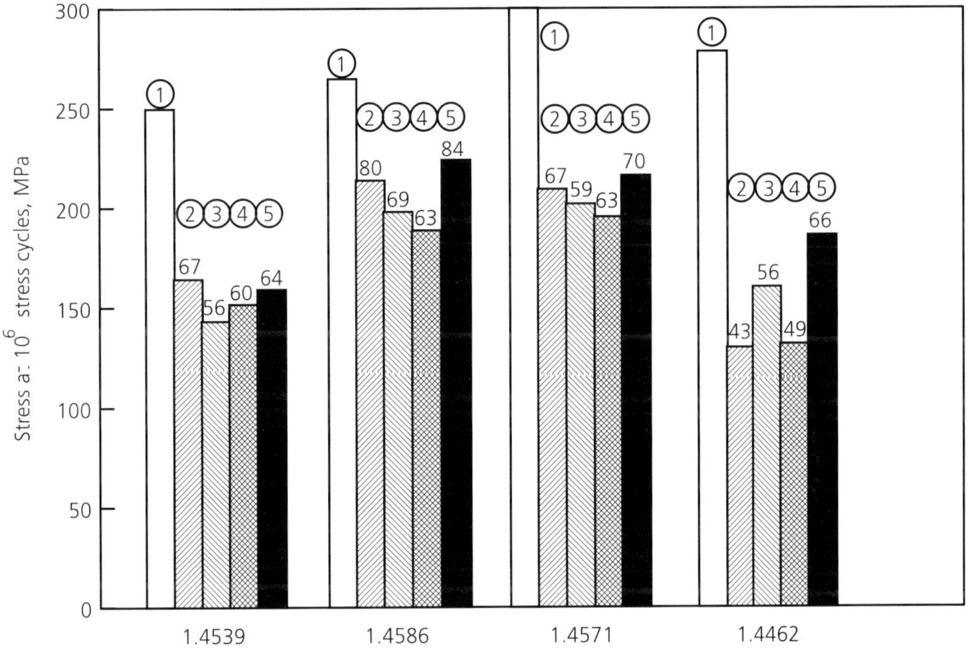

Figure 118: Fatigue limit of various steels in the rotating-bend fatigue test for 10^6 stress cycles in air and under corrosion in concentrated sulfuric acid at 353 K (80 °C) (numbers on the columns = % of the value in air) [217, 218]
① in air
② in 96 % H_2SO_4, dripped
③ in 90 % H_2SO_4, dripped
④ in 96 % H_2SO_4, immersed
⑤ in 96 % H_2SO_4, anodically protected

Weight loss determinations on the steel 1.4571 (X6CrNiMoTi17-12-2) have shown that in static and flowing sulfuric acid at the free corrosion potential in the concentration range from 67 to 98 % at temperatures above 353 K (80 °C), the corrosion rates decrease as the acid concentration increases and increase as the acid temperature increases. Practical experience, such as the information in Figure 72, also confirms that this steel can no longer be used with free corrosion under these conditions.

The corrosion rates in the temperature range from 373 to 413 K (100 to 140 °C) at sulfuric acid concentrations between 90 and 98 % depend on the potential applied. They fall with an increasingly positive potential, as can be seen from Figures 120 and 121. Above potentials of $U_H > +1{,}000$ mV, the corrosion rates are so low that corrosion damage is no longer to be expected. The corrosion rate falls with the acid concentration. A comparison of Figures 86 and 87 for the Mo-free steel with Figures 120 and 121 for the Mo-containing steel shows that, according to these studies, the Mo-free steel exhibits the more favorable corrosion behavior in slightly agitated sulfuric acid at a high concentration and temperature with anodic protection.

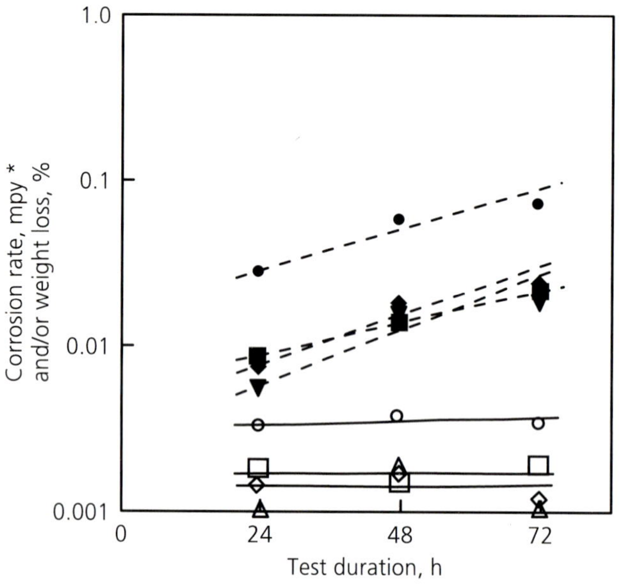

* 1 mpy = 0.0254 mm/a

Figure 119: Comparison of the corrosion behavior of the steel SAE 316 L (X2CrNiMo17-12-2) as a function of the treatment state in the Strauss test [187]
Values in:

mpy	%	
□	■	1,373 K (1,100 °C)/2 h/water
◇	◆	delivery state
△	▼	923 K (650 °C)/3 h/air
○	●	after LCF testing at 923 K (650 °C)

Figure 122 shows the results of weight loss measurements in flowing 98 % sulfuric acid (1 m/s). This graph shows that at potentials of U_H +900 mV, corrosion rates of below 0.1 mm/a (3.94 mpy) can be achieved. The optimum protection potential for these conditions is U_H = +1,350 mV with linear corrosion rates of below 0.01 mm/a (0.39 mpy). According to [168], the material 1.4541 (X6CrNiTi18-10) has a better corrosion behavior under the same test conditions than the Mo-containing steel 1.4571. The corrosion rates are lower by one power of ten.

Figure 88 shows a comparison of the two materials in static and flowing 98 % sulfuric acid at 413 K (140 °C). According to these studies [168], the Mo-free steel shows the better corrosion behavior with anodic protection. Table 54 contains electrochemical data from the anodic corrosion protection of steels, with the example of 1.4571, at 373 K (100 °C) [102].

The effect of anodic corrosion protection in concentrated sulfuric acid at a high temperature is summarized in Figure 123 for a steel containing 18 % Cr, 10 % Ni and 2 % Mo. As a result of this protectional measure, the steel can be regarded as being practically resistant in hot concentrated sulfuric acid. Whilst without this precaution, it would be destroyed with corrosion rates far in excess of 1 mm/a (39.4 mpy) [221].

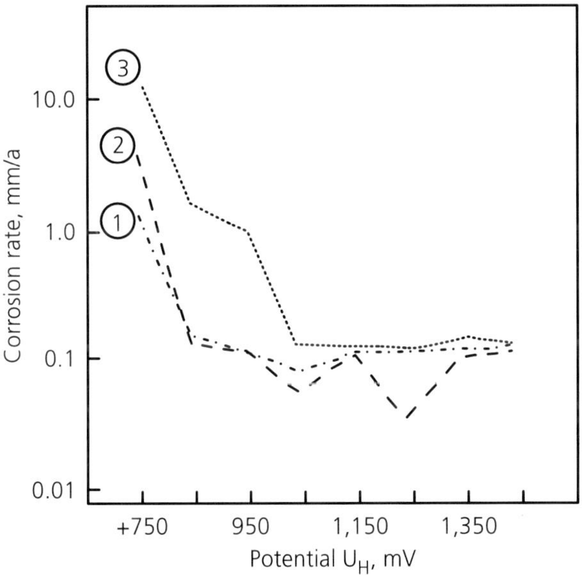

Figure 120: Corrosion rate of the CrNiMo-steel X6CrNiMoTi17-12-2 (1.4571, SAE 316 Ti) in aerated and gently agitated 90 % sulfuric acid as a function of the potential and temperature (test duration 5.3 d) [168]
① 373 K (100 °C)
② 393 K (120 °C)
③ 413 K (140 °C)

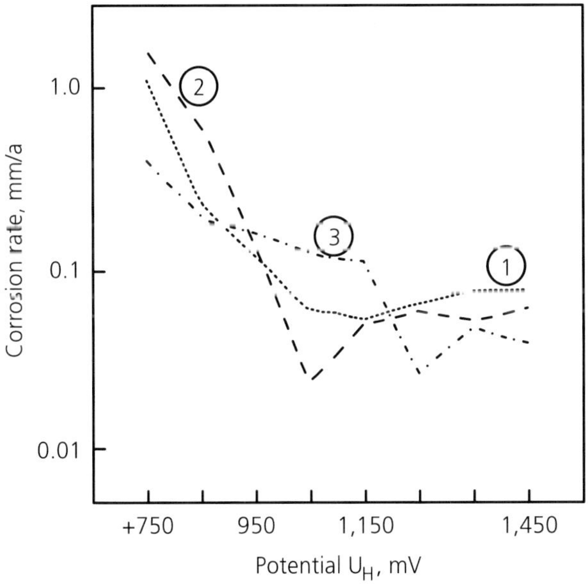

Figure 121: Corrosion rate of the CrNiMo-steel X6CrNiMoTi17-12-2 (1.4571, SAE 316 Ti) in aerated and gently agitated 97 to 98 % sulfuric acid as a function of the potential and temperature (test duration 5.3 d) [168]
① 373 K (100 °C)
② 393 K (120 °C)
③ 413 K (140 °C)

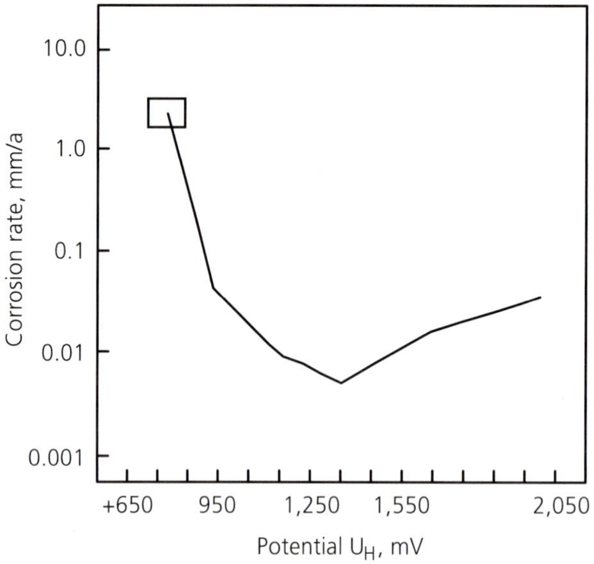

Figure 122: Corrosion rate of the CrNiMo-steel X6CrNiMoTi17-12-2 (1.4571, SAE 316 Ti) in flowing (1 m/s) 97 to 98 % sulfuric acid at 393 K (120 °C) as a function of the potential (test duration 10 d) [168]
– with anodic protection
□ without protection

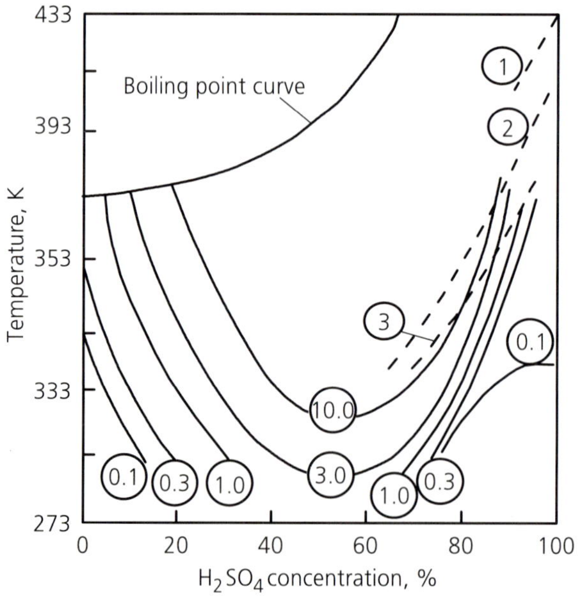

Figure 123: Isocorrosion curves of a CrNiMo-steel containing 18 % Cr, 10 % Ni and 2 % Mo in various concentrations of sulfuric acid with 1), 2), 3) and without anodic corrosion protection [221]
① 0.05 mm/a (1.97 mpy)
② 0.03 mm/a (1.18 mpy)
③ 0.01 mm/a (0.39 mpy)

Under conditions of corrosion fatigue (CF), anodic protection is not sufficient to achieve the fatigue limit in the case of the steel 1.4571 (see CF in this section). However, it helps to make the fatigue limit of the steels calculable on exposure to sulfuric acid (see also Figures 117 and 118) [217, 218]. A reliable passive behavior of the material is ensured by anodic protection, whereas without this protection measure it could be in either the passive or the active state. For cathodic corrosion protection, see Section Austenitic chromium-nickel steels.

– Fields of use and failure –

In 98.3 % sulfuric acid at a relatively low temperature, increased local corrosive attack leading to failure of plant components may occur as a consequence of downtime corrosion [65].

A remedy for severe sulfuric acid corrosion can be provided by using plastics, in particular fluoropolymers [223], or ceramic linings [224].

In the production of tubes for sausage casings by the viscose process, sodium cellulose xanthogenate (viscose) of a certain viscosity is forced through an annular die of a gut spinneret into a spinning bath which gently flows around the gut spinneret. The viscose is decomposed to hydrated cellulose and thus forms the sausage casing. The spinning bath consists of 235 to 245 g/l sulfuric acid, 240 to 245 g/l sodium sulfate, 115 to 120 g/l ammonium sulfate and 60 to 120 mg/l chlorides. The spinning bath solution was constantly circulated. The spinneret was 400 mm below the spinning bath surface, and the temperature was 303 K (30 °C), rising to a maximum of 313 K (40 °C). Sodium sulfate and hydrogen sulfide as well as small amounts of carbon disulfide are formed during the reaction of the viscose with the spinning bath. The spinnerets of X 8 CrNiMoTi 18 11 were cleaned irregularly during downtimes. The corrosion damage was due to structural heterogeneities and arose mainly as selective corrosion with crevice corrosion.

Gut spinnerets produced from the ultra-pure steel UR X3CrNiMo18-12-3 showed no corrosive attack after an operating time of 2 years. As well as this steel, higher-alloy Cu-containing steels or the alloy NiMo 30 have been proposed for possible use [225].

Adequate conditions for the appearance of different corrosion types, in the absence of mechanical stresses, are found, in the presence of electrolytes, in the region of the weld seam where differences in potential occur. An identical potential of the base material, welding material and heat-affected zone is desirable, but is difficult to achieve and can scarcely be realized without heat treatment of the weld seam. In pipes of X6CrNiMoTi17-12-2, welded in the same manner, in 30 % oleum in the temperature range between 308 and 423 K (35 and 150 °C), selective corrosion occurred in the weld seam after an operating time of 9 months, and also in the base material of a pipe bend. This is an indication that in highly corrosive media small differences in potential are sufficient to cause intense corrosion. In a nobler weld seam, a small cathode would be opposed by a relatively large anode. The corrosion current densities which develop under such conditions and the resulting corrosion rates are unacceptable in practice [226].

In a pipeline of 1.4571 conveying 96 % sulfuric acid, the weld seams of 1.4430 displayed local corrosion at room temperature in the region of the heat-affected zone. This can be prevented by using a higher-alloy filler material low in Mo- and C-content.

The formation of dilute sulfuric acid at digester circulating pumps in the paper industry from “sulfurous acid” and SO_2 at temperatures of 413 to 443 K (140 to 170 °C) may cause very intense corrosion in designs from CrNiMo-steel. A flue gas channel of 1.4571 (X6CrNiMoTi17-12-2) which was exposed to 0.3 % SO_2 and unknown quantities of SO_3 and sulfuric acid showed severe trough-shaped corrosive attack at the weld seams and on the free sheet metal surfaces after an operating time of 300 h, leading to rupture of the walls. The causes of this were the sulfuric acid condensates which formed at a concentration of 10 to 50 % when the temperature fell below the dew point [227].

The processability of sewage sludge is improved by dewatering. Its solids content is enriched to about 20 % by mechanical dewatering. It is heated up to 373 K (100 °C) in a reaction tank made of SAE 316 L (X2CrNiMo17-12-2) and sulfuric acid added (1.83 g H_2SO_4 per g solid). To keep corrosion of the material low, the sulfuric acid is partly neutralized by addition of ammonia. Corrosion rates of 1.15 mm/a (45.3 mpy) have been observed at a pH of about 0.6. At pH 1.6, the corrosion rate dropped to 0.002 mm/a (0.08 mpy). SAE 316 L can thus be used as a construction material at pH values of 1.6 or more [228].

Austenitic chromium-nickel steels with special alloying additions

Copper alone improves the corrosion resistance of CrNi-steels in cold sulfuric acid, as can be seen from the graph in Figure 66. The effect of copper is even more pronounced if molybdenum is present in the steel. Under these conditions, the resistance range at low concentrations in particular is extended. A simultaneous rise in the nickel content increases the corrosion resistance to the extent that the resistance gap in the range between about 30 and 80 % H_2SO_4 is closed, and a maximum corrosion rate of 0.1 mm/a (3.94 mpy) at 313 K (40 °C) can be expected in the entire concentration range of sulfuric acid [9, 63]. See also Figures 70, 95, 96 and 99 for the effect of Cu-additions on the resistance of CrNi-steels to sulfuric acid.

For the use ranges and resistance of the Cu-containing steels, see also Figures 15, 16, 45, 47, 48, 57, 65, 66, 70, 95, 96, 99, 101–103, 108–118 and Tables 6, 8 column 9, 34 (1.4505), 37, 39, 43, 44, 46, 48, 49, 53, 54, 63–65 and 68.

As well as high-silicon cast iron, steels with a Cu-, Mo- and increased Ni-content are amongst the most resistant ferrous materials in sulfuric acid.

The increase in the corrosion resistance as the Ni-content increases or as the alloying content itself increases can furthermore be seen from the isocorrosion curves in Figures 124 and 125.

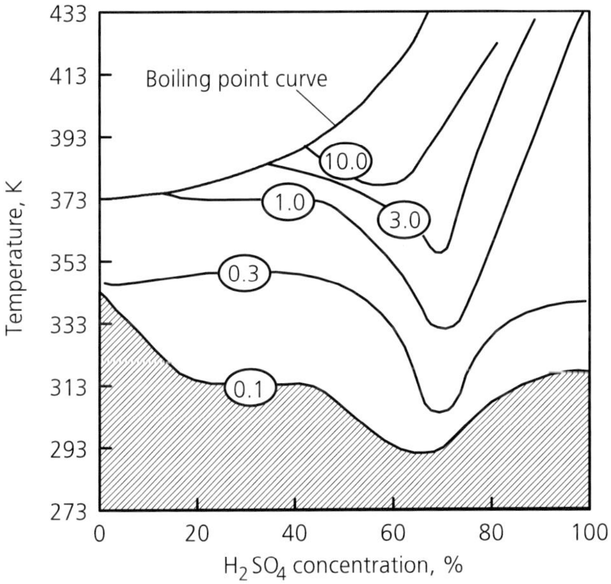

Figure 124: Isocorrosion curves (mm/a) of a CrNiMoCu-steel ((%) Fe-0.05C-18Cr-18Ni-2Mo-2Cu, Nb) in various concentrations of sulfuric acid [61]

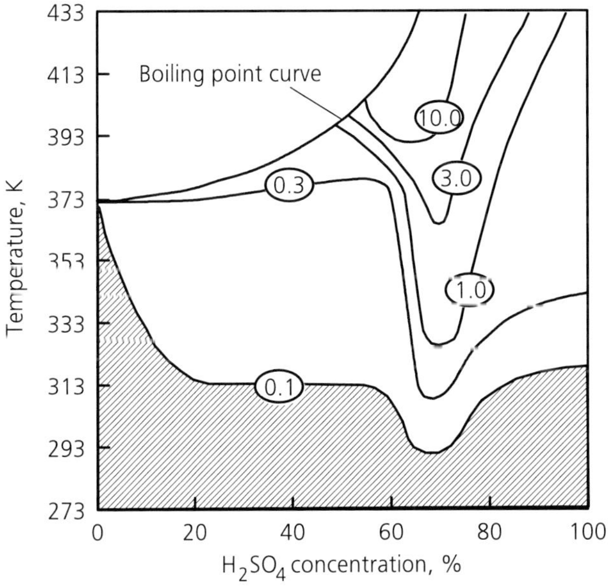

Figure 125: Isocorrosion curves (mm/a) of a CrNiMoCu-steel ((%) Fe-0.05C-25Ni-20Cr-3Mo-2Cu) in various concentrations of sulfuric acid [61]

The region up to a maximum corrosion rate of 0.3 mm/a (11.8 mpy) is raised significantly to 373 K (100 °C) in Figure 125 in comparison with Figure 124. The resistance gap is pushed back to the relatively narrow range between 60 and 80 % H_2SO_4 (compare also Figures 69, 97, 98) [61].

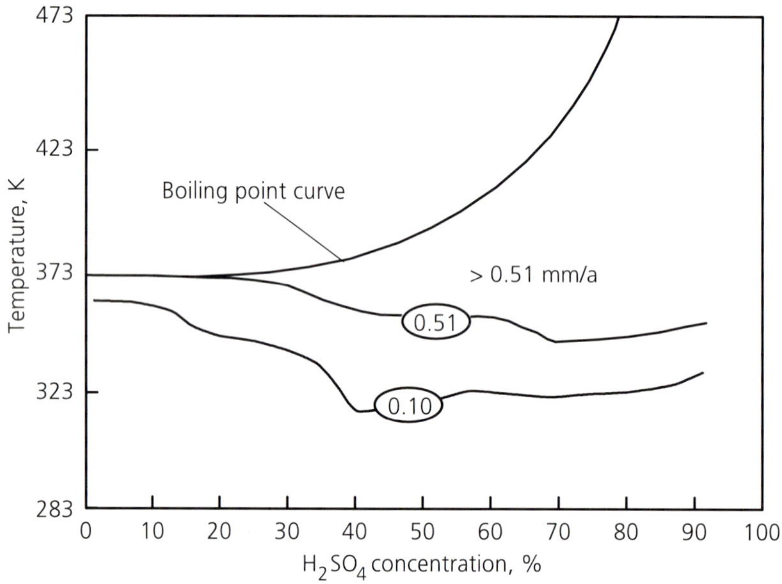

Figure 126: Isocorrosion curves (mm/a) and corrosion ranges of the steel 20Cb-3 (cf. 2.4660, NiCr20CuMo) in sulfuric acid of various concentrations (test duration 168 h) [229, 230]

By comparison, Figure 126 shows an even higher-alloy steel, the material 20Cb-3 (cf. 2.4660, NiCr20CuMo). In this graph, the resistance limit (0.1 mm/a (3.94 mpy)) is raised even more and the resistance gap has almost disappeared. However, the corrosion resistance at temperatures above 323 K (50 °C) is still not sufficient for long-term exposure, and the range in which corrosion rates of between 0.1 and 0.5 mm/a (3.94 and 19.7 mpy) are to be expected is only relatively narrow, so that corrosion rates of more than 0.5 mm/a (19.7 mpy) above about 323 K (50 °C) are still to be expected with this material [229, 230]. In addition to this high resistance to sulfuric acid, the high-alloy steels also have a substantially improved general resistance to other media and mixtures. In solutions containing chloride, they also have an increased resistance to pitting and crevice corrosion due to the raised content of Cr, Mo and Ni. Although there is no increased stress corrosion cracking resistance according to the $MgCl_2$ test, which is dubious in practice, they have an increased service life in comparison with the CrNiMo-steels 18 10. 1.5 to 2 % Cu is alloyed with the steel 1.4539 (X1NiCrMoCu25-20-5) containing 4.7 % Mo in order to increase its resistance towards reducing media, such as sulfuric acid and H_3PO_4. The adequate resistance is achieved by lowering the carbon content to below 0.025 % C. This results in hardly any problems during welding if the material is in the solution-annealed state. Higher Mo-contents of about 6 % or more lead (e.g. in this steel containing 25 % Ni and 20 % Cr) to precipitation of the χ-phase, to an associated poorer processability and, which is particularly important here, to a poorer corrosion resistance. This can be counteracted, on the other hand, by increasing the nitrogen content (see Figure 108). The following corrosion rates in Table 66 were determined for the steel 1.4539 in laboratory studies.

H_2SO_4 concentration %	*Temperature* K (°C)	*Corrosion rate* mm/a (mpy)
3	boiling temp.	0.13 (5.12)
30	333 (60)	0.38 (15.0)
50	353 (80)	0.26 (10.2)
60	358 (85)	1.52 (59.8)

Table 66: Corrosion behavior of 1.4539 in various concentrations of H_2SO_4 at various temperatures [231]

For the steel 1.4503 (X3NiCrCuMoTi27-23) or the Ni-base alloy 2.4858 (NiCr21Mo), corrosion rates of 0.36 mm/a (14.2 mpy) were measured in 80 % H_2SO_4 at 323 K (50 °C), and rates of 0.05 mm/a (1.97 mpy) in an aqueous solution of 140 g/l H_2SO_4 + 1,000 ppm Cl^- + 30 g/l Fe^{3+} at 353 K (80 °C) [231].

As the alloying contents increase, the pitting resistance equivalence (% Cr + 3.3 % Mo) of the materials, which is a criterion of their resistance to pitting and crevice corrosion, increases. The nitrogen content is also taken into account in the pitting resistance equivalence, although it is not present in all materials. The resistance of the materials to these types of corrosion is determined in $FeCl_3$ solution. The pitting resistance equivalence indicates a ranking of the resistance only in this medium, but if trials are to be performed, it is also a measure of the suitability or ranking of the materials in other media.

The steel Nicrofer® 3127 hMo (X1NiCrMoCu32-28-7, 1.4562), containing 31 % Ni, 27 % Cr, 6.5 % Mo, 1.2 % Cu, 0.20 % N and a maximum of 0.015 % C, displays an outstanding corrosion resistance to oxidizing media as well as to pitting and crevice corrosion in aqueous neutral and acid solutions. It is significantly superior to the known steels containing 6 % Mo, such as, for example, 1.4529 (X1NiCrMoCuN25-20-7) and the Ni-base alloy 2.4619 (NiCr22Mo7Cu), and is suitable for use in flue gas desulfurization plants and for concentration of dilute acids. The pitting resistance equivalence of this material is calculated from the above formula at the very high value of 47, or 53 if the nitrogen content is included. The addition of 1.2 % Cu, moreover, ensures a good resistance in reducing acids [232].

For the use of the steel 1.4539 (SAE 904 L) in sulfuric acid, see also [233–237] and Table 8, column 9. This steel, which was originally developed for the sulfuric acid industry, also exhibits the high corrosion resistance just described in other media. Figure 57 shows the isocorrosion curves for corrosion rates of 0.1 mm/a (3.94 mpy) in comparison with other steels. According to this graph, the steel CN-7M, which has a higher Ni-content and contains 29 % Ni, 20 % Cr, 3.5 % Mo and 2.5 % Cu, is even more suitable for use in sulfuric acid. Figure 127 is a graph of the resistance ranges of this material [238].

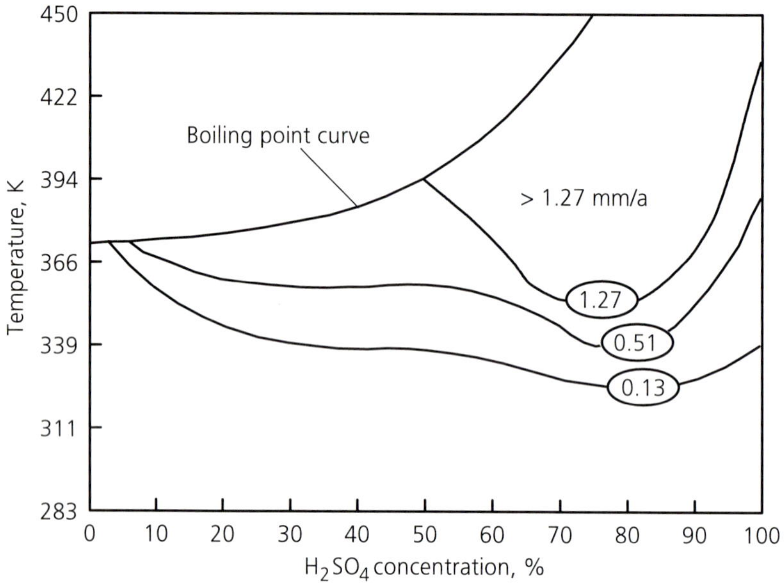

Figure 127: Isocorrosion curves (mm/a) and corrosion regions of the steels UNS N08700 (CN-7M) containing 29 % Ni, 20 % Cr, 3.5 % Mo and 2.5 % Cu in various concentrations of sulfuric acid [238]

Figures 128 and 129 show the corrosion resistance of the high-alloy steels 1.4539 (X1NiCrMoCu25-20-5) and 1.4563 (X1NiCrMoCu31-27-4) with additional isocorrosion curves with corrosion rates of 0.3 mm/a (11.8 mpy) [130]. The higher-alloy steel 1.4563 exhibits a visibly better resistance in sulfuric acid at higher temperatures, although it contains less molybdenum (3.5 %) and Cu (1.0 %) in comparison with 1.4539, which contains 4.5 and 1.5 % respectively [130]. Both steels were developed for use in the sulfuric acid industry.

Table 39 contains corrosion rates for the steel 06ChN28MDTL in various concentrations of sulfuric acid at 333 and 353 K (60 and 80 °C). The results are from test durations of 100 h [107].

Corrosion rates for the steel 1.4505 (X4NiCrMoCuNb20-18-2) in sulfuric acid and sulfuric acid mixtures are summarized in Table 34.

The corrosion behavior of various metallic materials in chamber acid (75.3 to 75.9 % H_2SO_4 + 0.035 to 0.045 % HNO_3) at various temperatures and flow rates is shown in Figure 15 for laboratory studies, and in Table 6 for field studies. Under these conditions, sharply increasing corrosion rates are recorded for the steels dealt with in this section (06ChN28MDT) at elevated temperatures, which means they have only limited use [23].

Corrosion studies in flowing media, simulated by means of a rotating disk, a rotating cylinder or a pipe through which the media flow, cannot simulate the conditions in pumps. It was found, in particular in pumps exposed to sulfuric acid, that considerable corrosion and erosion damage occurred under operating conditions if

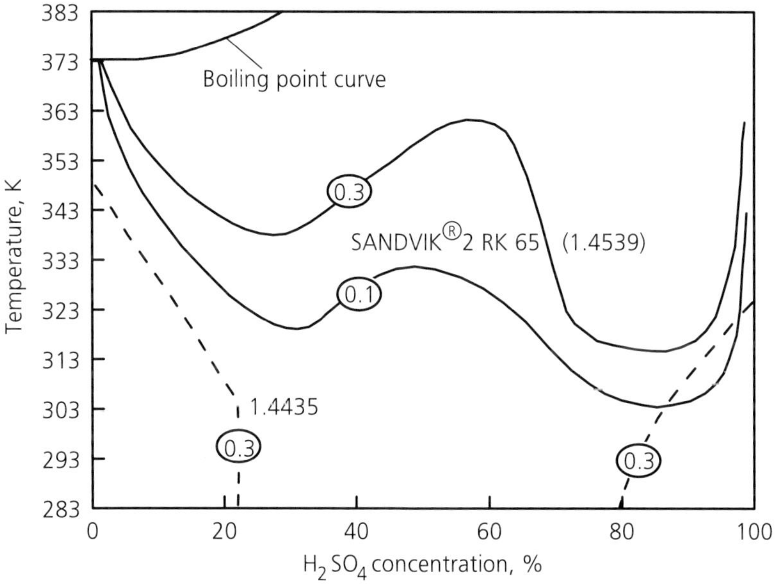

Figure 128: Isocorrosion curves (mm/a) of the CrNiMoCu-steel 1.4539 in various concentrations of sulfuric acid in comparison with the Cu-free steel 1.4435 [130]

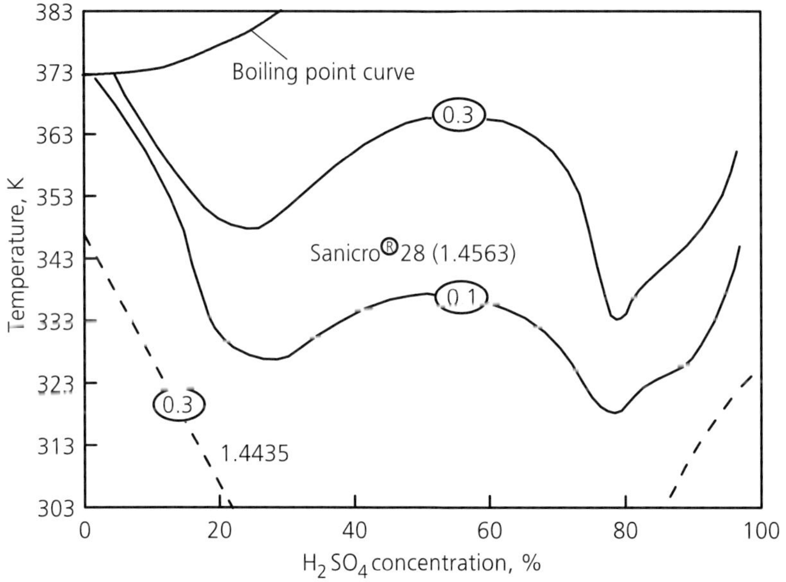

Figure 129: Isocorrosion curves (mm/a) of the CrNiMoCu-steel 1.4563 in various concentrations of sulfuric acid in comparison with the Cu-free steel 1.4435 [130]

the choice of material was based on results from studies using a rotating disk or pipe. These discrepancies are due to pump-specific flow conditions which are characterized by locally high flow rates as a result of separations and turbulences. Extreme conditions exist in particular in pumping media containing abrasive solids. Materials which form surface coatings are particularly at risk here. Pump impellers made of 1.4500 (GX7NiCrMoCuNb25-20) were completely destroyed by 25 % sulfuric acid at 318 K (45 °C) at a speed of rotation of 1,450 rpm after an operating time of 1,000 h, whereas the ferritic-austenitic steel G-X 3 CrNiMoCuN 24 6 showed no attack under the same conditions (see also Section Ferritic-austenitic steels with more than 12 % chromium). This was followed by corrosion studies under conditions close to those in practice on, amongst others, the steel 1.4500 with the following chemical composition: 0.015 % C, 19.12 % Cr, 24.73 % Ni, 3.29 % Mo, 1.97 % Cu, 0.27 % Nb, 1.12 % Si, 1.22 % Mn, 0.012 % P, 0.006 % S, balance Fe.
Orientating studies were performed in various concentrations of static H_2SO_4 flushed with air. These demonstrated that the corrosion rate can be very high at the start of the experiment, as a consequence of passivation occurring after a longer period of time. On the basis of this, the exposure times were standardized at 21 days. The corrosion data were determined from 5 specimens each time. Figure 130 shows the results of the continuous immersion test. The peculiarities shown in Figure 131 were found in 98.5 % sulfuric acid. Between 353 and 373 K (80 and 100 °C), the corrosion rate drops as the temperature rises. This is due to the formation of a continuous surface layer. As the temperature increases further to 423 K (150 °C), the corrosion rate increases to 1.3 mm/a (51.2 mpy), only to fall again to 0.5 mm/a (19.7 mpy) at 493 K (220 °C). This phenomenon is interpreted as better adhesion of the salt surface layer to the metal surface as the temperature of the sulfuric acid increases [133]. To supplement Figure 130, Figure 132 shows the corrosion rates on the cast steel GX7NiCrMoCuNb25-20 (CN-7M) as a function of the temperature for various sulfuric acid concentrations. Under these conditions, the corrosive attack on the high-alloy steel is virtually uniform in all cases.

Further studies were performed on static and rotating disks (20 mm ∅) in 10 to 98 % sulfuric acid solutions, flushed with air, at various temperatures over 21 days. The specimens were activated before the start of the experiments. The results are shown in Figure 133. These show the distinctly pronounced temperature-dependence of the corrosion on static and rotating disks (∅ 20 mm, 2,000 rpm) at all the acid concentrations. The corrosion losses on the rotating disks are about ten times greater than those on the static specimens. The corrosion was also uniform in all cases under these conditions. A linear relationship between the logarithm of the linear corrosion rate and the reciprocal of the absolute temperature was found within the temperature ranges shown in Figure 133. At the concentrations tested, a change in concentration has the effect of a parallel shift in the straight lines, which can be described by the equation $\log W_{lin} = -2{,}600 \times 1/T + b$. In this equation, b assumes the values listed in Table 67.

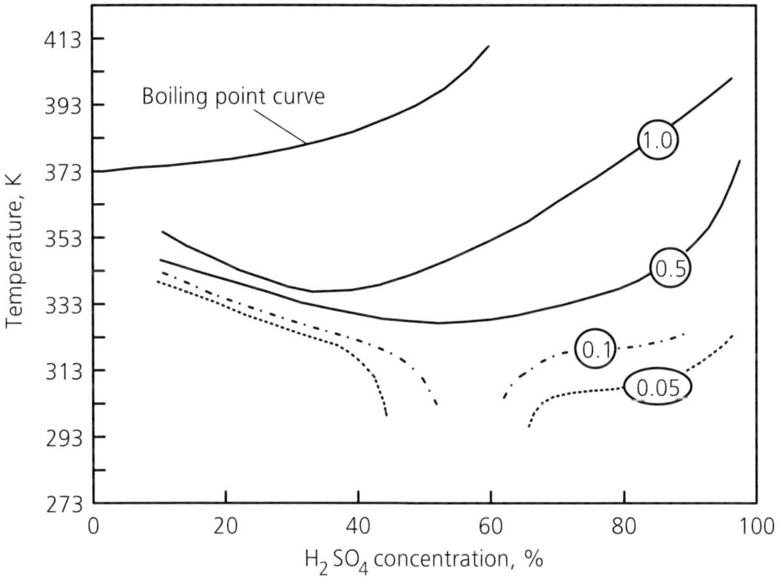

Figure 130: Isocorrosion curves (mm/a) of the steel GX7NiCrMoCuNb25-20 (CN-7M) in sulfuric acid (flushed with air, static, activated specimens, test duration 21 d) [133]

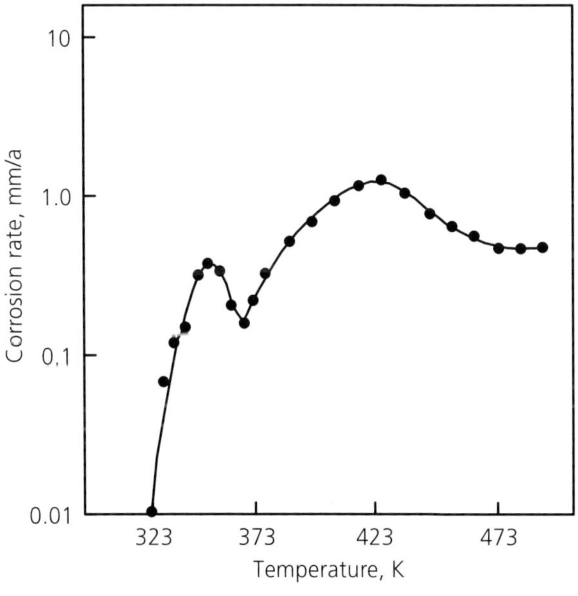

Figure 131: Corrosion of the steel GX7NiCrMoCuNb25-20 in 98.5 % sulfuric acid as a function of the temperature (flushed with air, test duration 21 d) [133]

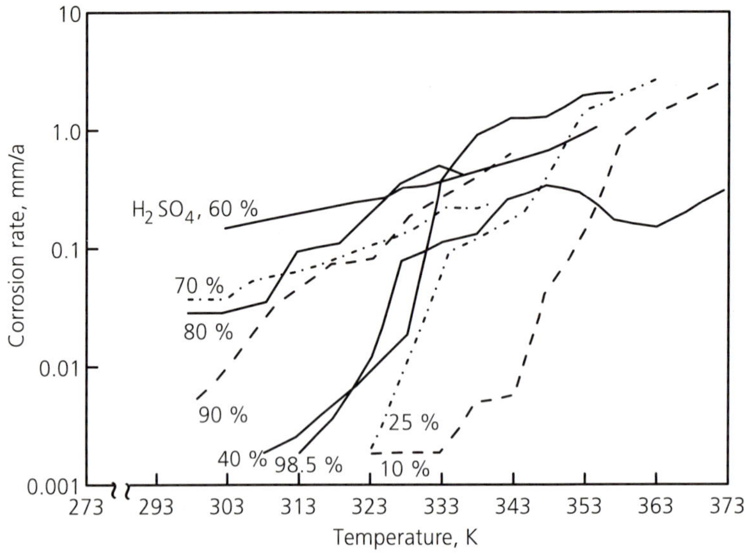

Figure 132: Corrosion of the steel GX7NiCrMoCuNb25-20 (CN-7M) in various concentrations of sulfuric acid as a function of the temperature (activated specimens, acid flushed with air, test duration 21 d) [133]

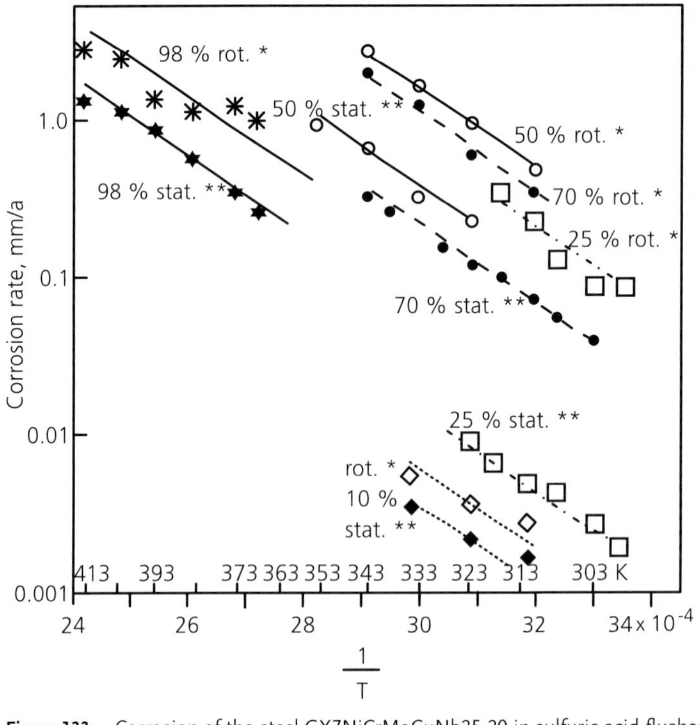

Figure 133: Corrosion of the steel GX7NiCrMoCuNb25-20 in sulfuric acid flushed with air (*rotating and **static) over a test duration of 21 d as a function of the temperature (Arrhenius plot log w = f(1/T)) [133]

H_2SO_4 concentration %	b	
	Static disk	Rotating disk n = 2,000 rpm
10	5.4	5.6
25	6.0	7.5
50	7.3	7.9
70	7.1	7.7
98	6.5	6.8

Table 67: Value b for determination of the straight line shift according to the above equation as a function of the H_2SO_4 concentration [133]

Sulfuric acid temperatures outside the ranges shown in Figure 133 cannot be calculated using the formula shown [133].

Pipe specimens 40 mm long with an internal diameter of 8 mm and an effective area of 10 cm^2 were used for studies with a pipe containing flowing medium. These were incorporated in a test zone in which the flow rate of the acid could be varied up to 21 m/s (description of the experiment and apparatus in [133, 144]).

The test durations were 42 days. Under the conditions in the pipe, the resistance range of the high-alloy austenitic steel and also the ferritic-austenitic steel (see Section Ferritic-austenitic steels with more than 12 % chromium) G-X 3 CrNiMoCuN 24 6 was extended. Of the groups of 5 specimens investigated, of the same melt, heat treatment, working and air passivity, some remained in the passive state with corrosion rates of below 0.01 mm/a (0.39 mpy), while others became active and exhibited corrosion rates of up to 8 mm/a (315 mpy). The temperature range in which the active and passive states may exist simultaneously is less pronounced with austenitic steel than with ferritic-austenitic steel, as can be seen from Figure 134. The steel GX7NiCrMoCuNb25-20 (CN-7M) is in the passive state at 323 K (50 °C) with corrosion rates of less than 0.01 mm/a (0.39 mpy), but corrodes at 333 K (60 °C) in flowing acid with a corrosion rate of up to 8 mm/a (315 mpy). Uniform general corrosion also occurred under these conditions [144].

Table 65 summarizes the corrosion behavior of CrNi-steels and the alloy NiCr21Mo in pure sulfuric acid and sulfuric acid contaminated with Cl^-- or Cu^{2+}-ions at various temperatures [213, 214].

The investigations were carried out in static acid in air. The test duration was 21 days. Intermediate weighings were carried out after 7 and 14 days at 323 K (50 °C) and after 10 days at 373 K (100 °C), the acid being renewed each time. As the corrosion rates on the steels 1.4539 and 1.4503 in Table 65 show, these steels are resistant at 323 K (50 °C) in 20 % sulfuric acid (0.01 and 0.04 mm/a (0.39 and 1.57 mpy) respectively). A sharp increase in the corrosion rates to 2.4 and 1.25 mm/a (94.5 and 49.2 mpy) respectively is observed at 373 K (100 °C). At 423 K (150 °C), they are destroyed under these conditions. Both steels are also resistant at 323 K (50 °C) in

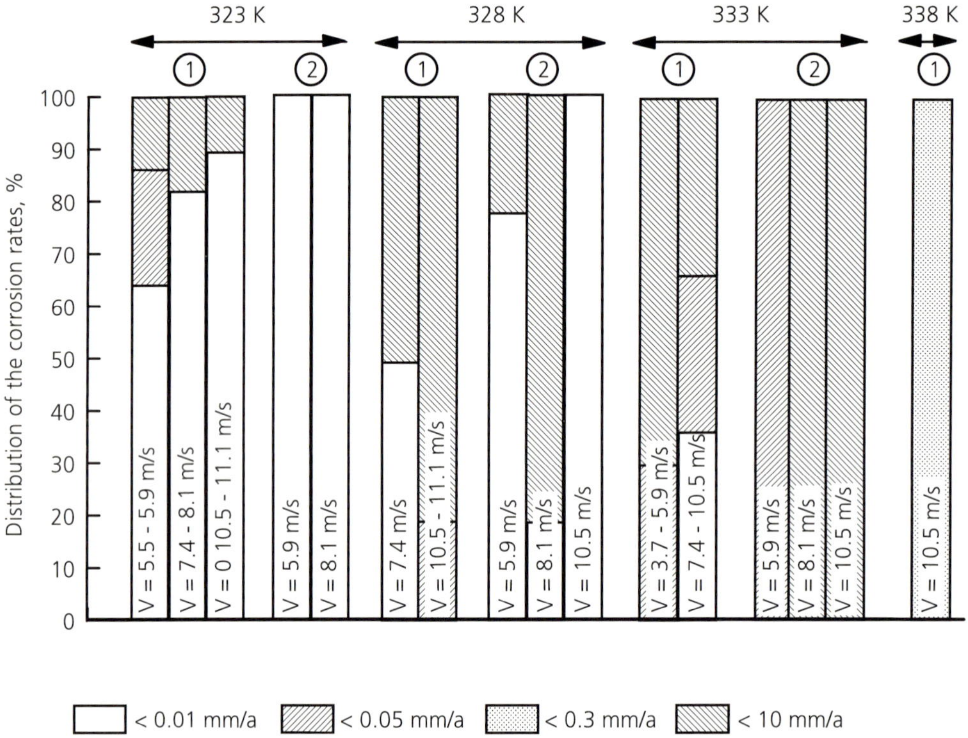

Figure 134: Percentage distribution of the corrosion rates determined in a series of experiments on pumps (impeller with tubular test specimens, 70 % H_2SO_4, flushed with air, test duration 42 d, 5 specimens each) at various temperatures [144]
1) G-X 3 CrNiMoCuN 24 6 (Noridur®)
2) GX7NiCrMoCuNb25-20 (CN-7M)

50 % sulfuric acid containing no chloride. The corrosion rate in this case is 0.05 mm/a (1.97 mpy). Addition of 500 ppm chloride to 20 % acid increases the corrosion rate at 323 K (50 °C) to 0.1 mm/a (3.94 mpy), but reduces it at 373 K (100 °C), in contrast to the pure acid, to 0.2 mm/a (7.87 mpy). This favorable influence of chloride ions can be explained by the formation of a thin, red-brown, protective layer of copper and Cu_2O by examination using a microprobe. On addition of Cu-ions to sulfuric acid, hardly any weight losses are to be found at 373 K (100 °C) in 20 % H_2SO_4, in contrast to acid without Cu. The steels are polarized into the stable passive state by the Cu-ions. Figures 110–115 show results of electrochemical studies. Cold forming of 20 % resulted in no change in the corrosion resistance of these steels. It was improved in 10, 20 and 50 % chloride-free acid by laboratory annealing (10 min at 1,353 K (1,080 °C) with subsequent quenching in water). In all other cases it was the same as the delivery state. Annealing of the steel 1.4539 had a favorable effect in chloride-containing acid solutions, but not in chloride-free solutions. Electroslag refining of the materials resulted in no significant improvement in the corrosion resistance of these steels [214].

There were no signs of pitting corrosion on any of the steels investigated in Table 65. This was in agreement with the electrochemical studies, since in all cases the repassivation potential of pitting corrosion was more positive than the free corrosion potential. The electrochemical interpretation of the experimental results is given in Section Austenitic chromium-nickel-molybdenum steels. According to these studies, the Cu-containing steels 1.4539 and 1.4503 are no longer adequately resistant in dilute sulfuric acid solutions at 423 K (150 °C) or above [213, 214].

The corrosion behavior of Mo- and Cu-containing high-alloy steels is shown, inter alia, in Figure 101 in comparison with other materials, and in Figures 130–132 for the steel GX7NiCrMoCuNb25-20. The resistance is not adequate at temperatures above 373 K (100 °C).

Figures 47 and 135 give further information on the corrosion behavior of the high-alloy steels 1.4539 (X1NiCrMoCu25-20-5), 1.4529 (X1NiCrMoCuN25-20-7), Uranus® SB 8 (25Cr, 25Ni, 4.9Mo, 1.5Cu, balance Fe; cf. 1.4537), 1.4503 (X3NiCrCuMoTi27-23) and 1.4563 (X1NiCrMoCu31-27-4) in hot 95 % sulfuric acid at 373, 398 and 423 K (100, 125 and 150 °C). While the corrosion rate of an Mo- and Cu-free CrNi-steel drops at a concentration of above 90 % H_2SO_4 at 373 K (100 °C) as the acid concentration increases, and reaches a value of only 0.03 mm/a (1.18 mpy) in approximately by 100 % H_2SO_4 (Figure 73), the high-alloy steel X1NiCrMoCu31-27-4 shows a different behavior, which can be seen in Figure 136. The drop in corrosion rate at 373 K (100 °C) occurs only after the rate has passed through a maximum of 2.61 mm/a (103 mpy) at approximately 93 % acid. The sharply decreasing corrosion rate as the water content decreases is due to diffusion polarization of the cathodic and anodic processes. Contamination of the acid by sulfates of Fe, Cr and Ni reduced the corrosion rates for the same reasons (compare with Figure 74) [102, 103].

No, or only insignificant, locally limited corrosion was found on the materials shown in Figures 47 and 135. The plots show that the steel 1.4563 (X1NiCrMoCu31-27-4) has the lowest corrosion rate in 95 % H_2SO_4 at all three temperatures investigated (Figure 135).

The steel 1.4503 (X3NiCrCuMoTi27-23) shows a comparable resistance at 373 and 398 K (100 and 125 °C) in 95 % acid with a corrosion rate of about 0.3 mm/a (11.8 mpy). In this range, it is the cheaper material with about the same corrosion rate, but with a corrosion rate of 0.8 mm/a (31.5 mpy) at 423 K (150 °C), drops off noticeably in comparison with the higher-alloy steel with a rate of 0.5 mm/a (19.7 mpy).

The steel 1.4575 (X1CrNiMoNb28-4-2) (see Section Ferritic chromium steels with more than 12 % chromium) behaves even better under these conditions, its corrosion rate being 0.1 mm/a (3.94 mpy) at 398 K (125 °C) and about 0.35 mm/a (13.8 mpy) at 423 K (150 °C). However, an active state may arise with this steel, causing it to dissolve. The Cu-free steel 1.4465 (X1CrNiMoN25-25-2) gives a similarly better corrosion behavior under these conditions.

An improvement in the corrosion behavior of the materials investigated occurs in oxidizing hot sulfuric acid with an increased Cr-content, in particular in the test range investigated from 373 to 423 K (100 to 150 °C). The influence of Mo is positive

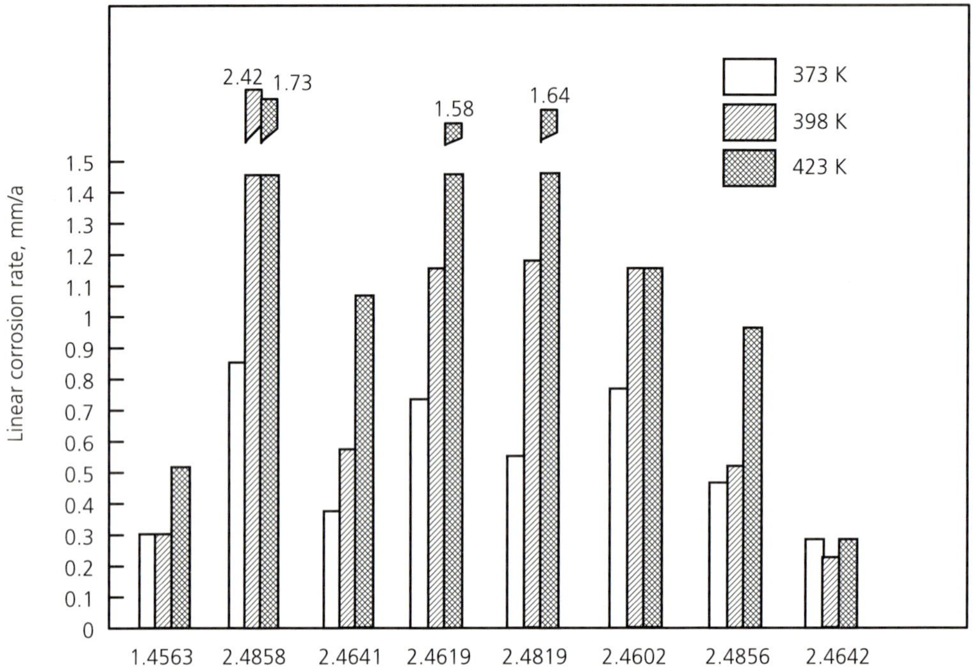

Figure 135: Comparison of the corrosion resistance of nickel alloys with a CrNi-steel in static 95 % sulfuric acid at various temperatures (test duration 21 d) [102, 103]

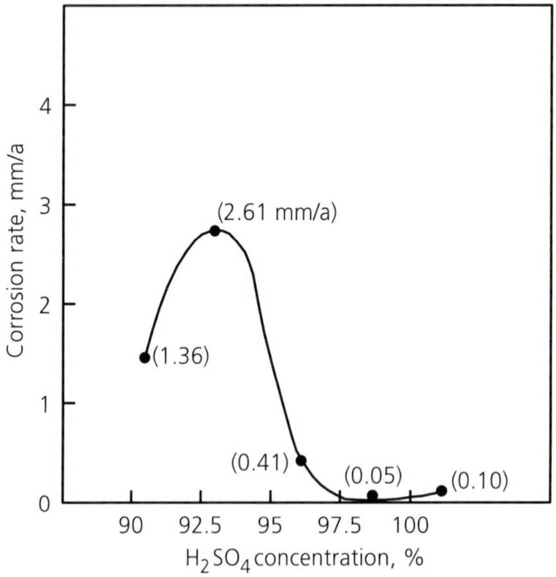

Figure 136: Corrosion rate of the steel X1NiCrMoCu31-27-4 (1.4563) in static non-aerated sulfuric acid at 373 K (100 °C) (test duration 21 d) [102, 103]

up to a level of 6 % at temperatures up to 398 K (125 °C). Higher Mo-contents in combination with high Ni-contents lead to increased corrosive attack (Figure 135). Cu-contents in the steel showed a corrosion-inhibiting action under reducing conditions as a result of incorporation or cementation of Cu into the surface layer. This is not possible under oxidizing conditions. The influence of the cathodic partial reaction, at least under static conditions, therefore seems to emerge as the positive, corrosion-inhibiting action of copper under oxidizing conditions [102].

The materials listed in Figures 47 and 135 show a stable-passive behavior in highly concentrated sulfuric acid above 333 K (60 °C). When a material-specific, critical temperature is reached, this behavior changes into a metastable state. As already mentioned in Section Austenitic chromium-nickel steels, the corrosion behavior of high-alloy austenitic steels can be described best with the aid of current density/ potential curves. The variations in potential found in [102] between $U_{R,a}$ and $U_{R,p}$ (free corrosiont potential in the active and passive region) are called potential oscillations. The corrosion rate established within the period of the experiment depends on the residence time at the particular corrosion potential, and the associated corrosion current density. In the case of the stainless steels and nickel alloys, an increase in corrosion current density in sulfuric acid of high concentration is associated with an increase in temperature. Material-specific threshold temperatures (potential oscillation temperatures), at which a metastable state of the material can be expected, are summarized for some CrNi-steels and Ni-alloys in Table 61.

This table shows that the Mo-free steels (X5CrNi18-10 and X1CrNiSi18-15-4) show no potential oscillations. The potential oscillation temperatures of the iron and nickel alloys increase as the alloying contents increase. The oscillation ranges of all the materials investigated overlap substantially. However, it cannot be assumed on principle that materials having high, or the highest, oscillation temperatures are the most corrosion-resistant. The reason for this is that at the same potential but different temperatures, the particular corrosion current densities differ [102]. In this context, compare also Figures 47 and 135. The potential oscillations can be suppressed by an anodic protection current (in this context, see – Electrochemical corrosion protection processes –).

– Corrosion behavior in sulfuric acid-acid mixtures –

The corrosion resistance of the Cu-containing CrNiMo-steels is also improved in mixed media in comparison with Cu-free steels, as can be seen from Tables 46, 48, 49, 63–65. As is also the case with Cu-free steels, the corrosion resistance is improved by oxidizing additions, such as, for example, nitric acid, as can be seen from Figures 65 and 102. Chlorides impair the corrosion resistance by eliminating the passivity. Although the effect of chlorides is less pronounced on these steels than on the Cu-free or Cu- and Mo-free steels (in this context, see Figure 108). For the effect of chlorides in sulfuric acid solutions on the steel X5NiCrMoCuNb22-18. As their content in the acid increases, they impair the resistance of the material. As a rule, they cause local corrosion, such as pitting corrosion, which mainly occurs at poorly accessible points and can lead to disastrous consequences. As corrosion studies in 25 to 35 %

H_2SO_4 containing 4 to 8 % HF (see Table 63) have shown, increased corrosion occurs at points of embossed markings on Cu-containing CrNi-steels, due to the stresses induced, but this cannot be described as stress corrosion cracking [16].

According to corrosion studies under simulated conditions in a chloride-containing, oxidizing solution of 23 % H_2SO_4, 1.2 % HCl, 1 % $FeCl_3$ and 1 % $CuCl_2$, similar to those found in pickling plants or during the concentration of waste sulfuric acid (ASTM G-28, method B), the steels 1.4529 (X1NiCrMoCuN25-20-7) and 1.4563 (X1NiCrMoCu31-27-4) are destroyed with corrosion rate of 80 mm/a (3,150 mpy) each at the boiling point. The corrosion rate of the steel Nicrofer® 3127 hMo (see Table 70) is 4.7 mm/a (185 mpy) [239]. The aggressiveness of phosphoric acid is intensified by impurities. Figure 103 shows the influence of sulfuric acid in H_3PO_4 [208]. See also Section Austenitic chromium-nickel-molybdenum steels.

The Cu-containing steel 0Ch23N28M3D3T cannot be used in a reaction mixture of about 25 % sulfuric acid, 7 % hydrochloric acid, sulfochloride, dichloroethane and chlorosulfonic acid under absolute pressures of 0.1 to 0.17 MPa and at temperatures of up to 383 K (110 °C), since corrosion rates of 35 mm/a (1,378 mpy) occur. A NiMo-alloy can be used under these conditions. The experiences during operation extend to a total period of 15,000 h [240].

– Corrosion behavior in sulfuric acid solutions –

Solutions which have approximately the following composition are obtained in the production of alkylsulfonates: 13.3 % alkylsulfonic acid, 29.6 % hydrocarbons, 3.7 % sulfuric acid, 2.1 % sulfur trioxide and 51.3 % water. The temperatures are about 393 K (120 °C). The high-alloy steel 0Ch23N28M3D3T can be employed under these conditions. In this reaction mixture, the steel corrodes at a rate of 0.09 mm/a (3.54 mpy) in argon and 0.01 mm/a (0.39 mpy) in air at 303 K (30 °C). At 363 K (90 °C), the maximum corrosion rate is 0.04 mm/a (1.57 mpy). No test durations were given [174].

The steel 06ChN28MDT (22–25 % Cr, 26–29 % Ni, 2.5–3.0 % Mo, 2.5–3.5 % Cu, balance Fe) proves to be practically resistant in the production of sodium hexafluorosilicate in sulfuric acid solutions by the sulfate process (Table 7) [24].

Titanium is particularly suitable for use under these conditions [173].

In the hydrolysis sulfuric acid obtained during titanium dioxide production (10 to 15 % H_2SO_4 + 0.8 to 1.4 % $TiOSO_4$ + 11 to 17 % $FeSO_4$) at 313 K (40 °C), the steel 0Ch23N28M3D3T (23 % Cr, 28 % Ni, 3 % Mo, 3 % Cu, balance Fe) corrodes at a rate of only 0.001 mm/a (0.04 mpy), whereas under production conditions after a test duration of 1,440 h the rate is 0.02 mm/a (0.79 mpy) [169].

Sulfuric acid solutions of thiourea, which were to be separated in centrifuges, contained, in addition to solids, 20 g/l H_2SO_4 and 100 g/l urea. At a temperature of 323 K (50 °C) and a pH of 0.7, a corrosion rate of 0.0026 mm/a (0.10 mpy) occurred on the material 0Ch23N28M3D3T after a test duration of 100 h. A considerably higher corrosion rate of 0.45 mm/a (17.7 mpy) occurred in the crevices after the same test duration as a result of crevice corrosion. This steel is therefore unsuitable for this application [172].

For the recovery of sulfur dioxide from sodium sulfite/sodium bisulfite, the steel Avesta® 254 SMO (20 % Cr, 18 % Ni, 6.1 % Mo, 0.2 % N, 0.7 % Cu, balance Fe; cf. 1.4547) proved to be a suitable container material for an aqueous solution of 20 % sodium sulfate, 0.5 to 3.5 % sulfuric acid and 0.8 to 1.5 % sulfur dioxide at pH 2 and a temperature of 353 to 368 K (80 to 95 °C). Material thicknesses of 0.6 to 0.8 mm are used for linings [241].

– Pitting corrosion in sulfuric acid and sulfuric acid solutions –

The corrosion resistance of the CrNi-steels is mainly achieved by their passivity in the corrosion medium. Local damage to these steels leads to pitting corrosion under certain circumstances. Predictions or tests in this context are therefore of particular interest for practical application. The temperature at which pitting corrosion is to be expected under the conditions specified, or at which pitting corrosion becomes detectable, is called the critical pitting corrosion temperature. Table 53 contains a summary of the critical pitting and crevice corrosion temperatures (the same applies to crevice corrosion as to pitting) [139]. According to this, the critical pitting or crevice corrosion temperatures in a solution of 0.01 mol/l H_2SO_4 + 0.1 % $Fe_2(SO_4)_3$ + 1,000 ppm Cl^- are 303 and 293 K (30 and 20 °C) respectively for the steel 20Cb-3 (NiCr20CuMo), and 313 and 318 K (40 and 45 °C) respectively for the steel SAE 904 L (X1NiCrMoCu25-20-5). These materials are more resistant to pitting corrosion than lower-alloy grades (compare also the pitting resistance equivalent).

While according to [209] the steel SAE 904 L has pitting and crevice corrosion temperatures of 303 and 293 K (30 and 20 °C) respectively in a solution of 7 percent by volume H_2SO_4 + 3 percent by volume HCl + 1 % $FeCl_3$ + 1 % $CuCl_2$, the figures are significantly higher for the steel Avesta® 254 SMO (see Figure 108), at 328 and 308 K (55 and 35 °C) (Table 68). A critical pitting corrosion temperature of 348 K (75 °C) was determined under these conditions for the recently developed steel Nicrofer® 3127 hMo (31 % Ni, 27 % Cr, 6.5 % Mo, 1.2 % Cu, 0.20 % N, max. 0.015 % C, balance Fe) (the test temperature was increased in steps of 2.5 K (2.5 °C) after every 24 h until the critical pitting corrosion temperature was reached) [239]. Temperatures of 353 K (80 °C) are possible in these studies. Thus values approaching those of the nickel alloy Alloy C-276 are obtained (compare Table 68 and Figure 109).

– Intercrystalline corrosion (IC) in sulfuric acid and sulfuric acid solutions –

The steel 1.4539 (X1NiCrMoCu25-20-5), which is widely used in the chemical industry, is distinguished by its resistance to intercrystalline corrosion as a consequence of its low C-content of about 0.02 %. Testing was performed in the Strauss test in accordance with DIN EN ISO 3651-2 (100 ml H_2SO_4, density 1.84, addition of 1 l H_2O + 110 g $CuSO_4 \times 5H_2O$; 50 g/l electrolytic copper are added to this solution in the form of filings before the start of the experiment, boiling test solution), by the intensified Strauss test using 37 % H_2SO_4, otherwise as DIN EN ISO 3651-2, and by the modified Streicher test (40 % H_2SO_4 + 25 g/l $Fe_2(SO_4)_3$) over a test duration of 24 h. No IC susceptibility was found on welded specimens with the necessary additional materials under the same test conditions. Table 69 contains a summary of the test results [231].

Material	Section	CPT[1] K (°C)	Material consumption rate at CPT[1] $g/m^2\ h$[3]	CCT[2] K (°C)	Material consumption rate at CCT[2] $g/m^2\ h$[3]
SAE 316	H	< 288 (< 15)	0.16 at 288 K (15 °C)	< 288 (< 15)	0.13 at 288 K (15 °C)
Alloy 904 L	I	303 (30)	0.011	293 (20)	0.003
Avesta® 254 SMO[4]	I	328 (55)	0.018	308 (35)	0.005
Alloy G		318 (45)	0.20	313 (40)	0.022
Alloy 625		338 (65)	2.7	318 (45)	0.006
Alloy C-276		> 373 (> 100)	0.36 at 373 K (100 °C)	343 (70)	0.036

[1] critical pitting corrosion temperature
[2] critical crevice corrosion temperature
[3] value × 0.36: density = mm/a
[4] see Table 44

H: see section Austenitic chromium-nickel-molybdenum steels
I: see section Austenitic chromium-nickel steels with special alloying additions

Table 68: Comparison of the critical pitting and crevice corrosion temperatures and the resulting material consumption rates on CrNiMo-steels and Ni-base alloys in a solution of 7 percent by volume H_2SO_4 + 3 percent by volume HCl + 1 % $FeCl_3$ + 1 % $CuCl_2$ (ASTM G 48), test duration 24 h (see also Figure 109) [209, 210]

		Strauss-Test DIN EN ISO 3651-2[1]	Intensified Strauss-Test[2]	Streicher-Test[3]	Material consumption rate $g/m^2\ d$	Corrosion rate mm/a (mpy)
Delivery state		+*	+	+	2.23	0.103 (4.06)
923 K (650 °C)	15 min	+	+	+	2.00	0.092 (3.62)
	30 min	+	+	+	1.62	0.075 (2.95)
	60 min	+	+	+	1.90	0.088 (3.46)
973 K (700 °C)	15 min	+	+	+	3.31	0.153 (6.02)
	30 min	+	+	+	3.28	0.151 (5.94)
	60 min	+	+	+	3.00	0.138 (5.43)
1,023 K (750 °C)	15 min	+	+	+	2.65	0.122 (4.80)
	30 min	+	+	+	2.51	0.116 (4.57)
	60 min	+	+	+	2.57	0.119 (4.69)

Test conditions:
[1] 100 ml H_2SO_4 (density 1.84) + 1 l H_2O + 110 g $CuSO_4 \times 5H_2O$ + 50 g/l electrolytic Cu-filings, boiling
[2] 37 % H_2SO_4, otherwise as DIN EN ISO 3651-2
[3] 40 % H_2SO_4 + 25 g/l $Fe_2(SO_4)_3$, 24 h, weight loss and IC after bending
* + no IC (intercrystalline corrosion)

Table 69: IC resistance of the steel 1.4539 (X1NiCrMoCu25-20-5) containing 0.018 % C in various test solutions, 4 mm sheet, solution-annealed, quenched, IIa [231]

The high-alloy steel Nicrofer® 3127 hMo was tested for intercrystalline corrosion susceptibility by the ASTM G-28 test, method A. 50 % H_2SO_4 with added $Fe_2(SO_4)_3$ at boiling point was used as the test medium. The test duration was 120 h. 5 mm-thick specimen sheets were annealed between 737 and 1,023 K (464 and 750 °C) over a period of 1 to 100 h and then quenched in water. As well as determination of the weight loss, which allows a conclusion as to whether sensitization exists in relation to the optimum solution-annealed starting state, the penetration depth of the intercrystalline corrosion was measured by metallography. In accordance with the steel-iron test sheet SEP 1877, a penetration depth of 0.050 mm is a criterion of intercrystalline corrosion. The investigations resulted in a typical TTS curve (time-temperature-sensitization) for this high-alloy steel with a relatively narrow susceptibility range, as is shown in Figure 137. From this graph, the shortest time for the occurrence of intercrystalline corrosion in this material is a treatment time of 2 h in the temperature range between 923 and 973 K (650 and 700 °C). According to [232], the low temperatures lead to the conclusion that the sensitization is a consequence of chromium carbide precipitates.

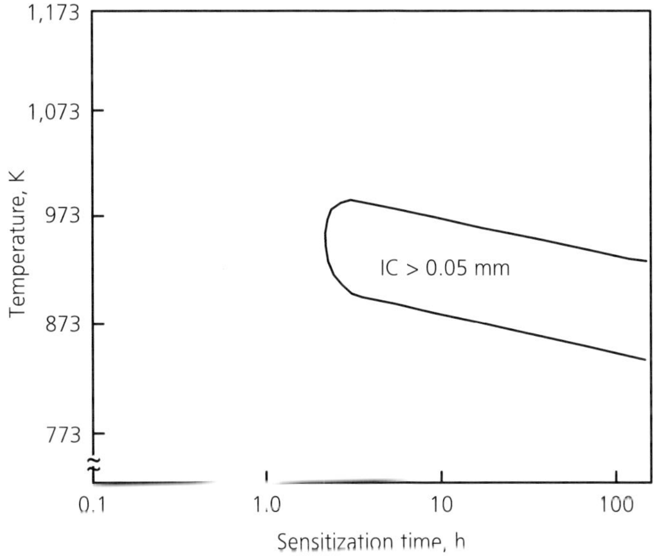

Figure 137: Time-temperature-sensitization diagram (TTS diagram) of Nicrofer® 3127 hMo ((%) Fe-31Ni-27Cr-6.5Mo-1.2Cu-<0.015C-0.020N) in boiling sulfuric acid (50 %) + addition of $Fe_2(SO_4)_3$ showing its susceptibility to intercrystalline corrosion (IC) (test duration 120 h (ASTM G-28 test, method A)) [232]

Further to this, studies were also performed on welded specimens, both in accordance with ASTM G-28, method A (120 h) and in the IC test according to Stahl-Eisen-Prüfblatt SEP 1877/method II (24 h). Table 70 shows the corrosion rates in comparison with two other high-alloy steels. With corrosion rates according to ASTM of about 0.2 mm/a (7.87 mpy) and of 0.05 mm/a (1.97 mpy) according to SEP, this material exhibits a low susceptibility to corrosion. The weld material, heat-affected zone and base material are free from intercrystalline corrosion. According to this work, the conclusions are very reliable, since they were determined on a laboratory batch having a C-content of 0.03 %, whereas maximum C-contents of 0.010 % can be established on production batches [232]. In comparison with the steel 1.4563 (X1NiCrMoCu31-27-4), Nicrofer® 3127 hMo has a 3 % higher Mo-content, a lower C-content of 0.015 % and an additional N-content of 0.20 %. This gives a pitting index of 47, or 53 if the nitrogen is included.

– Stress corrosion cracking in sulfuric acid solutions –

High-alloy steels which are used in ammonium sulfate saturation vessels in coking plants are exposed to a wide range of aggressive media under high mechanical and thermal stresses. Solutions with an excess of free H_2SO_4 of about 3 %, a pH of 1.5 and a Cl^--content of 1.2 g/l at a temperature of 367 to 371 K (94 to 98 °C) are formed during production of ammonium sulfate from ammonia vapors. The steels 1.4465 (X1CrNiMoN25-25-2), 1.4577 (X3CrNiMoTi25-25) and 1.4539 (X1NiCrMoCu25-20-5) were investigated for the purpose of material refinement in the design of ammonium sulfate saturation vessels. See also – Stress corrosion cracking – in Section Austenitic chromium-nickel-molybdenum steels. Transcrystalline cracks in the heat-affected zone of the weld seam and stress corrosion cracks on a reinforcing strut within an ammonium sulfate saturation vessel, probably caused by separation stresses during cutting on the guillotine, were the reasons for these studies. The material used was the steel 1.4577. Specimens of dimensions 15 × 15 × 8 mm ground to grain 600 were used for the electrochemical studies. Comparable conditions in respect to passivation were achieved by further chemical treatment (no indication in this reference) [216].

Figure 138 shows the current density/potential curves plotted in the above medium with the chloride content increased to 20 g/l. After a holding time of 20 h in the medium, the free corrosion potentials of the steels are between –460 and –515 mV_{SHE}.

The specimens were polarized starting from the free corrosion potential with a rate of change in the potential of less than 180 mV/h up to the region of activation by pitting corrosion. The passivation behavior of the materials is detectable by reversing the direction of the change in potential. Important aspects here are the repassivation current density, as a measure of metal dissolution in the active state, as well as the potential at which the current density is zero and complete passivation (repassivation) occurs. This potential must be more positive than the free corrosion potential, since only then can self-passivation after activation by pitting corrosion take place, or be expected, in the medium investigated. The copper-containing CrNiMo-steel 1.4539 has a pitting corrosion potential which is 300 mV more positive, but like the steel 1.4577 exhibits a high repassivation current density. According to Figure 138, the repassivation potentials of all three steels are about the same, lying between –340 and –380 mV_{SHE}, and are therefore more positive than the free corrosion potentials. Under these conditions, the copper-containing steel exhibits a less favorable repassivation behavior, which is why the copper-free steel 1.4465 has been given priority after these studies, as practical experiences over a period of 2 years have confirmed.

Material	*ASTM G-28, method A*[1)]		*SEP 1877/method II*[2)]	
	Base material	*Welded specimen*	*Base material*	*Welded specimen*
	Corrosion rate, mm/a (mpy)			
X1NiCrMoCuN25-20-7 (1.4529)	0.43; 0.37 (16.9; 14.6)	0.45; 0.58 (17.7; 22.8)	0.13 (5.12)	0.24 (9.45)
X1NiCrMoCu31-27-4 (1.4563)	0.18; 0.14 (7.09; 5.51)	0.20; 0.23 (7.87; 9.06)	0.03 (1.18)	0.08 (3.15)
Nicrofer® 3127 hMo[3)]	0.18; 0.13 (7.09; 5.12)	0.17; 0.20 (6.69; 7.87)	0.04 (1.57)	0.05 (1.97)

[1)]50 % $H_2SO_4 + Fe_2(SO_4)_3$, boiling, 120 h
[2)]method II: 40 % H_2SO_4 + 25 g/l Fe(III)-sulfate
[3)]31 % Ni, 27 % Cr, 6.5 % Mo, 1.2 % Cu, ≤ 0.015 % C, 0.20 % N, balance Fe

Table 70: Corrosion rates when testing for IC (intercrystalline corrosion) resistance on welded specimens (TIG hand welding) of high-alloy steels in accordance with ASTM and SEP (Stahl-Eisen-Prüfblatt), surface ratio of weld material: base material of the corrosion specimens 1:6 [232]

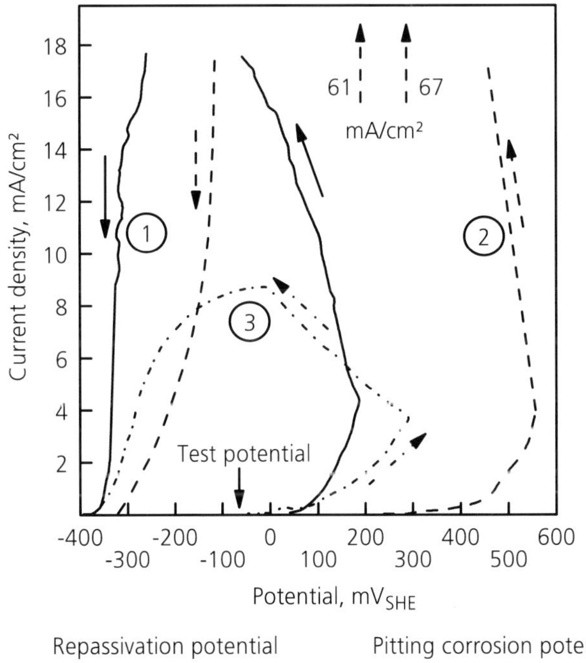

Figure 138: Current density/potential curves of the materials 1.4577, 1.4539 and 1.4465 in an operating medium of 3 % free H_2SO_4, 20 g/l Cl^-, pH 1.5, 367 to 371 K (94 to 98 °C) (repassivation analysis) [216]

– Corrosion fatigue in sulfuric acid –

Experiences in practice have shown that, in addition to other causes, corrosion fatigue is a culprit in damage to plant components exposed to sulfuric acid [217, 218]. For details on experimental procedure, see – Corrosion fatigue – in Section Austenitic chromium-nickel-molybdenum steels.

The results of these tests are shown in Figures 116–118. The fatigue curves in Figure 117 show that the bending fatigue strength achieved in air on the steels 1.4539 (X1NiCrMoCu25-20-5) and 1.4586 (X5NiCrMoCuNb22-18) is significantly reduced by dripping 96 % sulfuric acid at 353 K (80 °C) in the same way as on the other steels investigated in the figures. Under these conditions, these materials no longer have a fatigue strength. They have a corrosion fatigue limit of limited duration. The corrosion fatigue limit in 90 % H_2SO_4 at 353 K (80 °C) is still below that in 96 % acid. According to [217, 218], the higher dilution is the reason for an accelerated local attack on a micronotch which had been formed. Figure 118 contains a comparison of the corrosion fatigue limits after 10^6 load cycles with the corresponding fatigue limit in air. After 10^6 load cycles in 96 % H_2SO_4 at 353 K (80 °C), the steel 1.4586 achieves a significantly higher corrosion fatigue limit, with 80 %, than the steel 1.4539, with 67 %. As can be seen from Figure 116, the steels are passivatable in 90 % H_2SO_4 at 353 K (80 °C). The protection potential (anodic protection) chosen was +600 mV_{SHE}. Overall, it can be said that the corrosion fatigue limits of the steels in 96 % H_2SO_4 at 353 K (80 °C) are improved by anodic protection, since consumption corrosion is largely prevented in the passive state. Anodic protection is effective in particular on the materials which are most severely attacked by free corrosion (Figure 118). However, this protection is not sufficient to achieve the fatigue limit of the steels determined in air, although it is an effective means of rendering the corrosion fatigue on these steels calculable [218].

– Electrochemical corrosion protection processes –

In this context, see also Sections Austenitic chromium-nickel steels and Austenitic chromium-nickel-molybdenum steels. For the possibility of using anodic corrosion protection on plant components exposed to corrosion fatigue in hot concentrated sulfuric acid, see CF above. As already mentioned elsewhere, anodic protection is only possible if the material can become passive, or be passivated in the corrosion medium. This possibility exists in dilute sulfuric acid solutions. This is shown by the cyclic-dynamic current density/potential curves at 323 K (50 °C), for the media in Table 65, in Figures 113–115 for the materials 1.4539 and 1.4503. According to [214], the corrosion parameters of passivation current density and passive dissolution current density increase as the temperature rises. Anodic protection is evidently possible only for chloride-free sulfuric acid solutions or those at low temperatures. A possible use of anodic protection for these steels in dilute sulfuric acid solutions at higher temperatures can be clarified only in potentiostatic long-term holding tests in the passive range (if available for the medium to be investigated) [214]. In this context, compare also the free corrosion potential curves in Figures 110–112 and the text in Section Austenitic chromium-nickel-molybdenum steels.

Figures 47 and 135 show the corrosion behavior of high-alloy copper-containing CrNiMo-steels 1.4539, 1.4529, 1.4503, Uranus® SE8 and 1.4563 in 95 % H_2SO_4 at 373, 398 and 423 K (100, 125 and 150 °C) without anodic protection. Average protection potentials have been determined from current density potential curves. According to [102], Figures 139 and 140 show typical examples of such curves for the steel 1.4529, plotted at various rates of change in the potential in the original sulfuric acid of a zinc smelter. According to these studies and in agreement with the literature (see [102]), the best results in choosing the protection potentials are achieved if the latter have been determined from the plot of the curve of the so-called reverse scan. Table 54 contains a summary of the protection potentials of the steels investigated. The various corrosion rates with 98.7 % H_2SO_4 at 373 K (100 °C) of the steels with and without anodic protection and the current density measured in a semi-industrial pilot plant can also be seen from this table. The corrosion rates did not exceed 0.006 mm/a (0.24 mpy) on any of the steels, even on the lower-alloy 1.4541. Under these conditions, the use of such high-alloy steels is not absolutely essential. All the anodically protected materials in this table exhibited surfaces corresponding to the starting state. The results of these studies show that even materials with high molybdenum content could be given excellent anodic protection [102].

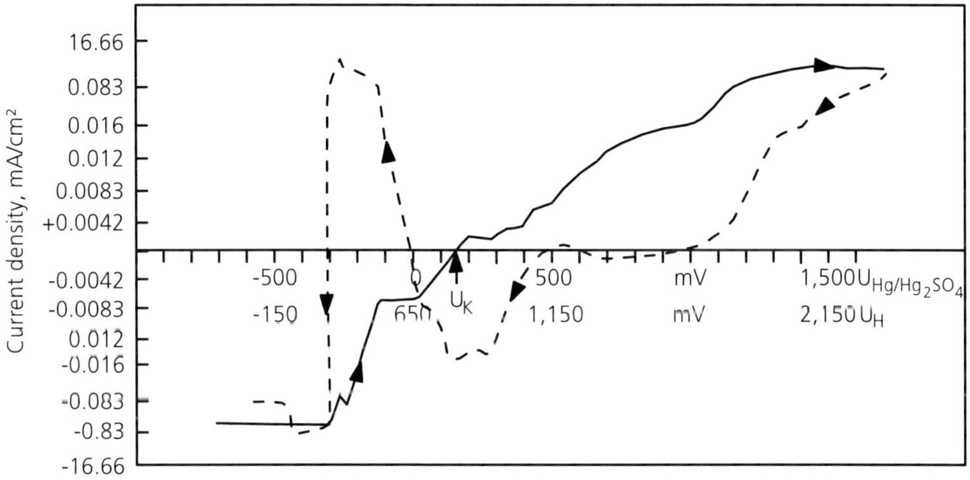

Figure 139: Current density/potential curve for the steel 1.4529 for dU/dt = 0.8 V/h at 373 K (100 °C) in 98.7 % H_2SO_4 [102]

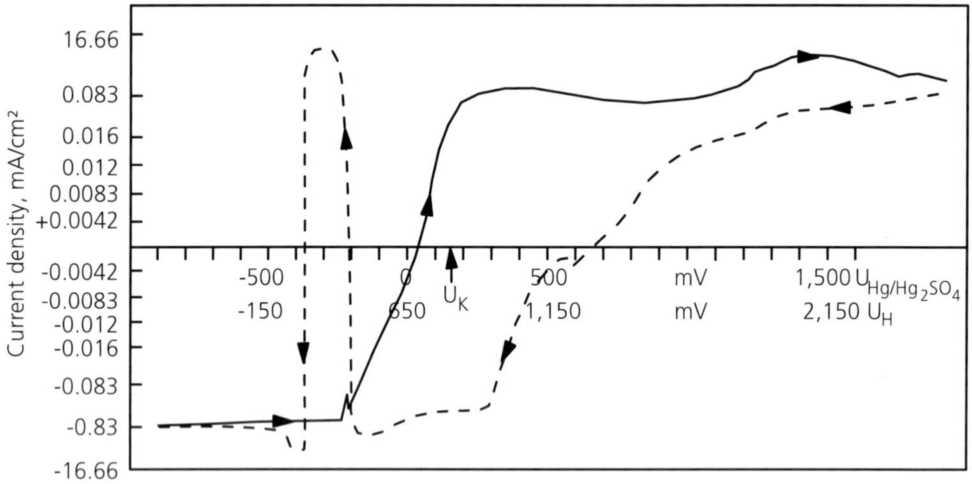

Figure 140: Current density/potential curve for the steel 1.4529 for dU/dt = 40 V/h at 373 K (100 °C) in 98.7 % H_2SO_4 [102]

In experiments the corrosion rate on the steel 1.4529 rose from 0.001 mm/a (0.04 mpy) at 373 K (100 °C) to 0.004 mm/a (0.16 mpy) at 393 K (120 °C) under anodic protection in each case, whereas it is attacked relatively severely with corrosion rates of 0.31 mm/a (12.2 mpy) at 373 K (100 °C) under these conditions, and very severely at 393 K (120 °C) at more than 10 mm/a (394 mpy) (Table 54) [102, 103].

– Fields of use and failure –

The steel 1.4539 (X1NiCrMoCu25-20-5) was developed for working with sulfuric acid [126, 208, 234, 236]. Figure 103 shows its better resistance in phosphoric acid contaminated with sulfuric acid in comparison with the Cu-free steel 1.4435. The two acids are similar in their corrosion properties, except that the corrosion rate according to [61] on Cu-containing CrNiMo-steels does not exceed 0.1 mm/a (3.94 mpy) in the entire concentration range up to 373 K (100 °C), even for a steel containing 0.05 % C, 18 % Cr, 18 % Ni, 2 % Mo, 2 % Cu, traces of Nb (Cb) and Fe as the balance.

This steel has proved to be a suitable material for pipelines for drying tower columns (74 % H_2SO_4 at 293 to 308 K (20 to 35 °C)), pickling liquor tanks (10 to 20 % H_2SO_4 at 363 K (90 °C)), chimney linings and in sulfuric acid contaminated with chlorides and fluorides [236].

Pipes of a heat exchanger made from 904 L SAE 904 L(1.4539) have proved suitable in gas condensates containing solid contents and up to 20 % H_2SO_4 at 323 K (50 °C). They have been used to replace unalloyed steel, which had a service life of less than one year at a pipe wall thickness of 3.9 mm. The high-alloy steel gave no cause for complaint after an operating life of about 4 years [234]. According to this work, the steel proved itself over a period of 5 years in 80 to 96 % H_2SO_4 at 303 to

313 K (30 to 40 °C), and in particular in an inert gas purification plant using sulfuric acid. The sulfuric acid was diluted on entry of water. The associated simultaneous increase in temperature led to the plant being destroyed. Erosion corrosion may be caused by sulfuric acid during separation of a two-phase mixture of phenol and sulfuric acid (55 to 60 %). Laboratory studies at 333 K (60 °C) showed corrosion rates of 0.08 mm/a (3.14 mpy) in 58 % H_2SO_4 containing solid contents of manganese dioxide and manganese sulfide. Corrosion studies were also performed in the actual corrosion solution at flow rates of up to 26 m/s. The steel 904 L showed no noticeable weight loss at 333 and 318 K (60 and 45 °C) and was completely resistant under these conditions. During production of plant-based oils, hydrolysis products are cleaved by means of sulfuric acid. This process continuously consumes H_2SO_4 at an initial concentration of 96 % and a process temperature of 363 K (90 °C). A material which is resistant over almost the entire concentration range of sulfuric acid is required here. After using the steel NU 904 L (1.4539), the corrosion problems were solved [234] – compare also Figures 108 and 128 and see – Corrosion behavior in sulfuric acid-acid mixtures –. An ion exchanger for the photographic industry is exposed to 10 % H_2SO_4 at an operating temperature of 353 K (80 °C). Nitrogen is passed through this medium. The plant also operated with no cause for complaints. In concentrated sulfuric acid (90 to 98 %), anodic protection is used successfully on cooling plants [234].

To purify the gas formed in the decomposition of fluorspar, which contains up to 80 % HF, 5 to 10 % H_2SO_4 and 2 to 3 % H_2O, the condensable constituents are removed. The condensates contain 18 to 40 % HF, 35 to 70 % H_2SO_4 and up to 40 % H_2O. Under the gas inlet conditions at 393 to 423 K (120 to 150 °C), the Cu-containing steel 0Ch23N28M3D3T (23 % Cr, 28 % Ni, 3 % Mo, 3 % Cu, balance Fe) is corroded at a rate of 0.84 mm/a (33.1 mpy) in the liquid phase (353 K (80 °C)) (test duration 1,680 h), and at a rate of 1.7 mm/a (66.9 mpy) in the gas phase (test duration 1,970 h), especially if the latter contains SO_2. At high gas speeds the corrosion rate can rise to 2.3 mm/a (90.6 mpy). However, the steel is considerably more corrosion-resistant under these conditions than the lower-alloy steels in Sections Austenitic chromium-nickel steels and Austenitic chromium-nickel-molybdenum steels [25].

A flue gas channel of 1.4571 (X6CrNiMoTi17-12-2, cf. SAE 316 Ti) installed in the paper industry between the dust scrubber (about 343 K (70 °C)) and the desulfurization in the Venturi washer exhibited severe corrosive attack in the form of wide pitting at the weld seams and wall surfaces after 300 h, leading to rupture of the walls. The flue gas contained 0.3 % SO_2, an unknown, probably larger amount of SO_3, and a resulting amount of H_2SO_4. The cause of the corrosion here was the temperature falling below the dew point and the formation and concentration of sulfuric acid. The steel 1.4505 (X4NiCrMoCuNb20-18-2) or a higher-alloy material and immediate washing out of the sulfuric acid have been recommended under these conditions [227].

Special iron-based alloys

Figure 66 shows the effect of various alloying elements in CrNi-steels on their resistance to sulfuric acid. Si-contents only slightly improve the corrosion resistance (Figure 66), and the resistance gap in sulfuric acid is narrowed [63].

Corrosion studies on CrNi-steels of varying Mo- and Si-content in respect of their corrosion resistance in sulfuric acid have demonstrated that if the chromium content is 18 %, about 20 % Ni must be present in the steel in order to retain an austenitic structure in the range up to 1 % Si and 0 to 10 % Mo. Table 71 shows the influence of Si- and Mo-content in 4 and 20 % sulfuric acid. The temperatures in the sulfuric acid in question at which transition from the passive into the active state takes place in the particular alloy are shown here as a function of the alloying content.

Alloys or CrNi-steels having Mo-contents of 10 % give only very poor results on forging and rolling. An alloying content of 5 % Mo is regarded as the optimum. In this case, the corrosion resistance is increased only slightly by Si-additions [163].

CrNi-Steel 18 20		*Sulfuric acid concentration*	
		4 %	*20 %*
Si, %	*Mo, %*	*Temperature of the passive-active transition, K (°C)*	
–	–	303 (30)	–
–	2.5	321 (48)	306 (33)
–	5.0	342 (69)	321 (48)
–	10.0	347 (74)	331 (58)
0.5	–	304 (31)	–
0.5	2.5	331 (58)	312 (39)
0.5	5.0	341 (68)	321 (48)
1.0	–	299 (26)	–
1.0	2.5	330 (57)	312 (39)
1.0	5.0	337 (64)	323 (50)
1.0	10.0	339 (66)	326 (53)

Table 71: Influence of the Mo- and Si-content on the temperature of the transition from the passive to the active state in CrNi-steel 18 20 in 4 and 20 % H_2SO_4 [163]

All or some of the nickel in austenitic CrNi-steels may be replaced by manganese, and eventually also by nitrogen. These steels include the types SAE 201 (UNS S20100) and SAE 202 (1.4371, X2CrMnNiN17-7-5). Their mechanical properties are significantly better than those of the CrNi-steels 18 10. However, their corrosion resistance is lower than that of the steels with higher carbon content. Figures 141 and 142 give a comparison of the corrosion resistance in sulfuric acid at concentrations up to 12 %. Molybdenum contents have a similar effect in Mn-alloyed steels as at the same level in austenitic CrNi-steels. More so than in these, molybdenum has the effect of stabilizing delta ferrite, which means that the steels SAE 201 and 203 mentioned already become austenitic-ferritic after customary heat treatment with Mo-contents of 1 % and above. A purely austenitic structure is obtained by compensating for Mo with higher Ni-contents. The grades SAE 216 (CrMnNiMo 18 10 8 2.5) and SAE 217 (CrMnNiMo 18 10 10 3.5) were developed analogously to the steels SAE 316 (X5CrNiMo17-12-2) and SAE 317 (X3CrNiMo18-12-3). They are more resistant than the Mo-free steels SAE 201 and 202, but less resistant than SAE 316 (Figure 142). The ferritic-austenitic steel SAE 202 containing 2.5 % Mo is more resistant than SAE 316 in sulfuric acid as can also be seen from Figure 142. Alongside these considerations, it should be remembered that these steels, for example, have a greater susceptibility to intercrystalline corrosion than the usual austenites, as a result of their readiness to undergo precipitation and their structural instability. They are of less significance for use in chemical engineering, although additional Cu-contents of about 1 % increase the corrosion resistance further.

In the annealed state, the austenitic CrMn-steel 18 18 achieves higher strengths than the steels SAE 304 (X5CrNi18-10) and SAE 316 (X5CrNiMo17-12-2). It has a better resistance than the CrNiMo-steel SAE 316 in dilute sulfuric acid (5 and 10 %), and is of comparable resistance in 25 % acid at room temperature. In oxidizing sulfuric acid however, the CrMn-steel 18 18 is less resistant than SAE 316, a corrosion rate of 0.31 mm/a (12.2 mpy) being regarded as resistant to only a certain extent, as can be seen from Table 72 [161].

See also Sections Ferritic chromium steels with more than 12 % chromium and Ferritic-austenitic steels with more than 12 % chromium as well as Figures 49 and 64 and Table 41 for the resistance of CrMn-steels. The corrosion behavior of FeCr-alloys can be seen in Figures 50 and 77 and Tables 42 and 43. A substantial increase in corrosion resistance in sulfuric acid can be achieved for these groups of materials by cathodic alloying, as can be seen also from the figures and tables just mentioned. For cathodic alloying, see Section Ferritic chromium steels with more than 12 % chromium.

An improvement in the corrosion resistance in sulfuric acid in the region of the "resistance gap" of metallic materials is achieved with the cast alloy Ni-Resist®, as can be seen from Figure 143. This cast material containing 3 % C, 17 % Ni, 7 % Cu, 2 % Cr, 2 % Si and as balance Fe is also non-resistant in sulfuric acid in the range between about 30 and 70 %, but exhibits lower corrosion rates than CrNi-steel (Figure 69) as well as CrNiMo-steels (Figure 97 and 98), in the region around 60 % H_2SO_4 at higher temperatures in the latter case. Table 73 contains a summary of the corrosion rates of the alloy Ni-Resist® 1 (cf. EN-JL3011) in sulfuric acid solutions of different concentrations and at various temperatures and states of agitation. Corrosion rates of other austenitic, alloyed, iron-based cast materials in 1 and 65 % H_2SO_4 are summarized in Table 74 [60]. If the alloys from this table are compared with lamellar and nodular graphite, it is found that the corrosion rate of the cast iron with a lamellar graphite construction in 1 % H_2SO_4 is considerably higher than that

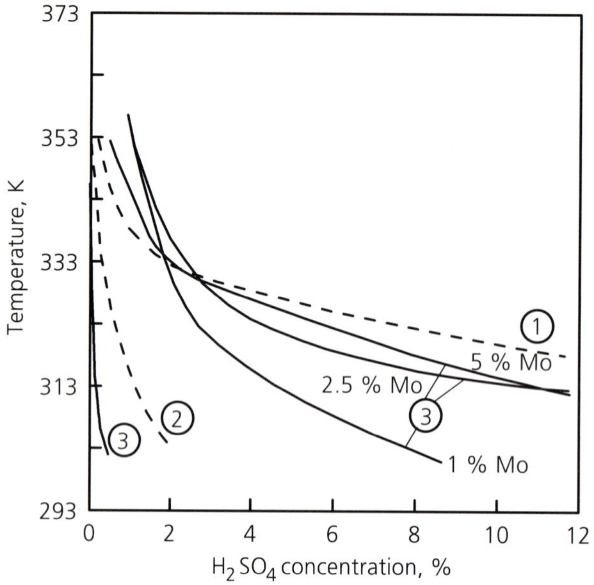

Figure 141: Isocorrosion curves (0.254 mm/a (10.0 mpy)) of CrNi-steels in comparison with CrMnNi-steels with and without Mo-additions in non-aerated H_2SO_4 [163, 242]
① SAE 316 (1.4401); ② SAE 302 (1.4300); ③ SAE 201 (CrNiMn-steel 17 6 4)

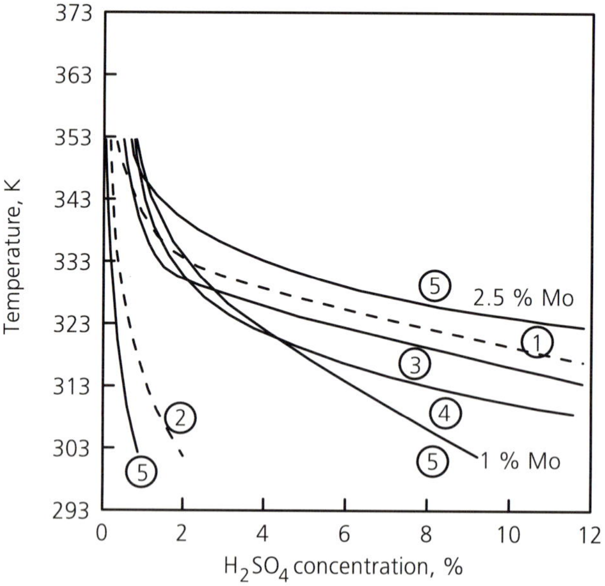

Figure 142: Isocorrosion curves (0.254 mm/a (10.0 mpy)) of CrNi-steels with and without Mo in comparison with CrNiMn-steels with and without Mo in non-aerated sulfuric acid [163, 242]
① SAE 316 (1.4401); ② SAE 302 (1.4300); ③ SAE 217 (CrNiMnMo-steel 18 10 10 3.5);
④ SAE 216 (CrNiMnMo-steel 18 10 8 2.5); ⑤ SAE 202 (1.4371)

containing nodular graphite. Similar circumstances are also to be found in 65 % sulfuric acid with the other alloy containing lamellar graphite, although not so obviously [60].

The steel 08Ch18G8N2T ((%) Fe-18Cr-8Mn-2Ni) is particularly suitable for use in chamber acid downstream of the condenser (for details, see Table 6). See – Corrosion behavior in sulfuric acid-acid mixtures – in Sections Austenitic chromium-nickel steels to Austenitic chromium-nickel steels with special alloying additions for the influence of nitric acid, which has an inhibiting effect on the corrosion of CrNi-steels.

Corrosion rates of between 0.12 and 1.22 mm/a (4.72 and 48.0 mpy) are quoted for the FeNiCr-alloy Incoloy® 800, 1.4876 (X10NiCrAlTi32-21) in sulfuric acid concentrations of 15 to 70 % at room temperature [243].

Laboratory studies on the alloy Incoloy® 800 over a period of 10 days resulted in a corrosion rate of 4.6 mm/a (181 mpy) in 5 % H_2SO_4 at 373 K (100 °C) and 7.5 mm/a (295 mpy) in 40 % H_2SO_4 at 333 K (60 °C). It is therefore less resistant than the CrNiMo-steel SAE 316 (X5CrNiMo17-12-2) [13].

Material	*H_2SO_4 concentration %*	*Temperature K (°C)*	*Corrosion rate mm/a (mpy)*
CrMn 18 18	5	353 (80)	0.04 (1.57)
SAE 304 (X5CrNi18-10)	5	353 (80)	2.11 (83.1)
SAE 316 (X5CrNiMo17-12-2)	5	353 (80)	1.18 (46.5)
CrMn 18 18	10	353 (80)	0.04 (1.57)
SAE 304	10	353 (80)	7.47 (294)
SAE 316	10	353 (80)	2.92 (115)
CrMn 18 18	25	room temperature	0.09 (3.54)
SAE 304	25	room temperature	0.75 (29.5)
SAE 316	25	room temperature	0.07 (2.76)
CrMn 18 18	96	323 (50)	0.96 (37.8)
SAE 304	96	323 (50)	0.48 (18.9)
SAE 316	96	323 (50)	0.31 (12.2)

Table 72: Comparison of the corrosion behavior of the austenitic CrMn-steel 18 18 with the CrNi-steel SAE 304 and the CrNiMo-steel SAE 316 in sulfuric acid [161]

H_2SO_4 concentration %	Temperature K (°C)	Corrosion rate mm/a (mpy)	Remarks
1	room temperature	0.13 to 0.20 (5.12 to 7.87)	duration 3 × 48 h
1	room temperature	0.26 (10.2)	
5	room temperature	0.37 (14.6)	
5	room temperature	0.5 (19.7)	no aeration
5	303 (30)	2.29 (90.2)	with agitation
5	363 (90)	9.5 (374)	
5	363 (90)	9.5 (374)	no aeration
7.8	368 to 373 (95 to 100)	10.2 (402)	
10	room temperature	0.20 to 0.25 (7.87 to 9.84)	duration 3 × 48 h
10	303 (30)	3.0 (118)	aerated
10	303 (30)	2.1 (82.7)	with agitation
10	303 (30)	0.5 (19.7)	no agitation
10	305 (32)	15.5 (610)	gently aerated
10	363 (90)	13.5 (531)	
20	room temperature	0.42 (16.5)	
25	room temperature	0.05 to 0.08 (1.97 to 3.15)	duration 3 × 48 h
25	333 (60)	0.2 (7.87)	acid slurry
30	300 (27)	1.25 (49.2)	no aeration
30	303 (30)	1.4 (55.1)	no agitation
30	303 (30)	1.52 (59.8)	with agitation
30	363 (90)	32.3 (1,272)	no aeration
30	363 (90)	32.8 (1,291)	
78	368 to 373 (95 to 100)	14.2 (559)	
80	303 (30)	0.5 (19.7)	no aeration
80	303 (30)	0.4 (15.7)	no agitation
80	303 (30)	0.5 (19.7)	with agitation
80	363 (90)	6.25 (246)	no aeration
80	363 (90)	6.25 (246)	

Table 73: Summary of corrosion rates of the cast iron alloy Ni-Resist® 1 (EN-JL3011) in sulfuric acid [60]

Table 73: Continued

H_2SO_4 concentration %	Temperature K (°C)	Corrosion rate mm/a (mpy)	Remarks
86	333 (60)	0.76 (30.0)	
95	353 (80)	< 0.125 (< 4.92)	
95	boiling temperature	< 0.125 (< 4.92)	
96	303 (30)	1.2 (47.2)	with agitation
96	303 (30)	0.13 (5.12)	no agitation

Table 73: Summary of corrosion rates of the cast iron alloy Ni-Resist® 1 (EN-JL3011) in sulfuric acid [60]

Material	H_2SO_4 concentration %	Temperature K (°C)	Corrosion rate mm/a (mpy)
GGL-NiCr 20 2 † (cf. UNS F41002)	1	333 (60)	2.4 (94.5)
	65	308 (35)	0.33 (13.0)
GGG-NiCr 20 2 † (EN-JS3011)	1	333 (60)	0.43 (16.9)
	65	308 (35)	0.33 (13.0)
GGG-NiCr 20 3 † (cf. UNS F43001)	1	333 (60)	0.50 (19.7)
	65	308 (35)	0.25 (9.84)
GGG-NiSiCr 20 4 2 †	1	333 (60)	0.25 (9.84)
	65	308 (35)	0.10 (3.94)
GGL-NiSiCr 20 4 3 †	1	333 (60)	0.90 (35.4)
	65	308 (35)	0.13 (5.12)

†: old designation

Table 74: Comparison of the corrosion behavior of austenitic cast iron alloys in 1 and 65 % sulfuric acid [60]

The high-alloy Si-steel 0Ch18N20S3M3D3B containing 18 % Cr, 20 % Ni, 3 % Si, 3 % Mo, 3 % Cu and Fe as balance displays a particularly good corrosion behavior in operating media in the production of caprolactam, with corrosion rates of sometimes far below 0.1 mm/a (3.94 mpy), and is thus, in some cases, superior to the CrMn-steel 10Ch14AG15 containing 14 % Cr and 15 % Mn as expected (Table 49).

An iron alloy containing 20 % Co, 14 % Cr, 2 % Mo and 2 % Cu is no longer usable in 10 % sulfuric acid even at room temperature. The corrosion rate after quenching from 1,323 K (1,050 °C) is 2.4 mm/a (94.5 mpy). After tempering, in each case for 4 hours at 773 and 923 K (500 and 650 °C) and cold working, the rate sometimes rises to values far above this [244].

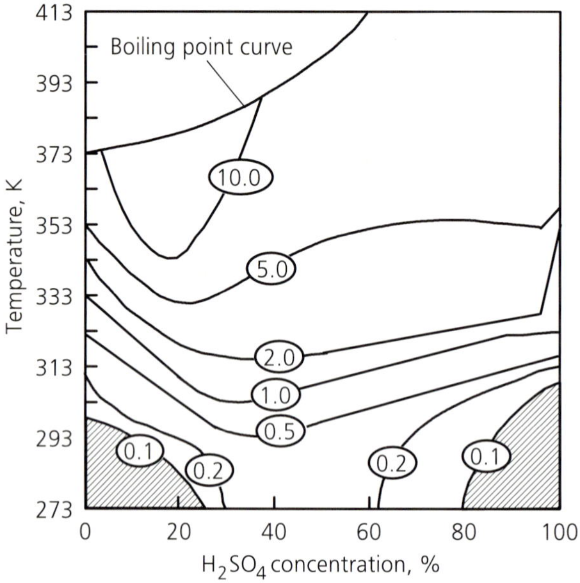

Figure 143: Isocorrosion curves (mm/a) of the cast iron alloy Ni-Resist® ((%) Fe-3C-17Ni-7Cu-2Cr-2Si) in various concentrations of sulfuric acid [61]

The corrosion behavior of the austenitic high-silicon steel 1.4361 (X1CrNiSi18-15-4) containing 4 % Si in dilute sulfuric acid solutions is comparable to that of the standard materials 1.4541 (X6CrNiTi18-10) and 1.4571 (X6CrNiMoTi17-12-2), whereas in hot concentrated sulfuric acid it is significantly superior. In this context, see Figures 47 and 135 for a comparison with nickel alloys. In static 95 % sulfuric acid at 373 K (100 °C) it has the best resistance of the materials shown in Figure 47 and 135 with a corrosion rate well below 0.1 mm/a (3.94 mpy). It is particularly suitable for use in 95 % sulfuric acid at 373 K (100 °C) for heat recovery if boiler feed water is using for cooling. Its resistance to pitting corrosion in artificial sea water is even lower than that of the steel X6CrNiMoTi17-12-2 [102, 245].

In hot concentrated H_2SO_4, the steel 1.4361 shows no potential oscillation and no passive-active transitions (see also Sections Austenitic chromium-nickel steels to Austenitic chromium-nickel steels with special alloying additions and Table 61).

The range of use in highly concentrated sulfuric acid is limited to certain concentrations, as the sometimes very high corrosion rates below 95 % H_2SO_4 demonstrate. For this reason and because of its higher resistance to pitting corrosion, the alloy Hastelloy® C-276 is used for heat exchangers.

– Stress corrosion cracking in sulfuric acid solutions –

Stress corrosion cracking may occur in metallic materials, especially steels, under the influence of atomic hydrogen (in this context, see Section Unalloyed steels and cast steel). For testing of the influence of hydrogen see Section Austenitic chromium-nickel steels. According to this graph, the materials 1.4841 (X15CrNiSi25-21, SAE 310), 1.4876 (X10NiCrAlTi32-21, UNS N08800) and 1.5662 (X8Ni9, cf. UNS K81340) in this section undergo substantial losses in elongation after fracture as a result of acid corrosion or possible H-exposure. The material X8Ni9 proves to be the worst, followed by X15CrNiSi25-21 and X10NiCrAlTi32-21 [183, 184]. In the case of X8Ni9, very short exposure times with hydrogen, in comparison with other materials, are sufficient to cause damage similar to that on pure iron.

Recommended ferrous materials

The choice of ferrous materials in sulfuric acid solution is not without problems, as can be seen from Figures 1, 16, 24, 39, 44, 48, 55, 66–70, 96–100, 104–108, 124–126, 128–130 and 143 amongst others. Resistance with relatively low corrosion rates is limited to a low or very low concentration range, or to a very high concentration range. In between, a resistance gap exists in the concentration range between about 20 and 80 %, with corrosion rates which rule out the use of many steels, especially at elevated temperatures. As a result of the evolution of hydrogen during acid corrosion, the mechanical properties of some materials may be impaired. Agitation and additions can reduce the aggressiveness of the acid solution. The presence of chlorides leads to a deterioration in the corrosion resistance (e.g. pitting corrosion) in all cases where the corrosion resistance of the steel is due to the formation of a passive layer.

Generally, the corrosion resistance of ferrous materials increases as the alloying content increases, as can be seen, for example, from Figure 70. High-Si cast iron exhibits extensive resistance in sulfuric acid, but can be used only to a limited degree because of its poor processability (see Figure 39). Materials shown in Figure 45, for example, can be used in its region of low resistance. See Figure 47 and 135 for the use of metallic materials in hot concentrated sulfuric acid.

The ferritic steel 1.4575 (X1CrNiMoNb28-4-2) is particularly resistant in hot 98.5 to 99 % sulfuric acid. The corrosion rates at a temperature of 448 K (175 °C) and flow rates of 1.3 m/s are below 0.01 mm/a (0.39 mpy). No critical corrosion phenomena were found on this material after more than 10,000 hours of operation [246].

When steels are used in sulfuric acid solutions, after the material has been chosen it is always advisable to consult the steel producer or to perform appropriate experiments corresponding to operational conditions or equivalent to field tests.

The application range of ferrous materials in sulfuric acid solutions can be extended by anodic or cathodic protection methods, and in the case of unalloyed and low-alloy steels with the aid of both processes. The usability of high-alloy steels can be improved by means of anodic protection. These steels cannot be protected cathodically in acids however, since the corrosion rate almost always increases in acid media at potentials which are more negative than the free corrosion potential.

Further possibilities for inhibiting corrosion in sulfuric acid solutions are the use of inhibitors or, if possible, mixing with oxidizing media.

Bibliography

[1] Barker, W.; Evans, T. E.; Williams, K. J.
Effect of alloying additions on the microstructure, corrosion resistance and mechanical properties of nickel-silicon alloys
Br. Corros. J. *5* (1970) March, p. 78

[2] Rabald, E.
Einiges über das Verhalten von Werkstoffen gegenüber Schwefelsäure (Some information about the behavior of materials towards sulfuric acid) (in German)
Werkst. Korros. *7* (1956) 11, p. 652

[3] Dechema-Werkstoff-Tabelle "Schwefelsäure"
(Corrosion data sheets "sulfuric acid") (in German), Dec. 1971
DECHEMA, D-6000 Frankfurt

[4] Groth, V. J.; Hafsten, R. J.
Corrosion of refinery equipment by sulfuric acid and sulfuric acid sludges
Corrosion *10* (1954) 11, p. 368

[5] Sander, U.; Rothe, U.; Gerken, R.
in: Winnacker – Küchler "Chemische Technologie – Anorganische Technologie I" (Chemical technology – inorganic technology I) (in German), 4th ed., vol. *2*, p. 35
Carl Hanser Verlag, München-Wien, 1982

[6] McDowell, D. W.
Sulfuric acid plants – materials of construction
Chem. Eng. Progr. *71* (1975) 3, p. 69

[7] McDowell, D. W.
Handling sulfuric acid
Chem. Eng. *81* (1974) 24, p. 118

[8] Nowakowsky, W. M.
Korrosion von Eisen in konzentrierter Schwefelsäure (Corrosion of iron in concentrated sulfuric acid) (in German)
Chem. Ing.-Tech. *25* (1953) 11, p. 690

[9] Tödt, F.
Korrosion und Korrosionsschutz (Corrosion and corrosion protection) (in German), 2nd ed.
Verlag Walter de Gruyter & Co., D-1000 Berlin, 1961

[10] van der Hoeven, H. W.
Strömungsgeschwindigkeit als besonderer Faktor bei der Schwefelsäurekorrosion (Flow rate, a special factor in sulfuric acid corrosion) (in German)
Werkst. Korros. *6* (1955) 2, p. 57

[11] Bartonichek, R.
Corrosion during heat exchange (in Russian)
Zashch. Met. *15* (1979) 3, p. 298

[12] Whitman, W. G.; Russel, R. P.
The acid corrosion of metals, effect of oxygen and velocity
Ind. Eng. Chem. *17* (1925) 4, p. 348

[13] Ellison, B. T.; Schmeal, W. R.
Corrosion of steel in concentrated sulfuric acid
J. Electrochem. Soc. *125* (1978) 4, p. 524

[14] Klyuchnikov, N. G.; Verizhskaya, E. V.
Corrosion of certain metals in sulfuric acid solutions passed through a magnetic field (in Russian)
Zashch. Met. *8* (1972) 6, p. 700

[15] Albrecht, J.; Bühler, H.-E.; Baumgartl, S.
Einfluß des Phosphors auf die Korrosionsgeschwindigkeit weicher, unlegierter Stähle in verdünnter Schwefelsäure
(Effect of phosphorus on the corrosion rate of mild, unalloyed steels in dilute sulfuric acid) (in German)
Arch. Eisenhüttenwes. *45* (1974) 8, p. 561

[16] Product Information
Korrosionsbeständigkeit nickelhaltiger Werkstoffe gegenüber Fluor, Fluorwasserstoff, Flußsäure und anderen Fluorverbindungen
(Corrosion resistance of nickel-containing materials to fluoride, hydrogen fluoride, hydrofluoric acid and other fluorine compounds) (in German), Information No. 59, 1st ed., March 1970
International Nickel Deutschland GmbH, D-4000 Düsseldorf 1

[17] Wolf, G. K.
Ionenbeschuß als Methode in der Korrosionsforschung und -verhütung (Ion bombardment – a method for investigating and preventing corrosion) (in German)
Werkst. Korros. *30* (1979) 12, p. 853

[18] Ferber, H.; Kasten, H.; Wolf, G. K.; Lorenz, W. J.; Schweickert, H.; Folger, H.
The influence of ion implantation on the corrosion behavior of iron in acid solution
Corros. Sci. *20* (1980) p. 117

[19] Lazurina, L. P.; Maltzeva, V. P.; Novikova, O. P.; Boikov, A. V.; Morozov, V. V.; Pichugin, N. M.
Corrosion resistance of steel in a (leather) pickling solution (in Russian)
Kozh. -Obuv. Prom. (1983) 9, p. 40

[20] Heitz, E.; v. Meysenbug, C. M.
Die Korrosion von Eisen, Nickel, Kupfer und Aluminium in organischen Lösungsmitteln mit geringem Mineralsäuregehalt
(The corrosion of iron, nickel, copper and aluminium in organic solvents containing a small amount of mineral acid)
(in German)
Werkst. Korros. *16* (1965) 7, p. 578

[21] Koebel, M.; Elsener, M.
Korrosion von ölgefeuerten Zentralheizungskesseln
(Corrosion of oil-fired domestic boilers)
(in German)
Werkst. Korros. *40* (1989) 5, p. 285

[22] Rabald, E.
Corrosion Guide, 2nd revised edition
Elsevier Publishing Company, Amsterdam-London-New York, 1968

[23] Poluboyartseva, L. A.; Reifer, A. A.
Corrosion resistance of steels and alloys in sulfuric acid production (in Russian)
Khim. Prom. (1980) 8, p. 481

[24] Yurlova, L. N.; Lyashenko, A. Z.
The corrosion behavior of steels in the production of sodium silicofluoride
(in Russian)
Khim. Prom. (1983) 7, p. 447

[25] Zotikov, V. S.; Bakhmutova, G. B.; Bocharova, N. A.
The use of apparatus made of the steel OCh23N28M3D3T in the production of hydrogen fluoride (in Russian)
Khim. i. Neft. Mash. (1973) 5, p. 21

[26] Product Information
Corrosion tables for stainless steels and titanium, ed. 1979
Avesta GmbH, D-4000 Düsseldorf

[27] Prötzl, M.
Werkstoffauswahl im Chemieanlagenbau
(Selection of materials in chemical plant construction) (in German)
Erdöl und Kohle – Erdgas – Petrochemie vereinigt mit Brennstoff-Chemie *31* (1978) 12, p. 572

[28] Mayer, K. H.; König. H.
Spannungsrißkorrosionsuntersuchungen an warmfesten Schraubenwerkstoffen für Dampfturbinen in turbinenspezifischen Lösungen
(Investigations into stress corrosion cracking of heat-resistant screw stock for steam turbines in turbine-specific solutions) (in German)
Final report FE-Project-No. O-B 2.3/5, Dec. 1984
DECHEMA, D-6000 Frankfurt

[29] Mayer, K. H.; König. H.
Spannungsrißkorrosionsuntersuchungen an warmfesten Schraubenwerkstoffen für Dampfturbinen in turbinenspezifischen Lösungen
(Investigations into stress corrosion cracking of heat-resistant screw stock for steam turbines in turbine-specific solutions)
(in German)
Werkst. Korros. *36* (1985) 11, p. 524

[30] Schmitt-Thomas, Kh. G.; Stengel, W.
Möglichkeiten zur Früherkennung von Wasserstoffschädigungen in metallischen Werkstoffen durch Anwendung der Schallemissionsanalyse
(Feasibility of early detection of hydrogen damage in metals by acoustic emission analysis) (in German)
Werkst. Korros. *34* (1983) 1, p. 7

[31] Grubitsch, H.; Künne, P.; Hilbert, F.
Über den Einfluß von Thioharnstoff und Thioharnstoffderivaten als Beizinhibitoren auf die Wasserstoffaufnahme, die Inhibitorwirksamkeit und die Wechselbiegefestigkeit von Kohlenstoffstahl bei der Säurebeizung
(Influence of thiourea and its derivatives as pickling inhibitors on hydrogen pickup, degree of inhibition and fatigue load of carbon steel during pickling in acids) (in German)
Werkst. Korros. *31* (1980) 8, p. 626

[32] Riecke, E.; Johnen, B.; Grabke, H. J.
Einflüsse von Legierungselementen auf die Korrosion und Wasserstoffaufnahme von Eisen in Schwefelsäure – Teil I: Permeation, Diffusion und Löslichkeit von Wasserstoff in binären Eisenlegierungen
(Effects of alloying elements on corrosion and hydrogen uptake of iron in sulfuric acid – Part I: Permeation, diffusion and solubility of hydrogen in binary iron alloys) (in German)
Werkst. Korros. *36* (1985) 10, p. 435

[33] Riecke, E.; Möller, R.; Johnen, B.; Grabke, H. J.
Einflüsse von Legierungselementen auf die Korrosion und Wasserstoffaufnahme von Eisen in Schwefelsäure – Teil II: Korrosion und Deckschichtbildung
(Effects of alloying elements on corrosion and hydrogen uptake of iron in sulfuric acid – Part II: Corrosion and formation of surface layers) (in German)
Werkst. Korros. *36* (1985) 10, p. 447

[34] Riecke, E.; Johnen, B.; Grabke, H. J.
Einflüsse von Legierungselementen auf die Korrosion und Wasserstoffaufnahme von Eisen in Schwefelsäure – Teil III: Kinetik der Abscheidung und Aufnahme von Wasserstoff an binären Eisenlegierungen
(Effects of alloying elements on corrosion and hydrogen uptake of iron in sulfuric acid – Part III: Kinetics of proton discharge and hydrogen uptake in binary iron alloys) (in German)
Werkst. Korros. *36* (1985) 10, p. 455

[35] Heitz, E.; Henkhaus, R.; Rahmel, A.
Korrosionskunde im Experiment. Untersuchungsverfahren – Meßtechnik – Aussagen
(Experimental corrosion science. Methods of investigation – measuring technique – results) (in German)
Verlag Chemie, Weinheim-Deerfield Beach (Florida)-Basel, 1983

[36] Dayal, R. K.; Grabke, H. J.
Hydrogen induced stress corrosion cracking in low and high strength ferritic steels of different phosphorus content in acid media
Werkst. Korros. *38* (1987) 8, p. 409

[37] Herbsleb, G; Pöpperling, R.; Schwenk, W.
Potentialabhängigkeit der H-induzierten Korrosion (HIC) und H-induzierten Spannungsrißkorrosion (HSCC) bei Röhrenstählen in schwach sauren und neutralen Medien
(Influence of potential on hydrogen-induced cracking (HIC) and hydrogen-induced stress corrosion cracking (HSCC) of pipeline steels in weak acid and neutral environments) (in German)
Werkst. Korros. *31* (1980) 2, p. 97

[38] Riecke, E.; Johnen, B.
Wasserstoffinduzierte Spannungsrißkorrosion an unverzinkten und verzinkten Baustählen
(Hydrogen-induced stress corrosion cracking of non-galvanized and galvanized construction steels) (in German)
Werkst. Korros. *37* (1986) 6, p. 310

[39] Kuzub, V. S.; Novitski, V. S.; Moysa, V. G.; Kemen, D.; Makerov, V. A.; Kuzub, L. G.; Gnezdilova, V. I.
Anodic protection of sulfuric acid stores at the Tissa chemical plant (in Russian)
Zashch. Met. *19* (1983) 1, p. 18

[40] Babej, Yu. I.; Maksimishin, M. D.
Protective efficiency of the inhibitor Kh0SP-10 on steel under conditions of low frequency stress due to vibrational corrosion (in Russian)
Fiz.-Khim. Mekh. Mater. *12* (1976) 5, p. 82

[41] Bartsch, E.; Danninger, H.
Über die Inhibitorwirkung der Sorbinsäure
(The inhibitory effect of sorbic acid) (in German)
Werkst. Korros. *38* (1987) 2, p. 73

[42] Kuron, D.; Gräfen, H.; Rother, H.-J.
Anwendung von Beizinhibitoren
(Application of pickling inhibitors) (in German)
Werkst. Korros. *37* (1986) 5, p. 223

[43] Hudson, R. M.; Butler, T. J.; Warning. C. J.
The effect of pyrrole-halide mixtures in inhibiting the dissolution of low-carbon steel in sulphuric acid
Corros. Sci. *17* (1977) 7, p. 571

[44] Przewlocka, H.; Bala, H.
Hemmung der sauren Korrosion unterschiedlicher Kohlenstoffstähle unter Verwendung von Thioharnstoff
(Inhibition of the acid corrosion of various carbon steels) (in German)
Werkst. Korros. *32* (1981) 4, p. 443

[45] Kuron, D. et al.
Wasserstoff und Korrosion
(Hydrogen and corrosion) (in German),
Bonner Studien Reihe, p. 275
Verlag Irene Kuron, Bonn, 1986

[46] Chen, N. G., Kiryukha, A. S.; Chen, L. N.; Sokolyan, L. N.; Panfilova, Z. V.
Inhibitors of acid corrosion of metals EK and EK-1 (in Russian)
Zashch. Met. 9 (1973) 2, p. 211

[47] Kiriluk, S. S.; Titakova, I. K.; Korsunskaya, A. L.; Miskidzhyan, S. P.
Anticorrosion action of some urotropin derivatives (in Russian)
Zashch. Met. *16* (1980) 2, p. 180

[48] Abd El Rehim, S. S.; Tohamy, F. M.; Select, M. M.
Effect of some polyamino polycarboxylic acids on the corrosion of steel in sulphate solutions
Surf. Tech. *21* (1984) 2, p. 169

[49] Schmitt, G.
Application of inhibitors for acid media
Br. Corros. J. *19* (1984) 4, p. 165

[50] Dhirendra, Thesis, Ph. D.
Korrosion von Eisen und Eisenlegierungen in Schwefelsäure unter schwierigen Betriebsbedingungen und ihre Abschwächung
(Corrosion of iron and iron alloys in sulfuric acid under difficult operating conditions and its inhibition) (in German)
Korrosion *17* (1986) 2, p. 63

[51] Schmitt, G.
Beizinhibitoren – Hemmstoffe oder Promotoren für die Wasserstoffschädigung
(Pickling inhibitors – inhibitors or promoters of hydrogen damage)
(in German)
Oberfläche Surf. *28* (1987) 1/2, p. 8

[52] Hommel, G. et al.
Handbuch der gefährlichen Güter,
4. Auflage: Merkblatt 174 "Rauchende Schwefelsäure"; Merkblatt 183 "Schwefelsäure"; Merkblatt 183a "Batterie-Säure" (Schwefelsäure mit nicht mehr als 51 % Säure); Merkblatt 184 "Schwefeltrioxid" (stabilisiert)
(Handbook of dangerous goods, 4th edition; sheet 174 "fuming sulfuric acid"; sheet 183 "sulfuric acid"; sheet 183a "battery acid" (sulfuric acid with not more than 51 % acid), sheet 184 "sulfur trioxide" (stabilized)) (in German)
Springer Verlag, Berlin-Heidelberg-New York-London-Paris-Tokyo, 1987

[53] Burbridge, J. F.
Corrosion-resistant reinforced polyester for process plant
Anticorros. Methods Mater. *23* (1976) 2, p. 7

[54] Fyfe, D.; Vanderland, R.; Rodda, J.
Sulfuric acid plant operations: Corrosion in sulfuric acid storage tanks
Chem. Eng. Progr. *73* (1977) 3, p. 65

[55] Zhuravleva, L. V.; Kasinskaya, L. L.; Nosivets, L. A.
Corrosion resistance of the steel 08Ch18G8N2T in sulfuric acid (in Russian)
Khim. Neft. Mashinostr. (1979) 12, p. 21

[56] Ruge, J.; Trarbach, K.
Plasmaauftragschweißen mit Heißdrahtelektrode von Sonderwerkstoffen (Plasma deposition welding of special materials using a hot wire electrode) (in German)
Schweissen Schneiden *34* (1982) 8, p. 369

[57] Fichtl, W.
Über neue Erkenntnisse auf dem Gebiet des Oberflächenborierens
(On recent findings in the field of surface boriding) (in German)
Oberflächentechnik *51* (1974) 12, p. 535

[58] Sander, U.; Rothe, U.; Kola, R.
in: Ullmanns Encyklopädie der technischen Chemie
(Ullmann's encyclopedia of industrial chemistry) (in German), 4th ed., vol. *21*, p. 118
Verlag Chemie, Weinheim-Deerfield Beach (Florida)-Basel, 1982

[59] Corrosion Data Survey, Metals Section, 6th ed., 1985
National Association of Corrosion Engineers, Houston (USA)
[60] Katz, W.
Gießen für die Chemie – Das Verhalten metallischer Gußwerkstoffe unter korrosiver Beanspruchung
(Casting in chemistry – the behavior of metallic cast materials under corrosive attack) (in German)
VDI-Verlag GmbH, D-4000 Düsseldorf
[61] Berg. F. F.
Korrosionsschaubilder
(Corrosion diagrams) (in German)
VDI-Verlag GmbH, D-4000 Düsseldorf, 1969
[62] Product Information
Western zirconium – Corrosion resistance of zirconium and other metals, 1980
Robert Zapp, D-4000 Düsseldorf
[63] Bünger, J.
Die Korrosion durch Schwefelsäure (The corrosion caused by sulfuric acid) (in German)
Werkst. Korros. *7* (1956) 6, p. 322
[64] Podschus, R.
in: Ullmanns Encyklopädie der technischen Chemie (Ullmann's encyclopedia of industrial chemistry) (in German), 3rd ed., vol. *15*, p. 970
[65] Kuzyukov, A. N.; Khanzaddev, I. V.; Pishchev, Yu. S.
Corrosion of plants in sulfuric acid production (in Russian)
Khim. Prom. (1984) 4, p. 232
[66] Katz, W.
Die Beständigkeit von Gußeisen in Schwefelsäure
(The resistance of cast iron in sulfuric acid) (in German)
Konstruieren und Gießen (1976) 2, p. 28
[67] Roll, F.
Beitrag zum Verhalten von Temperguß unter dem Einfluß von Korrosion
(Contribution to the behavior of malleable cast iron under the action of corrosion) (in German)
Werkst. Korros. *12* (1961) 4, p. 209, 368
[68] Maahn, E.
Corrosion of cast iron in concentrated sulphuric acid under potentiostatic conditions
Br. Corros. J. *1* (1966) Nov., p. 350
[69] Gramberg. U.; Günther, T.
Untersuchungen an einem durch Oleum geschädigten Rührwerksbehälter aus Gußeisen mit Lamellengraphit
(Examination of a lamellar graphite cast iron mixing chamber damaged by oleum) (in German)
Prakt. Metallogr. *10* (1973) 12, p. 682
[70] v. Plessen, H.; Vogt, H.
Korrosion von grauem Gußeisen in konzentrierter Schwefelsäure
(Corrosion of gray cast iron in concentrated sulfuric acid) (in German)
Z. Werkstofftech. *14* (1983) 5, p. 141
[71] Salminkeit, V.; v. Plessen, H.; Vollmüller, H.
Die Korrosion der Gußeisenkessel beim Pauling-Verfahren
(Corrosion of the cast iron boiler in the Pauling process) (in German)
Chem.-Ing.-Tech. *53* (1981) 10, p. 822
[72] Donaldson, E. G.
Der Einsatz von Meehanite®-Gußeisen in Chemieanlagen
(The use of Meehanite® cast iron in chemical plants) (in German)
Chemie-Technik *4* (1975) 4, p. 127
[73] Dilthey, U.; Wanke, R.
Hochleistungs-Schweißplattierverfahren für den Chemie-Apparatebau
(The high-performance cladding process in chemical apparatus engineering) (in German)
Chem.-Ing.-Tech. *46* (1974) 11, p. 467
[74] Ashby, W. A.; Evans, L. S.; Shepherd, W.
Anodic protection of cast iron in boiling sulphuric acid under industrial conditions
Br. Corros. J. *13* (1978) 2, p. 85
[75] Rabald, E.
in: Ullmanns Encyklopädie der technischen Chemie
(Ullmann's encyclopedia of industrial chemistry) (in German), 3rd ed., vol. *1*, p. 920
Urban & Schwarzenberg, München-Berlin, 1951

[76] Tödt, F.
Korrosion und Korrosionsschutz
(Corrosion and corrosion protection)
(in German)
Verlag Walter de Gruyter & Co., D-1000
Berlin, 1955

[77] Ford, E.
Tantiron® – Eigenschaften und
Anwendung bei der Förderung aggressiver
Flüssigkeiten
(Tantiron® – Properties and use in the
extraction of aggressive liquids)
(in German)
Oberflächentechnik *50* (1973), Metall-
Journal 2/73, p. A3

[78] Wiegand, H.
Werkstoff – Konstruktion – Fertigung
(Material – design – production)
(in German)
Z. Werkstofftech. *8* (1977) 6, p. 209

[79] Droscha, H.
Erfahrungen mit Guß im Chemiebau
(Experiences with cast iron in chemical
plant engineering) (in German)
Chem.-Anl. + Verfahren (1973) 4, p. 79

[80] Zorin, A. I.; Dergunov, B. D.;
Nikitenko, E. A.; Saprykin, G. D.
Choice of new corrosion-resistant materials
for the production of sulfuric acid
(in Russian)
Zashch. Met. *7* (1971) 5, p. 579

[81] Rabald, E.
Werkstoffe, physikalische Eigenschaften
und Korrosion
(Materials, physical properties and
corrosion) (in German), vol. *1*
Verlag von Otto Spamer, Leipzig, 1931

[82] Saldanha, B. J.; Streicher, M. A.
Effect of silicon on the corrosion resistance
of iron in sulfuric acid
Mater. Performance *25* (1986) 1, p. 37

[83] Saldanha, B. J.; Streicher, M.
Effect of silicon on the corrosion resistance
of iron in sulfuric acid Corrosion in
sulfuric acid. (Proc. Conf.), Boston, March
1985, p. 41
National Association of Corrosion
Engineers, Houston (Tx/USA)

[84] NACE Publication
Corrosion in sulfuric acid
Proceedings of the Corrosion '85,
Symposium on Corrosion in Sulfuric Acid,
Catalog Card Number: 85-61573
National Association of Corrosion
Engineers, Houston (Texas/USA), 1985

[85] Uhlig. H.
Korrosion und Korrosionsschutz
(Corrosion and corrosion protection)
(in German)
Akademie-Verlag, Berlin, 1970

[86] Shreir, L.L.
Corrosion, metal/environment reactions,
vol. *1*
Newnes-Butterworths, London-Boston,
1976

[87] Uhlig, H. H.
Korrosion und Korrosionsschutz
(Corrosion and corrosion protection)
(in German)
Akademie-Verlag, D-1000 Berlin, 1970

[88] Swandby, R. K.
Corrosion charts: Guides to materials
selection
Chem. Eng. *69* (1962) Nov. 12, p. 186

[89] Product Information
Zirkonium im Säuretest, Ausgabe zur
ACHEMA 1988 (Zirconium in acid testing,
publication for ACHEMA 1988)
(in German)
TISTO Titan u. Sonderlegierungen GmbH,
D-4000 Düsseldorf

[90] Frank, K.
Die Anwendung von 65–70 %igem
Ferrosilizium als Säureschutz im
chemischen Apparatebau
(The use of 65 %–70 % ferrosilicon for acid
protection in chemical apparatus
engineering) (in German)
Dechema Monographien *13* (1943) p. 71

[91] Novitski, V. S.; Kuzub, L. G.; Kuzub, V. S.;
Odarchenko, V. V.; Gnezdilova, V. I.;
Kuzyukov, A. N.
Corrosion and electrochemical behavior of
modified silicon pig iron in 40 % sulfuric
acid (in Russian)
Zashch. Met. *18* (1982) 6, p. 924

[92] Gräfen, H. et al.
Die Praxis des Korrosionsschutzes, Kontakt und Studium,
(Corrosion protection in practice; dealings with and study of) (in German), vol. *64*
expert verlag, D-7031 Grafenau, 1981

[93] Product Information
Chemische Beständigkeit der nichtrostenden REMANIT®-Stähle
(Chemical resistance of the stainless Remanit® steels) (in German),
Information No. 1127/2, Dec. 1985
Thyssen Edelstahlwerke AG, D-4150 Krefeld

[94] Skuin, K.; Kreissing, T.
Herstellung und Eigenschaften des ökonomisch mit Nickel legierten, höherfesten, austenitischen rost- und säurebeständigen Stahls X 5 CrNiN 19 7
(Manufacture and properties of the economically Ni-alloyed, high-strength, austenitic, rust and acid-resistant steel X 5 CrNiN 19 7) (in German)
Die Technik *30* (1975) 2, p. 107

[95] Paulekat, F.; Kassat, H.
Anwendung des anodischen Schutzes bei der Kühlung von Schwefelsäure
(Application of anodic protection in the cooling of sulfuric acid) (in German)
vt Verfahrenstechnik *13* (1979) 9, p. 697

[96] Hicks, P. D.; Robinson, F. P. A.
The aqueous corrosion behaviour of an ion implanted 12 % chromium steel
Corros. Sci. *24* (1984) 10, p. 885

[97] Hubler, G. K.; Singer, I. L.; Clayton, C. R.
Mechanical and chemical properties of tantalum-implanted steels
Mater. Sci. Eng. *69* (1985) p. 203

[98] Singh, R. P.; Modi, O. P.; Mungole, M. N.; Singh, K. P.
Corrosion of 2.25 Cr-1 Mo ferritic steel in sulphuric acid and sea water
Br. Corros. J. *20* (1985) 1, p. 28

[99] Ergang. R.; Rockel, M. B.
Die Korrosionsbeständigkeit der nichtrostenden Stähle an der Atmosphäre – Auswertung von Versuchen bis zu 10-jähriger Auslagerung
(The corrosion resistance of stainless steels in the atmosphere – analysis of trials with an ageing period of up to 10 years) (in German)
Werkst. Korros. *26* (1975) 1, p. 36

[100] Diekmann, H.
Hochlegierte nichtrostende Stähle und Nickelbasislegierungen im Chemie-Apparatebau
(High alloy stainless steels and nickel-base alloys in chemical equipment construction) (in German)
Werkst. Korros. *37* (1986) 3, p. 130

[101] Anonymous
X 1 CrNiMoNb 28 4 2 (Remanit® 4575) – Korrosionsbeständigkeit
(X 1 CrNiMoNb 28 4 2 (Remanit® 4575) – corrosion resistance) (in German)
Thyssen Edelstahl Tech. Ber. *5* (1979) 1, p. 48

[102] Renner, M.
Hochlegierte Austenite für die Anwendung in ruhender und bewegter hochkonzentrierter Schwefelsäure (+ 95 %)
(High-alloy austenites for use in static and agitated sulfuric acid of high concentration (+ 95 %)) (in German), Final report FE-project No. 0-d 3.9/1, Sept. 1986
DECHEMA, D-6000 Frankfurt

[103] Renner, M.
Hochlegierte Austenite für die Anwendung in ruhender und bewegter, hochkonzentrierter Schwefelsäure (+ 95 %)
(High-alloy austenites for use in static and agitated sulfuric acid of high concentration (+ 95 %)) (in German)
Werkst. Korros. *38* (1987) 4, p. 191

[104] Lennartz, G.; Kiesheyer, H.
Korrosionsverhalten von hochchromhaltigen, ferritischen Stählen
(Corrosion behavior of high-chromium, ferritic steels) (in German)
DEW-Tech. Ber. *11* (1971) 4, p. 230

[105] Kiesheyer, H.; Lennartz, G.; Brandis, H.
Korrosionsverhalten hochchromhaltiger, ferritischer, chemisch beständiger Stähle
(Corrosion behavior of high-chromium, ferritic, chemically resistant steels) (in German)
Werkst. Korros. *27* (1976) p. 416

[106] Charbonnier, J.-C.
Influence du molybdène sur la résistance des aciers inoxydables à différents types de corrosion, en présence de solutions aqueuses minérales ou de milieux organiques (étude bibliographique)
(Effect of molybdenum on the resistance of stainless steels towards various kinds of corrosion in inorganic solutions or organic media) (in French)
Métaux-Corros.-Ind. (1975) No. 598, p. 201

[107] Chernova, G. P.; Bogdashkina, N. L.; Kalinichenko, V. A.; Dobroljubov, V. V.; Tomashov, N. D.
Corrosion and electrochemical behavior of the steel 90Cr28MFTAL in sulfuric acid (in Russian)
Zashch. Met. *18* (1982) 2, p. 224

[108] Chernova, G. P.; Bogdashkina, N. L.; Tomashov, N. D.; Kalinichenko, V. A.; Dobrolyubov, V. V.; Grushina, V. A.; Kuznetsov, V. l.; Porshneva, A. l.; Shvets, V.A.
Effect of plastic deformation on the dissolution of iron in sulfuric acid (in Russian)
Zashch. Met. *19* (1983) 4, p. 534

[109] Gräfen, H.; Kuron, D.
Neuentwicklung von Bleilegierungen aufgrund elektrochemischer Untersuchungen über das Korrosionsverhalten in Schwefelsäure
(New development of lead alloys based on electrochemical tests on the corrosion behavior in sulfuric acid) (in German)
Werkst. Korros. *20* (1969) 9, p. 749

[110] Tomashov, N. D.; Chernova, G. P.; Tschigirinskaja, L. A.; German, M. F.
Einfluß der Legierungselemente auf die Passivierung und die Korrosionsbeständigkeit der Legierungen auf Eisen-Chrom-Basis
(The effect of alloying elements on the passivation and corrosion resistance of iron-chromium based alloys) (in German)
Werkst. Korros. *27* (1976) 9, p. 636

[111] Tomashov, N. D.
Untersuchungen zur Passivität und Korrosionsbeständigkeit von Legierungen
(Investigations into the passivity and corrosion resistance of alloys) (in German)
Chem. Tech. *30* (1978) 1, p. 6

[112] Tomashov, N. D.; Chernova, G. P.; Kupriyanova, L. S.
Corrosion resistance of palladium-alloyed cast chrome steel (in Russian)
Zashch. Met. *6* (1970) 5, p. 533

[113] Tomashov, N. D.; Chernova, G. P.
New corrosion-resistant alloys based on titanium and high chromium steel with additions of Pd (in Russian)
Zashch. Met. *11* (1975) 4, p. 403

[114] Tomashov, N. D.; Lakomskij, V. I.; Chernova, G. P.; Torkhov, G. F.; Chigirinskaya, L. A.; Slyshankova, V. A.
Corrosion stability of austenitic stainless steels alloyed with palladium and with a high nitrogen content (in Russian)
Zashch. Met. *13* (1977) 1, p. 10

[115] Anonymous
Korrosionsbeständiger Automatenstahl 1802
(Corrosion-resistant free cutting steel 1802) (in German)
Draht *33* (1982) 4, p. 188

[116] Brandis, H.; Kiesheyer, H.
Betrieblich hergestellte nichtrostende Automatenstähle mit Titanzusatz
(Commercially manufactured free cutting stainless steels with titanium additions) (in German)
Thyssen Edelst. Tech. Ber. *11* (1985) 1, p. 45

[117] Brandis, H.; Kiesheyer, H.; Lennartz, G.
Einfluß von Titanzusätzen auf die Korrosionsbeständigkeit nichtrostender Automatenstähle
(Influence of additions of titanium on the corrosion resistance of free cutting stainless steels) (in German)
Thyssen Edelstahl Tech. Ber. *8* (1982) 2, p. 135

[118] Steigerwald, R. F.
New molybdenum stainless steels for corrosion resistance. a review of recent developments
Mater. Performance *13* (1974) 9, p. 9

[119] Davison, R. M. (Product Information)
An 18Cr-2Mo ferritic stainless steel
Climax Molybdenum GmbH, D-4000 Düsseldorf

[120] Smallwood, R. E.
Heat exchanger tubing reliability
Mater. Performance *16* (1977) 2, p. 27

[121] Coutsouradis, D.
The corrosion resistance of some stainless steels alloyed with cobalt
Cobalt (1959) 5, p. 3

[122] Heimann, W.; Kiesheyer, H.
Ein weichmagnetischer Stahl für höchste korrosionschemische Beanspruchung: X 1 CrNiMoNb 28 4 2 (Remanit® 4575)
(A soft magnetic steel for highest corrosion-chemical stress: X 1 CrNiMoNb 28 4 2 (Remanit® 4575)) (in German)
Thyssen Edelst. Tech. Ber. *3* (1977) 1, p. 23

[123] Dillon, C. P.
Use of low interstitial 26Cr-1Mo stainless steel in chemical plants
Mater. Performance *14* (1975) 8, p. 36

[124] Süry, P.
Schwefelsäurekorrosion metallischer Werkstoffe unter extremen Bedingungen
(Corrosion of metallic materials by sulfuric acid under extreme conditions)
(in German)
Chemie-Technik *6* (1977) 10, p. 415

[125] Süry, P.; Brezina, P.
Kurzzeitprüfmethoden zur Untersuchung des Einflusses von Wärmebehandlung und chemischer Zusammensetzung auf die Korrosionsbeständigkeit martensitischer Chrom-Nickel-(Molybdän)-Stähle mit tiefem Kohlenstoffgehalt
(Short-duration test for evaluating the influence of heat-treatment and chemical composition on the corrosion resistance of low carbon martensitic chromium-nickel (molybdenum) steels) (in German)
Werkst. Korros. *30* (1979) 5, p. 341

[126] Gräfen, H.
Entwicklungstendenzen metallischer Werkstoffe aus der Sicht der Verwendung im Chemieanlagenbau
(Developments of metallic materials in view of their use for chemical plants)
(in German)
Chem.-Ing.-Tech. *54* (1982) 2, p. 108

[127] Vermilyea, D. A.
Susceptibility of iron- and nickel-base alloys to SCC in pH 2.5 H_2SO_4 at 289 °C
Corrosion *31* (1975) 12, p. 421

[128] Herbsleb, G.; Stoffels, H.
Zur Prüfung von Schweißplattierungen mit austenitischen Bandelektroden der Art X 2 CrNiNb 24 13 auf Beständigkeit gegen interkristalline Korrosion
(Testing of weld-claddings with austenitic band electrodes of the type X 2 CrNiNb 24 13 for resistance to intercrystalline corrosion) (in German)
Werkst. Korros. *29* (1978) 9, p. 576

[129] Herbsleb, G.; Schwaab, P.
Die Ausscheidungen von intermetallischen Phasen, Nitriden und Carbiden in AF 22 Duplex-Stahl und ihr Einfluß auf das Korrosionsverhalten in Säuren
(Precipitation of intermetallic compounds, nitrides and carbides in AF Duplex steel and their influence on corrosion behavior in acids) (in German)
Mannesmann Forschungsberichte (1983) No. 957, p. 1

[130] Hummel, O.H.
Werkstoffe im Behälter- und Apparatebau
(Materials in vessel construction and apparatus engineering) (in German)
Chemie-Technik *11* (1982) 8, p. 930

[131] Product Information
Ferralium® Alloy 255-3SC. Cast duplex stainless steel, April 1987
Deutsche Langley Alloys GmbH, D-6000 Frankfurt

[132] Product Information
Fermanel® alloy – ein hochlegierter Edelstahl einer neuen Generation, der noch höhere Maßstäbe für Korrosionsbeständigkeit bei hoher Festigkeit setzt
(Fermanel® alloy – a new generation cast high-alloy stainless steel which sets even higher standards for corrosion resistance combined with high strength) (in German), Aug. 1986
Deutsche Langley Alloys GmbH, D-6000 Frankfurt

[133] Horn, E.-M.; Diekmann, H.; Kilian, R.; Kratzer, A.; Stiepel, H.; Tischner, H.
Praxisnahe Prüfung korrosions- und erosionsbeständiger Pumpenwerkstoffe
(Corrosion- and erosion-resistant pump material – investigation under flow conditions close to practice) (in German)
Z. Werkstofftech. *14* (1983) 9, p. 311

[134] Tacke, G.; Köhler, H. J.
Untersuchungen zum Einfluß höherer Mo-Gehalte auf das Korrosionsverhalten von höherfesten nichtrostenden ferritisch-austenitischen Stählen
(Effect of molybdenum on corrosion behavior of high strength ferritic-austenitic stainless steels) (in German)
steel research *58* (1987) 3, p. 129

[135] Koren, M.; Hochörtler, G.
Eigenschaften des ferritisch-austenitischen Stahles X 3 CrMnNiMoN 25 6 4
(Properties of the ferritic-austenitic steel X 3 CrMnNiMoN 25 6 4) (in German)
Stahl Eisen *102* (1982) 10, p. 509

[136] Product Information
Beständigkeit des Gußstahls 9.4462 (G-X 3 CrNiMoCuN 26 6 3) in Schwefelsäure
(Resistance of the cast steel 9.4462 (G-X 3 CrNiMoCuN 26 6 3) in sulfuric acid) (in German), Investigation report, Dec. 1986
Eisen- und Stahlwerk Pleissner GmbH, D-3420 Herzberg

[137] Product Information
Langley corrosion resistance, GA 8692/3M/481
Langley Alloys Ltd., Slough, Berks (GB)

[138] Product Information
Ferritisch austenitischer Stahlguß GX-3 CrNiMoCuN 26 6 3
(Ferritic-austenitic cast steel GX-3 CrNiMoCuN 26 6 3) (in German), Test reports 1987
Eisen- und Stahlwerk Pleissner GmbH, D-3420 Herzberg

[139] Manning. P. E.; Tuff, W. F.; Zordan, R. D.; Schuur, P. D.
Evaluation of high performance alloys used in the pulp and paper industry
Mater. Performance *23* (1984) 1, p. 19

[140] Product Information
Ferralium® Alloy 255 – Hochfester korrosionsbeständiger Edelstahl
(Ferralium® Alloy 255 – High-strength corrosion-resistant steel) (in German), 4th. ed., 1987
Deutsche Langley Alloys GmbH, D-6000 Frankfurt

[141] Katz, W.
ACHEMA '79 – Teil 2: Kurzer Überblick über Werkstoffe und Werkstoffprobleme
(ACHEMA '79 – Part 2: A short survey of materials and material problems) (in German)
Werkst. Korros. *30* (1979) 9, p. 651

[142] Kratzer, A.; Heumann, A.; Wittekindt, W.
Tauchmotorpumpen aus korrosions- und verschleißbeständigen Werkstoffen
(Submersible motor pumps made of corrosion and wear-resistant materials) (in German)
Chemie-Technik *13* (1984) 12, p. 56

[143] Horn, E.-M.; Mollenkopf, G.
Praxisnahe Prüfung von korrosions- und erosionsbeständigen Pumpenwerkstoffen
(Corrosion- and erosion-resistant pump material – investigation under flow conditions close to practice) (in German)
Werkst. Korros. *34* (1983) 3, p. 117

[144] Horn, E.-M.; Diekmann, H.; Kilian, R.; Kratzer, A.; Stiepel, H.; Tischner, H.
Praxisnahe Prüfung korrosions- und erosionsbeständiger Pumpenwerkstoffe
(Investigation of corrosion and erosion resistant pump materials under flow conditions close to practice) (in German)
Z. Werkstofftech. *14* (1983) 10, p. 350

[145] Kuzub, V. S.; Krikun, B. P.; Novitski, V. S.; Khrustenko, T. A.
Effect of speed and temperature of 20 % sulfuric acid on the anodic behavior of stainless steels (in Russian)
Zashch. Met. *18* (1982) 6, p. 920

[146] Kiselev, V. D.; Dobrolyubov, V. V.
The corrosion behavior of stainless steels in hot wet process phosphoric acid (in Russian)
Khim. Prom. (1980) 12, p. 737

[147] Vasileva, V. A.; Dzutsev, V. T.; Moskvichev, I. F., Uljanin, E. A.; Feldgandler, E. G.; Lobova, M. V.
Effect of nitric acid on the corrosion resistance of steels in dilute mixtures of sulfuric and nitric acid (in Russian)
Khim. Prom. (1982) 3, p. 165

[148] Faingold, L. L.; Shatova, T. Yu.
Effect of nitric acid additions on stainless steel corrosion in sulfuric acid (in Russian)
Zashch. Met. *17* (1981) 3, p. 312

[149] Kiselev, V. D.; Dobrolyubov, V. V.
Corrosion resistance of stainless steels in the phosphoric-sulfuric-fluoric acid system (in Russian)
Khim. Prom. (1979) 3, p. 160

[150] Dvali, T. M.
Corrosion resistance of steels in caprolactam production media (in Russian)
Zashch. Met. *9* (1973) 1, p. 58

[151] Poluboyartseva, L. A.; Reifer, A. A.
The corrosion resistance of the steel 0Ch22N5T in oleum (in Russian)
Khim. Prom. (1976) 2, p. 158

[152] Rockel, M.B.; Weidemann, R.
Untersuchung über die Schwingungsrißkorrosion einiger nichtrostender Stähle in Schwefelsäure hoher Konzentration
(Study of the vibration corrosion cracking of some stainless steels in highly concentrated sulfuric acid) (in German)
Werkst. Korros. *33* (1982) 1, p. 40

[153] Altpeter, E.; Rockel, M. B.
Untersuchung über die Schwingungsrißkorrosion rost- und säurebeständiger Stähle in Schwefelsäure hoher Konzentration
(Investigation into the corrosion fatigue of stainless steels in highly concentrated sulfuric acid) (in German)
Werkst. Korros. *33* (1982) 9, p. 483

[154] Spähn, R.; Blümmel, G.; Speckhardt, H.
Beanspruchungssynchrone elektrochemische Messungen an SwRK-beanspruchten Proben aus rost- und säurebeständigem Stahl
(Stress synchron electrochemical measurement with stainless steel in corrosion fatigue) (in German)
Z. Werkstofftech. *17* (1986) 4, p. 154

[155] Süry, P.
Korrosionseigenschaften gegossener Chrom-Nickel- und Chrom-Nickel-Molybdänstähle in aggressiven Medien
(Corrosion properties of cast chromium-nickel and chromium-nickel-molybdenum steels in aggressive media) (in German)
Chemie-Technik *4* (1975) 7, p. 241

[156] Anonymous
Tantalum coatings protect chemical equipment at lower cost
Mater. Eng. *75* (1972) 1, p. 36

[157] Anonymous
Verschleißfest und abriebbeständig: LP-Legierungen Tribaloy® für Gußteile und zum Beschichten
(Wear-resistant and abrasion-resistant: LP-alloys Tribaloy® for cast parts and coating) (in German)
Werkstoffe und ihre Veredlung *1* (1979) 2, p. 24

[158] Team of authors
Werkstoffeinsatz und Korrosionsschutz in der chemischen Industrie
(Material application and corrosion protection in the chemical industry) (in German)
VEB Deutscher Verlag für Grundstoffindustrie, Leipzig, 1986

[159] Beyer, B.
Entwicklungsrichtungen bei Bleiwerkstoffen
(Trends in the development of lead materials) (in German)
Neue Hütte *17* (1972) 2, p. 65

[160] Oppenheim, R.
Nichtrostende Stähle: Kennzeichnung, Eigenschaften und Verwendung
(Stainless steels: characterization, properties and use) (in German)
DEW-Tech. Ber. *14* (1974) 1, p. 5

[161] Anonymous
Austenitic stainless steel is strong, resists corrosion
Mater. Eng. *84* (1976) 5, p. 47

[162] McDowell jr., D. W.
Choosing materials for sulfuric-acid services
Chem. Eng. *84* (1977) 14, p. 137 (1B)

[163] Colombier, L. (Product Information)
Molybdän in rost- und säurebeständigen Stählen und Legierungen
(Molybdenum in stainless and acid-resistant steels and alloys) (in German)
Climax Molybdenum GmbH, D-4000 Düsseldorf-Oberkassel

[164] Lamb, B. R.
Sulphuric acid – technology moves platewards
Processing *27* (1981) 10, p. 56

[165] Tsetjtlin, Kh. L.; Faingold, L. L.; Shapiro, M. B.; Shuratova, G. P.; Sidlin, Z. A.
Comparative chemical resistance of the stainless steels 000Cr18Ni8 11 and Cr18Ni10T in dilute sulfuric and hydrochloric acid (in Russian)
Zashch. Met. 9 (1973) 5, p. 571

[166] Riedel, G.; Voigt, C.; Werner, H.; Günzel, M.; Erkel, K.-P.
Zum Einfluß von Mangan auf die Korrosionseigenschaften von austenitischen 18.10-CrNi-Stählen (Influence of manganese on the corrosion properties of austenitic 18.10 CrNi stainless steels) (in German)
Werkst. Korros. *37* (1986) 10, p. 519

[167] Chekhovskij, S. V.; Smirnova, I. S.; Dobrolyubov, V. V.; Dubinin, V. G.
Effect of saturation of sulfuric acid with sulfur dioxide on the corrosion of stainless steel (in Russian)
Zashch. Met. *8* (1972) 6, p. 695

[168] Kuron, D.; Paulekat, F.; Gräfen, H.; Horn, E.-M.
Werkstoffauswahl für schwefelsäureführende Rohrleitungen bei Anwendung des anodischen Korrosionsschutzes
(Material selection for sulfuric acid pipes under conditions of anodic corrosion protection) (in German)
Werkst. Korros. *36* (1985) 11, p. 489

[169] Chuvilova, V. A.; Emelyanova, V. P.; Borovskaya, K. I.; Kolesnikova, E. A.; Dobrolskij, I. P.; Gerasimov, D. A.
Corrosion of titanium VTi-1 and certain steels in hydrolyzed sulfuric acid in titanium dioxide production (in Russian)
Zashch. Met. *10* (1974) 6, p. 711

[170] Kryuchek, V. G.; Shapovalova, O. M.; Vazhenin, S. F.; Mineeva, I. K.; Sinyavskaya, M. K.; Skalskaya, L. P.; Vulikhman, A. A.
Selection of materials for plants in the production of tartaric acid (in Russian)
Vinodel. Vinograd. SSSR *32* (1972) 2, p. 49; (C. A. 77 (1972) 23931 h)

[171] Dvali, T. M.
Corrosion resistance of steels in caprolactam production media (in Russian)
Zashch. Met. 9 (1973) 1, p. 58

[172] Kotlyar, N. Z.; Konstantinova, E. V.; Khokhlova, Z. S.
Selection of materials for plants in contact with solutions of sulfuric acid with thiourea (in Russian)
Khim. Prom. (1975) 7, p. 524

[173] Sapak, L. N.; Denisova, R. I.
Corrosion resistance of titanium in media used in the production of iodine (in Russian)
Khim. i. Neft. Mashinostr. (1980) 10, p. 16

[174] Riskin, I. V.; Orlova, F. A.; Balakirev, E. S.
Corrosion resistance of metals in the production of alkyl sulfonates (in Russian)
Khim. Prom. *48* (1972) 10, p. 748

[175] Vajpeyi, M.; Dhirendra, S. G.; Pandey, G. N.
Corrosion of stainless steel (AlSl 304) in H_2SO_4 contaminated with HCl and HNO_3
Corros. Prev. Control *32* (1985) 5, p. 102

[176] Dhirendra, M.; Panadey, G. N.; Sanyal, B.
Comparative corrosion of unalloyed and alloyed steels in single and mixed mineral acids
Corros. Prev. Control *28* (1981) 1, p. 19

[177] Bozin, N. A.; Gulyaev, V. E.; Bozin, S. N.
Effect of fluoride additives on steel corrosion in sulfuric acid (in Russian)
Zashch. Met. *17* (1981) 4, p. 437

[178] Poluboyartseva, L. A.; Reifer, A. A.; Yurlova, L. N.; Shiskin, B. I.; Lukinykh, V. M.
The corrosion resistance of the steel 1Ch18N10T in oleum (in Russian)
Khim. Prom. (1973) 5, p. 372

[179] Wolf, G. K.
Zur Anwendung von durch Ionenimplantation erzeugten Oberflächenlegierungen in der Korrosionsforschung und -verhütung
(The application of surface alloys produced by ion-implantation in corrosion research and prevention) (in German)
Werkst. Korros. *35* (1984) 6, p. 273

[180] Wolf, K.; Munn, P.; Meger, A.; Ferber, H.; Ensinger, W.; Dressler, J.
Zur Anwendung von durch Ionenimplantation erzeugten Oberflächenlegierungen in der Korrosionsforschung und -verhütung
(The application of surface alloys produced by ion-implantation in corrosion research and prevention) (in German)
Final report FE-Projekt-No. 2.5/9, Universität Heidelberg, Institut für Physikalische Chemie

[181] Wolf, G. K.
Ionenstrahlen in der Metallforschung
(Ionic beams in metal research) (in German)
Metall *38* (1984) 5, p. 402

[182] Cragnolino, G.; Macdonald, D. D.
Intergranular stress corrosion cracking of austenitic stainless steel at temperatures below 100 °C – a review
Corrosion *38* (1982) 8, p. 406

[183] Erdmann-Jesnitzer, F.
Untersuchungen zur Wasserstoffversprödung und Spannungsrißkorrosion austenitischer Stähle unter besonderer Berücksichtigung des Einflusses einer thermomechanischen Behandlung (HERF)
(Investigation of the hydrogen embrittlement and stress corrosion cracking of austenitic steels with special consideration of the influence of thermomechanical treatment) (in German)
Werkst. Korros. *33* (1982) 4, p. 221

[184] Erdmann-Jesnitzer, F.; Wessel, A.
Untersuchungen zur Wasserstoffversprödung einiger kubisch-flächenzentrierter Werkstoffe
(Investigation into the hydrogen embrittlement of some cubic face-centered materials) (in German)
Arch. Eisenhüttenwesen *52* (1981) 2, p. 77

[185] Herbsleb, G.; Westerfeld, K.-J.
Der Einfluß von Stickstoff auf die korrosionschemischen Eigenschaften lösungsgeglühter und angelassener 18/10 Chrom-Nickel- und 18/12 Chrom-Nickel-Molybdän-Stähle
(The effect of nitrogen on the chemical corrosion properties of solution-annealed and tempered 18/10 chromium-nickel steels and 18/12 chromium-nickel-molybdenum steels) (in German)
Werkst. Korros. *27* (1976) p. 404

[186] Dorn, L.; Anik, S.; Günaltan, A.
Untersuchungen zur Beständigkeit widerstandspunktgeschweißter Verbindungen aus X 12 CrNi 18 8 gegenüber interkristalliner Korrosion
(Investigations relating to the intergranular corrosion resistance of resistance spot welded joints in X 12 CrNi 18 8) (in German)
Schweissen Schneiden *32* (1980) 1, p. 5

[187] Hoffman, C.; McEvily, A. J.
The effect of high temperature low cycle fatigue on the corrosion resistance of austenitic stainless steels
Metallurg. Transactions *13A* (1982) 5, p. 923

[188] Herbsleb, G.; Hildebrand, H.; Schwenk, W.
Die Lochkorrosion austenitischer Chrom-Nickel- und Chrom-Nickel-Molybdän-Stähle in schwefelsaurer Bromidlösung und ihre Inhibition durch Nitrationen
(The pitting corrosion of austenitic chromium-nickel steels and chromium-nickel-molybdenum steels in a bromide solution containing sulfuric acid and its inhibition by nitrate ions) (in German)
Werkst. Korros. *27* (1976) p. 618

[189] v. Makarov, A.; Zarubin, P. I.; Mironov, Yu. M.; Poluboryartseva, L. A.; Yurlova, L. N.; Sidelnikova, E. A.
Investigation and modelling of anodic protection of heat exchangers made from the stainless steel Ch18N10T (in Russian)
Zashch. Met. *13* (1977) 2, p. 181

[190] Horn, E.-M.; Kuron, D.; Paulekat, F.
Werkstoffliche Auslegung von schwefelsäureführenden Rohrleitungen unter den Bedingungen des anodischen Korrosionsschutzes
Ergebnisse des Forschungs- und Entwicklungsprogramms "Korrosion und Korrosionsschutz", Kurzberichte abgeschlossener FE-Vorhaben und Nullprojekte
(Choice of material for sulfuric acid pipelines under conditions of anodic corrosion protection) (in German) Results of the research and development program "Corrosion and corrosion protection", Brief reports of finished FE-projects and pilot projects, vol. *3* (1981/1983)
DECHEMA, D-6000 Frankfurt 97

[191] Gupta, S.; Pandey, G. N.; Sanyal, B.
Effect of potassium iodide on the attack of stainless steel by sulfuric acid
Met. Finish. *80* (1982) 8, p. 51

[192] Karner, W.; Nazmy, M. Y.; Arfaj, A.
Phase instability and corrosion behaviour of AISI 316 P/M stainless steel
Werkst. Korros. *31* (1980) 6, p. 446

[193] Itzhak, D.; Harush, S.
The effect of Sn addition on the corrosion behaviour of sintered stainless steel in H_2SO_4
Corros. Sci. *25* (1985) 10, p. 883

[194] Schäfer, G.
Korrosionsverhalten des Stahls X 5 CrNiMoCuTi 18 18 in Schwefelsäure
(Corrosion behavior of the steel X 5 CrNiMoCuTi 18 18 in sulfuric acid) (in German)
Schweisstechnik *22* (1972) 3, p. 118

[195] Swales, G. L.; Todd, B.
Stainless steel for chemical tankers, Information No. 15, July 1971
International Nickel Ltd., London (GB)

[196] Moniz, B.J.
Inco checks out stainless steels
Canadian Chemical Processing *59* (1975) 3, p. 28

[197] Anonymous
Rostfreie Sonderstähle aus Schweden
(Special stainless steels from Sweden) (in German)
Technica, Basel *30* (1981) 25, p. 2353

[198] Hopkins, D. J.
Korrosionsbeständige Nickellegierungen und deren Anwendungsmöglichkeiten
(Corrosion-resistant nickel-alloys and their possible applications) (in German)
Chem. Anl. + Verfahren (CAV) (1979) 9, p. 110

[199] Anonymous
New stainless steels to resist stress-corrosion cracking and pitting
Anticorr. Methods Mater. *28* (1981) 2, p. 10

[200] Kohl, H.; Hochörtler, G.; Kriszt, K.; Koren, M.
Sonderstähle für den Apparatebau mit sehr guter Korrosionsbeständigkeit und erhöhter Festigkeit
(Special steels with superior corrosion resistance and strength for chemical equipment manufacture) (in German)
Werkst. Korros. *34* (1983) 1, p. 1

[201] Schäfer, G.
Korrosionsverhalten des Stahls X 5 CrNiMoCuTi 18 18 in Schwefelsäure
(Corrosion behavior of the steel X 5 CrNiMoCuTi 18 18 in sulfuric acid) (in German)
Schweisstechnik *22* (1972) 3, p. 118

[202] McClain, G. E.; Mueller, W. A.
Corrosion problems in acid flow control
Chem. Eng. Progr. (CEP) *78* (1982) 2, p. 48

[203] Product Information
Wiggin korrosionsbeständige Legierungen
(Wiggin corrosion-resistant alloys) (in German), edition Dec. 1977
Henry Wiggin & Company Ltd., Hereford (GB)

[204] Clayton, jr. C. H.; Johnson, T. E.
Engineering materials for pumps and valves
Chem. Eng. Progr. *74* (1978) 9, p. 54

[205] Niendorf, K.; Peuker, R.
Zum Einfluß des Nitrosegehaltes konzentrierter Schwefelsäure auf das Korrosionsverhalten austenitischer Chrom-Nickel-Stähle bei hohen Temperaturen
(The effect of the nitrous content of concentrated sulfuric acid on the corrosion behavior of austenitic chromium-nickel steels at high temperatures) (in German)
Chem. Tech. *28* (1976) 4, p. 236

[206] Product Information
Korrosionsbeständiger Edelstahl und Edelstahlguß
(Corrosion-resistant alloy steel and alloy steel casting) (in German), 12/1983
Schmidt + Clemens Edelstahlwerk, D-5253 Lindlar

[207] Product Information
Korrosionsbeständigkeit nickelhaltiger Werkstoffe gegenüber Phosphorsäure und Phosphaten
(Corrosion resistance of nickel-containing materials to phosphoric acid and phosphates) (in German)
Information No. 61, March 1970
International Nickel Deutschland GmbH, D-4000 Düsseldorf

[208] Nordin, St.
Rostfreie Sonderstähle für die chemische Industrie
(Special stainless steels for the chemical industry) (in German)
Chemie-Technik (1975) 4, p. 123

[209] Wallen, B.; Liljas, M.; Olsson, J. (Product Information)
Avesta® 254 SMO – Performance of a high molybdenum stainless steel in gas cleaning systems with particular reference to the pulp and paper industry, January 1982
Avesta Jernverks AB, Avesta (Sweden)

[210] Wallen, B.; Liljas, M.; Olsson, J.
Performance of a high molybdenum stainless steel in gas cleaning systems with particular reference to the pulp and paper industry
Mater. Performance *21* (1982) 6, p. 40

[211] Gräfen, H.
Bedeutung, Eigenschaften und Entwicklungsstand der nichtrostenden Stähle für Chemieanlagen
(Importance, properties and state of development of stainless steels for chemical plants) (in German),
Lecture given on the congress "Werkstoffe im Chemieanlagenbau", D-4300 Essen, June 1989

[212] Heubner, U.; Heimann, W.; Kirchheiner, R.
Korrosionsbeständige metallische Werkstoffe für Rauchgasentschwefelungsanlagen
(Corrosion-resistant metallic materials for flue gas desulfurization plants) (in German)
Werkst. Korros. *38* (1987) 12, p. 746

[213] Rockel, M. B.; Weidemann, R.
Untersuchung über den Einfluß der Vorbehandlung auf das Korrosionsverhalten hochlegierter Sonderstähle in reiner und chloridhaltiger Schwefelsäure
(Study of the effect of pretreatment on the corrosion behavior of highly alloyed special steels in pure sulfuric acid and sulfuric acid containing chloride) (in German)
Werkst. Korros. *33* (1982) 1, p. 42

[214] Altpeter, E.; Rockel, M. B.
Untersuchungen über den Einfluß des Gefügezustandes auf das Korrosionsverhalten hochlegierter, rost- und säurebeständiger Stähle in reiner sowie in chloridhaltiger Schwefelsäure
(Corrosion properties of high alloyed stainless steels in pure as well as in chloride containing sulfuric acid) (in German)
Werkst. Korros. *34* (1983) 10, p. 505

[215] Bandy, R.; van Rooyen, D.
Properties of nitrogen-containing stainless alloy designed for high resistance to pitting
Corrosion *41* (1985) 4, p. 228

[216] Stewen, W.; Jörgens, H.; Remke, M.; Litzkendorf, M.
Korrosion hochlegierter Stähle in Ammoniumsulfatsättigern auf Kokereien
(Corrosion of high-alloy steels in ammonium sulfate saturators in coking plants) (in German)
Glückauf *121* (1985) 13, p. 1030

[217] Rockel, M. B.; Weidemann, R.
Untersuchung über die Schwingungsrißkorrosion einiger nichtrostender Stähle in Schwefelsäure hoher Konzentration
(Study of the vibration corrosion cracking of some stainless steels in highly concentrated sulfuric acid) (in German)
Werkst. Korros. *33* (1982) 1, p. 40

[218] Altpeter, E.; Rockel, M. B.
Untersuchung über die Schwingungsrißkorrosion rost- und säurebeständiger Stähle in Schwefelsäure hoher Konzentration
(Investigation into the corrosion fatigue of stainless steels in highly concentrated sulfuric acid) (in German)
Werkst. Korros. *33* (1982) 9, p. 483

[219] Spähn, R.; Blümmel, G.; Speckhardt, H.
Beanspruchungssynchrone elektrochemische Messungen an SwRK-beanspruchten Proben aus rost- und säurebeständigem Stahl
(Stress synchron electrochemical measurement with stainless steel in corrosion fatigue) (in German)
Z. Werkstofftech. *17* (1986) 4, p. 154

[220] Kuzub, V. S.; Tokarenko, A. G.; Kuzub, L. G.; Makarov, V. A.; Mironov, Yu. M.; et al.
Anodic protection of tube bundle coolers (in Russian)
Khim. Prom. (1979) 4, p. 224

[221] Paulekat, F.; Gräfen, H.; Kuron, D.
Anodic corrosion protection of sulfuric acid plants with regard to the recovery of heat
Werkst. Korros. *33* (1982) 5, p. 254

[222] Locke, C. E.
Status of anodic protection twenty-five years old
International Congress on Metallic Corrosion, Toronto (Canada), 1984 (Proc. Conf.) vol. *1*, p. 316

[223] Pavlenko, V. F.
The use of polymer coatings for the protection of plate heat exchangers in the production of sulfuric acid (in Russian)
Khim. Neft. Mashinostr. (1986) 8, p. 21

[224] Pierce, R. R.; Semler, C. E.
Ceramic and refractory linings for acid condensation – part I
Chem. Eng. *90* (1983) 25, p. 81

[225] Frömberg. M.
Selektive Korrosion an Darmspinndüsen aus rost- und säurebeständigem Stahl
(Selective corrosion of gut spinnerets made from acid-resistant stainless steel) (in German)
Korrosion *10* (1979) 4, p. 217

[226] Gramberg. U.
Betriebsschäden an Schweißnähten – Erfahrungen aus der Chemietechnik; Schäden durch korrosive Beanspruchung und Häufigkeit der Schadensarten
(Damage on weld joints in operation – Experiences from chemical technology; damage by corrosive stress and frequency of damage type) (in German)
Schweissen Schneiden *33* (1981) 8, p. 357

[227] Forchhammer, P.
Beispiele für Korrosionsschäden in der Papierindustrie und Hinweise zur Schadensverhütung
(Examples of corrosion damage in the paper industry and advice on damage prevention) (in German)
Wochenblatt für Papierfabrikation *109* (1981) 22, p. 871

[228] Barber, J. C.; Moeller, M. G.
Incorporation of sewage sludge in commercial fertilizers
Ind. Eng. Chem. Prod. Res. Dev. *20* (1981) 3, p. 409

[229] Crum, J. R.; Adkins, M. E.
Comparison of electrochemical behavior and corrosion rate of alloy 625 in sulfuric acid
Mater. Performance *25* (1986) 2, p. 27

[230] NACE Publication
Corrosion in sulfuric acid
Proceedings of the Corrosion /85, Symposium on Corrosion in Sulfuric Acid, Catalog Card Number: 85-61573
National Association of Corrosion Engineers, Houston (Texas/USA), 1985

[231] Rockel, M. B.
Einsatz hochlegierter Stähle und Nickelbasislegierungen im chemischen Anlagenbau
(The application of high-alloy steels and nickel-based alloys in chemical plant construction) (in German)
Werkstoffe und ihre Veredlung *4* (1982) 4/5, p. 153

[232] Heubner, U.; Rockel, M.; Wallis, E.
Ein neuer hochlegierter Nickel-Chrom-Molybdän-Stahl für den Chemie-Apparatebau
(A new high-alloyed nickel-chromium-molybdenum steel for the chemical process industry) (in German)
Werkst. Korros. *40* (1989) 7, p. 418

[233] Katz, W.
ACHEMA '79 – Teil 2: Kurzer Überblick über Werkstoffe und Werkstoffprobleme (ACHEMA '79 – Part 2: A short survey of materials and material problems) (in German)
Werkst. Korros. *30* (1979) 9, p. 651

[234] Nordin, S.
Werkstoffe für den Chemieanlagenbau, NU Stainless 904L
(Materials for the construction of chemical plants: NU stainless 904L) (in German)
Blech – Rohre – Profile *28* (1981) 11, p. 540

[235] Nolan, P. D.
How CIL handles sulfuric acid corrosion
Can. Chem. Process. *61* (1977) 4, p. 35

[236] Product Information (Nyby Uddeholm GmbH, D-4000 Düsseldorf)
Nichtrostender Stahl mit hohen Korrosionseigenschaften, NU Stainless 904L
(Stainless steel with excellent corrosion properties, NU stainless 904L) (in German)
Blech – Rohre – Profile *28* (1981) 12, p. 600

[237] Nordin, S.
Rostfreie Spezialstähle für die chemische Industrie. Teil 1: Korrosionseigenschaften und Anwendungsbereiche
(Special stainless steels for the chemical industry. Part 1: Corrosion properties and fields of application) (in German)
Chem.-Anl. + Verfahren (CAV) (1979) 5, p. 67

[238] NACE Publication
Corrosion in sulfuric acid
Proceedings of the Corrosion '85, Symposium on Corrosion in Sulfuric Acid, Catalog Card Number: 85-61573
National Association of Corrosion Engineers, Houston (Texas/USA), 1985

[239] Heubner, U.; Rockel, M.; Wallis, E.
Ein neuer hochlegierter Nickel-Chrom-Molybdän-Stahl für den Chemie-Apparatebau
(A new high-alloyed nickel-chromium-molybdenum steel for the chemical process industry) (in German)
Werkst. Korros. *40* (1989) 7, p. 418

[240] Nazarov, V. N.; Boiko, A. Z.; Demyanenko, N. S.; Elizov, A. M.; Zatulin, V. D.; Tishchenko, G. A.
Testing on the use of apparatus made of the nickel-molybdenum alloy N70M27F in the Chemical Combine Cherkassk (in Russian)
Khim. Prom. (1975) 8, p. 593

[241] Olsson, J.; Grützner, H.
Experiences with a high-alloyed stainless steel under highly corrosive conditions
Werkst. Korros. *40* (1989) 5, p. 279

[242] Nair, F.; Semchyshen, M.
Corrosion resistance of molybdenum-modified Cr-Ni-Mn austenitic stainless steels
Corrosion *19* (1963) 6, p. 210t

[243] Anonymous
Fertigungs- und Standardlieferprogramm M3-M56 Schrauben, Muttern, Zubehör nach DIN, ISO und EURONORM
(Manufacturing and standard delivery program M3-M56 screws, nuts and equipment according to DIN, ISO and EURONORM) (in German)
Werkstoffe und ihre Veredlung *2* (1980) 5, p. 250

[244] Uhlig. H. H.; Asphahani, A. I.
Corrosion behavior of cobalt base alloys in aqueous media
Mater. Performance *18* (1979) 11, p. 9

[245] Horn, E. M.; Kügler, A.
Entwicklung. Eigenschaften, Verarbeitung und Einsatz des hochsiliciumhaltigen austenitischen Stahls X 2 CrNiSi 18 15
(Development, properties, processing and applications of the high-silicon steel grade X 2 CrNiSi 18 15) (in German)
Z. Werkstofftech. *8* (1977) 12, p. 410

[246] Wallis, E.; Brücken, V.
Strömungsabhängige Korrosion – Untersuchungen zur Korrosions- und Erosionsbeständigkeit metallischer Guß- und Knetwerkstoffe in ruhender und strömender hochkonzentrierter Schwefelsäure
(Flow-dependent corrosion – investigations into the corrosion and erosion resistance of metallic cast materials and forgeable materials in static and flowing sulfuric acid of high concentration) (in German)
Werkst. Korros. *41* (1990) 6, p. 351

Key to materials compositions

Table 1: Chemical compositions of alloys according to German and other standards

German Standard		***Materials Compositions***	***US-Standard***
Mat.-No.	***DIN-Design***	***Percent in Weight***	***SAE/ASTM/UNS***
0.6015	EN-JL1020	–	A 48 (25B)
0.6025	EN-JL1040	–	A 48 (40B)
0.6655	GGL-NiCuCr 15 6 2	Fe-max. 3.0C-1.0-2.8Si-0.5-1.5Mn-13.5-17.5Ni-1.0-2.5Cr-5.5-7.5Cu	A 436 Type 1
0.6656	GGL-NiCuCr 15 6 3	Fe-max. 3.0C-1.0-2.8Si-0.5-1.5Mn-13.5-17.5Ni-2.5-3.5Cr-5.5-7.5Cu	A 436 Type 1b
0.6660	GGL-NiCr 20 2	Fe-max. 3.0C-1.0-2.8Si-0.5-1.5Mn-18.0-22.0Ni-1.0-2.5Cr	A 436 Type 2
0.6661	GGL-NiCr 20 3	Fe-max. 3.0C-1.0-2.8Si-0.5-1.5Mn-18.0-22.0Ni-2.5-3.5Cr	A 436 Type 2b
0.6667	GGL-NiSiCr 20 5 3	Fe-≤2.5C-3.5-5.5Si-0.5-1.5Mn-1.5-4.5Cr-18.0-22.0Ni	
0.6676	GGL-NiCr 30 3	Fe-max. 2.5C-1.0-2.8Si-0.5-0.8Mn-2.5-3.5Cr-28.0-32.0Ni	A 436 Type 3
0.6680	GGL-NiSiCr 30 5 5	Fe-≤2.5C-5.0-6.0Si-0.5-1.5Mn-4.5-5.5Cr-29.0-32.0Ni	A 436 Type 4
0.7040	EN-JS1030	–	A 536 (60 40 18)
0.7660	EN-GJSA-XNiCr20-2	Fe-≤3.0C-1.5-3.0Si-0.5-1.5Mn-≤0.080P-1.0-3.5Cr-≤0.50Cu-18.0-22.0Ni	A 439 Type D-2
0.7661	GGG-NiCr 20 3	Fe-≤3.0C-1.5-3.0Si-0.5-1.5Mn-≤0.080P-2.5-3.5Cr-18.0-22.0Ni	A 439 Type D-2B
0.7665	GGG-NiSiCr 20 5 2	Fe-≤3.0C-4.5-5.5Si-0.5-1.5Mn-≤0.080P-1.0-2.5Cr-18.0-22.0Ni	A 439 Type D-2C
0.7670	EN-GJSA-XNi22	Fe-≤3.0C-1.0-3.0Si-1.5-2.5Mn-≤0.080P-≤0.5Cr-≤0.5Cu-21.0-24.0Ni	A 439 Type D-2C
0.7679	GGG-NiSiCr 30 5 2	Fe-≤2.6C-4.0-6.0Si-0.5-1.5Mn-≤0.08P-29.0-32.0Ni	
0.7680	EN-GJSA-XNiSiCr30-5-5	Fe-≤2.6C-5.0-6.0Si-0.5-1.5Mn-≤0.080P-4.5-5.5Cr-≤0.5Cu-28.0-32.0Ni	A 439 Type D-4

German Standard		Materials Compositions	US-Standard
Mat.-No.	**DIN-Design**	**Percent in Weight**	**SAE/ASTM/UNS**
0.7688	EN-GJSA-XNiSiCr35-5-2; GGG-NiSiCr 35 5 2	Fe-≤2.0C-4.0-6.0Si-0.5-1.5Mn-≤0.08P-1.5-2.5Cr-≤0.5Cu-34.0-36.0Ni	A 439 (Type D-5S)
0.9625	EN-GJN-HV550	Fe-≤3.0-3.6C-≤0.080Si-≤0.080Mn-≤0.1P-≤0.1S-1.5-3.0Cr-≤0.5Mo-3.0-5.5Ni	A 532 (IA NiCr-HC)
0.9635	EN-JN3029	Fe-max. 1.8-2.4C-1.0Si-0.5-1.5Mn-0.08P-0.08S-14.0-18.0Cr-3.0Mo-2.0Ni	A 532
0.9640	EN-GJN-HV600(XCr14)	Fe-≤1.8-2.4C-≤1.0Si-≤0.5-1.5Mn-≤0.080P-≤0.080S-14.0-18.0Cr-≤3.0Mo-≤2.0Ni	A 532
0.9650	EN-JN3049	Fe-max. 2.4-3.2C-1.0Si-0.5-1.5Mn-0.08P-0.08S-23.0-28.0Cr-3.0Mo-2.0Ni	A 532
1.0030	St 00	Fe-≤0.30C-≤0.30Si-0.20-0.50Mn-≤0.08P-≤0.05S	
1.0032	St 34-2; (S205GT)	Fe-≤0.15C-≤0.3Si2.0-0.5Mn-0.05P-0.05S-0.007N	1010 (SAE)
1.0035	St 33; S 185		
1.0036	S235JRG1; USt 37-2; USt 37-2 G; (S235JRG1+CR)	Fe-≤0.17C-≤1.4Mn-≤0.0045P-≤0.045S-≤0.007N	K02502 (UNS)
1.0037	St 37-2; S235JR	Fe-≤0.17C-≤0.3Si-≤1.4Mn-≤0.045P-≤0.045S-≤0.009N	A 283; SAE 1015
1.0038	RSt 37-2; S235JR	Fe-≤0.17C-≤1.4Mn-≤0.045P-≤0.045S-≤0.009N	UNS K02502
1.0040	USt 42-2	Fe-≤0.25C-≤0.2-0.5Mn-≤0.05P-≤0.05S-≤0.007N	
1.0044	S275JR; St 44-2	Fe-≤0.21C-≤1.50Mn-≤0.045P-≤0.045S-≤0.009N	UNS K03000 A 853 (1020) SAE 1020
1.0050	E295; St 50-2	Fe-≤0.045P-≤0.045S-≤0.009N	
1.0070	E360; St 70-2	Fe-≤0.045P-≤0.045S-≤0.009N	
1.0114	S235J0	Fe-≤0.17C-≤1.40Mn-≤0.040P-≤0.040S-≤0.009N	
1.0116	S235J2G3; St 37-3 N	Fe-max. 0.17C-1.4Mn-0.035P-0.035S	UNS K02001
1.0120	S235JRC	Fe-≤0.17C-≤1.40Mn-≤0.045P-≤0.045S-≤0.009N	
1.0204	UQSt 36	Fe-≤0.14C-≤0.25-0.50Mn-≤0.040P-0.040S	SAE 1008
1.0208	RSt 35-2; (C10G2)	Fe-0.06-0.12C-≤0.25Si-0.40-0.60Mn-≤0.035P-≤0.035S-≤0.25Cu-≤0.012N	
1.0253	USt 37.0	Fe-≤0.2C-≤0.55Si-≤1.6Mn-≤0.04P-≤0.04S-≤0.007N	
1.0254	P235TR1, St 37.0	Fe-max. 0.16C-0.35Si-1.2Mn-0.025P-0.020S-0.30Cr-0.30Cu-0.08Mo-0.010Nb-0.30Ni-0.04Ti-0.02V	UNS K02501
1.0256	St 44.0	Fe-≤0.21C-≤0.55Si-≤1.60Mn-≤0.040P-≤0.040S-≤0.009N	A 106

German Standard		Materials Compositions	US-Standard
Mat.-No.	DIN-Design	Percent in Weight	SAE/ASTM/UNS
1.0301	C10	Fe-≤0.07-0.13C-≤0.4Si-≤0.3-0.6Mn-≤0.045P-≤0.045S	SAE 1010
1.0305	St 35.8	Fe-≤0.17C-≤0.10-0.35Si-≤0.40-0.80Mn-≤0.040P-≤0.040S	UNS K01200
1.0308	E 235	Fe-≤0.17C-≤0.35Si-0.4Mn-≤0.05P-≤0.05S-≤0.007N	SAE 1010
1.0309	DX55D	Fe-≤0.16C-0.17-0.40Si-0.35-0.65Mn-≤0.05P-≤0.050S-≤0.30Cr-≤0.30Ni-≤0.30Cu	UNS K02501
1.0330	St 12; DC01 + ZN	Fe-≤0.12C-≤0.60Mn-≤0.045P-≤0.045S	A 366 (C)
1.0333	USt 13	Fe-≤0.08C-≤0.007N	
1.0336	USt 4	Fe-≤0.09C-≤0.25-0.50Mn-≤0.030P-≤0.030S-≤0.007N	
1.0338	DC04; St 14	Fe-≤0.08C-≤0.40Mn-≤0.03P-≤0.030S	
1.0345	P235GH; H I	Fe-≤0.16C-≤0.35Si-≤0.40-1.20Mn-≤0.030P-≤0.025S-0.02Al-≤0.30Cr-≤0.30Cu-≤0.08Mo-≤0.010Nb-≤0.30Ni-≤0.03Ti-≤0.02V	A 285; A 414
1.0346	H220G1	Fe-≤0.04C-≤0.40Mn-≤0.03P-≤0.02S-≤0.01-0.04Ti	UNS K02202 A 516 (55)(380)
1.0356	TTSt35 N	Fe-≤0.18C-≤0.13-0.45Si-≤0.55-0.98Mn-≤0.035P-≤0.04S	UNS K03000 A 524 (I,II)
1.0375	TH57; T 57	Fe-≤0.1C-Traces Si-0.25-0.45Mn-≤0.04P-≤0.04S-0.007N	
1.0401	C 15	Fe-≤0.12-0.18C-≤0.40Si-≤0.3-0.6Mn-≤0.045P-≤0.045S	SAE 1015
1.0402	C 22	Fe-≤0.17-0.24C-≤0.4Si-≤0.4-0.7Mn-≤0.045P-≤0.045S-≤0.4Cr-≤0.1Mo-≤0.4Ni	SAE 1020
1.0405	St 45.8	Fe-≤0.21C-≤0.10-0.35Si-≤0.40-1.20Mn-≤0.040P-≤0.040S	A 106
1.0408	St 45	Fe-≤0.25C-≤0.035Si-0.40Mn-≤0.050P-≤0.050S	A 108; SAE 1020
1.0414	C20D; D 20-2	Fe-≤0.18-0.23C-≤0.30Si-≤0.3-0.6Mn-≤0.035P-≤0.035S-≤0.01Al-≤0.2Cr-≤0.3Cu-≤0.05Mo-≤0.25Ni	UNS G10200; SAE 1020
1.0425	P265GH; H II	Fe-≤0.20C-≤0.4Si-≤0.5-1.4Mn-≤0.030P-≤0.025S-0.02Al-≤0.3Cr-≤0.3Cu-≤0.08Mo-≤0.01Nb-≤0.3Ni-≤0.03Ti-≤0.02V	UNS K01701
1.0426	P280GH	Fe-≤0.08-0.20C-≤0.4Si-≤0.9-1.5Mn-≤0.025P-≤0.015S-≤0.30Cr	A 662 (A)
1.0461	StE 255	Fe-≤0.18C-≤0.40Si-≤0.50-1.30Mn-≤0.035P-≤0.030S-≥0.02Al-≤0.30Cr-≤0.20Cu-≤0.08Mo-≤0.02N-≤0.03Nb-≤0.30Ni	UNS K02202 A 516 (55)(380)
1.0473	P355GH; 19 Mn 6	Fe-≤0.1-0.22C-≤0.6Si-≤1.0-1.7Mn-≤0.03P-≤0.025S-≤0.30Cr-≤0.3Cu-≤0.08Mo-≤0.3Ni	A 299

German Standard		Materials Compositions	US-Standard
Mat.-No.	DIN-Design	Percent in Weight	SAE/ASTM/UNS
1.0481	17 Mn 4; P 295 GH	Fe≤0.08-0.20C-≤0.4Si-≤0.90-1.50Mn-≤0.030P-≤0.025S-0.02Al-≤0.30Cr-≤0.30Cu-≤0.08Mo-≤0.010Nb-≤0.3Ni-≤0.03Ti-≤0.02V	A 414, 515
1.0482	19 Mn 5	Fe-≤0.17-0.22C-≤0.30-0.60Si-≤1.00-1.30Mn-≤0.045P-≤0.045S-≤0.30Cr	UNS K12437; A 537
1.0490	S275N	Fe-≤0.18C-≤0.40Si-0.50-1.40Mn-≤0.035P-≤0.030S-≥0.0200Al-≤0.30Cr-≤0.35Cu-≤0.10Mo-≤0.015N-≤0.050Nb-≤0.30Ni-≤0.03Ti-≤0.05V	UNS K03000
1.0501	C 35	Fe-≤0.32-0.39C-≤0.4Si-≤0.5-0.8Mn-≤0.045P-≤0.045S-≤0.4Cr-≤0.1Mo-≤0.4Ni	SAE 1035
1.0503	C 45	Fe-≤0.42-0.50C-≤0.4Si≤0.5-0.8Mn-≤0.045P-≤0.045S-≤0.4Cr-≤0.1Mo-≤0.4Ni	SAE 1045
1.0505	StE 315	Fe-≤0.18C-≤0.45Si-≤0.70-1.50Mn-≤0.035P-≤0.030S-≤0.30Cr-0.020Al-≤0.20Cu-≤0.020N-≤0.03Nb-≤0.08Mo-0.30Ni	A 573
1.0528	C 30	Fe-≤0.27-0.34C-≤0.4Si-≤0.5-0.8Mn-≤0.045P-0.045S-≤0.4Cr-≤0.1Mo≤0.4Ni	SAE 1030
1.0540	C50	Fe-0.47-0.55C-≤0.40Si-0.60-0.90Mn-≤0.045P-≤0.045S-≤0.40Cr-≤0.10Mo-≤0.40Ni	A 689 (1050) ASTM A 866 (1050) ASTM
1.0545	S355N	Fe-≤0.20C-≤0.50Si-≤0.90-1.65Mn-≤0.035P-0.030S-0.02Al-≤0.30Cr-≤0.35Cu-≤0.10Mo-≤0.015N-≤0.050Nb-≤0.50Ni-≤0.03Ti-≤0.12V	UNS K12709
1.0553	S355J0	Fe-≤0.20C-≤0.55Si-≤1.60Mn-≤0.04P-≤0.04S-≤0.009N	
1.0562	StE 355; P 355 N	Fe-≤0.20C-≤0.50Si-≤0.90-1.70Mn-≤0.03P-≤0.025S-0.02Al-≤0.30Cu-≤0.30Cr-≤0.08Mo-≤0.02N-≤0.05Nb-≤0.50Ni-0.03Ti	A 633 (C)
1.0564	N-80	Fe-≤0.030P-≤0.030S	—
1.0570	St 52-3 N; S 355 J2G3	Fe-≤0.20C-≤0.55Si-≤1.60Mn-≤0.035P-≤0.035S	SAE 1024
1.0580	E 355	Fe-≤0.22C-≤0.55Si-≤1.60Mn-≤0.025P-≤0.025S	A 513 (1024) (ASTM)
1.0589	GL-E 36	Fe-≤0.18C-≤0.50Si-0.90-1.60Mn-≤0.040P-≤0.040S-≥0.0200Al-≤0.20Cr-≤0.35Cu-≤0.08Mo-0.020-0.050Nb-≤0.40Ni-0.05-0.10V	UNS K11852
1.0601	C60	Fe-0.57-0.65C-≤0.40Si-0.60-0.90Mn-≤0.045P-≤0.045S-≤0.40Cr-≤0.10Mo-≤0.40Ni	A 830 (1060) ASTM A 713 (1060) ASTM
1.0605	C 75	Fe-≤0.7-0.8C-≤0.15-0.35Si-≤0.6-0.8Mn-≤0.045P-≤0.045S	SAE 1074
1.0616	C86D	Fe-≤0.83-0.88C-≤0.10-0.30Si-≤0.50-0.80Mn-≤0.035P-≤0.035S-≤0.01Al-≤0.15Cr-≤0.25Cu-≤0.05Mo-≤0.20Ni	SAE 1086

German Standard		*Materials Compositions*	*US-Standard*
Mat.-No.	***DIN-Design***	***Percent in Weight***	***SAE/ASTM/UNS***
1.0619	GP240GH	Fe-≤0.18-0.23C-≤0.60Si-≤0.50-1.20Mn-≤0.03P-≤0.02S	A 216
1.0664	St 160/180	Fe-≤0.80C-≤0.20Si-≤0.70Mn-≤0.04P-≤0.04S	
1.0670	P-105	Fe-≤0.70C-≤0.03-0.30Si-1.0Mn-≤0.04P-≤0.04S-≤0.007N	—
1.0721	10S20	Fe-0.07-0.13C-≤0.40Si-0.70-1.10Mn-≤0.060P-0.150-0.250S	SAE 1109
1.0854	M125-35P	consult producer	
1.0912	46Mn7	Fe-≤0.42-0.50C-≤0.15-0.35Si-≤1.6-1.9Mn-≤0.05P-≤0.05S-≤0.007N	SAE 1345
1.1013	RFe 100	Fe-≤0.05C-≤0.10Si-≤0.20-0.35Mn-≤0.03P-≤0.035S-≤0.04-0.10Al	
1.1104	EStE 285; P275NL2	Fe-≤0.16C-≤0.4Si-≤0.5-1.5Mn-≤0.025P-≤0.015S-0.02Al-≤0.30Cr-≤0.30Cu-≤0.02N-≤0.5Ni	P275NL2
1.1106	P355NL2; EStE 355	Fe-≤0.18C-≤0.5Si-≤0.9-1.7Mn-≤0.025P-≤0.015S-≤0.3Cr-≤0.3Cu-≤0.3Mo-0.5Ni-≤0.02N	A 707
1.1121	C10E; Ck 10	Fe-≤0.07-0.13C-≤0.40Si-≤0.30-0.60Mn-≤0.035P-≤0.035S	SAE 1010
1.1127	36Mn6	Fe-≤0.34-0.42C-≤0.15-0.35Si-≤1.4-1.65Mn-≤0.035P-≤0.035S	
1.1136	G24Mn4	Fe-0.20-0.28C-0.30-0.60Si-0.90-1.20Mn-≤0.035P-≤0.035S	
1.1151	Ck 22; C22E	Fe-≤0.17-0.24C-≤0.4Si-≤0.4-0.7Mn-≤0.035P-≤0.035S-≤0.4Cr-≤0.1Mo-≤0.4Ni	SAE 1023
1.1166	34Mn5	Fe-0.30-0.37C-0.15-0.30Si-1.20-1.50Mn-≤0.035P-≤0.035S	G15360 (UNS) A 711 (1536) (ASTM) 1536 (SAE)
1.1176	G36Mn5	Fe-≤0.32-0.40C-≤0.15-0.35Si-≤1.20-1.50Mn-≤0.035P ≤0.035S	UNS H10380, A 830, SAE 1038
1.1186	C40E; Ck 40	Fe-≤0.37-0.44C-≤0.4Si-≤0.5-0.8Mn-≤0.035P-≤0.035S-≤0.4Cr-≤0.1Mo-≤0.4Ni	SAE 1040
1.1191	C45E; Ck 45	Fe-≤0.42-0.50C-≤0.4Si-≤0.50-0.80Mn-≤0.035P-≤0.035S-≤0.40Cr-≤0.10Mo-≤0.4Ni	SAE 1045
1.1520	C70U	Fe-≤0.65-0.74C-≤0.10-0.30Si-≤0.10-0.35Mn-≤0.030P-≤0.030S	
1.1525	C80U; C80W1	Fe-≤0.75-0.85C-≤0.10-0.25Si-≤0.10-0.25Mn-≤0.020P-≤0.020S	SAE W 108
1.1545	C105U; C105W1	Fe-≤1.0-1.1C-≤0.10-0.25Si-≤0.10-0.25Mn-≤0.020P-≤0.020S	SAE W 110
1.1730	C45U; C45W	Fe-≤0.42-0.50C-≤0.15-0.40Si-≤0.60-0.80Mn-≤0.03P-≤0.03S	A 830 (1045); SAE 1045

German Standard		Materials Compositions	US-Standard
Mat.-No.	DIN-Design	Percent in Weight	SAE/ASTM/UNS
1.2210	115CrV3	Fe-1.10-1.25C-0.15-0.30Si-0.20-0.40Mn-≤0.030P-≤0.030S-0.50-0.80Cr-0.07-0.12V	L 2 SAE A 681 (L2) ASTM
1.2311	40CrMnMo7	Fe-0.35-0.45C-0.20-0.40Si-1.30-1.60Mn-≤0.035P-≤0.035S-1.80-2.10Cr-0.15-0.25Mo	
1.2343	X37CrMoV5-1; X38CrMoV5-1	Fe-0.33-0.41C-0.80-1.20Si-0.25-0.50Mn-≤0.030P-≤0.020S-4.80-5.50Cr-1.10-1.50Mo-0.30-0.50V	H 11 SAE A 681 (H 11) ASTM T 20811 UNS
1.2344	X40CrMoV5-1	Fe-0.35-0.42C-0.80-1.20Si-0.25-0.50Mn-≤0.030P-≤0.020S-4.80-5.50Cr-1.20-1.50Mo-0.85-1.15V	H 13 SAE T 20813 UNS A 681 (H13) ASTM
1.2365	32CrMoV12-28 X32CrMoV33	Fe-≤0.28-0.35C-≤0.10-0.40Si-≤0.15-0.45Mn-≤0.030P-≤0.030S-≤2.70-3.20Cr-≤2.60-3.00Mo-≤0.40-0.70V	SAE H 10
1.2550	60WCrV8; 60WCrV7	Fe-0.55-0.65C-0.70-1.00Si-0.15-0.45Mn-≤0.030P-≤0.030S-0.90-1.20Cr-0.10-0.20V-1.70-2.20W	
1.2567	30WCrV17-2	Fe-0.25-0.35C-0.15-0.30Si-0.20-0.40Mn-≤0.035P-≤0.035S-2.20-2.50Cr-0.50-0.70V-4.00-4.50W	
1.2787	X23CrNi17	Fe-≤0.10-0.25C-≤1.00Si-≤1.00Mn-≤0.035P-≤0.035S≤15.5-18.0Cr-≤1.0-2.5Ni	
1.2823	70Si7	Fe-≤0.65-0.75C-≤1.50-1.80Si-≤0.60-0.80Mn-≤0.03P-≤0.03S	
1.2842	90MnCrV8	Fe-≤0.85-0.95C-≤0.10-0.40Si-≤1.80-2.20Mn-≤0.030P-≤0.030S-≤0.20-0.50Cr-≤0.05-0.20V	UNS T31502; SAE O2; A 681 (O2); SAE O2
1.3247	HS2-9-1-8 S2-10-1-8	Fe-1.05-1.15C-≤0.70Si-≤0.40Mn-≤0.030P-≤0.030S-7.50-8.50Co-3.50-4.50Cr-9.00-10.00Mo-0.90-1.30V-1.20-1.90W	UNS T11342
1.3355	HS 18-0-1	Fe-≤0.70-0.78C-≤0.45Si-≤0.4Mn-≤0.030P-≤0.030S-≤3.8-4.5Cr-≤1.0-1.2V-≤17.5-18.5W	A 600
1.3505	100Cr6	Fe-≤0.93-1.05C-≤0.15-0.35Si-≤0.25-0.45Mn-≤0.025P-≤0.015S-≤0.05Al-≤1.35-1.60Cr-≤0.30Cu-≤0.10Mo	SAE 52100; A 29
1.3551	80MoCrV42-16	Fe-≤0.77-0.85C-≤0.40Si-≤0.15-0.35Mn-≤0.025P-≤0.015S-≤3.90-4.30Cr-≤0.30Cu-≤4.00-4.50Mo-≤0.90-1.10V-≤0.25W	SAE M50; A 600 (M50)
1.3728	AlNiCo 9/5	Fe-≤11.0-13.0Al-5.0Co-2.0-4.0Cu-≤57.0Fe-21.0-28.0Ni-≤1.0Ti	
1.3813	X40MnCrN19	Fe-≤0.30-0.50C-≤0.80Si-≤17.0-19.0Mn-≤0.10P-≤0.030S-≤3.0-5.0Cr-≤0.08-0.12N	
1.3817	X40MnCr18	Fe-0.30-0.50C-≤1.00Si-17.00-19.00Mn-≤0.060P-≤0.030S-≤3.00-5.00Cr-≤0.100N-≤1.00Ni	

German Standard		*Materials Compositions*	*US-Standard*
Mat.-No.	*DIN-Design*	*Percent in Weight*	*SAE/ASTM/UNS*
1.3914	X2CrNiMnMoNNb21-15-7-3	Fe-≤0.03C-≤0.75Si-6.00-8.00Mn-≤0.025P-≤0.010S-20.00-22.00Cr-3.00-3.50Mo-0.350-0.500N-0.100-0.250Nb-14.00-16.00Ni	
1.3940	GX2CrNiN18-13	Fe-≤0.03C-≤1.00Si-≤2.00Mn-≤0.035P-≤0.020S-16.50-18.50Cr-0.100-0.200N-12.00-14.00Ni	
1.3951	G-X 4 CrNiMoN 22 15	Fe-≤0.05C-0.80-1.10Si-0.50-1.00Mn-≤0.030P-≤0.030S-22.00-23.50Cr-1.10-1.30Mo-0.150-0.250N-17.00-18.00Ni	
1.3952	X2CrNiMoN18-14-3	Fe-≤0.03C-≤1.00Si-≤2.00Mn-≤0.045P-≤0.015S-16.50-18.50Cr-2.50-3.00Mo-0.150-0.250N-13.00-15.00Ni	
1.3955	GX12CrNi18-11	Fe-≤0.15C-≤1.00Si-≤2.00Mn-≤0.045P-≤0.030S-16.50-18.50Cr-≤0.75Mo-10.00-12.00Ni	
1.3964	X2CrNiMnMoNNb21-16-5-3	Fe-≤0.03C-≤1.00Si-4.00-6.00Mn-≤0.025P-≤0.010S-20.00-21.50Cr-3.00-3.50Mo-0.200-0.350N-≤0.250Nb-15.00-17.00Ni	NITRONIC 50
1.3974	X2CrNiMnMoNNb23-17-6-3	Fe-≤0.03C-≤1.00Si-≤4.50-6.50Mn-≤0.025P-≤0.01S-≤21.00-24.50Cr-≤2.80-3.40Mo-≤0.30-0.50N-≤0.10-0.30Nb-≤15.50-18.00Ni	
1.3981	NiCo 29 18; (X3NiCo29-18)	Fe-≤0.050C-≤0.30Si-≤0.50Mn-≤17.0-18.0Co-≤28.0-30.0Ni	UNS K94610
1.4000	X6Cr13	Fe-≤0.08C-≤1.0Si-≤1.0Mn-≤0.04P-≤0.015S-≤12.0-14.0Cr	SAE 403, 410S
1.4001	X 7 Cr 14	Fe-≤0.08C-≤1.00Si-≤1.00Mn-≤0.045P-≤0.030S-≤13.00-15.00Cr	429 (SAE) A240 (410S) (ASTM) 410S (SAE)
1.4002	X6CrAl13	Fe-≤0.08C-≤1.0Si-≤1.0Mn-≤0.04P-≤0.015S-≤0.10-0.3Al-≤12-14Cr	SAE 405
1.4003	X2CrNi12	Fe-≤0.03C-≤1.0Si-≤1.5Mn-≤0.04P-≤0.015S-≤10.5-12.5Cr-≤0.03N-≤0.3-1.0Ni	UNS S40977
1.4005	X12CrS13	Fe-≤0.08-0.15C-≤1.0Si-≤1.5Mn-≤0.04P-≤0.15-0.35S-≤12.0-14.0Cr-≤0.6Mo	SAE 416
1.4006	X12Cr13	Fe-≤0.08-0.15C-≤1.00Si-≤1.5Mn-≤0.04P-≤0.015S-≤11.0-13.5Cr-≤0.75Ni	SAE 410
1.4008	GX7CrNiMo12-1	Fe-≤0.10C-1.0Si-1.0Mn-≤0.035P-≤0.025S-12.0-13.5Cr-0.20-0.50Mo-1.0-2.0Ni	UNS J91150
1.4015	X8Cr18	Fe-≤0.10C-≤1.50Si-≤1.50Mn-≤0.030P-≤0.030S-≤16.50-18.50Cr	S43080 (UNS)
1.4016	X6Cr17	Fe-≤0.08C-≤1.0Si-≤1.0Mn-≤0.04P-≤0.015S-≤16.0-18.0Cr	SAE 430
1.4021	X20Cr13	Fe-≤0.16-0.25C-≤1.0Si-≤1.5Mn-≤0.04P-≤0.015S-≤12.0-14.0Cr	SAE 420

German Standard		Materials Compositions	US-Standard
Mat.-No.	DIN-Design	Percent in Weight	SAE/ASTM/UNS
1.4024	X15Cr13	Fe-≤0.12-0.17C-≤1.0Si-≤1.0Mn-≤0.045P-≤0.03S-≤12.0-14.0Cr	SAE 420
1.4028	X30Cr13	Fe-≤0.26-0.35C-≤1.0Si-≤1.5Mn-≤0.04P-≤0.015S-≤12.0-14.0Cr	A 743; UNS J91153
1.4031	X39Cr13	Fe-≤0.36-0.42C-≤1.00Si-≤1.00Mn-≤0.04P-≤0.015S-≤12.5-14.5Cr	UNS S42080
1.4034	X46Cr13	Fe-≤0.43-0.5C-≤1.0Si-≤1.0Mn-≤0.04P-≤0.015S-≤12.5-14.5Cr	
1.4057	X17CrNi16-2	Fe-≤0.12-0.22C-≤1.0Si-≤1.5Mn-≤0.04P-≤0.015S-≤15.0-17.0Cr-≤1.5-2.5Ni	SAE 431
1.4085	GX70Cr29	Fe-≤0.50-0.90C-2.0Si-1.0Mn-≤0.045P-≤0.030S-27.0-30.0Cr	
1.4104	X14CrMoS17	Fe-≤0.10-0.17C-≤1.0Si-≤1.5Mn-≤0.04P-≤0.15-0.35S-≤15.5-17.5Cr-≤0.2-0.6Mo	SAE 430 F
1.4105	X6CrMoS17	Fe-≤0.08C-≤1.50Si-≤1.50Mn-≤0.040P-≤0.15-0.35S-≤16.00-18.00Cr-≤0.20-0.60Mo	
1.4110	X55CrMo14	Fe-≤0.48-0.60C-≤1.0Si-≤1.0Mn-≤0.04P-≤0.015S-≤13.0-15.0Cr-≤0.5-0.8Mo-≤0.15V	
1.4112	X90CrMoV18	Fe-≤0.85-0.95C-≤1.0Si-≤1.0Mn-≤0.04P-≤0.015S-≤17.0-19.0Cr-≤0.9-1.3Mo-≤0.07-0.12V	SAE 440 B
1.4113	X6CrMo17-1	Fe-≤0.80C-≤1.0Si-≤1.00Mn-≤0.04P-≤0.03S-≤16.0-18.0Cr-≤0.90-1.40Mo	SAE 434
1.4116	X50CrMoV15	Fe-≤0.45-0.55C-1.0Si-1.0Mn-≤0.040P-≤0.015S-14.0-15.0Cr-0.5-0.8Mo-0.10-0.20V	
1.4117	X38CrMoV15	Fe-≤0.35-0.40C-1.0Si-1.0Mn-≤0.045P-≤0.030S-14.0-15.0Cr-0.4-0.6Mo-0.10-0.15V	
1.4120	X20CrMo13	Fe-≤0.17-0.22C-≤1.0Si-≤1.0Mn-≤0.04P-≤0.015S-≤12-14Cr-≤0.9-1.3Mo-≤1.0Ni	
1.4122	X39CrMo17-1	Fe-≤0.33-0.45C-≤1.0Si-≤1.5Mn-≤0.04P-≤0.015S-≤15.5-17.5Cr-≤0.8-1.3Mo-≤1.0Ni	
1.4125	X105CrMo17	Fe-≤0.95-1.20C-1.0Si-1.0Mn-≤0.040P-≤0.015S-16.0-18.0Cr-0.4-0.8Mo	SAE 617
1.4126	X 110 CrMo 13	Fe-≤1.05-1.15C-≤1.0Si-≤1.0Mn-≤0.040P-≤0.030S- ≤17.0-18.0Cr-≤0.8-1.0Mo	
1.4131	X 1 CrMo 26 1	Fe-≤0.010C-≤0.40Si-≤0.40Mn-≤0.020P-≤0.020S-≤0.015N-≤25.0-27.5Cr-≤0.75-1.50Mo-≤0.50Ni	
1.4133		Fe-≤0.01C-1.0Si-1.0Mn-≤0.030P-≤0.030S-26.0-30.0Cr-1.7-2.3Mo-0.010N	
1.4136	GX70CrMo29-2	Fe-≤0.50-0.90C-2.0Si-1.0Mn-≤0.045P-≤0.030S-27.0-30.0Cr-2.0-2.5Mo	
1.4138	GX120CrMo29-2	Fe-≤0.90-1.30C-2.0Si-1.0Mn-≤0.045P-≤0.030S-27.0-30.0Cr-2.0-2.5Mo	

German Standard		Materials Compositions	US-Standard
Mat.-No.	**DIN-Design**	**Percent in Weight**	**SAE/ASTM/UNS**
1.4153	X80CrVMo13-2	Fe-≤0.76-0.86C-≤1.00Si-≤1.00Mn-≤0.045P-≤0.030S-≤12.00-14.00Cr-≤0.40-0.60Mo-≤1.50-2.10V	
1.4300	X 12 CrNi 18 8	Fe-≤0.12C-≤1.0Si-≤2.0Mn-≤0.045P-≤0.030S-≤17.0-19.0Cr-≤8.0-10.0Ni	
1.4301	X5CrNi18-10	Fe-≤0.07C-≤1.0Si-≤2.0Mn-≤0.045P-≤0.015S≤17-19.5Cr-≤0.11N-≤8.0-10.5Ni	SAE 304
1.4302	X5CrNi19-9	Fe-≤0.05C-≤1.40Si-≤1.90Mn-≤0.025P-≤0.015S-≤18.2-19.8Cr-≤8.70-10.30Ni	UNS S30888
1.4303	X4CrNi18-12	Fe-≤0.06C-≤1.0Si-≤2.0Mn-≤0.045P-≤0.015S-≤17-19Cr-≤0.11N-≤11.0-13.0Ni	SAE 305/308
1.4304	X5CrNi18-12E	Fe-≤0.12C-≤1.0Si-≤2.0Mn-≤0.045P-≤0.030S-≤17-19Cr-≤8.0-10.0Ni	
1.4305	X8CrNiS18-9; X10CrNiS189	Fe-≤0.10C-≤1.0Si-≤2.0Mn-≤0.045P-≤0.15-0.35S-≤17-19Cr-≤1.0Cu-≤0.11N-≤8.0-10.0Ni	SAE 303
1.4306	X2CrNi19-11	Fe-≤0.03C-≤1.0Si-≤2.0Mn-≤0.045P-≤0.015S-≤18-20Cr-≤0.11N-≤10.0-12.0Ni	SAE 304 L
1.4307	X2CrNi18-9	Fe-≤0.03C-≤1.0Si-≤2.0Mn-≤0.045P-≤0.015S-≤17.5-19.5Cr-≤0.11N-≤8.0-10.0Ni	
1.4308	GX5CrNi19-10	Fe-≤0.07C-≤1.5Si-≤1.5Mn-≤0.04P-≤0.030S-≤18.0-20.0Cr-≤8.0-11.0Ni	SAE 304 H
1.4310	X10CrNi18-8	Fe-≤0.05-0.15C-≤2.0Si-≤2.0Mn-≤0.045P-≤0.015S-≤16.0-19.0Cr-≤0.8Mo-≤0.11N≤6.0-9.5Ni	SAE 301
1.4311	X2CrNiN18-10	Fe-≤0.03C-≤1.0Si-≤2.0Mn-≤0.045P-≤0.015S-≤17.0-19.5Cr-≤8.5-11.5Ni-≤0.12-0.22N	SAE 304 LN
1.4312	GX10CrNi18-8	Fe-≤0.12C-≤2.0Si-≤1.5Mn-≤0.045P-≤0.03S-≤17.0-19.5Cr-≤8.0-10.0Ni	A 743
1.4313	X3CrNiMo13-4	Fe-≤0.05C-≤0.7Si-≤1.5Mn-≤0.04P-≤0.015S-≤12.0-14.0Cr-≤0.3-0.7Mo-≤0.02N-≤3.5 4.5Ni	UNS J91540
1.4315	X5CrNiN19-9	Fe-≤0.06C-≤1.0Si-≤2.0Mn-≤0.045P-≤0.03S-≤18.0-20.0Cr-≤0.12-0.22N-≤8.0-11.0Ni	
1.4317	GX4CrNi13-4	Fe-≤0.06C-≤1.00Si-≤1.00Mn-≤0.035P-≤0.025S-≤12.00-13.50Cr-≤0.70Mo-≤3.50-5.00Ni	UNS J91540 A 743 (CA-6NM)
1.4318	X2CrNiN18-7	Fe-≤0.03C-≤1.0Si-≤2.0Mn-≤0.045P-≤0.015S-≤16.5-18.5Cr-≤0.1-0.2N-≤6.0-8.0Ni	
1.4319	X3CrNiN17-8	Fe-≤0.05C-≤1.00Si-≤2.00Mn-≤0.045P-≤0.015S-≤16.00-18.00Cr-≤0.04-0.08N-≤7.00-8.00Ni	
1.4330	X 2 CrNi 25 20	Fe-≤0.03C-≤1.0Si-≤1.5Mn-≤0.045P-≤0.035S-≤18.0-22.0Cr-≤23.0-27.0Ni	

German Standard		Materials Compositions	US-Standard
Mat.-No.	DIN-Design	Percent in Weight	SAE/ASTM/UNS
1.4333	X 5 NiCr 32 21	Fe-≤0.07C-≤1.40Si-≤2.40Mn-≤0.045P-≤0.030S-≤19.00-22.00Cr-≤30.00-34.00Ni	S33200 (UNS) S33200 (SAE) B 710 (N 08330) (ASTM)
1.4335	X1CrNi25-21	Fe-≤0.02C-≤0.25Si-≤2.0Mn-≤0.025P-≤0.010S-≤24.0-26.0Cr-≤0.20Mo-≤0.110N-≤20.0-22.0Ni	
1.4340	GX40CrNi27-4	Fe-≤0.30-0.50C-≤2.00Si-≤1.50Mn-≤0.045P-≤0.030S-≤26.0-28.0Cr-≤3.5-5.5Ni	A 743
1.4347	GX6CrNiN26-7	Fe-≤0.08C-≤1.50Si-≤1.50Mn-≤0.035P-≤0.020S-25.00-27.00Cr-0.100-0.200N-5.50-7.50Ni	
1.4361	X1CrNiSi18-15-4	Fe-≤0.015C-≤3.70-4.50Si-≤2.00Mn-≤0.025P-≤0.010S-≤16.5-18.5Cr-≤0.20Mo-≤0.110N-≤14.0-16.0Ni	A 336
1.4362	X2CrNiN23-4	Fe-≤0.03C-≤1.00Si-≤2.00Mn-≤0.035P-≤0.015S-≤22.0-24.0Cr-≤0.10-0.60Cu-≤0.10-0.60Mo-≤0.050-0.200N-≤3.5-5.5Ni	
1.4371	X2CrMnNiN17-7-5	Fe-≤0.30C-≤1.0Si-≤6.0-8.0Mn-≤0.045P-≤0.015S-≤16.0-17.0Cr-≤0.15-0.20N-≤3.5-5.5Ni	SAE 202
1.4401	X5CrNiMo17-12-2	Fe-≤0.07C-≤1.0Si-≤2.0Mn-≤0.045P-≤0.015S-≤16.5-18.5Cr-≤2.0-2.5Mo-≤0.110N-≤10.0-13.0Ni	SAE 316
1.4404	X2CrNiMo17-12-2; X2 CrNiMo 17 12 2	Fe-≤0.03C-≤1.0Si-≤2.0Mn-≤0.045P-≤0.015S-≤16.5-18.5Cr-≤2.0-2.5Mo-≤10.0-13.0Ni≤0.110N	SAE 316 L
1.4405	X 5 CrNiMo 16 5; GX4CrNiMo16-5-1	Fe-≤0.07C-≤1.0Si-≤1.0Mn-≤0.035P-≤0.025S-≤15.0-16.5Cr-≤0.5-2.0Mo-≤4.5-6.0Ni	
1.4406	X2CrNiMoN17-11-2; X2 CrNiMoN 17 12 2	Fe-≤0.03C-≤1.0Si-≤2.0Mn-≤0.045P-≤0.015S-≤16.5-18.5Cr-≤2.0-2.5Mo-≤0.12-0.22N-≤10.0-12.0Ni	SAE 316 LN
1.4408	GX5CrNiMo19-11-2	Fe-≤0.07C-≤1.50Si-≤1.50Mn-≤0.04P-≤0.030S-≤18.0-20.0Cr-≤2.0-2.5Mo-≤9.0-12.0Ni	CF-8M
1.4410	X2CrNiMoN25-7-4	Fe-≤0.03C-≤1.0Si-≤2.0Mn-≤0.035P-0.015S-≤24.0-26.0Cr-≤3.0-4.5Mo-≤0.200-0.350N-≤6.0-8.0Ni	2507; A 182
1.4413	X3CrNiMo13-4	Fe-≤0.05C-≤0.30-0.60Si-≤0.50-1.00Mn-≤0.03P-≤0.02S-≤12.00-14.00Cr-≤0.30-0.70Mo-≤3.50-4.00Ni	UNS S42400 A 988 (S 41500) SAE S41500
1.4417	GX2CrNiMoN25-7-3	Fe-≤0.03C-≤1.0Si-≤1.5Mn-≤0.030P-≤0.020S-≤24.0-26.0Cr-≤1.0Cu-≤3.0-4.0Mo-≤0.150-0.250N-≤6.0-8.5Ni-≤1.0W	3RE60; A 789

German Standard Mat.-No.	DIN-Design	Materials Compositions Percent in Weight	US-Standard SAE/ASTM/UNS
1.4418	X4CrNiMo16-5-1	Fe-≤0.06C-≤0.7Si-≤1.5Mn-≤0.040P-≤0.015S-≤15.0-17.0Cr-≤0.8-1.5Mo-≤0.020N-≤4.0-6.0Ni	
1.4424	X2CrNiMoSi18-5-3	Fe-≤0.03C-1.40-2.00Si-1.20-2.00Mn-≤0.035P-≤0.015S-18.00-19.00Cr-2.50-3.00Mo-0.050-1.00N-4.50-5.20Ni	
1.4427	X12CrNiMoS18-11	Fe-≤0.12C-≤1.0Si-≤2.0Mn-≤0.06P-≤0.15-0.35S-≤16.5-18.5Cr-≤2-2.5Mo-≤10.5-13.5Ni	
1.4429	X2CrNiMoN17-13-3	Fe-≤0.03C-≤1.0Si-≤2.0Mn-≤0.045P-≤0.015S-≤16.5-18.5Cr-≤2.5-3.0Mo-≤11.0-14.0Ni-≤0.12-0.22N	SAE 316 LN
1.4430	X2CrNiMo19-12	Fe-≤0.02C-≤1.40Si-≤1.90Mn-≤0.025P-≤0.015S-17.20-19.80Cr-2.50-3.00Mo-10.70-13.30Ni	S 31688 UNS S 31683 UNS
1.4434	X2CrNiMoN18-12-4	Fe-≤0.03C-≤1.00Si-≤2.00Mn-≤0.045P-≤0.015S-≤16.50-19.50Cr-≤3.00-4.00Mo-≤0.10-0.20N-≤10.50-14.00Ni	
1.4435	X2CrNiMo18-14-3; X2 CrNiMo 18 4 3	Fe-≤0.03C-≤1.0Si-≤2.0Mn-≤0.045P-≤0.015S-≤17.0-19.0Cr-≤2.5-3.0Mo-≤0.110N-≤12.5-15.0Ni	SAE 316 L
1.4436	X3CrNiMo17-13-3; X5 CrNiMo 17 13 3	Fe-≤0.05C-≤1.0Si-≤2.0Mn-≤0.045P-≤0.015S-≤16.5-18.5Cr-≤2.5-3.0Mo-≤0.110N-≤10.5-13.0Ni	SAE 316
1.4438	X2CrNiMo18-15-4	Fe-≤0.03C-≤1.0Si-≤2.0Mn-≤0.045P-≤0.015S-≤17.5-19.5Cr-≤3.0-4.0Mo-≤0.110N-≤13.0-16.0Ni	SAE 317 L
1.4439	X2CrNiMoN17-13-5	Fe-≤0.030C-≤1.00Si-≤2.00Mn-≤0.045P-≤0.015S-≤16.5-18.5Cr-≤4.00-5.00Mo-≤12.5-14.5Ni-≤0.12-0.22N	
1.4440	X2CrNiMo18-16-5	Fe-≤0.03C-≤1.00Si-2.00-3.00Mn-≤0.025P-≤0.025S-17.00-20.00Cr-4.00-5.00Mo-16.00-19.00Ni	SAE 317 L
1.4441	X2CrNiMo18-15-3	Fe-≤0.03C-≤1.00Si-≤2.00Mn-≤0.025P-≤0.01S-≤17.00-19.00Cr-≤0.50Cu-≤2.50-3.20Mo-≤0.10N-≤13.00-15.50Ni	
1.4442	X2CrNiMoN18-15-4	Fe-≤0.03C-1.0Si-2.0Mn-≤0.025P-≤0.010S-17.0-18.5Cr-3.7-4.2Mo-0.1-0.2N-14.0-16.0Ni	SAE 317 LN
1.4447	X5CrNiMo18-13	Fe-≤0.06C-1.5Si-2.0Mn-≤0.025P-≤0.020S-17.0-19.0Cr-4.0-5.0Mo-12.5-15.5Ni	
1.4449	X3CrNiMo18-12-3	Fe-≤0.035C-≤1.00Si-≤2.00Mn-≤0.045P-≤0.015S-≤17.0-18.2Cr-≤1.0Cu-≤2.25-2.75Mo-≤0.08N-≤11.5-12.5Ni	SAE 317
1.4457	X8CrNiMo17-5-3	Fe-≤0.07-0.11C-≤0.50Si-≤0.50-1.25Mn-≤16.00-17.00Cr-≤2.50-3.25Mo-≤4.00-5.00Ni	

German Standard		Materials Compositions	US-Standard
Mat.-No.	DIN-Design	Percent in Weight	SAE/ASTM/UNS
1.4460	X3CrNiMoN27-5-2	Fe-≤0.05C-≤1.0Si-≤2.0Mn-≤0.035P-≤0.015S-≤25.0-28.0Cr-≤1.3-2.0Mo-≤0.05-0.2N-≤4.5-6.5Ni	SAE 329
1.4462	X2CrNiMoN22-5-3	Fe-≤0.03C-≤1.00Si-≤2.00Mn-≤0.035P-≤0.015S-≤21.0-23.0Cr-≤2.50-3.50Mo-≤4.50-6.50Ni-≤0.1-0.22N	2205; A 182
1.4463	GX6CrNiMo24-8-2	Fe-≤0.07C-1.5Si-1.5Mn-≤0.045P-≤0.030S-23.0-25.0Cr-2.0-2.5Mo-7.0-8.5Ni	
1.4464	GX40CrNiMo27-5	Fe-0.30-0.50C-≤2.00Si-≤1.50Mn-≤0.045P-≤0.030S-26.00-28.00Cr-2.00-2.50Mo-4.00-6.00Ni	
1.4465	X1CrNiMoN25-25-2	Fe-≤0.02C-≤0.70Si-≤2.0Mn-≤0.020P-≤0.015S-≤24.0-26.0Cr-≤2.0-2.5Mo-≤22.0-25.0Ni-≤0.08-0.16N	SAE 310 MoLN
1.4466	X1CrNiMoN25-22-2	Fe-≤0.02C-≤0.7Si-≤2.0Mn-≤0.025P-≤0.010S-≤24.0-26.0Cr-≤2.0-2.5Mo-≤21.0-23.0Ni-≤0.1-0.16N	
1.4467	X2CrMnNiMoN26-5-4	Fe-≤0.03C-≤0.8Si-≤4.0-6.0Mn-≤0.03P-≤0.015S-≤24.5-26.5Cr-≤2.0-3.0Mo-≤0.3-0.45N-≤3.5-4.5Ni	
1.4469	GX2CrNiMoN26-7-4	Fe-≤0.03C-1.0Si-1.0Mn-≤0.035P-≤0.025S-25.0-27.0Cr-≤1.30Cu-3.0-5.0Mo-0.12-0.22N-6.0-8.0Ni	UNS J93404
1.4492	X 8 CrNiMoN 17 5	Fe-≤0.07-0.11C-≤0.5Si-≤0.5-1.25Mn-≤0.04P-≤0.03S-≤16.0-17.0Cr-≤2.5-3.25Mo-≤4.0-5.0Ni	
1.4500	GX7NiCrMoCuNb25-20	Fe-≤0.08C-≤1.50Si-≤2.0Mn-≤0.045P-≤0.03S-≤19.0-21.0Cr-≤1.5-2.5Cu-≤2.5-3.5Mo-≤24.0-26.0Ni	A 351
1.4501	X2CrNiMoCuWN25-7-4	Fe-≤0.03C-≤1.0Si-≤1.0Mn-≤0.035P-≤0.015S-≤24.0-26.0Cr-≤0.5-1.0Cu-≤3.0-4.0Mo-≤0.2-0.3N-≤6.0-8.0Ni-≤0.5-1.0W	UNS S32760
1.4502	X8CrTi18	Fe-≤0.09C-≤1.40Si-≤1.40Mn-≤0.030P-≤0.020S-16.70-18.30Cr-≤0.060N-0.35-0.65Ti	
1.4503	X3NiCrCuMoTi27-23	Fe-≤0.04C-0.75Si-0.75Mn-≤0.030P-≤0.015S-22.0-24.0Cr-2.5-3.5Cu-2.5-3.0Mo-26.0-28.0Ni-0.4-0.7Ti	
1.4504		Fe-≤0.09C-≤0.50Si-≤1.00Mn-≤0.025P-≤0.025S-≤0.75-1.25Al-≤16.00-17.25Cr-≤6.50-7.75Ni	UNS S17780 SAE S17700 A 705 (631)
1.4505	X4NiCrMoCuNb20-18-2	Fe-≤0.05C-≤1.0Si-≤2.0Mn-≤0.045P-≤0.015S-≤16.5-18.5Cr-≤2.0-2.5Mo-≤19.0-21.0Ni-≤1.8-2.2Cu	

German Standard		Materials Compositions	US-Standard
Mat.-No.	DIN-Design	Percent in Weight	SAE/ASTM/UNS
1.4506	X5NiCrMoCuTi20-18	Fe-≤0.07C-≤1.0Si-≤2.0Mn-≤0.045P-≤0.030S-≤16.5-18.5Cr-≤2.0-2.5Mo-≤19.0-21.0Ni-≤1.8-2.2Cu	
1.4507	X2CrNiMoCuN25-6-3	Fe-≤0.03C-≤0.7Si-≤2.0Mn-≤0.035P-≤0.015S-≤24.0-26.0Cr-≤1.0-2.5Cu-≤0.15-0.30N-≤5.5-7.5Ni-≤2.7-4.0Mo	UNS S43940
1.4509	X2CrTiNb18	Fe-≤0.03C-≤1.0Si-≤1.0Mn-≤0.040P-≤0.015S-≤17.5-18.5Cr-≤0.10-0.60Ti	
1.4510	X6CrTi17	Fe-≤0.05C-≤1.0Si-≤1.0Mn-≤0.040P-≤0.015S-≤16.0-18.0Cr	SAE 430 Ti
1.4511	X3CrNb17; X6CrNb17	Fe-≤0.05C-≤1.0Si-≤1.0Mn-≤0.040P-≤0.015S-≤16.0-18.0Cr	
1.4512	X2CrTi12	Fe-≤0.03C-≤1.0Si-≤1.0Mn-≤0.040P-≤0.015S-≤10.5-12.5Cr	SAE 409
1.4515	GX2CrNiMoCuN26-6-3	Fe-≤0.03C-≤1.0Si-≤2.0Mn-≤0.030P-≤0.020S-≤0.12-0.25N-≤24.5-26.5Cr-≤2.5-3.5Mo-≤5.5-7.0Ni-≤0.8-1.3Cu	
1.4517	GX2CrNiMoCuN25-6-3-3	Fe-max. 0.03C-1.0Si-1.5Mn-0.035P-0.025S-24.5-26.5Cr-2.75-3.50Cu-2.50-3.50Mo-0.12-0.22N-5.0-7.0Ni	UNS J93372
1.4519	X2CrNiMoCu20-25	Fe-≤0.02C-≤1.40Si-≤2.10-4.90Mn-≤0.025P-≤0.02S-≤19.20-21.80Cr-≤0.90-1.90Cu-≤4.10-5.90Mo-≤24.30-26.70Ni	
1.4520	X2CrTi17	Fe-≤0.025C-≤0.5Si-≤0.5Mn-≤0.04P-≤0.015S-≤16.0-18.0Cr-≤0.015N-≤0.3-0.6Ti	
1.4521	X2CrMoTi18-2	Fe-≤0.025C-≤1.0Si-≤0.040P-≤0.015S-≤1.0Mn-≤17.0-20.0Cr-≤1.8-2.5Mo-≤0.030N	SAE 444
1.4522	X2CrMoNb18-2	Fe-≤0.025C-≤1.0Si-≤1.0Mn-≤0.040P-≤0.015S-≤17.0-19.0Cr-≤1.8-2.3Mo-≤0.25Ni	SAE 443
1.4523	X2CrMoTiS18-2	Fe-≤0.03C-≤1.0Si-≤0.5Mn-≤0.040P-≤0.15-0.35S-≤17.5-19.0Cr-≤2.0-2.5Mo-≤0.30 0.80Ti	
1.4525	GX5CrNiCu16-4; GX4CrNi-CuNb16-4	Fe-≤0.07C-≤0.80Si-≤1.00Mn-≤0.035P-≤0.025S-≤15.00-17.00Cr-≤2.50-4.00Cu-≤0.80Mo-≤0.050N-≤0.350Nb-≤3.50-5.50Ni	
1.4527	GX4NiCrCuMo30-20-4	Fe-≤0.06C-≤1.50Si-≤1.50Mn-≤0.040P-≤0.030S-19.00-22.00Cr-3.00-4.00Cu-2.00-3.00Mo-27.50-30.50Ni	
1.4528	X105CrCoMo18-2	Fe-≤1.0-1.1C-≤1.0Si-≤1.0Mn-≤0.045P-≤0.030S-≤16.5-18.5Cr-≤1.0-1.5Mo-≤0.30-0.80Ti-≤1.3-1.8Co≤0.07-0.12V	
1.4529	X1NiCrMoCuN25-20-7	Fe-≤0.02C-≤0.50Si-≤1.00Mn-≤0.030P-≤0.010S-≤19.0-21.0Cr-≤6.0-7.0Mo-≤0.5-1.5Cu-≤0.15-0.25N-≤24.0-26-0Ni	A 249; ASTM N08926

German Standard		Materials Compositions	US-Standard
Mat.-No.	DIN-Design	Percent in Weight	SAE/ASTM/UNS
1.4530	X1CrNiMoAlTi12-9	Fe-≤0.015C-0.10Si-0.10Mn-≤0.020P-0.6-1.0Al-11.0-12.5Cr-1.5-2.5Mo-8.5-10.5Ni-0.25-0.40Ti	
1.4532	X8CrNiMoAl15-7-2	Fe-≤0.1C≤0.7Si≤1.2Mn≤0.04P≤0.015S-0.7-1.5Al-14.0-16.0Cr-2.0-3.0Mo-6.5-7.8Ni	UNS S15700
1.4533	X6CrNiTi18-10S	Fe-≤0.06C-≤1.00Si-≤2.00Mn-≤0.035P-≤0.015S-≤0.20Co-≤17.00-19.00Cr-≤9.00-12.00Ni	
1.4534	X3CrNiMoAl13-8-2	Fe-≤0.05C-≤0.10Si-≤0.10Mn-≤0.010P-≤0.008S-≤0.90-1.20Al-≤12.25-13.25Cr-≤2.0-2.50Mo-≤0.01N-≤7.50-8.50Ni	S 13800
1.4536	GX2NiCrMoCuN25-20	Fe-≤0.03C≤1.0Si≤1.0Mn≤0.035P≤0.01S-19.0-21.0Cr-1.5-2.0Cu-2.5-3.5Mo-24.0-26.0Ni	UNS J94650
1.4537	X1CrNiMoCuN25-25-5	Fe-≤0.02C-≤0.70Si-≤2.00Mn-≤0.030P-≤0.010S-24.00-26.00Cr-1.00-2.00Cu-4.70-5.70Mo-0.170-0.250N-24.00-27.00Ni	
1.4539	X1NiCrMoCu25-20-5	Fe-≤0.02C-≤0.70Si-≤2.0Mn-≤0.030P-≤0.010S-≤19.0-21.0Cr-≤4.0-5.0Mo-≤24.0-26.0Ni-≤1.2-2.0Cu-≤0.150N	SAE 904 L
1.4540		Fe-≤0.06C-1.0Si-1.0Mn-15.0-17.0Cr-2.5-4.0Cu-0.050N-0.15-0.40Nb-3.5-5.0Ni	UNS J92130
1.4541	X6CrNiTi18-10	Fe-≤0.08C-≤1.0Si-≤2.0Mn-≤0.045P-≤0.015S-≤17.0-19.0Cr-≤9.0-12.0Ni	SAE 321
1.4542	X5CrNiCuNb16-4	Fe-≤0.07C-≤0.70Si-≤1.5Mn-≤0.040P-≤0.015S-≤15.0-17.0Cr-≤3.0-5.0Ni-≤3.0-5.0Cu-≤0.60Mo	SAE 630; 17-4 PH
1.4544	X 10 CrNiMnTi 18 10	Fe-≤0.08C-≤1.0Si-≤2.0Mn-≤0.035P-≤0.025S-≤17.0-19.0Cr-≤9.0-11.5Ni	SAE 321; UNS J92630
1.4545		Fe-≤0.07C-≤1.00Si-≤1.00Mn-≤0.03P-≤0.015S-≤14.00-15.50Cr-≤2.50-4.50Cu-≤0.50Mo-≤3.50-5.50Ni	UNS S15500 SAE S15500 A 705 (XM-12)
1.4546	X5CrNiNb18-10	Fe-≤0.08C-≤1.0Si-≤2.0Mn-≤0.045P-≤0.030S-≤17.0-19.0Cr-≤9.0-11.5Ni	SAE 347
1.4547	X1CrNiMoCuN20-18-7	Fe-≤0.02C-≤0.70Si-≤1.0Mn-≤0.030P-≤0.010S-≤19.5-20.5Cr-≤0.5-1.0Cu-≤6.00-7.00Mo-≤0.18-0.25N-≤17.5-18.5Ni	254 SMO; A 182
1.4548	X5CrNiCuNb17-4-4	Fe-≤0.07C-≤1.0Si-≤1.0Mn-≤0.025P-≤0.025S-≤15.0-17.5Cr-≤3.0-5.0Cu-≤0.15-0.45Nb-≤3.00-5.00Ni	17-4 PH; SAE 630
1.4550	X6CrNiNb18-10	Fe-≤0.08C-≤1.0Si-≤2.0Mn-≤0.045P-≤0.015S-≤17.0-19.0Cr≤9.0-12.0Ni	SAE 347

German Standard		Materials Compositions	US-Standard
Mat.-No.	DIN-Design	Percent in Weight	SAE/ASTM/UNS
1.4551	X5CrNiNb19-9	Fe-≤0.06C-≤1.40Si-≤1.90Mn-≤0.025P-≤0.015S-≤18.20-19.80Cr-≤8.20-9.80Ni	UNS S34780 S34781 S34788
1.4552	GX5CrNiNb19-11	Fe-≤0.07C≤1.5Si≤1.5Mn≤0.04P≤0.03S-18.0-20.0Cr-9.0-12.0Ni	UNS J92710
1.4557	GX2CrNiMoCuN20-18-6	Fe-≤0.025C-≤1.00Si-≤1.20Mn-≤0.03P-≤0.01S-≤19.50-20.50Cr-≤0.50-1.00Cu-≤6.00-7.00Mo-≤0.18-0.24N-≤17.50-19.50Ni	
1.4558	X2NiCrAlTi32-20	Fe-≤0.03C-≤0.70Si-≤1.0Mn-≤0.020P-≤0.015S-≤0.15-0.45Al≤20.0-23.0Cr-≤32.0-35.0Ni	
1.4561	X1CrNiMoTi18-13-2	Fe-≤0.02C-≤0.50Si-≤2.0Mn-≤0.035P-≤0.015S-≤17.0-18.5Cr-≤2.0-2.5Mo-≤11.5-13.5Ni-≤0.4-0.6Ti	
1.4562	X1NiCrMoCu32-28-7	Fe-≤0.015C-≤0.30Si-≤2.0Mn-≤0.020P-≤0.010S-≤26.0-28-0Cr-≤1.0-1.4Cu-≤6.0-7.0Mo-≤0.15-0.25N-≤30.0-32.0Ni	Alloy 31
1.4563	X1NiCrMoCu31-27-4	Fe-≤0.02C-≤0.70Si-≤2.0Mn-≤0.030P-≤0.010S-≤26.0-28.0Cr-≤3.0-4.0Mo-≤30.0-32.0Ni-≤0.70-1.5Cu-≤0.11N	B 668
1.4564		Fe-≤0.09C-≤1.00Si-≤1.00Mn-≤0.04P-≤0.03S-≤0.75-1.50Al-≤16.00-18.00Cr-≤0.50Cu-≤6.50-7.75Ni	UNS 17700 SAE S17700 A 705 (631)
1.4565	X2CrNiMnMoNbN25-18-5-4	Fe-≤0.03C-≤1.0Si-≤3.5-6.5Mn-≤0.030P-≤0.015S-≤23.0-26.0Cr-≤3.0-5.0Mo-≤0.3-0.5N-≤0.15Nb-≤16.0-19.0Ni	UNS S34565
1.4566	X3CrNiMnMoCuNbN 23-17-5-3	Fe-≤0.04C≤1.0Si-4.5-6.5Mn≤0.03P-≤0.015S-21.0-25.0Cr-0.3-1.0Cu-3.0-4.5Mo-15.0-18.0Ni-0.1-0.3Nb	
1.4567	X3CrNiCu18-9-4; X 3 CrNiCu 18 9	Fe-≤0.04C-≤1.00Si-≤2.00Mn-≤0.045P-≤0.015S-≤17.00-19.00Cr-≤3.00-4.00Cu-≤0.110N-≤8.50-10.50Ni	304 Cu (SAE) S 30430 (UNS) A 493 (S 30430) (ASTM)
1.4568	X7CrNiMoAl17-7	Fe-≤0.09C-≤0.70Si-≤1.0Mn-≤0.040P-0.015S-≤16.0-18.0Cr-≤6.5-7.80Ni-≤0.70-1.5Al	17-7 PH; SAE 631
1.4571	X6CrNiMoTi17-12-2	Fe-≤0.08C-≤1.0Si-≤2.0Mn-≤0.045P-0.015S-≤16.5-18.5Cr-≤2.0-2.5Mo-≤10.5-13.5Ni	SAE 316 Ti
1.4573	GX3CrNiMoCuN24-6-5	Fe-≤0.40C-≤1.0Si-≤1.0Mn-≤0.030P-≤0.020S-≤22.0-25.0Cr-≤1.5-2.5Cu-≤4.5-6.0Mo-≤0.15-0.25N-≤4.5-6.5Ni	SAE 316 Ti
1.4574		Fe-≤0.09C-≤1.00Si-≤1.00Mn-≤0.040P-≤0.030S-≤0.75-1.50Al-≤14.00-16.00Cr-≤2.00-3.00Mo-≤6.50-7.75Ni	S 15700 (SAE) A 579 (63) (ASTM) S 15700 (UNS)

German Standard		Materials Compositions	US-Standard
Mat.-No.	DIN-Design	Percent in Weight	SAE/ASTM/UNS
1.4575	X1CrNiMoNb28-4-2	Fe-≤0.015C-≤1.0Si-≤1.0Mn-≤0.025P-≤0.015S-≤26.0-30.0Cr-≤1.8-2.5Mo-≤0.035N-≤3.0-4.5Ni	25-4-4; A 176
1.4577	X3CrNiMoTi25-25	Fe-≤0.04C-≤0.50Si-≤2.0Mn-≤0.030P-≤0.015S-≤24.0-26.0Cr-≤2.0-2.5Mo-≤24.0-26.0Ni	
1.4580	X6CrNiMoNb17-12-2	Fe-≤0.08C-≤1.0Si-≤2.0Mn-≤0.045P-≤0.015S-≤16.5-18.5Cr-≤2.0-2.5Mo-≤10.5-13.5Ni	SAE 316 Cb UNS J92971
1.4581	GX5CrNiMoNb19-11-2	Fe-≤0.07C≤1.5Si≤1.5Mn≤0.04P≤0.03S-18.0-20.0Cr≤2.0-2.5Mo-9.0-12.0Ni	
1.4582	X4CrNiMoNb25-7	Fe-≤0.06C-≤1.00Si-≤2.00Mn-≤0.045P-≤0.030S-≤24.00-26.00Cr-≤1.30-2.00Mo-≤6.50-7.50Ni	
1.4583	X10CrNiMoNb18-12	Fe-≤0.10C-≤1.00Si-≤2.00Mn-≤0.045P-≤0.030S-≤16.5-18.5Cr-≤2.5-3.0Mo-≤12.0-14.5Ni	318 (Spec)
1.4585	GX7CrNiMoCuNb1818	Fe-≤0.080C-≤1.50Si-≤2.0Mn-≤0.045P-≤0.030S-≤16.5-18.5Cr-≤2.0-2.5Mo-≤19.0-21.0Ni-≤1.8-2.4Cu	
1.4586	X5NiCrMoCuNb22-18	Fe-≤0.07C-1.0Si-2.0Mn-≤0.045P-≤0.030S-16.5-18.5Cr-1.5-2.0Cu-3.0-3.5Mo-21.5-23.5Ni	
1.4589	X5CrNiMoTi15-2	Fe-≤0.080C-≤1.0Si-≤1.0Mn-≤0.045P-≤0.030S-≤13.0-15.5Cr-≤0.2-1.2Mo-≤1.0-2.5Ni-≤0.3-0.5Ti	UNS S42035
1.4591	X1CrNiMoCuN33-32-1	Fe-≤0.015C-≤0.5Si-≤2.0Mn-≤0.020P-≤0.010S-≤31.0-35.0Cr-≤0.3-1.2Cu-≤0.5-2.0Mo-≤0.35-0.6N-≤30.0-33.0Ni	
1.4592	X1CrMoTi29-4	Fe-≤0.025C-1.0Si-1.0Mn-≤0.030P-≤0.010S-28.0-30.0Cr-3.5-4.5Mo-0.045N	
1.4593	GX3CrNiMoCuN24-6-2-3	Fe-≤0.04C-1.5Si-1.5Mn-≤0.030P-≤0.020S-23.0-26.0Cr-2.75-3.5Cu-2.0-3.0Mo-0.1-0.2N-5.0-8.0Ni	
1.4603	X1CrTi17	Fe-≤0.02C-≤1.00Si-≤1.00Mn-≤0.040P-≤0.015S-≤16.00-18.00Cr	
1.4604	X2CrTi20	Fe-≤0.03C-≤1.00Si-≤1.00Mn-≤0.040P-≤0.015S-≤19.00-21.00Cr-≤0.40-0.80Ti	
1.4652	X1CrNiMoCuN24-22-8	Fe-≤0.02C-≤0.5Si-≤2.0-4.0Mn-≤0.03P-≤0.005S-≤23.0-25.0Cr-≤0.3-0.6Cu-≤7.0-8.0Mo-≤0.45-0.55N-≤21.0-23.0Ni	
1.4712	X10CrSi6	Fe-≤0.12C-≤2.00-2.50Si-≤1.00Mn-≤0.045P-≤0.030S-≤5.50-6.50Cr	

German Standard		*Materials Compositions*	*US-Standard*
Mat.-No.	***DIN-Design***	***Percent in Weight***	***SAE/ASTM/UNS***
1.4713	X10CrAl7; X10CrAlSi7	Fe-≤0.12C-≤0.50-1.00Si-≤1.00Mn-≤0.040P-≤0.015S-≤0.5-1.0Al-≤6.00-8.00Cr	
1.4718	X45CrSi9-3	Fe-0.4-0.5C-2.7-3.3Si-≤0.6Mn-≤0.04P-≤0.03S-8.0-10.0Cr-≤0.5Ni	S65007 (UNS)
1.4720	X7CrTi12	Fe-≤0.08C-1.0Si-1.0Mn-≤0.040P-≤0.030S-10.5-12.5Cr	SAE 409
1.4722	X 10 CrSi 13	Fe-≤0.12C-≤1.90-2.40Si-≤1.00Mn-≤0.045P-≤0.030S-≤12.0-14.0Cr	
1.4724	X10CrAl13; X10CrAlSi13	Fe-≤0.12C-≤0.70-1.40Si-≤1.00Mn-≤0.040P-≤0.015S-≤0.70-1.20Al-≤12.0-14.0Cr	
1.4725	CrAl 14 4; (X8CrAl14-4)	Fe-≤0.1C-≤0.5Si-≤1.0Mn-≤0.045P-≤0.03S-3.5-5.0Al-13.0-15.0Cr	K91670 (UNS)
1.4742	X10CrAlSi18; X10CrAl18	Fe-≤0.12C-≤0.70-1.40Si-≤1.00Mn-≤0.040P-≤0.015S-≤0.70-1.20Al-≤17.00-19.00Cr	
1.4749	X18CrN28	Fe-0.15-0.20C-1.0Si-1.0Mn-≤0.040P-≤0.015S-26.0-29.0Cr-0.15-0.25N	
1.4762	X10 CrAl 24; X10CrAlSi25	Fe-≤0.12C-≤0.70-1.40Si-≤1.00Mn-≤0.040P-≤0.015S-≤1.20-1.70Al-≤23.0-26.0Cr	SAE 446
1.4765	CrAl 25 5; (X8CrAl25-5)	Fe-≤0.10C-≤1.00Si-≤0.60Mn-≤0.045P-≤0.030S-≤4.50-6.00Al-≤22.00-25.00Cr	
1.4773	X8Cr30	Fe-≤0.09C-≤1.90Si-≤1.40Mn-≤0.030P-≤0.025S-≤28.80-31.20Cr-≤2.00Ni	
1.4776	GX40CrSi28	Fe-0.30-0.50C-1.0-2.5Si-1.0Mn-≤0.040P-≤0.030S-27.0-30.0Cr-0.50Mo-1.0Ni	UNS J92605
1.4777	GX130CrSi29	Fe-1.20-1.40C-1.0-2.5Si-0.5-1.0Mn-≤0.035P-≤0.030S-27.0-30.0Cr-0.50Mo-1.0Ni	
1.4821	X15CrNiSi25-4; X20 CrNiSi 25 4	Fe-0.1-0.2C-0.8-1.5Si-≤2.0Mn-≤0.04P-≤0.015S-24.5-26.5Cr-≤0.11N-3.5-5.5Ni	
1.4828	X15CrNiSi20-12	Fe-≤0.20C-≤1.50-2.50Si-≤2.0Mn-≤0.045P-≤0.015S-≤19.0-21.0Cr-≤0.11N-≤11.0-13.0Ni	SAE 309
1.4829	X12CrNi22 12	Fe-≤0.14C-0.90-1.90Si-1.90Mn-≤0.025P-≤0.015S-20.8-23.2Cr-10.2-12.8Ni	UNS S30980
1.4833	X7CrNi23 14; X12CrNi23-14	Fe-≤0.15C-≤1.00Si-≤2.00Mn-≤0.045P-≤0.015S-≤22.0-24.0Cr-≤0.11N-≤12.0-14.0Ni	SAE 309 S
1.4835	X9CrNiSiNCe21-11-2	Fe-≤0.05-0.12C-≤1.4-2.5Si-≤1.0Mn-≤0.045P-≤0.015S-≤0.030-0.080Ce-≤20.0-22.0Cr-≤0.12-0.20N-≤10.0-12.0Ni	253 MA; A 182
1.4841	X15CrNiSi25-20; X15CrNiSi25-21	Fe-≤0.20C-≤1.50-2.50Si-≤2.00Mn-≤0.045P-≤0.015S-≤24.0-26.0Cr-≤0.11N-≤19.0-22.0Ni	3RE60; SAE 310; SAE 314
1.4845	X8CrNi25-21; X12CrNi25-21	Fe-≤0.15C-≤1.50Si-≤2.00Mn-≤0.045P-≤0.015S-≤24.0-26.0Cr-≤0.11N-≤19.0-22.0Ni	SAE 310 S

German Standard		Materials Compositions	US-Standard
Mat.-No.	DIN-Design	Percent in Weight	SAE/ASTM/UNS
1.4847	X8CrNiAlTi20-20	Fe-≤0.08C-≤1.0Si-≤1.0Mn-≤0.030P-≤0.015S-≤0.6Al-18.0-22.0Cr-18.0-22.0Ni-0.6Ti	334 (SAE)
1.4848	GX40CrNiSi25-20	Fe-≤0.30-0.50C-≤1.00-2.50Si-≤1.50Mn-≤0.035P-≤0.030S-≤24.0-26.0Cr-≤19.0-21.0Ni	A 297 (HK)
1.4856	GX40NiCrSiNbTi35-25	Fe-≤0.35-0.45C-≤1.00-1.50Si-≤0.5-1.50Mn-≤0.035P-≤0.030S-≤23.0-27.0Cr-≤0.9-1.5Nb-≤33.0-37.0Ni-≤0.10-0.25Ti	
1.4857	GX40NiCrSi35-25	Fe-≤0.30-0.50C-≤1.00-2.50Si-≤1.50Mn-≤0.035P-≤0.030S-≤24.0-26.0Cr-≤34.0-36.0Ni	A 297 (HP)
1.4862	X8NiCrSi38-18	Fe-≤0.1C-1.5-2.5Si-0.8-1.5Mn-≤0.03P-≤0.03S-17.0-19.0Cr-≤0.5Cu-35.0-39.0Ni-≤0.2Ti	
1.4864	X12NiCrSi35-16; X12 NiCrSi 36 16	Fe-≤0.15C-≤1.0-2.0Si-≤2.0Mn-≤0.045P-≤0.015S-≤15.0-17.0Cr-≤0.11N-≤33.0-37.0Ni	SAE 330
1.4871	X53CrMnNiN21-9	Fe-≤0.48-0.58C-≤0.25Si-≤8.00-10.00Mn-≤0.045P-≤0.030S-≤20.00-22.00Cr-≤0.35-0.50N-≤3.25-4.50Ni	S 63008 (UNS) EV 8 (SAE)
1.4873	X45CrNiW18-9	Fe-≤0.40-0.50C-≤2.00-3.00Si-≤0.80-1.50Mn-≤0.045P-≤0.030S-≤17.00-19.00Cr-≤8.00-10.00Ni-≤0.80-1.20W	
1.4875	X55CrMnNiN20-8	Fe-≤0.50-0.60C-≤0.25Si-≤7.00-10.00Mn-≤0.045P-≤0.030S-≤19.50-21.50Cr-≤0.20-0.40N-≤1.50-2.75Ni	S 63012 (UNS) EV 12 (SAE)
1.4876	X10NiCrAlTi32-21; X10 NiCrAlTi 32 20	Fe-≤0.12C-≤1.00Si-≤2.00Mn-≤0.030P-≤0.015S-≤0.15-0.60Al-≤19.00-23.00Cr-≤30.00-34.00Ni-≤0.15-060Ti	N 08332 (UNS) B 366 (N08332) (ASTM) N 8810 (SAE)
1.4877	X6NiCrNbCe32-27	Fe-0.04-0.08C-≤0.3Si-≤1.0Mn-≤0.02P-≤0.01S-≤0.025Al-0.05-0.1Ce-26.0-28.0Cr-0.11N-0.6-1.0Nb-31.0-33.0Ni	S33228 (UNS)
1.4878	X10CrNiTi18-10; X12 CrNiTi 18 9	Fe-≤0.10C-≤1.0Si-≤2.0Mn-≤0.045P-0.015S-≤17.0-19.0Cr-≤9.0-12.0Ni	
1.4903	X10CrMoVNb9-1	Fe-≤0.08-0.12C-≤0.20-0.50Si-≤0.30-0.60Mn-≤0.020P-≤0.010S-≤0.04Al-≤8.0-9.5Cr-≤0.85-1.05Mo-≤0.030-0.070N-≤0.06-0.1Nb-≤0.4Ni-≤0.18-0.25V	A 182
1.4913	X19CrMoNbVN11-1	Fe-0.17-0.23C-0.50Si-0.40-0.90Mn-≤0.025P-≤0.015S-≤0.02Al-≤0.0015B-10.0-11.5Cr-0.5-0.8Mo-0.05-0.1N-0.25-0.55Nb-0.20-0.60Ni-0.10-0.30V	

German Standard		Materials Compositions	US-Standard
Mat.-No.	DIN-Design	Percent in Weight	SAE/ASTM/UNS
1.4919	X6CrNiMo17-13	Fe-≤0.04-0.08C-≤0.75Si-≤2.0Mn-≤0.035P-≤0.015S-≤0.0015-0.0050B-≤16.0-18.0Cr-≤2.0-2.5Mo-≤0.11N-≤12.0-14.0Ni	SAE 316 H
1.4922	X20CrMoV11-1	Fe-≤0.17-0.23C-≤0.50Si-≤1.00Mn-≤0.030P-≤0.030S-≤10.0-12.5Cr-≤0.80-1.20Mo-≤0.30-0.80Ni-≤0.25-0.35V	
1.4943	X4NiCrTi25-15	Fe-≤0.06C-≤1.0Si-≤2.0Mn-≤0.025P-≤0.015S-≤0.35Al-≤0.003-0.01B-≤13.5-16Cr-≤1.0-1.50Mo-≤24.00-27.00Ni-≤1.70-2.00Ti-≤0.10-050V	SAE HEV 7 UNS S66545 ASI S 66286 A 891
1.4944		Fe-≤0.08C-≤1.0Si-≤2.0Mn-≤0.025P-≤0.015S-≤0.35Al-≤0.003-0.01B-≤13.50-16.0Cr-≤1.0-1.50Mo-≤24.00-27.00Ni-≤1.90-2.30Ti-≤0.10-0.50V	UNS S66286; ASI 660; A 638
1.4947		Fe-≤0.07C-0.50-1.2Si-1.5-2.0Mn-≤0.035P-≤0.025S-22.0-23.0Cr-≤0.3Cu-≤0.75Mo-9.5-10.5Ni	UNS J93001
1.4948	X6CrNi18-10	Fe-≤0.04-0.08C-≤0.75Si-≤2.0Mn-≤0.035P-≤0.015S-≤17.0-19.0Cr-≤10.0-12.0Ni	SAE 304 H
1.4958	X5NiCrAlTi31-20	Fe-0.03-0.08C-≤0.7Si-≤1.5Mn-0.015P-≤0.01S-0.2-0.5Al-0.5Co-19.0-22.0Cr-≤0.5Cu-≤0.03N-≤0.1Nb-30.0-32.5Ni-0.2-0.5Ti	N08810 (UNS)
1.4959	X8NiCrAlTi32-21	Fe-0.05-0.1C-≤0.7Si-≤1.5Mn-≤0.015P-≤0.01S-0.2-0.65Al-≤0.5Co-19.0-22.0Cr-≤0.5Cu-≤0.03N-30.0-34.0Ni-0.25-0.65Ti	N08811 (UNS)
1.4961	X8CrNiNb16-13	Fe-≤0.04-0.1C-≤0.3-0.6Si-≤1.5Mn-≤0.035P-≤0.015S-≤15.0-17.0Cr-≤12.0-14.0Ni	
1.4970	X 10 NiCrMoTiB 15 15	Fe-0.08-0.12C-0.25-0.45Si-1.6-2.0Mn-≤0.03P-≤0.015S-0.003-0.006B-14.5-15.5Cr-1.05-1.25Mo-15.0-16.0Ni-0.35-0.55Ti	
1.4971	X12CrCoNi21-20	Fe-≤0.08-0.16C-≤1.00Si-≤2.00Mn-≤0.035P-≤0.015S-≤18.50-21.00Co ≤20.00-22.50Cr-≤2.50-3.50Mo-≤0.10-0.20N-≤0.75-1.25Nb-≤19.00-21.00Ni-≤2.00-3.00W	HEV 1 (SAE) 661 (SAE) R 30155 (UNS)
1.4977	X 40 CoCrNi 20 20	Fe-0.35-0.45C-≤1.00Si-≤1.50Mn-≤0.045P-≤0.030S-19.00-21.00Co-19.00-21.00Cr-3.50-4.50Mo-3.50-4.50Nb-19.00-21.00Ni-3.50-4.50W	R 30590 UNS
1.4980	X6NiCrTiMoVB25-15-2; X5NiCrTi26-15	Fe-0.03-0.08C-≤1.0Si-1.0-2.0Mn-≤0.025P-≤0.015S-≤0.35Al-0.003-0.01B-13.5-16.0Cr-1.0-1.5Mo-24.0-27.0Ni-1.9-2.3Ti-0.1-0.5V	663 (SAE)
1.4981	X8CrNiMoNb16-16	Fe-≤0.04-0.10C-≤0.30-0.60Si-≤1.50Mn-≤0.035P-≤0.015S-≤15.5-17.5Cr-≤1.60-2.00Mo-≤15.5-17.5Ni	

German Standard		Materials Compositions	US-Standard
Mat.-No.	DIN-Design	Percent in Weight	SAE/ASTM/UNS
1.4982	X10CrNiMoMnNbVB15-10-1	Fe-≤0.07-0.13C-≤1.00Si-≤5.50-7.00Mn-≤0.040P-≤0.030S-≤0.003-0.009B-≤14.00-16.00Cr-≤0.80-1.20Mo-≤0.110N-≤0.75-1.25Nb-≤9.00-11.00Ni-≤0.15-0.40V	
1.4986	X8CrNiMoBNb16-16	Fe-≤0.04-0.1C-≤0.3-0.6Si-≤1.5Mn-≤0.045P-≤0.030S-≤0.05-0.1B-≤15.5-17.5Cr-≤1.6-2.0Mo-≤15.5-17.5Ni	
1.4988	X8CrNiMoVNb16-13	Fe-≤0.04-0.1C-≤0.3-0.6Si-≤1.5Mn-≤0.035P-≤0.015S-≤15.5-17.5Cr-≤1.1-1.5Mo-≤12.5-14.5Ni-≤0.60-0.85V-≤0.06-0.14N	
1.5069	36Mn7	Fe-≤0.35C≤0.5Si≤1.6Mn≤0.025P≤0.025S	UNS H13400
1.5094	38MnS6	Fe-≤0.35-0.40C-≤0.20-0.65Si-≤1.30-1.60Mn-≤0.045P-≤0.045-0.065S-≤0.01-0.05Al-≤0.10-0.20Cr-≤0.015-0.020N	
1.5122	37MnSi5	Fe-≤0.33-0.41C-≤1.1-1.4Si-≤1.1-1.4Mn-≤0.035P-≤0.035S	
1.5219	41MnV5	Fe-0.38-0.44C-0.1-0.4Si-1.1-1.3Mn≤0.035P≤0.035S-0.1-0.15V	
1.5415	15 Mo 3; 16Mo3	Fe-≤0.12-0.2C-≤0.35Si-≤0.4-0.9Mn-≤0.030P-≤0.025S-≤0.30Cr-≤0.30Cu-≤0.25-0.35Mo-≤0.30Ni	A 204 (A)
1.5431	G12MnMo7-4	Fe-0.08-0.15C≤0.6Si-1.5-1.8Mn≤0.02P≤0.015S≤0.2Cr-0.3-0.4Mo≤0.05Nb≤0.1V	
1.5511	35B2	Fe-≤0.32-0.39C-≤0.4Si-≤0.5-0.8Mn-≤0.035P-≤0.035S-≤0.02Al-≤0.0008-0.005B	
1.5662	X8Ni9	Fe-≤0.10C-0.35Si-0.30-0.80Mn-≤0.020P-≤0.010S-≤0.10Mo-8.5-10.0Ni-0.05V	UNS K71340
1.5680	X12Ni5; 12 Ni 19	Fe-≤0.15C-≤0.35Si-0.3-0.8Mn-≤0.02P-≤0.01S-4.75-5.25Ni-≤0.05V-≤0.5Cr-≤0.5Mo-≤0.5Cu	A 2515 (SAE)
1.5736	36NiCr10	Fe-max. 0.32-0.40C-0.15-0.35Si-0.40-0.80Mn-0.035P-0.035S-0.55-0.95Cr-2.25-2.75Ni	SAE 3435
1.6354		Fe-≤0.03C-≤0.10Si-≤0.10Mn-≤0.01P-≤0.01S-≤0.05-0.15Al-≤8.00-9.50Co-≤4.60-5.20Mo-≤17.00-19.00Ni-≤0.60-0.90Ti	UNS J93150
1.6511	36CrNiMo4	Fe-≤0.32-0.40C-≤0.4Si-≤0.5-0.8Mn-≤0.035P-≤0.035S-≤0.9-1.2Cr-≤0.15-0.3Mo-≤0.9-1.2Ni	SAE 9840
1.6545	30NiCrMo2-2	Fe-≤0.27-0.34C-≤0.15-0.4Si-≤0.7-1.0Mn-≤0.035P-≤0.035S-≤0.4-0.6Cr-≤0.15-0.3Mo-≤0.4-0.7Ni	SAE 8630

German Standard		Materials Compositions	US-Standard
Mat.-No.	**DIN-Design**	**Percent in Weight**	**SAE/ASTM/UNS**
1.6562	40 NiCrMo 8 4	Fe-≤0.37-0.44C-≤0.20-0.35Si-≤0.70-0.90Mn-≤0.02P-≤0.015S-≤0.005-0.05Al-≤0.70-0.95Cr-≤0.30-0.40Mo-≤1.65-2.00Ni	UNS G43406 SAE 4340 UNS H 43406 A 829 SAE E 4340 H
1.6565	40NiCrMo6	Fe-0.35-0.45C-≤0.15-0.35Si-≤0.50-0.70Mn-≤0.035P-≤0.035S-≤0.90-1.4Cr-≤0.20-0.30Mo-≤1.4-1.7Ni	SAE 4340
1.6580	30CrNiMo8	Fe-≤0.26-0.34C-≤0.4Si-≤0.3-0.6Mn-≤0.035P-≤0.035S-≤1.8-2.2Cr-≤0.3-0.5Mo-≤1.8-2.2Ni	
1.6582	34CrNiMo6	Fe-≤0.3-0.38C-≤0.4Si-≤0.5-0.8Mn-≤0.035P-≤0.035S-≤1.3-1.7Cr-≤0.15-0.3Mo-≤1.3-1.7Ni	
1.6751	22NiMoCr3-7	Fe-≤0.17-0.25C-≤0.35Si-≤0.5-1.0Mn-≤0.02P-≤0.02S-≤0.05Al-≤0.3-0.5Cr-≤0.18Cu-≤0.5-0.8Mo-≤0.6-1.2Ni-≤0.03V	A 508
1.6900	X 12 CrNi 18 9	Fe-≤0.12C-≤1.00Si-≤2.00Mn-≤0.045P-≤0.030S-≤17.00-19.00Cr-≤0.5Mo-≤8.00-10.00Ni	UNS J92801
1.6903	X 10 CrNiTi 18 10	Fe-≤0.10C-≤1.00Si-≤2.00Mn-≤0.045P-≤0.030S-≤17.0-19.0Cr-≤0.5Mo-≤10.0-12.0Ni	
1.6906	X 5 CrNi 18 10	Fe-≤0.07C-≤1.0Si-≤2.0Mn-≤0.045P-≤0.030S-≤17.0-19.0Cr-≤0.50Mo-≤9.0-11.5Ni	
1.6932	28NiCrMoV8-5	Fe-0.24-0.32C≤0.4Si-0.15-0.4Mn≤0.035P≤0.035S-1.0-1.5Cr-0.35-0.55Mo-1.8-2.1Ni-0.05-0.15V	
1.6944		Fe-≤0.35-0.40C-≤0.15-0.35Si-≤0.50-0.80Mn-≤0.015P-≤0.01S-≤0.65-0.90Cr-≤0.30-0.40Mo-≤1.65-2.00Ni-≤0.08-0.15V	
1.6948	27NiCrMoV11-6; 26NiCr-MoV11 5	Fe-≤0.22-0.32C-≤0.15Si-≤0.15-0.40Mn-≤0.010P-≤0.007S-≤1.20-1.80Cr-≤0.25-0.45Mo-≤2.40-3.10Ni-≤0.05-0.15V	
1.6952	24NiCrMoV14-6	Fe-≤0.20-0.28C-≤0.15-0.40Si-≤0.30-0.60Mn-≤0.035P-≤0.035S-≤1.20-1.80Cr-≤0.35-0.55Mo-≤3.00-3.80Ni-≤0.04-0.12V	K 42885 (UNS) A 649 (6, 7, 8) (ASTM) A 470 (5, 6, 7) (ASTM)
1.6956	33NiCrMoV14-5; 33NiCrMo14-5	Fe-≤0.28-0.38C-≤0.40Si-≤0.15-0.40Mn-≤0.035P-≤0.035S-≤1.00-1.70Cr-≤0.30-0.60Mo-≤2.90-3.80Ni-≤0.08-0.25V	
1.6957	27NiCrMoV15-6	Fe-0.22-0.32C≤0.15Si-0.15-0.4Mn≤0.01P≤0.007S-1.2-1.8Cr-0.25-0.45Mo-3.4-4.0Ni-0.05-0.15V	ASTM A 470
1.7005	45Cr2	Fe-≤0.42-0.48C-≤0.15-0.40Si-≤0.50-0.80Mn-≤0.025P-≤0.035S-≤0.40-0.60Cr	
1.7033	34Cr4	Fe-≤0.3-0.37C-≤0.4Si-≤0.6-0.9Mn-≤0.035P-≤0.035S-≤0.9-1.2Cr	UNS G51320

German Standard		Materials Compositions	US-Standard
Mat.-No.	**DIN-Design**	**Percent in Weight**	**SAE/ASTM/UNS**
1.7035	41Cr4	Fe-≤0.38-0.45C-≤0.4Si-≤0.6-0.9Mn-≤0.035P-≤0.035S-≤0.9-1.2Cr	SAE 5140; UNS H51400
1.7120		Fe-≤0.1C-≤0.25Si-≤0.45Mn-≤0.16Cu-≤0.07Ni-≤0.05Cr-≤0.035P-≤0.035S	
1.7131	16MnCr5	Fe-0.14-0.19C-≤0.40Si-1.00-1.30Mn-≤0.035P-≤0.035S-0.80-1.10Cr	G 51170 UNS A 711 (5115) ASTM
1.7147	20MnCr5	Fe-≤0.17-0.22C-≤0.40Si-≤1.10-1.40Mn-≤0.035P-≤0.035S-≤1.00-1.30Cr	UNS H51200 A 752 (5120) SAE 5120H
1.7214		Fe-≤0.22-0.29C-≤0.15-0.35Si-≤0.5-0.8Mn-≤0.02P-0.015S-≤0.90-1.20Cr-≤0.15-0.20Mo-0.30Ni	
1.7218	25CrMo4	Fe-≤0.22-0.29C-≤0.40Si-≤0.60-0.90Mn-≤0.035P-≤0.035S-≤0.90-1.20Cr-≤0.15-0.30Mo	SAE 4130
1.7219	26 CrMo 4; 26CrMo4-2	Fe-≤0.22-0.29C-≤0.35Si-≤0.5-0.8Mn-≤0.03P-≤0.025S-≤0.9-1.2Cr-≤0.15-0.30Mo	A 372
1.7220	34CrMo4	Fe-≤0.3-0.37C-≤0.4Si-≤0.6-0.9Mn-≤0.035P-≤0.035S-≤0.9-1.2Cr-≤0.15-0.30Mo	SAE 4130
1.7225	42CrMo4	Fe-≤0.38-0.45C-≤0.40Si-≤0.60-0.90Mn-≤0.035P-≤0.035S-≤0.90-1.20Cr-≤0.15-0.30Mo	UNS G41400 A 866 (4140) SAE 4140 RH
1.7242	16CrMo4	Fe-≤0.13-0.20C-≤0.15-0.35Si-≤0.50-0.80Mn-≤0.035P-≤0.035S-≤0.90-1.20Cr-≤0.20-0.30Mo-≤0.40Ni	
1.7259	26CrMo7	Fe-≤0.22-0.30C-≤0.15-0.35Si-≤0.50-0.70Mn-≤0.035P-≤0.035S-≤1.50-1.80Cr-≤0.20-0.25Mo	
1.7273	24CrMo10	Fe-≤0.20-0.28C-≤0.15-0.35Si-≤0.50-0.80Mn-≤0.035P-≤0.035S-≤2.30-2.60Cr-≤0.20-0.30Mo-≤0.80Ni	
1.7276	10CrMo11	Fe-≤0.08-0.12C-≤0.15-0.35Si-≤0.30-0.50Mn-≤0.035P-≤0.035S-≤2.70-3.00Cr-≤0.20-0.30Mo	
1.7279	17 MnCrMo 3 3	Fe-≤0.20C-0.50-0.90Si-0.70-1.10Mn-≤0.035P-≤0.035S-0.60-1.00Cr-0.20-0.60Mo-0.06-0.12V-0.06-0.12Zr	
1.7281	16CrMo9-3	Fe-≤0.12-0.20C-≤0.15-0.35Si-≤0.30-0.50Mn-≤0.035P-≤0.035S-≤2.00-2.50Cr-≤0.30-0.40Mo	
1.7335	13 CrMo 4 4; 13CrMo4-5	Fe-≤0.08-0.18C-≤0.35Si-≤0.4-1.0Mn-≤0.030P-≤0.025S-≤0.7-1.15Cr-≤0.3Cu-≤0.4-0.6Mo	A 182

German Standard		*Materials Compositions*	*US-Standard*
Mat.-No.	***DIN-Design***	***Percent in Weight***	***SAE/ASTM/UNS***
1.7357	G17CrMo5-5	Fe-≤0.15-0.20C-≤0.60Si-≤0.50-1.0Mn-≤0.020P-≤0.020S-≤1.00-1.50Cr-≤0.45-0.65Mo	A 217; UNS J11872
1.7362	X12CrMo5	Fe-≤0.08-0.15C-≤0.50Si-≤0.30-0.60Mn-≤0.025P-≤0.020S-≤4.00-6.00Cr-≤0.45-0.65Mo	SAE 501
1.7375	12CrMo9-10	Fe-≤0.10-0.15C-≤0.30Si-≤0.30-0.80Mn-≤0.015P-≤0.010S-≤0.01-0.04Al-≤2.00-2.50Cr-≤0.20Cu-≤0.9-1.10Mo-≤0.012N-≤0.30Ni	UNS K21590
1.7380	10CrMo9-10	Fe-≤0.08-0.14C-≤0.50Si-≤0.40-0.80Mn-≤0.030P-≤0.025S-≤2.00-2.50Cr-≤0.30Cu-≤0.90-1.10Mo	A 182 (F22); UNS J21890
1.7383	11CrMo9-10	Fe-≤0.08-0.15C-≤0.50Si-≤0.40-0.80Mn-≤0.030P-≤0.025S-≤2.00-2.50Cr-≤0.30Cu-≤0.90-1.10Mo	
1.7386	X12CrMo9-1	Fe-≤0.07-0.15C-≤0.25-1.0Si-≤0.30-0.60Mn-≤0.025P-≤0.020S-≤8.0-10.0Cr-≤0.90-1.1Mo	SAE 504; UNS S50488
1.7388	X7CrMo9-1	Fe-≤0.04-0.09C-≤0.45-0.75Si-≤0.43-0.72Mn-≤0.015P-≤0.015S-≤8.60-9.90Cr-≤0.90-1.10Mo	S 50480 (UNS)
1.7707	30CrMoV9	Fe-0.26-0.34C-≤0.4Si-0.4-0.7Mn-≤0.035P-≤0.035S-2.3-2.7Cr-≤0.25Mo-≤0.6Ni-0.1-0.2V	G43406 (UNS)
1.7711	40CrMoV4-6; 40CrMoV4-7	Fe-≤0.36-0.44C-≤0.40Si-≤0.45-0.85Mn-≤0.03P-≤0.03S-≤0.015Al-≤0.90-1.20Cr-≤0.50-0.65Mo-≤0.25-0.35V	A 437 (B4D)
1.7715	14MoV6-3	Fe-≤0.1-0.18C-≤0.1-0.35Si-≤0.4-0.7Mn-≤0.035P-≤0.035S-≤0.3-0.6Cr-≤0.5-0.7Mo-≤0.22-0.32V	UNS K11591
1.7734		Fe-≤0.12-0.18C-≤0.20Si-≤0.80-1.10Mn-≤0.02P-≤0.015S-≤1.25-1.50Cr-≤0.80-1.00Mo-≤0.20-0.30V	
1.7766	17CrMoV10	Fe-≤0.15-0.20C-≤0.15-0.35Si-≤0.30-0.50Mn-≤0.035P-≤0.035S-≤2.70-3.00Cr-≤0.20-0.30Mo-≤0.10-0.20V	
1.7779	20 CrMoV 13 5; 20CrMoV13-5-5	Fe-≤0.17-0.23C-≤0.15-0.35Si-≤0.30-0.50Mn-≤0.025P-≤0.020S-≤3.00-3.30Cr-≤0.50-0.60Mo-≤0.45-0.55V	
1.7783	X41CrMoV5-1	Fe-≤0.38-0.43C-≤0.80-1.0Si-≤0.20-0.40Mn-≤0.015P-≤0.010S-≤4.75-5.25Cr-≤1.2-1.4Mo-≤0.40-0.60V	SAE 610
1.8070	21CrMoV5-11	Fe-≤0.17-0.25C-≤0.30-0.60Si-≤0.30-0.60Mn-≤0.035P-≤0.035S-≤1.20-1.50Cr-≤1.00-1.20Mo-≤0.60Ni-≤0.25-0.35V	

German Standard		Materials Compositions	US-Standard
Mat.-No.	**DIN-Design**	**Percent in Weight**	**SAE/ASTM/UNS**
1.8075	10CrSiMoV7	Fe-≤0.12C-≤0.9-1.2Si-≤0.35-0.75Mn-≤0.035P-≤0.035S-≤1.6-2Cr-≤0.25-0.35Mo-≤0.25-0.35V	
1.8159	51CrV4; 50 CrV 4	Fe-≤0.47-0.55C-≤0.40Si-≤0.70-1.10Mn-≤0.035P-≤0.035S-≤0.90-1.20Cr-≤0.10-0.25V	UNS G61500 A 866 (6150) SAE 6150H
1.8719	15MnCrMo3-2	Fe-0.10-0.20C-0.15-0.35Si-0.60-1.00Mn-≤0.025P-≤0.025S-0.0005B-0.40-0.65Cr-0.15-0.50Cu-0.40-0.60Mo-0.70-1.00Ni-0.03-0.08V	
1.8812	18MnMoV5-2	Fe-≤0.20C-0.20-0.50Si-1.00-1.50Mn-≤0.030P-≤0.025S-0.10-0.30Mo-≤0.02N-0.05-0.10V	A 202 (A) (ASTM) A 202 (B) (ASTM) A 302 (A) (ASTM)
1.8850	S460MLH	Fe-≤0.16C-≤0.60Si-≤1.70Mn-≤0.030P-≤0.025S-0.02Al-≤0.20Mo-≤0.025N-≤0.050Nb-≤0.30Ni-≤0.05Ti-≤0.12V	A 514 (F) (ASTM) A 517 (F) (ASTM) A 592 (F) (ASTM)
1.8901	S460N	Fe-≤0.2C≤0.6Si-1.0-1.7Mn-≤0.035P-≤0.03S-≤0.3Cr-≤0.7Cu-≤0.1Mo-≤0.05Nb-≤0.8Ni-≤0.03Ti≤0.2V	ASTM A 572
1.8905	P460N; StE 460	Fe-≤0.20C-≤0.60Si-≤1.00-1.70Mn-≤0.030P-≤0.025S-≤0.02Al-≤0.30Cr-≤0.70Cu-≤0.10Mo-≤0.025N-≤0.050Nb-≤0.80Ni-≤0.03Ti-≤0.20V	A 225 (C), A 633 (E)
1.8907	StE 500	Fe-≤0.21C-≤0.1-0.6Si-≤1.0-1.7Mn-≤0.035P-≤0.03S-≤0.02Al-≤0.30Cr-≤0.70Cu-≤0.10Mo-≤0.020N-≤0.05Nb-≤1.0Ni-≤0.2Ti-≤0.22V	6386 B; UNS K02001
1.8912	S420NL; TStE 420	Fe-≤0.2C-≤0.6Si-≤1.0-1.7Mn-≤0.03P-≤0.025S-≤0.02Al-≤0.3Cr-≤0.7Cu-≤0.1Mo-≤0.025N-≤0.050Nb-≤0.8Ni-≤0.03Ti-≤0.2V	A 737: UNS K02002
1.8924	S500Q; StE 500V	Fe-≤0.2C-≤0.8Si-≤1.7Mn-≤0.025P-≤0.015S-≤1.5Cr-≤0.5Cu-≤0.7Mo-≤0.06Nb-≤2.0Ni-≤0.05Ti-≤0.15Zr	
1.8931	S690Q; StE 690V	Fe-≤0.2C-≤0.8Si-≤1.7Mn-≤0.025P-≤0.015S-≤1.5Cr-≤0.5Cu-≤0.7Mo-≤0.06Nb-≤2.0Ni-≤0.05Ti-≤0.15Zr	
1.8935	WstE 460; P460NH	Fe-≤0.20C-≤0.60Si-≤1.0-1.70Mn-≤0.030P-≤0.025S-≤0.02Al-≤0.30Cr-≤0.70Cu-≤0.10Mo-≤0.025N-≤0.050Nb-≤0.8Ni-≤0.03Ti-≤0.2V	A 350; UNS K02900
1.8940	S890Q	Fe-≤0.2C-≤0.8Si-≤1.7Mn-≤0.025P-≤0.015S-≤1.5Cr-≤0.5Cu-≤0.7Mo-≤0.06Nb-≤2.0Ni-≤0.05Ti-≤0.15Zr	
1.8946	S355J2WP	Fe-≤0.12C-≤0.75Si-≤1.0Mn-0.06-0.15P-≤0.035S-0.30-1.25Cr-0.25-0.55Cu-≤0.009N-≤0.65Ni	K02601 (UNS)

German Standard		*Materials Compositions*	*US-Standard*
Mat.-No.	*DIN-Design*	*Percent in Weight*	*SAE/ASTM/UNS*
1.8952	L450QB	Fe-≤0.16C-≤0.45Si-≤1.60Mn-≤0.025P-≤0.020S-≤0.015-0.06Al-≤0.30Cr-≤0.25Cu-≤0.10Mo-≤0.012N-≤0.05Nb-≤0.30Ni-≤0.06Ti-≤0.09V	
1.8961	S235J2W; WTSt 37-3	Fe-≤0.13C-≤0.4Si-≤0.2-0.6Mn-≤0.040P-≤0.035S-≤0.02Al-≤0.4-0.8Cr-≤0.25-0.55Cu-≤0.015-0.060Nb-≤0.65Ni-≤0.02-0.10Ti-≤0.02-0.10V	
1.8962	9CrNiCuP3-2-4	Fe-≤0.12C-≤0.25-0.75Si-≤0.2-0.5Mn-≤0.07-0.15P-≤0.035S-≤0.5-1.25Cr-≤0.25-0.55Cu-≤0.65Ni	A 242; UNS K11430
1.8963	S355J2G1W; WTSt 52-3	Fe-≤0.16C-≤0.50Si-≤0.50-1.5Mn-≤0.035P-≤0.035S-≤0.02Al-≤0.40-0.80Cr-≤0.25-0.55Cu-≤0.3Mo-≤0.015-0.060Nb-≤0.65Ni-≤0.02-0.10Ti-≤0.02-0.12V-≤0.15Zr	A 588 (A)
1.8972	L415NB	Fe-≤0.21C-≤0.45Si-≤1.60Mn-≤0.025P-≤0.020S-≤0.015-0.060Al-≤0.30Cr-≤0.25Cu-≤0.10Mo-≤0.012N-≤0.050Nb-≤0.30Ni-≤0.04Ti-≤0.15V	API 5LX 60 (API)
1.8975	L450MB; StE 445.7	Fe-≤0.16C-≤0.45Si-≤1.6Mn-≤0.025P-≤0.02S-0.015-0.06Al-≤0.3Cr-≤0.25Cu-≤0.1Mo-≤0.05Nb-≤0.3Ni-≤0.06Ti	API 5LX65
1.8977	L485MB; StE 480.7	Fe-≤0.16C-≤0.45Si-≤1.70Mn-≤0.025P-≤0.020S-≤0.015-0.06Al-≤0.30Cr-≤0.25Cu-≤0.10Mo-≤0.012N-≤0.06Nb-≤0.30Ni-≤0.06Ti-≤0.10V	API 5LX70
2.4060	Ni 99,6	≤99.60Ni-≤0.08C-≤0.15Si-≤0.35Mn-≤0.005S-≤0.15Cu-≤0.25Fe-≤0.15Mg-≤0.10Ti	UNS N02200
2.4061	LC-Ni 99,6	Fe-≤0.02C-0.15Si-0.35Mn-≤0.005S-<0.15Cu-≤0.25Fe-≤0.15Mg-99.6Ni-≤0.10Ti	UNS N02201
2.4066	Ni 99,2; S-Ni 99,2	≤99.20Ni-≤0.10C-≤0.25Si-≤0.35Mn-≤0.005S-≤0.25Cu-≤0.40Fe-≤0.15Mg-≤0.10Ti	UNS N02200
2.4068	LC-Ni 99	≤99.0Ni-≤0.02C-≤0.25Si-≤0.35Mn-≤0.005S-≤0.25Cu-≤0.40Fe-≤0.15Mg-≤0.10Ti	UNS N02201
2.4360	NiCu 30 Fe	≤63.0Ni-≤0.15C-≤0.50Si-≤2.0Mn-≤0.020S-≤0.5Al-≤28.0-34.0Cu-≤1.0-2.5Fe-≤0.3Ti	UNS N04400
2.4361	LC-NiCu 30 Fe	Fe-≤0.04C-≤0.3Si-≤2.0Mn-≤0.02S-≤0.5Al-28.0-34.0Cu-1.0-2.5Fe-63.0Ni-≤0.3Ti	N04402 (UNS)
2.4363	NiCu30Fe5	≤0.30C-≤0.50Si-≤2.0Mn-0.025-0.60S-≤0.50Al-28.00-34.00Cu-≤2.50Fe-63.00-70.00Ni-≤0.30Ti	
2.4365	G-NiCu 30 Nb	Ni-≤0.15C-0.5-1.5-0.5-1.5Mn-≤0.5Al-26.0-33.0Cu-1.0-2.5-1.0-1.5Nb	UNS N24130

German Standard		Materials Compositions	US-Standard
Mat.-No.	***DIN-Design***	***Percent in Weight***	***SAE/ASTM/UNS***
2.4366	EL-NiCu 30 Mn	≤62.0Ni-≤0.15C-≤1.0Si-≤1.0-4.0Mn-≤0.030P≤0.015S-≤0.5Al-≤27.0-34.0Cu-≤0.5-2.5Fe-≤1.0Nb≤1.0Ti	B 127-98
2.4368	G-NiCu 30Si4	Fe-≤0.25C-3.5-4.5Si-0.5-1.5Mn-27.0-31.0Cu-1.0-2.5Fe-60.0-68.0Ni	UNS N10665
2.4374	NiCu30Al	≤0.25C-≤1.00Si-≤1.50Mn-≤0.010S-2.00-4.00Al-27.00-34.00Cu-≤2.00Fe-≥63.00Ni-0.25-1.00Ti	
2.4375	NiCu 30 Al	Ni-≤0.20C-≤0.50Si-≤1.5Mn-≤0.015S-≤2.2-3.5Al-≤27.0-34.0Cu-≤0.5-2.0Fe-≤63.0Ni-≤0.3-1.0Ti	UNS N05500
2.4566	ACN 17	Ni-≤0.12C-≤10Si-≤1.2Mn-≤4Co-≤3Cu	
2.4600	NiMo29Cr	Ni-≤0.01C-≤0.1Si-≤3.0Mn-≤0.025P-≤0.015S-≤0.1-0.5Al-≤3.0Co-≤0.5-3.0Cr-≤0.5Cu-≤1.0-6.0Fe-≤26.0-32.0Mo-≤0.4Nb-≤0.2Ti-≤0.2V-≤3.0W	
2.4602	NiCr21Mo14W	Ni-≤0.01C-≤0.08Si-≤0.5Mn-≤0.025P-≤0.010S-≤2.5Co-≤20.0-22.5Cr-≤2.0-6.0Fe-≤12.5-14.5Mo-≤0.35V-≤2.5-3.5W	UNS N06022
2.4603	NiCr30FeMo	Ni-≤0.03C-≤0.08Si-≤2.0Mn-≤0.04P-≤0.02S-≤5.0Co-≤28.0-31.5Cr-≤1.0-2.4Cu-13.0-17.0Fe-≤4.0-6.0Mo-≤0.3-1.5Nb-≤1.5-4.0W	UNS N06002
2.4605	NiCr23Mo16Al	Ni-≤0.01C-≤0.10Si-≤0.5Mn-≤0.025P-≤0.015S-≤0.1-0.4Al-≤0.3Co-≤22.0-24.0Cr-≤0.5Cu-≤1.5Fe-≤15.0-16.5Mo	UNS N06059
2.4606	NiCr21Mo16W	Ni-≤0.01C-≤0.08Si-≤0.75Mn-≤0.025P-≤0.015S-≤0.5Al-≤1.0Co-≤19.0-23.0Cr-≤2.0Fe-≤15.0-17.0Mo-≤0.02-0.25Ti-≤0.2V-≤3.0-4.0W	UNS N06686
2.4607	SG-NiCr23Mo16	Ni-≤0.015C-≤0.08Si-≤0.50Mn-≤0.02P-≤0.015S-≤0.1-0.4Al-≤0.3Co-≤22.0-24.0Cr-≤1.5Fe-≤15.0-16.5Mo	UNS N06059
2.4608	NiCr26MoW	Fe-0.03-0.08C-0.7-1.5Si-≤2.0Mn-≤0.03P-≤0.015S-2.0-4.0Cu-24.0-26.0Cr-≤0.5Cu-2.5-4.0Mo-44.0-47.0Ni-2.5-4.0W	N06333 (UNS)
2.4610	NiMo16Cr16Ti	Ni-≤0.01C-≤0.08Si-≤1.0Mn-≤0.025P-≤0.015S-≤2.0Co-≤14.0-18.0Cr-≤0.5Cu-≤3.0Fe-≤14.0-18.0Mo-≤0.7Ti	UNS N06455
2.4612	EL-NiMo15Cr15Ti	≤0.02C-≤0.20Si-≤1.00Mn-≤0.015S-≤2.00Co-≤14.00-18.00Cr-≤3.00Fe-≤14.00-17.00Mo-at least56Ni	
2.4615	SG-NiMo27	Fe-≤0.02C-≤0.10Si-≤1.0Mn-≤0.015S-≤1.0Cr-≤2.0Fe-26.0-30.0Mo-64.0Ni	UNS N10665

German Standard		Materials Compositions	US-Standard
Mat.-No.	**DIN-Design**	**Percent in Weight**	**SAE/ASTM/UNS**
2.4617	NiMo28	Ni-≤0.01C-≤0.08Si-≤1.0Mn-≤0.025P-≤0.015S-≤1.0Co-≤1.0Cr-≤0.5Cu-≤2.0Fe-≤26.0-30.0Mo	UNS N10665
2.4618	NiCr22Mo6Cu	Ni-≤0.05C-≤1.0Si-≤1.0-2.0Mn-≤0.025P-≤0.015S-≤2.5Co-≤21.0-23.5Cr-≤1.5-2.5Cu-≤18.0-21.0Fe-≤5.5-7.5Mo-≤1.75-2.5Nb-≤1.0W	UNS N06007
2.4619	NiCr22Mo7Cu	Ni-≤0.015C-≤1.0Si-≤1.0Mn-≤0.025P-≤0.015S-≤5.0Co-≤21.0-23.5Cr-≤1.5-2.5Cu-≤18.0-21.0Fe-≤6.0-8.0Mo-≤0.5Nb-≤1.5W	UNS N06985
2.4621	EL-NiCr20Mo9Nb	Ni-≤0.1C≤0.8Si≤2.0Mn≤0.4Al-20.0-23.0Cr ≤0.5Cu≤6.0Fe-8.0-10.0Mo-2.0-4.0Nb≤0.4Ti	
2.4623	EL-NiCr23Mo7Cu	Ni-≤0.02C≤1.0Si≤1.0Mn≤0.04P ≤0.03≤5.0Co-21.0-23.5Cr-1.5-2.5Cu-18.0-21.0Fe-6.0-8.0Mo≤0.5Nb≤0.5Ta≤1.5W	
2.4627	SG-NiCr22Co12Mo	Fe-≤0.1C-≤0.5Si-≤1.0Mn-≤0.015S-0.8-1.5Al-10.0-14.0Co-20.0-24.0Cr-≤0.5Cu-≤1.0Fe-8.0-10.0Mo-50.0Ni-≤0.6Ti	N06617 (UNS)
2.4630	NiCr20Ti	Fe-0.08-0.15C-≤1.0Si-≤1.0Mn-18.0-21.0Cr-≤0.5Cu-≤5.0Fe-0.2-0.6Ti	N06075 (UNS)
2.4631	NiCr20TiAl	Ni-0.04-0.1C≤1.0Si≤1.0Mn≤0.03P≤0.015S-1.0-1.8Al≤2.0Co-18.0-21.0Cr≤0.2Cu≤1.5Fe-1.8-2.7Ti	UNS N07080
2.4632	NiCr20Co18Ti	Ni-≤0.13C≤1.0Si≤1.0Mn≤0.02P≤0.015S-1.0-2.0Al-15.0-21.0Co-18.0-21.0Cr≤0.2Cu ≤1.5Fe-2.0-3.0Ti	UNS N07090
2.4633	NiCr25FeAlY	Fe-0.15-0.25C-≤0.5Si-≤0.5Mn-≤0.02P-≤0.01S-1.8-2.4Al-24.0-26.0Cr-≤0.1Cu-8.0-11.0Fe-0.1-0.2Ti-0.05-0.12Y-0.01-0.1Zr	
2.4634	NiCo20Cr15MoAlTi	Ni-0.12-0.17C-≤1.0Si-≤1.0Mn-≤0.045P-≤0.015S-4.5-4.9Al-18.0-22.0Co-14.0-15.7Cr-≤0.2Cu-≤1.0Fe-4.5-.5.5Mo-0.9-1.5Ti	UNS N13021
2.4636	NiCo15Cr15MoAlTi	Ni-0.12-0.2C≤1.0Si≤1.0Mn≤0.045P≤0.03S-4.5-5.5Al-13.0-17.0Co-14.0-16.0Cr≤0.2Cu ≤1.0Fe-3.0-.5.0Mo-3.5-4.5Ti	NIMONIC alloy 115
2.4641	NiCr21Mo6Cu	Fe-≤0.025C-≤0.50Si-≤1.0Mn-≤0.025P-≤0.015S-≤0.2Al-≤1.0Co-20.0-23.0Cr-1.5-3.0Cu-5.5-7.0Mo-39.0-46.0Ni-0.6-1.0Ti	UNS N08042
2.4642	NiCr29Fe	≤58.0Ni-≤0.05C-≤0.5Si-≤0.5Mn-≤0.020P-≤0.015S-≤0.5Al-≤27.0-31.0Cr-≤0.5Cu-≤7.0-11.0Fe-≤0.5Ti	UNS N06690
2.4650	NiCo20Cr20MoTi	Ni-≤0.04-0.08C-≤0.4Si-≤0.6Mn-≤0.007S-≤0.3-0.6Al-≤0.005B-≤19.0-21.0Co-≤19.0-21.0Cr-≤0.2Cu-≤0.7Fe-≤5.6-6.1Mo-≤1.9-2.4Ti	UNS N07263

German Standard		Materials Compositions	US-Standard
Mat.-No.	***DIN-Design***	***Percent in Weight***	***SAE/ASTM/UNS***
2.4652	EL-NiCr26Mo	≤37.0-42.0Ni-≤0.03C-≤0.7Si-≤1.0-3.0Mn-≤0.015S-≤0.1Al-≤23.0-27.0Cr-≤1.5-3.0Cu-≤30.0Fe-≤3.5-7.5Mo-≤37.0-42.00Ni-≤1.0Ti	UNS S32654
2.4654	NiCr20Co13Mo4Ti3Al; NiCr19Co14Mo4Ti	Fe-0.02-0.2C-≤0.15Si-≤0.1Mn-≤0.015P-≤0.015S-1.2-1.6Al-0.003-0.01B-12.0-15.0Co-18.0-21.0Cr-≤0.1Cu-≤2.0Fe-3.5-5.0Mo-2.8-3.3Ti-0.02-0.08Zr	N07001 (UNS)
2.4658	NiCr7030; NiCr 70 30	Fe-≤0.1C-0.5-2.0Si-≤1.0Mn-≤0.02P-≤0.15S-≤0.3Al-≤1.0Co-≤29.0-32.0Cr-≤0.5Cu-≤5.0Fe-60.0Ni	N06008 (UNS)
2.4660	NiCr20CuMo	≤32.0-38.0Ni-≤0.07C-≤1.0Si-≤2.0Mn-≤0.025P-≤0.015S-≤19.0-21.0Cr-≤3.0-4.0Cu-≤2.0-3.0Mo	UNS N08020
2.4662	NiCr13Mo6Ti3	Fe-0.02-0.06C-≤0.40Si-≤0.50Mn-≤0.020P-≤0.020S-≤0.35Al-0.01-0.02B-≤1.0Co-11.0-14.0Cr-≤0.04Cu-5.0-6.5Mo-40.0-45.0Ni-2.8-3.1Ti	UNS N09901
2.4663	NiCr23Co12Mo	Ni-≤0.05-0.10C-≤0.2Si-≤0.2Mn-≤0.01P-≤0.01S-≤0.7-1.4Al≤0.006B-≤11.0-14.0Co-≤20.0-23.0Cr-≤0.5Cu-≤2.0Fe-≤8.5-10.0Mo-≤0.2-0.6Ti	UNS N06617
2.4665	NiCr22Fe18Mo	Fe-0.05-0.15C-≤1.0Si-≤1Mn-≤0.02P-≤0.015S-≤0.5Al-0.01-0.1B-0.5-2.5Co-20.5-23.0Cr-≤0.5Cu-17.0-20.0Fe-8.0-10.0Mo-0.2-1.0W	680 (SAE)
2.4667	SG-NiCr19NbMoTi	Fe-≤0.08C-≤0.40Si-≤0.40Mn-≤0.015S-0.2-0.8Al-≤0.006B-17.0-21.0Cr-≤0.30Cu-≤22.0Fe-2.8-3.3Mo-4.8-5.5Nb-50.0Ni-0.60-1.20Ti	
2.4668	NiCr19Fe19Nb5Mo3	Fe-0.02-0.08C-≤0.35Si-≤0.35Mn-≤0.015P-≤0.015S-0.3-0.7Al-0.006B-≤1.0Co-17.0-21.0Cr-≤0.30Cu-2.8-3.3Mo-4.7-5.5Nb-50.0-55.0Ni-0.60-1.20Ti	UNS N07718
2.4669	NiCr15Fe7TiAl; NiCr15Fe7Ti2Al	Ni-≤0.08C-≤0.5Si-≤1.0Mn-≤0.02P-≤0.015S-≤0.4-1.0Al≤1.0Co-≤14.0-17.0Cr-≤0.5Cu-≤5.0-9.0Fe-≤0.7-1.2Nb-≤2.25-2.75Ti	UNS N07750
2.4670	G-NiCr13Al6MoNb	Ni-≤0.08-0.20C-≤0.50Si-≤0.25Mn-≤0.015P-≤0.015S-≤5.50-6.50Al-≤0.005-0.15B-≤1.00Co-≤12.00-14.00Cr-≤0.50Cu-≤3.80-5.20Mo-≤1.50-2.50Nb-≤0.40-1.00Ti-≤0.05-0.15Zr	UNS N07713
2.4675	NiCr23Mo16Cu	Ni-≤0.01C-≤0.08Si-≤0.5Mn-≤0.025P-≤0.015S-≤0.5Al≤2.0Co-≤22.0-24.0Cr-≤1.3-1.9Cu-≤3.0Fe-≤15.0-17.0Mo	

German Standard		Materials Compositions	US-Standard
Mat.-No.	**DIN-Design**	**Percent in Weight**	**SAE/ASTM/UNS**
2.4679	G-NiCr35	Ni-≤0.10C-≤1.00Si-≤0.30Mn-≤34.00-36.00Cr-≤1.00Fe-≤0.30N	
2.4680	G-NiCr50Nb	Ni-≤0.10C-≤1.00Si-≤0.50Mn-≤0.02P-≤0.02S-≤48.00-52.00Cr-≤1.00Fe-≤0.50Mo-≤0.16N-≤1.00-1.80Nb	
2.4681	CoCr26Ni9Mo5W	Ni-≤1.0C-≤1.0Si-≤1.5Mn-≤23.5-27.5Cr-≤1.0-3.0Fe-≤4.0-6.0Mo-≤0.12N-≤7.0-11.0Ni-≤1.0-3.0W	
2.4683	CoCr22NiW	Fe-0.05-0.15C-0.2-0.5Si-≤1.25Mn-≤0.02P-≤0.015S-20.0-24.0Cr-≤3.0Fe-0.02-0.12La-20.0-24.0Ni-13.0-16.0W	R30188 (UNS)
2.4686	G-NiMo 17 Cr	Ni-≤0.03C≤0.5Si≤1.0Mn≤2.5Co-15.5-17.5Cr≤7.0Fe-16.0-18.0Mo	
2.4694	NiCr16Fe7TiAl	Fe-≤0.08C-≤0.5Si-≤0.5Mn-≤0.015P-≤0.01S-0.8-1.6Al-14.0-17.0Cr-≤0.5Cu-5.0-9.0Fe-0.7-1.2Nb-70.0Ni-2.0-2.6Ti	N07031 (UNS)
2.4800	S-NiMo 30	–≤60.0Ni-≤0.05C-≤1.0Si-≤1.0Mn-≤0.045P-≤0.025S-≤2.5Co-≤1.0Cr-≤4.0-7.0Fe-≤26.0-30.0Mo-≤0.2-0.4V	UNS N10001
2.4810	NiMo 30	≤62.0Ni-≤0.05C-≤0.5Si-≤1.0Mn-≤0.030P-≤0.015S-≤2.5Co-≤1.0Cr-≤0.5Cu-≤4.0-7.0Fe-≤26.0-30.0Mo-≤0.6V	UNS N10001
2.4811		Fe-≤0.03C-≤0.05Si-≤0.80Mn-≤0.030P-≤0.015S-19.0-21.0Cr-≤0.50Cu-≤2.5Fe-14.0-17.0Mo-58.0Ni-0.35V	
2.4816	NiCr15Fe	≤72.0Ni-≤0.05-0.1C-≤0.5Si-≤1.0Mn-≤0.020P-≤0.015S-≤0.3Al-≤0.0060B-≤1.0Co-≤14.0-17.0Cr-≤0.5Cu-≤6.0-10.0Fe-≤0.3Ti	UNS N06600
2.4817	LC-NiCr15Fe	≤72.0Ni-≤0.025C-≤0.50Si-≤1.0Mn-≤0.020P-≤0.015S-≤0.3Al-≤0.0060B-≤1.0Co-≤14.0-17.0-≤0.5Cu-≤6.0-10.0Fe-≤0.3Ti	
2.4819	NiMo16Cr15W	Ni-≤0.01C-≤0.08Si-≤1.0Mn-≤0.025P-≤0.015S-≤2.5Co-≤14.5-16.5Cr-≤0.5Cu-≤4.0-7.0Fe-≤15.0-17.0Mo-≤0.35V-≤3.0-4.5W	UNS N10276
2.4831	SG-NiCr21Mo9Nb	Fe-≤0.10C-≤0.50Si-≤0.50Mn-≤0.015S-≤0.4Al-20.0-23.0Cr-≤0.50Cu-≤5.0Fe-8.0-10.0Mo-3.0-4.5Nb-60.0Ni-≤0.40Ti	UNS N06625
2.4851	NiCr23Fe	Fe-0.03-0.1C-≤0.5Si-≤1.0Mn-≤0.02P-≤0.015S-1.0-1.7Al-≤0.006B-21.0-25.0Cr-≤0.5Cu-≤18.0Fe-58.0-63.0Ni-≤0.5Ti	N06601 (UNS)
2.4856	NiCr22Mo9Nb	Ni-≤0.03-0.10C-≤0.5Si-≤0.5Mn-≤0.020P-≤0.015S-≤0.4Al-≤1.0Co-≤20.0-23.0Cr-≤0.5Cu-≤5.0Fe-≤8.0-10.0Mo-≤3.15-4.15Nb-≤0.4Ti	UNS N06625

German Standard		*Materials Compositions*	*US-Standard*
Mat.-No.	***DIN-Design***	***Percent in Weight***	***SAE/ASTM/UNS***
2.4858	NiCr21Mo	≤38.0-46.0Ni-≤0.025C-≤0.50Si-≤1.0Mn-≤0.025P-≤0.015S-≤0.20Al-≤1.0Co-≤19.5-23.5Cr-≤1.5-3.0Cu-≤2.5-3.5Mo-≤0.6-1.2Ti	B 163 UNS N08825
2.4869	NiCr80-20	Fe-≤0.15C-0.50-2.0Si-≤1.0Mn-≤0.020P-≤0.015S-≤0.3Al-≤1.0Co-19.0-21.0Cr-≤0.50Cu-≤1.0Fe-75.0Ni	UNS N06003
2.4882		Fe-≤0.12C-0.50-2.0Si-≤1.0Mn-≤0.020P-≤0.015S-≤0.3Al-≤1.0Co-19.0-21.0Cr-≤0.50Cu-≤1.0Fe-75.0Ni	UNS N10001
2.4883		Fe-≤0.03C-≤0.50Si-≤1.0Mn-≤2.50Co-15.50-17.50Cr-≤7.0Fe-16.0-18.0Mo	UNS N10002
2.4886	SG-NiMo16Cr16W; UP-NiMo16Cr16W	≤50.0Ni-≤0.02C-≤0.08Si-≤1.0Mn-≤0.015S-≤14.5-16.5Cr-≤4.0-7.0Fe-≤15.0-17.0Mo-≤0.4V-≤3-4.5W	UNS N10276
2.4887	EL-NiMo15Cr15W	≤50.0Ni-≤0.02C-≤0.20Si-≤1.0Mn-≤0.015S-≤14.5-16.5Cr-≤4.0-7.0Fe-≤15.0-17.0Mo-≤0.4V-≤3-4.5W	
2.4951	NiCr20Ti	Ni-≤0.08-0.15C-≤1.0Si-≤1.0Mn-≤0.020P-≤0.015S-≤0.3Al-≤0.0060B-≤5.0Co-≤18.0-21.0Cr-≤0.5Cu-≤5.0Fe-≤0.2-0.6Ti	UNS N06075
2.4952	NiCr20FeMo3TiCuAl	Fe-≤0.03C-≤0.5Si-≤1.0Mn-≤0.03P-≤0.03S-0.1-0.5Al-19.5-22.5Cr-1.5-3.0Cu-≤22.0Fe-2.5-3.5Mo-≤0.5Nb-42.0-46.0Ni-1.9-2.4Ti	
2.4964	CoCr20W15Ni	≤9.0-11.0Ni-≤0.05-0.15C-≤0.4Si-≤2.0Mn-≤0.020P-≤0.015S-≤19.0-21.0Cr-≤3.0Fe-≤9.0-11.0Ni-≤14.0-16.0W	UNS R30605
2.4973	NiCr19CoMo	Ni-≤0.12C-≤0.50Si-≤0.10Mn-≤1.40-1.80Al-≤10.00-12.00Co-≤18.00-20.00Cr-≤5.00Fe-≤9.00-10.50Mo-≤2.80-3.30Ti	N 07041 (UNS) 683 (SAE)
2.4975	NiFeCr12Mo	Fe-≤0.10C-≤0.60Si-≤2.00Mn-≤0.020P-≤0.010S-≤0.350Al-≤1.00Co-11.00-14.00Cr-5.00-7.00Mo-40.00-45.00Ni-2.35-3.10Ti	
2.4976	NiCr20Mo	Ni-≤0.10C-≤1.00Si-≤1.00Mn-≤0.020P-≤0.010S-0.50-1.80Al-≤2.00Co-18.00-21.00Cr-≤5.00Fe-4.00-5.00Mo-1.80-2.70Ti	
2.4983	NiCr18Co	Ni-≤0.15C-≤0.50Si-≤1.00Mn-≤0.02P-≤0.01S-≤2.50-3.20Al-≤17.00-20.00Co-≤17.00-20.00Cr-≤4.00Fe-≤3.00-5.00Mo-≤2.50-3.20Ti	UNS N07500 ASTM B637 (N07500)(864)
2.4999	MP35N	35.0Ni-≤0.01C-≤20.0Cr-≤9.5Mo	

Table 2: Chemical compositions of different American, CIS, Bulgarian and other steels

Steel	*Materials Compositions, Percent in Weight*	*Note*
000Ch16N13M2	Fe-≤0.07C-≤1.0Si-≤2.0Mn-16.5-18.5Cr-2-2.5Mo-10-13Ni-≤0.045P-≤0.015S-≤0.11N	CIS, formerly USSR, identical with SAE 316
000Ch16N13M3	Fe-≤0.07C-≤1.0Si-≤2.0Mn-16.5-18.5Cr-2-2.5Mo-10-13Ni-≤0.045P-≤0.015S-≤0.11N	CIS, formerly USSR, identical with SAE 316
000Ch16N16M4	Fe-≤0.07C-≤1.0Si-≤2.0Mn-16.5-18.5Cr-2-2.5Mo-10-13Ni-≤0.045P-≤0.015S-≤0.11N	CIS, formerly USSR, identical with SAE 316
000Ch18N10	Fe-≤0.03C-≤0.8Si-≤2.0Mn-17-19Cr-≤0.3Mo-9-11Ni	CIS, formerly USSR/Bulg.
000Ch18N11	Fe-≤0.03C-≤1.0Si-≤2.0Mn-18-20Cr-10-12Ni-≤0.045P-≤0.015S-≤0.11N	Bulg., comparable with 1.4306
000Ch20N20	Fe-≤0.03C-≤18.57Cr-≤19.40Ni-≤0.71Mn-≤0.26Si	CIS, formerly USSR
000Ch21N6M2	Fe-≤0.036C-≤21.1Cr-6.5Ni-2.4Mo	CIS, formerly USSR
000Ch21N10M2	Fe-≤0.02C-≤19.8Cr-≤10.5Ni-≤2.1Mo	CIS, formerly USSR
000Ch21N21M4B	Fe-≤0.03C-≤20-22Cr-20-21Ni-3.4-3.7Mo-≤0.6Mn-≤0.6Si-≤0.03P-≤0.02S-0.45-0.8Nb	CIS, formerly USSR
005Ch25B	Fe-0.005C-0.007N-25Cr	CIS, formerly USSR
00Ch18G8N2T	Fe-≤0.08C-≤0.8Si-7-9Mn-17-19Cr-≤0.3Mo-1.8-2.8Ni-≤0.2W-≤0.3Cu-≤0.2Ti-0.2-0.5Al	CIS, formerly USSR
00Ch18N10	Fe-≤0.015C-≤0.7Si-≤1.7Mn-≤17.3Cr-≤10.4Ni	CIS, formerly USSR
02Ch12N10S5	Fe-0.02C-12Cr-10Ni-5Si, Nb-stabilized	CIS, formerly USSR
02Ch12N10S5B	Fe-0.02C-12Cr-10Ni-5Si, Nb-stabilized	CIS, formerly USSR
02Ch12N10S5T	Fe-0.02C-12Cr-10Ni-5Si, Nb-stabilized	CIS, formerly USSR
02Ch17NS6	Fe-0.02C-4-6.5Si-0.43-0.52Mn-16.3-18Cr-10.5-18.2Ni-0.005-0.008S-0.012-0.014P	CIS, formerly USSR
02Ch8N22S6	Fe-≤0.02C-≤5.4-6.7Si-0.6Mn-≤0.030P-≤0.020S-7.5-10Cr-0.3Mo-21-23Ni-≤0.2Ti-≤0.20W	CIS, formerly USSR
02Ch8N22S6B	Fe-0.02C-8Cr-122Ni-6Si, Nb-stabilized	CIS, formerly USSR
02Ch8N22T	Fe-0.02C-8Cr-122Ni-6Si, Ti-stabilized	CIS, formerly USSR
03Ch16N15M3	Fe-≤0.03C-15.0-17.0Cr-14.0-16.0Ni-2.5-3.0Mo-Ti	CIS, formerly USSR
03Ch18N11	Fe-0.03C-≤0.80Si-≤0.70-2.0Mn-≤0.035P-0.020S-17.0-19.0Cr-≤0.10Mo-10.5-12.5Ni-≤0.20W-0.30Cu-≤0.50Ti	CIS, formerly USSR, comparable with DIN Mat.No. 1.4306
03Ch18N14	Fe-0.03C-18Cr-14Ni	CIS, formerly USSR
03Ch21N21M4GB	Fe-≤0.03C-≤0.60Si-1.8-2.5Mn-≤0.030P-≤0.020S-20-22Cr-3.4-3.7Mo-20-22Ni-≤0.2W-≤0.3Cu-≤0.2Ti, Nb 15 x C-0.80	CIS, formerly USSR
03Ch23N6	Fe-≤0.03C-≤0.40Si-1.0-2.0Mn-≤0.035P-≤0.020S-22-24Cr-≤5.3-6.3Ni	CIS, formerly USSR
03Ch25	Fe-about 0.03C-25Cr-0.6Ni	CIS, formerly USSR
03ChN28MDT	Fe-≤0.03C-≤0.80Si-≤0.80Mn-≤0.035P-≤0.020S-22.0-25.0Cr-2.5-3.0Mo-26.0-29.0Ni-0.50-0.90Ti-2.5-3.5Cu	CIS, formerly USSR
04Ch18N10	Fe-≤0.04C-≤0.8Si-≤2.0Mn-17-19Cr-≤0.030P-≤0.02S-≤0.3Mo-9-11Ni-≤0.2W-≤0.3Cu-≤0.2Ti	CIS, formerly USSR

Steel	Materials Compositions, Percent in Weight	Note
04Ch18N10T	Fe-≤0.04C-≤0.8Si-≤2.0Mn-17-19Cr-≤0.3Mo-9-11Ni-≤0.2W-≤0.3Cu-≤0.2Ti	CIS, formerly USSR
05Ch16N15M3	Fe-0.05C-16Cr-15Ni-3Mo	CIS, formerly USSR
06Ch17G15NAB	Fe-0.05C-18.36Cr-16.5Mn-1.6Ni-0.31Nb-ß.12Si-0.01Ce-0.017P-0.014S	CIS, formerly USSR
06Ch23N28M3D3T	Fe-≤0.06C-≤0.8Si-≤2.0Mn-22-25Cr-≤2.4-3Mo-26-29Ni-0.5-0.9Ti-2.5-3.5Cu	CIS, formerly USSR/Bulg.
06Ch28MDT	Fe-≤0.06C-≤0.8Si-≤0.80Mn-22.0-25.0Cr-2.5-3.0Mo-26.0-29.0Ni-0.50-0.90Ti-2.5-3.5Cu	CIS, formerly USSR
06ChN28MDT	Fe-≤0.06C-≤0.8Si-≤0.8Mn-≤0.035P-≤0.02S-22-25Cr-2.5-3.0Mo-26-29Ni-0.5-0.9Ti-2.5-3.5Cu	CIS, formerly USSR
06ChN40B	Fe-0.055C-17.01Cr-39.04Ni-1.99Mn-0.50Nb-0.60Si-0.013S-0.022P	CIS, formerly USSR
07Ch13AG20	Fe-≤0.07C-≤0.60Si-≤19-22Mn-≤0.035P-≤0.025S-≤0.0030B-≤0.1Ca-≤0.1Ce -12.-14.8Cr-≤0.30Cu≤0.1Mg-≤0.30Mo≤0.08-0.18N-≤1.0Ni-≤0.20W≤0.20Ti	CIS, formerly USSR
07Ch16N4B	Fe-0.05-0.10C-≤0.60Si-≤0.2-0.5Mn-≤0.025P-≤0.020S-15-16.5Cr-≤0.30Cu-≤0.30Mo-0.2-0.4Nb-3.5-4.5Ni-≤0.20W	CIS, formerly USSR
07Ch17G15NAB	Fe-0.05C-18.4Cr-16.5Mn-1.6Ni-0.01Ce-0.005B-0.32N	CIS, formerly USSR
07Ch17G17DAMB	Fe-0.06C-17.6Cr-15.2Mn-0.43Mo-0.3Nb-0.005B-0.38N	CIS, formerly USSR
08Ch17N5M3	Fe-0.06C-0.10C-≤0.80Si-≤0.80Mn-≤0.035P-≤0.020S-16.0-17.5Cr-3.0-3.5Mo-4.5-5.5Ni-≤0.20W-≤0.30Cu-≤0.20Ti	CIS, formerly USSR
08Ch17N15M3B	Fe-≤0.08C-16.0-18.0Cr-14.0-16.0Ni-3.0-4.0Mo-Ti	CIS, formerly USSR
08Ch17N15M3T	Fe-≤0.08C-≤0.80Si-≤2.0Mn-≤0.35P-≤0.020S-16.0-18.0Cr-3.00-4.00Mo-14.0-16.0Ni≤0.20W-0.30Cu-0.30-0.60Ti	CIS, formerly USSR
08Ch17T	Fe-≤0.08C-≤0.80Si-≤0.80Mn-≤0.035P-≤0.025S-16.0-18.0Cr-≤0.6Ni-≤0.30Cu	CIS, formerly USSR
08Ch18G8N2M2T	Fe-0.08C-18.2Cr-3.42Ni-8.9Mn-2.32Mo-0.22Ti	CIS, formerly USSR
08Ch18G8N2T	Fe-≤0.08C-≤0.80Si-7.0-9.0Mn-≤0.035P-17.0-19.0Cr-≤0.30Mo-1.80-2.80Ni-≤0.30Cu-≤0.035P-≤0.025S-≤0.20-0.50Ti-≤0.20W	CIS, formerly USSR
08Ch18N10	Fe-≤0.08C-≤0.8Si-≤2.0Mn-≤0.035P-≤0.020S-17-19Cr-0.3Mo-9.0-11.0Ni-≤0.2W-≤0.3Cu	CIS, formerly USSR
08Ch18N10T	Fe-≤0.08C-≤0.8Si-≤2.0Mn-≤0.035P-≤0.020S-17-19Cr-0.5Mo-9.0-11.0Ni-0.5Ti-≤0.2W-≤0.3Cu	CIS, formerly USSR
08Ch21N6M2T	Fe-≤0.08C-≤0.8Si-≤0.8Mn-≤0.035P-≤0.025S-20-22Cr-1.8-2.5Mo-5.5-6.5Ni-0.2-0.4Ti-≤0.2W-≤0.3Cu	CIS, formerly USSR
08Ch22N6M2T	Fe-≤0.08C-≤0.80Si-≤0.80Mn-20.0-22.0Cr-1.80-2.50Mo-5.50-6.50Ni-≤50.20W-≤0.30Cu-≤0.035P-≤0.025S-0.20-0.40Ti	CIS, formerly USSR

Steel	*Materials Compositions, Percent in Weight*	*Note*
08Ch22N6T	Fe-≤0.08C-≤0.8Si-≤0.8Mn-≤0.035P–≤0.025S-21-23Cr-≤0.3Mo-5.3-6.3Ni-≤0.2W-≤0.3Cu, 5x% C max. 0.65Ti	CIS, formerly USSR
08ChP	Fe-0.25Cr-0.25Ni-0.25Mo	CIS, formerly USSR
09Ch16N15M3B	Fe-≤0.09C-≤0.80Si-≤0.80Mn-≤0.035P-≤0.020S-15.0-17.0Cr-≤0.30Cu-2.5-3.0Mo-0.6-0.9Nb-14.0-16.0Ni-≤0.20Ti-≤0.20W	CIS, formerly USSR
09G2S	Fe-≤0.12C-0.5-0.8Si-1.3-1.7Mn-≤0.035P-≤0.035S-≤0.3Cr-≤0.3Ni-≤0.3Cu	Bulg.
0Ch17N16M3T	Fe-≤0.080C-≤0.8Si-≤2.00Mn-16.0-18.0Cr-3.00-4.00Mo-14.0 16.0Ni-< 0.035P-≤0.025S-0.30-0.60Ti	CIS, formerty USSR
0Ch18G8N3M2T	Fe-about 18 Cr-8Mn-3Ni-2Mo, Ti	CIS, formerly USSR
0Ch18N10T	Fe-≤0.08C-≤1.0Si-≤2.0Mn-≤0.045P-≤0.015S-17-19Cr-9-12Ni, Ti 5xC-0.70	CIS, formerly USSR/Bulg.
0Ch18N12B	Fe-≤0.08C-≤1.0Si-≤2.0Mn-≤0.045P-≤0.015S-17-19Cr-≤9.0-12Ni, Nb 10xC-1.00	CIS, formerly USSR/Bulg., comparable with DIN-Mat. No. 1.4550
0Ch20N14S2	Fe- about 20Cr-14Ni-2Si	CIS, formerly USSR
0Ch21N5T	Fe-≤0.08Cr-21Cr-5Ni, Ti	CIS, formerly USSR s. text HNO3
0Ch23N18	Fe-≤0.20C-≤1.00Si-1.50Mn-22.0-25.0Cr-≤0.30Mo-17.0-20.0Ni-≤0.035P-≤0.025S	CIS, formerly USSR
0Ch23N28M3D3T	Fe-≤0.06C-≤0.8Si-≤2Mn-22-25Cr-2.4-3Mo-26-29Ni-0.5-0.9Ti-2.5-3.5Cu	CIS, formerly USSR/Bulg.
0Ch25T	Fe-≤0.01C-25Cr, Ti-stabilized	CIS, formerly USSR
0H17N12M2T	Fe-≤0.05C-≤1.0Si-≤2.0Mn-16-18Cr-2-3Mo-11-14Ni-≤0.045P-≤0.030S, Ti 5xC-0.60	Poland
10Ch13 (1Ch13)	Fe-0.08-0.15C-≤1.0Si-≤1.5Mn-≤0.040P-≤0.015S-11.5-13.5Cr-≤0.75Ni	CIS, formerly USSR/Bulg.
10Ch14AG15	Fe-≤0.10C-≤0.80Si-14.5-16.5Mn-≤0.045P-≤0.030S-13.0-15.0Cr-<0.60Ni-≤0.60Cu-≤0.20Ti-0.15-0.25N	CIS, formerly USSR
10Ch14G14N4T	Fe-≤0.10C-≤0.80Si-13.0-15.0Mn-13.0 15.0Cr-≤0.30Mo-2.80-4.50Ni-≤0.20W-≤0.30Cu-≤0.035P-≤0.020S-5x% C max. 0.60Ti	CIS, formerly USSR
10Ch17	Fe-0.10C-17Cr	CIS, formerly USSR
10Ch17N13M2T	Fe-≤0.10C-≤0.80Si-≤2.0Mn-≤0.035P-≤0.020S-16.0-18.0Cr-≤0.30Cu-2.0-3.0Mo-12.0-14.0Ni-≤0.20W, Ti>-5x% C	USA, comparable with DIN-Mat. No. 1.4571
10Ch17N13M3T	Fe-≤0.10C-≤0.80Si-≤2.0Mn-≤0.035P-≤0.020S-16.0-18.0Cr-≤0.30Cu-3.0-4.0Mo-12.0-14.0Ni-≤0.20W-≤0.7Ti	USA, comparable with DIN-Mat. No. 1.4573
10Ch18N9T(Ch18N9T)	Fe-0.08C-1.0Si-≤2Mn-≤0.045P-≤0.015S-17-19Cr-≤0.3Mo-9-12Ni-Ti5xC-0.70	CIS, formerly USSR/Bulg.
10Ch18N10M2T	Fe-≤0.10C-18Cr-10Ni-2Mo, Ti stabilized	CIS, formerly USSR

Steel	Materials Compositions, Percent in Weight	Note
10Ch18N10T	Fe-≤0.10C-≤0.80Si-≤1.0-2.0Mn-≤0.035P-≤0.020S-≤17.0-19.0Cr-≤10.0-11.0Ni	CIS, formerly USSR/Bulg.
12Ch13G18D	Fe-0.12C-13Cr-18Mn-Cu	CIS, formerly USSR
12Ch17G9AN4	Fe-≤0.12C-≤0.80Si-8.0-10.5Mn-≤0.035P-≤0.020S-16.0-18.0Cr-≤0.30Mo-3.5-4.5Ni-≤0.20W-≤0.30Cu-≤0.20Ti-0.15-0.25N	CIS, formerly USSR
12Ch18N9T	Fe-≤0.12C-≤0.80Si-≤2.0Mn-≤0.035P-≤0.020S-17.0-19.0Cr-≤0.50Mo-8.0-9.5Ni-≤0.20W-≤0.30Cu-Ti = 5x % C	CIS, formerly USSR
12Ch18N10T	Fe-≤0.12C-≤0.8Si-≤2.0Mn-≤0.025P-≤0.020S-17.0-19.0Cr-≤0.30Cu-≤0.50Mo-9.0-11.0Ni-≤0.20W-≤0.70Ti	CIS, formerly USSR; comparable with DIN-Mat. No. 1.4878
12Ch2M1	Fe-0.12C-2Cr-1Mo	CIS, formerly USSR/Bulg., comparable with DIN-Mat. No. 1.7380, A 182, F 22, B.S. 1501-622
12Ch2N4A	Fe-0.09-0.15C-0.17-0.37Si-0.30-0.60Mn-≤0.025P-≤0.025S-1.25-1.65Cr-3.25-3.65Ni-≤0.30Cu-≤0.15Mo-≤0.03Ti-≤0.05V-≤0.12W	CIS, formerly USSR
12Ch21N5T	Fe-0.09-0.14C-≤0.80Si-≤0.80Mn-≤0.035P-≤0.025S-20.0-22.0Cr-≤0.30Mo-4.80-5.80Ni-≤0.20W-≤0.30Cu-0.25-0.50Ti-≤0.08Al	CIS, formerly USSR
12ChN2	Fe-0.09-0.16C-0.17-0.37Si-0.30-0.60Mn-≤0.035P-≤0.035S-0.60-0.90Cr-1.50-1.90Ni-≤0.30Cu-≤0.15Mo-≤0.03Ti-≤0.05V-≤0.20W	CIS, formerly USSR/Bulg.
13-4-1	Fe-0.043C-12.7Cr-3.9Ni-1.5Mo-0.68Mn-0.39Si-0.009P-0.013S-0.030N	CIS, formerly USSR
14Ch17N2	Fe-≤0.11-0.17C-≤0.80Si-≤0.80Mn-16.0-18.0Cr-≤0.30Mo-1.50-2.50Ni-≤0.20W-≤0.30Cu-≤0.030P-≤0.025S-≤0.20Ti	CIS, formerly USSR
15Ch17N2	Fe-0.13C-0.49Si-0.52Mn-17.17Cr-1.75Ni-0.012P-0.09S	CIS, formerly USSR
15Ch25T	Fe-≤0.15C-≤1.00Si-≤0.80Mn-≤0.035P-≤0.025S-24.0-27.0Cr-≤0.30Cu-≤1.00Ni-0.09Ti	CIS, formerly USSR
15Ch28	Fe-≤0.15C-≤1.00Si-≤0.80Mn-27.0-30.0Cr-0.60Ni-≤0.30Cu-≤0.035P-≤0.025S-≤1.0Ni-≤0.20Ti	CIS, formerly USSR
15Ch2M2FBS	Fe-about 0.15C-2Cr-2Mo-V-Nb-Si	CIS, formerly USSR
15Ch5M	Fe-≤0.15C-≤0.50Si-≤0.50Mn-4.50-6.00Cr-0.40-0.60Mo-≤0.60Ni-≤0.03Ti-≤0.030P-≤0.025S-≤0.20Cu-≤0.05V-≤0.30W	CIS, formerly USSR
16GS	Fe-≤0.12-0.18C-≤0.40-0.70Si-0.90-1.20Mn-≤0.30Cr- ≤0.30Ni-≤0.30Cu-≤0.035P-≤0.040S-≤0.05Al-≤0.08As-≤0.012N-≤0.03Ti	CIS, formerly USSR/Bulg., comparable with DIN-Mat. No. 1.0481, A 414, A 515, A 516
18/8-CrNi-steel	Fe-≤0.12C-1.0Si-≤2.0Mn-≤0.045P-≤0.030S-≤17.0-19.0Cr≤-8.0-10.0Ni	–

Steel	Materials Compositions, Percent in Weight	Note
1815-LCSi	Fe-0.006C-18.3Cr-15.1Ni-1.5Mn-4.1Si-0.005S-0.010P-0.010N	UNS S30600, comparable with DIN-Mat. No. 1.4361
18-18-2	Fe-≤0.08C-1.5-2.5Si-≤2.0Mn-≤0.030P-≤0.030S-17.0-19.0Cr-17.5-18.5Ni	USA
18G2A	Fe-≤0.20C-≤0.50Si-≤0.9-1.7Mn-≤0.025P-≤0.020S-≤0.0200Al-≤0.30Cr-≤0.50Ni-≤0.30Cu.-≤0.08Mo-≤0.020N-≤0.050Nb-≤0.03Ti-≤0.1V	Poland
20Ch13 (2Ch13)	Fe-0.16-0.25C-≤0.8Si-≤0.8Mn-12-14Cr-≤0.6Ni-≤0.030P-≤0.025S-≤0.30Cu-≤0.20Ti	CIS, formerly USSR/Bulg., comparable with DIN-Mat. No. 1.4021, SAE 420, 420 S 29
20Ch23N18	Fe-≤0.2C-≤1.0Si-≤2.0Mn-22-25Cr-≤0.3Mo-17-20Ni-≤0.2W-≤0.3Cu-≤0.2Ti-≤0.035P-≤0.025S	CIS, formerly USSR
20Ch2G2SR	Fe-0.16-0.26C-0.75-1.55Si-1.4-1.8Mn-≤0.040P-≤0.040S-1.4-1.8Cr-≤0.30Ni-≤0.30Cu-0.02-0.08Ti-0.015-0.050Al-0.001-0.007B	CIS, formerly USSR
23Ch2G2T	Fe-0.19-0.26C-0.40-0.70Si-1.4-1.7Mn-≤0.045P-≤0.045S-1.35-1.70Cr-≤0.30Ni-≤0.30Cu-0.02-0.08Ti-0.015-0.05Al	CIS, formerly USSR
36NChTJu	Fe-≤0.05C-≤0.3-0.7Si-≤0.8-1.2Mn-≤0.020P-≤0.020S-11.5-13Cr-35-37Ni-0.9-1.2Al-2.7-3.2Ti	CIS, formerly USSR
40Ch13	Fe-0.36-0.45C-≤0.80Si-≤0.80Mn-≤0.030P-≤0.025S-12.0-14.0Cr-≤0.30Cu-≤0.60Ni-≤0.20Ti	CIS, formerly USSR, comparable with DIN-Mat. No. 1.4031
45 G2	Fe-0.41-0.49C-0.17-0.37Si-1.4-1.8Mn-0.035P-≤0.035S-≤0.30Cr-≤0.30Cu-≤0.15Mo-≤0.30Ni-≤0.03Ti-≤0.05V-≤0.20W	CIS, formerly USSR, comparable with DIN-Mat. No. 1.0912
50Ch	Fe-0.46-0.54C-0. 17-0.37Si-0.50-0.80Mn-≤0.035P-≤0.035S-0.80-1.10Cr-≤0.30Ni-≤0.30Cu-≤0.15Mo-≤0.03Ti-≤0.05V-≤0.20W	CIS, formerly USSR
70G	Fe-0.67-0.75C-0.17-0.37Si-0.90-1.20Mn-≤0.035P-<0.035S-≤0.25Cr-≤0.25Ni- ≤0.20Cu	CIS, formerly USSR
80S	Fe-0.74-0.82C-0.60-1.10Si-0.50-0.90Mn-≤0.040P-≤0.045S-≤0.30Cr-≤0.30Ni-≤0.30Cu-0.015-0.040Ti	CIS, formerly USSR
2320	Fe-≤0.08-1.0Si-1.0Mn-≤0.040P-≤0.030S-16.0-18.0Cr-≤1.0Ni	Sweden, comparable with DIN-Mat. No. 1.4016; SAE 430, 10Ch17T
ASTM A-159	Fe-3.1-3.4C-1.9-2.3Si-0.6-0.9Mn-0.15S-0.15P	USA
ASTM A-516 Gr. 70	Fe-0.27C-0.13-0.45Si-0.79-1.30Mn-≤0.035P-≤0.040S	USA, comparable with DIN-Mat. No. 1.0050 and No. 1.0481
ASTM XM-27	Fe-≤0.01C-≤0.40Si-≤0.40Mn-≤0.020P-≤0.020S-25.0-27.5Cr-≤0.20Cu-≤0.75-1.50Mo-≤0.015N-≤0.050-0.2Nb-0.50Ni, Ni+Cu≤0.50	USA, comparable with SAE XM-27

Steel	*Materials Compositions, Percent in Weight*	*Note*
C 1204	Fe-0.20C-0.35Si-0.50Mn-0.050P-0.050S-0.30Cr	Yugoslavia, comparable with DIN-Mat. No. 1.0425, B.S. 1501 Gr. 161-400, 164-350, 164-400; 16 K
C 90	Fe-0.85-0.94C-≤-0.35Si-≤0.35Mn-≤0.03P-≤0.03S	Italy
Carpenter 20 Cb-3	Fe-≤0.06C-≤1.00Si-≤2.00Mn-19.0-21.0Cr-2.0-3.0Mo-32.5-35Ni-3.0-4.0Cu-≤0.035P-≤0.035S	USA
Ch12M	Fe-1.45-1.65C-0.15-0.35Si-0.15-0.4Mn-11-12.5Cr-0.4-0.6Mo-≤0.35Ni-15-0.3V-≤0.2W≤0.3Cu-≤0.03Ti	CIS, formerly USSR
Ch14N40SB	Fe-0.034C-4.0Si-0.05Mn-14.4Cr-38.9Ni-0.63Nb	CIS, formerly USSR
Ch15T	Fe-≤0.1C-≤0.8Si-≤0.8Mn-14-16Cr-≤0.3Mo-≤0.6Ni, 5x%C≤Ti≤0.8	CIS, formerly USSR/Bulg.
Ch17	Fe-≤0.08C-≤1Si-≤1Mn-≤0.040P-≤0.015S-16.0-18.0Cr	Bulg., comparable with DIN-Mat. No. 1.4016, SAE 430, 12Ch17T, X6Cr17
Ch17N2	Fe-17Cr-2Ni	CIS, formerly USSR
Ch17N12M3T	Fe-≤0.12C-≤1.5Si-≤2.0Mn-16-19Cr-3-4Mo-11-13Ni-0.3-0.6Ti	CIS, formerly USSR/Bulg.
Ch17N18M2T	Fe-0.09C-0.6Si-1.4Mn-16.9Cr-1.9Mo-12.3Ni, Ti stab. (p.a.)	CIS, formerly USSR/Bulg.
Ch17T	Fe-≤0.05C-≤1Si-≤1Mn-≤0.040P-≤0-0.15S-16.0-18.0Cr, Ti4x(C+N)+0.15-0.80	Bulg., comparable with DIN-Mat. No. 1.4510, 08Ch17T, X3CrTi17
Ch18AG14	Fe-18Cr-14Mg-0.5N	
Ch18N9T Ch18N10T	Fe-≤0.08C-≤1Si-≤2Mn-≤0.045P-≤0.015S-17.0-19.0Cr-9.0-12.0Ni, Ti 5xC-0.70	CIS, formerly USSR/Bulg., comparable with DIN-Mat. No. 1.4541, X6CrNiTi18-10
Ch18N10	Fe-0.08C-18.4Cr-10.2Ni-1.08Mn-0.3Si-0.005P-0.014S-0.005N	
Ch18N12M2T	Fe-≤0.15C-≤5 1.5Si-≤2Mn-17-19Cr-2-2.5Mo-11-13Ni, 4x%C≤Ti≤0.8	CIS, formerly USSR/Bulg.
Ch18N12T	Fe-0.08C-1.0Si-≤2.0Mn-≤0.045P-0.015S-17.0-19.0Cr-9.0-12.0Ni-Ti = 5x%C≥0.70	CIS, formerly USSR/Bulg.
Ch18N14	Fe-0.035C-18.8Cr-14.6Ni-0.35Mn-0.75Si-0.005P-0.03S-0.004N	
Ch18N40T	Fe-<0.08C-<1.0Si-<2.0Mn-<0.045P-<0.015S-17-19Cr-9-12Ni-<0.7Ti	Comparable with DIN-Mat. No. 1.4541, SAE 321
Ch20N20	Fe-0.004-0.015C-19.4-21.8Cr-19.3-20.8Ni-0.05-5.40Si-0.002-0.1P	CIS, formerly USSR/Bulg
Ch22N5	Fe-0.07C-21.54Cr-5.73Ni	CIS, formerly USSR
Ch23N18	Fe-≤0.20C-≤1.00Si-≤1.50Mn-22.0-25.0Cr-≤0.30Mo-17.0-20.0Ni-≤0.035P-≤0.025S	CIS, formerly USSR/Bulg.
Ch23N27M2T	Fe-27Ni-23Cr-2Mo-Ti	CIS, formerly USSR/Bulg.
Ch23N28M3D3T	Fe-28Ni-23Cr-3Mo-3Cu-Ti	CIS, formerly USSR

Steel	Materials Compositions, Percent in Weight	Note
Ch25T	Fe-≤0.15C-≤1.00Si-≤0.80Mn-24.0-27.0Cr-≤0.30Mo-≤0.60Ni-≤0.035P-≤025S-5x%C≤Ti≤0.9	CIS, formerly USSR/Bulg.
Ch28N18	Fe-0.16C-1.7Mn-1.1Si-22.6Cr-18Ni-0.4Ti (p.a.)	CIS, formerly USSR/Bulg.
ChN28MDT	Fe-0.03-0.046C-22.2-23.5Cr-26.55-27.88Ni-2.55-3.06Mo-2.68-3.38Cu-0.54-0.76Ti-0.15-0.30Mn-0.39-0.69Si-0.021-0.43P-0.008-0.017S	CIS, formerly USS
ChN40B	Fe-0.032C-18.2Cr-40.4Ni-0.08Si-0.05Mn-0.49Nb	CIS, formerly USSR
ChN40S	Fe-0.031C-20.0Cr-38.9Ni-4.2Si-0.05Mn-0.13Nb	CIS, formerly USSR
ChN40SB	Fe-0.04C-18.8Cr-39.4Ni-4.3Si-0.06Mn-0.63Nb	CIS, formerly USSR
ChN58W	Ni-0.03C-14.5-16.5Cr-15-17Mo-3.0-4.5W-1.5Fe-1.0Mn-<0.12Si-0.02S-0.025P	CIS, formerly USSR
ChN60V	Fe-0.01-0.02C-0.09N-max.0.05Zr-0.1Ti-0.015Ce-max.1.7Nb-max0.009B	CIS, formerly USSR
ChN77TJuR	Fe-≤0.07C-≤0.60Si-≤0.40Mn-≤0.015P-≤0.007S-0.6-1.0Al-≤0.01B-≤0.02Ce-19.0-22.0Cr-≤1.0Fe-≤0.001Pb-2.4-2.8Ti	CIS, formerly USSR
FC 20	Fe-3.92C-1.12Si-0.63Mn-0.072P-0.012S	Japan
JS 700	Fe-0.04C-≤1.0Si-≤2.0Mn-≤0.040P-≤0.030S-19-23Cr-4.3-5.0Mo-24-26Ni, Nb≥8xC≤0.40	USA
OZL-17u	Fe-0.04C-0.32Si-1.5Mn-23.2Cr-0.2Mo-29.4Ni-0.01P-0.01S-0.1Ti	CIS, formerly USSR
SAE 1008	Fe-0.10C-0.30-0.50Mn-≤0.030P-≤0.050S	USA, comparable with DIN-Mat. No. 1.0204
SAE 1018	Fe-0.15-0.20C-0.60-0.90Mn-≤0.030P-≤0.050S	USA
SAE C-1018	Fe-0.20C-0.25Si-0.58Mn-0.16Cr-0.04Mo-0.012-0.014P-0.02S	USA
S35C	Fe-0.32-0.38C-0.15-0.35Si-0.60-0.90Mn-≤0.030P-≤0.035S-≤0.20Cr-≤0.30Cu-≤0.20Ni	Japan, comparable with DIN-Mat. No. 1.0501
SIS 2333	Fe-≤0.05C-1.0Si-2.0Mn-≤0.045P-≤0.030S-17.0-19.0Cr-8.0-11.0Ni	Sweden, comparable with DIN-Mat. No. 1.4303; SAE 304, 03Ch18N11
SKH 2	Fe-0.73-0.83C-0.45Si-0.40Mn-≤0.030P-≤0.030S-3.8-4.5Cr-≤0.25Cu-17.0-19.0W-1.0-1.2V	Japan, comparable with DIN 1.3355
SKH-4A	Fe-0.80C-0.29Si-0.31Mn-4.16Cr-17.64W-9.3Co-1.1V	Japan
SKH-9	Fe-0.85C-0.16Si-0.31Mn-4.14Cr-4.97Mo-6.03W-1.88V	Japan
SS41	Fe-≤0.050P-≤0.050S	Japan, comparable with DIN-Mat. No. 1.0040
St35b-2	Fe-≤0.16C-0.17-0.40Si-0.35-0.65Mn-≤0.30Cr-≤0.30Ni-≤0.30Cu-0.050P-0.050S	Germany, formerly GDR, comparable with DIN-Mat. No. 1.0309
St35hb	Fe-≤0.18C-≤0.17Si-0.35-0.65Mn-0.050P-0.050S	Germany, formerly GDR
St38	Fe-≤0.20C-≤0.080-≤0.060S	Germany, formerly GDR, comparable with DIN-Mat. No. 1.0037

Steel	Materials Compositions, Percent in Weight	Note
St38b-2	Fe-0.12-0.20C-0.17-0.37Si-0.40-0.65Mn-≤0.045P-≤0.050S-Cr+Cu+Ni≤0.70	Germany, formerly GDR, comparable with DIN-Mat. No. 1.0038, BS 4360-40C and A 570 Gr. 36
St5	Fe-≤0.045P-≤0.045S-≤0.009N	Poland, comparable with DIN-Mat. No. 1.0050
SUS 304	Fe-≤0.08C-≤1.0Si-≤2.0Mn-≤0.045P-≤0.030S-18.0-20.0Cr-8-10.5Ni	Japan, comparable with DIN-Mat. No. 1.4301
SUS 430	Fe-≤0.12C-≤0.75Si-≤1.00Mn-16.0-18.0Cr-≤0.60Ni-≤0.040P-≤0.030S	Japan, comparable with DIN-Mat. No. 1.4016, SAE 430, 430 S 15, 12Ch17
Sv-08	Fe-≤0.10C-≤0.03Si-0.35-0.60Mn-≤0.040P-≤0.040S-≤0.015Cr-≤0.30Ni-≤0.01Al	CIS, formerly USSR
TsL-17	Fe-0.1C-5Cr-1Mo	CIS, formerly USSR
TsL-9	Fe-0.07C-1.00Si-2.3Mn-24.1Cr-12.9Ni-1.1Nb-0.03P-0.02S	CIS, formerly USSR
U7A	Fe-0.65-0.75C-0.10-0.30Si-0.10-0.40Mn-≤0.030P-≤0.030S	CIS, formerly USSR, comparable with DIN-Mat. No. 1.1520
U8A	Fe-0.75-0.85C-0.10-0.30Si-0.10-0.40Mn-≤0.030P-≤0.030S	CIS, formerly USSR, comparable with DIN-Mat. No. 1.1525
U10A	Fe-0.96-1.03C-0.17-0.33Si-0.17-0.28Mn-≤0.025P-≤0.018S-≤0.20Cr-≤0.20Ni-≤0.20Cu	CIS, formerly USSR, comparable with DIN-Mat.No. 1.1545, SAE W 1 10
X5CrNiMoCuTi18-18	Fe-≤0.07C-≤0.80Si-≤2.0Mn-≤0.045P-≤0.030S-16.5-18.5Cr-2.0-2.5Mo-19.0-21.0Ni-1.8-2.2Cu-Ti≥7x%C	Germany, formerly GDR, comparable with DIN-Mat. No. 1.4506
X8CrNiTi18-10	Fe-≤0.10C-≤1.0Si-≤2.0Mn-≤0.045P-≤0.015S-17.0-19.0Cr-9.0-12.0Ni-Ti≥5 x C≤0.80	Germany, formerly GDR, comparable with DIN-Mat. No. 1.4541

Index of materials

z

Subject index

n

o

p

q

r